International Academic Symposium of Social Science 2022

International Academic Symposium of Social Science 2022

Editors

Mohamad Rahimi Mohamad Rosman
Nor Erlissa Abd Aziz
Mohd Nasir Ismail
Ghazali Osman
Khalid Abdul Wahid

MDPI • Basel • Beijing • Wuhan • Barcelona • Belgrade

Editors
Mohamad Rahimi Mohamad Rosman
Universiti Teknologi MARA
Kelantan Branch
Malaysia

Nor Erlissa Abd Aziz
Universiti Teknologi MARA
Kelantan Branch
Malaysia

Mohd Nasir Ismail
Universiti Teknologi MARA
Kelantan Branch
Malaysia

Ghazali Osman
Universiti Teknologi MARA
Kelantan Branch
Malaysia

Khalid Abdul Wahid
Universiti Teknologi MARA
Kelantan Branch
Malaysia

Editorial Office
MDPI
St. Alban-Anlage 66
4052 Basel, Switzerland

This is a reprint of Proceedings published online in the open access journal *Proceedings* (ISSN 2504-3900) in 2022 (available at: https://www.mdpi.com/2504-3900/82/1).

For citation purposes, cite each article independently as indicated on the article page online and as indicated below:

LastName, A.A.; LastName, B.B.; LastName, C.C. Article Title. *Journal Name* **Year**, *Article Number*, Page Range.

Volume 1
ISBN 978-3-0365-6093-9 (Hbk)
ISBN 978-3-0365-6094-6 (PDF)

Volume 1-2
ISBN 978-3-0365-6017-5 (Hbk)
ISBN 978-3-0365-6018-2 (PDF)

Contents

proceedings

MDPI

Editorial

Statement of Peer Review †

Mohamad Rahimi Mohamad Rosman *, Nor Erlissa Abd Aziz, Mohd Nasir Ismail, Ghazali Osman and Khalid Abdul Wahid

Faculty of Information Management, Universiti Teknologi MARA Kelantan Branch, Kelantan 18500, Malaysia
* Correspondence: rahimimr@uitm.edu.my

† Presented at the International Academic Symposium of Social Science 2022, Kota Bharu, Malaysia, 3 July 2022.

When submitting conference proceedings to the journal *Proceedings*, the volume's editors notify the publisher that they carried out a peer review of all published papers. Reviews were conducted by expert referees while upholding all the professional and scientific standards expected of the *Proceedings* journal.

- Type of peer review: single-blind.
- Conference submission management system: EasyChair.
- Number of submissions sent for review: 186.
- Number of submissions accepted: 114.
- Acceptance rate (number of submissions accepted/number of submissions received): 61%.
- Average number of reviews per paper: 3.
- Total number of reviewers involved: 114.
- Any additional information on the review process: not applicable.

This publication gathered the proceedings of the International Academic Symposium of Social Science 2022 (IASSC2022) held on 3 July 2022 in Kota Bharu, Kelantan, Malaysia. The conference was jointly organized by the Faculty of Information Management of Universiti Teknologi MARA Kelantan Branch, Malaysia; the Digital Information Interest Group (DIGIT), Malaysia; the University of Malaya, Malaysia; Universitas Pembangunan Nasional Veteran Jakarta, Indonesia; Universitas Ngudi Waluyo, Indonesia; Camarines Sur Polytechnic Colleges, Philippines; and UCSI University, Malaysia. Featuring experienced keynotes speakers from Malaysia, Australia, and England, these proceedings provided an opportunity for researchers, postgraduate students, and industry practitioners to gain knowledge and an understanding of advanced topics concerning digital transformations in the perspective of social sciences and information systems, focusing on issues, challenges, impacts, and theoretical foundations. These proceeding could be helpful in shaping the future of academia and the industry by compiling state-of-the-art works, as well as future trends concerning the digital transformation of social science and the information systems field. It is also considered an interactive platform allowing for the gathering of academicians, practitioners, and students from various institutions and industries. A total of 186 manuscripts was received for consideration; however, only 114 manuscripts were accepted for presentation and publication, comprising of participants from Malaysia, Australia, Indonesia, Jordan, Morocco, Nigeria, Pakistan, Philippines, Sudan, Thailand, and the United Kingdom.

International Advisory Board

Prof. Datuk Ts. Dr. Roziah Mohd Janor, *Universiti Teknologi MARA, Malaysia.*
Prof. Dr. Mohd Sazili Shahibi, *Universiti Teknologi MARA, Malaysia.*
Prof. Dr. Haji Abdol Samad Nawi, *Universiti Teknologi MARA, Malaysia.*
Dr. Charlito P. Cadag, *Camarines Sur Polytechnic Colleges, Philippines.*
Prof. Engr. Harold Jan Terano, *Camarines Sur Polytechnic Colleges, Philippines.*
Assoc. Prof. Filmor J. Murillo, *Camarines Sur Polytechnic Colleges, Philippines.*

check for updates

Citation: Rosman, M.R.M.; Aziz, N.E.A.; Ismail, M.N.; Osman, G.; Wahid, K.A. Statement of Peer Review. *Proceedings* **2022**, *82*, 1. https://doi.org/10.3390/proceedings2022082001

Published: 6 September 2022

Prof. Dr. Khalid Mahmood, *University of the Punjab, Pakistan.*
Prof. Dr. Subyantoro, M.Hum, *Universitas Ngudi Waluyo, Indonesia.*
Meida Rachmawati, S.E,.M.M,.MH, *Universitas Ngudi Waluyo, Indonesia.*
Assoc. Prof. Dr. Keoy Kay Hooi, *UCSI University, Malaysia.*
Prof. Datuk Dr. Rohana Binti Yusof, *UCSI University, Malaysia.*
Prof. Dr. Garry Tan Wei Han, *UCSI University, Malaysia.*
Prof. Dr. Chan Nee, *UCSI University, Malaysia.*

General Chairman

Dr. Nor Erlissa Abdul Aziz, *Universiti Teknologi MARA, Malaysia.*

General Co-Chairman

Ts. Inv. Dr. Mohamad Rahimi Mohamad Rosman, *Universiti Teknologi MARA, Malaysia.*

Keynote Speakers

Prof. Ts. Dr. Mohamad Noorman Masrek, *Universiti Teknologi MARA, Malaysia.*
Assoc. Prof. Dr. Sherah Kurnia, *University of Melbourne, Australia.*
Dr. Susanta Udagedara, *University of Salford, United Kingdom.*

Moderators

Assoc. Prof. Filmor J. Murillo, *Camarines Sur Polytechnic Colleges, Philippines.*
Assoc. Prof. Ts. Dr. Mohd Nasir Ismail, *Universiti Teknologi MARA, Malaysia.*
Assoc. Prof. Dr. Masitah Ahmad, *Universiti Teknologi MARA, Malaysia.*
Dr. Rudsada Kaewsaeng-On, *Prince of Songkla University, Thailand.*
Dr. Zubaidah Iberahim, *Universiti Putra Malaysia, Malaysia.*
Dr. Hafizah Mat Nawi, *Universiti Pertahanan Nasional, Malaysia.*
Dr. Suryani Awang, *Universiti Teknologi MARA, Malaysia.*
Dr. Nurazleena Ismail, *Universiti Teknologi MARA, Malaysia.*

Technical Committee

Assoc. Prof. Ts. Dr. Mohd Nasir Ismail, *Universiti Teknologi MARA, Malaysia.*
Assoc. Prof. Ts. Dr. Ghazali Osman, *Universiti Teknologi MARA, Malaysia.*
Assoc. Prof. Dr. Marziana Madah Marzuki, *Universiti Teknologi MARA, Malaysia.*
Assoc. Prof. Dr. Masitah Ahmad, *Universiti Teknologi MARA, Malaysia.*
Asst. Prof. Ts Chit Su Mon, *UCSI University, Malaysia.*
Ho Meng Chuan, *UCSI University, Malaysia.*
Asst. Prof. Dr Shyala Islam, *UCSI University, Malaysia.*
Dr Tanaraj Krishna, *Quest International University, Malaysia.*
Dr. Maria Teresa J. Manzano, *Camarines Sur Polytechnic Colleges, Philippines.*
Dr. Jocelyn O. Jintalan, *Camarines Sur Polytechnic Colleges, Philippines.*
Dr. Norel Peter M. Illo, *Camarines Sur Polytechnic Colleges, Philippines.*
Dr. Crezel B. Obrero, *Camarines Sur Polytechnic Colleges, Philippines.*
Dr. Challiz D. Omorog, *Camarines Sur Polytechnic Colleges, Philippines.*
Dr. Ian P. Benitez, *Camarines Sur Polytechnic Colleges, Philippines.*
Henki Bayu Seta, *Universitas Pembangunan Nasional Veteran Jakarta, Indonesia.*
Tjahjanto, *Universitas Pembangunan Nasional Veteran Jakarta, Indonesia.*
Erly Krisnanik, *Universitas Pembangunan Nasional Veteran Jakarta, Indonesia.*
M. Octaviano Pratama, *Universitas Pembangunan Nasional Veteran Jakarta, Indonesia.*
Mayanda Mega Santoni, *Universitas Pembangunan Nasional Veteran Jakarta, Indonesia.*

Editors

Ts. Inv. Dr. Mohamad Rahimi Mohamad Rosman, *Universiti Teknologi MARA, Malaysia.*
Dr. Nor Erlissa Abdul Aziz, *Universiti Teknologi MARA, Malaysia.*
Assoc. Prof. Ts. Dr. Mohd Nasir Ismail, *Universiti Teknologi MARA, Malaysia.*

Assoc. Prof. Ts. Dr. Ghazali Osman, *Universiti Teknologi MARA, Malaysia.*
Ts. Dr. Khalid Abdul Wahid, *Universiti Teknologi MARA, Malaysia.*

Web Chair

Faizal Haini Fadzil, *Universiti Teknologi MARA, Malaysia.*

Parallel Session Chair

Izzatil Husna Arshad, *Universiti Teknologi MARA, Malaysia.*

Registration Chair

Khadijah Abdul Rahman, *Universiti Teknologi MARA, Malaysia.*

Corporate Office Chair

Salliza Md Radzi, *Universiti Teknologi MARA, Malaysia.*

Secretariat Chair

Siti Aishah Mokhtar, *Universiti Teknologi MARA, Malaysia.*

Proceedings Compilation Team

Ts. Inv. Dr. Mohamad Rahimi Mohamad Rosman, *Universiti Teknologi MARA, Malaysia.*
Ainatul Mardhiah Mat Nawi, *Universiti Teknologi MARA, Malaysia.*
Amira Idayu Mohd Shukry, *Universiti Teknologi MARA, Malaysia.*
Fatin Nur Zulkipli, *Universiti Teknologi MARA, Malaysia.*
Haslinda Abdul Hamid, *Universiti Teknologi MARA, Malaysia.*
Mohamad Syauqi Mohamad Arifin, *Universiti Teknologi MARA, Malaysia.*
Nik Nur Izzati Nik Rosli, *Universiti Teknologi MARA, Malaysia.*
Noor Arina Md Arifin, *Universiti Teknologi MARA, Malaysia.*
Noor Azreen Alimin, *Universiti Teknologi MARA, Malaysia.*
Nur Shaliza Sapial, *Universiti Teknologi MARA, Malaysia.*
Nurfatihah S Baharuddin, *Universiti Teknologi MARA, Malaysia.*
Siti Aishah Mokhtar, *Universiti Teknologi MARA, Malaysia.*
Zaila Idris, *Universiti Teknologi MARA, Malaysia.*
Noor Masliana Razlan, *Universiti Teknologi MARA, Malaysia.*
Hadhifah Fadhlina bt Ismail, *Universiti Teknologi MARA, Malaysia.*

proceedings

Proceeding Paper

Gadgets and Their Impact on Child Development †

Zarina Mohd Zain *, Fatin Nur Najidah Jasmani, Nurul Hadirah Haris and Suzei Mat Nurudin

Faculty of Administrative Science and Policy Studies, Universiti Teknologi MARA, Seremban 70300, Malaysia
* Correspondence: zarina752@uitm.edu.my
† Presented at the International Academic Symposium of Social Science 2022, Kota Bharu, Malaysia, 3 July 2022.

Abstract: Having gadgets for children can be beneficial, as children can be creative through mobile games or with a stimulus for their senses and imagination through some creative applications. However, the overuse of gadgets can make the children rely on them, and it also can lead to addiction if they use them without proper guidance from their parents. This paper aims to visualize the use of gadgets among children on their acceptance and the impacts on their development. A total of fifteen informants were selected among children, caretakers, parents, and teachers via face-to-face interviews. The findings show that the children accept the use of gadgets due to their usefulness and easiness. However, there were some negative impacts of gadget usage in terms of social life, health, speech delay, and children's cognitive skills, which also could affect their education in the long term.

Keywords: the gadget; child development

check for updates

Citation: Zain, Z.M.; Jasmani, F.N.N.; Haris, N.H.; Nurudin, S.M. Gadgets and Their Impact on Child Development. *Proceedings* **2022**, *82*, 6. https://doi.org/10.3390/proceedings2022082006

Academic Editor: Mohamad Rahimi Mohamad Rosman

Published: 7 September 2022

1. Introduction

Children nowadays or those who can be referred to as Generation Z, are born during the digital world and can also be described as Digital Natives [1]. Members of Generation Z are more tech savvy, and they are the ones who are going to explore new applications as if they are part of their lives. The author stated that members of Generation Z could not live a day without gadgets, as if that is one of their basic needs to get through the day. As time goes by, Generation Alpha has the potential to be occupied with gadgets since they are being influenced by the immersion of technology since birth. Hence, gadgets have been used by parents as a medium to control their children and get them to behave well in public. There are few factors that would lead to an excessive usage of gadgets such as the place where the children are growing up from either urban or rural areas. Parents always established the idea of having gadgets for their children as an obligation, parallel with modern life. However, parents forget about their responsibility to weigh the pros and cons of giving gadgets to their children as it can be harmful or beneficial depending on how they use them. Having gadgets for children can be beneficial as the children can be creative through mobile games or with a stimulus for their senses and imagination through some creative applications [2]. However, the overuse of gadgets can make their children rely on them and it also can lead to addiction as if they use them without proper guidance from the parents.

2. Past Studies on Related Subjects

This part begins with the overview of gadget usage among children, followed by discussion on effects of technology used among children.

2.1. Issue of Gadget Use among Children

Gadget usage among children is worrisome because the Malaysian Communications and Multimedia Commission (MCMC) found that 83.2 percent of Internet users are children between the ages of 5 to 17 years old [3]. To illustrate, 93 percent of Internet users are children using smartphones to access WhatsApp, Telegram, and any other applications that have a communication element. Furthermore, studies found that 75 percent of children do not get enough sleep in their daily activities, which would affect their development due to the impact of the use of gadgets [4]. In addition, findings from the Michael Cohen Group, which specializes in children's education, stated that 60 percent of parents that has a child aged 12 years old are also occupied with gadgets. To elaborate, 30 percent of them are using gadgets concurrently, while 36 percent of them did not know the long-term implication of using gadgets too much [5]. Referring to the scenario, this study aims to clarify the impact of gadgets on children's development and to identify parents' initiatives in controlling their children's usage of gadgets.

2.2. Technology and the Use of Gadgets among Children

Technology is unavoidable, and everyone must adapt to it. Technology is now the main tool in not only administration but also in education, the health sector, and even trade and businesses. In the 21st century, gadgets and the Internet are not taboo in our society. The behavior relating to the use of gadgets among children can be explained through several theories, such as the Technology Acceptance Model (TAM) and the Theory of Reasoned Action (TRA). In TAM, two elements predict the acceptance of technology, which are perceived ease of use (PEOU) and perceived usefulness (PU) [6]. PEOU is the ratio of individuals who believe that using technology will make it easy for them to do any task. PU is a comparison of individuals who believe that using technology can increase their productivity. These two elements reflect the ideas in the TRA that explain the behavior of using technology. This theory explained that a person's attitude and subjective norms influenced the intensity of the behavior of using technology. All of these lead to the level of acceptance or usage regarding the human behavior of using technology in a certain context.

Many studies related to technology focus on the use of a system approach in dealing with daily work, but few studies touch on how the use of devices affects child development. The first study, which was conducted by the original scholar for this model, Davis featured lab experiments with emails and graphics and identified that with these technologies, work can be done very quickly, easily, and effectively and lead to high job performance [7]. In addition, emails and graphics are also easy to learn, controllable, flexible, and understandable [8]. A lab experiment with a word processor, which was conducted by Davis, Bagozzi, Warshaw, and Venkatesh, found that using the processor improved performance and enhanced the effectiveness of completing the task given [9,10]. The process of learning to operate the processor is also easy and allows an individual to become skillful in handling the system. Heijden in his study on the use of website technology, identified that the information on sites is very interesting and adds value to the performance [11]. It is easy to navigate around sites, and users can obtain information very quickly. The system is also easy to learn, flexible, clear, and understandable. Another study, which involved voicemail, also identified that this technology is very important when it comes to job performance and effective decision-making. It is easy to get voice mails to do what individuals want them to do, and they are very comfortable to use. To the best of the author's knowledge, no studies in which children's use of gadgets was utilized to measure PU and PEOU have been conducted, and all the previous studies discussed the usefulness and ease of the technology, but not the impact of gadget use on children.

2.3. The Future of Gadgets for Children

Chiu [12] found that gadgets are most frequently used by people between 20 and 25 years old. However, as technology advances, the use of electronic gadgets by children is becoming a more important research topic, as some parents are very open with their children regarding the usage of gadgets. There are some benefits of gadget usage among children since technically it can improve their education because there is infinite information at their fingertips. However, gadget dependency can also arise when parents start to introduce gadgets to their children due to the rapid development of technology. Children use gadgets to do several things, such as listening to songs, online chatting, video gaming, and browsing through the Internet. In other words, children rely on gadgets to be the source of their entertainment, use them to explore information on the Internet, and spend most of their time on gadgets [13]. Therefore, if the children are left without the supervision of any adults, they tend to self-neglect, causing unwanted consequences to their vision and health.

During the Covid-19 lockdown, the usage of gadgets among children has increased [14]. Children spend hours on gadgets, watching YouTube videos and movies, listening to music, and playing games. Hence, when children use gadgets for a long period of time, they tend to be aggressive, furious, and disrespectful most of the time because they are attached to the gadgets since they did not know what else to do while being in lockdown. Nor Azah Abdul Aziz stated that children as young as one year old are being surrounded by gadgets that become their incentive to behave well in public [15]. Parents are introducing their children to gadgets, and because children as young as one year old are full of curiosity, as time goes by, they get addicted. According to Naquiah Nahar, children in Malaysia spend 19 h a week on gadgets, which is abnormal for children without commitments such as working from home [16]. In Korea, as stated by Joo and Sang, most of the Korean people use gadgets because of the features provided by the developers, and they feel that the applications in the gadgets suit their needs [17]. This leads to more gadget addiction.

2.4. *The Impacts of Excessive Usage of Gadgets among Children*

The excessive usage of gadgets among children may affect their social skills. For example, they might become introverted and prefer being alone instead of talking to their friends. Moreover, when gadgets become part of their routine, it might also affect their health, causing joint pain, back pain, and eyestrain. Gadgets also might affect children's speech development if they just focus on videos rather than communicating with their peers verbally. On a side note, the excessive usage of gadgets could also affect the way children learn, write, and read, depending on their cognitive skills. When children are occupied with gadgets, they are reluctant to learn and to understand basic knowledge in school, which in turn can affect their cognitive skills. Chiu agrees that the effects of gadget addiction can lead to emotional stress and poor international relationships, as people who are addicted to gadgets tend to be very passive and alienated from their surroundings. Besides, Andreesen et al. also discussed that those who are addicted to gadgets, especially when it comes to video games, have increased stress and self-perception problems [18]. If they win the game, then they will continue to play, but if they lose, they become angry and rebellious, which affects their communication with other people. Jap said that those who always spend their time on gadgets suffer from a lack of sleep, escapism, mood disorders, aggression, physical injuries, and addiction [19]. Commonly, the effects can be categorized into social, health, communication, and knowledge effects.

2.4.1. Effects on Socialization

According to Mildayani Suhana, the excessive usage of gadgets leads children to become anti-social and lack emotional management, so they tend to have tantrums in public or in their own home [20]. Hence, they may become introverted, have lower self-esteem, and keep away from their peers. Even during family gatherings, they keep themselves occupied with gadgets. The researcher also emphasized the importance of parents managing their children's behavior through limiting the usage of gadgets. A child who has unrestricted screen time is more likely to have trouble communicating with their peers and even family [21]. This is because when they are too occupied with gadgets, they have a limited vocabulary; hence, it may lead to stuttering because they did not practice speaking in real life with actual people [22].

2.4.2. Effects on Health

Moreover, gadget dependency among children can cause hazardous health problems. The most obvious effect is that children who depend on gadgets tend to feel lazy and weak and suffer malnutrition compared to those who do not. This is because children tend to sacrifice necessities such as sleep and food just to have their alone time playing with gadgets. As time goes by, this can affect their sleeping habits to the point where they only sleep for 3–4 h a day and can lead to insomnia and headaches. Consequently, children who do not undergo proper physical development are more prone to diseases involving the eyes, head, bones, joints, or back, in which they will experience tiredness and feel weak. Children who are too focused on their gadgets tend to suffer from sleep deprivation due to their addiction; hence, without them, they cannot go through their days like normal people. Moreover, children who are occupied with gadgets are more likely to have difficulty concentrating, which could affect their education while they are learning in school [23]. In addition, using gadgets can cause eye problems, and the blue light might negatively affect the brain and body.

2.4.3. Effects on Speech Development

A child should be able to speak starting from as young as one year old, even if it is just babbling with the help of their parents. A child that is occupied with gadgets tends to respond slowly to any questions given, as if they cannot digest the question quickly. This also may affect their vocabulary, as they only know a few words in conjunction with repetitive words in cartoons and video games. Children who spend most of their time on electronic gadgets such as smartphones, tablets, and any other handheld devices are most likely to have delays in expressive speech. To elaborate, a study conducted by a pediatrician in Toronto, Canada found that children who play with gadgets for an additional 30 min each time could face the risk of speech delay by 49%. To adduce, when a child cannot express their needs verbally, they tend to express their frustration through movement, so they tend to get violent. This can be seen when a parent refuses to give a gadget to them, as they will throw a tantrum even though it is in public, which shows that a child who is too dependent on gadgets has no control over their emotions.

2.4.4. Effects on Cognitive Skills

Cognitive skills are important for children to survive in education. If a child lacks in any cognitive skills, they will not be able to learn along with their peers. Cognitive skills can be referred to as how a child learns, writes, and gathers information. Children who are too dependent on gadgets are facing cognitive delays, as they do not communicate with their parents as often as needed [24]. A child needs their parents to practice communication skills, not gadgets; hence, this can affect their cognitive development, as their communication skills become stunted. Moreover, a child who depends on gadgets might face anxiety in which they worry about future events because they are not well-prepared and they may experience increased impulsivity [25]. The excessive use of gadgets may also lead to tantrums, fast heart rate, and shakiness. When children become too occupied with gadgets, they often cannot focus or multitask, which can cause lack of proactivity.

3. Methodology

The purpose of this research is to study the issue of gadget dependency among children as well as the impact of the excessive usage of gadgets and whether it gives a positive or negative effect towards children's development. Hence, a qualitative method was used to achieve the objectives of this study. Informants were selected and interviewed through an open-ended questions session regarding the usage of gadgets among their children or children whom they take care of. To finalize the number of informants, the saturation points were identified where the trend of responses received was almost similar, so a total of fifteen informants were selected for this study. The informants consist of people who are the parents, caretakers, and teachers of children aged 10 to 14 years old who are being exposed to gadgets and live in the Seremban 3 area. To examine the responses of the informants, this study followed the suggestions made by Creswell and Poth on the data collecting techniques for interviews [26]. The data was analyzed using thematic analysis, in which the initial step is to read the interview answers from respondents one by one. Before developing a code list, the researchers went through the text parts in the interview and the answer text from respondents. Then, texts in which respondents agreed with a code were circled. Lastly, the themes were developed. From the findings, the researchers developed the themes into levels of use, the benefits, and the impacts and finally provided the subthemes, as stated in Table 1 and Figure 1.

Table 1. The use and the impacts on gadget among the children.

Informants	The Use of Gadget *	The Benefits	The Impacts
I1 (caretaker)	Useful to get info, help to learn	Flexible	Speech delay
I2 (caretaker)	Can control playing time of the children	Controllable	Social, physical, and mental growth problem
I3 (caretaker)	Easy to monitor their study	Controllable and easy	Not being able to socialize with peers
I4 (teacher)	Easy for T&L, easy to explain	Easy to navigate the lesson plan	Attention disorder, slow cognitive development
I5 (teacher)	Help better understanding	Clear and understandable	Hinder to have high thinking skills
I6 (teacher)	Fast in sharing the info and pass the work	Easy to be used	Lazy to learns, write, and gather info
I7 (parents)	Speed up learning process	Easy to remember on how to perform the task given	Throw tantrums when separated from gadgets
I8 (parents)	Bait to control children's behavior	No cost to get info	Lazy, weak and suffer malnutrition
I9 (parents)	Can control the children seen they are close to them	Easy to monitor	Keep themselves away from peoples
I10(parents)	Easy to learn and very useful to get info	Comfortable for children	Rebellious
I11 (children)	Helpful to search info, easy to study, for leisure time	Easy to use, get info quick	Lead to insomnia and headache
I12 (children)	Easy to expose to new games or software	Not complicated and comfortable	Do not want to socialize with others
I13 (children)	Easy to do homework	Clear instruction and understandable	Has sleep deprivation
I14 (children)	Can explore more information and help to improve my quality of work	Easy to learn on how to be used or play with the gadgets	Difficult to concentrate to one thing
I15 (children)	Increase patience to learn something new via the searching	Flexible and easy to interact with	Cause addition, lessened the curiosity

* The gadget refers to handphones, tablets, laptops and personal computers.

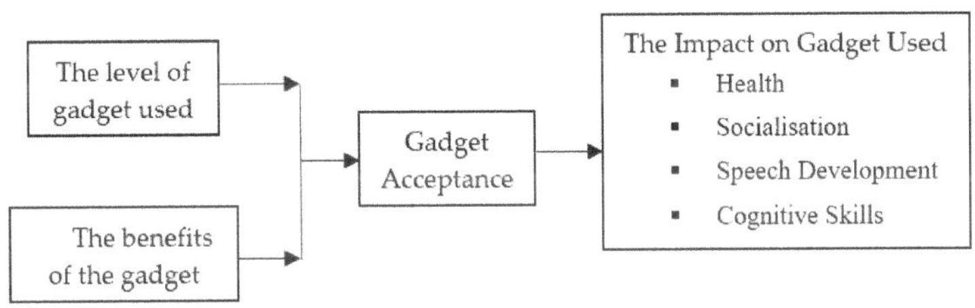

Figure 1. The framework on the gadget acceptance and its impact.

4. Results

Due to the pandemic, the whole system changed, including education and social life. The students from all levels (primary, secondary, and tertiary) who began studying online and their parents who began working from home used devices (gadgets such as handphone, tablet, laptop and personal computer) to perform their tasks. At this time, no one could resist this change. The participants' feedback showed that gadgets help them so much when it comes to their daily activities. However, most of them were aware that the use of gadgets can have a negative impact on children's development. The parents, teachers and caretakers noticed that children tended to behave aggressively if they could not play with gadgets in a day.

Our findings show that children use gadgets as their main tools not only for studying but also for leisure. They believe that these gadgets benefit them greatly, especially during the pandemic, which increased their level of gadget acceptance and caused gadgets to become important items that are always by their sides. Losing or being away from their devices can make children lose their direction. However, the response from the informants also agreed that the overuse of devices can lead to several negative implications in terms of health, social life, speech development, and cognitive skills.

It has been proven that technology has a significant impact on social life. The use of technology helps facilitate most matters, including not only administration but also education, especially during the pandemic. This was shown that most students, especially children, used gadgets for learning. The technology is easy to use, and they can explore more information in just a second. The technology is also flexible, controllable, comfortable, and understandable. However, one must consider the consequences of technology addiction when it begins to harm children's health, leading to text claw, an unhealthy level of radiation, phantom vibration syndrome, speech delay, and the poor development of cognitive skills. These could be temporary, but if parents or those who are close to children just ignore these signs, they may cause permanent effects and ruin their lives.

5. Conclusions

Technology and gadgets make life easy. This study found the concentration and impact of gadgets used among children. Basically, children use gadgets for learning activities and entertainment because they are useful and convenient. However, the impacts of gadget usage must be considered. Even though children use gadgets for online learning or to gather information, the negative impacts still apply, especially in cases of continuous use. These impacts can be seen in terms of health, social, physical, and mental development as well as cognitive skills. It is very important for parents, caretakers, and even teacher to reduce gadget dependency among children by limiting their screen time or enrolling them in extra classes so they can do other social activities. This can enhance their cognitive, affective, and psychomotor domains. At the same time, it is the responsibility of the parents and guardians who are very close to them to monitor the children to ensure they are not too occupied with gadgets. It is time for the government to re-plan and re-frame the policies related to the issue of gadget dependency among children by having a focused discussion with related ministries such as the Ministry of Education and the Ministry of Women, Family, and Community.

Author Contributions: Conceptualization, Z.M.Z., F.N.N.J. and N.H.H.; methodology, Z.M.Z. and S.M.N.; validation, Z.M.Z. and S.M.N.; formal analysis, Z.M.Z.; investigation, Z.M.Z., F.N.N.J., N.H.H. and S.M.N.; resources, Z.M.Z., F.N.N.J., N.H.H. and S.M.N.; data curation, Z.M.Z., F.N.N.J., N.H.H. and S.M.N.; writing—original draft preparation, Z.M.Z., F.N.N.J., N.H.H. and S.M.N.; writing—review and editing, Z.M.Z. and S.M.N. All authors have read and agreed to the published version of the manuscript.

Funding: This research received no external funding.

Institutional Review Board Statement: Not applicable.

Informed Consent Statement: Not applicable.

Data Availability Statement: Not applicable.

Acknowledgments: This study was supported by the Faculty of Administrative Science and Policy Studies, Universiti Teknologi Mara, Malaysia). The appraisal was carried out by the Research, Innovation, Entrepreneurship, Industry, Community and Alumni Network (RICAEN), FSPPP. The authors appreciate the valuable comments.

Conflicts of Interest: The authors declare no conflict of interest.

References

1. Che Had, M.Z.; Ab Rashid, R. A Review of Digital Skills of Malaysian English Language Teachers. *Int. J. Emerg. Technol. Learn.* **2019**, *14*, 139–145. [CrossRef]
2. Sarvananthan, D.R. *Kids and Their Gadget: What Do You Do When Your Children Have an All-Consuming "Relationship" with Their Gadgets?* Press Reader: Richmond, BC, Canada, 2015. Available online: https://www.pressreader.com/ (accessed on 17 November 2021).
3. Malaysian Communications and Multimedia Commission (MCMC). Communications Multimedia: Facts Figures. 2017. Available online: https://www.mcmc.gov.my/en/resources/statistics/ (accessed on 29 September 2020).
4. Esther, F.A. Parental Influence on Social Anxiety in Children and Adolescents: Its Assessment and Management Using Psychodrama. *Psychology* **2013**, *4*, 246–253.
5. Aziz, F. Gajet: Membantu Atau Merosakkan Perkembangan Anak Anda? Article in Hello Doktor. 2020. Available online: https://hellodoktor.com/keibubapaan/tips-keibubapaan/gajet-membantu-atau-merosakkan-perkembangananak-anda/#gref (accessed on 17 November 2021).
6. Gefen, D.; Straub, D. The relative importance of perceived ease-of-use in is adoption: A study of e-commerce adoption. *J. Assoc. Inf. Syst.* **2000**, *1*, 8. [CrossRef]
7. Davis, F.D. Perceived Usefulness, Perceived Ease of Use, And User Acceptance. *MIS Q.* **1989**, *13*, 319–340. [CrossRef]
8. Sundus, J. The impact of using gadgets on children. *J. Depress. Anxiety* **2018**, *7*, 1–3. [CrossRef]
9. Davis, F.D.; Venkatesh, V. A Critical Assessment of Potential measurement Biases in the technology Acceptance Model: Three Experiment. *Int. J. Hum.-Comput. Stud.* **1996**, *45*, 19–45. [CrossRef]
10. Davis, F.D.; Bagozzi, R.P.; Warshaw, P.R. User Acceptance of Computer technology: A Comparison of Two. *Manag. Sci.* **1989**, *35*, 982–1001. [CrossRef]
11. Heijden, H. *Using the Technology Acceptance Model to Predict Website Usage: Extensions and Empirical Test.* VU University Amsterdam, Faculty of Economics, Business Administration and Econometrics, Serie Research Memoranda. Technical Report, No. 0025. 2000. Available online: https://ideas.repec.org/p/vua/wpaper/2000-25.html (accessed on 17 November 2021).
12. Chiu, S.I. The relationship between life stress and smartphone addiction on taiwanese university student: A mediation model of learning self-Efficacy and social self-Efficacy. *Comput. Human Behav.* **2014**, *34*, 49–57. [CrossRef]
13. Daria, J.K. Internet addiction and problematic Internet use: A systematic review of clinical research. *World J. Psychiatry* **2016**, *6*, 143–176.
14. Nadeem, K.; Ahmed, N. Persistent Use of Gadgets and Internet in Lockdown Endangers Childhood. *Electron. Res. J. Soc. Sci. Humanit.* **2020**, *2*, 16–22.
15. Abdul Aziz, N.A. Gerakan Jari Yang Digunakan Untuk Aplikasi Kanak-Kanak Pada Skrin Sesentuh iPad. *Natl. Consum. Disput. Redressal Comm.* **2013**, *2*, 101–104.
16. Nahar, N. Impak Negatif Teknologi Moden Dalam Kehidupan Dan Perkembangan Kanak-Kanak Hingga Usia Remaja. *Int. J. Islamic Civiliz. Stud.* **2017**, *5*. [CrossRef]
17. Joo, J.; Sang, Y. Exploring Koreans' smartphone usage: An integrated model of the technology acceptance model and uses and gratifications theory. *Comput. Human Behav.* **2013**, *29*, 2512–2518. [CrossRef]
18. Andreassen, C.S.; Billieux, J.; Griffiths, M.D.; Kuss, D.J.; Demetrovics, Z.; Mazzoni, E.; Pallesen, S. The relationship between addictive use of social media and video games and symptoms of psychiatric disorders: A large scale cross-sectional study. *Psychol. Addict. Behav.* **2016**, *30*, 252–262. [CrossRef] [PubMed]
19. Jap, T.; Tiatri, S.; Jaya, E.S.; Suteja, M.S. The Development of Indonesian Online Game Addiction Questionaire. *PLoS ONE* **2013**, *8*, e61098.
20. Suhana, M. Influence of Gadget Usage on Children's Social-Emotional Development. *Adv. Soc. Sci. Educ. Humanit. Res.* **2017**, 224–227. [CrossRef]
21. Sodik, M.A.; Afdila, I. Pengaruh Penggunaan Gadget Pada Tumbuh Kembang Anak Usia Dini. OSF Preprints. 2018. Available online: https://www.researchgate.net/publication/328409190_Pengaruh_Penggunaan_Gadget_Pada_Tumbuh_Kembang_Anak_Usia_Dini (accessed on 17 November 2021).
22. Sari, D.N. An Analysis of the Impact of the Use of Gadget on Children's Language and Social Development. *Adv. Soc. Sci. Educ. Humanit. Res.* **2019**, *449*, 201.
23. Mappapoleonro, A.M. The Effect of Gadget Toward Early Childhood Speaking Ability. *Indones. J. Early Child. Educ. Stud.* **2018**, *2*, 86.
24. Pertiwi, L. The Use of Technology in the Cognitive Development of Early Children. *J. Psychol.* **2021**, *6*, 49. [CrossRef]
25. Sinta, S.; Ali, M.; Halida, H. Pengaruh Gadget Terhadap Perkembangan Sosial Anak Di Tk Aisyiyah Bustanul Athfal Vi. *J. Pendidik. Dan Pembelajaran Khatulistiwa* **2018**, *7*. [CrossRef]
26. Creswell, J.W.; Poth, C.N. *Qualitative Inquiry and Research Design Choosing among Five Approaches*, 4th ed.; SAGE Publications, Inc.: Thousand Oaks, CA, USA, 2018.

proceedings

Proceeding Paper

Barriers Faced by Teachers in Acclimatizing to Online Teaching †

Nurul Ashikin Izhar [1], **Yahya M. Al-dheleai** [1], **Rozniza Zaharudin** [1,*] and **Si Na Kew** [2]

1 School of Educational Studies, Universiti Sains Malaysia, USM 11800, Penang, Malaysia
2 Language Academy, Faculty of Social Science and Humanities, Universiti Teknologi Malaysia, Skudai 81310, Johor, Malaysia
* Correspondence: roz@usm.my

† Presented at the International Academic Symposium of Social Science 2022, Kota Bharu, Malaysia, 3 July 2022.

Abstract: Centered on the transformation of the norm of education from face-to-face teaching to online platforms, this article explores the intrinsic and extrinsic factors that contribute to the challenges teachers faced in delivering online lessons during the pandemic. The study adapts thematic deductive qualitative analysis approaches by using an open-ended questionnaire where the data were collected via Internet survey. The results portrayed intrinsic factors to exhibit the biggest barriers in comparison to extrinsic factors. Hence, teachers' pedagogical needs should be considered, with priority given to teachers' technological access by the education authority to improve the quality of the online teaching. This study provides insights for education authorities in addressing the challenges of online teaching, especially in terms of technological access.

Keywords: challenges; COVID-19; educational technology; online education; online teaching/learning; remote teaching; secondary school; teachers; technology in education

Citation: Izhar, N.A.; Al-dheleai, Y.M.; Zaharudin, R.; Kew, S.N. Barriers Faced by Teachers in Acclimatizing to Online Teaching. *Proceedings* **2022**, *82*, 11. https://doi.org/10.3390/proceedings2022082011

Academic Editors: Mohamad Rahimi Mohamad Rosman

Published: 7 September 2022

1. Introduction

The outbreak of the COVID-19 pandemic has had global implications for education, whereby most lessons which were traditionally conducted through face-to-face teaching are now delivered through online platforms [1]. Not only do governments worldwide face the health threats of the contagious virus, but they also need to develop practical strategies to ensure that school classes do not cease [2]. For examples, Malaysia's neighboring countries such as Indonesia, Brunei, Singapore, Timor-Leste, and the Philippines have adopted online teaching and home-based learning as their strategies to maintain the continuation of education during the pandemic. However, due to the differences in local and economic backgrounds among citizens, the countries have made changes to their online education which include adapting the curriculum, learning materials and online delivery modes [3].

The announcement of the Movement Control Order (MCO) in Malaysia has put pressure on teachers, as they need to be prepared to adapt to the online teaching and learning mode immediately. Online teaching and learning readiness are the state of teachers' preparation for online teaching and learning practices [4]. During the shift from physical classroom interaction to this practice, teachers needed to plan and deliver online lessons. Past research pointed out that the sudden and massive global adoption of online teaching and learning practices has challenged teachers' readiness for online teaching like never before [5]. However, the severity of the COVID-19 crisis and its widespread outbreak may indicate that online teaching and learning practices will soon come into the picture of the global education scenario.

2. Significance of the Study

The observation of Malaysians' education experience with the use of technology shows that blended learning is being used at a limited scale. For instance, [6] it was found that only 0.57% to 4.69% of teachers in Malaysia have used the Learning Management

System (LMS) for blended learning. This suggests that a significant number of teachers never or rarely used blended learning technology in their teaching before COVID-19 mandated online teaching and learning practices. Although online teaching and learning practices have existed in the educational context for a long time, it is still perceived as something novel when teachers have never encountered this technology in their teaching [7]. Teaching through an online teaching and learning platform becomes more difficult when teachers face infrastructure problems, limited bandwidth, unstable internet access and time challenges [8].

Nevertheless, previous research shows that technologies such as online platform teaching and learning practices were only used as a tool for teaching, unlike the current scenario during the pandemic where the entire teaching and learning process takes place online [9]. Therefore, teachers who had never used online teaching and learning practices before the school closure need to learn the basic skills to implement it. Teachers who have had experience with online teaching and learning practices need to improve their skills for more successful implementation.

Thus, there are a variety of challenges arising from the new norm of online education during the outbreak of COVID-19. Therefore, this study is essential in exploring the challenges faced by secondary schools due to the sudden shift from face-to-face teaching to online teaching and learning. The findings of this study will inform teachers, researchers, and education policy makers around the world, particularly in Malaysia, of the actions that are needed to be taken in order to address the challenges and improve the infrastructures for online teaching and learning practices and teacher readiness, not only during the pandemic but also in terms of future education plans.

3. Review of Related Studies

The current research suggests that there are several intrinsic and extrinsic factors that challenge teachers' readiness to deliver content online [10]. Intrinsic factors refer to factors derived from the teachers that challenged their readiness to conduct online teaching [11]. To put it in another way, it has to do with the fundamental and personal characteristics that are ingrained in their belief about education and technologies [12]. Concomitantly, the Self-Determination Theory [13] highlights the fact that the fundamental elements that affect the way a teacher engaged in a certain behavior were founded based on the need for relatedness, autonomy, and the need for competence. Noh et al. [10] found that the intrinsic factors that challenge teachers' willingness to teach online are teachers' innovation, information technology-specific innovation, computer self-efficacy, and technological knowledge and skills. In the context of professional development, when teachers understood the importance of intrinsic motivation, it engaged both their teaching process and their students' learning [14]. Therefore, the beliefs and thoughts of teachers influence their pedagogy. The lack of confidence has been found by Al-Maroof et al. [11] to be the factor that disconnect teachers from implementing online teaching and learning practices. This will be detrimental, as students need more support from teachers in adjusting from physical face-to-face class sessions into cyberspace. In order to cultivate self-efficacy in teachers to adapt to online teaching and learning practices, there is a need for external support such as teachers' training in adequate skills needed to teach online by the administrative [15], and the availability and stability of internet connections [9].

In this study, the external factors influencing the teachers were acknowledged as extrinsic factors. Extrinsic factors were related to external sources that affect teachers in their online teaching and learning practices. In the context of online teaching and learning practices, Wang [12] suggests extrinsic factors as elements in technology preparation to be integrated in pedagogy. Meanwhile, Rogers [7] emphasized in the theory of Diffusion of Innovation that technology does not have to take the form of a tangible product. Technology is an innovation that materializes in the form of hardware such as a laptop, software such as Google Meet, concepts, and practices such as online teaching and learning itself, or any combination of these things. Some of the extrinsic factors that affect the readiness for online teaching and learning practices are identified as administrative support [15], the role of change agents [7], and school and training institutions [12]. Botham [14] found that the policies that were developed by those in positions of higher authority (such as the Ministry of Education) ultimately led to the implementation of those policies into effective procedures (i.e., the implementation of online teaching as an alternative to ensure education continuation during school closure).

Therefore, the intrinsic and extrinsic factors affecting teachers' readiness should not be considered as a separate entity, but instead as side-by-side factors. In this research, intrinsic factors are those originating from teachers themselves that challenge their readiness to implement online teaching. Meanwhile, extrinsic factors are external factors that influence teachers' willingness.

The objective of the current study is to find out the intrinsic and extrinsic factors that affect teachers' willingness to conduct online teaching and learning practices and therefore raises the following question: What are the intrinsic and extrinsic factors that affect teachers' willingness to conduct online teaching and learning?

4. Population and Sample

The population of this study is public school teachers that have conducted online teaching and learning processes. The teachers who participated in this study were recruited on a voluntary basis, and anyone who received the link to it could answer the questionnaire. A non-probability volunteer type of sampling was suggested by Cohen, Manion, and Morrison [16] when access to a sample is difficult. The adaptation of volunteer sampling and internet survey in the current study is a feasible and appropriate alternative method due to limited freedom of movement and health aspects to be considered amidst a pandemic. Internet-based surveys have the potential to act as representative data because of its ability to collect responses from a greater number of people compared to traditional methods.

Statistical Techniques Used in the Present Study

The researchers developed an open-ended questionnaire as the research instrument to collect feedback from teachers on the challenges they faced when teaching online. The items of the instrument were validated for language and content by two experts in the field. Subsequently, the Google Form questionnaire was distributed via social media platforms such as Facebook, WhatsApp, and Telegram. A descriptive analysis of distribution was then used to present the findings derived from the samples. Consequently, frequency and percentage were then used as data presentations.

5. Data Analysis and Interpretation

At the first screening, the data were filtered by three types of sources, which were (i) sources from teachers, (ii) sources from students, and (iii) sources from parents. The data source from teachers was then further analyzed in this study. A total of 136 valid responses were collected from 116 secondary school teachers. The data were then sifted to remove unrelated information and were prepared for further analysis.

The data were analyzed using the deductive thematic analysis method in order to determine the frequency with which each particular theme emerged in the study that was being discussed. In carrying out deductive theme analysis, data were analyzed by employing a structure or framework that has been determined in advance. A deductive thematic analysis was adopted as it helps organize the data into categories and with the evaluation of the themes retrieved from literature that emerged in the scenario studied. After all the data were classified according to their respective themes, the next process was to categorize the data into intrinsic or extrinsic factors. The frequencies of occurrence and their respective percentages were then recorded.

116 teachers responded to the questionnaire, highlighting issues with facilitating online teaching, interaction and providing feedback. Of the eight issues, 33.1% are intrinsic factors and 66.9% are extrinsic factors. Among the factors, technological access is the

most prominent, with 62.5% of mentions, followed by communication self-efficacy (14%). In comparison, online lesson planning and technological skills each account for 5.9%, knowledge for 5.1%, and time, students' lesson comprehension and environment for 2.2% each. Figure 1 shows the number of occurrences and the respective percentage in the current research.

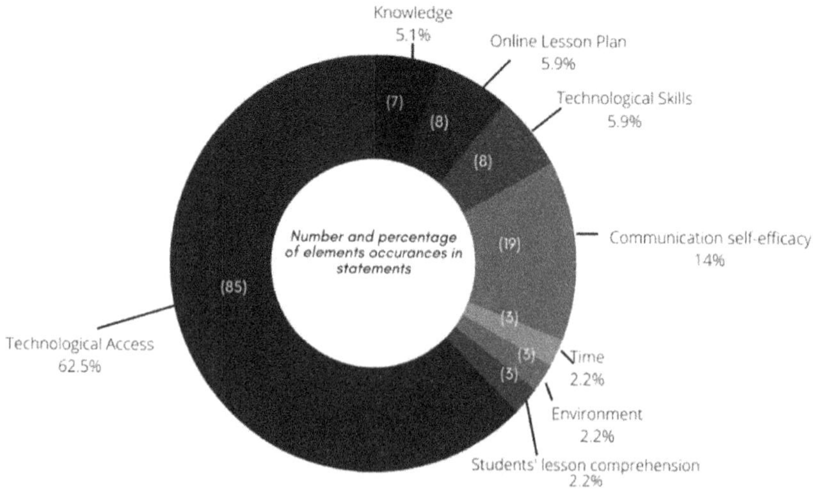

Figure 1. Number and percentage of element occurrences in statements.

6. Study Variables and Novelty

This study has a unique focus within the factors included in this study. One of the investigated intrinsic factors is that the lesson plan has not been widely investigated in previous research. The authors of this study believe that the lesson plan prepared prior to teaching is particular and important to promote the success of online teaching, as the conducted online class requires a different method and skills of preparation than that of face-to-face lesson planning. Moreover, this study investigated the extrinsic factor components, which are the teachers' working time, students online learning comprehension and the teacher's working environment. To the authors' knowledge, those components have not been widely investigated in the context of online learning. However, the extrinsic factor focus on the working environment of the teachers itself, as the class was mostly conducted in the teachers' homes hence, making it easier for the teachers to feel distracted as they are bound to their own personal responsibilities and the need to balance work-personal life may lead to the lack of time they needed to achieve work-related task and their own personal tasks.

The themes and the type of factor, sources of adaptations and the number of statements that arose in the current study are shown in Table 1.

Table 1. Pre-determined and emerging themes and their number of statements.

No	Theme	Type of Factor	Adaptation	Number of Statements
1	Communication Self-Efficacy		Martin, Budhrani, Kumar, and Ritzhaupt, (2019) [17] and Watkins et al., (2004) [18]	19
2	Online Lesson Plan		Martin, et al., (2019) [17]	8
3	Technological Skills	Intrinsic	Martin, et al., (2019) [17] and Watkins et al., (2004) [18]	8
4	Knowledge		Adapted from Pamuk, Ergun, Cakir, Yilmaz, and Ayas, (2015) [19], and Martin, et al., (2019) [18]	7
5	Technological Access	Extrinsic	Watkins, et al., (2004), [18]	85
6	Time		Martin, et al., (2019) [17]	3
7	Students' Lesson Comprehension			3
8	Environment			3

6.1. Intrinsic Factors

The intrinsic factors are Knowledge, Online Lesson Plan, Technological Skills, Communication Self-Efficacy and Time. Figure 2 shows some examples expressed by the respondents in the current study.

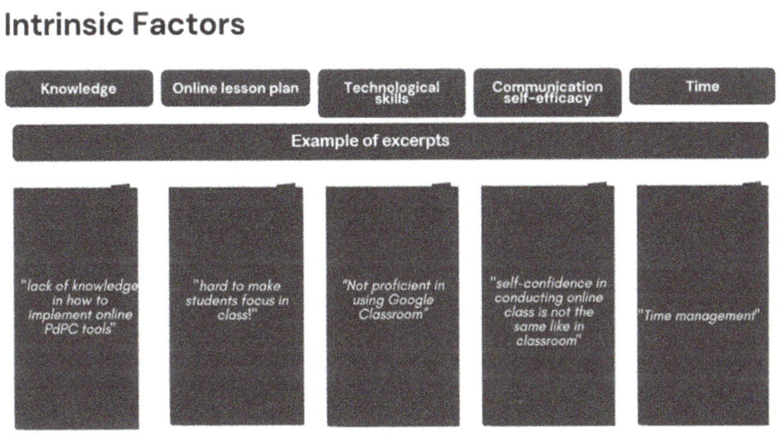

Figure 2. Intrinsic Factor.

6.1.1. Knowledge and Technological Skills

The theme of Knowledge is labelled when a teacher knows how to use online instruction, develop instructional materials, assess students' learning, and use available online resources to teach on school days. In other words, knowledge means knowing what is and how to conduct online instruction [17]. For example, Teacher 4 emphasized that it was confusing to keep track of students' attendance and assignments. This was so because in the past, teachers could ask class leaders to monitor the attendance of respective classes early in the morning and teachers would update the attendance software system upon being informed by class leaders. For online courses, it is then difficult to track students' attendance. Teacher 35 stated that she did not know how to use the online teaching tools because using the online software was new to her. According to Rogers [7], a technology, practice or idea that has existed but that no one has ever been exposed to is perceived as a new innovation.

Teaching using technology, such as online teaching, requires technological knowledge, pedagogical knowledge, content knowledge, technological pedagogical knowledge, technological content knowledge and pedagogical content knowledge [20]. Teacher 109, for example, admitted that she is not very skilled in using technology or online teaching tools. She expressed that she only knows how to use social media such as WhatsApp and Telegram for social interaction (chatting).

6.1.2. Online Lesson Plan

Online Lesson Plan demand the teachers to design the way a lesson should be conducted through an online medium [17]. Online teaching differs from physical face-face class session as the change in the medium to conduct teaching and learning call for different approach to ensure the students' focus in class. For example, Teacher 72 expressed her difficulties in "... making sure that students are ready for the lesson they are planning, excited about the lesson, and participate in the lesson until the end". Teachers need to use their knowledge and skills in online teaching to prepare a lesson that will practically make students participate in the lesson and stay until the end, taking into account the students' screen time and attention spans. Teacher 102 also mentioned that it is difficult for her to ensure students' concentration throughout the lesson. With online teaching, teachers lose some of their aptitude to manage the class, such as controlling students' behaviors. Hence, unlike physical classrooms where natural interactions can happen, teachers cannot plan or monitor students' acts when teaching online.

For this reason, the Ministry of Education (MoE) and United Nation Children's Fund (UNICEF) Malaysia, in collaboration with the Ministry of Education, has established the Komuniti Guru Digital Learning platform (Teacher Digital Learning Community), which consists of learning modules for teachers to learn and develop their skills in preparing and delivering online lessons [21]. In addition, the Ministry of Education has prepared a teaching and learning module called the PdPR module, which focuses on mitigating lessons during school closure [22]. Nevertheless, there is no recent study on the effectiveness of Komuniti Guru Digital Learning and the PdPR module for teachers' professional development.

6.1.3. Communication Self-Efficacy

The current research views Communication Self-Efficacy as teachers' self-confidence in online teaching, their expressiveness in words, voice and video, and their well-being [23]. Teachers 38 and 39 both indicated that they lacked confidence in conducting online classes because they could only use their smartphones for social communication and not for teaching purposes. Teacher 36 preferred to use pre-recorded videos for his lessons, as he had little time available for live video streaming.

6.1.4. Time

In the current research, the term time is understood as the allocation of time to complete online classes and assignments. The problems were highlighted by Teacher 114, while Teacher 87 emphasized that she is happy about students being responsible for their education. She shared that "the students did not ask much during class, but afterwards they keep texting and asking about the lesson, no matter how late it is". This highlights the need for teachers to set up a class setting in the first week of online teaching that is sustainable and supportive for both teachers and students. It can be concluded that when teaching online, teachers need to plan their time thoroughly so that their online teaching will not be disrupted.

6.2. Extrinsic Factors

The extrinsic factors are Technological Access, Students' Lesson Comprehension, and Environment, as shown in Figure 3. Figure 3 also shows some excerpts received in the current study.

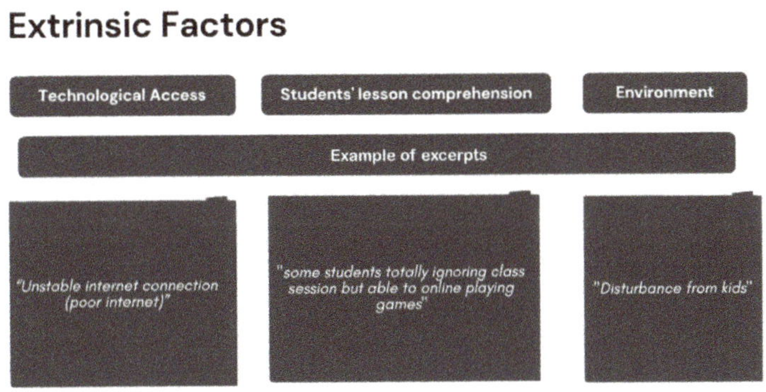

Figure 3. Extrinsic Factors.

6.2.1. Technological Access

Technological access is at the heart of the successful implementation of online teaching. Other factors, such as knowledge, online lesson planning, etc., only emerge when the access to technology is available. Thus, technological access contributes to a larger part of the problems in implementing online teaching. Online teaching requires sufficient data coverage to conduct live streaming and share teaching materials. Furthermore, sufficient data coverage also contributes to the stability and the strength of internet connection, and most importantly, the availability of devices like laptops, smartphones, and tablets for online classes. For example, Teachers 2, 11, 12, 15, 17 and 29 mentioned that they had problems with internet connections that kept disconnecting during lessons.

The slow internet coverage is not only due to the internet capacity subscribed by the teachers prior to their lessons, but is also influenced by geographical location. Zhang, Wang, Yang, and Wang [24] mentioned that some of the problems in implementing online teaching in China are due to geographical location, whereby rural and sub-rural areas have little to no internet access compared to urban areas. In the current study, Teacher 67 said, "I live in a fairly rural area. The internet coverage here is slow with only one to two bars (the indicator of internet coverage) and I have to find a better place to get internet". In addition, teachers who depend on mobile phone data coverage shoulder a bigger financial burden than those who use home Wi-Fi. Teacher 114 mentioned having to upgrade his mobile data coverage subscription to be able to do live streaming.

If the teachers have excellent technological access but the students do not have the same, it also makes online teaching difficult to conduct. Teacher 90 reported that her students only have a limited number of devices which they have to share among their siblings. Some even sacrificed their learning opportunities because other siblings were sitting for important exams and had to attend online classes to prepare for exams. Teacher 26 also mentioned that one of his students did not have access to the lessons because the only devices available in the household belonged to the parents, and his parents had to use them for their work. Therefore, technological access is a problem that needs to be solved by both teachers and students. It also shows that online teaching is not something that can be mastered quickly, especially when technological access plays a key role in its implementation.

6.2.2. Students' Lesson Comprehension

The primary purpose of a teacher's delivery of learning is his or her students' understanding of the subject matter [20]. Teacher 14 indicated that they were concerned whether students understand the materials he taught in his online classes. In face-to-face physical classes, teachers can observe their students' facial expressions and body languages. However, according to Teacher 97 and Teacher 112, students tend to not turn on their cameras during online classes. Some students do so because they are uncomfortable, and some do not have functional webcams on their devices (Teacher 113).

However, students' understanding of the subject matter may also be influenced by their behaviors. Teacher 69 expressed his disappointment; some of the students have openly said that they do not want to participate in class. This shows that the students do not take their education seriously. At the same time, teacher 81 mentioned that "students did not attend classes because they forgot, even though the timetable was announced beforehand on different platforms".

6.2.3. Environment

During Malaysian Movement Control Order (MCO), everyone was instructed to stay at their own respective home as a measure to control the spreading of COVID -19. Therefore, apart from teaching online, teachers are occupied with chores and the preparation of online teaching materials. Teacher 92, who has young children, expressed her concern with regard to concentrating during class while her children need her attention at the same time. Nonetheless, the distractions in online classes do not only come from the teacher's home environment, but also that of students. Teacher 24 mentioned that she and the other students could hear noises from one of the students during their live streaming session, which distracted the rest of the class. However, these problems can be overcome if the teacher uses the software skillfully, e.g., the mute function or good self-efficacy in communication, as mentioned earlier.

7. Recommendations

The current study only considered teachers' views on the challenges affecting their readiness and not the views of students or parents. In addition, individuals tend to emphasize the negative over the positive when presenting challenges. Therefore, it is recommended that suggestions for improvement of online education be developed from the perspective of teachers, students, parents, and other organizations such as telecommunication providers and devices (e.g., laptops, smartphones), manufacturers, etc. It is possible to overcome extrinsic issues by giving funding, sufficient training, technological help, and support within the teacher community [12].

The government could establish a task force group made up of representatives from teachers, students, and parents that reports transparently on strategies implemented to improve online teaching over time. Setting up a task force was also suggested by United Nation Educational, Scientific and Cultural Organization (UNESCO) [25], which would better facilitate online teaching during the crisis. This article also identified the critical factors that should be considered if the government intends to integrate online teaching into its strategies of using technology in education for future education plans.

It is recommended that education authority involved should provide more teachers' training regarding the use of the online teaching with the focus on technological knowledge, technological pedagogical knowledge, technological content knowledge and the communication self-efficacy as they are the necessary aspects to improve the quality of online teaching. The education authority and the telecommunication companies could work together to improve the condition of the technological access to online teaching and learning for both the teachers and students in terms of the devices and reliable, affordable internet.

Future Research Recommendations

This study was conducted qualitatively using an open-ended questionnaire. It is recommended that future researchers study the significance of the obtained factors regarding their study quantitatively, within the actual usage of online teaching and learning practice among teachers as the higher number of responses would be able to predict the situation in a more accurate way.

8. Conclusions

It can be concluded that there are intrinsic and extrinsic issues and challenges that challenge teachers' readiness to deliver online content. In the context of this study, technological access is the factor with the highest degree of occurrence as barriers faced by teachers in acclimatizing to online teaching. Despite this, the number of intrinsic factors, which include knowledge, an online lesson plan, technological skills, communication self-efficacy, and time, has been found to outnumber the number of extrinsic factors, which include only three factors (technological access, student lesson comprehension, and time). It can also be seen that even though technological access is categorized as the extrinsic factor having the highest occurrence, the intrinsic factors still form the utmost barriers. Therefore, technological access and teacher self-efficacy are essentials for the online teaching and learning implementation. Moreover, the government could establish a task force from teachers, students, and parents in addition to education experts that reports transparently on strategies implemented to improve online teaching over time.

Hence, in an educational system that is structured and centralized, concern should still be focused on teachers' pedagogical needs in order to gain a better understanding of the updated scenario of technology integration in education. The only constant in technology is advancement, hence there will be several areas that will keep on changing and challenges that will need to be overcome in education. This article also identified the critical factors that should be considered if the government intends to integrate online teaching into its strategies of using technology in education for future education plans.

Author Contributions: Conceptualization, N.A.I., Y.M.A.-d. and R.Z.; methodology, N.A.I.; data curation, N.A.I. and Y.M.A.-d.; writing—original draft preparation, N.A.I., Y.M.A.-d. and R.Z.; writing—review and editing, S.N.K.; visualization, N.A.I.; supervision, Y.M.A.-d.; project administration, N.A.I. All authors have read and agreed to the published version of the manuscript.

Funding: This research received no external funding.

Institutional Review Board Statement: The study was conducted in accordance with the Declaration of Helsinki and approved by the Human Research Ethics Committee USM (HREC) of Universiti Sains Malaysia (USM) (USM/JEPeM/COVID19-19 and date of approval 17 May 2020).

Informed Consent Statement: Informed consent was obtained from all subjects involved in the study.

Data Availability Statement: Not applicable.

Acknowledgments: The authors would like to acknowledge Human Research Ethics Committee USM (HREC) for the ethical clearance for the study and all participants that were volunteering in current research.

Conflicts of Interest: The authors declare that they have no conflict of interest.

References

1. Izhar, N.A.; Al-Dheleai, Y.M.; Ishak, N.A. Education continuation strategies during COVID-19 in Malaysia. *Int. J. Acad. Res. Bus. Soc. Sci.* **2021**, *11*, 1423–1436. [CrossRef]
2. Al Salman, S.; Alkathiri, M.; Bawaneh, A. School off, learning on: Identification of preference and challenges among school students towards distance learning during COVID19 outbreak. *Int. J. Lifelong Educ.* **2021**, *40*, 53–71. [CrossRef]
3. UNESCO. COVID-19 education response policy. *Int. Organ.* **2020**, *4*, 326–327.
4. Martin, F.; Budhrani, K.; Wang, C. Examining faculty perception of their readiness to teach online. *Online Learn. J* **2019**, *23*, 97–119. [CrossRef]
5. Cutri, R.M.; Mena, J.; Whiting, E.F. Faculty readiness for online crisis teaching: Transitioning to online teaching during the COVID-19 Pandemic. *Eur. J. Teach. Educ.* **2020**, *43*, 523–541. [CrossRef]
6. Awang, H.; Aji, Z.M.; Yaakob, M.F.M.; Osman, W.R.S.; Mukminin, A.; Habibi, A. Teachers' intention to continue using virtual learning environment (VLE): Malaysian context. *J. Technol. Sci. Educ.* **2018**, *8*, 439–452. [CrossRef]
7. Rogers, E.M. *Diffusion of Innovation*, 3rd ed.; Free Press: New York, NY, USA, 1983.
8. Ramli, N.; Saleh, S. Frogvle application in science teaching in secondary schools in North Malaysia: Teachers' Perspective. *Educ. Sci.* **2019**, *9*, 262. [CrossRef]
9. Izhar, N.A.; Al-Dheleai, Y.M.; Na, K.S. Teaching in the Time of COVID -19: The Challenges Faced by Teachers in Initiating Online Class Sessions. *Int. J. Acad. Res. Bus. Soc. Sci.* **2021**, *1*, 1294–1306. [CrossRef]

10. Noh, N.M.; Abdullah, N.; Teck, W.K.; Hamzah, M. Cultivating blended learning in teaching and learning: Teachers' intrinsic and extrinsic readiness in Malaysia. *Int. J. Acad. Res. Progress. Educ. Dev.* **2019**, *8*, 257–265. [CrossRef]
11. Al-Maroof, R.S.; Alhumaid, K.; Salloum, S. The continuous intention to use E-Learning, from two different perspectives. *Educ. Sci.* **2021**, *11*, 6. [CrossRef]
12. Wang, Y. In-service teachers' perceptions of technology integration and practices in a Japanese university context. *JALT Call J.* **2021**, *17*, 45–71. [CrossRef]
13. Ryan, R.M.; Deci, E.L. Self-determination theory and the facilitation of intrinsic motivation, social development, and well-being. *Am. Psychol.* **2000**, *55*, 68–78. [CrossRef]
14. Botham, K.A. An analysis of the factors that affect engagement of higher education teachers with an institutional professional development scheme. *Innov. Educ. Teach. Int.* **2017**, *55*, 176–189. [CrossRef]
15. Paetsch, J.; Drechsel, B. Factors influencing pre-service teachers' intention to use digital learning materials: A study conducted during the COVID-19 pandemic in Germany. *Front. Psychol.* **2021**, *12*, 1–12. [CrossRef] [PubMed]
16. Cohen, L.; Manion, L.; Morrison, K. *Research Methods in Education*, 8th ed.; Routledge, Taylor & Francis Group: New York, NY, USA, 2018.
17. Martin, F.; Budhrani, K.; Kumar, S.; Ritzhaupt, A. Award-winning faculty online teaching practices: Roles and competencies. *Online Learn. J.* **2019**, *23*, 184–205. [CrossRef]
18. Watkins, R.; Leigh, D.; Triner, D. Assessing readiness for E-Learning. *Perform. Improv. E-Learn.* **2004**, *17*, 66–79. [CrossRef]
19. Pamuk, S.; Ergun, M.; Cakir, R.; Yilmaz, H.B.; Ayas, C. Exploring relationships among TPACK components and development of the TPACK instrument. *Educ. Inf. Technol.* **2015**, *20*, 241–263. [CrossRef]
20. Mishra, P.; Koehler, M.J. Technological Pedagogical Content Knowledge: A Framework for Teacher Knowledge. *Teach. Coll. Rec.* **2006**, *108*, 1017–1054. [CrossRef]
21. Digital Learning Teachers Community with MOE (Komuniti Guru Digitial Learning Bersama KPM). Available online: https://sites.google.com/moe-dl.edu.my/ldp-digital-learning/home?authuser=1 (accessed on 4 May 2020).
22. Teaching & Learning at Home Manual. Available online: https://www.moe.gov.my/pemberitahuan/pengumuman/manual-pdpdr (accessed on 18 January 2021).
23. Hung, M.L. Teacher readiness for online learning: Scale development and teacher perceptions. *Comput. Educ.* **2016**, *94*, 120–133. [CrossRef]
24. Zhang, W.; Wang, Y.; Yang, L.; Wang, C. Suspending classes without stopping learning: China's education emergency management policy in the COVID-19 outbreak. *J. Risk Financ. Manag.* **2020**, *13*, 55. [CrossRef]
25. UNESCO. *UNESCO COVID-19 Education Response Education Sector Issue Notes: Distance Learning Strategies in Response to COVID-19 School Closures*; UNESCO: Paris, France, 2020; pp. 1–8.

proceedings

Proceeding Paper

TPACK Readiness among English-Language Lecturers for Open Distance Learning (ODL) Adoption in a Malaysian Public University †

Azlan Ab Aziz [1], Ameiruel Azwan Ab Aziz [2,*], Nur Aqilah Norwahi [2], Nor Afifa Nordin [2], Nuramirah Zaini [2], Sumayyah Shaidin [2] and Sheik Badrul Hisham Jamil Azhar [2]

1 Fakulti Sains Komputer & Matematik, Universiti Teknologi MARA Cawangan Melaka Kampus Jasin, Melaka 77300, Malaysia
2 Akademi Pengajian Bahasa, Universiti Teknologi MARA Cawangan Melaka Kampus Alor Gajah, Melaka 78000, Malaysia
* Correspondence: ameirul@uitm.edu.my
† Presented at the International Academic Symposium of Social Science 2022, Kota Bharu, Malaysia, 3 July 2022.

Abstract: In the wake of the coronavirus disease 2019 (COVID-19) crisis, the education domain all around the globe suddenly shifted to online/open distance learning (ODL) platforms. Despite decades of technological advancement in the fabric of education, it is progressively important to understand English-language lecturers' technological pedagogical content knowledge (TPACK) readiness to adopt ODL. Generally, language-learning classes are conducted in the traditional physical setting. This is of great importance to investigate lecturers' ability to integrate technology into teaching and learning, as it is a significant factor that affects online learning success. This quantitative case study aims to explore the TPACK readiness of English-language lecturers in one public university in Malaysia. Based on the TPACK model, an online survey was designed and administered to collect data among 143 English-language lecturers during the movement control order. Despite an immediate compulsory ODL deployment, the results indicate that the English-language lecturers' TPACK readiness is progressive, and they are acquiescent and fairly receptive towards the ODL with efforts to overcome ODL challenges and stay positive. This, in turn, contributed to students' positive performance. It is hoped that such information can offer some insights for ensuring positive impacts in tertiary educational teaching and learning practice.

Keywords: COVID-19; English lecturers; open distance learning; self-efficacy; technological pedagogical content knowledge

Citation: Aziz, A.A.; Aziz, A.A.A.; Norwahi, N.A.; Nordin, N.A.; Zaini, N.; Shaidin, S.; Azhar, S.B.H.J. TPACK Readiness among English- Language Lecturers for Open Distance Learning (ODL) Adoption in a Malaysian Public University. *Proceedings* **2022**, *82*, 12. https://doi.org/10.3390/proceedings2022082012

Academic Editor: Mohamad Rahimi Mohamad Rosman

Published: 8 September 2022

1. Introduction

The outbreak of the COVID-19 pandemic has presented unprecedented challenges to educational practices all around the world while severely crippling people's routines and devastating the worldwide economy. The sudden outbreak left many teachers, academicians, and lecturers with no choice but to hastily accept, embrace, and adopt online/open distance learning (ODL) as the solitary method for teaching and learning. English-language lecturers who previously enjoyed and engaged in face-to-face (F2F) interactions like other language classes generally had to move with the tide. On 18 March 2020, the Malaysian government issued an initiative in the form of the movement control order (MCO) to prevent the spread of the outbreak in the nation [1]. In response to the MCO, which closed down retail and stunted the economy, the F2F interactions in schools and higher education institutions were suspended, and teachers were instructed that all these educational practices and activities to be conducted online.

More than half of the education population worldwide has been adversely affected by this sudden technological transition [2], and the COVID-19 pandemic propelled the whole world into unexpected online teaching and learning (T&L) dramatically. The transition from F2F to a heavily relied-on technology was no longer a matter of choice but rather imposed and the only viable alternative for future education. Thus, the teachers, lecturers, and educators alike needed to abruptly adapt, adopt, and equip themselves with information and communication technology (ICT) literacy to conduct classes through ODL. ICT application in higher education has remained a major concern at the global level for decades [3]. The shift from traditionally proposed lessons is now revolutionised to accommodate the new online approach. The day's main course is no longer facilitated by solely the beliefs and skills of English-language lecturers on T&L before but also includes unique online student responses, learner–instructor relationships, online dynamics and management, and suitable online teaching approaches on top of the online T&L activities. This has become a critical factor in meeting the needs of the learner [4] as well as ensuring the online learning process is still as exciting and motivational as the F2F [3] and continues to be a significant predictor of the learners' academic accomplishment [5].

Implementing effective ODL is vital to ensure the attainment of educational and institutional goals in a higher education setting. The emergency ODL has varied challenges for both the lecturers and the learners alike. Emergency ODL implementation posed a myriad of challenges such as lack of online technologies exposure, adequate facilities, online resources, and limited technological knowledge in utilising the appropriate educational technologies available, which may affect the learners' learning process [6]. Many previous studies on online learning touched on ever so many dimensions such as readiness, pedagogy, technology, support, faculty, ethics, planning, evaluation, management, and institution [7]. Online learning readiness is considered one of the essential dimensions being studied.

Past research has indicated that learners usually show better academic performance online than in traditional settings [8]. However, the same cannot be said for lecturers. Lecturers' digital proficiencies are found to be inadequate and more so in the lesson plan formulation [9] even when, in the normal sense, they are digitally literate and can conduct online classes but are concomitantly unsuccessful at delivering online content efficiently [10]. In the same context, this certainly raises the need to evaluate lecturers' technological abilities to remain successful pedagogically in the ODL setting.

TPACK model is regarded as a useful framework for describing and understanding the goals for technology use in the T&L delivery. Technology competency encompasses all critical components, including technological and pedagogical, content knowledge, skills, and attitudes [11]. Previous studies have evaluated lecturers' technological competencies focusing on their knowledge, beliefs, and adaptation [12] and investigated the technological competency regarding other TPACK determinants in various countries [13–15]. It is believed that to deliver effective instructions, English-language lecturers must acquire sufficient technological competency. Hence, this study focuses on examining the English-language lecturers' TPACK readiness for ODL adoption.

Several works of literature have examined how English as a second language (ESL) university lecturers adopted, adapted, endured, and persisted with the ODL delivery of lessons. However, only several studies have surfaced in the context of Malaysian university language lecturers. This study was designed to investigate the English language lecturers' readiness on the related technology competency via the TPACK perspective for online T&L delivery during the COVID-19 pandemic. The results of this study are estimated to contribute to understanding these university lecturers' acquired ICT literacy to deliver effective ODL and provide some insight for others to emulate. The English-language lecturers were considered for the following research questions:

1. What are the levels of TPACK of faculty members (English-language lecturers) in three southern region branches of a public university in Malaysia?

2. Is there any significant difference between faculty members' TPACK and their age and/or teaching experience?

2. Literature Review

The transition of F2F education to ODL delivery fundamentally concerns the teaching practices and administrative support central to sustaining important student engagement [4]. This phenomenon is rooted in the subject discipline and the learner characteristics with different teaching approaches and learning resources. What is almost seamlessly accomplished during the F2F setting needs careful consideration when carried out in ODL. It requires the lecturers to be adept at the available technologies, which may have been uncommon in their practice. From a common perspective, the shift from F2F to ODL is not without challenges. Far from enjoying a smooth transition, the shift from F2F to ODL, which can be attested from the various studies, was perceived as aggressive, disastrous, disruptive, and unwelcome to certain quarters [16]. Nonetheless, by understanding the challenges or unwillingness of lecturers, this study hoped better to assist the lecturers in the technology-supported pedagogical activities because they are vital to supporting and establishing change within the educational process [17].

2.1. *Technological Pedagogical Content Knowledge (TPACK) Model*

The technological pedagogical content knowledge model (TPACK) model is a valuable framework for encapsulating today's lecturers' knowledge and skill demands. It is used to explore the lecturers' effective teaching practices in a digital setting. Mishra and Koehler's TPACK framework reiterates that effective instructors have the subject knowledge, adequate teaching skills, and technology capabilities to assist their students in achieving learning goals. TPACK is seen as a framework for measuring and enhancing teachers' knowledge of integrating technology into learning and instructing processes, which adds relationship and complexity to fundamental teaching knowledge [18].

The TPACK model has been constructed in the education domain regarding the central focus of an educator's knowledge on pedagogy, content, and technology for effective teaching [19]. Pedagogical knowledge (PK) relates to the lecturers' knowledge of teaching and learning methods, practices, and strategies. Content knowledge (CK) involves the lecturers' knowledge of the subject matter, while technological knowledge (TK) involves the lecturers' knowledge of modern information, communication technology, and the Internet. In addition, the integration between pedagogy and content knowledge creates an overlapping area of pedagogical content knowledge (PCK) on the idea of strategies for teaching specific content. Integrating pedagogy and technology knowledge forms technological pedagogical knowledge (TPK). This is about the technology that supports pedagogical goals. The content and technology knowledge overlap with technological content knowledge (TCK), referring to the transformation of the subject matter through technology. Lastly, the integrated knowledge of pedagogy, content, and technology, including the PCK, TPK, and TCK, is formalised as technological pedagogical content knowledge (TPACK). Within the TPACK context, the lecturers understand the use of related technologies in effectively delivering effective content for attaining different pedagogical goals. Most of the general technology integration studies from the TPACK perspective have been carried out regarding the F2F learning setting [18]. Indeed, it is exciting and perhaps insightful to examine the technology integration and readiness involving English-language lecturers in the ODL context.

2.2. *Self-Efficacy*

Self-efficacy is considered an effective tool to predict the behaviours of individuals in performing specific tasks. It concerns how a person views his abilities and capabilities through self-reflection, internalisation, and actions. The self-efficacy theory suggests that the uncertainty or the lack of competency in oneself within his environment can transform into resistance to change [20]. An individual who performs at high levels has high self-efficacy and engages and participates more willingly than a low self-efficacy individual [20]. High self-efficacy individuals are ready to put in more exertion as necessary to fulfil a task and spend valuable time working out challenges.

Concerning this study, self-efficacy is viewed as a belief in the English-language lecturers' context reflected in their readiness and confidence in their professional capability to deliver educational activities using the appropriate technological tools, which will be an influential factor in the achievement of the educational outcomes. Knowledge of technology increases the self-efficacy of language lecturers about technology integration, and such readiness and belief are important to impact or impede the lecturers' ability to create an effective technology-friendly learning environment [17,21–23].

The literature suggests that educators' self-belief in integrating technology effectively is a significant factor in determining its use and implementation in their teaching practice [24]. Even though self-efficacy of technology integration is seen as a credible indicator of the educators' ability and willingness to adopt technology, other research discoveries also showed that technology integration for impactful instructional activities remains among the greatest challenges facing educators of today [3,17,23,24]. As a result, there is a need to explore the relationship between the educators' self-efficacy relating to technology and their ability to use and integrate technology into their teaching practice effectively.

Lastly, the researchers consider self-efficacy as a significant controlling factor that affects the adjustment of an individual's behaviour and thinking. Furthermore, it is hoped that the development of professional self-efficacy is a promising way to stimulate and motivate educators to accomplish effective educational activities with the support of appropriate technology.

3. Methodology

This study employed a quantitative approach using a survey instrument in Google Form that was developed and distributed online. The data collection method used 5-point Likert scale items to measure the four self-efficacy constructs of "Technology Knowledge" (TK), "Technological Content Knowledge" (TCK), "Technological Pedagogical Knowledge" (TPK), and "Technological Pedagogical Content Knowledge" (TPACK). The instrument consisted of three demographic items (age, position, and years of teaching experience). The other four sections contained eleven items on the construct TK, five items on the construct TCK, ten items on the construct TPK, and seven items on the construct TPACK, with the 5-point scale being 1 = strongly disagree, 2 = disagree, 3 = neutral, 4 = agree, and 5 = strongly agree. These items were developed in the frame of "can do" statements in which self-efficacy is perceived as a capability.

This study intends to measure the readiness of English-language lecturers to implement ODL during the COVID-19 pandemic condition. A total of 143 English-language lecturers were involved, and they were from three branch campuses representing three different states in the southern regions. They have a varied online experience; some stated that this was their first exposure to online teaching, while some had some exposure to online teaching. The respondents' demographic is shown in Table 1.

A descriptor for each scale or the alpha (α) value is where an α value of >0.9 is considered excellent, an α value of >0.8 is considered strong, and α value of >0.7 is considered acceptable, and an α value of >0.6 is considered reasonable [25]. Cronbach's alpha coefficient value was used to evaluate the internal consistency reliability of all the TPACK constructs. Table 2 shows that the Cronbach or the coefficient value of each construct was greater than 0.8 or 80%, indicating a solid internal consistency and reliability.

Table 1. Respondent Demographics (n = 143).

Age	Frequency	Percentage (%)	Years of Teaching	Frequency	Percentage (%)
30 or younger	12	8.4	5 or less	26	18.2
31–40	76	53.1	6 to 10	46	32.2
41–50	42	29.4	11 to 15	34	23.8
51–60	13	9.1	16 to 20	19	13.3
			21 or more	18	12.6
Position					
Lecturer	62	43.4			
Senior Lecturer	76	53.1			
Associate Professor	5	3.5			

Table 2. Reliability evaluation.

Constructs	No. of Items	CA Value	M	SD
Technology Knowledge (TK)	11	0.921	3.54	0.65
Technological Content Knowledge (TCK)	5	0.878	3.89	0.53
Technological Pedagogical Knowledge (TPK)	10	0.927	3.91	0.58
Technological Pedagogical Content Knowledge (TPACK)	7	0.914	3.85	0.59

4. Results and Discussion

The survey instrument was intended to examine the English-language lecturers' readiness to adopt the ODL approach from the TPACK point of view. Various descriptive and inferential statistical tests were used to analyse the collected data. Based on the four (4) constructs of TPACK, the English-language lecturers scored the highest in the TPK dimension (M = 3.91 SD = 0.58), followed by TCK (M = 3.89, SD = 0.53), and followed by TPACK (M = 3.85, SD = 0.59). The lowest score is the TK (M = 3.57, SD = 0.65), indicating that their age is not a significant factor contributing to the limited exposure to technological knowledge, as their focus or area of expertise is on the English language. Since the Cronbach's alpha value on all the constructs of TPACK was larger than 0.80, implying that the scales had strong reliability, it is believed that the values of the descriptive and inferential statistics for the four constructs, as shown in Table 2, are reliable and valid.

It is safe to say that these educators have minimal issues with technology fundamentals that enhance teaching and learning delivery, thus showing a progressive level of TPACK readiness for ODL. Table 3, which involves the TPACK construct, indicates that these English-language lecturers can relate and profess to know about selecting, combining, strategising, and utilising the appropriate technology for enhancing their T&L delivery in ODL. They can select and choose the suitable technology to teach the students (M = 3.98, SD = 0.64). They also are able to strategise and combine (M = 3.95, SD = 0.67) both the technologies and teaching approach as well as appropriately teach a particular lesson using the combined technology and teaching approach (M = 3.88, SD = 0.70).

However, a point worth noting in Table 4 is that even though the TK construct is the lowest, there is still the belief by the English language lecturers that they can learn technology easily (M = 3.88, SD = 0.76), and they may possess the necessary knowledge in solving their technical problems (M = 3.78, SD = 0.80). This relates to operating online ODL technologies in delivering the teaching and learning content during the pandemic. At the same time, these English-language lecturers understand the need to keep abreast with the new technologies being used or introduced in the environment (M = 3.76, SD = 0.77). This means that these lecturers are accommodating to the introduction of technologies into the T&L delivery.

Table 3. Statistics on Technological Pedagogical Content Knowledge (TPACK) construct for all respondents.

Items on Technological Pedagogical Content Knowledge (TPACK) Construct	M	SD
1. I can teach lessons that appropriately combine (the particular content), technologies and teaching approaches.	3.88	0.70
2. I can select technologies to use in my classroom that enhance what I teach, how I teach and what students learn.	3.98	0.64
3. I can use strategies that combine (the particular content), technologies and teaching approaches that I learned about in my coursework in my classroom.	3.95	0.67
4. I can display leadership in helping others to coordinate the use of (the particular content), technologies and teaching approaches at my school and/or district.	3.67	0.87
5. I can choose technologies that enhance the learning of (the particular content) for a lesson.	3.98	0.66
6. I can evaluate and select new information resources and technological innovations based on their appropriateness to specific tasks in (the particular content).	3.85	0.73
7. I can use (the particular content)-specific tools (e.g., software, simulation, environmental probes, graphing calculators, exploratory environments, Web tools) to support learning and research.	3.67	0.80

Table 4. Statistics on Technology Knowledge (TK) construct for all respondents.

Items on Technology Knowledge (TK) Construct	M	SD
1. I know how to solve my own technical problems.	3.78	0.80
2. I can learn technology easily.	3.88	0.76
3. I keep up with important new technologies.	3.76	0.77
4. I frequently play around the technology.	3.63	0.87
5. I know about a lot of different technologies.	3.34	0.94
6. I have the technical skills I need to use technology.	3.72	0.78
7. I have had sufficient opportunities to work with different technologies.	3.58	0.85
8. I can use technology tools to process data and report results.	3.72	0.84
9. I can use technology in the development of strategies for solving problems in the real world.	3.51	0.86
10. I have the ability to design web pages and use authoring software.	2.45	1.05
11. I understand the legal, ethical, cultural, and societal issues related to technology.	3.53	0.94

Meanwhile, the results in Table 5 show that the respondents in the age level of 31–40 scored the highest in all the TPACK constructs, i.e., TK (M = 3.69, SD = 0.58), TCK (M = 3.97, SD = 0.51), and TPK (M = 3.96, SD = 0.54) except TPACK. In the TPACK construct, the 13 respondents in the 51–60 age group scored the highest with M = 3.91 and SD = 0.52. This could mean that even though these very senior respondents are not that well-versed with the TK, with the lowest score of M = 3.25 and SD = 0.42, they can still relate to the technological inclusion in the other constructs (TCK, TPK, TPACK). In short, this could mean that they make up in the other areas where they lacked in TK and reach a balance.

Table 5. *t*-test Analysis by the age of Respondents on TK, TCK, TPK, and TPACK constructs (n = 143).

Technology Knowledge (TK)	N	M	SD
30 or younger	12	3.57	0.81
31–40	76	3.69	0.58
41–50	42	3.34	0.70
51–60	13	3.25	0.42
Technological Content Knowledge (TCK)	**N**	**M**	**SD**
30 or younger	12	3.96	0.51
31–40	76	3.97	0.51
41–50	42	3.78	0.55
51–60	13	3.70	0.60
Technological Pedagogical Knowledge (TPK)	**N**	**M**	**SD**
30 or younger	12	3.93	0.50
31–40	76	3.96	0.54
41–50	42	3.82	0.68
51–60	13	3.90	0.52
Technological Pedagogical Content Knowledge (TPACK)	**N**	**M**	**SD**
30 or younger	12	3.84	0.63
31–40	76	3.89	0.57
41–50	42	3.75	0.63
51–60	13	3.91	0.52

As Table 6 shows, the cumulative teaching experience or the years of teaching did not significantly influence the English-language lecturers' level of TK, TCK, TPK, and TPACK. However, the teaching experience had a significant bearing on the younger lecturers with 15 years or less of teaching experience. The lecturers with five or fewer years of teaching scored the highest (M = 3.72, SD = 0.70) in the TK construct. Again, the same group of lecturers scored the highest with M = 4.01 and SD = 0.58 in the TCK construct, while the lecturers with six to ten years of teaching scored the highest (M = 4.00, SD = 0.45) in the TPK construct. For the TPACK construct, the English-language lecturers attained the highest score with 11 to 15 years of teaching experience (M = 3.91, SD = 0.58). None of the English-language lecturers with the most years of teaching experience (16 to 20, 21, or more) achieved a high score in any of the constructs. This indicates that the years of teaching experience do not relate to technological knowledge and competency. This means that the more senior lecturers accept technology but are slightly unfamiliar if they are not exposed to or engage in technology.

Pearson's correlation was conducted to examine the relationship between other constructs and the TK and the lecturers' readiness to carry out the ODL activities effectively. The findings are shown in Table 7. Based on the Pearson's correlation figures, there was a significantly strong positive correlation between the lecturers' self-perceived TK and their PCK (rs (143) = 0.753, $p < 0.05$). This is followed by the second-highest score with a significantly positive correlation between the lecturers' self-perceived TK and TCK (rs (143) = 0.731, $p < 0.05$). The third-highest score with a significantly moderate positive relationship existed between lecturers' self-perceived TK and the TPK (rs (143) = 0.713, $p < 0.05$).

It is also beneficial to identify the construct that most significantly influenced the respondents' readiness to conduct ODL activities by conducting a multiple linear regression analysis. Table 8 shows the model of the summary. All of the independent constructs showed a mere 4.2% contribution that would influence these lecturers' readiness to be well-prepared to handle related language ODL activities. This value indicates that 4.2% of the variance in "technology readiness and adoption" (self-efficacy) can be predicted from the TPACK constructs, but that does not mean the contribution was insignificant.

Table 6. *t*-test Analysis by Teaching Experience on TK, TCK, TPK, TPACK) constructs (n = 143).

Technology Knowledge (TK)	N	M	SD
5 or less	26	3.72	0.70
6–10	46	3.64	0.56
11–15	34	3.58	0.74
16–20	19	3.45	0.41
21 or more	18	3.05	0.61
Technology Content Knowledge (TCK)	**N**	**M**	**SD**
5 or less	26	4.01	0.58
6–10	46	3.93	0.38
11–15	34	3.98	0.50
16–20	19	3.83	0.37
21 or more	18	3.50	0.81
Technology Pedagogical Knowledge (TPK)	**N**	**M**	**SD**
5 or less	26	3.90	0.57
6–10	46	4.00	0.45
11–15	34	3.95	0.62
16–20	19	3.76	0.40
21 or more	18	3.77	0.89
Technology Pedagogical Content Knowledge (TPACK)	**N**	**M**	**SD**
5 or less	26	3.89	0.66
6–10	46	3.87	0.55
11–15	34	3.91	0.58
16–20	19	3.76	0.37
21 or more	18	3.70	0.79

Table 7. Correlation between the TCK, TPK, and TPCK Constructs and the TK Construct.

		TCK	TPK	TPCK
Technology Knowledge	Pearson Correlation	0.731 **	0.713 **	0.753 **
	Sig. (2-tailed)	0.000	0.000	0.000

** Correlation is significant at the 0.01 level (2-tailed).

Table 8. Test Significance of Multiple Linear Regression Model.

R	R Square	Adjusted R Square	Std. Error of the Estimate	Sig. F Change
0.206 [a]	0.42	0.036	0.52259	0.014

[a] Dependent Variable: Overall TPACK.

Lastly, a multiple correlation analysis was conducted between all the constructs in the TPACK model, which are the TK, TCK, TPK, and TPACK. The statistics in Table 9 indicates that the value (F (1141) = 6.228, $p < 0.5$, R2 = 0.42) was significant at 0.05 level of significance. This indicated that the English-language lecturers' self-efficacy on TPACK constructs showed they are generally ready and on the right track for carrying out pedagogical activities through ODL.

Table 9. ANOVA for multiple correlations between TK, TCK, TPK, and TPACK.

Model	Sum of Squares	df	Mean Square	F	Sig.
Regression	1.701	1	1.701	6.228	0.014 [b]
Residual	38.507	141	0.273		
Total	40.208	142			

[b] Predictors: (Constant), Years of Teaching.

5. Limitation of Study

The study identified some limitations. Firstly, the English-language lecturers' responses on their technological competencies may be biased, as they were self-reported. It is useful for future studies to undertake different available approaches to discover and understand the lecturers' TPACK competencies. Secondly, this study concentrated only on the TPACK model to gauge the technological proficiencies of the English-language lecturers to understand their technological readiness and knowledge. Therefore, it is also significant to reinforce the findings of this study by assessing other indicators to examine the English language lecturers' competencies in their actual technology-supported teaching. Lastly, evaluating these lecturers' online technological, content, and pedagogical competencies is beneficial over time, as this study was carried out during the COVID-19 pandemic outbreak and the initial digital transition. Over time, changes may need to be observed; thus, a longitudinal study can be conducted to strengthen the findings further. Such a move can also provide an avenue critical to the elevation and enhancement of online instructional practices.

6. Conclusions

As the COVID-19 pandemic terrified the world, the education system worldwide was not spared from the pandemic's rage. The devastating virus suspended all the F2F interactions of educational institutions and forced the pedagogical practices online. However, this involuntary and unexpected move somehow affected the continuous delivery of F2F lessons. The shift to ODL is undoubtedly not without concerns and challenges, as many works of literature demonstrate. This quantitative study examined the impact of TPACK readiness on the English language lecturers' technology competencies to handle the ODL instructional activities. In today's technology advancement era, language lecturers must be competent in digital delivery and related technologies to ensure an engaging and uninterrupted lesson, albeit digitally. The findings showed that the level of TPACK readiness among English-language lectures is progressive, as they are becoming more willing and receptive to the ODL with efforts to overcome any ODL challenges and stay positive. It is also noted that age is not a barrier to the use of technology, as older lecturers can relate to the technology. The more senior lecturers also were not apprehensive about technology if they were exposed to it properly. The teaching experience years did not influence the rejection of technology, but it was the other way around, showing that these more senior lecturers are open and willing to technology with the proper exposure. Overall, the English-language lecturers are considered on the right track regarding their readiness and continued effort towards delivering their pedagogical activities on the ODL platform.

Author Contributions: Methodology, A.A.A. and A.A.A.A.; conceived and designed the analysis, A.A.A.; writing—original draft preparation, A.A.A.; conceptualisation, A.A.A.A.; writing—reviewing and editing, A.A.A.A.; data collection and curation, N.A.N. (Nur Aqilah Norwahi), N.A.N. (Nor Afifa Nordin), N.Z. and S.S.; visualisation, S.B.H.J.A.; investigation, S.B.H.J.A. All authors have read and agreed to the published version of the manuscript.

Funding: This research was funded by Skim Geran Dalaman TEJA (GDT2022/1-2).

Institutional Review Board Statement: Not applicable.

Informed Consent Statement: Informed consent was obtained from all subjects involved in the study.

Data Availability Statement: Not applicable.

Acknowledgments: The researchers would like to thank the Research and Industrial Linkages Department of Universiti Teknologi MARA Cawangan Melaka, which provided funding under the Skim Geran Dalaman TEJA (GDT2022/1-2).

Conflicts of Interest: The authors declare no conflict of interest.

References

1. Bunyan, J. PM: Malaysia under Movement Control Order from Wed until March 31, All Shops Closed Except for Essential Services. Available online: https://www.malaymail.com/news/malaysia/2020/03/16/pm-malaysia-in-lockdown-from-wed-until-march-31-all-shops-closed-except-for/1847204 (accessed on 3 February 2022).
2. UNESCO. UNESCO Building Peace in the Minds of Men and Women. Available online: https://en.unesco.org/70years/building_peace (accessed on 3 February 2022).
3. Daniela, L.; Visvizi, A.; Gutierrez-Braojos, C.; Lytras, M. Sustainable higher education and technology-enhanced learning (TEL). *Sustainability* **2018**, *10*, 3883. [CrossRef]
4. Martins, J.; Branco, F.; Goncalves, R.; Au-Yong-Oliveira, M.; Oliveira, T.; Naranjo-Zolotov, M.; Cruz-Jesus, F. Assessing the success behind the use of education management information systems in higher education. *Telemat. Inf.* **2019**, *38*, 182–193. [CrossRef]
5. Xu, Z.; Yuan, H.; Liu, Q. Student performance prediction based on blended learning. *IEEE Trans. Educ.* **2020**, *64*, 6–73. [CrossRef]
6. Lynch, D.; Smith, R.; Provost, S.; Yeigh, T.; Turner, D. The correlation between 'teacher readiness' and student learning improvement. *Int. J. Innov. Creat. Chang.* **2017**, *3*, 1–12.
7. Torun, E.D. Online distance learning in higher education: E-learning readiness as a predictor of academic achievement. *Open Prax.* **2020**, *12*, 191–208. [CrossRef]
8. Shehzadi, S.; Nisar, Q.A.; Hussain, M.S.; Basheer, M.; Hameed, W.U.; Chaudhry, N.I. The role of digital learning toward students' satisfaction and university brand image at educational institutes of Pakistan: A post-effect of COVID-19. *Asian Educ. Dev. Stud.* **2020**, *10*, 276–294. [CrossRef]
9. Farid, S.; Ahmad, R.; Niaz, I.; Arif, M.; Shamshirband, S.; Khattak, M. Identification and prioritisation of critical issues for the promotion of e-learning in Pakistan. *Comput. Hum. Behav.* **2015**, *51*, 161–171. [CrossRef]
10. Adnan, M.; Anwar, K. Online learning amid the COVID-19 pandemic: Students' perspectives. *Online Submiss.* **2020**, *2*, 45–51. [CrossRef]
11. Voogt, J.; Fisser, P.; Good, J.; Mishra, P.; Yadav, A. Computational thinking in compulsory education: Towards an agenda for research and practice. *Educ. Inf. Technol.* **2015**, *20*, 715–728. [CrossRef]
12. Kim, M.; Xie, K.; Cheng, S.L. Building teacher competency for digital content evaluation. *Teach. Teach. Educ.* **2017**, *66*, 309–324. [CrossRef]
13. Acikgul, K.; Aslaner, R. Effects of Geogebra supported micro-teaching applications and technological pedagogical content knowledge (TPACK) game practices and the TPACK levels of prospective teachers. *Educ. Inf. Technol.* **2020**, *25*, 2023–2047. [CrossRef]
14. Castera, J.; Marre, C.; Yok, M.; Sherab, K.; Impedovo, M.; Sarapuu, T.; Armand, H. Self-reported TPACK of teacher educators. *Educ. Inf. Technol.* **2020**, *25*, 3003–3019. [CrossRef]
15. Ortega-Sanchez, D.; Gomez-Trigueros, I.M. MOOCs and NOOCs in the training of future geography and history teachers: A comparative cross-sectional study based on the TPACK model. *IEEEE Access* **2019**, *8*, 4035–4042. [CrossRef]
16. Rapanta, C.; Botturi, L.; Goodyear, P.; Guardia, L.; Koole, M. Balancing technology, pedagogy and the new normal: Postpandemic challenges for higher education. *Postdigital Sci. Educ.* **2021**, *3*, 715–742. [CrossRef]
17. Williams-Buffonge, N.A.G. COVID-19 and education: The untold story of the barriers to technology adoption from a tertiary viewpoint. *J. Educ. Pract.* **2021**, *12*, 8–18.
18. Rahmadi, I. Teachers' technology integration and distance learning adoption amidst the COVID-19 crisis: A reflection for the optimistic future. *Turk. Online J. Distance Educ.* **2021**, *22*, 26–41. [CrossRef]
19. Mishra, P.; Koehler, M.J. Technological pedagogical content knowledge: A framework for teacher knowledge. *Teach. Coll. Rec.* **2006**, *108*, 1017–1054. [CrossRef]
20. Bandura, A. Guide for constructing self-efficacy scales, in self-efficacy beliefs of adolescents, Connecticut. *Inf. Age Publ.* **2006**, *5*, 307–337.
21. Akram, H.; Yingxiu, Y.; Al-Adwan, A.; Alkhalifah, A. Technology integration in higher education during COVID-19: An assessment of online teaching competencies through Technological Pedagogical Content Knowledge model. *Front. Psycholog.* **2021**, *12*, 1–11. [CrossRef]
22. Drajati, N.; Rakerda, H.; Sulistyawati, H.; Nurkamto, J. Investigating the adoption of TPACK-21CL by English pre-service teachers in a COVID-19 teaching practicum. *Indones. J. Appl. Linguist.* **2021**, *11*, 124–133. [CrossRef]
23. Saienko, N.; Lavrysh, Y.; Lukianenko, V. The impact of educational technologies on university teachers' self-efficacy. *Int. J. Learn. Teach. Educ. Res.* **2020**, *19*, 323–336. [CrossRef]

24. Gomez, F.; Trespalacios, J.; Hsu, Y.C.; Yang, D. Exploring teachers' technology integration self-efficacy through the 2017 ISTE standards. *TechTrends* **2021**, *66*, 159–171. [CrossRef] [PubMed]
25. Tavalok, M.; Dennick, R. Making sense of Cronbach's Alpha. *Int. J. Med. Educ.* **2011**, *2*, 53–55.

proceedings

MDPI

Proceeding Paper

Measuring Willingness to Communicate in English among Malaysian Language Learners through Domains of Language Use †

Fatin Nabila Abd Razak [1,*], Ain Nadzimah Abdullah [2], Vahid Nimehchisalem [3] and Nur Ida Aniza Rusli [4]

1 Academy of Language Studies, Kampus Kuala Pilah, University Teknologi Mara Cawangan Negeri Sembilan, Kuala Pilah 72000, Malaysia
2 Faculty of Social Science and Leisure Management, School of Education, Taylor's University, Subang Jaya 47500, Malaysia
3 Department of English, Faculty of Modern Languages and Communication, Universiti Putra Malaysia, Serdang 43400, Malaysia
4 Faculty of Mathematics and Computer Science, Kampus Kuala Pilah, University Teknologi Mara Cawangan Negeri Sembilan, Kuala Pilah 72000, Malaysia
* Correspondence: fatinnrazak@uitm.edu.my
† Presented at the International Academic Symposium of Social Science 2022, Kota Bharu, Malaysia, 3 July 2022.

Abstract: In the field of second language acquisition (SLA) and language teaching, willingness to communicate (WTC), a construct of oral communication, has been extensively researched as it is considered a facilitative factor for language development. Most studies examine this construct using the quantitative method. There are fewer studies that have examined how languages are codeswitched and used interchangeably across different social domains, a common practice among Malaysian English language users. The purpose of this research was to develop and validate a WTC measuring tool for Malaysian English language learners. In the questionnaire, WTC in English was examined and determined via four language use domains—education, friendship, transaction and family. The validity of the four domain factors was tested using the two-stage approach factor analysis. The results suggest that WTC can be seen as a domain-based construct where learner social domains are contextualized. This paper aims to briefly introduce the study and presents its validation results.

Keywords: willingness to communicate; domains of language use; SLA

check for updates

Citation: Abd Razak, F.N.; Abdullah, A.N.; Nimehchisalem, V.; Rusli, N.I.A. Measuring Willingness to Communicate in English among Malaysian Language Learners through Domains of Language Use. *Proceedings* **2022**, *82*, 16. https://doi.org/10.3390/proceedings2022082016

Academic Editors: Mohamad Rahimi Mohamad Rosman

Published: 8 September 2022

Publisher's Note: MDPI stays neutral with regard to jurisdictional claims in published maps and institutional affiliations.

1. Introduction

Willingness to communicate (WTC) plays a crucial role in facilitating oral interaction among speakers. The term was coined [1] and defined as the inclination to participate in an interaction when there is an opportunity for it. This idea was shaped by other studies [2,3] who investigated the reluctance of native speakers to engage in communication.

Although various variables that influence a native speaker's WTC communication competence have been identified (for example, communication anxiety, introversion, self-efficacy, and cultural diversities), early studies on WTC of native speakers primarily focused on its link to personality attributes. Native speaker WTC has also been examined across several different communication settings as well as involving different types of recipients [4]. As a result of these personality-based studies, WTC is seen as a construct that remains mostly constant across various communicative situations. This native speaker view, or the L1 view, of WTC, however, is in contradiction to the second language speaker view, or the L2 view, posited by [5].

In the L2 context, WTC is defined as the psychological preparedness to interact in the target language when the opportunity arises. The manifestation of WTC among L2 speakers is different from L1 speakers as they generally have high oral competency and therefore the situation is rather simple, without any issues [5]. This however is not the case for L2 speakers whose oral competency levels could range from the lowest to the highest. Authors [5] believe that WTC among L2 speakers is dependent on the context as their predisposition to interact seems to be contingent on the situational condition that is different in each context. Given this, [6] included situation-bound contextual variables such as the topic, interactants, magnitude of the communicative group, and cultural setting to facilitate his research on WTC of Korean L2 speakers.

In addition, [5] also points out that there are many intergroup issues in the L2 context that have social and political ramifications. In Malaysia, for example, to ensure the unity of its multiethnic and multilingual population, Malay, the country's official language, is to be used as a language of unity. English, on the other hand, is seen as a global language—a language of international business, diplomacy, knowledge, technology etc. [7]. Because of the complex linguistic landscape of the country, an exploration of this crucial communication construct is warranted to avoid the over-generalisation of WTC in the L2 setting.

1.1. The WTC Scale

To identify the WTC levels of L2 learners, an adapted version of the WTC scale developed by [8] is usually employed. The instrument determines the WTC of language learners through their scores in four context-based situations: (1) group discussion, (2) meetings, (3) interpersonal interaction, and (4) public speaking. Additionally, it measures WTC in terms of the scores of the recipients which include strangers, acquaintances and friends. Most of the studies that have employed this instrument have been conducted in Western countries, particularly America and Canada [9–11]. Over in Asia, similar studies have been conducted in Japan [12–14] and China [4,15].

As noted by [16], these L2 studies view WTC from a monolithic perspective, the East versus the West perspective. She argues that given the pluralistic nature of Malaysia, WTC should be approached from a pluralistic viewpoint as it would be more representative of its society. Following this argument, the present study considered two variables—socio-cultural and psychological—in its investigation of WTC. These factors are particularly pertinent to an ethnically and linguistically diverse country such as Malaysia.

This study measured the WTC in English of Malaysian undergraduates in four language use domains—education, family, friendship, and transaction—as outlined by [17]. This selection was also guided by specifications stated in government documents that touched on language policy and use such as the Rahman Talib Report and Tenth Malaysia Plan. Besides official documents, media also play a role in influencing the way Malaysians think and what they express in certain language domains [18].

1.2. Language Use Domain

Studies on language use domains describe the language choices of speakers, which are determined by the individuals they are conversing with, the conversation topic, and the location of the conversation. Their language choice is further confined by the cultural norms and social expectations of their society [17,19].

Discourse on language use is of particular relevance to Malaysia given its linguistically diverse society where it is common to code switch and use more than one language when communicating in some social domains [17,20]. Researchers such as [21–25] feel that in the context of a multilingual society, the decision on which language to use in a particular language domain is determined largely by the interactants, their relationships, the discussion topic and the setting.

The domains of language use are also discussed in relation to the familiarity among the interactants. Typical interactions between typical participants in typical settings create a domain [26]. A study by Platt links the domain to the continuum of formality [27]. In tandem with this view, it is stated that the degree of formality in a domain is dependent on the outcome intended by the interactants [28]. For example, if the intention is to create an air of elitism, a formal code might be preferred while a more casual less formal code might be preferred to foster a sense of friendliness and kinship.

Another factor that affects language use choices is the social organization (the family, community, educational institution, workplace, etc.) that individuals are in [29]. Comparatively, the educational and workplace settings are likely to be more formal than the friendship and family settings.

Additionally, the government language policies also influence what language would be used in an institution and/or organisation. This is relevant for the Malaysian context where English is taught as a subject in schools. In other words, the elements of language use construct such as the discussion topic, relationship, setting, and formality are crucial to an understanding of the language choices that individuals made. For these reasons, this study aimed to focus on the education, transactional, friendship and family domains to study WTC in English among Malaysian undergraduates.

2. Method

The respondents were 540 undergraduates from a public university in Malaysia selected using proportionate quota sampling. The Statistical Package for Social Sciences Program (SPSS) version 17 (SPSS Inc., Chicago, IL, USA) was used to analyse the data collected. After the data were analysed and assessed, structural equation modelling (SEM) was utilised. The second-order model presented in Figure 1 was validated using the two-stage approach in SmartPLS 2.0.M3 software (SmartPLS2., Hamburg, Germany). Each item in the questionnaire is a reflective indicator of its domain in the model while each domain is the formative indicator of WTC in English. Hence, a complete measurement model of WTC in English for Malaysian language learners was developed.

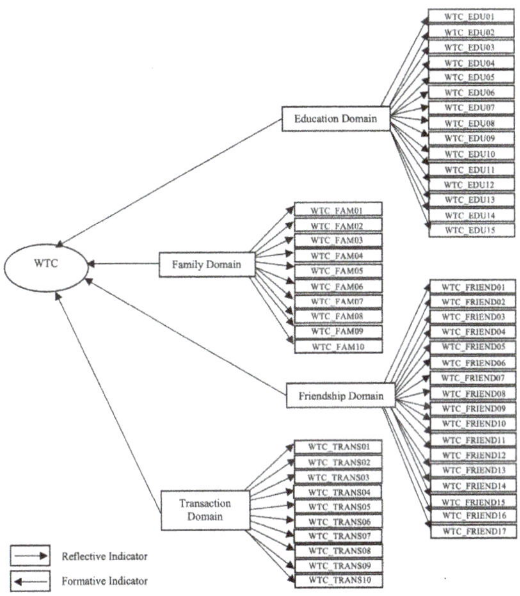

Figure 1. The measurement model of WTC.

3. Results

The measurement model of WTC in English was validated statistically. First, factor analysis of the reflective-formative hierarchical component model (HCM) of the WTC construct was conducted using a two-stage approach. For the first stage, the first or lower order components of the reflective measurement model were evaluated whereas the second stage involved the evaluation of the second or higher order components of the formative measurement model.

3.1. Reflective Measurement Model

The reflective measurement model determines the validity and reliability of the questionnaire items or indicators. For this study, two types of validity assessment were conducted: (1) convergent validity and (2) discriminant validity.

"Convergent validity is the degree to which indicators of a specific construct converge or share a high proportion of variance in common" [30]. Both factor loadings and average variance extracted (AVE) were used to assess the convergent validity of these indicators [31]. The indicator loadings, AVE, and composite reliability (CR) of the reflective construct of WTC in the four domains of language use are presented in Table 1.

Table 1. WTC in English, reflective measurement model.

Construct	Item	Loadings	AVE	CR
WTC in English in the Education Domain	WTC_EDU1	0.735	0.606	0.939
	WTC_EDU2	0.790		
	WTC_EDU3	0.741		
	WTC_EDU4	0.825		
	WTC_EDU6	0.848		
	WTC_EDU8	0.833		
	WTC_EDU9	0.715		
	WTC_EDU10	0.830		
	WTC_EDU14	0.755		
	WTC_EDU15	0.692		
WTC in English in the Family Domain	WTC_FAM1	0.905	0.825	0.979
	WTC_FAM2	0.875		
	WTC_FAM3	0.893		
	WTC_FAM4	0.922		
	WTC_FAM5	0.918		
	WTC_FAM6	0.917		
	WTC_FAM7	0.911		
	WTC_FAM8	0.913		
	WTC_FAM9	0.920		
	WTC_FAM10	0.909		
WTC in English in the Friendship Domain	WTC_FRIEND1	0.790	0.656	
	WTC_FRIEND2	0.750		
	WTC_FRIEND3	0.792		
	WTC_FRIEND4	0.818		
	WTC_FRIEND5	0.798		
	WTC_FRIEND6	0.747		
	WTC_FRIEND7	0.832		
	WTC_FRIEND8	0.850		
	WTC_FRIEND9	0.826		
	WTC_FRIEND10	0.836		
	WTC_FRIEND11	0.815		
	WTC_FRIEND12	0.783		
	WTC_FRIEND13	0.825		
	WTC_FRIEND14	0.840		
	WTC_FRIEND15	0.814		
	WTC_FRIEND16	0.829		
	WTC_FRIEND17	0.813		
WTC in English in the Transaction Domain	WTC_TRANS1	0.904	0.754	
	WTC_TRANS2	0.910		
	WTC_TRANS3	0.906		
	WTC_TRANS4	0.897		
	WTC_TRANS5	0.901		
	WTC_TRANS6	0.898		
	WTC_TRANS7	0.846		
	WTC_TRANS8	0.737		
	WTC_TRANS9	0.806		
	WTC_TRANS10	0.859		

Note: WTC_EDU5, WTC_EDU7, WTC_EDU11, WTC_EDU12, WTC_EDU13, and WTC_EDU16 were deleted due to low loadings.

Items with loadings higher than 0.708, as suggested by [31], were retained while the others were omitted. These low loading items, WTC_EDU5, 7, 12, 13 and 16, were from the education construct and are not depicted in Table 1. With their omission, the loading for WTC_EDU15 fell from 0.705 to 0.692. Therefore, this item remained since the total loading scores are high with its inclusion. Furthermore, the AVE score is more than 0.5, which is acceptable [32]. The four constructs also attained the threshold values for CR and AVE. CR scores for all four domains are greater than 0.7 while the AVE scores are greater than 0.5 after the item deletion [31]. It can be concluded that the requirements for reliability and convergent validity of the four domains of the WTC construct have been attained.

To determine the discriminant validity of the model, the Fornell–Larcker indicators were used (see Table 2 for the results). According to [33], the indicators should load more strongly within the same construct compared to the other constructs of the model. Furthermore, the average variance of each construct and its measure should be more than the variance shared between the construct and other constructs. As indicated in Table 2, all the constructs attained satisfactory discriminant validity [33], that is, the square root of AVE (diagonal) is larger than the correlations (off-diagonal) of the four reflective constructs.

Table 2. Discriminant validity using the Fornell-Larcker criterion.

	WTC_EDU	WTC_FAMILIY	WTC_FRIEND	WTC_TRANS
WTC_EDU	**0.778**			
WTC_FAMILIY	0.503	**0.908**		
WTC_FRIEND	0.758	0.689	**0.810**	
WTC_TRANS	0.676	0.612	0.752	**0.868**

Based on the results of the Fornell–Lacker discriminant validity test, it can be concluded that the construct WTC in English in the education, friendship, transaction, and family domains met the requirement of discriminant validity for the reflective measurement model.

3.2. *Formative Measurement Model*

A three-step approach that involved determining the convergent validity, addressing collinearity issues, and assessing the significance and relevance of the formative indicators was used to establish the validity of the formative measurement model for this HCM. The measurement properties of the formative construct of WTC are indicated in Table 3.

Table 3. Measurement properties of the formative construct WTC.

Construct	Item	Convergent Validity	Weight	VIF	*t*-Value Weight	*p*
WTC	Education	0.728	0.288	2.619	4.096	0.000
	Friendship		0.501	3.849	6.747	0.000
	Transaction		0.384	2.512	5.947	0.000
	Family		−0.102	2.043	1.951	0.051

Results of the redundancy analysis indicate that the path coefficient of 0.728 is bigger than 0.7. This means that the WTC formative construct has a satisfactory level of convergent validity [34]. Besides, the Varian Inflation Factor (VIF) values are all lower than the threshold value of 5 [31]. In addition, the collinearity does not reach a critical level in any of the formative construct indicators and therefore it can be used for estimating the Partial Least Square (PLS) path model.

As to the significance level, three of the WTC constructs (education, friendship and transaction) were found to be significant whereas the family domain was insignificant. By employing Hair et al.'s [31] absolute contribution method, the loading value was 0.640 and the *t*-value was 15.414. Thus, the family domain was retained. In conclusion, the results proved the validity and reliability of the reflective-formative measurement model for the WTC in English.

4. Conclusions

This research on WTC in English examines the interchangeable use of languages in the daily lives of Malaysians within and across the four language use domains of education, friendship, transaction, and family. It is quite usual for Malaysians to be able to converse in two or more languages and to codeswitch between them in their daily interaction. Furthermore, language policies (Malay as a language of unity and public domains; English as a global language and a language of knowledge and technology) dictate that one language is used more than another in specific domains. The validated WTC scale for Malaysian English language learners, which is reflective of the unique local social and linguistic landscape, is a valuable contribution to research in the area.

Author Contributions: Conceptualization, F.N.A.R. and A.N.A.; methodology, F.N.A.R.; software, F.N.A.R. and N.I.A.R.; validation, F.N.A.R. and V.N.; formal analysis, F.N.A.R.; writing—original draft preparation, F.N.A.R.; writing—review and editing, A.N.A. and V.N.; supervision, A.N.A. and V.N.; funding acquisition, F.N.A.R. and N.I.A.R. All authors have read and agreed to the published version of the manuscript.

Funding: This research was funded by Universiti Putra Malaysia under the Putra Grant—Putra Graduate Initiative (GP-IPS/2018/9631200) and UiTM Cawangan Negeri Sembilan.

Institutional Review Board Statement: Not applicable.

Informed Consent Statement: Not applicable.

Data Availability Statement: Not applicable.

Acknowledgments: Paper presented at the International Academic Symposium of Social Science 2022 by the first author.

Conflicts of Interest: The authors declare no conflict of interest.

References

1. McCroskey, J.C.; Baer, J.E. Willingness to communicate: The construct and its measurement. In Proceedings of the Annual Convention of the Speech Communication Association, Denver, CO, USA, 7–10 November 1985.
2. Burgoon, J.K. The unwillingness-to-communicate scale: Development and validation. *Commun. Monogr.* **1976**, *43*, 60–69. [CrossRef]
3. Mortensen, C.D.; Arntson, P.H.; Lustig, M. The measurement of verbal predispositions: Scale development and application. *Hum. Commun. Res.* **1977**, *3*, 146–158. [CrossRef]
4. Xie, Q.M. Willingness to Communicate in English among Secondary School Students in the Rural Chinese English as a Foreign Language (EFL) Classroom. Doctoral Dissertation, AUT University, Auckland, New Zealand, 2011.
5. MacIntyre, P.D.; Clément, R.; Dörnyei, Z.; Noels, K.A. Conceptualizing willingness to communicate in a L2: A situational model of L2 confidence and affiliation. *Mod. Lang. J.* **1998**, *82*, 545–562. [CrossRef]
6. Kang, S.J. Dynamic emergence of situational willingness to communicate in a second language. *System* **2005**, *33*, 277–292. [CrossRef]
7. Gill, S.K. Language policy in Malaysia: Reversing direction. *Lang. Policy* **2005**, *4*, 241–260. [CrossRef]
8. McCroskey, J.C. Reliability and Validity of the Willingness to Communicate Scale. *Commun. Q.* **1992**, *40*, 16–25. [CrossRef]
9. Clément, R.; Baker, S.C.; MacIntyre, P.D. Willingness to communicate in a second language. *J. Lang. Soc. Psychol.* **2003**, *22*, 190–209. [CrossRef]
10. MacIntyre, P.; Baker, S.; Clément, R.; Conrod, S. Willingness to Communicate, Social Support, and Language-learning Orientations of Immersion Students. *Stud. Second Lang. Acquis.* **2001**, *23*, 369–388. [CrossRef]
11. MacIntyre, P.; Baker, S.; Clément, R.; Donovan, L. Talking in order to Learn: Willingness to Communicate and Intensive Language Programs. *Can. Mod. Lang. Rev.* **2003**, *59*, 589–608. [CrossRef]
12. Hashimoto, Y. Motivation and Willingness to communicate as predictors of L2 use. *Second Lang. Stud.* **2002**, *20*, 29–70.
13. Yashima, T. Willingness to communicate in a second language: The Japanese EFL context. *Mod. Lang. J.* **2002**, *86*, 54–66. [CrossRef]
14. Yashima, T.; Zenuk-Nishide, L.; Shimizu, K. The influence of attitudes and affect on willingness to communicate and second language communication. *Lang. Learn.* **2004**, *54*, 119–152. [CrossRef]
15. Peng, J.E. Willingness to communicate in an L2 and integrative, motivation among college students in an intensive English language program in China. *Univ. Syd. Pap. TESOL* **2007**, *2*, 33–59.

16. Saidi, S.B. Willingness to Communicate in English among Malaysian Undergraduates: An Identity-Based Motivation Perspective. Doctoral Dissertation, University of York, Heslington, UK, 2018.
17. Fishman, J.A. *Language in Sociocultural Change*; Stanford University Press: Redwood City, CA, USA, 1972.
18. Abd Razak, F.N.; Nimehchisalem, V.; Nadzimah, A. The relationship between ethnic group affiliation (EGA) and willingness to communicate (WTC) in English among undergraduates in a public university in Malaysia. *Int. J. Appl. Linguist. Engl. Lit.* **2018**, *7*, 207–213. [CrossRef]
19. Granhemat, M. Language Choice and Use among Undergraduates at a Malaysian Public University. Doctoral Dissertation, Universiti Putra Malaysia, Seri Kembangan, Malaysia, 2015.
20. Giles, H.; Smith, P.M. *Accommodation Theory: Optimal Levels of Convergence*; Giles, H.S., Clair, R.N., Eds.; Peter Lang: New York, NY, USA, 1979; pp. 45–65.
21. Rahman, A.R.M.; Chan, S.H.; Abdullah, A.N. What determine the choice of language with friends and neighbors? The case of Malaysian university undergraduate. *Lang. India* **2008**, *8*, 1–16.
22. Lim, C.C. Language Choices of Malaysian Youth: A Case Study. Doctoral Dissertation, University of Malaya, Kuala Lumpur, Malaysia, 2008.
23. Burhanudeen, H. *Language and Social Behavior: Voices from the Malay World*; Penerbit Universiti Kebangsaan Malaysia: Bangi, Malaysia, 2006.
24. Yeh, H.; Chan, H.; Cheng, Y. Language Use in Taiwan: Language Proficiency and Domain Analysis. *J. Taiwan Norm. Univ. Humanit. Soc. Sci.* **2004**, *49*, 75–108.
25. Parasher, S.N. Mother-Tongue-English Diglossia: A Case Study of Educated Indian Bilinguals' Language Use. *Anthropol. Linguistics.* **1980**, *22*, 151–168.
26. Holmes, J. *An Introduction to Sociolinguistics*, 2nd ed.; Longman: London, UK, 2001.
27. Platt, J.T. A Model for Polyglossia and Multilingualism (With Special Reference to Singapore and Malaysia). *Lang. Soc.* **1977**, *6*, 361–378. [CrossRef]
28. Gasiorek, J.; Giles, H. Accommodating the Interactional Dynamics of Conflict Management. *Iran. J. Soc. Cult. Lang.* **2013**, *1*, 1–12.
29. Giddens, A. *Sociology*; Blackwell Publishers: Hoboken, NJ, USA, 1989.
30. Hair, J.F.; Black, W.C.; Babin, B.J.; Anderson, R.E. *Multivariate Data Analysis*, 7th ed.; Pearson: London, UK, 2010.
31. Hair, J.F.J.; Hult, G.T.M.; Ringle, C.M.; Sarstedt, M. *A Primer on Partial Least Squares Structural Equation Modeling (PLS-SEM)*, 2nd ed.; Sage: Thousand Oaks, CA, USA; London, UK, 2017.
32. Byrne, B.M. *Structural Equation Modelling with AMOS: Basic Concepts, Applications, and Programming*, 3rd ed.; Routledge: New York, NY, USA, 2016.
33. Fornell, C.; Larcker, D.F. Evaluating Structural Equation Models with Unobservable Variables and Measurement Error. *J. Mark. Res.* **1981**, *18*, 39–50. [CrossRef]
34. Klassen, R.D.; Whybark, D.C. The Impact of Environmental Technologies on Manufacturing Performance. *Acad. Manag. J.* **1999**, *42*, 599–615.

proceedings

MDPI

Proceeding Paper

Demystifying the Specific Roles and Challenges of Educational Audiologists: A Narrative Review †

Mohd Fadzil Nor Rashid [1], Tian Kar Quar [2], Nashrah Maamor [2], Foong Yen Chong [2], Mohd Normani Zakaria [1], Mazlina Che Mustafa [3,4] and Hasrul Hosshan [3,4,*]

1 Audiology Programme, School of Health Sciences, Health Campus, Universiti Sains Malaysia, Kubang Kerian 16150, Malaysia
2 Audiology Programme, Center for Rehabilitation & Special Needs, Faculty of Health Sciences, Universiti Kebangsaan Malaysia, Kuala Lumpur 50300, Malaysia
3 National Child Development Research Centre, Universiti Pendidikan Sultan Idris, Tanjung Malim 35900, Malaysia
4 Faculty of Human Development, Universiti Pendidikan Sultan Idris, Tanjung Malim 35900, Malaysia
* Correspondence: hasrul.hosshan@fpm.upsi.edu.my; Tel.: +60-1548797642

† Presented at the International Academic Symposium of Social Science 2022, Kota Bharu, Malaysia, 3 July 2022.

Abstract: Educational audiology is a critical subspecialty in the field of audiology. An educational audiologist is responsible for providing services to hearing-impaired children in educational settings. Despite their responsibilities in performing their roles, they also encounter challenges. Therefore, the aim of this literature review was to demystify the specific roles of educational audiologists and challenges faced by them in real school environments. Materials and Methods: A search of the Scopus and Web of Science (WoS) databases was conducted in February 2020, and 17 relevant articles were identified. The inclusion criteria were educational audiology studies conducted in all countries or regions, and articles written in English. Results: The review identified six main themes concerning the roles of educational audiologists, and five main themes concerning the challenges that are faced by them. The findings from this review provide essential information on current educational practices in the audiology field. Conclusions: School-age hearing-impaired children have specific needs and require specific services to be provided to them by educational audiologists. In this review, the knowledge gaps in the roles and the challenges faced by educational audiologists are revealed accordingly. As a result, several recommendations are highlighted based on the review results. Notably, school-based audiological data for evidence-based practice in school settings are required and should be the focus of future research.

Keywords: educational audiologist; roles and challenges; literature review; narrative review

Citation: Rashid, M.F.N.; Quar, T.K.; Maamor, N.; Chong, F.Y.; Zakaria, M.N.; Mustafa, M.C.; Hosshan, H. Demystifying the Specific Roles and Challenges of Educational Audiologists: A Narrative Review. *Proceedings* **2022**, *82*, 17. https://doi.org/10.3390/proceedings2022082017

Academic Editor: Mohamad Rahimi Mohamad Rosman

Published: 8 September 2022

1. Introduction

To increase the likelihood of having adequate speech and language development among infants and young children with hearing impairments, hearing assessments and respective interventions must take place as early as possible. This is further facilitated by the advancements in technology (e.g., sophisticated hearing aids, cochlear implants etc.) and the availability of services related to hearing, speech, and language specialties. Hearing-impaired children who have been appropriately receiving intervention by audiologists and speech language pathologists (SLPs) may then enroll in specific educational institutions.

Provided with adequate academic background and appropriate clinical competency licenses, audiologists are able to work in various settings, including hospitals, private practices, hearing aid companies, universities, schools, and others. Educational audiologists who work in educational settings may play important roles in supporting the educational management of students with hearing impairments. As defined by the Educational Audiology Association (EAA), educational audiologists are a member of the school multidisciplinary team who deliver a full spectrum of hearing services to all children, particularly those in educational settings [1]. To provide guidelines related to the roles of educational audiologists, suggestions have been made by three prominent associations, including the EAA [1], the British Association of Educational Audiologists (BAEA) [2], and the American Speech-Language-Hearing Association (ASHA) [3]. The roles of educational audiologists provided by these associations are summarized in Table A1. It is worth noting that there are several acts and laws in the United States that support audiological services in schools such as the Disabilities Education Act (IDEA, 2004), Rehabilitation Act 1973 (No Child Left Behind Act, 2001), and Americans with Disabilities Act (1990) [1].

It is imperative to have educational audiologists to provide essential services to hearing-impaired school children so that their educational aims can be achieved. Nevertheless, there are many factors to be considered with regard to this specialized service delivery. As such, information on the specific roles of educational audiologists and challenges faced by them in real school settings should be made available. For example, the number of certified educational audiologists must be sufficient to provide optimal services to school children. Herein, the EAA stated that an educational audiologist should provide services to 10000 registered students at local education agencies [4]. In line with this, in a survey study conducted by Richburg and Smiley, the full-time educational audiologist to the general student population ratio in the United States ranged from 1:10,000 to 1:15,000 [5]. This "good" ratio, nevertheless, may not be achieved in other countries due to a limited number of audiologists available in both clinical and educational settings.

As shown in Table A1, there are some disagreements between the three associations, in which some roles of educational audiologists are not emphasized by the others. As such, it is essential to ascertain the suggested roles and identify the main elements based on the research evidence. The challenges faced by educational audiologists in the actual school environments should be unveiled to provide essential information on improving the existing services. In this paper, we performed a literature review to demystify the specific roles and challenges faced by educational audiologists. See Table A1 here.

2. Methods

A literature review was performed using the two main journal databases: Scopus and Web of Science (WoS). Scopus is the main database of peer-reviewed literature that supports the Elsevier Research Intelligence and has been used by more than 5000 corporate, academic, and government institutions worldwide. The second database used in the review was WoS and has been established by Clarivate Analytics Company. The authors conducted the selection process focusing on a narrative review, not a systematic literature review. In line with this, Green et al. suggested that the literature search should include at least two databases related to the study in order to provide a reasonable scientific discussion [6]. Furthermore, grey literature searches through an internet web engine such as Google and Google Scholar were also conducted to identify any related publications, with the first five pages of results examined.

The review process was conducted in February 2020, and the identified keywords were used in the search process. Based on the previous studies and thesaurus, keywords which are similar and related to educational audiologist, roles, and challenges were used (Table A2). The selection process was carried out based on the method suggested by Moher et al. [7]. The inclusion criteria were educational audiology research conducted in all countries or regions, and articles in English. The exclusion criteria were articles not relevant to the topic, review articles, and those with unavailable abstract or full text. With regard to the timeline, the period between 2000 and 2020 (20 years) was selected and considered appropriate to gather related publications and information on the changes of scope of practices. Table A2 here.

The first stage of the review process was the identification. That is, 136 articles were found based on the keywords used in the search process. Of these, four duplicate articles were removed (based on EndNote X7 software, Clarivate Analytics, Philadelphia,

US). In the second phase (i.e., screening), the 132 articles were screened by the authors. Subsequently, 102 articles were excluded as some of them did not focus on educational audiology, they were non-English articles, as well as systematic or scoping review articles. In the third stage (i.e., eligibility), the full text of 30 articles was thoroughly examined by the authors. Of these, 17 articles were found to be eligible and included in this review (Figure A1). Based on the categories, a content analysis was conducted to identify specific themes. A qualitative content analysis is typically carried out for evaluating the semantic content in the text of the data [8,9]. Figure A1 here.

3. Result and Discussion

The review of the 17 articles resulted in two main categories: educational audiologists' roles and the challenges faced by educational audiologists. Within these categories, specific themes were identified. Therefore, the subsequent discussion is based on these two categories and their specific themes supported by related articles.

3.1. Roles of Educational Audiologists

Findings from the previous studies are useful to provide the essential information with regard to the roles of educational audiologists in the actual school environment. This review resulted in six main themes related to the roles of educational audiologists. These were collaboration, (re)habilitation, support personnel, audiological assessments, managing hearing instruments, and monitoring classroom acoustics. Each theme is discussed accordingly in the subsequent paragraphs.

3.1.1. Developing Professional Collaboration

The dynamic nature of the educational audiologist role requires strong collaboration with other professionals and paraprofessionals in order to deliver the optimum (evidence-based) service to hearing-impaired students in school settings [10]. In line with this, Welling and Ukstins stated that the medical and healthcare service providers and the school-based professionals should develop a bidirectional relationship [11,12]. For example, guidelines for service provision to students with otitis media with effusion (OME) must include school nurses, teachers, classroom paraprofessional staff members, speech–language pathologists, and educational audiologists. Furthermore, audiologists in clinical settings and school-based audiologists who are able to work collaboratively on behalf of students will create an efficient partnership to fulfil students' needs [11–13].

Richburg and Knickelbein conducted a study to determine whether school-based speech–language pathologists (SLPs) had access to the services of educational audiologists [14,15]. This study used a 36-item survey titled "How Can Educational Audiologists Assist Speech–Language Pathologists and Special Educators" to measure collaboration outcomes. It consists of four sections: (i) demographics; (ii) basic knowledge of audiological practices; (iii) access, benefit, and responsibility; and (iv) collaboration with educational audiologists. The findings from this study revealed that more than half of the SLPs (61.5%, n = 126) reported that they had access to audiologists, and 113 of these SLPs (89.7%) answered that they did receive the benefits. The authors then concluded that a notable avenue for building collaborative efforts was the involvement of educational audiologists in the individualized educational program (IEP) teams of students who need an auditory (re)habilitation.

Teachers are the other professionals who are involved in assisting hearing-impaired students in academic settings. In this regard, educational audiologists must find ways to shift their workload to include more time for collaboration with the classroom teachers [11,13]. In fact, around 79% of 110 special educators reported that they had benefited from collaboration with educational audiologists [16]. Likewise, Richburg and Goldberg shared their opinions regarding this collaboration and its impact on managing students with minimal hearing loss (MHL) [17]. Apart from providing the essential information about MHL and its educational consequences, they also found that the teachers felt more confident in dealing with difficulties experienced by students with MHL when working with the educational audiologists [17].

In addition to the collaborations mentioned above, educational audiologists should also collaborate with architects to ensure that new building projects comply with their acoustic specifications [14]. In fact, to ensure students with hearing problems receive appropriate interventions, the educational audiologist should also be involved with the school multidisciplinary team to design clinical and educational programs [18]. In line with this, Welling and Ukstins added that in school settings, recommendations for educational modifications and accommodations should be achieved in collaboration with the SLPs, educational audiologists, and teachers [12].

3.1.2. Providing Audiological (re)Habilitation

Loss of hearing sensitivity is the most common consequence of auditory disorders. In this regard, it is useful to have educational audiologists for providing aural (re)habilitation to school-age children, particularly if the hearing impairment cannot be medically treated. The aim of (re)habilitation is to help and ameliorate the effect of hearing loss on communication, psychological, and social aspects [19].

According to Soman and Nevins, hearing aid function assessment, appropriate hearing aid fitting, and periodic speech perception testing (using sounds, words and/or sentences) should be conducted by audiologists [20]. They emphasized the evidence-based principles of listening and spoken language (LSL) intervention and the involvement of practitioners such as SLPs, educational audiologists, and teachers when dealing with hearing-impaired school-age children. The proposed LSL interventions were: (i) learning through listening; (ii) language and literacy development; (iii) individualized, systematic, and multidimensional; (iv) interprofessional practice, and (v) family involvement [20]. They also stated the importance of collaboration and consultation with educational audiologists to maximize the auditory ability in all learning environments and the listening needs of students.

Meanwhile, according to Stach, the (re)habilitation should be carried out based on a patient-centered approach which includes communication needs, self, and family assessment of disability, selection of goals, and non-auditory needs assessment (physical abilities, psychosocial status, and financial status) [20–23]. In this regard, educational audiologists may consider this approach as a guideline for providing the respective (re)habilitation.

3.1.3. Managing Hearing Instruments

The hearing instrument technology has made a great deal of headway in the audiological field. The benefits of having hearing instruments for hearing-impaired children are obvious. By using appropriately prescribed hearing aids, assistive listening devices, or cochlear implants, they are able to improve their hearing and communication skills. Therefore, the selection and management of appropriate hearing instruments, and linking hearing instruments with the school's technological facilities are important roles for educational audiologists [13]. In this regard, the educational audiologists must perform validation assessment to ensure the effectiveness of the amplification in the actual classroom setting.

Salathiel et al. suggested that educational audiologists need to provide in-service training for hearing amplification and should be aware of students' high-tech needs [14]. The teacher's responsibilities will be more challenging when hearing-impaired students attend his/her class. Therefore, comprehensive hands-on demonstrations of hearing instrument usage by an educational audiologist would be beneficial to teachers. In line with the advancement of technology, educational audiologists have to be knowledgeable in hearing technology to meet students' communication needs. For example, high-tech-literate students may request to build connectivity between the hearing instruments and other devices such as computers, iPods, cell phones, and recreational devices. Thus, educational audiologists should have the supplementary essential information regarding frequency modulated systems and class settings, compared to audiologists working in clinical settings [24–26].

3.1.4. Monitoring Classroom Acoustics

Given the nature of room acoustics, the variables that can interfere with speech perception are background noise, signal to noise ratio, reverberation time, the distance between the talker and the listeners, and interactions among these variables [27–29]. For classroom acoustics, educational audiologists are the competent personnel to utilize these concepts and maximize learning adaptations to classroom environments [17]. This view is in agreement with those of Smaldino et al. [30], Johnson et al. [13], and Johnson [31] who suggested that educational audiologists were often the first professionals to assess classroom acoustics and student performance.

Providing the "best" room acoustics for conducive learning environments requires specific financial planning. The optimal listening in the classroom and educational audiologists' responsibilities in financial planning were discussed in-depth by Salathiel et al. [17]. For example, educational audiologists must be aware of the financial issues and work cooperatively with the local education agency staff to update technology and improve listening facilities in classrooms (including the application of acoustic modifications of the rooms). It would also be advisable to have proper planning in consultation with educational audiologists, contractors, architects, teachers, and school management team before the construction of the building.

3.2. Challenges Faced by Educational Audiologists

The educational audiologists may encounter clinical or non-clinical challenges while performing their roles in the educational settings. Their ongoing challenge is to maintain flexibility without sacrificing their professional ethics and the standard of care delivered to the children, youth, and families they serve [27]. Five themes have emerged from this review aiming to demystify the challenges faced by educational audiologists, namely the personnel shortage, high workload, limited understanding by school personnel, financial constraints, and limited collaboration. Each theme is discussed accordingly in the subsequent paragraphs.

3.2.1. Personnel Shortage

As mentioned earlier, the full-time educational audiologist to the general student population ratio in the United States ranges from 1:10,000 to 1:15,000 [5], which is in line with the EAA recommendations [4]. Nevertheless, Johnson et al. stated that most educational audiologists need to travel to multiple school districts and do not see students in a single location because of personnel shortage [13]. The issue of personnel shortage is perhaps more prominent in developing and underdeveloped countries, and research in this area is greatly warranted.

Fitzpatrick and Olds conducted a qualitative study using a semi-structured focus group interview with 28 professionals to gain understanding of the functioning of school-age children who use cochlear implants [32]. The interview findings were divided into two categories of professionals' perspectives on the functioning of school-age children and how to support the needs of children with cochlear implants. Under the category of supporting the needs of children, three themes emerged, i.e., specialized support and integration of school-based services, parental and family support, and service provider recommendations. They also found that some professionals reported that it would be an advantage to have educational audiologists because none of the participating school districts had educational audiologists in place.

3.2.2. High Workload

According to the school survey by The American Speech-Language-Hearing Association (ASHA) in 2018, the workload approach was based on all activities required and performed by educational audiologists [33]. The response rate for this survey was 41.3%, and 61.6% of respondents reported a high workload. ASHA also produced a school survey report regarding the trend in educational audiology from 2010 to 2018 (every two-year survey) and found that high workload was reported by 42% to 51% of the respondents [33]. In line with this, Richburg and Smiley shared their concerns regarding the workload of educational audiologists (which was too high) because many states did not have enough full time equivalent (FTE) audiologists working in school settings [5].

Several factors that contributed to high workloads among the educational audiologists were thoughtfully discussed by Johnson et al. [13]. Firstly, the educational audiologists need to spend more time educating the school personnel in managing students with multiple disabilities and complex communications needs. Secondly, the use of the hearing instruments that require specialized service (lacks the ease of plug and play). Thirdly, the educational audiologists need to travel to several districts to meet their students, and the travel time is a contributing factor that increases their workload. Finally, due to the success of early hearing detection and intervention (EHDI) programs and inclusion agendas, the educational audiologists are also required to support students in the general education settings.

While many studies found the workload of educational audiologists to be high due to several factors [5,13,34], a contradictory outcome was reported by Blood and colleagues [34]. That is, in their survey study involving 332 members of the EAA, job burnout was assessed using the Maslach Burnout Inventory (MBI) [34]. It was then found that the job burnout among educational audiologists was low compared to other educators' normative samples [35]. In particular, only 16% of the participants reported high burnout (the other 84% scored in the average and low burnout ranges) [34]. Nevertheless, the authors stated that their findings must be interpreted with caution due to a very low response rate. Other than this, factors such as the backgrounds of samples, the FTE ratio of audiologist to general student population, the type of questionnaire used, and the availability of adequate support personnel may also influence the study outcomes. Further studies are warranted to shed light on this area of educational audiology.

3.2.3. Limited Understanding by School Personnel and Financial Constraints

In this review, only one article emphasized that some school personnel believed that educational audiologists were diagnosticians [13]. Consequently, this may affect the educational audiologists' roles, especially for collaboration and (re)habilitation tasks. In the worst situation, this may impede the effort to add more positions for educational audiologists and school personnel may suggest eliminating a few tasks such as consultation, collaboration, and counselling. Furthermore, this article also stated that financial constraints may cause the elimination of an educational audiologist position when he/she retires or leaves the position. That would not only affect the career path of educational audiologists, but the students' outcomes may also be jeopardized.

3.2.4. Limited Collaboration

As mentioned earlier, collaboration plays an important role in expanding audiological services in school settings. In the survey study carried out by Knickelbein and Richburg involving SLPs, collaboration was not widely achieved [16]. Some respondents expressed a lack of satisfaction with the services provided by educational audiologists because the information provided was already known. Other respondents reported that the information provided was insufficient, particularly in the preparation of the student's IEP [13].

4. Limitation of the Review

In this paper, a narrative review was employed to understand the roles of educational audiologists and the challenges faced by them in academic settings. Since different countries have different educational regulations and laws, educational audiology practice may differ across the countries. This factor, nevertheless, was not highlighted in this review. Additionally, this review included eligible articles regardless of the study type. Perhaps better review outcomes would be obtained if the study type is categorized with more specific discussions.

5. Conclusions

Audiologists should provide continuous support to hearing-impaired children. As such, school-age hearing-impaired children have specific needs and require specific services from educational audiologists. This paper provides a literature review of the specific roles of educational audiologists and the challenges faced by them in real educational settings. Six main themes related to roles, including collaboration, (re)habilitation, support personnel, audiological assessments, managing hearing instruments, and monitoring classroom acoustics, were identified and discussed accordingly. This information may serve as a guideline for audiologists working in academic settings. For challenges faced by educational audiologists, five themes (personnel shortage, high workload, limited understanding by school personnel, financial constraints, and limited collaboration) were identified and discussed as intended.

Several recommendations are highlighted based on the review results. Notably, school-based audiological data for evidence-based practice in school settings are required and should be the focus of future research.

Author Contributions: Conceptualization, M.F.N.R., T.K.Q. and N.M.; methodology, M.F.N.R.; T.K.Q. and F.Y.C.; formal analysis, M.F.N.R. and T.K.Q.; investigation, M.F.N.R.; resources, M.F.N.R. and H.H.; data preparation, M.F.N.R.; writing—original draft preparation, M.F.N.R.; writing—review and editing, M.F.N.R., T.K.Q., M.N.Z., M.C.M. and H.H.; visualization, H.H.; supervision, T.K.Q. and N.M.; project administration, M.F.N.R. All authors have read and agreed to the published version of the manuscript.

Funding: This research received no external funding.

Institutional Review Board Statement: Not applicable.

Informed Consent Statement: Not applicable.

Data Availability Statement: Not applicable.

Acknowledgments: Special thanks to School of Health Sciences, Universiti Sains Malaysia for the financial assistance.

Conflicts of Interest: The authors declare no conflict of interest.

Appendix A

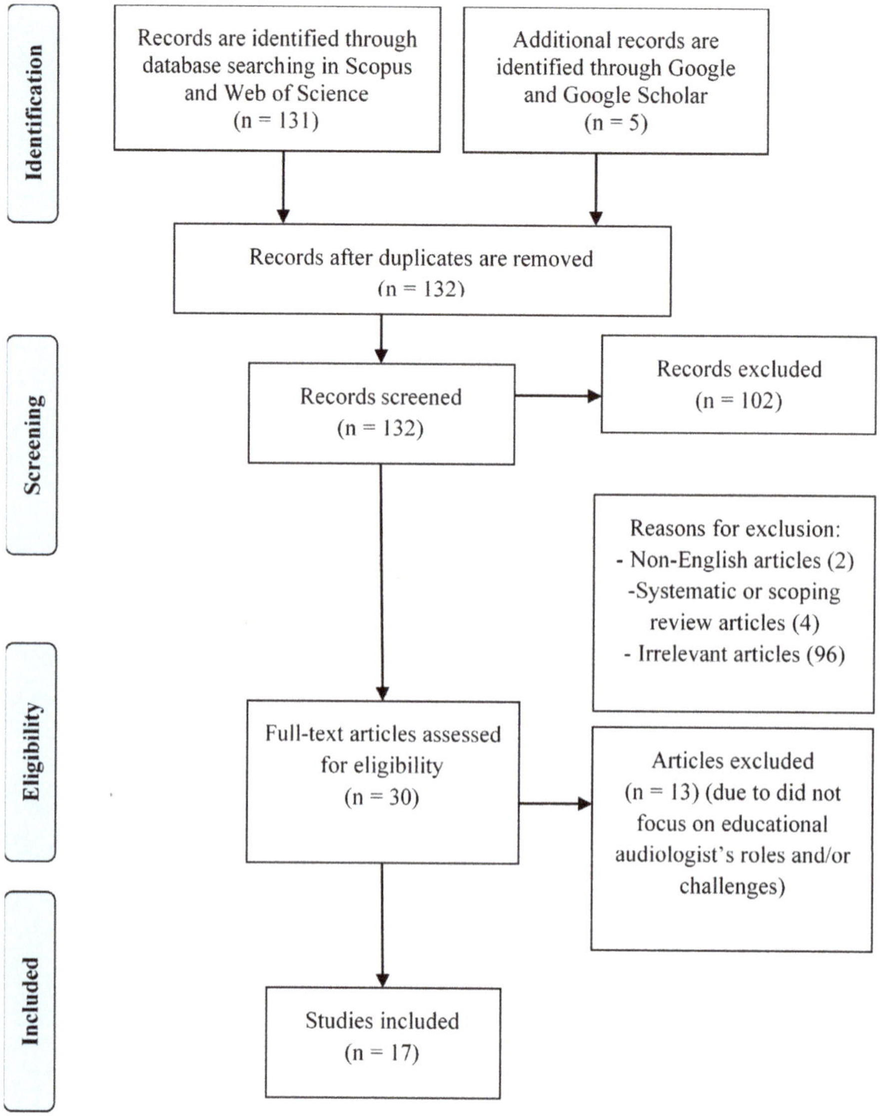

Figure A1. The flow chart of the literature review.

Appendix B

Table A1. A summary of educational audiologists' roles according to Educational Audiology Association (EAA, 2019), American Speech-Language-Hearing Association (ASHA, 2002) and British Association of Educational Audiologists (BAEA, 2016).

Educational Audiologists' Role	EAA	ASHA	BAEA
Identification of hearing problems and referrals	√	√	
Hearing assessment and classroom acoustic measurement	√	√	√
Evaluating, fitting and managing amplification	√	√	√
Facilitating and/or providing support for (re)habilitation	√	√	√
Providing counselling and training to teachers, parents and students	√	√	√
Organising program for hearing loss prevention	√	√	
Education management (collaboration, consultation etc)	√	√	√
Demonstrate understanding of education law		√	
Involvement in early hearing detection and intervention (EHDI) program		√	
Serve as advocate and community resource liaison	√	√	√
Comply with continuing education requirement		√	√

Table A2. The search string used for the literature review process.

Databases	Keywords Used
Scopus (117)	TITLE-ABS-KEY (educational AND audiology) OR TITLE-(audiology AND in AND school) OR (educational AND audiologist) OR (audiologist AND in AND school) OR (pediatric AND audiologist) AND (role) OR (task) OR (responsibility) OR (function) OR (duty) OR (job) OR (contribution) AND (challenge) OR (problem) OR (trouble) OR (obstacle) OR (issue) OR (difficulty) OR (trouble) AND DOCTYPE (ar OR re) 2000–2020
Web of Science (14)	(TS = ((“educational audiologist”) OR (“audiology in school”) OR (“educational audiologist”) OR (“audiologist in school”) OR (“pediatric audiologist”) AND (role OR task OR responsibility OR function OR duty OR job OR contribution) AND (challenge OR problem OR trouble OR obstacle OR issue OR difficulty OR trouble)) AND Language: (English) AND Document Types: (Article) Indexes = SCI-EXPANDED, SSCI, A&HCI, CPCI-S, CPCI-SSH, BKCI-S, BKCI-SSH, ESCI Timespan = 2000–2020

References

1. Educational Audiology Association. Educational Audiology Scope of Practice. Available online: www.edaud.org/pdf/scope-of-practice.pdf (accessed on 26 February 2020).
2. Webster, G. The Role of the Educational Audiologist. Available online: http://www.educational-audiologists.org.uk/index.php (accessed on 25 February 2020).
3. American Speech Language Hearing Association. Guidelines for Audiology Service Provision in and for Schools (Guidelines). Available online: www.asha.org/policy (accessed on 15 February 2020).
4. Educational Audiology Association. Recommended Professional Practices for Educational Audiology. Available online: https://edaud.org/educational-audiologist-role-defined/ (accessed on 26 February 2020).
5. Richburg, C.; Smiley, D.F. The “State” of Educational Audiology Revisited. *J. Educ. Audiol.* **2009**, *15*, 63–73.
6. Green, B.N.; Johnson, C.D.; Adams, A. Writing Narrative Literature Reviews For Peer-Reviewed Journals: Secrets Of The Trade. *J. Chiropr. Med.* **2006**, *5*, 101–117. [CrossRef]
7. Moher, D.; Liberati, A.; Tetzlaff, J.; Altman, D.G. Preferred Reporting Items for Systematic Reviews and Meta-Analyses: The Prisma Statement. *Ann. Intern. Med.* **2009**, *151*, 264–269. [CrossRef] [PubMed]
8. Knudsen, L.V.; Laplante-Levesque, A.; Jones, L.; Preminger, J.E.; Nielsen, C.; Lunner, T.; Hickson, L.; Naylor, G.; Kramer, S.E. Conducting Qualitative research in audiology: A tutorial. *Int. J. Audiol.* **2012**, *51*, 83–92. [CrossRef] [PubMed]
9. Hosshan, H. Perspectives of Teachers on Supporting the Participation of Students with Learning Disabilities in Inclusive Secondary Schools: A Qualitative Study. *Asia Pac. J. Educ. Educ.* **2022**, *37*, 47–60.

10. Lee, W.Y.; Tan, J.T.A.; Kok, K. The Educational Experiences of Deaf Students in Ipoh, Malaysia. *J. Pendidik. Bitara UPSI* **2021**, *14*, 9–17.
11. Masnan, A.H.; Mustafa, M.C.; Hosshan, H. New preschool teachers and implementation of inclusive classes issues in Malaysia. *Int. J. Acad. Res. Bus. Soc. Sci.* **2017**, *7*, 644–652. [CrossRef]
12. Welling, D.R.; Ukstins, C.A. Otitis Media: Beyond the Examining Room. *Pediatr. Clin.* **2018**, *65*, 105–123.
13. Johnson, C.D.; Cannon, L.; Oyler, A.; Seaton, J.; Smiley, D.; Spangler, C. Shift Happens: Evolving Practices in School-Based Audiology. *J. Educ. Audio.* **2014**, *20*, 48–62.
14. Salathiel, K.; Steele, J.; Edwards, D. Advocating for Optimal Listening in the Classroom. *Sem. Hear.* **2010**, *31*, 241–251. [CrossRef]
15. Richburg, C.M.; Knickelbein, B.A. Educational Audiologists: Their Access, Benefit, and Collaborative Assistance to Speech-Language Pathologists in Schools. *Lang Speech Hear. Serv. Sch.* **2011**, *42*, 444–460. [CrossRef]
16. Knickelbein, B.A.; Richburg, C.M. Special Educators' Perspectives on the Services and Benefits of Educational Audiologists. *Commun. Disord. Q* **2012**, *34*, 17–28. [CrossRef]
17. Richburg, C.M.; Goldberg, L.R. Teachers' Perceptions About Minimal Hearing Loss: A Role for Educational Audiologists. *Commun. Disord. Q* **2005**, *27*, 4–19. [CrossRef]
18. Schery, T.K.; Peters, M.L. Developing Auditory Learning in Children with Cochlear Implants. *Top Lang Disord.* **2003**, *23*, 4–15. [CrossRef]
19. Hull, R.H. *Introduction to Aural Rehabilitation: Serving Children and Adults with Hearing Loss*; Plural Publishing: San Diego, CA, USA, 2019.
20. Soman, U.; Nevins, M.E. Guiding Principles and Essential Practices of Listening and Spoken Language Intervention in the School-Age Years. *Top Lang Disord.* **2018**, *38*, 202–224. [CrossRef]
21. Ibrahim, H.; Mokhshein, S.E.; Anal, A.; Jabar, S.A. Assessment for children with special educational needs. *J. Pendidik. Bitara UPSI* **2019**, *7*, 9–15.
22. Wai Leng, A.P.; Hui- Shen, C.L.; Dhamotharan, M.; Mustafa, M.C. Preschool teachers' beliefs and classroom practices of child-centred learning at private preschools in central region, Malaysia. *Southeast Asia Early Child. J.* **2021**, *10*, 69–83.
23. Stach, B. *Clinical Audiology: An Introduction*; Nelson Education: San Diego, CA, USA, 2008.
24. Mckay, S. Management of Young Children with Unilateral Hearing Loss. *Volta Rev.* **2006**, *106*, 299–319. [CrossRef]
25. Richburg, C.M.; Imhoff, L. Survey of Hearing Screeners: Training and Protocols Used in Two Distinct School Systems. *J. Educ. Audio.* **2007**, *14*, 31–46.
26. Berg, A.L.; Ip, S.C.; Hurst, M.; Herb, A. Cochlear Implants in Young Children: Informed Consent as a Process and Current Practices. *Am. J. Audiol.* **2007**, *16*, 13–28. [CrossRef]
27. Johnson, C.D.; Seaton, J.B. *Educational Audiology Handbook*, 2nd ed.; Delmar Cengage Learning: Clifton Park, NY, USA, 2012.
28. Tang, T.P.; Mcpherson, B.; Yuen, K.C.; Wong, L.L.; Lee, J.S. Auditory Neuropathy/Auditory Dys-Synchrony in School Children with Hearing Loss: Frequency of Occurrence. *Int. J. Pediatr. Otorhinolaryngol.* **2003**, *68*, 175–183. [CrossRef] [PubMed]
29. Smaldino, J.; Crandell, C.; Kreisman, B.; John, A.; Kreisman, N. Room Acoustics and Auditory Rehabilitation Technology. In *Handbook of Clinical Audiology*; Katz, J., Ed.; Lippincott Williams & Wilkins: Philadelphia, PA, USA, 2009.
30. Smaldino, J.J.; Doggett, F.; Thunder, T. The Complimentary Roles of Audiologists and Acoustic Consultants in Solving Classroom Acoustic Problems. *Sem. Hear.* **2004**, *25*, 179–188. [CrossRef]
31. Johnson, C.D. Making a Case for Classroom Listening Assessment. *Sem. Hear.* **2010**, *31*, 177–187. [CrossRef]
32. Fitzpatrick, E.M.; Olds, J. Practitioners' Perspectives on the Functioning of School-Age Children with Cochlear Implants. *Cochlear Implant. Int.* **2015**, *16*, 9–23. [CrossRef]
33. American Speech Language Hearing Association. Schools Survey Report: Trends in Educational Audiology, 2010–2018. Available online: www.asha.org/research/memberdata/schoolssurvey/ (accessed on 23 February 2020).
34. Blood, I.M.; Cohen, L.; Blood, G.W. Job Burnout, Geographic Location, and Social Interaction among Educational Audiologists. *Percept Mot Ski.* **2007**, *105*, 1203–1208. [CrossRef] [PubMed]
35. Maslach, C.; Jackson, S.E.; Leiter, M.P. *Maslach Burnout Inventory*, 3rd ed.; Consulting Psychologists Press: Palo Alto, CA, USA, 1996.

proceedings

Proceeding Paper

Adaptation of Teleaudiology Approach in Undergraduate Clinical Examinations: Lesson Learned †

Mohd Fadzil Nor Rashid *, Wan Najibah Wan Mohamad, Mahamad Almyzan Awang and Mohd Normani Zakaria

Audiology Programme, School of Health Sciences, Health Campus, Universiti Sains Malaysia, Kota Bharu 16150, Malaysia
* Correspondence: fadzilnor@usm.my; Tel.: +60-97677692
† Presented at the International Academic Symposium of Social Science 2022, Kota Bharu, Malaysia, 3 July 2022.

Abstract: USM's undergraduate audiology program began in 2005 and lasted eight semesters. Advances in information and communications technology (ICT) and the global COVID-19 epidemic are promoting telehealth in clinical teaching. This study used Krumm's teleaudiology model for eight pediatric face-to-face clinical audiology examinations with final-year clinical students. Observation, Internet connectivity, and audiologist satisfaction confirmed the model's feasibility. Lack of ICT resources and staff caused technical issues in most sessions. Internet speed was faster than early estimates. Live observation was more satisfying than recordings. We successfully implemented a trial version of an adapted teleaudiology approach that may be used to observe audiology clinical examinations. With minor changes, this approach can also be used for clinical observation in the future, particularly during the COVID-19 pandemic.

Keywords: teleaudiology; audiology; COVID-19; clinical examination

Citation: Rashid, M.F.N.; Mohamad, W.N.W.; Awang, M.A.; Zakaria, M.N. Adaptation of Teleaudiology Approach in Undergraduate Clinical Examinations: Lesson Learned. *Proceedings* **2022**, *82*, 18. https://doi.org/10.3390/proceedings2022082018

Academic Editors: Mohamad Rahimi Mohamad Rosman

Published: 8 September 2022

1. Introduction

Audiology is a health discipline that investigates hearing, balance, and related diseases. It is dedicated to the care of impaired people of all ages and backgrounds. According to Katz, audiology is concerned with the intersection of science and art [1]. Aside from the objective evaluation of tasks, audiology provides emotional and social support to people impacted. An audiologist is a professional that specializes in identifying, evaluating, and treating people with hearing loss. The hearing care professional is also involved in a multidisciplinary team of speech therapists, ear, nose and throat specialists, deaf instructors, engineers, pediatricians, psychologists, and occupational therapists.

A career in audiology is related to academic expectations, similar to the expansion of the audiology field. First, most audiologists hold a Bachelor's degree before pursuing a Master's or Ph.D. Universiti Kebangsaan Malaysia, Universiti Islam Antarabangsa Malaysia, and Universiti Sains Malaysia are the only three universities that provide undergraduate audiology programs in Malaysia [2]. A clinical audiologist in a private hospital or clinic and a dispensing audiologist in various hearing aid centers are possible job options for audiology graduates. Despite the increase in audiology graduates, audiology services require more attention in special education than other sectors such as clinical or private hearing aid centers [3]. Moreover, the significant number of special education students can lead to inadequate audiological care.

Since students must apply their knowledge when caring for patients, the connection between the lecturer's room and the clinical scene is critical. It covered clinical teaching foundations, effective clinical teaching tactics, adjusting clinical teaching techniques to individual situations, and clinical evaluation for audiology students. Aside from continuous clinical assessments, audiology students must complete all clinical hours and pass the audiological clinical examination to demonstrate clinical proficiency. For example, the

Malaysian Qualifications Agency stated that a Bachelor's degree in audiology required 350 clinical hours with at least 200 direct student–patient contacts [4]. Meanwhile, final year students are evaluated on their clinical competency in dealing with new or follow-up patients in pediatrics and adults.

Following the expansion of ICT and the global COVID-19 pandemic, clinical training in higher education has lately changed to conform to new normal and standard operating procedures (SOP). As a result, telemedicine and telehealth approaches are becoming increasingly popular in clinical teaching [5–7]. Telehealth as a clinical teaching approach at universities has many benefits. Novak can also be used for professional collaboration, as the students learned excellent communication skills and expanded their expertise in diverse teams through this project [8].

The word teleaudiology has been widely used in research papers and publications to describe audiological services delivered via telehealth. Teleaudiology uses ICT to provide audiological services and information to clients [9], and it is classified into screening, diagnostic, and intervention services [10]. American Speech-Language-Hearing Association (ASHA) identified three teleaudiology service delivery methods: synchronous, asynchronous, and hybrid [11]. The synchronous method entails an audiologist and a client communicating through video or audio. Asynchronous refers to sending images or data to an audiologist for interpretation. A hybrid is a mix of these two approaches. These approaches are implemented directly with clients or with facilitator support.

Teleaudiology was incorporated into clinical instruction for undergraduate audiology students at the Universiti Sains Malaysia (USM) Health Campus in 2021. First, we provide online clinical knowledge and theories for lectures, clinical case studies, and problem-based learning. During the COVID-19 epidemic, clinical students did not attend clinical practice as anticipated at the start of the semester due to a rigorous movement restriction order (MCO) and decreased patient numbers at USM Audiology Clinic, notably for small children and seniors. To guarantee that clinical students could apply their knowledge, they were required to perform pure tone audiometry using the Audsim Flex audiometer simulator (audstudent.com, Hollywood, FL, USA) on the clinic's personal computer (PC). Alternatively, the students were taught how to use Chrome Remote Desktop Version 1.5 (Google, Mountain View, CA, USA) to control the audiology clinic PC from their hostel or house.

As previously stated, clinical evaluation is critical in measuring clinical student competence. For example, we administered an online Oriented Structured Clinical Exam (OSCE) to examine their knowledge and clinical abilities. We recently conducted face-to-face clinical audiology exams for our final year clinical student, modifying Krumm's Teleaudiology Model [12]. As such, the purpose of this study was to determine the feasibility of an adapted teleaudiology model for pediatric clinical assessment with the following specific objectives:

i. To design a trial version of the adapted teleaudiology approach for observation in clinical audiology examinations.
ii. To measure Internet connectivity and make comparisons between two distinct types of connections and geographical areas.
iii. To assess clinical audiologists' satisfaction with the audio-visual quality of recorded sessions.

2. Materials and Methods

Efforts are currently being made to maximize the USM Audiological Clinic's audiological resources, include teleaudiology in clinical education, and establish its practical feasibility. Instead, the teleaudiology model was modified to limit the testing room to one student and one patient. Moreover, each clinical step is visible to other lecturers and clinicians. We were able to link the audiological clinical examination to the principal investigator (PI), who was around 17 km away from the clinic. In total, four laptops and a personal computer running Windows 10 were employed in this study: the Acer Aspire A515-56 for Laptop A and B (Acer Inc., New Taipei City, Taiwan), the Asus ZenBook Flip UX360UAK for Laptop C (ASUSTek Computer Inc., Taipei, Taiwan), the Asus X450C Series for Laptop D (ASUSTek Computer Inc., Taipei, Taiwan), and the Lenovo c40 for PC (Lenovo Group Limited, Hong Kong, China).

2.1. Description of ICT Setup and Background of the Study

Equipment and ICT stability are crucial for long-term teleaudiology services feasibility, as this approach aims to deliver a reliable clinical service similar to face-to-face. This study optimized clinic facilities without upgrading ICT or audiology systems. The clinical coordinator's Cisco Webex Meeting (https://usm-cmr.webex.com (accessed on 25 July 2021), Ver. 41.7.7.9) was shared with participants of laptops A, B, and D. The student clinical (tester) has to log into Webex using laptop A using their USM email. Technical support utilized TeamViewer Version 15.19.5 (TeamViewer AG, Göppingen, Germany) to remotely control the Panasonic HC-V550 video camera (Panasonic Holdings Corporation, Osaka, Japan) and digital web camera from a PC in the observation room (free version, 64 bit). The technical support will change the camera input for Webex as asked by PI. We employed a video camera to record the candidates' threshold searching on the audiometer (audiogram). An external digital web camera focused on the tympanometer, supporting the examiner in qualitative and quantitative tympanogram interpretation. Laptop A's built-in camera also records candidates' interactions with the child's parents.

Additionally, laptop B is located in the testing room to monitor the child's response during the audiological examinations. In general, hearing evaluations in pediatric cases may vary according to developmental age and may include distraction testing, visual reinforcement audiometry, and play audiometry. As a result, it is critical to monitor the child's response during the audiological exams. Moreover, this laptop is connected via TeamViewer to laptop D (PI) to assist the investigator in determining the Internet connection. The clinical coordinator records all sessions and uploads them to the cloud for educational purposes and to accomplish the third objective of this project. Furthermore, as noted previously, the personal computer (observation room) was connected to laptop A (test room), allowing additional examiners or lecturers to observe all clinical procedures and outcomes as displayed on laptop A (mirror screen concept). As illustrated in Figure A1, the PI was outside the USM Audiology Clinic, located within the USM Speech-Language Clinic (Day 1) and his home (Day 2).

2.2. Internet Connection and Measurements

As previously stated, we observed eight sessions of audiological tests using four laptops (A, B, C, and D) and a PC. Laptops A and C and the PC are essential for camera input, administrator access to the Webex application, and observation. As a result, these three laptops were connected to the Internet via a local area network (LAN) to ensure robust connectivity. Meanwhile, laptop B (testing room) and laptop D (day 1) were linked to USMSecure WiFi because of the room's insufficient LAN port. On Day 2, the PI was at home and connected to the Internet via a mobile hotspot for this investigation. Unifi's unlimited 2 h mobile hotspot service was subscribed to to ensure that each session's Internet connection was uninterrupted.

Speedtest.net (Ookla, Seattle, WA, USA) and BlazeMeter (Perforce Software, Inc., Minneapolis, MN, USA) were used to measure Internet connectivity. The Internet connection speed of three laptops (A, B, and D) was measured regularly before, during, and after clinical assessment sessions. Ookla has maintained this website since 2006, reporting over 35 billion tests that meet the wide-angle contact measurement requirements [13,14]. It has three main features: it can upload, download, and ping. Second, Webex's performance with four participants was rated using BlazeMeter. This open-source load testing tool for mobile apps, databases, online services, and websites [15]. However, we did not restrict the license or scalability during the test. The test results were automatically sent to the PI after completion. Maximum users, average throughput, faults, and 50% response time were provided. The average bandwidth could not be tracked due to the testing credit restriction.

2.3. *Live Observations and Satisfaction Level*

During these two-day clinical tests, the principal investigator must observe all live sessions. It was also necessary to record all data acquired during the observation, including Internet speed, audio-visual quality, and comments on technical challenges. Because this is a trial version of the adapted teleaudiology approach, the PI serves as an external examiner located outside the clinic or in another state. For this study, the principal investigator did not evaluate the candidates' performance throughout their clinical examinations. The course coordinator recorded every session and sent the URL to the PI. Because the recorded sessions contain private and confidential information, only experienced audiologists at USM were invited to evaluate the audio-visual quality. Quality and Impairment Scales of the International Telecommunication Union-R chose the grading structure (ITU, 2015). It was decided that the assessment process would be relevant to the recorded sessions. Each section covers a different aspect of the exam (Appendix C): history taking (HT), otoscopic examination (OE), and tympanometry (tymp), as well as providing feedback (PF) and other relevant tests (OT). This process requires four experienced audiologists. During the assessments, the recorded sessions were shown on laptop B and a Panasonic television.

3. Results

All candidates for the audiological clinical examinations were familiar with the testing room's facilities and audiological equipment, having spent nearly two years practicing in the USM Audiology Clinic. As illustrated in Figure A1, we implemented only a few ICT facilities and incorporated comments from students, technologists, clinicians, and examiners to ensure that they do not interfere with or distract during clinical assessments. Technically, we optimized all existing infrastructure and had a few issues connecting laptop A to the exterior digital web camera due to a short cable. As a result, we were prompted to connect the camera via a three-meter USB extension wire, which resulted in signal loss. As a consequence, one of three web camera models was chosen following several trials. In general, the trial version of the adapted teleaudiology approach was successful in observing clinical examinations, and further details about the results are presented below.

3.1. *Internet Connection*

Ookla and BlazeMeter were used to test Internet speed for eight clinical sessions. However, in laptops B and D, the Internet speed cannot be determined during the 'middle' and 'after' sessions of Session 8 due to the PI's laptop's poor Internet connection. Because the data are significantly deviant from normal, all non-parametric tests were chosen to analyze ping, download, and upload Internet features. For this study, we compared the following: (i) the Internet speed properties of the same laptop, measured three times for each session; (ii) the Internet speed properties of different types of connectivity in different laptops, and (iii) the Internet speed properties of USMSecure Wireless (within USM) and unlimited Unifi Hotspot (at home) connections using laptop D.

Friedman's analysis revealed no statistically significant change in the ping (ms), download (Mbps), and upload (Mbps) values for each session, which were measured three times (before, middle, and after) ($p > 0.05$). As a result, the repeated values for online properties were merged, and the mean values were calculated, as illustrated in Figure A2. A Kruskal–Wallis H test revealed a statistically significant difference between the three laptops' ping, download, and upload times (ms) (Table A1). A post hoc analysis using Conover yielded a significance level of $p < 0.05$. Between laptop B and laptop D, there were no significant differences in ping and upload speeds. However, statistically significant differences in download speeds were observed across laptops A, B, and C. Additionally, a Mann—Whitney test revealed no statistically significant difference between the USMSecure Wireless and the unrestricted Unifi Hotspot in all online properties.

Webex's performance during audiology clinical examinations was evaluated using BlazeMeter. According to the test findings, the maximum number of virtual users that could be tested was 20, and the Webex load capabilities were kept to a minimum during the test period, as only four people participated. Overall, the average throughput, error rate, and response time were 16.50 hits per second, 99 percent, and 1215.88 ms, respectively. These values indicated that the Webex on the laptop D was operating at a deficient level of performance.

3.2. Live Observation Report

The live observation occurred in the USM Speech-Language Clinic on Day 1 and the PI's home on Day 2. The PI reported on various points, and each observation criterion was divided into six segments (Table A2). Seven clinical sessions used play audiometry to examine behavior, while one session applied visual reinforcement audiometry. These approaches were chosen following age-appropriate diagnostic audiology procedures, which required candidates to select differential diagnostic techniques that were developmentally suitable for the kid.

Due to the limits of the audiology equipment, the findings for otoscopic examination and distortion product otoacoustic emissions cannot be noticed during live observation. Overall, Session 8 had the most reported issues, at four, while Sessions 6 and 7 each had one. Additionally, the most frequently occurring recurring difficulties occurred only in two distinct sessions, and the most frequently occurring problems were noted in part for providing feedback (not related to the limitation of the equipment).

3.3. Level of Satisfaction among the Principal Investigator and Experienced Audiologists

Live observation by the PI established the level of satisfaction, and four experienced audiologists analyzed the recorded Webex sessions. The assessors, two male and three female audiologists, had an average of 5.36–15.8 years of experience. Participants were required to rate the audio-visual quality in at least five segments during each session. Respondents were generally satisfied with the audio quality (M = 3.02, SD = 1.14), with behavioral testing scoring the highest (M = 3.50, SD = 0.76) and history taking scoring the lowest (M = 2.63, SD = 1.06) during the live observation. Similarly, the audio quality of behavioral testing in recorded sessions suggested the greatest degree of satisfaction (M = 1.63, SD = 0.66), but otoscopic examination indicated the lowest level of satisfaction (M = 1.26, SD = 0.44). Meanwhile, the highest degree of visual quality was reported for audiograms performed under live observation (M = 4.75, SD = 0.71), while the lowest level of satisfaction was reported for recorded tympanometry testing (M = 2.13, SD = 1.06). Additionally, we averaged all areas for each session to compare satisfaction levels across live and recorded sessions. For audio (U = 867.50, p = 0.001) and visual (U = 4448.50, p = 0.001), satisfaction with live observation was significantly higher than satisfaction with recorded sessions. Other testing results were removed since the task was not included in all clinical sessions. Figure A3 summarizes the degree of satisfaction.

4. Discussion

The essential premise of a basic model of teleaudiology for adult clients was that the services should be comparable to those provided in conventional clinical settings and that they could be used in a variety of telecommunications studies utilizing synchronous, asynchronous, or hybrid technology [14]. In this study, we adapted this model to create a trial version for use during audiology clinical evaluations. Thus, the principal investigator acted as an examiner in this study, evaluating the live audiology clinical examination via ICT applications from a distance. This approach may have a substantial impact, particularly during the COVID-19 pandemic, due to the restriction of interstate travel orders, particularly for an external examiner. As a result, it may minimize the cost of travel [16,17]. Additionally, live observation via Webex can be used for undergraduate or postgraduate audiology clinical practice by domestic or international students. This may be advantageous for students who remained at home or in a hostel during the epidemic COVID-19. Additionally, as demonstrated by a previous study, it can foster health professional teamwork, who successfully collaborated with nursing and audiology students to fit hearing aids using teleaudiology [8].

In Malaysia, the first research concentrating on teleaudiology was conducted to ascertain audiologist attitudes about the field [18] and conduct remote hearing assessments for deaf and hard-of-hearing school children [3]. Because this approach is novel in Malaysia, we maximized the use of existing audiology equipment in the clinic, which was not PC-based. In this case, more ICT facilities are required for communication and recording of all audiological findings. For audio-visual communication, prior researchers have described the usage of Cisco Webex Meeting [19,20], and other studies have successfully adopted a nearly identical technique for cochlear implant users [21–23]. However, several researchers used a variety of platforms in their investigations, including AudioProConnect (AudioProConnect Company, Bobigny, France) [24], the Polycom System (Plantronics, Inc., 500 Series, CA, USA) [21,25], and Skype (Microsoft Corp., Luxembourg City, Luxembourg) [26]. Second, this study employed TeamViewer to control the laptop in the testing room remotely, comparable to earlier teleaudiology research [27–29]. Although this study's general design blended prior studies, its primary objective was distinct and did not even include remote hearing assessment. As a result, it is unlikely that this design may be used in the future to accomplish a similar purpose as the prior study.

The summary of the live observation report (Table A2) shows that most sessions encountered difficulties due to a lack of ICT resources and technical support staff. These two critical components should be highlighted in the teleaudiology approach planning checklist [20,30–32]. For instance, in our study, we allocated only existing staff members who are not ICT experts and must execute another duty during the clinical examination. As a result of the shortage of ICT technical support employees, specific technical issues occurred, such as camera selection input and monitoring Webex recording sessions. Additionally, because this study was conducted during a clinical assessment, all technological difficulties encountered inside the testing room were resolved remotely, as we did not intend to disrupt the sessions. In comparison, if we were to use the teleaudiology approach in daily clinical practice, any technical difficulties that arose could be resolved directly by technical support.

As a guideline to assure the proper functioning of this teleaudiology service, several researchers have established a minimum upload and download Internet connection speed of between 0.1 Mbps and 0.38 Mbps [27,28,33–35]. In comparison to this study, the Internet speed was significantly faster. The download speed was recorded as being between 2.77 Mbps and 86.21 Mbps, while the upload speed was reported as being between 3.01 Mbps and 92 Mbps. Meanwhile, researchers observed ping values for Internet speed was 16.75 ms to 76.32 ms, which were found to be greater than those reported in a study conducted by Penteado, which ranged between 58.5 ms and 7.3 ms [36]. However, no specific ping values have been established by prior studies in the teleaudiology approach.

Additionally, while the TeamViewer application utilized in this investigation was similar to those used in prior studies to operate the laptop at the testing site [27,28], the application's performance is reliant on the Internet connection. According to their website, the minimal criteria for Internet speed are 6 Mbps for download and 1 Mbps for upload [37]. Only laptop B exceeded the recommended download speed (Figure A1). This finding is consistent with PI's experience, during which TeamViewer regularly disconnected and one of the sessions encountered connectivity issues while using TeamViewer to test the Internet speed. Apart from that, PI observed a reduction in the audio-visual quality on laptop B after connecting to this application, which could be attributed to the volume of Internet traffic on laptop B, which was connected only via USMSecure WiFi. Additionally, we discovered that the wireless connection between USMSecure and Unifi Mobile was comparable in ping and upload speeds. This circumstance may have a beneficial influence on the use of mobile networks such as 3G cellular networks [38] or 3G Vodafone mobile hotspots [27] in areas with restricted Internet access, mainly rural areas.

Dharmar conducted a teleaudiology study for infant hearing assessment and discovered that audiologist satisfaction with live testing was 5.9 and 6.7, respectively, on a 7-point Likert scale [38]. These findings indicated that satisfaction accounted for more than half of the entire score, which corresponds to this study's findings for live observation, where the total satisfaction score for audio-visual was more significant than 2.5 on a 5-point Likert scale. However, the total score for the recorded sessions' audio quality was less than half (M = 1.39, SD = 0.55). As we are aware, the sound quality is poor because we can only use the built-in microphone on laptop A or the video camera, which is fixed in that position. Additionally, the recorded sessions do not focus on the screen that was selected during the live observation but rather on the four primary screens of the Webex participants, which include the clinical coordinator (for recording purposes), the principal investigator (for live observation), clinical students (for testing purposes), and technical staff (patient's response). As a result, this may affect the evaluations of the recorded sessions.

5. Conclusions and Recommendations for the Future

Students' abilities are routinely examined in a variety of ways during pediatric clinical assessment. However, these factors can be classified into two broad categories: communication abilities and hands-on skills. The following objectives for implementing a comprehensive observation system for audiology clinical examinations are provided together with their associated challenges:

- During clinical evaluations, establishing valid assessments of communication and hands-on skills.

Challenges: Communication abilities are contingent upon the adequacy of ICT infrastructure and audio-visual quality. We employed an existing communication device in the clinic (Figure A1) and fitted it to their position in this study. As a result, it may impair the audio-visual quality, mainly when working with pediatric patients who actively move throughout sessions. Meanwhile, clinical abilities are contingent on the ability to utilize audiology tools to examine the hearing status and be adaptable when dealing with patients. As previously stated, all audiology devices are not operated via a personal computer, necessitating an external camera (video camera or web camera) to watch all clinical stages while operating the machines. In general, these two concerns can be addressed by updating equipment to more portable, stable audio-visual, and PC-based devices capable of screen sharing and remote control.

- Creating a steady Internet connection in terms of network traffic or Internet speed.

Challenges: Due to the limitations of the existing Internet infrastructure, the testing room was equipped with only one LAN port for connection to laptop A (audiology device), while laptop B (patient reaction) was wirelessly connected. Thus, this may result in a significant difference in Internet speed between these two laptops with similar specifications and may affect the level of satisfaction with audio-visual quality. Additionally, using a Mobile Hotspot (Day2) may provide a more accurate forecast of the teleaudiology approach that can be used in areas with restricted Internet connectivity. However, network provider collaboration is critical to ensuring the stability of the Internet connection.

- Forming a team of individuals who are technically adept and educated about the teleaudiology program.

Challenges: We utilized all available employees throughout this clinical evaluation and assigned only one person to provide technological support, specifically for camera input choices. However, this staff member had a relationship with his regular employment, and as a result, the camera selection for specific clinical sessions was not as planned. Thus, the team should include skilled personnel dedicated solely to technological support, particularly during the teleaudiology session.

Overall, we were successful in implementing a trial version of an adapted teleaudiology approach that can be utilized to observe audiology clinical examinations and this strategy can also be used in the future for clinical observation, particularly during pandemic COVID-19, when the total number of students in the clinic had to be limited. Additionally, the observation can be used to expose nonclinical professionals or students who work with special needs children, such as teachers, policymakers, and parents. With specific changes and upgrades to the teleaudiology equipment, it is possible to conduct hearing tests from a distance using the same approach.

Author Contributions: Conceptualization, M.F.N.R. and W.N.W.M.; methodology, M.F.N.R.; formal analysis, M.F.N.R. and M.N.Z.; investigation, M.F.N.R.; resources, M.A.A.; writing—original draft preparation, M.F.N.R.; writing—review and editing, M.N.Z., W.N.W.M. and M.A.A.; project administration, M.F.N.R. All authors have read and agreed to the published version of the manuscript.

Funding: This research received no external funding.

Institutional Review Board Statement: Not applicable.

Informed Consent Statement: Not applicable.

Data Availability Statement: Not applicable.

Acknowledgments: We would like to express our gratitude to Suzana Mansor, Mohd Khary Hussain, Rozazipah Ahmad, and Halimah Mahmud for their participation and constructive remarks on this work. Additionally, we appreciate Mohd Alamin Mohd's willingness to serve as technical assistant during the audiology clinical examinations and the entire team of the USM audiology program. Special thanks to School of Health Sciences, Universiti Sains Malaysia for the financial assistance.

Conflicts of Interest: The authors declare no conflict of interest.

Appendix A

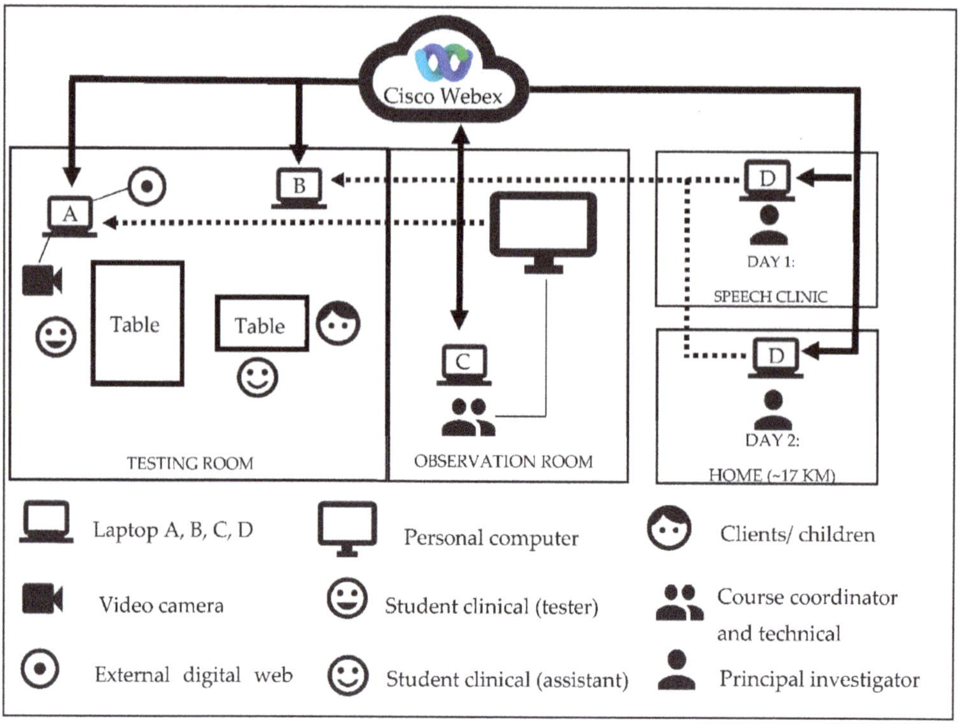

Figure A1. Audiology clinical examination setup.

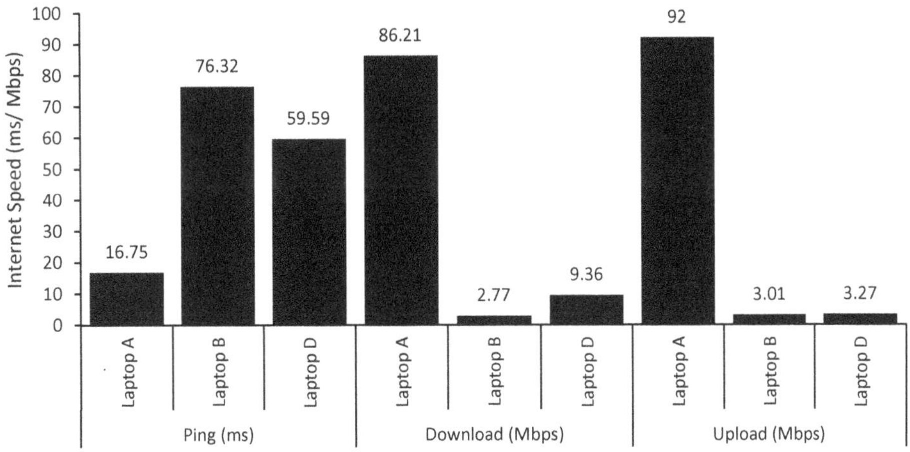

Figure A2. Mean values for Internet speed measurements using three different laptops.

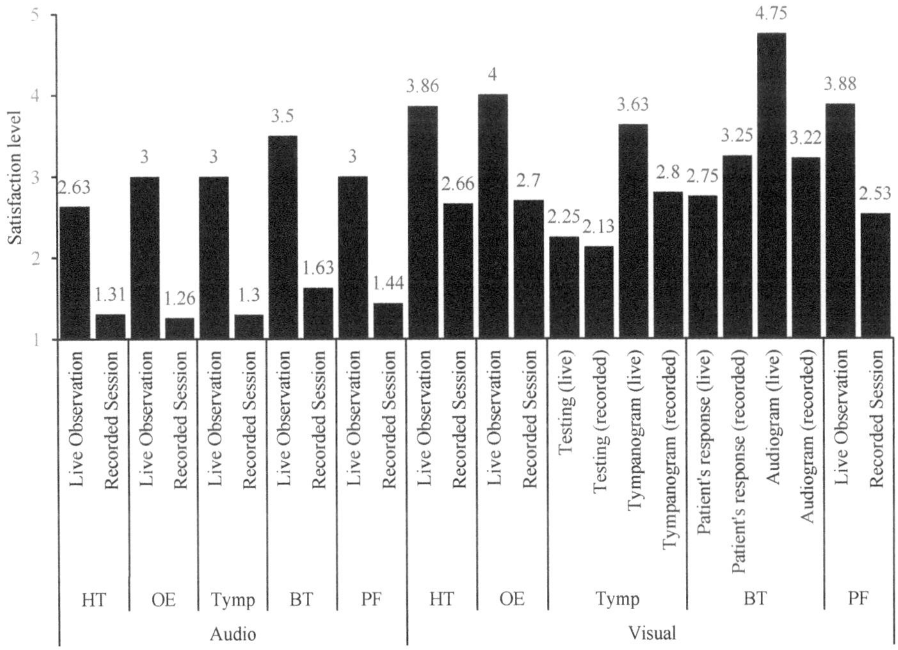

Figure A3. The satisfaction level of audio-visual quality for history taking (HT), otoscopic examination (OE), tympanometry (Tymp), behavioral testing (BT), and providing feedback (PF).

Appendix B

Table A1. Internet properties in the different type of connectivity in different laptops.

Internet Properties	Laptop	Type of Connection	Kruskal–Wallis H	*p* Value	Conover Post-Hoc Analysis	
					Average Rank	Different ($p < 0.05$)
Ping (ms)	Laptop A	LAN	29.20	0.001	17	Laptop B and D
	Laptop B	Wireless [1]			43.66	Laptop A
	Laptop D	Wireless [2]			44.43	Laptop A
Download (Mbps)	Laptop A	LAN	51.79	0.001	56.50	Laptop B and D
	Laptop B	Wireless [1]			15.27	Laptop A and D
	Laptop D	Wireless [2]			29.73	Laptop A and B
Upload (Mbps)	Laptop A	LAN	56.50	0.001	56.50	Laptop B and D
	Laptop B	Wireless [1]			21.23	Laptop A
	Laptop D	Wireless [2]			23.77	Laptop A

Wireless [1] = USMSecure, Wireless [2] = Unifi Mobile.

Table A2. Summary of the live observation.

Segments		Descriptions of the Problems	Sessions
History taking (HT)	a.	Mother's voice is too soft	8
	b.	Unable to hear due to microphone malfunctions	7
	c.	Mothers are far from the mic	4
Otoscopic examination (OE)	a.	Unable to observe at least in one ear	5 and 2
	b.	Unable to observe as the patient sit outside the focus point of the camera	3
	c.	Unable to judge the ear canal status and tympanic membrane conditions	All
Tympanometry (Tymp)	a.	Unable to observe probe insertion	8
	b.	Unable to observe tympanometer screen due to wrong camera selection input	
	c.	Unable to hear the instruction	7 and 1
	d.	Blurred screen	6 and 3
Behavioral testing (BT, Play = 7, VRA =1) *	a.	The audiometer screen blocked by candidates	8 and 1
	b.	Unable to observe the patient's response as the patient sit outside of the camera focus point	3
Other testing (OT)	a.	Unable to observe screen for distortion product otoacoustic emissions	All
Providing Feedback (PF)	a.	Intermittent sound quality and unable to understand	8 and 2
	b.	Unable to hear as the caregiver sit far away from the microphone	7 and 5
	c.	Unable to observe the interaction between candidates and parents due to wrong camera selection	4 and 3

* Play = Play Audiometry, VRA = Visual Reinforcement Audiometry.

Appendix C

Name : ……………………………………………………………………………………
Position : ……………………………………………………………………………………
Working experience : …………… years
Clinical session : ……………………….

Instructions:
This assessment form was designed to evaluate the quality of audio-visual recorded sessions during audiology clinical examination. Kindly rate for each clinical session as below:

i. Please rate the quality of audio-visual on a scale of 'Bad', 'Poor', 'Fair', 'Good', or 'Excellent'. This grading process was recommended by International Telecommunication Union (ITU, 2015). The details of these rating scales are as below:
   a. Bad: Very annoying or unable to listen (audio) or observe (visual) the recorded session
   b. Poor: Annoying or able to listen (audio) or observe (visual) the recorded session, but with distortion/disconnected for more than five times
   c. Fair: Slightly annoying or able to listen (audio) or observe (visual) the recorded session, but with distortion/disconnected for less than/ equal to five times
   d. Good: Not annoying or able to listen (audio) or observe (visual) without any distortion/disconnected
   e. Excellent: Able to listen (audio) or observe (visual) clearly without any distortion/disconnected

ii. Each session will be divided into five main sessions, which are 'History Taking', 'Otoscopic Examination', 'Tympanometry', 'Behavioral audiology testing', 'others testing' and 'Giving Feedback'

Section	Assessment	Bad	Poor	Fair	Good	Excellent	Comments
History taking	Audio						
	Visual (interaction)						
Otoscopic examination	Audio (instruction)						
	Visual (perform)						
Tympanometry	Audio (instruction)						
	Visual (perform)						
	Visual (tympanogram)						
Behavioral testing	Audio (instruction and reinforcement)						
	Visual (patient's response)						
	Visual (audiogram)						
Others testing	Audio (instruction)						
	Visual (patient's response)						
Giving feedback	Audio						
	Visual (interaction)						
Overall comments *(if any)*							
……							

References

1. Katz, J.; Chasin, M.; English, K.M.; Hood, L.J.; Tillery, K.L. *Handbook of Clinical Audiology*; Wolters Kluwer Health: Philadelphia, PA, USA, 2015; Volume 7, p. 3.
2. Sistem Pengiktirafan Kelayakan Perkhidmatan Awan (eSisraf). Available online: https://www2.mqa.gov.my/esisraf/cariankelayakan.cfm (accessed on 15 July 2022).
3. Rashid, M.F.N. Needs and implementation of teleaudiology services in special education service centres. *Disabil. Soc.* **2021**, *36*, 848. [CrossRef]
4. Programme Standards: Medical and Health Sciences. Available online: https://www2.mqa.gov.my/qad/v2/garispanduan/2019/July/PS%20Medical%20BI.pdf (accessed on 15 July 2022).
5. Abraham, H.N.; Opara, I.N.; Dwaihy, R.L.; Acuff, C.; Brauer, B.; Nabaty, R.; Levine, D.L. Engaging Third-Year Medical Students on Their Internal Medicine Clerkship in Telehealth During COVID-19. *Cureus* **2020**, *12*, e8791. [CrossRef] [PubMed]
6. He, X.; Shelden, D.; Kraftson, A.; Else, T.; Auchus, R.J. A virtual teaching clinic for virtual care during the COVID-19 pandemic. *Clin. Diabetes Endocrinol.* **2020**, *6*, 25. [CrossRef] [PubMed]
7. Kopp, A.R.; Rikin, S.; Cassese, T.; Berger, M.A.; Raff, A.C.; Gendlina, I. Medical student remote eConsult participation during the COVID-19 pandemic. *BMC Med. Educ.* **2021**, *21*, 120. [CrossRef]
8. Novak, R.E.; Cantu, A.G.; Zappler, A.; Coco, L.; Champlin, C.A. The Future of Healthcare Delivery: IPE/IPP Audiology and Nursing Student/Faculty Collaboration to Deliver Hearing Aids to Vulnerable Adults via Telehealth. *J. Nurs. Interprofessional Leadersh. Qual. Saf.* **2016**, *1*, 1.
9. Jacobs, P.G.; Saunders, G.H. New opportunities and challenges for teleaudiology within Department of Veterans Affairs. *J. Rehabil. Res. Dev.* **2014**, *51*, 7–12. [CrossRef]
10. Swanepoel, D.W.; Hall III, J.W. A Systematic Review of Telehealth Applications in Audiology. *Telemed. e-Health* **2010**, *16*, 181–200. [CrossRef]
11. American Speech-Language-Hearing Association. Key Issues in Telepractice. 2017. Available online: https://www.asha.org/practice-portal/professional-issues/telepractice/#collapse_1 (accessed on 28 August 2021).
12. Krumm, M. Teleaudiology Model Considerations. *SIG 18 Perspect. Telepractice* **2014**, *4*, 4–10. [CrossRef]
13. Ookla 2019. Available online: https://www.speedtest.net/about (accessed on 12 August 2021).
14. Bauer, S.; Clark, D.D.; Lehr, W. Understanding Broadband Speed Measurements. In Proceedings of the TPRC 2010: Telecommunications Policy Research Conference, Arlington, VA, USA, 1–3 October 2010.
15. Memon, P.; Hafiz, T.; Bhatti, S.; Qureshi, S.S. Comparative study of testing tools Blazemeter and Apache JMeter. *Sukkur IBA J. Comput. Math. Sci.* **2018**, *2*, 70–76.
16. Govender, S.; Mars, M. The use of telehealth services to facilitate audiological management for children: A scoping review and content analysis. *J. Tlemedicine Telecare* **2017**, *23*, 392–401. [CrossRef]
17. Kokesh, J.; Ferguson, A.S.; Patricoski, C.; LeMaster, B. Traveling an audiologist to provide otolaryngology care using store-and-forward telemedicine. *Telemed. e-Health* **2009**, *15*, 758–763. [CrossRef] [PubMed]
18. Rashid, M.F.N.B.; Quar, T.K.; Chong, F.Y.; Maamor, N. Are we ready for teleaudiology?: Data from Malaysia. *Speech Lang. Hear.* **2019**, *23*, 146–157. [CrossRef]
19. Rushbrooke, E.; Houston, K.T. *Telepractice in Audiology*; Plural Publishing: San Diego, CA, USA, 2016; pp. 153–188.
20. Telehealth for Audiological Management. Available online: www.entdev.uct.ac.za (accessed on 28 December 2020).

21. Sevier, J.D.; Choi, S.; Hughes, M.L. Use of Direct-Connect for Remote Speech-Perception Testing in Cochlear Implants. *Ear Hear.* **2019**, *40*, 1162–1173. [CrossRef]
22. Hughes, M.L.; Goehring, J.L.; Sevier, J.D.; Choi, S. Measuring Sound-Processor Thresholds for Pediatric Cochlear Implant Recipients Using Visual Reinforcement Audiometry via Telepractice. *J. Speech Lang. Hear. Res.* **2018**, *61*, 2115–2125. [CrossRef]
23. Goehring, J.L.; Hughes, M.L. Measuring sound-processor threshold levels for pediatric cochlear implant recipients using conditioned play audiometry via telepractice. *J. Speech Lang. Hear. Res.* **2017**, *60*, 732–740. [CrossRef] [PubMed]
24. Venail, F.; Picot, M.C.; Marin, G.; Falinower, S.; Samson, J.; Cizeron, G.; Balcon, M.; Blanc, D.; Bricaud, J.; Lorenzi, A.; et al. Speech perception, real-ear measurements and self-perceived hearing impairment after remote and face-to-face programming of hearing aids: A randomized single-blind agreement study. *J. Telemed. Telecare* **2019**, *27*, 409–423. [CrossRef] [PubMed]
25. Fletcher, K.T.; Dicken, F.W.; Adkins, M.M.; Cline, T.A.; McNulty, B.N.; Shinn, J.B.; Bush, M.L. Audiology telemedicine evaluations: Potential expanded applications. *Otolaryngol. Head Neck Surg.* **2019**, *161*, 63–66. [CrossRef]
26. Constantinescu, G.; Waite, M.; Dornan, D.; Rushbrooke, E.; Brown, J.; McGovern, J.; Ryan, M.; Hill, A. A pilot study of telepractice delivery for teaching listening and spoken language to children with hearing loss. *J. Telemed. Telecare* **2014**, *20*, 135–140. [CrossRef]
27. Monica, S.D.; Ramkumar, V.; Krumm, M.; Raman, N.; Nagarajan, R.; Venkatesh, L. School entry level tele-hearing screening in a town in South India—Lessons learnt. *Int. J. Pediatric Otorhinolaryngol.* **2017**, *92*, 130–135. [CrossRef]
28. Ramkumar, V.; Nagarajan, R.; Kumaravelu, S.; Hall, J.W. Providing tele ABR in rural India. *Perspect. Telepractice* **2014**, *4*, 30–36. [CrossRef]
29. Tomaszewska-Hert, I.; Skarzynsk, P.H.; Ludwikowski, M. Audiology measurement using telemedical solution in Central Asia. *J. Int. Soc. Telemed. Ehealth* **2017**, *5*, e56.
30. American Speech-Language-Hearing Association. Telepractice. Available online: https://www2.asha.org/PRPSpecificTopic.aspx?folderid=8589934956&section=Key_Issues (accessed on 24 August 2021).
31. Critical Steps in Establishing a Teleaudiology Practice. Hearing Review. Available online: https://hearingreview.com/practice-building/office-services/telehealth/critical-steps-establishing-teleaudiology-practice (accessed on 24 August 2021).
32. Hayes, D.; Eclavea, E.; Dreith, S.; Habte, B. From colorado to guam: Infant diagnostic audiological evaluations by telepractice. *Volta Rev.* **2012**, *112*, 243–253. [CrossRef]
33. Campos, P.D.; Ferrari, D.V. Teleaudiology: Evaluation of teleconsultation efficacy for hearing aid fitting. *J. Da Soc. Bras. De Fonoaudiol.* **2012**, *24*, 301–308. [CrossRef] [PubMed]
34. Gladden, C. The nuts and bolts of preparing for audiology telepractice. *ASHA Lead.* **2017**, *22*, 28–29. [CrossRef]
35. Penteado, S.P.; Bento, R.F.; Battistella, L.R.; Silva, S.M.; Sooful, P. Use of the satisfaction with amplification in daily life questionnaire to assess patient satisfaction following remote hearing aid adjustments (telefitting). *JMIR Med. Inform.* **2014**, *2*, e18. [CrossRef] [PubMed]
36. How to Get Your Remote Support to the Next Level. Available online: https://community.teamviewer.com/English/discussion/4819/how-to-get-your-remote-support-to-the-next-level (accessed on 28 August 2021).
37. Visagie, A.; Swanepoel, D.W.; Eikelboom, R.H. Accuracy of remote hearing assessment in a rural community. *Telemed. J. E-Health Off. J. Am. Telemed. Assoc.* **2015**, *21*, 930–937. [CrossRef]
38. Dharmar, M.; Simon, A.; Sadorra, C.; Friedland, G.; Sherwood, J.; Morrow, H.; Deines, D.; Nickell, D.; Lucatorta, D.; Marcin, J.P. Reducing loss to follow-up with tele-audiology diagnostic evaluations. *Telemed. e-Health* **2016**, *22*, 159–164. [CrossRef]

proceedings

MDPI

Proceeding Paper

Moral Values Application in Islamic Education Teaching and Learning through the 21CE Activities †

Norfarahi Zulkifli *, Zuraidah Ramdzan, Wan Ali Akbar Wan Abdullah, Mohd Isa Hamzah, Khadijah Abdul Razak and Hafizhah Zulkifli [1]

Faculty of Education, Universiti Kebangsaan Malaysia, Bangi 43600, Selangor, Malaysia
* Correspondence: farahijpa@gmail.com
† Presented at the International Academic Symposium of Social Science 2022, Kota Bharu, Malaysia, 3 July 2022.

Abstract: This study aims to explore the values that can be applied through the 21st Century Education (21CE) implementation as well as explore creative methods that can be used to apply moral values to students during the 21CE activities. Online interviews were conducted with five Islamic education teachers (IET) and the data obtained were subsequently transcribed and analyzed. The results highlighted three main themes to answer the first objective, namely: religious appreciation, daily interaction, and leadership. Meanwhile, five main themes emerged to answer the second objective, namely role models, rewards and punishments, advice, social media use, and innovation production. The findings of this study are expected to provide ideas for IET to apply moral values creatively in the implementation of 21CE activities.

Keywords: moral values; creative method; Islamic education

check for updates

Citation: Zulkifli, N.; Ramdzan, Z.; Abdullah, W.A.A.W.; Hamzah, M.I.; Razak, K.A.; Zulkifli, H. Moral Values Application in Islamic Education Teaching and Learning through the 21CE Activities. *Proceedings* **2022**, *82*, 22. https://doi.org/10.3390/proceedings2022082022

Academic Editor: Mohamad Rahimi Mohamad Rosman

Published: 9 September 2022

1. Introduction

Morality is one of the Islamic teaching aspects related to a Muslim's behavior in daily life. Morality is also a measure of a Muslim's personality excellence. Based on Al-Ghazali's view, one of the purposes of Islamic education is to inculcate an excellent Muslim personality as a perfect human being (*Al-Insān al-Kāmil*) through the application of morality in the teaching process by using certain methods [1]. This also means that education not only aims to produce good citizens but also good, moral individuals. This matter is emphasized in the National Education Philosophy, which is the main aspiration and goal in the education system in Malaysia that emphasizes physical, emotional, spiritual, and intellectual balance [2]. The same thing is also emphasized in the Philosophy of Islamic Education (FPI) on the application of moral values in the educational process. In the context of education today, values and ethics are highly accentuated in the Malaysian Education Development Plan 2013–2025 through the implementation of 21st-century learning (21CE) in teaching (Pelan Pembangunan Pendidikan, 2013–2025).

The 21st-century student-centered learning is implemented based on five basic standards, namely, communication, creativity, critical thinking, and collaboration as well as values and ethics, which has encouraged teachers to progressively implement teaching and learning towards achieving educational excellence, realizing the aspirations of the National Education Philosophy in general and the Philosophy of Islamic Education in particular, which focuses on the moral formation to be a good Muslim. The application of values and ethics in teaching through the 21CE activities develops a high sense of identity. This is because teachers are the closest to the students when they are in school. Therefore, teachers' efforts to inculcate positive values in students during the T&L session can meet the students' aspirations [3]. In addition, the application of values and ethics emphasized in the 21CE is one of the main purposes of education because the goal is to inculcate noble morals and develop human beings to be able to perform the most important responsibility of performing the duties of God's vicegerent (caliph) on earth according to the Divine law [4]. Therefore, the application of the values in T&L activities based on the 21CE elements will reveal the noble values practiced in daily life.

Educators are an important factor that influences students in the teaching and learning process. To ensure that good moral values can be applied to students, educators should implement appropriate methods. From Imam Al-Ghazali's perspective, the teacher or educator's role is not just to impart knowledge but as a *muaddib* [5]. In addition to knowledge, teachers need to have pedagogical skills, professional skills, personality skills, and social skills. Al-Ghazali also reminded teachers to prepare themselves in terms of knowledge, competence, ability, skills, attitude change, and readiness in education. Education in Islam is a process that emphasizes morals and relies on the teacher or educator's role to emulate good morals for students. Hence, the objective of this study is to explore the values that can be applied through the implementation of 21CE activities and to explore the creative methods that can be used to apply moral values to students during 21CE activities.

2. Methodology

This study used a qualitative approach by conducting interviews with five study participants. The design of this study is a single-case study that uses a constant comparison method to analyze the data. Data collection was done through in-depth interviews. The raw data were analyzed using Atlas.ti version 9 software to make comparisons, find similarities and differences between datasets through triangulation of interview data, observation, and document analysis. Atlas ti.9 is one of the established research tool for qualitative data and details on the features and capabilities of the software can be found from the free manual and video tutorials available at http://atlasti.com (accessed on 20 May 2022). The interview sessions were conducted online and the data obtained were then transcribed. Study participants were selected based on the purposive sampling technique and started by identifying the main selection criteria of the participants [6]. Therefore, the study participants' selection was based on the following criteria: (1) teaching Islamic education; (2) having more than 10 years of experience in teaching Islamic education; and (3) being approachable and cooperative. Five study participants were involved. They are given the codes G1 to G5. The demographics of the study participants were as follows (Table 1):

Table 1. Demography of study participants.

Study Participants	Code	Teaching Experience
Study participant 1	G1	16 years
Study participant 2	G2	15 years
Study participant 3	G3	12 years
Study participant 4	G4	12 years
Study participant 5	G5	22 years

Table 1 shows the coding and years of teaching experience of the five participants. All participants have more than 10 years of teaching experience The analysis covered all types of data from all study participants. Findings are reported thematically according to the research questions. For the purpose of preserving the validity and reliability of the data, the researcher practiced: (1) triangulation, (2) review of study participants, (3) researcher bias description, (4) peer review, (5) in-depth description, and (6) audit trail.

3. Findings

The students' outcomes can be developed through the strength of Islamic appreciation in their lives. Therefore, 21st-century learning can enhance religious appreciation through *aqidah* (faith/belief). As stated by G1, the practice of the values in the 21CE brings students closer and instills their love of Allah SWT. G1 said, "when we make and practice 21CE in the Islamic way, then these pupils will be closer to Allah because the things with human beings will eventually reflect on his belief in Allah". Consequently, the Muslim personality of the student can be inculcated towards obedience to Allah. G2 said, "This concept if the teacher really does it, we can shape our students to be God's servants who follow the true path". In addition, 21CE can also improve moral practice among students. For example, maintaining cleanliness and tidiness while doing work. G1 said, "*Ok, we have to* do work neatly, God likes tidiness, we have to emphasize our values. That, I think, ease me. They did that".

The 21CE also improves students' daily interactions, such as respecting their friends (G3) and teachers (G2). G1 further explained the interaction by giving an example of the greeting practice, he said, "In life, they can evaluate the situation whether it is good or bad. For example, greetings ... It is according to the Prophet's sunnah. So, these students make it a practice. Isn't it parallel with the 21CE?". They also practiced working together and celebrating the differences between them. G1 explained, "They have a kind of value for them to be can work together ... The pupils who are not very smart, those who are weak, those who are smart, those who are moderate can blend with the 21CE".

In addition, 21CE stimulates students to become leaders when planning and organizing activities, interacting with friends as well as respecting opinion differences. G3 said, "For example (students) are trained to lead in groups and have the opportunity to present their work, students will usually think before making a decision and respect the findings of other friends". Pupils also become courageous to go forward and express their ideas. G2 explained, "21CE when applied in students, they dare to give speeches, ideas, and they are courageous to go forward". Consequently, students can be "independent" (G1). For example, doing a stage performance without guidance from the teacher. G2 described, "All our students make their own presentations without any guidance from the teacher and I really salute the students that they can come out with the idea that even the teacher doesn't teach them like that. I think that's the effect that the student got". Next, five main themes emerge from the creative methods that the IET used to inculcate values among students during 21CE activities, namely role model, reward and punishment, advice, use of social media, and innovation.

3.1. Role Model

Study participants agreed on the method of modeling values in teaching, especially by giving an example through the project implementation and the teachers' manners during the T&L delivery. G1 said, "Creative approach should be more effective in conveying the moral values. Either through the project, mentioned earlier, modeling the project through examples. How to appoint a group leader, obey the leader's instructions." G2 provides a more detailed example, "If we look at the modeling method, it can still be used ... even if it is online ... Although students do not look at us, our way when we say something, in terms of relationships and communication ... even communication via WhatsApp and so on, students can see our value". This method requires Islamic education teachers to set a good example for students because students can evaluate the teachers in any situation. G2 said, "In terms of delivery, our words ... Even online, we have to take care of that things, the manners and so on. So, when students look at us, even in an online class at home, sloppily dressed, for example, people look at that thing. Look at the students, for example, when they are on camera, we can actually see it".

3.2. Rewards and Punishments

Reward and punishment methods were also mentioned by the study participants among the methods that helped the application of the values. G1 said, "In terms of rewards, right? If you fail to follow the plan, it may be in the form of punishment. That's the application of moral values right". Punishment methods can be applied to students if they fail to comply with something set by the teacher, whereas, rewards are given by teachers based on the need of students to complete a task. G2 explained, "We can also use the reward. Rewards and punishments. For example, a student is supposed to be in class at 9 o'clock, but the pupil enters the class at 10 o'clock, that is a problem, right? So maybe we can use our power to cut the marks. If we see the pupil do his homework, it's okay, if we see his attendance record is good, we can give him a bonus, right"?

3.3. Advice

Giving advice is also one way to inculcate moral values. Islamic education teachers need to advise students in various situations, either inside or outside the classroom, during face-to-face or non-face-to-face teaching. G1 said, "The input of the application of the moral value is actually more effective than in a traditional class. So, those are the values that we want to discuss in class. Most students are touched by this". G2 added, "Another method is giving advice. If in a face-to-face class, we can directly reprimand them if they did wrong. Even during online classes, if we can, ask the student to open his camera. Sometimes the student sleeps, or lies down, isn't it? We can see where he is. If we see something wrong, we can still advise them. But in a good manner. Not reprimand them in front of their friends." In addition, teachers can advise students either at the beginning of the lesson or at the end. The method of advising students can also be done by making reflections and conclusions. G2 said, "When the class is ending, maybe we can make a conclusion. That time is also suitable to give advice. Usually, if we stop before we close the class with a prayer or something, that's even more beautiful, right? But before that, we can also advise on what is appropriate. We can also ask students what they want to improve in our class, and what they want to do, and we can ask the student's opinions. Or we can ask the student to respond to our T&L as well. Like a reflection ... ".

3.4. Social Media Use

Social media as a delivery platform is mentioned as one of the methods to instill values in students. For G3, he sees that moral values can be inculcated in teaching by using social media as a medium of delivery. G3 explained, "Looking at the current situation, there are various mediums or platforms to convey the T&L process and inculcate values. So, use various mediums. Medium of technology through social media. So, the students are more creative by using the medium of technology".

3.5. Innovation

To ensure that the 21CE activities are focused on values when developing innovations, the emphasis on values has become one of the important T&L objectives. G5 explained, "When pupils' morals are formed, for example, they keep up prayers, obey their parents, respect their teacher, what they learn, in the thick textbook, automatically they can get it. But why is it hard for students to study today? They didn't even want to come to school. They slept and did not pray. They are against their parents. If the parent who gave birth to them, are against them, how about we, teachers? Small in their eyes. That's why my methods are, I think that innovation is something that ... really helps the students' inner self". Therefore, the focus of innovation is given to topics that can have a direct impact on character development, such as prayer (G4). The emphasis on the prayer topic is made because of the belief in the Divine decree that prayer will prevent abominable and evil things. Although students can pray, the quality and appreciation of students during prayer are problematic. This is explained by G4, "Belief that, *"Innas salata tanha 'anil fahshai wal munkar"* ... When I did the prayer chapter, that's my effort to make the students and my child closer, to prevent them from falling into sin. So, we did a study on prayer. They apparently can pray. They prayed, and perform the prayer, but when you check back, apparently there is a problem with the prayer. That's why we made the prayer puzzle, we made the meaning of the prayer recitation ... That's what I see that I can help the children".

In line with the views of G4, G3 also focuses on the production of prayer innovations. The intention is to produce students of strong faith even when they enter the career world later. When there is faith, then students will be trusted at work. It all starts with prayer. Cakna Solat was produced. G3 explained, "We have to provide a strong base for them, in which when they finish school, after this, even when they become mechanics, the become a Faithful mechanics. A Faithful cleaner. The one with integrity. How? We make sure their basis is strong ... They care of their prayer" (G3). G1, on the other hand, wants to educate his students to use good language. Thus, he produced Army Yell. The goal is to replace "their language ... It is not suitable for the Islamic Education class" (G1) to "So if it's used like that, they will say when they are grateful, fun, they say, "*Allahu akbar*!", Right? "*Alhamdulillah*!" (G1). Through Army Yell as well, students are educated to "share their knowledge ... Let's say, they got the knowledge but refused to share it with friends. But when we taught them the concept, they wanted to share because I said, what we learn, when we teach our friends, is a charity. Ha, so, it is one form of jihad, I said. So, there seems to be a little bit (of improvement). *Alhamdulillah*" (G1).

In addition, G5 stated that by using Trademark, a voice-based innovation, students' hearts will be touched. As a result, any reprimand and advice will be heard by the student (G5). As for G2, any content in Islamic education has an implicit element of da'wah (preaching) (G2). Therefore, when formula-based innovation is used, although it helps students remember the content, it is also a form of reminder for students to take lessons from what they learn. G2 explained, "For me, these formulas have an element of da'wah. For example, in the atom formula that I made, the destruction of the past people reminds us of how the people of Luth were destroyed. So, when they hear this, oh, this is the consequence if I do that. So, these students (think), they know that homosexual relationship is a great sin, and will be punished by Allah. Need to learn from Luth's people." In addition, the use of singing innovation about the life of a figure can evoke an appreciation for their struggle and inspire students. G2 explained, "If the song of the figure is on their struggle, need to be strong, to do jihad. If we do not fight and become lazy, we will not be successful. That's the *da'wah*, for them to study hard" (G2).

4. Discussion

The findings have identified three themes of values that can be applied through the 21CE activities, namely religious appreciation, daily interaction, and leadership. Religious appreciation plays an important role in shaping students' beliefs and personal values. The results showed that the 21CE activities can cultivate a sense of love and affection for God when the teachers apply 21CE activities such as critical questioning, group discussion, and games. The study of IET understanding of critical thinking activities, one of the 21CE elements, found that teachers understand that thinking activities are an act of worship and a tool to strengthen faith while wisely spreading Islamic teachings [7]. Indirectly, religious appreciation can have a good impact on the daily interaction of students based on the implementation of 21CE activities carried out by the teachers. This can be seen through the contribution of consistently applied knowledge in the teaching process with the elements of values and ethics, which has opened space for students to practice social manners such as greeting, cooperating, and helping each other in daily life. Therefore, the application of good and positive values should be inculcated to form students who are leaders with good morals to have a positive impact on their fellow students and the surrounding community. Effective teaching activities play an important role in producing individuals who fear God [8] and inculcate positive moral values while stimulating students' cognitive thinking and cultivating leadership [9].

Additionally, there are three themes on creative methods that can be used by the IET to inculcate values during 21CE activities, namely role model, reward and punishment, advice, use of social media, and innovation production. The creative method to cultivate values is also in line with the views of al-Ghazali in *Ihya 'Ulum al-Din* [4,10], such as: (1) loving and sympathetic, loving students like their own children; (2) sincerity, teaching with the sole intention of seeking the pleasure of Allah S.W.T; (3) giving advice, the teacher always advises the student through good words; (4) reprimanding mistakes wisely; (5) gradual teaching according to the ability of the student's intellect; and (6) practice with knowledge. Islamic education strongly emphasizes moral education and teachers play an important role in ensuring that this can be achieved. Creative methods to apply moral values can be practiced by the IET because students always need guidance and advice in any T&L situation and activities so that students are spared from moral and social problems. Educators are responsible for playing a role in 'touching the hearts' of students and trying to limit their involvement in social problems by giving advice, reprimand, guidance, and being a good example and role model to students while implementing the T&L [11]. In addition, Islamic values can also be incorporated into teaching through innovations produced by the IET. There is a change in students' willingness to accept teaching innovations and Islamic values, and this can help produce students who excel academically and have outstanding personalities [12].

5. Conclusions

A summary of the theme for the first research question can be seen in Table 2:

Table 2. The theme of values applied through the implementation of 21CE activities.

Theme	Sub-Theme	Sub-Sub-Theme
Religious appreciation	*Akidah* (Belief) *Akhlak* (Values)	
Daily interaction	Respecting friends and teachers	Greeting by giving *'salam'* Work together Celebrate the difference
Religious appreciation	Plan, organize, and interact with friends Courageous to come up with ideas Respect differences of opinion	

A summary of the themes for the second research question can be seen in the following table (Table 3):

Table 3. The theme of creative methods in applying moral values during 21CE activities.

Theme	Sub-Theme	Sub-Sub-Theme
Role model	Project implementation Teacher's manners	
Rewards and punishments	Task completion Evaluation	
Advice	The beginning or end of T&L In the form of reflection or summary	
Use of social media		
Innovation	Focus on specific topics such as prayer and manners The development of innovation according to the nature of the Islamic education subject	

Table 2 shows the summary of the theme for values applied through the implementation of 21CE activities namely religious appreciation, daily interaction and religious appreciation. Table 3 shows the summary of the themes of creative methods in applying moral values during 21CE activies. The themes are role model, rewards and punishmnents, advice, use of social media and innovation. Therefore, various methods to apply moral values can be done in the Islamic education teaching and learning process. However, its implementation requires a tremendous effort and commitment from teachers to equip themselves with various skills to ensure that the T&L runs smoothly and strives to educate students in line with current developments. IET not only has a role to convey the content but also has a greater role and responsibility to achieve the student outcome as an excellent human being with noble morals. The findings of this study are expected to provide ideas to IET to apply moral values creatively in the 21CE activities.

Author Contributions: Conceptualization, W.A.A.W.A. and N.Z.; methodology, W.A.A.W.A.; software, W.A.A.W.A. and Z.R.; validation, Z.R.; formal analysis, N.Z.; investigation, W.A.A.W.A., Z.R. and N.Z.; resources, W.A.A.W.A., Z.R. and N.Z.; data curation, W.A.A.W.A., Z.R. and N.Z.; writing—original draft preparation, N.Z.; writing—review and editing, W.A.A.W.A. and N.Z.; visualization, W.A.A.W.A., Z.R. and N.Z.; supervision, M.I.H., K.A.R. and H.Z.; project administration, M.I.H., K.A.R. and H.Z. All authors have read and agreed to the published version of the manuscript.

Funding: This research received no external funding.

Institutional Review Board Statement: Not applicable.

Informed Consent Statement: Not applicable.

Data Availability Statement: Not applicable.

Acknowledgments: We are grateful to all of those with whom we have had the pleasure to work during this research and also to all the team members for the supports and encouragement.

Conflicts of Interest: The authors declare no conflict of interest.

References

1. Suryadarma, Y.; Haq, A.H. Pendidikan akhlak menurut Imam Al-Ghazali. *At-Ta'dib* **2015**, *10*, 362–381.
2. Sabilan, S.; Mohamed Lip, S.; Ishak, M.F.; Zdainal Abidin, S.; Sulaiman, S.H. Konsep penerapan dan penghayatan nilai-nilai murni berasaskan Falsafah Pendidikan Kebangsaan (FPK). In Proceedings of the International Conference on Moslem Society, Kolej Universiti Islam Antarabangsa Selangor, Bangi, Malaysia, 24 October 2016; pp. 1–13.
3. Ismail, S.N.F.S.; Maat, S.M. Pembelajaran Abad Ke 21: Pendekatan Kontekstual Untuk Pembelajaran Bermakna dalam Pengajaran dan Pembelajaran Matematik. In Proceedings of the Seminar Pendidikan Serantau ke-VIII, Universiti Kebangsaan Malaysia, Bangi, Malaysia, 7 September 2017; pp. 75–89.
4. Mohamad, W.M.W.; Teh, K.S.M. Pembelajaran abad ke-21 dalam Al-Quran: Satu manual pendidikan istimewa daripada Allah. *J. Educ. Res. Indig. Stud.* **2019**, *2*, 192–201.
5. Noor, N.E.M.; Kasim, T.S.A.T.; Yusoff, Y.M. Peranan guru dalam pelaksanaan e-pembelajaran pendidikan islam menurut perspektif Al-Ghazali: The role of teachers in the implementation of e-learning in islamic education according to Al-Ghazali's Perspective. *J. Islamic Educ. Res.* **2021**, *6*, 52–63.
6. Merriam, S.B. *Qualitative Research: A Guide to Design and Implementation*; Jossey-Bass: San Francisco, CA, USA, 2009.
7. Zhaffar, N.M. Kefahaman dan Amalan Penerapan Pemikiran Kritis Dalam Pengajaran Pendidikan Islam Sekolah Menengah. Ph.D. Thesis, Universiti Kebangsaan Malaysia, Bangi, Malaysia, 2017.
8. Othman, M.S.; Kassim, A.Y. Kesahan dan kebolehpercayaan instrumen amalan pengajaran guru pendidikan islam sekolah rendah di dalam mengintegrasikan Kemahiran Berfikir Aras Tinggi (Kbat) menerusi pengajaran akidah. *O-JIE Online J. Islamic Educ.* **2019**, *7*, 15–26.
9. Kamal, S.K.M.; Hamzah, M.I. Penerapan elemen nilai murni dan etika dalam pembelajaran abad ke-21 dalam kalangan guru pendidikan islam sekolah rendah. *Int. J. Adv. Res. Islamic Stud. Educ. (ARISE)* **2021**, *1*, 16–31.
10. Khadijah, I. Etika guru dan murid dalam pendidikan perspektif Imam Al-Ghazali. *MENDIDIK J. Kaji. Pendidik. Dan Pengajaran* **2019**, *5*, 89–102.
11. Noor, A.F.M.; Jasmi, K.A.; Shukor, K.A. Masalah sosial dalam kalangan pelajar dan peranan pensyarah Pendidikan Islam dalam pengajaran dan pembelajaran. *e-BANGI* **2016**, *11*, 50–61.
12. Khalil, A.B.A.; Othman, M.K.B.; Saidon, M.K.B. Memacu pendidikan di era Revolusi Industri 4.0: Penerapan nilai-nilai islam dan inovasi dalam pengajaran di institusi pengajian tinggi. *Islamiyyat* **2020**, *42*, 13–20.

proceedings

Proceeding Paper

Integrating Design Features for E-Learning Platforms †

Lilian Lee Shiau Gee

Academy of Arts and Creative Technology, Universiti Malaysia Sabah, Sabah 88450, Malaysia; lilian@ums.edu.my

† Presented at the International Academic Symposium of Social Science 2022, Kota Bharu, Malaysia, 3 July 2022.

Abstract: The reliance on online learning platforms to continue teaching and learning with COVID-19 issues and implications emphasises the necessity and preparedness of e-learning platforms as a main means of providing students with quality education. Supporting web platforms for higher education requires an awareness of content design elements for online learning sustainability and optimisation. This study proposes that e-learning platforms must prioritise usability and utility from a design standpoint to meet future needs. This article presents an overview of the primary features necessary in an online learning platform and conceptual work linked to design feature variables that may be developed to produce a good online learning platform and fulfil local students' preferences. Sharing, search, networking, and organisation enhance distance learning. This study examines new e-learning platform features to improve educational technology.

Keywords: e-learning; higher education; design platform

Citation: Gee, L.L.S. Integrating Design Features for E-Learning Platforms. *Proceedings* **2022**, *82*, 23. https://doi.org/10.3390/proceedings2022082023

Academic Editor: Mohamad Rahimi Mohamad Rosman

Published: 13 September 2022

1. Introduction

Students are no longer unfamiliar with e-learning platform utilisation. The sudden change in daily activities brought on by the COVID-19 pandemic increased the importance of web-based learning, especially the shift from face-to-face to online learning. Clearly, e-learning allows individuals to learn in a variety of ways; however, creating a high-quality e-learning platform is often a challenge for higher education. The evolution of web design and the results of a survey conducted at Universiti Malaysia Sabah indicate a pressing need to rethink and revamp the concept of an e-learning platform that not only provides "easy to access" content, but also stimulates a sensual learning experience, ease of communication, and a "multi-functional" context [1].

Therefore, the e-learning platform approach should be reconsidered in light of a more strategic, multifunctional, and experience-oriented design [2,3] for characterising students in relation to learning domains (cognitive, psychomotor, and affective). The design of the online learning procedure should incorporate technical (mechanism), content (design), and visual considerations (attractiveness). The creation of a straightforward e-learning design necessitates appropriate design characterization [4] and parallel technical and visual implementation. The requirement for didactic and effective content extends the knowledge transfer to the formation of an effective, integrated education.

This study aims to develop a conceptual framework for e-learning platform design characterization in order to assist researchers, educators, and designers in developing an appropriate platform for learning in higher education. This framework intends to shed light on the classification of key e-learning platform features, the nature and structure of existing dashboard displays, and the application of necessary component aspects for online learning. This overview aims to maximise the use and functionality of e-learning platforms, so that local students can continue learning remotely and maintain contact with instructors and peers.

2. Education, Online Learning, and Web Platforms

In the design of e-learning platforms that support lecturers, students, and other stakeholders in teaching and learning activities, a multi-functional awareness is often emphasized [5]. This specification must investigate the use of integrated design features to define screen interfaces for e-learning that support learning activities, synchronous interaction, asynchronous interaction, and sensory engagement. Consider the current interest in how e-learning should be shaped and the implications of the required categorization for e-future learning. Dias et al. [6] identify communication tools, administration tools, and course delivery tools as crucial design principles for online learning platforms. According to Al Ajlan's [7] classification, in order for an e-learning platform to be highly functional, it must incorporate three main features: learning tools, support tools, and technical tools. Regarding the capabilities of e-learning platforms developed for academic institutions, Garrote Jurado et al. [8] assert that the most appropriate features for higher education are distribution tools, communication tools, interactive tools, and course administration tools to support the platform's design structure, and establish its relationship to blended and distance learning. In addition, Reitano's [9] study on students' need for e-learning platforms reveals that students are more concerned with the variety of communication tools and the adaptability of collaboration tools (such as Wiki and Google docs) when evaluating e-learning platforms.

Therefore, a deeper understanding of design features allows educational institutions, such as universities, to optimise the use of e-learning platforms and increase the likelihood of integrating institutional and consumer culture into the digital environment. All of these factors can ultimately result in a deeper understanding of how engagement with e-learning platforms impacts learning activities and motivational factors for e-learning. The platform is closely associated with the rising propensity of young people to engage in multiple tasks and learning activities. Given that "multi-function" is viewed as the platform's ability to communicate with the environment to deliver learning-specific information, the potential of the platform in continuous learning by students has necessitated features (integration of tools) that meet the need to maintain continuous use for addressing current learning problems, particularly in a crucial and necessary higher education environment.

3. Online Platform Design Features Support

Each learning platform's characteristics are highly dependent on its intended purpose and principle of application within the field of education. Thus, instructional design research begins to consider the requirements and consequences of e-learning platforms for education, with an emphasis on the development of higher learning component features.

Facilitated file sharing, media exchange (images, audio, and video), and instant discussion increased student participation in online discussions, thereby enhancing the platform's use in online education [10]. The ability to send real-time messages to individuals or groups of friends simultaneously, low cost, and privacy are characteristics that increase the popularity of e-learning use [11], and these characteristics strengthen student acceptance of the platform in addition to indirectly influencing students' attitudes and experiences. This suggests that the online learning platform must incorporate the following features in order to gain widespread student acceptance.

Richness of media influences the experience of lecturers and students as a learning tool, and the positive experiences that result increase students' assignment-completion productivity. This media richness [12,13] refers to the possibility of design principles supporting social interaction (lecturers and students), information processing, usability, and motivation. Similarly, the description of education platform development should avoid monotonous interface backgrounds, multifunctional and cluttered systems, and excessively formal design arrangements [14].

Users benefit from the availability of collaborative knowledge sharing, social connections, simple accessibility, engaging representations, and multiple functions [12,14]. Editing, annotating, and sharing information digitally and instantaneously are essential features for easily accessible instructional content, and efficiently facilitate students' search for information during independent study. In addition, educators and students benefited from the absence of time constraints and barriers in the use of online platforms in the context of teaching and learning. Therefore, platform pages that address privacy and security are required to increase users' confidence in the platforms presented.

This demonstrates that the platform principle must be implemented with appropriate design features to improve online learning, where these features should contribute to student activation, ease of browsing, collaboration, expanded social interaction, active discussion, peer support, instant feedback, an efficient information provider, a large number of members, and achievement motivation. As a network of educational resources and support, it is crucial to benefit online classrooms and engage students and teachers. In addition, there is a requirement for an instantaneous alert function that includes announcements to encourage interaction and the sharing of course materials, assignments, tutorials, and class sessions.

4. Tools Integration in the Platform

The development of a successful platform requires more than the simple combination of teaching content and online capabilities; it should also include a combination of diverse features with four primary aspects: sharing, searching, networking, and organising to meet demand, flexibility, prompt delivery, and an efficient learning environment. This study examines in greater detail the integration of tools into the e-learning platform. These tools are divided into the following four categories:

- Sharing tools:
- A sharing tool that allows lecturers and students to upload text documents and various types of media files;
- Search tools:
- Lecturers and students have easy access to information exploration tools during learning activities;
- Networking tools:
- Tools that permit interaction with additional rewards, reactions, and immediate feedback could encourage student activity and collaboration, thus, enhancing the learning experience;
- Organizing tools:
- Course organizing tools used to monitor and record the educational process. The tool also enables course evaluation and course management by lecturers, as well as the ability to customise screen layouts according to student preferences and instructional needs.

With a focus on robust design features, and a greater emphasis on tool integration, it is crucial that e-learning platforms include the most essential components in each tool category. The categories of e-learning-platform-required tools and their constituent parts are depicted in Figure 1.

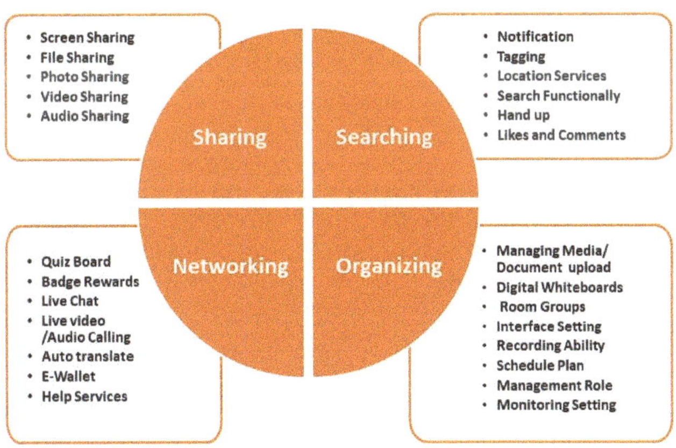

Figure 1. The four tool categories in e-learning platform.

4.1. Sharing Tools

The speed of sharing is a benefit, while online learning is established. Sharing tools must include a method of obtaining media, links, and files instantly in live chat, unlimited media size, and must ensure that shared files, including meeting (class) recordings, are accessible on live chat throughout the semester. A new functionality proposed for sharing is a cloud-based folder where students and lecturers can access all collected media and files with a single click.

4.2. Search Tools

Depending on the responses of students and educators, the searching segment could expand to include features such as notifications, tagging, location services, search functionality, hand up, likes, and comments for enhanced distance learning or online learning.

4.2.1. Notification

Post alerts directly from the platform as a friendly reminder to students and instructors enrolled in scheduled classes. To join the online class, simply click on the pop-up notification.

4.2.2. Tagging

The tag feature enables students or lecturers to rapidly disseminate knowledge and pertinent learning topics. Tag dependency is a classroom benefit to increase student interest in mutual learning keyword topics.

4.2.3. Location Services

Enable shared locations, which could be an excellent way for instructors to monitor their students' conditions.

4.2.4. Search Functionally

A search feature is a design element that allows students to easily access content by inputting specific keywords or phrases.

4.2.5. Hand Up

The handheld function permits students to pose questions or request anything from an instructor or a friend.

4.2.6. Likes and Comments

The ability to like and comment in live chat as a contact source specification for individual posts in order to enhance individual participation and feedback in the learning process.

4.2.7. Hashtag

The inclusion of the symbol "#" in a post with the intent of outlining shared knowledge topics and facilitating the search for information by keywords asserts knowledge sharing.

These search features contribute to the value of knowledge in online education, such as the 'like' button, which measures the majority of students' consent and opinions. The comments feature enables user discussions, thereby promoting the dissemination of new information. The tag and hashtag features are the most recent means of facilitating communication between users. The capability of instant tagging is useful for alerting other users to important information or activities, so they do not miss out.

4.3. Networking Tools

Concerning student engagement and immersion, it is essential to use networking tools such as quiz boards to maximise online classroom interactivity. The quiz board is a multiple-choice knowledge evaluation game that can be played in real time. In addition, quiz boards can be used for formative assessments, such as voting and student opinion polls. In an educational platform, badge rewards are essential for student motivation. These badges may be rewarded with stickers, GIF emoticons, or emojis. This is another simple method for promoting positive attitudes among online learners. Transparency and the speed of the live chat tool, audio, and video are essential in online classrooms for maintaining positive emotions and attitudes among students and instructors, and for continuing the teaching and learning process. Cultural differences can lead to varying interpretations of the use of particular sentences and terms; therefore, automatic translation is advantageous for international students, allowing them to communicate effectively and comprehend learning themes more thoroughly. The subtitling capabilities of an automatic translation tool enhances learning, assignments, and conversations.

The pandemic in Malaysia requires the integration of e-wallets into the platform. The e-wallet system allows students to manage their finances and quickly reload their phone data. Long-term, the e-wallet on the platform enables students, particularly those at a distance, to make cashless and secure payments. Help services are a component of a networking tool intended to serve as a resource and aid for students. The aid service is responsible for providing IT support for platform-related technical issues, as well as troubleshooting and institution-related services.

4.4. Organizing Tools

Support services for students and instructors are systematically organised. In accordance with the recommendations of Pireva et al. [15], the standard of e-learning platforms should be enhanced in terms of student control over the screen interface, in order to increase student interaction and motivation by optimising the use of allocated features. The availability of digital whiteboards improves the responsiveness of online learning and facilitates the presentation by lecturers and students in classes, especially in tutorial classes and practical training, such as art classes, digital art classes, and courses involving skill acquisition. Digital whiteboards promote mutual comprehension by facilitating collaboration, idea development, and editing between students and instructors. Group rooms facilitate small-group interactions among students, making it easier for instructors to supervise and monitor student tasks, particularly group assignments and group projects. The group room also serves as a waiting area where students can engage in preliminary preparations and conversations prior to presenting their assignments or displaying their work.

The interface settings allow students to customise the layout of the display to their preference. By allowing users to select screen style, screen colour, display size, text size, font style, background blurring, and screen effects, interface options can emphasise student comfort and confidentiality.

Changing educators' reliance on recording online classes on e-learning platforms allows students in the classroom to create the most effective remote learning environment possible. The ability for anyone to take recordings increases the visibility of instructional content for all users, and the recorded content can be distributed instantly to the chat room of the platform.

Adding new tools such as schedule plans facilitates time and schedule management for instructors and students. Announcements and reminders regarding class sessions, events, and discussions can be scheduled by instructors. The schedule plan serves as an alert system for the public. Schedules allow instructors to be informed of invitations to instruct, reviews of student work, the return of formative assessment, and grade reports. The schedule plan aids students in submitting assignments, exam dates, and daily agenda reminders on time.

The intent of management tools is to help lecturers and students optimise the flow of lessons. The access button enables the instructor to designate the group leader, presenter, and accessor. In chat messages, attendees can include the status of a class post. Accessors can regulate who can join their sessions. Not only instructors, but also students, can record and send recordings automatically in chats. Each student has the ability to mute a classmate's microphone using a simple "mute" button. Students could invite their friends to enrol in the class via a notification menu. Using the "role swap" option, both educators and students can share screens for discussions using the "share" function. The monitoring setting is a distinguishing feature that makes the job of online educators easier. Using the monitoring setting tool, lecturers have greater control over the management of their teaching courses by configuring reminders for class times, replacement classes, assignment submission schedules, late assignment submissions, re-submissions of student assignments, and additional class events. Activating reminder alerts enables lecturers to ensure that all students can effectively follow online classes.

5. Conclusions

To overcome this obstacle, the instructional feature framework of this built-in e-platform aims to upgrade and provide more social features in the interface to enhance online learning in accordance with current social platform image trends. By optimising the necessary tools, it improves the platform requirements that meet the needs of local students and expands the advantages of functionalization in online learning. The framework also aims to bridge the gap between learning outcomes and the implications of online learning tools by developing ideal practises for the formation of a variety of effective learning methods.

This feature design framework is intended to provide educators, designers, developers, researchers, and organisations with recommendations for developing a more beneficial and effective online learning platform. It is hoped that future research will be expanded by testing the design framework of instructional features of this e-platform in order to gain a better understanding of the effectiveness of each of the features in the implementation function of online learning, and to develop a more knowledgeable iteration model in instructional design to provide advantageous instruction. Through the study of the structure of learning, technology becomes a critical part in the creation of innovative and high-tech online classroom instruction.

Funding: This research was funded by SKIM DANA KHAS, grant number SDK0284-2020 and the APC was funded by Universiti Malaysia Sabah.

Institutional Review Board Statement: Not applicable.

Informed Consent Statement: Not applicable.

Data Availability Statement: Not applicable.

Acknowledgments: Sincere gratitude is extended to the Ministry of Higher Education (MOHE) and Universiti Malaysia Sabah (UMS), both of which encouraged and supported this study as part of a Grant Project research endeavour. In addition, many thanks to all organisations for their great assistance, and to the reviewer for their thoughtful comments.

Conflicts of Interest: The authors declare no conflict of interest.

References

1. Gee, L.L.S.; Pangayan, V.; Ee, C.Y.; Dolah, J. Effective social media features as strategy platform teaching and learning during the movement control order (MCO) period. *Turk. J. Physiother. Rehabil.* **2021**, *32*, 7193–7199.
2. Alkhasawnh, S.; Alqahtani, M.A.M. Fostering students' self-regulated learning through using a learning management system to enhance academic outcomes at the University of Bisha. *TEM J.* **2019**, *8*, 662–669. [CrossRef]
3. Kehrwald, B.A.; Parker, B. Implementing online learning: Stories from the field. *J. Univ. Teach. Learn. Pract.* **2019**, *16*, 1. [CrossRef]
4. Bhatia, R.P. Features and effectiveness of e-learning tools. *Glob. J. Bus. Manag. Inf. Technol.* **2011**, *1*, 1–7.
5. Borrellia, L.; Perrella, S. User interface design for e-learning platform and institutional portal of University of Foggia. In Proceedings of the Second Workshop on Technology Enhanced Learning Environments for Blended Education, Foggia, Italy, 5–6 October 2021.
6. Dias, S.B.; Dinis, J.A. Towards an enhanced learning in higher education incorporating distinct learner's profiles. *Educ. Technol. Soc.* **2014**, *17*, 307–319.
7. Al-Ajlan, A.S. A comparative study between e-learning features. In *Methodologies, Tools and New Developments for E-Learning [Internet]*; Pontes, E., Silva, A., Guelfi, A., Kofuji, S.T., Eds.; IntechOpen: London, UK, 2012.

8. Garrote Jurado, R.; Pettersson, T.; Regueiro Gomez, A.; Scheja, M. Classification of the features in learning management systems. In Proceedings of the XVII Scientific Convention on Engineering and Architecture, Havana City, Cuba, 24–28 November 2014.
9. Reitano, R. Criteria for an eLearning platform. In Proceedings of the Quality in Blended Learning Conference, Wiener Neustadt, Austria, 20–22 February 2014.
10. Salloum, S.A.; Alhamad, A.Q.; Al-Emran, M.; Monem, A.A.; Shaalan, K. Exploring students' acceptance of e-learning through the development of a comprehensive technology acceptance model. *IEEE Access* **2019**, *7*, 128445–128462. [CrossRef]
11. Cetinkaya, L. The impact of whatsapp use on success in education process. *Int. Rev. Res. Open Distrib. Learn.* **2017**, *18*, 7. [CrossRef]
12. Baehr, C. Incorporating user appropriation, media richness, and collaborative knowledge sharing into blended e-learning training tutorial. *IEEE Trans. Prof. Commun.* **2012**, *55*, 175–184. [CrossRef]
13. Gyamfi, A. Media richness perspective of social media usage for learning: Perception of cocoa researchers in Ghana. In *Evaluating Media Richness in Organizational Learning*; Gyamfi, A., Idongesit, W., Eds.; IGI Global: Hershey, PA, USA, 2018; pp. 1–5. [CrossRef]
14. Mahmood, J.; Dahlan, H.M.; Hussin, A.R.C. Enhancement of e-learning system by using social network features. In Proceedings of the 2013 IEEE Conference on e-Learning, e-Management and e-Services, Kuching, Malaysia, 2–4 December 2013; pp. 24–29. [CrossRef]
15. Pireva, K.; Nuqi, F.; Hajrizi, E. The impact of LMSs in learners'. In Proceedings of the Quality in Blended Learning Conference, Wiener Neustadt, Austria, 20–22 February 2014.

proceedings

MDPI

Proceeding Paper

Major Selection Tendency among Creative Arts Students, Academy of Arts and Creative Technology, UMS †

Victor Pangayan

Academy of Arts and Creative Technology, Universiti Malaysia Sabah, Kota Kinabalu 88450, Malaysia; victorpangayan@ums.edu.my

† Presented at the International Academic Symposium of Social Science 2022, Kota Bharu, Malaysia, 3 July 2022.

Abstract: Both teachers and students are concerned about choosing majors. The gap between initial student expectations and external circumstances might lead to inferior academic performance and higher dropout rates. The purpose of this study was to determine the majors chosen by university freshmen and the factors that influenced their selection of creative arts majors. They filled out surveys about ASTiF's Creative Arts majors. Results show that desire, family, lecturer, and course difficulty influence student major selection. According to the survey, students have their own perceptions of majors before pursuing these specialisations. Academic accomplishment, personal and social growth, and retention must be outlined in advance. Better preparation and communication between teachers, students, and parents are needed to identify more appropriate majors for each student. Universities could perhaps offer courses geared toward the aspects of academic choices, attitudes, and social skills. This should be a process, not an event, and should include peer-mentoring and staff—student engagement to choose an ASTiF Creative Arts major.

Keywords: creative arts; higher education; student perceptions

check for updates

Citation: Pangayan, V. Major Selection Tendency among Creative Arts Students, Academy of Arts and Creative Technology, UMS. *Proceedings* **2022**, *82*, 26. https://doi.org/10.3390/proceedings2022082026

Academic Editors: Mohamad Rahimi Mohamad Rosman

Published: 13 September 2022

1. Introduction

The Academy of Arts and Creative Technology (ASTiF) is a new faculty of Universiti Malaysia Sabah created in 2021 [1]. ASTiF was founded in accordance with the reorganisation of the Social Sciences and Arts cluster, which had been united under the Faculty of Humanities, Arts, and Heritage since its establishment in 2014 [2]. The existence of ASTiF implicitly recalls the existence of the School of Arts (*Sekolah Pengajian Seni*), a former body at Universiti Malaysia Sabah that offered arts-focused study programmes such as the Creative Arts Programme, the Music Arts Programme, and the Visual Arts Technology Programme [3]. Compared to the School of Arts, ASTiF is an academy that concentrates on subjects attributed with the creative sector by offering three main programmes, namely the Creative Arts Programme, the Music Arts Programme, and the Visual Arts Technology Programme, as well as a forthcoming programme, the Cinematography Programme. Creative Arts is one of these programmes that frequently has an imbalance of students between the various majors. This makes the programme so unique that it merits thorough observation.

2. Creative Arts Programme

The Creative Arts Programme (UH2162002) is a programme that channels the specialisation of students into three branches: dance, theatre, and writing. This three-year, six-semester programme awards a Bachelor of Arts with Honours upon completion (Creative Arts). This programme offers both theoretical and practical courses, including Dance Art Analysis, Art Criticism, and Comparative Art Analysis, as well as Dance Art Fundamentals, Acting Fundamentals, Direction, Composition, and Choreography, New Media Writing, and Short Story Writing, and others [4]. However, the distribution of students according to their chosen major is unequal and follows a consistent percentage trend year after year. This percentage is displayed in Table 1, which depicts the number of students per strip from 2019 to 2021.

Table 1. Number of Creative Arts Students Based on Specialisation/Strip from 2019 to 2021.

Year / Strip	2019	2020	2021
Writing	74 students	67 students	64 students
Theatre	37 students	40 students	31 students
Dance	20 students	6 students	10 students

Based on the distribution of students indicated in Table 1, we can conclude that, on average, only 14 percent of students opted to concentrate on the dance strip, 29 percent chose the theatre segment, and the remaining students chose the writing segment. This has increased the significance of the study examining the primary selection parameters used by students in the Creative Arts. This study's objective was to assess the factors that influence students' choices in dance, theatre, and writing. Understanding students' attitudes, trends, and interests is crucial, especially for guiding and assisting students to define their talents at the beginning of the semester and choosing a field that matches their abilities, which is essential for their success in their professional careers.

3. Research Methodology

The purpose of this study was to evaluate student major selection by requesting responses to six essential factors that may influence their major choice. Using a random selection technique to obtain a sample, this study remained only focused on Year 1 Creative Arts Program students. This study was conducted between December 2020 and February 2021. Using Google Forms, questionnaires were distributed. Each respondent was given a questionnaire containing a total of 24 items. The level of respondents' agreement was measured using a Likert scale, with 1 representing "strongly disagree" and 5 representing "strongly agree." According to this study's findings, a total of 78 valid questionnaires were collected.

The first section evaluates respondents' feedback on preference tendency factors, the second section measures the influence of respondents' friends and family on major selection, the third section focuses on aspects of impact from lecturers and mentors, the fourth section focuses on the effect of course difficulty level, the fifth section assesses curriculum factors, and the final section focuses on career factors on student major selection. All obtained data will be evaluated using statistical packages for social science programmes (SPSS), with frequency analysis and percentage analysis serving as the fundamental analyses.

4. Results

4.1. Student Demographics

According to the number of students who responded to this survey, 59 (75.6%) are female students and 19 (24.4%) are male students. Sabah accounted for 82.1% of the students, followed by Peninsular Malaysia with 10.2% and Sarawak with 7.7%. A total of 65.4% (51 individuals) of students are from rural areas, while 34.6% (27 individuals) are from urban areas. This indirectly indicates that the majority of student households belong to the low-income B40 category, which is documented at 70.5%. (55 individuals). In terms of student entrance routes to the Creative Arts Program, the UPU offer, which registered 64 students, accounts for the majority of students. Meanwhile, 14 more individuals have passed through the appeals process. However, no student recruitment through talent has been documented as yet. This is evident from Table 2, which displays the proportion of students' personal information.

Table 2. Student Personality Information.

Student Personality	Percentage
Gender	
• Male • Female	24.4% 75.6%
Residential area	
• Urban Area • Rural Area	34.6% 65.4%
Household income	
• B40 • Others	70.5% 29.5%
Methods of entering the Creative Arts Program	
• UPU System • Appeal • Entry through talent	82.1% 17.9% 0.0%

When asked about their selection within the Creative Arts Program, the majority of students (60.3%, or 47 individuals) chose writing. A total of 22 students, 28.2%, chose theatre, , while 9 students, 11.5%, chose dance . Again, this percentage implies an inconsistent selection pattern for specialisations. This is shown in Figure 1.

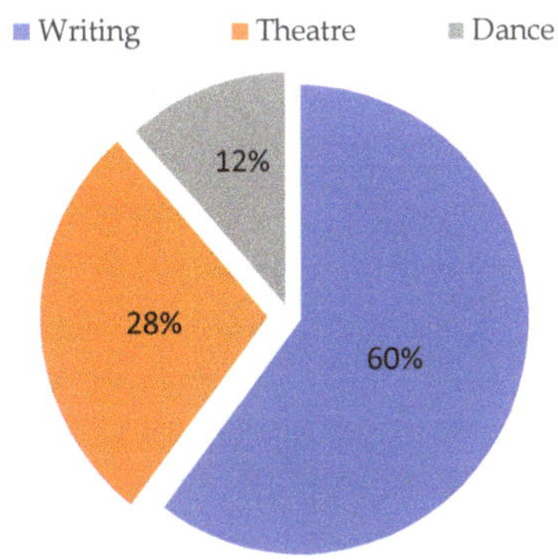

Figure 1. Student Choice Specialisation.

4.2. Tendencies and Preferences

Students in the Creative Arts Program base their choice of specialisation firstly on their own inclinations or interests, uninfluenced by any third party. This is the factor that academics are most interested in observing, as it has the most influence on students' choice of academic direction [5]. To examine this factor, three questions have been outlined in order to precisely determine the tendencies of students. On the first question regarding their level of interest in their chosen major, 70.5% responded "strongly agree," while 25.6% responded "agree" and 3.8% responded "quite agree." Regarding the second question, which inquired about the target major they had chosen since the beginning of the course, 62.8% of the students responded "strongly agree," 25.6% responded "agree," and 11.5% responded "quite agree." The same answer was also displayed on the answer scale for question three, which examined the relationship between students' major selection and their ambitions, with a score of 60.3% "strongly agree", 32.1% "agree", and 7.7% "quite agree". On the basis of the percentage of students' responses, it can be concluded that the majority of Creative Arts Program students are very clear and confident in their personal decisions. This is clearly displayed in Table 3.

Table 3. Tendencies and Preferences.

No.	Tendencies and Preferences	SA	A	QA	D	SD
1.	I am interested in my preferred major.	70.5	25.6	3.8	0.0	0.0
2.	I have been aiming to follow this major since the beginning of my studies.	62.8	25.6	11.5	0.0	0.0
3.	I chose this option major to achieve my ambitions.	60.3	32.1	7.7	0.0	0.0

4.3. *Influence of Society and Media*

Peer and family influence is also one of the variables that influences students' decisions and choices regarding their academic choices [6]. "Strongly agree" was selected by 26.9% on social media, followed by "agree" with 35.9%, "quite agree" with 33.3%, "disagree" with 3.8 percent, and "strongly disagree" with no responses. The influence of family members resulted in the second highest percentage for "strongly agree" with 25.6%, followed by 23.1% for "agree", 26.9% for "quite agree", 15.4% for "disagree", and 9% for "strongly disagree". Peer influence resulted in the lowest percentages of "strongly agree" with a score of 3.8%, "agree" with a level of 6.4%, "quite agree" with a score of 26.9%, "disagree" with a score of 15.4%, and "strongly disagree" with a score of 28.2%. This is evident from Table 4, which provides percentage breakdowns for each question in this section. Through the total percentages shown in Table 4, it is clear that major selection among Creative Arts students is not influenced by their peers, but more by the influence of social media as well as their family members.

Table 4. Influence of Society and Media.

No.	Influence of Society and Media	SA	A	QA	D	SD
1.	Peer influence is one of the main factors I take into account in my selection of major.	3.8	6.4	19.2	42.3	28.2
2.	My family exerted a huge influence in my selection of major.	25.6	23.1	26.9	15.4	9.0
3.	Major selection was due to advice and recommendations from my Senior.	11.5	26.9	34.6	16.7	10.3
4.	Looking at the success of my family members in this field, has helped me choose a major.	21.8	26.9	15.4	28.2	7.7
5.	I got input from social media that helped me with major selection.	26.9	35.9	33.3	3.8	0.0

4.4. *Influence of Mentor/Lecturer*

As shown in Table 5, the third factor employed as a measurement in this survey is the influence of mentors or programme lecturers. This is significant because students are more inclined to select their fields of study based on their idols [5]. Five questions are outlined for measuring the influential factors of mentors or lecturers, and the highest percentage indicates that mentors or lecturers are the primary source of information for students to explain the objectives, content, and assessment methods in the academic programme they will take. A total of 62.8% of students selected "strongly agree", 30.8% selected "agree", 6.4% selected "quite agree", and no student selected "disagree". In addition, 60.3% of students selected "strongly agree" (60.3%), 35.9% selected "agree" and 3.8% selected "quite agree" regarding the influence of mentors and lecturers in guiding students decisions. Further, none disagreed. By omitting the percentages for "disagree" and "strongly disagree," the distribution of student responses in Table 5 reveals that the vast majority of items were answered positively by students.

Table 5. Influence of Mentor/Lecturer.

No.	Mentor/Lecturer	SA	A	QA	D	SD
1.	My mentor/lecturer had a huge influence on my selection of major.	23.1	35.9	23.1	10.3	7.7
2.	My mentor/lecturer facilitates communication between students and students and lecturers.	41.0	48.7	10.3	0.0	0.0
3.	My mentor/lecturer gave an overview of the beginning and end of the course we studied in class.	53.8	33.3	12.8	0.0	0.0
4.	My mentor/lecturer maintains an objective and respectful nature towards students.	60.3	35.9	3.8	0.0	0.0
5.	My mentor/lecturer provides clear information on the objectives, content, and methods of assessment in the subject curriculum.	62.8	30.8	6.4	0.0	0.0

4.5. Course Intensity

In addition to the three previously discussed factors, there is a fourth factor that must be considered: the level of difficulty of the Creative Arts Program courses. This is accomplished by outlining the three questions listed in Table 6. Approximately 56.4% responded they "strongly agree" in response to the second of the three questions answered. The response rate for the third question was 50.0%, while the response rate for the first question was 44.9%. On the "quite agree" scale, the percentages for the three questions were comparable, with the first question garnering the greatest percentage (15.4%), followed by the third question (10.3%), and the second question (5.1%). Only the third question on the "disagree" scale recorded a percentage of 1.3%, while the other two questions did not record a percentage figure. The same holds true for the "strongly disagree" scale, which had no graphical representation percentages for the next three items. Table 6 makes this explicitly clear.

Table 6. Creative Arts Program Course Intensity.

No.	Course Intensity	SA	A	QA	D	SD
1.	The courses offered in my preferred major are areas in which I am proficient and I can get excellent marks.	44.9	39.7	15.4	0.0	0.0
2.	I can demonstrate and polish my abilities as well as talents in my chosen major through activities and coursework.	56.4	38.5	5.1	0.0	0.0
3.	I am happy to face the challenges I face in my chosen major.	50.0	38.5	10.3	1.3	0.0

4.6. Curriculum Factors

Through this poll, it was also found that curriculum factors in a programme had a significant impact on students' decisions regarding their specialisation or major. Table 7 indicates that, as a whole, all 78 students agreed that curriculum variables play a major role in the selection of their profession and major. This is shown by the average percentage score of 56.4% for the "strongly agree" scale across the five question segments. According to Table 7, just 1.3% of respondents who responded negatively to the first question indicated disagreement. Based on the responses to these five questions, the tutor support and assistance factor received the highest percentage (62.8%), followed by the effective learning module delivery factor (61.5%), the description method at the beginning of the course (57.7%), and the flexibility in teaching delivery for selected majors (48.0%). This is clearly illustrated in Table 7 below.

Table 7. Curriculum Factors.

No.	Curriculum Factors	SA	A	QA	D	SD
1.	My major choice is having flexibility in classroom teaching delivery.	48.7	42.3	7.7	1.3	0.0
2.	The modules in my choice major are delivered effectively and help me improve my skills.	61.5	33.3	5.1	0.0	0.0
3.	Each course is delivered regularly and in accordance with the activities of students being active in the major of choice.	51.3	41.0	7.7	0.0	0.0
4.	Assessment and teaching methods are well explained at the beginning of the course.	57.7	35.9	6.4	0.0	0.0
5.	The tutors are very helpful in carrying out assignments and facilitating learning.	62.8	32.1	5.1	0.0	0.0

4.7. Career Factors

The final factor identified in this research is the career factor, which should be evaluated while choosing a particular major [7]. This is evident by the fact that the overall average percentage of student responses for this career category was 83.3%. In order to measure this aspect, it is divided into three questions, the first of which is a combination of the percentage scales of "strongly agree" and "agree" (89.7%). While the second impact reported 83.3% (strongly agree + agree) who considered career security in major options and 76.9% (strongly agree + agree) on the relationship between the application of IR 4.0 and major options. This conclusively shows that students agree that their choice of major is influenced by criteria related to their future employment.

5. Conclusions

According to the findings of this study, four factors—inclination, family, lecturer, and course difficulty—play a crucial influence in the primary selection of UH2612002 students majoring in creative arts at the Academy of Arts and Creative Technology (ASTiF). The statistics presented in Tables 4–8 provide clear evidence for this conclusion, as can be seen above. The difficulty of the course that will be studied comes in second with a record of 89.33%, followed by the influence of mentors and lecturers with a percentage of 85.12%, and a career factor with a record of 83.3%. Among the factors that have a big impact on student major selection is the co-curricular factor of the course—the content of the course that will be taken by students. The influence of this factor was rated as 93.32%. This alone is sufficient evidence to show that each of the three majors or specialised paths presents a compelling variety of co-curriculum and various complexity levels, as determined by the viewpoints of the students. This highest percentage may also be influenced by the backgrounds of undergraduates in creative arts, the majority of whom have a Malaysian Higher Certificate of Education (Malaysia Higher Certificate of Education) (STPM).

Table 8. Career Factors.

No.	Career Factors	SA	A	QA	D	SD
1.	I chose this major because of the career opportunities I see in the future.	56.4	33.3	10.3	0.0	0.0
2.	I chose this major because of the guaranteed career opportunities in this field.	43.6	39.7	16.7	0.0	0.0
3.	I chose this major because it relates to IR 4.0 applications in this field.	28.2	48.7	20.5	28.2	0.0

Moreover, significant alternatives have direct ties to the teaching and learning (T&L) of existing courses. Particularly, well-structured classes will offer favourable outcomes for students' decision making. It is noticeable that a learning module that prioritises concept development that facilitates students' learning skills, flexibility, conceptual knowledge, and skill is chosen by students. Regarding curriculum aspects, leadership (stakeholders) has a crucial role in the enhancement of current majors through various approaches. The dean employs leadership to support a number of initiatives implemented by lecturers to enhance the quality of classroom instruction. In addition, programmes of study can ensure that information, techniques, and resources positively impact the educational environment of students, thereby influencing their major choices. This study's findings recommend a greater emphasis on the pleasant interaction between teachers and parents of students in order to comprehend students' learning experiences and, consequently, the selection of acceptable major subjects. In addition, the preparation of workshops for future careers will also encourage students to strive and show good performance in the faculty.

Funding: This research was funded by SKIM DANA KHAS, and the APC was funded by SDK0284-2020 Universiti Malaysia Sabah.

Institutional Review Board Statement: The data reported in this research are obtainable from the corresponding author upon request.

Informed Consent Statement: Informed consent was obtained from all subjects involved in this study.

Data Availability Statement: Not applicable.

Acknowledgments: A huge thank you is extended to the students from the Universiti Malaysia Sabah who took part in the online survey for this research. Due to their receptivity, this study was able to obtain the necessary data for conducting this research. The authors would also like to express their gratitude to the Ministry of Higher Education (MOHE) and Universiti Malaysia Sabah (UMS) for their encouragement and support for this project.

Conflicts of Interest: The authors declare no conflict of interest.

References

1. BPA. Akademi Seni dan Teknologi Kreatif (ASTiF). UMS. 2021. Available online: http://bpa.ums.edu.my/images/dokumen/Prospektus/2021/ASTiF/PENGENALAN.pdf (accessed on 20 February 2022).
2. BPA. Fakulti Kemanusiaan, Seni dan Warisan (FKSW). UMS. 2019. Available online: https://bpa.ums.edu.my/index.php/ms/prospektus-prasiswazah/2019-2020/fakulti/fksw (accessed on 13 March 2022).
3. Han, Y.S. Penubuhan ASTiF Perkukuh Industri Kreatif Sabah. Utusan Borneo. 2021. Available online: https://www.utusanborneo.com.my/2021/10/23/penubuhan-astif-perkukuh-industri-kreatif-sabah (accessed on 14 March 2022).
4. BPA. Struktur Krusus Seni Kreatif 2021/2022. Universiti Malaysia Sabah. 2021. Available online: https://bpa.ums.edu.my/index.php/en/undergraduate-prospectus/2021-2022/fakulti/31- (accessed on 14 March 2022).
5. Mohamad, N.; Md. Tap, A.O. Keutamaan pemilihan bidang dan tempat pengajian: Pendekatan konjoin kabur. *J. Teknol. Mklm. Dan Sains Kuantitatif* **2002**, *4*, 67.
6. Awang, A.H.; Ramli, Z.; Ibrahim, I. Faktor tarikan pelajar siswazah antarabangsa ke universiti penyelidikan di Malaysia. *Geogr. OnlineTM Malays. J. Soc. Space* **2012**, *8*, 32–41.
7. Hussin, M.H.N.; Mujahdeen, H.A. Faktor-faktor pendorong pemilihan kursus di institusipengajian tinggi-satu ulasan. *J. Humanit. Lang. Cult. Bus.* **2017**, *1*, 139–146.

proceedings

MDPI

Proceeding Paper

Students' Attitudes and Behavior towards Academic Dishonesty during Online Learning †

Amirun Hasri, Rafidah Supar *, Nurul Dizyana Nor Azman, Hairenanorashikin Sharip and Lyana Shahirah Mohamad Yamin

Faculty of Health Sciences, Universiti Teknologi MARA, Bandar Puncak Alam 42300, Malaysia
* Correspondence: rafidah5575@uitm.edu.my
† Presented at the International Academic Symposium of Social Science 2022, Kota Bharu, Malaysia, 3 July 2022.

Abstract: Online learning has been an integral part of the educational process in universities, particularly during the COVID-19 pandemic. Despite the popularity of online learning, concerns exist over their level of academic integrity. The aim of this study is to investigate students' attitudes and behavior towards academic dishonesty during online learning. In total, 319 undergraduate health sciences students at a public university took part in the survey. The online self-administered questionnaire was distributed through a social media platform. Data collected were analyzed using the Statistical Package for Social Sciences (SPSS) Version 25.0. Majority of the respondents perceived the indicated behavior as serious cheating. However, most respondents (86.2%) self-report that they have engaged in academically dishonest behaviour at least once for the past one year. Furthermore, approximately 77% (n = 246) of respondent has witnessed act of academic dishonesty among their friends for the past one year. Spearman correlation test revealed no association between students' attitudes and behavior towards academic dishonesty during online learning. The result of this study, in summary, is that students perceive the indicated behaviors as serious cheating and have engaged in academically dishonest behaviors less frequently.

Keywords: academic dishonesty; academic integrity; attitude and behavior

Citation: Hasri, A.; Supar, R.; Azman, N.D.N.; Sharip, H.; Yamin, L.S.M. Students' Attitudes and Behavior towards Academic Dishonesty during Online Learning. *Proceedings* **2022**, *82*, 36. https://doi.org/10.3390/proceedings2022082036

Academic Editor: Mohamad Rahimi Mohamad Rosman

Published: 13 September 2022

1. Introduction

Academic integrity is an integral part of education that should be upheld by every member of an academic community to instill a good learning environment, allowing success and growth. Within tertiary education, acts of academic dishonesty are often used as a measure of lacking academic integrity. Academic dishonesty among students can be defined as academic behavior that does not conform with the university's policies, whereby the students perform acts of dishonesty to gain unjustified advantage in their assessment [1]. In addition, academic dishonesty is also defined as behavior such as plagiarism, unauthorized collaboration, violating examination's rules, cheating during examination and attending an exam in place of another individual [2–4].

Academic dishonesty is not a new phenomenon, as it has been a long-standing challenge, existing since the beginning of traditional learning. Various studies have reported the occurrence of academic dishonesty, indicating that it can occur in various events such as academic assignments, tests or final examinations [5]. It was reported that 75% of students had engaged in one form or another of academic dishonesty [6]. Additionally, it was reported that the prevalence of academic dishonesty among students in medical school ranges from 0% to 58% [7]. Similarly, it was found that about 20% of medical and health sciences students cheated at least once during their undergraduate studies [8]. In recent years, the prevalence of students that self-report on acts of academic dishonesty has increased tremendously, especially in regards to cheating during tests [9,10].

With the advancement of technology, many institutions have shifted from traditional on-campus education to hybrid (online and face-to-face) or entirely online. Online learning has become more prominent especially during the COVID-19 pandemic, in that most universities have shifted to online learning to ensure continuation of the learning process. In spite of its great advantages, online learning has further raised concern, particularly regarding new methods of academic dishonesty. One study found that the majority of students believe that cheating is easier in online learning as compared to traditional learning [11].

To effectively address academic dishonesty, understanding of various factors such as the cause of engaging in such activity is needed. Students' perceptions of what are acceptable and unacceptable acts related to dishonest practices might affect their behaviors [12].

Therefore, to further understand academic dishonesty among health sciences students, this study aimed to examine students' attitudes and behavior towards academic dishonesty.

2. Methods

2.1. Study Design and Data Collection

This cross-sectional survey study was conducted among undergraduate health sciences students at a public university. A total of 319 respondents were recruited in this study by using convenience sampling. The questionnaire was distributed online. Respondents were assured of the data confidentiality and their participation were kept anonymous.

2.2. Instrument

This survey study used self-administered questionnaires. The questionnaire consisted of 75 questions that was adapted from previous studies [13,14]. The questionnaire included three sections. Section A consisted of demographic information. Section B was used to identify students' attitude towards academic dishonesty. It consisted of 23 questions, which asked the respondents to identify how serious they believe each of the behaviors to be by using a scale of "Not Cheating", "Trivial Cheating", "Moderate Cheating" and "Serious Cheating". Section C was used to identify students' behaviors towards academic dishonesty. It consisted of 23 questions, which asked the respondents to indicate how often within the past year they (own self) and their friends (friends/classmates) had engaged in a set of behaviors.

2.3. Scoring of Instrument's Items

The responses to Section B and C were scored individually per respondent. For Section B, the responses were scored as 1 (Not Cheating), 2 (Trivial Cheating), 3 (Moderate Cheating) and 4 (Serious Cheating). An average score for each respondent was calculated with a value ranging from 1 to 4. Lower average value suggested that the respondent does not perceive the indicated behaviors as cheating, whereas higher average value indicated that the respondent perceived the behaviors as serious cheating. For Section C, responses were scored as 1 for "I never did it", scored 2 for "I've done it once" and scored 3 for "I have done it more than once". Responses of "Not Relevant" were scored as zero (0). An average score for each respondent was calculated with a value ranging from 1 to 3. Higher average value suggested that the students had engaged in academically dishonest behaviors more frequently.

2.4. Reliability

A total of 30 students were recruited in a pilot study for questionnaire reliability testing. Reliability test was conducted using test–retest reliability method. Cohen's kappa coefficient was used to determine the reliability of the questionnaire using SPSS, and it yielded a value of 0.917, which reflected a substantial agreement of reliability.

2.5. Statistical Analysis

Both descriptive and inferential data analysis were performed using IBM SPSS Statistics for Windows, version 25.0, Armonk, NY, USA: IBM Corp, with a value of $p < 0.05$ being considered statistically significant. Normality test was performed, which showed not-normally distributed data. Hence, Mann–Whitney, Kruskal–Wallis and Spearman's correlation tests were performed for data analysis.

3. Results

3.1. Demographic of Study Population

A total of 260 (81.5%) female and 59 (18.5%) male students participated in this study. The age of the respondents ranges from 19 to 26 years old, with 69% (N = 220) of the respondents being bachelor's degree students. A higher number of semester two students participated (N = 87, 27.3%), while semester six students recorded the least participation (N = 60, 19.1%). According to the findings, the majority of the respondents have CGPA of '3.01 to 3.50' (N = 167, 52.4%) whereas only 2 (0.6%) respondents are from those with CGPA of '2.00 to 2.50'.

3.2. Prevalence of Academic Dishonesty

A majority of respondent (86.2%, n = 264) self-report that they have committed academic dishonesty behaviour at least once for the past one year. Furthermore, study revealed that about 89.8% (n = 53) male students self-report that they've performed academic misconduct at least once for the past one year; which is higher than female students (85.4%, n = 260). In addition, approximately 77.1% (n = 246) of respondent has witnessed act of academic dishonesty among their friends at least once for the past one year.

3.3. Student's Attitude and Behaviour towards Academic Dishonesty

Table 1 reveals the scores for students' attitudes and behavior towards academic dishonesty. For attitude, this study found a higher average mean score, which means that the respondent perceives the indicated behaviors as serious cheating. In addition, for behavior of own self and friends, a lower mean score shows that the students and their friends have engaged in academically dishonest behaviors less frequently.

Table 1. Students' attitudes and behavior towards academic dishonesty.

	N	Min.	Max.	Mean	SD
Attitude	319	1.00	4.00	2.76	1.06
Behavior (own self)	319	0.61	3.00	1.31	0.32
Behavior (friends)	319	0.87	3.00	1.45	0.45

3.4. Association of Demographic Factors and Student's Attitude and Behaviour towards Academic Dishonesty

In addition, Table 2 shows that the score for attitude towards academic dishonesty during online learning for female students (3.22) is higher compared to male students (2.96), whereas the scores for behavior toward academic dishonesty are higher among male students. These scores show that female students perceived the indicated behavior more seriously whereas male students engaged in dishonest behavior more frequently compared to female. A Mann–Whitney test was conducted, and it was found that there is no association between gender and attitude towards academic dishonesty. However, it was found that there is an association between gender and behavior towards academic dishonesty (p value < 0.01).

Table 2. Association between gender and students' attitudes and behavior towards academic dishonesty.

Variable	Male Median (IQR)	Female Median (IQR)	Z Statistics [a]	p Value [a]
Attitude	2.96 (1.57)	3.22 (2.26)	−0.863	0.388
Behavior (own self)	1.39 (0.65)	1.22 (0.26)	−3.600	0.000 [a]*

[a] Mann–Whitney test. * Statistically significant, $p < 0.01$.

Furthermore, it was found that the scores of students' attitudes toward academic dishonesty are highest among students aged 'more than 25 years old' (Table 3). These scores show that older students perceived the indicated behavior more seriously but had engaged in academic dishonesty more frequently. A Kruskal–Wallis test was conducted and reported no association between age and attitude toward academic dishonesty. However, there is an association between age and behaviors toward academic dishonesty (p-value < 0.01).

Table 3. Association between age and students' attitudes and behavior towards academic dishonesty during online learning.

Variable	Age	N	Median (IQR)	X2 Statistic (df) [b]	p Value [b]
Attitude	19–21	145	3.17 (2.20)	0.470 (2)	0.791
	22–24	141	3.17 (2.26)		
	>25	33	3.30 (2.04)		
Behavior (own self)	19–21	145	1.17 (0.28)	12.737 (2)	0.002 [b]*
	22–24	141	1.26 (0.41)		
	>25	33	1.48 (0.57)		

[b] Kruskal–Wallis test. * Statistically significant, $p < 0.01$.

Table 4 shows that the scores of students' attitudes and behavior toward academic dishonesty are the highest among students in semester 4 and 8 respectively. These scores show that second year students perceived the indicated behavior more seriously whereas final year students had engaged in academic dishonesty more frequently compared to others. A Kruskal–Wallis test was conducted and showed no association between year of study and attitude towards academic dishonesty. Nevertheless, there is an association between year of study and behavior towards academic dishonesty (p-value < 0.01).

Table 4. Results for students' attitudes and behavior towards academic dishonesty during online learning among respondents in association with semester.

Variable	Semester	N	Median (IQR)	X2 Statistic (df) [b]	p Value [b]
Attitude	2	87	3.00 (2.35)	5.534 (3)	0.137
	4	84	3.35 (0.95)		
	6	61	3.17 (2.39)		
	8	87	3.30 (2.04)		
Behavior (own self)	2	87	1.17 (0.30)	24.915 (3)	0.000 [b]*
	4	84	1.21 (0.34)		
	6	61	1.26 (0.35)		
	8	87	1.35 (0.52)		

[b] Kruskal–Wallis test. * Statistically significant, $p < 0.01$.

In addition, Table 5 shows that the scores for students' attitudes towards academic dishonesty during online learning are higher among students with CGPA of '2.00–2.50' as compared to others. On the other hand, for behavior of own self towards academic dishonesty, students with CGPA of '3.51 to 4.00' recorded higher scores compared to others. These scores show that students with lower CGPA perceived the indicated behavior more seriously and had engaged in academic dishonesty less frequently. A Kruskal–Wallis test was conducted, which showed that there is an association between CGPA and attitude towards academic dishonesty (p-value < 0.05). However, there is no association between CGPA and behavior towards academic dishonesty.

Table 5. Results for students' attitudes and behavior towards academic dishonesty during online learning in association with CGPA.

Variable	CGPA	N	Median (IQR)	X2 Statistic (df) [b]	p Value [b]
Attitude	2.00–2.50	2	3.65 (0.00)	9.254 (3)	0.026 [b*]
	2.51–3.00	45	2.39 (2.41)		
	3.01–3.50	167	3.30 (2.00)		
	3.51–4.00	105	3.13 (1.70)		
Behavior (own self)	2.00–2.50	2	1.02 (0.00)	7.732 (3)	0.052
	2.51–3.00	45	1.17 (0.30)		
	3.01–3.50	167	1.22 (0.43)		
	3.51–4.00	105	1.26 (0.35)		

[b] Kruskal–Wallis test. * Statistically significant, $p < 0.05$.

3.5. Association between Student's Attitude and Behaviour towards Academic Dishonesty

Spearman's correlation test was conducted to determine association between students' attitudes and behavior of own self towards academic dishonesty. However, it was found that there is no association between students' attitudes and behavior towards academic dishonesty, as shown in Table 6.

Table 6. Results of association between students' attitudes and behavior (own self) towards academic dishonesty during online learning.

Independent Variable	rs	p Value
Attitude	0.047	0.406
Behavior (own self)	0.047	0.406

Spearman's rho.

4. Discussion

The results of this study reflect a positive attitude of health sciences students towards academic dishonesty during online learning amid COVID-19 pandemic. Higher average scores show that the respondents perceived the indicated behaviors as serious cheating. Respondent perceived cheating behaviours such as turning in work done by someone else, copying from other students during test; and copying other student's homework to be a serious cheating behaviour. This finding is in agreement with previous studies [15,16]. However, despite their positive perception, a majority (86.2%) of the respondents reported that they have engaged in cheating behaviour at least once for the past one year. In other words, despite the fact that students perceive academic dishonesty as serious problem and unethical, they couldn't dissuade from doing it. Various researches have reported cheating behaviours among medical and health sciences student [7,8,13]. Besides, 77% of students reported that they witnessed their friends engaged in academically dishonest behaviour. This is in line with previous study who reported that students cheat because they believe their friends do the same [17]. Additionally, they discover that their fellow friends engage in cheating behaviour more frequently than they do, which is viewed as "a justification for their behaviour" [18].

In addition, it is found that female students have higher scores of attitudes as compared to male students. However, statistically, the results of this study indicate that gender does not influence students' attitudes towards online learning. This is in contrast with study that reported significant difference in male and female students with female students perceive academic cheating behaviours as more serious [15]. The irony in this situation is that, despite the fact that most indicated behaviours were perceive by female students as serious, the study revealed that they got engaged in dishonest behaviour (85%) similar like male student. Although the scores on behaviour is low for both male and female respondents (less frequent of engaging in dishonest behaviour), male respondent reported to engage more frequently in dishonest behaviour compared to female respondent. This is similar with previous study which found that male students were more likely to believe a cheating culture is more prevalent [13]. Besides, it is found that there is an association between gender and cheating behaviour. This is in contrast with various studies whereby most researchers have found that gender does not have a significant impact on a student's decision to engage in acts of academic dishonesty [8,14,19]. The possible cause of outcome variation may be the difference in the number of male and female students in this research compared to other studies.

This study also shows that there's no statistically significant difference in student's attitudes towards academic dishonesty based on study level. However, it was found that their study level might have impact on the likelihood of engaging in the academically dishonest behaviour, with final year students engaged more frequently compared to other students. Previous studies stated that student's focus on academic is likely to diminish as they get older, which may affect how they perceive academic dishonesty and how likely they are to do it [14,19]. Furthermore, this study found that CGPA may have impact on the attitude but not likelihood of academic dishonesty. Interestingly, students with lower CGPA (2.00 to 2.50) perceive the indicated behaviour as serious cheating compared to higher achievers. This is in contrast with previous literature [16,19]. Besides, this study shows that students with CGPA of "3.51 to 4.00" engaged in dishonest act more frequently compared to other students. This result also suggests that as a student's cumulative GPA increases, their self-reported cheating behavior increased. This might be due to the fact that these students might feel the need to maintain their academic performance or for scholarship purposes. However, this is contradicted with study that suggested as student's CGPA increases, their self-reported cheating behaviour decreased [14,20]. The possible cause of outcome variation may be the difference in the number of respondents among lower and higher achievers in this research compared to other studies.

In addition, this study found that there is no association between students' attitude and behaviour towards academic dishonesty. It suggested that though students perceive the indicated behaviours as serious dishonesty act, it does not have an impact on their self-reported behaviour. In contrast, previous studies reported significant correlation between students' attitudes towards academic dishonesty and their self-reported cheating behaviours [14,21]. The absence of significant results in relation to individual components of sociodemographic data suggests that additional study into other possible elements, such as moral growth and institutional culture, that influence students' attitudes and behaviors is required.

This study presented some limitations. First, since students were asked to recollect actions that may have been conducted or observed for the past one year, recall bias may have affected the data. Besides, the data are relied on self-reports, which could make them vulnerable to social response bias due to sensitive subject matter. However, it was mitigated because the researcher guaranteed complete anonymity and stressing the importance of honest responses to the questions. Despite the limitation, this study has provided some insight into how university students perceived and act with regards to academic dishonesty.

5. Conclusions

Finally, this study revealed students' attitudes and behavior regarding academic dishonesty during online learning. The researcher achieved the overall goal of this study, which was to evaluate students' attitudes and behavior regarding academic dishonesty during online learning. Furthermore, the particular aim of determining the relationship between sociodemographic data and attitudes and behavior towards academic dishonesty during online learning has been effectively achieved.

A descriptive statistic has been administered to define the frequency of students' attitudes and behavior. Hence, it is helpful to determine the total students who experienced different attitudes and behavior based on this study. The analysis revealed that there is no association regarding both attitudes and behavior.

Lastly, the results of this study are, in summary, that the majority of students perceived the indicated behaviors as serious cheating and have engaged in academically dishonest behaviors less frequently. Serious attention and effective mechanism should be continuously implemented and monitored to ensure quality education as cheating can give unfair advantage to the cheater and it may falsify data about what students have truly learned.

Author Contributions: Conceptualization, A.H., R.S. and N.D.N.A.; methodology, A.H., R.S. and H.S.; formal analysis, A.H., R.S. and L.S.M.Y.; investigation, A.H.; resources, R.S.; data curation, R.S.; writing—original draft preparation, A.H.; writing—review and editing, R.S.; supervision, R.S.; project administration, R.S. and N.D.N.A. All authors have read and agreed to the published version of the manuscript.

Funding: This research received no external funding.

Institutional Review Board Statement: The study was conducted in accordance with the Declaration of Helsinki, and approved by the UiTM Research Ethics Committee (REC/12/2021 (UG/MR/995, 14/12/2021).

Informed Consent Statement: Not applicable.

Data Availability Statement: Not applicable.

Conflicts of Interest: The authors declare no conflict of interest.

References

1. Guthrie, C.L. Plagiarism and Cheating: A Mixed Methods Study of Student Academic Dishonesty. Ph.D. Thesis, The University of Waikato, Hamilton, New Zealand, 2009.
2. Aluede, O.; Omoregie, E.O.; Osa-Edoh, G.I. Academic dishonesty as a contemporary problem in higher education: How academic advisers can help. *Read. Improv.* **2006**, *43*, 97–107.
3. Brimble, M.; Stevenson-Clarke, P. Perceptions of the prevalence and seriousness of academic dishonesty in Australian universities. *Aust. Educ. Res.* **2005**, *32*, 19–44. [CrossRef]
4. Richards, D.; Saddiqui, S.; White, F.; McGuigan, N.; Homewood, J. A theory of change for student-led academic integrity. *Qual. High. Educ.* **2016**, *22*, 242–259. [CrossRef]
5. Herdian, H.; Mildaeni, I.N.; Wahidah, F.R. "There are always ways to cheat" academic dishonesty strategies during online learning. *J. Learn. Theory Methodol.* **2021**, *2*, 60–67. [CrossRef]
6. Chace, W.M. Cover story: A question of honor: Cheating on campus undermines the reputation of our universities and the value of their degrees. Now is the time for students themselves to stop it. *Am. Sch.* **2012**, *81*, 20–32.
7. Kusnoor, A.V.; Falik, R. Cheating in medical school: The unacknowledged ailment. *South. Med. J.* **2013**, *106*, 479–483. [CrossRef] [PubMed]
8. Desalegn, A.A.; Berhan, A. Cheating on examinations and its predictors among undergraduate students at Hawassa University College of Medicine and Health Science, Hawassa, Ethiopia. *BMC Med. Educ.* **2014**, *14*, 89. [CrossRef] [PubMed]
9. Parnther, C. Academic misconduct in higher education: A comprehensive review. *J. High. Educ. Policy Leadersh. Stud.* **2020**, *1*, 25–45. [CrossRef]
10. McCabe, D.L.; Butterfield, K.D.; Trevino, L.K. *Cheating in College: Why Students Do It and What Educators Can Do about It*; JHU Press: Baltimore, MD, USA, 2012.
11. King, C.G.; Guyette, R.W., Jr.; Piotrowski, C. Online exams and cheating: An empirical analysis of business students' views. *J. Educ. Online* **2009**, *6*, 1. [CrossRef]
12. Harper, M.G. High tech cheating. *Nurse Educ. Today* **2006**, *26*, 672–679. [CrossRef] [PubMed]
13. Maring, J.; Vail, M.; Wright, K.A.; Tebbenhoff, B.; Canova, K.; Costello, E. Attitudes toward academic dishonesty in health profession students. *J. Allied. Health* **2018**, *47*, e97–e103.
14. Pearson, K.A. *Student Attitudes and Behaviors Regarding Academic Dishonesty*; University of South Florida: Tampa, FL, USA, 2019.
15. Chala, W.D. Perceived seriousness of academic cheating behaviors among undergraduate students: An Ethiopian experience. *Int. J. Educ. Integr.* **2021**, *17*, 2. [CrossRef]
16. McCabe, D.L.; Treviño, L.K.; Butterfield, K.D. Cheating in academic institutions: A decade of research. *Ethics. Behav.* **2001**, *11*, 219–232. [CrossRef]
17. Engler, J.N.; Landau, J.D. Source is important when developing a social norms campaign to combat academic dishonesty. *Teach. Psychol.* **2011**, *38*, 46–48. [CrossRef]
18. Engler, J.N.; Landau, J.D.; Epstein, M. Keeping up with the Joneses: Students' perceptions of academically dishonest behavior. *Teach. Psychol.* **2008**, *35*, 99–102. [CrossRef]
19. Soroya, M.S.; Hashmi, M.A.; Soroya, S.H. Academic Integrity: Effects of Demographic Variables on Students' Conduct. *South Asian Stud.* **2016**, *31*, 27–43.

20. Teodorescu, D.; Andrei, T. Faculty and peer influences on academic integrity: College cheating in Romania. *High. Educ.* **2009**, *57*, 267–282. [CrossRef]
21. McCabe, D.L.; Trevino, L.K. Academic dishonesty: Honor codes and other contextual influences. *J. High. Educ.* **1993**, *64*, 522–538. [CrossRef]

proceedings

MDPI

Proceeding Paper

High School Students' Motivation Needs and Their Intentions to Pursue Tertiary Education †

Mazlina Mahdzar [1,*], Aini Jaapar [2] and Wan Zuraida Wan Mohd Zain [3]

1 Faculty of Hotel & Tourism Management, Universiti Teknologi MARA, Puncak Alam 42300, Selangor, Malaysia
2 Faculty of Architecture, Planning and Surveying, Universiti Teknologi MARA, Shah Alam 40450, Selangor, Malaysia
3 Faculty of Plantation and Agrotechnology, Universiti Teknologi MARA, Shah Alam 40450, Selangor, Malaysia
* Correspondence: nina@uitm.edu.my
† Presented at the International Academic Symposium of Social Science 2022, Kota Bharu, Malaysia, 3 July 2022.

Abstract: The study investigated the motivational needs of secondary school students and their intentions to pursue higher education. The descriptive survey method was utilised. This study's sample was comprised of 137 students in grades four and five in secondary schools in Langkawi Island, Malaysia. For data analysis, descriptive and inferential statistics were utilised. The results indicated that these students have high educational aspirations concerning physiological needs, safety needs, social needs, self-esteem needs, and self-improvement needs. The survey also revealed that the need for self-improvement influences the educational ambitions of students pursuing tertiary education. Other motivational elements, including physiological needs, safety needs, social needs, and self-esteem needs, have no statistically significant relationship with students' plans to pursue postsecondary education.

Keywords: motivation needs; secondary school students; tertiary education

Citation: Mahdzar, M.; Jaapar, A.; Zain, W.Z.W.M. High School Students' Motivation Needs and Their Intentions to Pursue Tertiary Education. *Proceedings* **2022**, *82*, 37. https://doi.org/10.3390/proceedings2022082037

Academic Editor: Mohamad Rahimi Mohamad Rosman

Published: 13 September 2022

1. Introduction

In Malaysia, it is exceedingly difficult for a student to acquire admission to post-secondary education, particularly in public higher education institutions. The dilemma arises because the current tertiary education options are insufficiently able to meet societal demands and the country's expanding needs [1]. Furthermore, studies on access to higher education institutions are critical in assisting underprivileged students in improving their academic performance and social position. Secondary school students must be encouraged to aspire to higher levels of study and advancement.

Motivation is the engine that drives achievement. Students perform their best academically when they are motivated. According to [2], motivation reinforces and sustains classroom activities. Motivation is viewed as satisfying human wants or demands in order to attain individual primary objectives or organisational objectives [3]. Motivation stimulates, activates, and propels conduct [4]. Students' motivational demands are those things that they require in order to exert greater effort in their academic work. They may have physical, emotional, social, or material needs, among others. Motivation can be internal (intrinsic) or external (extrinsic). Intrinsic motivation refers to all components of motivation that might boost students' self-determination to see the need to thrive in their academic pursuits. According to [5], intrinsic motivation places control in the hands of the individual who controls learning based on his or her own preferences. A learner who is intrinsically driven maintains a quality of academic endeavour because his or her motivation for success originates from within. Extrinsic motivation refers to motivation that comes from external sources. It takes external variables to provide such incentive. External variables that are required for extrinsic motivation include the students' teachers, parents, families' peers, and counsellors. Extrinsic motivation might come in the form of prizes, tokens, or money. Motivation enables pupils to push themselves in order to earn high grades [6]. Studies have indicated, for instance, that students from poor socioeconomic status households have low aspirations to continue their education [7,8]. They are therefore at risk of dropping out of higher education institutions.

A higher education institution (HLI) is primarily responsible for educating students according to specific prerequisites at a higher level, with an emphasis on practical and technical skill sets [9]. Education is a tool for reforming a country's economy, raising people's living standards, and reducing poverty issues [10]. Previous research has demonstrated that students are motivated and choose to enrol in universities because of a promising future profession [11,12]. In addition, research indicates that education helps students improve or requalify themselves as their future living circumstances change [7,13–16].

The theory focuses mostly on motivation. The study's central topic is Maslow's hierarchy of needs. According to the theory [17], individuals attempt to meet their wants in accordance with the hierarchy of those requirements. Needs are met in accordance with their relative importance. He ordered these requirements from lowest to highest priority. The order is as follows: physiological needs, safety needs, belonging and love needs, esteem needs, and self-actualization wants. Maslow stated that these requirements determine an individual's conduct, since behaviour stems from an individual's efforts to satisfy their needs. According to this idea, physiological needs relate to insufficiencies and necessities, while higher-order needs relate to human growth. The physiological demands include food, water, sleep, and sex. Safety needs relate to protection from physical and mental threats, as well as security of life, property, and employment. The purposes of the needs for belonging and love are to be accepted by others, to be liked, to enjoy a social circle of friendship, and to belong to a group. Respect is linked to self-esteem, accomplishment, position, and acknowledgment. Self-actualization demands relate to an individual's creativity, potential, and talent utilisation. Maslow recognised that these wants cannot be totally satisfied, and that when they are, they cease to inspire the individual.

Maslow's hierarchy of requirements postulated that individuals have needs and that these needs are structured in a hierarchy, beginning with the most basic or physiological needs which no one can live without. Such needs include food and shelter, among others. When the physiological needs are met, other needs, such as the need for safety, belonging, love, self-esteem, and self-actualization, will be satisfied. These needs must be addressed in order for students to be motivated to study; however, if these needs are not met, students will not be driven to learn. If these demands are addressed, however, they will move towards self-actualization. Therefore, students must be provided with an environment that encourages them to excel in their studies, thereby fostering their self-actualization.

Academic achievement [18], college experience [19], and demographic traits have been the subject of a number of earlier studies focusing on students [8,20]. Numerous studies on academic achievement, college experience, and demographic features of secondary school students have been conducted in western nations. Unfortunately, fewer studies have been undertaken, particularly in developing nations, on the motivations or variables that drive secondary school students, particularly those from rural areas, to enrol in higher education institutions. Based on Maslow's theory of the hierarchy of needs, the purpose of this study was to determine the motivation level of secondary school students in terms of their physiological, safety, social, esteem, and self-actualization needs, and to examine the ranking of their needs. This study also was conducted to examine the relationship between the influencing factors of secondary students' intentions to study at a higher education institution.

2. Material and Methods

Secondary school students in Malaysia comprised the population of this study, with its sampling frame comprising of students of two secondary schools in Langkawi. They include Sekolah Menengah Kebangsaan Pulau Tuba, which is a school in a rural island, and

Sekolah Menengah Kebangsaan Tunku Putra, Langkawi. Langkawi schools were selected because they are part of the requirements for the research grant obtained for the research to be conducted in Langkawi and the sample is believed to represent the population. The Pejabat Program Pra Pendidikan Tinggi, Universiti Teknologi MARA collected data from students over the course of three days in November 2021 as part of its corporate social responsibility (CSR) initiatives to help and provide a second chance to underprivileged youths of Malays, Sarawak origin descendant, Sabah origin descendant, and Orang Asli to enrol in institutions of higher education. The objective of this programme is to provide the B40 students with the option to pursue higher education through its pre-diploma programmes [21]. The B40 category is comprised of families with gross monthly incomes of less than MYR4849 [22]. There were 137 surveys distributed and analysed. For this study, fifth- and fourth-grade students from both schools who attended the motivational talk and programme were asked to complete the questionnaire. As the primary source of primary data for the study, a self-administered survey was utilised. The questions are composed of two portions and written in Bahasa Melayu. They include Part A (demographic portfolio) and Part B (students' perceptions on motivation factors to further their studies at HLI). Five elements were adopted from [23] for the demographic profile's questions. According to [23], motivation consists of five factors: physiological needs, safety needs, social needs, self-esteem needs, and self-improvement needs. Five items were utilised to measure each of the five motivational elements, for a total of 25 items. As the ultimate dependent construct, this study included two-item measures to examine secondary school students' behavioural intentions to continue their education at a higher education institution. Descriptive analysis, correlation analysis, and regression analysis were used to investigate the strength of the link between five motivational factors and students' plans to pursue higher education.

3. Results

3.1. Descriptive Analysis

Table 1 shows the total mean value for each motivation factor. Based on Table 1, the overall mean score for motivation is high, m = 4.42 (a mean score between 4.01–5.00, according to [24]). The highest total mean value for motivation is for safety needs (m = 4.60), followed by physiological needs (m = 4.60), self-esteem needs (m = 4.49), and self-improvement needs (m = 4.40), respectively. The least important factor is social needs (m = 3.99).

Table 1. Mean, SD, and Rank for Maslow Hierarchy of Needs.

Factor	Mean	Standard Deviation (SD)	Rank	N
Physiological needs	4.60	0.424	2	137
Safety needs	4.63	0.419	1	137
Social needs	3.99	0.635	5	137
Self-esteem needs	4.49	0.489	3	137
Self-improvement needs	4.40	0.483	4	137
Total	**4.42**	**0.490**		

This reveals that the students have the greatest motivation for safety needs. The hierarchical order of the remaining needs is psychological needs, self-esteem needs, self-improvement needs, and social needs.

Table 2 shows the important aspects for each of the motivation factors for the students in deciding to pursue their intentions to study at a higher education institution. Based on Table 2, the highest mean value in motivating secondary school children for physiological needs is to help family in improving living standards, m = 4.81. For the security needs factor, the most important aspect is to avoid being troublesome to others (m = 4.72), while for the social needs aspect it is the ability to meet new friends (m = 4.38). In the self-esteem factor, obtaining knowledge (m = 4.74) and the self-improvement needs factor of fulfilling dreams and aspirations (m = 4.71) are the most important parts for the secondary students in motivating them to pursue their studies at a higher learning institution. According to these group of students, the least important factor is social needs (wanting to be noticed by other friends).

Table 2. Motivation Aspects of Students in Secondary School in Langkawi.

Items	Mean	Standard Deviation	N
Physiological (to get a better job)	4.66	0.474	137
Physiological (to buy food and clothing necessities)	4.37	0.642	137
Physiological (to help families achieve a better standard of living)	4.81	0.446	137
Physiological (learning and continuing learning is important)	4.61	0.560	137
Physiological (to get a high paying job)	4.54	0.630	137
Safety (to obtain a permanent job after graduation)	4.66	0.585	137
Safety (to obtain a more secure life in the future)	4.74	0.470	137
Safety (to be a good example to other siblings)	4.62	0.516	137
Safety (to avoid being left behind by a friend who constantly achieves success)	4.61	0.670	137
Safety (to avoid bothering others)	4.72	0.499	137
Social (to avoid a boring life)	4.21	0.790	137
Social (To participate in various activities of the association)	4.07	0.720	137
Social (meet more friends)	4.38	0.677	137
Social (to get the attention of the teacher)	3.77	0.923	137
Social (want to be noticed by other friends)	3.53	0.932	137
Self-esteem (to increase self -confidence)	4.61	0.518	137
Self-esteem (to acquire knowledge)	4.74	0.442	137
Self-esteem (appreciated by others)	4.18	0.813	137
Self-esteem (respected by others)	4.28	0.764	137
Self-esteem (to earn an academic excellence award)	4.63	0.582	137
Self-improvement needs (to develop one's potential)	4.58	0.525	137
Self-improvement needs (to fulfill dreams and achieve personal ambitions)	4.71	0.487	137
Self-improvement needs (to understand your own weaknesses and strengths)	4.53	0.570	137
Self-improvement needs (to improve social status)	4.10	0.750	137
Self-improvement needs (to get more wealth in the future)	4.08	0.849	137

3.2. Correlation Analysis

Table 3 exhibits the Pearson Correlation Matrix for all the independent variables of this study, namely motivation factors such as physiological needs, safety needs, social needs, self-esteem needs, and self-improvement needs. The results of the correlations between the variables are statistically significant. There is a moderate to high association between all the elements of motivation among secondary school children ($r = 0.540$ to $r = 0.750$, $p < 0.01$), which also means that there is a positive relationship between the variables. Among all the elements, the physiological and safety needs element achieved the highest range ($r = 0.70$, $p < 0.01$). However, there is a low association between physiological needs and social needs ($r = 0.39$, $p > 0.01$), since the r-values are less than 0.40.

Table 3. Pearson Correlation Matrix.

	Physiological Needs	Safety Needs	Social Needs	Self-Esteem Needs	Self-Improvement Needs
Physiological needs	1	0.750	0.399	0.547	0.593
Safety needs	0.750	1	0.540	0.709	0.693
Social needs	0.399	0.540	1	0.662	0.684
Self-esteem needs	0.547	0.709	0.662	1	0.727
Self-improvement needs	0.593	0.693	0.684	0.727	1

3.3. *Regression Analysis*

To see whether there is a significant relationship between secondary school children and their motivation intentions to further their studies at HLI, a single step multiple regression was conducted. The 25 items in the independent (motivation) dimension and two items in the dependent dimension (behavioural intention) were collapsed and entered into the equation. The outcomes demonstrated that there is a significant relationship between secondary school children's motivation towards behavioural intentions. The secondary school children's perception of motivation was able to explain 40% ($R^2 = 0.400$, F—change = 17.481, $p < 0.000$) of the variance in behavioural intentions to pursue studies at a higher education institution. It evidently shows that secondary students' perception of motivation is found to significantly and positively influence their intentions to pursue tertiary education. Besides that, Table 4 shows the value of beta coefficients for this study. In [24] it is stated that the beta coefficient indicates which variable(s) are important to the model by looking at the largest value while ignoring the negative signs. This study suggests that, among the factors of motivation, self-improvement needs (0.332) is the strongest influence on the behavioural intention to continue studying. However, other factors of motivation, such as physiological needs, safety needs, social needs, and self-esteem needs, have statistically insignificant relationships towards students' behavioural intentions.

Table 4. Coefficient Behavioural Intentions.

Model	Unstandardized B	Std. Error	Standardized Beta	t	Sig.
(Constant)	0.508	0.547		−0.929	0.355
Physiological needs	0.157	0.166	0.098	0.945	0.346
Safety needs	0.138	0.201	0.086	0.688	0.493
Social needs	0.044	0.105	0.042	0.420	0.675
Self-esteem needs	0.230	0.158	0.166	1.455	0.148
Self-improvement needs	0.466	0.163	0.332	2.858	0.005

4. Discussion and Conclusions

This study's findings indicate that the physiological needs element is the most essential factor for respondents' motivation to pursue studies at a higher education institution (HLI), while the social needs factor is the least important factor. The results of the regression analysis reveal that the independent variable, students' perceptions of their motivational needs, has a substantial impact on their intention to pursue higher education. Among all motivational elements, the need for self-improvement has the greatest influence on their behavioural intention.

Students' desires to continue their education at HLI is primarily motivated by their physiological requirements. After graduation, the desire for a better career is the primary physiological needs factor. These findings confirm the findings of [12], which discovered that students' motivations for pursuing related subjects closely resemble their desire to obtain employment after earning a degree. In addition, the desire to improve the family's level of living is a big motivation for students to continue their education at colleges. This finding is consistent with the majority of findings from prior studies indicating that students intend to continue their education at colleges and universities to help their family better their living conditions and escape poverty [7,14–16]. In addition, security standards are the second most significant criterion that inspires students to continue their education at an institution of higher learning. The desire for a more secure life in the future is intimately tied to students' safety demands. The findings indicate that many pupils desire a permanent position following graduation. It is not surprising given that a survey of the relevant literature revealed that students wish to continue their studies at HLI in order to secure a job. The results of this study indirectly corroborate the findings of [13], which concluded that students will not attend college if they believe it will not improve their lives in the future.

The need for self-esteem is also an incentive for first-generation students to continue their studies at HLI. According to [17], once individuals have satisfied their basic needs and been accepted by others, they desire to be liked, respected, and appreciated by others. This study demonstrates that first-generation students are advancing their studies to HLI in order to receive academic excellence awards while continuing their education at HLI. They are doing so in order to gain respect and appreciation for their elevation. These findings are consistent with those of [25], which found that children want their parents to be proud of them. This demonstrates that the students' incentives to continue their education at HLI are closely tied to their desire to be praised by others, particularly their parents.

Additionally, students choose to continue their education at HLI due to the urge to achieve self-actualization. Students desire to attain their objectives and ambitions in order to satisfy their need for self-fulfilment. This finding is consistent with the findings of [12], which discovered that the primary reason students enrol in college is to attain personal and career goals, in particular, to get a degree. Additionally, [15,16] discovered that a student desires to attend college to increase social mobility.

Social needs, which is the final aspect that drives students to continue their studies at HLI, has the lowest mean scores. New friendships are not the driving force behind their decision to continue their studies at HLI. This is likely due to the fact that children already have numerous friends in secondary school. Consequently, they will not prioritise this factor when pursuing higher education at HLI in the future.

This study's findings also indicate that secondary students' perception of motivation was found to significantly and positively influence their intentions to pursue tertiary education. It was concluded that all five factors of needs, which are based on Maslow's theory of motivation, were considered motivators by secondary school students, invariably. It exhibited that there was a ladder in the motivational needs of secondary school children in Langkawi. The need for self-improvement was the greatest motivation for these students, which was followed by physiological needs, safety needs, social needs, and self-esteem needs.

This research will be useful and serve as a guide for authorities in the ministry of education and schools, as well as school counsellors, in recognizing the needs of their students. For these students' desires to continue their education at HLI to be achieved, the school must continually motivate or encourage them. For this reason, the school's counselling unit must concentrate on programmes that improve students' motivation and desire to continue their education at HLI. For instance, school counsellors are advised to arrange motivational programmes and invite universities such as the CSR teams from Pejabat Program Pra Pendidikan Tinggi (PPT), UITM to their individual schools. The purpose of PPT's programme is to expose students to the educational and employment opportunities available at HLI. Students in rural areas, such as Pulau Tuba, are only exposed to parental employment and other potential employment opportunities in their environment. Furthermore, it can be one of the effective techniques to help students succeed in their studies, particularly those from the B40 group, and provide them with the possibility to increase their family's future income [26]. In addition, with the career exposure programme for these students, it is hoped that these students' futures would be realised and secured by obtaining desired and better professions in the future.

Author Contributions: Conceptualization, A.J., M.M. and W.Z.W.M.Z.; methodology, M.M.; formal analysis, M.M.; investigation, A.J., M.M. and W.Z.W.M.Z.; resources, W.Z.W.M.Z.; data curation, M.M.; writing—original draft preparation, M.M.; writing—review and editing, M.M.; supervision, A.J.; project administration, M.M.; funding acquisition, W.Z.W.M.Z. All authors have read and agreed to the published version of the manuscript.

Funding: This research is funded through the research grant obtained from Universiti Teknologi MARA (UiTM)-SDG Pulau Tuba (Grant No: 600-RMC/LESTARI SDG-T 5/3 (117/2019)).

Institutional Review Board Statement: Not applicable.

Informed Consent Statement: Not applicable.

Data Availability Statement: Not applicable.

Conflicts of Interest: The authors declare no conflict of interest.

References

1. Ishak, Y. Analisis perbandingan institusi pengajian tinggi awam dengan institusi pengajian tinggi swasta di Malaysia. In *Pembangunan Sumber Manusia Dalam Era K-Ekonomi*; Rahmah Ismail, Ed.; Penerbit Universiti Kebangsaan Malaysia: Bangi, Malaysia, 2017; pp. 129–148.
2. Lee, W.; Reeve, J. Teachers' estimate of their students' motivation and engagement: Being in synch with students. *Educ. Psychol.* **2012**, *32*, 727–747. [CrossRef]
3. Steel, P. *Motivation: Theory and Applied*; Pearson Learning Solutions: Boston, MA, USA, 2012.
4. Ayub, N. Effect of Intrinsic and Extrinsic Motivation on Academic Performance. Available online: https://www.academia.edu/2599275/Effect_of_intrinsic_and_extrinsic_motivation_on_academic_performance (accessed on 22 January 2022).
5. Mazumder, Q. Student motivation and learning strategies of students from USA, China and Bangladesh. *Int. J. Eval. Res. Educ.* **2014**, *3*, 205–210. [CrossRef]
6. Ullah, I.M.; Sagheer, A.; Sattar, T.; Khan, S. Factors influencing students; Motivation to learn in Bahuddin Zakariya University, Multan. *Int. J. Hum. Resour. Stud.* **2013**, *3*, 2162–3058. [CrossRef]
7. Saenz, V.B.; Hurtado, S.; Barrera, D.; Wolf, D.; Yeung, F. *First in My Family: A Profile of Forst-Generation College Students at Four-Year Institutions Since 1971*; The Foundation for Independent Higher Education: Washington, DC, USA, 2017.
8. Terenzini, P.T.; Cabrera, A.; Bernal, E. *Swimming Against the Tide: The Poor in American Higher Education*; College Board: Princeton, NJ, USA, 2001.
9. Walchshofer. Strategy of Higher Educatio. Available online: http://www.appear.at/fileadmin/icm/appear/docs/ada/strategy_higher_education.pdf (accessed on 1 August 2019).
10. Annie, W.M.N.; Hamali, J. Higher education and employment in Malaysia. *Int. J. Bus. Soc.* **2006**, *7*, 66–76.
11. Tavares, O.; Ferreira, J.B. Choices and motivations: The why and how of Portuguese students' enrolment choices. *Eur. J. Educ.* **2012**, *47*, 310–326. [CrossRef]
12. Pope, M.L.; Fermin, B. The perceptions of college students regarding the factors most influential in their decision to attend postsecondary education. *Coll. Univ.* **2013**, *78*, 19–25.
13. Belonozhko, M.L.; Khitu, E.F. On the characteristics of higher education for adults. *Russ. Educ. Soc.* **2008**, *50*, 57–63. [CrossRef]
14. Gofen, A. Family capital: How first-generation higher education students break the intergenerational cycle. *Fam. Relat.* **2009**, *58*, 104–120. [CrossRef]
15. Smith, M.J. Right direction, wrong maps: Understanding the involvement of low-SES African American parents to enlist them as partners in college choice. *Educ. Urban Soc.* **2009**, *41*, 171–196. [CrossRef]
16. Smith, M.J. College choice process of first generation black female students: Encouraged to what end? *Negro Educ. Rev.* **2008**, *59*, 147–161.
17. Maslow, A.H. *Motivation and Personality*, 2nd ed.; Harper & Row Publishers: New York, NY, USA, 1970.
18. Riehl, R. The academic preparation, aspirations, and first-year performance of first-generation students. *Coll. Univ.* **1994**, *70*, 14–19.
19. Pike, G.; Kuh, G. First–and second generation college students: A comparison of their engagement and intellectual development. *J. High. Educ.* **2005**, *76*, 276–300. [CrossRef]
20. Bui, K.V. First-generation college students at a four-year university: Background characteristics, reasons for purposing higher education, and first-year experiences. *Coll. Stud. J.* **2002**, *36*, 3–11.
21. Pejabat Program PPT. Anjung PPT 2022. Available online: https://pppt.uitm.edu.my/index.php/ms/anjung/ppt (accessed on 26 January 2022).
22. Kumpulan Isi Rumah B40, M40, T20 Terbahagi Pada 10 Kategori, Berita Harian. Bernama. Available online: https://www.bharian.com.my/berita/nasional/2020/07/709371/kumpulan-isirumah-b40-m40-t20-terbahagi-pada-10-kategori (accessed on 10 July 2020).

23. Abdul Rahim, A.H.; Azman, N.; Mohd Majzub, R. Pelajar generasi pertama sebagai sumber manusia: Motivasi dan aspirasi melanjutkan pelajaran ke Institusi Pengajian Tinggi (IPT). *Malays. J. Learn. Instr.* **2013**, *10*, 203–232. [CrossRef]
24. Pallant, J. *SPSS Survival Manual a Step by Step Guide to Data Analysis Using SPSS Program*, 6th ed.; UK McGraw-Hill Education: London, UK, 2016.
25. Sheek, J.L. Rural Students, Aspiration, Motivation, and Music Education: Why don't more Rural Students Make it to College? Ph.D. Thesis, University of North Carolina, Greensboro, NC, USA, 2017.
26. Mahdzar, M.; Jaapar, A.; Zain, W.Z.W.M.; Ismail, I.; Razak, R.A. Tracing UiTM pre-diploma students' academic performance with the Faculty of Hotel and Tourism. *J. Tour. Hosp. Culin. Arts* **2021**, *13*, 134–141.

proceedings

MDPI

Proceeding Paper

The Impact of Social Media on the Teaching and Learning of EFL Speaking Skills during the COVID-19 Pandemic †

Latha Ravindran [1], Ikhram Ridzuan [2] and Bee Eng Wong [2,*]

1 Faculty of Education, Language & Psychology, Segi University, Jalan Teknologi, Kota Damansara, Petaling Jaya 47810, Malaysia
2 Faculty of Social Sciences and Liberal Art, UCSI University, UCSI Heights, 1, Jalan Puncak Menara Gading, Taman Connaught, Cheras, Kuala Lumpur 56000, Malaysia
* Correspondence: wongbe@ucsiuniversity.edu.my
† Presented at the International Academic Symposium of Social Science 2022, Kota Bharu, Malaysia, 3 July 2022.

Abstract: The shift to teaching online during the COVID-19 pandemic has resulted in the extensive use of social media in the English language classroom. The current study investigated the impact of social media on the speaking skills of tertiary-level EFL students during the pandemic. Adopting a qualitative framework, data was collected by using in-depth interviews with both students and lecturers. Findings generally point to the benefits of using social media in the speaking classroom. However, accounting for students' cultural differences is the most important aspect during the lessons. Implications on the pedagogical awareness of lecturers are underscored.

Keywords: COVID-19; EFL students; online teaching; social media; speaking skills

Citation: Ravindran, L.; Ridzuan, I.; Wong, B.E. The Impact of Social Media on the Teaching and Learning of EFL Speaking Skills during the COVID-19 Pandemic. *Proceedings* **2022**, *82*, 38. https://doi.org/10.3390/proceedings2022082038

Academic Editor: Mohamad Rahimi Mohamad Rosman

Published: 13 September 2022

1. Introduction

The English language is significant to the world community, especially in the 21st century. This is because English language proficiency not only encourages communication but also cooperation in aspects of everyday life, education, and employment, especially in a society that consists of members of various cultural backgrounds. In view of this, learning, understanding, and appreciating English is a long-term commitment in meeting social, employment, and educational needs, as well as individual needs [1]. English is a lingua franca as it is the most widely used language of communication in the world [2]. The importance of English has resulted in many students of other first languages learning English as a second language or as a foreign language. Learning English is a challenging process, and because of that, it requires a collaborative, massive and extraordinary effort that not only involves the students but also the educators [3]. The importance of English, especially during this era of globalization has resulted in the teaching of English in Malaysian higher education to increasingly shift from grammar translation to a communicative approach [4]. This approach has enabled students to be actively involved in the classroom and helped them improve their English, especially speaking and listening skills. There is no denying that the ability to listen, read, and write is important, but speaking competence is more important than the other skills because communication via speech often occurs in one's daily life, and in various settings.

Education plays a crucial role in the survival of a country especially in terms of economic and socio-economic aspects, as it is through education that people can be more productive. In the past two and a half years, however, the COVID-19 pandemic has affected the education system in Southeast Asia, especially in Indonesia and Malaysia [5]. This is because millions of students were not able to attend school physically as the schools had to be closed. By the end of March 2020, more than one billion students worldwide had been affected.

During the pandemic, schools were closed in late March 2020, and this disrupted learning for all students in the country. However, to enable them to learn from home, the Ministry of Education launched an online learning platform nationwide, called the Teacher Digital Learning Community (Komuniti Guru Digital Learning). This was established with support from UNICEF. The nation-wide platform enabled three million children to learn during physical closures of schools until 24 June 2020. The schools were gradually opened after that. The approach used to teach the students in schools then was blended—both face-to-face and online learning [6]. However, with periods of total lockdown in 2021, schools were again closed, and learning continued to be conducted online exclusively.

During this period, every educational institution helped educators by introducing online learning models, and providing laptops for students, teachers, and lecturers. Despite this, there were teachers and lecturers, especially in the less developed rural areas, who were still not adequately skilled in using e-learning for the teaching and learning process. At the same time, there were also students who had a negative attitude and did not want to accept the concept of e-learning although this is one of the best options during the pandemic era [7].

As stated earlier, due to the abrupt closure of educational institutions as a result of the spread of the COVID-19 virus, students faced various obstacles in the process of learning. In particular, these events had a negative impact on language learning, including the learning of the English language. Generally, the majority of students who study English in schools have tried to enhance their speaking abilities during this trying period.

In the same vein, universities and colleges in Malaysia adopted the online mode to teach courses. Lecturers had to be creative in their delivery of these courses. The teaching of speaking skills in English was a case in point. English, being the second-most important language in Malaysia after Bahasa Malaysia, is taught in all schools locally. Due to its status, we used to refer to the teaching of English as a second language (ESL) rather than a foreign language (EFL) in the Malaysian classroom. However, EFL describes the study of English by non-native speakers in countries where English is not a native language and where it is not used as a medium of instruction [8]. Thus, based on this definition, Malaysia is now more of an EFL country rather than an ESL country. Further, the students in our classroom were from China, where the status of English is definitely that of a foreign language.

In view of the situation, it was important to find potential solutions to overcome the challenges of EFL students and improve their speaking abilities to enable them to engage in meaningful communication outside the classroom. Ideas and techniques include incorporating social media such as Facebook, WhatsApp, and YouTube in the classroom. This paper explores the concept and impact of e-learning during the COVID-19 pandemic. Specifically, it investigates the challenges faced by teachers and EFL students in the teaching and learning of speaking skills respectively, and the use of social media platforms to assist them in facing the challenges of the COVID-19 pandemic era.

2. The Teaching of Speaking Skills in the ESL Classroom

The British Council gives ten reasons why we should learn the English language. Among the reasons that are important to students at the tertiary level are the facts that many universities teach courses in English, scientific papers are written in English, English is the language used most widely by the media industry and the Internet, and international businesses conduct meetings in English [9]. In view of these reasons, it is clearly advantageous for tertiary-level students to master the language to a proficient enough level to function adequately in their future workplace.

Of the four language skills, speaking is generally deemed a difficult skill to learn by most learners as it is an overtly active skill. In fact, one of the key differences between the spoken language and written language that makes the speaking skill more difficult to learn is speech is constructed spontaneously [10] (p. 4) while the written form can be revised and edited over and over. In fact, Ur [11] went further to say that speaking is the most important skill to master among the other skills. She further states that problems in speaking were due to the students' lack of confidence and their fear of making mistakes while using the language.

According to Goh and Burn [12] (p. 53), the fundamental components of second-language (L2) speaking competence are knowledge of language and discourse, essential speaking skills, and communication strategies. Burns [10] (p. 3) elaborates that knowledge of language and discourse requires the knowledge of the phonetics of the language, the grammar and vocabulary, and knowledge of how connected speech is organized. She further states that core speaking skills enable learners to comprehend the speech quickly, and to negotiate and manage the flow of it, while communication strategies are the cognitive and metacognitive strategies learners develop to compensate for their lack in linguistic knowledge. Pedagogically, she advocates for speaking lessons to be organized as "structured and supported learning opportunities that develop these various components of speaking competence" (p. 3). It is with this knowledge of what is required to aid the L2 learner toward a proficient level of speaking competence that we believe social media platforms are the best avenue for us to teach speaking skills in the Malaysian EFL classroom during the COVID-19 pandemic, especially during the periods of lockdown.

3. Theoretical Perspectives

Teaching and learning collectively is a social activity (see e.g., [13]), in particular if the environment is student-centered. Vygotsky's theory stresses the role of social interaction in the development of cognition, and language plays an important role in this development. His belief in the central role that community plays in the process of making meaning makes a great deal of sense when we anchor our study on this perspective. In such a student-centered learning situation, the teacher is a facilitator and the experience of learning on the students' part is more social constructivist in nature. In this environment, interaction and collaboration among the students is not only encouraged but a necessity. A resourceful teacher will capitalize on the students' interaction on social media platforms to aid them in the classroom. The teacher is able to disseminate the required and important information of the subject matter and tasks to be accomplished via these platforms. More crucially, students can interact, share information, learn from and discuss with each other, as well as collaborate via these social media platforms. As these platforms generally allow both speech and the written form to be communicated, the teacher could make it mandatory for students to only use the spoken form for their activities. Here, the teacher will scaffold the students' spoken language during the activities. In this way, the more proficient students will model the teacher's behavior, and the teacher could instruct the more proficient students to scaffold their less proficient friends via the social media platform(s) being used.

4. Social Media in the EFL Speaking Classroom

Social media tools enable their users to distribute and share knowledge over the Internet [14]. These tools or platforms can range from low social presence to frequent users [15]. These platforms, which include Facebook, WhatsApp, Twitter, Tik Tok, and Instagram, allow students to connect with each other and to be able to communicate with others. The features that are of concern to us are affordances of social interaction, sharing of content, and content creation of these platforms. Crucially, students can work on their tasks outside the class which would enable them to practice English-speaking skills. This is the ideal situation during the pandemic as students were not able to meet face to face either with their teacher or with any of their course mates. The researchers view this avenue as positive because schools and higher institutions of learning were hard-pressed to go on with their job in the online mode. It was also generally acknowledged as a mode that makes language learning more engaging and perhaps even motivating (see e.g., [16]).

In fact, some past studies have used the various social media platforms to teach speaking skills and the findings have generally been positive. However, it has been found that e-learning in general can create high levels of anxiety for students due to factors such as fear of being assessed, educator teaching techniques, and student culture [17,18].

Specifically, factors that could contribute to speaking problems in an EFL classroom include lack of confidence and fear of making mistakes [11]. In a study, Al-Sobhi and Preece [19] highlighted the lack of exposure and limited knowledge of English, lack of motivation, anxiety, and lack of confidence on the part of the learners, and inefficient teaching methods on the part of the teacher, as some of the key issues that are often encountered in the EFL-speaking classroom. Yen and Mohamad [2] also found that the main problems that hinder the speaking ability of EFL students include lack of desire and self-confidence, anxiety, and lack of vocabulary skills.

According to Burns [10], speaking is related to "social and functional motivation." In speaking, a distinction is made between the notion of interpersonally and pragmatically motivated speech [20]. The latter involves exchanging information or goods or services while the former, also known as interactional speech, is used in creating and maintaining social relationships. Speaking for interaction and communication makes this activity in an EFL classroom a daunting prospect for students. Furthermore, speaking is an overtly productive skill and the language produced is here and gone. Unlike the written form, students would not have the opportunity to correct what had been said. In writing, students could revise and edit what has been written as the writing task progresses. As such, teachers or lecturers of speaking skills in an EFL classroom have a great challenge, in that they have to incorporate activities by using various techniques that would involve students in interactions and communication in order to produce the language. It was during the pandemic that we found the application of social media to be most fruitful in the teaching of speaking skills in the EFL classroom.

Generally, studies have shown that improvement in speaking skills are possible if there is a decline in speaking anxiety. Teachers are able to integrate social media platforms, such as WhatsApp, Facebook, Tik Tok, YouTube, Instagram, WeChat, and Telegram, and others to give their students an opportunity to practice beyond the confines of the physical classroom. The studies that investigated the use of WhatsApp have shown that the application was able to increase communication away from the physical classroom [21,22]. Instagram, a popular social media platform among the students, was also shown to be useful for improving speaking even with students who are not proficient in English [23,24]. Overall, the studies have yielded positive results in that students do improve in their speaking skills and with higher confidence and increased motivation. A key element that contributed to improvement is social interaction [25,26]. In the latter study, Sevy–Biloon and Chroman [26] showed that chat functions in the video led to students practicing their speaking skills in authentic and meaningful communicative activities. Yet another study revealed the positive effect of social media integration (Telegram and You Tube) for teachers to make amends for the insufficient teaching time in the class and for its affordance in out-of-class practice in the teaching of pronunciation [27]. Such integration of social media seemed to have enhanced the teaching procedures in the online mode.

However, as EFL teachers, we have to be judicious in the introduction of these platforms in our speaking classroom. If students are too young, for example, they could be inexperienced, and these tools may have more of an adverse effect than what was intended [28]. In the study by Teoh and Yunus [28], the primary school participants were young and lacked the experience in using the social media platforms to learn English, although they had access to and the competence to use the tools. The participants in our study are young adult university students, so they would not have any issue incorporating social media in the learning of EFL speaking skills. However, the lecturer would need to structure her lessons and activities in such a way that these platforms could be used optimally.

5. Methodology

This study adopted a qualitative research design. The aim of the study is to investigate the impact of social media in the teaching and learning of speaking skills of EFL students during the pandemic. The study examines how social media was utilized by lecturers to enhance the speaking skills of EFL students. It also elicited the perspectives of students on how social media was useful to enhance their speaking skills and the difficulties faced by them. The paper also highlights the implications of using social media to enhance EFL students' speaking skills. The focus is to explore and give descriptive details of the data collected, and hence adopting a qualitative method is deemed appropriate. As Lancy [29] observed, "every aspect of one's work as a qualitative researcher demands more writing than would be the case for a quantitative scholar. Writing is to qualitative research what mathematics is to quantitative researcher" (p. 234).

The researchers selected three lecturers teaching EFL students at the intermediate level. These lecturers were selected based on the criteria that they have been teaching EFL courses for more than five years. This criterion was important as they are experienced in teaching EFL students. In-depth interviews were conducted with these lecturers and the interviews were recorded and transcribed verbatim. The participants were informed of the analysis to confirm the data (member-checking). The interview followed a protocol [30]. The researcher confirmed the findings with them, and further interviews were conducted for clarifications. In determining the number of participants, the researcher adopts the view contended by Patton [30] who recommends that we identify a minimum sample size "based on expected reasonable coverage of the phenomenon given the purpose of the study" (p. 186).

For the selection criteria of students, a total number of 20 of them were identified to be interviewed. These students were selected because they were EFL students and because they had been using social media for their intermediate English programs. The interviews focused on the speaking classes, as that is the only component being investigated. The students did a total of 10 h per week of speaking classes over a period of seven weeks. The interviews were conducted to get their perceptions on the difficulties they encountered in speaking classes, and their views on the problems encountered on the use of social media in EFL classes. The researchers conducted further follow-up interviews with the students for clarification and confirmed the findings with them. The results are discussed in the following section.

6. Result and Discussion

6.1. Cultural Differences as a Major Factor

In response to the difficulties they faced in learning speaking skills, one main theme that emerged in the discussion pertains to the cultural differences that accounted for the difficulties in mastering speaking skills. The spread of social media is beyond one culture and hence accounting for the cultural differences is deemed as one important finding of this study. Students posited that in China, eye contact is avoided as much as possible, whereas in classes they were required to have eye contact to establish the connection with the other listeners. In this class, students also had to pass the speaking test at the end of the seven weeks for their assessment. Topics were given to them by the lecturers, and they had to produce a speech outline and submit it to the lecturer. After the feedback given to them, they had to improve on the speech outline and deliver the speech as a form of assessment. Another factor relates to the learning environment, the learning context as well as the perseverance of the students. One student, Fan, whose band level is 5.5, said that, "English speaking skills are very important and we need to learn many words". However, "some of my group mates are not willing to communicate in English and hence it is difficult for us to practice".

Some students opine that it is good to learn English for 1 to 2 h a day. It is convenient to have conversations with lecturers via social media and students like platforms such as WeChat and Zoom, which make it convenient for them to practice more often. Hence, the lecturer must be able to understand their hesitancy to participate in class as cultural awareness is related to language learning. As Tomalin and Stempleski [31] posit, sensitivity to the impact of culture on language use and communication is important as they are interwoven. Moreover, they argued that language could not be taught without a reflection of the associated culture. In response to the concept of cultural mediation propounded by Vygotsky, the lecturers used topics related to China for an effective learning environment as the students would be more familiar with the context. Topics such as places of interest and food were picked up for speaking practice for a more rewarding experience. Some students also cited the benefits of the choice of social media platform. A majority of the students loved using Tik Tok videos for practice sessions.

Nevertheless, the aspect of good communication between the lecturer and the student cannot be overlooked. Sharing videos with others in class was beneficial as they had the opportunity to listen to others and correct their own pronunciation. Past studies have also indicated the positive aspects of using social media for teaching activities related to speaking skills. Ainun et al. [32] also point to the optimistic views on using WhatsApp to improve students' oral skills.

6.2. *Usage of Social Media for Speaking Classes*

Most of the students interviewed opined that the use of social media enhanced their speaking skills. Zoom and MS Teams were the most used platforms. Sometimes they were also given YouTube videos to watch, and the students then have a group discussion based on the given topic. A majority of the students preferred watching YouTube and interacted well on the Zoom and Teams platforms. Students viewed Zoom and Teams as useful platforms for class participation, as they liked to interact with other students on the chat groups and they also liked the break-up room functions in the Teams platform. However, there are some students who have negative perceptions in terms of participating in the activities. This hinders their progress, as it leads to lack of practice sessions. As another student, Wang, said "We are very conscious of having to speak in English and the other students listening to us. While we like to the use of social media and sometimes, we use it for entertainment but shy away from speaking in class".

This inhibition on the part of the students stems from their own lack of vocabulary and knowledge of the content related to the topic. Ilyas and Putri [33] reinforce that using YouTube to teach speaking skills lends a positive learning attitude among the learners. A majority of the students indicated an increased confidence as it allowed them to view the same video multiple times, enabling them to enhance their memory of a given word in the process. Using Facebook was not an option, as most of the students in China do not have access to it. As Boyd [34] postulates, the unprecedented spread of social media around the world indicates its importance in helping students to improve fluency and understanding in the target language. Social media can have beneficial impact on students' language skills, teaching practice, and student learning when used appropriately [35].

6.3. *Willingness of Students in Class to Coomunicate*

As speaking requires interactions with other classmates and developing social skills, lecturers were interviewed on how they ensured participation in class and structured the activities to ensure effective participation in class. The lecturers cited that creating interesting teaching activities stimulated the interest of the students. According to them, they tried to create a language environment that approximated the actual lives of the students, such that the target language plays a crucial role in their interaction [13]. Based on this constructivist theory, which contends that learning takes place while engaging in social experiences, students were given opportunities to work as a team to internalize their speaking skills. To this effect, role plays were incorporated during online classes and the situational context was given to them. This helped to a certain extent to train their communicative ability. The lecturer Tand said, "The activities were used as a starter to get the students talking. They would indicate their topic in the WhatsApp group so that no students repeated the same topics".

The other lecturers also opined that they ensured participation in class by integrating social media together with their main text. Some lecturers awarded participation marks to ensure students participated. The feedback from the students was that they really enjoyed the class, and the shy students were also keen to participate in the activities. Though such integration was beneficial to students, the lecturers' role in structuring the lessons is deemed important. Reinders and Wattana [36] also support integrating technology in the teaching and learning process, as this encourages learners to use the target language freely. Otherwise, the objective of the lesson could be easily lost with random talk. On what facilitated participation in class, the lecturers viewed that the classroom was a friendlier place and that was a great feat because online classes tended to be more impersonal. In this case, the classes were conducted in a hybrid mode. The effectiveness of participation in class also was based on the good choice of the materials. This finding also points to the need to have strong pedagogical knowledge and the capacity to deliver the content effectively on the part of the lecturers. Students thought that setting expectations and boundaries in terms of class participation ensured that everyone had a fair chance to speak on the given topic. They were in favor of this practice adopted by the lecturers. This clearly points out the strength of the pedagogy of the lecturers and their awareness of how the activities should be structured. Lecturers should be proficient to ensure the efficiency and the interest of the class. It requires a lot of coordination on the part of the lecturer to create a good learning environment. This finding further supports the idea of "structured and supported learning opportunities" that develop the various components of oral skills, as advocated by Burns [10].

6.4. Effects of Using Social Media to Enhance Speaking Skills

The lecturers were interviewed about the effects of using social media to enhance the speaking skills, and they responded that it had both positive and negative aspects to it. Using social media was popular with students as they liked engaging with the technology. It is already a norm for every student to have laptops owing to the online or hybrid classes. The emergence of the COVID-19 pandemic and the subsequent shift to the online platform has encouraged educators to turn to social media to motivate learners to participate in speaking classes to compensate for the lack of physical classes. Students who were shy to speak in conventional classes were readily participating in online platforms. As mentioned earlier, studies conducted by Sevy–Biloon and Chroman [26] also point out the chat functions that act as a platform to provide authentic and meaningful communication which are purposeful for the students. However, the lecturer points out that the negative aspect is that some of the students were not willing to switch on the cameras while they spoke, which made it difficult for the lecturer to ascertain if they were referring to some notes or a Google translation. As the teacher Ling said, "the positive aspect is that we were able to watch the videos we posted, and that helped them a lot with the pronunciation". The students were able to practice and listen repeatedly and thus gained confidence to speak. Some students preferred to use social media in the classroom as they perceived the environment to be friendlier whereas some others preferred online class as it was less threatening. This is also highlighted by Ghoneim and Elghotmy [25], who find that social interaction helps to improve speaking skills. They opined that with the online platform, the pressure to perform on the spot was less intense, as they could refer to other resources for vocabulary and sentence structures. Using social media enhanced their speaking skills, as they received materials from the instructor in advance, and this enabled them to go through the materials before the lesson. Students related positively to the fact that participation by all students was ensured by the lecturer and each student was responsible for their own learning. Some group discussions ensured that every member had watched the videos posted and that was an avenue as a starting point for the discussion to take place.

6.5. Implications of Using Social Media for Speaking Classes

The findings indicated that the students gained a lot of confidence and motivation by using social media for the speaking classes. The materials used were considered less formal, as most students perceived books to be formal whereas videos and other materials were less formal and hence less threatening. They could relate to the materials and the content and hence participated enthusiastically. This implies that the lecturers should be selective in choosing the content for the lesson and also choose materials that will engage the students in meaningful interactions. The implications according to the lecturers also point to the versatility of the use of social media and its immediate reach with the students. Using social media just makes it convenient for us to connect with other like-minded people in a harmonious manner [37]. It helped to develop the interpersonal skills of the students, and they were able to connect with their classmates. The usage of social media enabled a relaxed environment for the students, and they were not apprehensive of making mistakes in the class. On the negative aspects relating to social media, students cited connectivity as one of the issues during the online learning process. Some students did not really engage with the activities, and they only spoke for a few minutes. This duration was more of a restraint as the lecturer could only give feedback based on the input. The lecturers also noted that students were still shy and produced limited utterances, thus limiting the speech. One of the lecturers commented, "No matter which medium I used, the students only spoke for a few minutes, and it was quite difficult to motivate them to speak up". The solution might be that students require more scaffolding, and the lecturers also need to strengthen their pedagogical approaches towards online teaching and using social media. As the lecturer Jia noted, "we try some of the approaches we are familiar with. However, a proper training on pedagogy to use social media will be beneficial in the long run".

7. Conclusions and Recommendations

This study points to the positive impact and the benefits of using social media for speaking classes despite the challenges posed by the COVID-19 pandemic. However, the study underscores the implication on the training needs of the lecturers and the pedagogical awareness using social media to teach speaking skills. The students also will have to learn to utilize social media platforms for a specific purpose other than entertainment purposes. The limitation of the current study is that it only investigated a small sample from one private university, and further studies should be carried out with a bigger sample to consolidate the results of this study. Lecturers should also be able to incorporate text materials that cater to the cultural context of the students. Providing them with appropriate reading materials will stimulate their interest and participation in class. The implication that emanates from this is that lecturers should be given enough time to plan and to integrate social media in language teaching. It requires a considerable amount of time to plan according to each level of the students as well as to diversify their use of social media and not just resort to one platform.

Author Contributions: Conceptualization, L.R., I.R. and B.E.W.; methodology, L.R. and B.E.W.; validation, L.R., I.R. and B.E.W.; formal analysis, L.R.; investigation, B.E.W.; resources, L.R., I.R. and B.E.W.; data curation, L.R., I.R. and B.E.W.; writing—original draft preparation, L.R., I.R. and B.E.W.; writing—review and editing, L.R., I.R. and B.E.W.; visualization, L.R.; supervision, L.R.; project administration, I.R.; funding acquisition, L.R., I.R. and B.E.W. All authors have read and agreed to the published version of the manuscript.

Funding: This research received no external funding.

Institutional Review Board Statement: Not applicable.

Informed Consent Statement: Informed consent was obtained from all subjects involved in the study.

Data Availability Statement: Not applicable.

Acknowledgments: The authors would like to acknowledge the participants in this study.

Conflicts of Interest: The authors declare no conflict of interest.

References

1. Kukulska-Hulme, A.; Lee, H.; Norris, L. Mobile Learning Revolution: Implications for language pedagogy. In *The Handbook of Technology and Second Language Teaching and Learning*; Chapelle, C.A., Sauro, S., Eds.; Wiley & Sons: Hoboken, NJ, USA, 2017; pp. 217–233.
2. Yen, E.L.Y.; Mohamad, M. Utilising E-Learning to assist primary school ESL pupils in learning to spell during COVID-19 pandemic: A Literature Review. *Creat. Educ.* **2020**, *11*, 1223–1230. [CrossRef]
3. Khasbani, I. Revealing teachers' motivational strategy in Indonesia EFL classrooms. *Eur. J. Engl. Lang. Teach.* **2018**, *3*, 1–14.
4. Zakaria, A.H.; Shah, P.M. Communicative Language Teaching (CLT): Its implementation in teaching English to Malaysian ESL primary learners. *Int. J. Sci. Eng. Res.* **2019**, *10*, 785–799.
5. Pandemic Exploited to Normalise Mass Surveillances? The Asean Post. Available online: https://theaseanpost.com/article/pandemic-exploited-normalise-mass-surveillance (accessed on 22 May 2022).
6. UNICEF Education COVID-19 Case Study: Malaysia-Empowering Teachers to Deliver Blended Learning after School Reopening. Available online: https://www.unicef.org/evaluation/documents/unicef-education-COVID-19-case-study-malaysia-empowering-teachers-deliver-blended (accessed on 20 May 2022).
7. Wargadinata, W.; Maimunah, I.; Dewi, E.; Rofiq, Z. Student's responses on learning in the early COVID-19 pandemic. *Tadris J. Educ. Teach. Train.* **2020**, *5*, 141–153. [CrossRef]
8. English as a Foreign Language (EFL). Available online: https://www.thoughtco.com/english-as-a-foreign-language-efl-1690597 (accessed on 20 May 2022).
9. British Council. The Importance of the English Language: 10 Reasons to Learn English. Available online: https://www.british-study.com/en/importance-of-english-language (accessed on 21 May 2022).
10. Burns, A. Concept for teaching speaking in the English Language classroom. *Lang. Educ. Acquis. Res. Netw. J.* **2019**, *12*, 1–11.
11. Ur, P. *A Course in Language Teaching Practice and Theory*; Cambridge University Press: Harvard, UK, 1996.
12. Goh, C.C.M.; Burns, A. *Teaching Speaking: A Holistic Approach*; Cambridge University Press: New York, NY, USA, 2012.
13. Vygotsky, L.S. *Mind in Society: The Development of Higher Psychological Processes*; Harvard University Press: Cambridge, MA, USA, 1978.
14. Robbins, S.P.; Singer, J.B. From the editor-The medium is the message: Integrating social media and social work education. *J. Soc. Work Educ.* **2014**, *50*, 387–390. [CrossRef]
15. Kaplan, A.M.; Haenlein, M. Users of the world, unite! The challenges and opportunities of social media. *Bus. Horiz.* **2010**, *53*, 59–68. [CrossRef]
16. Khan, I.U.; Ayaz, M.; Khan, S.; Khan, M.F. Effect of social media on enhancement of English learning proficiency at university level in Khyber Paktunkhwa. *IASET J. Humanit. Soc. Sci.* **2016**, *2*, 71–83.
17. Savasci, M. Why are some students reluctant to use L2 in EFL Speaking Classes? An action research at tertiary level. *Procedia-Soc. Behav. Sci.* **2014**, *116*, 2682–2686. [CrossRef]
18. Idrus, H.; Salleh, H.I. Perceived self-efficacy of malaysian esl engineering and technology students on their speaking ability and its pedagogical implications. *Engl. Teach.* **2007**, *XXXVII*, 61–75.
19. Al-Sobhi, B.M.S.; Preece, A.S. Teaching English speaking skills to the Arab students in the Saudi school in Kuala Lumpur: Problems and solutions. *Int. J. Educ. Lit. Stud.* **2018**, *6*, 1–11. [CrossRef]
20. Brown, G.; Yule, G. *Discourse Analysis*; Cambridge University Press: Cambridge, UK, 1983.
21. Minalla, A.A. The effect of WhatsApp chat group in enhancing EFL learners' verbal interaction outside classroom contexts. *Engl. Lang. Teach.* **2018**, *11*, 1–7. [CrossRef]
22. Noni, N.; Basri, M. WhatsApp audio and video chat-based in stimulating students' self-confidence and motivation to speak English. *Asian EFL J.* **2019**, *23*, 181–203.
23. Rakhmanina, L.; Yuneva, Y. The application of Instagram activity to improve students' motivation in English speaking. *Edu-Ling J. Engl. Educ. Linguist.* **2018**, *2*, 49–59. [CrossRef]
24. Rahmah, R.E. Sharing photographs on Instagram boosts students' self-confidence in speaking English. *Pedagog. J. Engl. Lang. Teach.* **2020**, *6*, 148–158. [CrossRef]
25. Ghoneim, N.M.M.; Elghotmy, H.E.A. Using voice thread to develop EFL pre-service teachers' speaking skills. *Int. J. Engl. Lang. Teach.* **2016**, *4*, 13–31.
26. Sevy-Biloon, J.; Chroman, T. Authentic use of technology to improve EFL communication and motivation through international language exchange video chat. *Teach. Engl. Technol.* **2019**, *19*, 44–58.
27. Xodabande, I. The effectiveness of social media network telegram in teaching English language pronunciation to Iranian EFL learners. *Cogent Educ.* **2017**, *4*, 1347081. [CrossRef]
28. Teoh, C.X.; Yunus, M.M. Primary pupils use of social media to learn ESL. *Int. J. Innov. Technol. Explor. Eng.* **2019**, *8*, 2076–2080.
29. Lancy, D.F. *Qualitative Research In Education: An Introduction To The Major Traditions*; Longman Pub Group: Harlow, UK, 1993.
30. Patton, M.Q. *Qualitative Research and Evaluation Methods*, 3rd ed.; Sage Publications: Thousand Oaks, CA, USA, 2002.
31. Tomalin, B.; Stempleski, S. *Cultural Awareness-Resource Books for Teachers*; Oxford University Press: Oxford, UK, 2013.
32. Ainun, N.H.; Nurweni, A.; Sholihah, L. The students' perception of voice chatting through WhatsApp in speaking class. In Proceedings of the Second International Conference on Progressive Education (ICOPE), Bandar Lampung, Indonesia, 16–17 October 2020.

33. Ilyas, M.; Putri, M.E. YouTube channel: An alternative social media to enhance EFL students' speaking skill. *J-SHMIC J. Engl. Acad.* **2020**, *7*, 77–87. [CrossRef]
34. Boyd, D. Social media: A phenomenon to be analyzed. *Soc. Media Soc.* **2015**, *1*, 2056305115580148. [CrossRef]
35. Lederer, K. Pros and cons of social media in the classroom. *Campus Technol.* **2012**, *2*, 1–2.
36. Reinders, H.; Wattana, S. Can I say something? The effects of digital game play on willingness to communicate. *Lang. Learn. Technol.* **2014**, *18*, 101–123.
37. Nasri, M.; Biria, R.; Karimi, M. Projecting gender identity in argumentative written discourse. *Int. J. Appl. Linguist.* **2018**, *7*, 201–205. [CrossRef]

proceedings

MDPI

Proceeding Paper

Intention to Choose Education Course in UiTM Using Theory of Planned Behaviour (TPB) †

Muhammad Saiful Anuar Yusoff [1], Wan Nazihah Wan Mohamed [1,*], Zulkifli Mohamed [2] and Norhaiyati Abdul Muin [3]

1 Akademi Pengajian Bahasa, Universiti Teknologi MARA (UiTM) Cawangan Kelantan, Machang 18500, Malaysia
2 Faculty of Business Management, Universiti Teknologi MARA (UiTM) Cawangan Kelantan, Kota Bharu 15050, Malaysia
3 Independent Researcher, Machang 18500, Malaysia
* Correspondence: wnazihah@uitm.edu.my

† Presented at the International Academic Symposium of Social Science 2022, Kota Bharu, Malaysia, 3 July 2022.

Abstract: In line with the offering of a Master's of Education as a new course in UiTM Kelantan, this study was conducted to investigate the willingness of the community to enrol into the course based on the Theory of Planned Behaviour (TPB). The influences of attitude, subjective norms and perceived behavioural control towards intention were studied. A total of 347 data elements were obtained from employees working in both public and private sectors in Kelantan and Terengganu. The data were analysed using SmartPLS. The findings reveal a significant influence of attitudes, subjective norms and perceived behavioural control on the intention of the respondents to enrol into an education course offered at UiTM. The findings of this study are expected to contribute to identify the factors that influence the intentions of the community members, in particular public servants, to choose courses offered at UiTM Kelantan.

Keywords: attitude; subjective norms; perceived behavioural control

Citation: Yusoff, M.S.A.; Wan Mohamed, W.N.; Mohamed, Z.; Abdul Muin, N. Intention to Choose Education Course in UiTM Using Theory of Planned Behaviour (TPB). *Proceedings* **2022**, *82*, 40. https://doi.org/10.3390/proceedings2022082040

Academic Editor: Mohamad Rahimi Mohamad Rosman

Published: 14 September 2022

1. Introduction

The importance of education in increasing the productivity of individuals, workplaces and countries is undeniable since education is an ongoing process throughout a person's life. The learning process is not confined to only children and the working groups, but it also covers other community members including housewives, retirees, the disabled and the elderly. The ultimate goal of education is to improve the life quality of individuals and society; thus, the country is in dire need of knowledgeable human capital. Knowledge needs to be updated in order to produce human beings that are relevant to the progress and development of the nation [1]. A study by the United Nations Educational, Scientific and Cultural Organization (UNESCO) found that educational programs provided to all community members in the developing countries has fostered progress in the respective countries' development. As such, lifelong education has been recognized by the United Nations (UN) as the key to human resource development. Subsequently, the government has taken proactive measures in the education system to produce high quality human capital that can move towards the global economy era of knowledge-based communities [2]. All of these measures are designed to fulfil the country's aspirations to produce balanced, high quality and knowledgeable human capital from all angles.

The rapid growth of education has indirectly created many opportunities for school students and workers in both public and private sectors to pursue studies into higher education. As a result, the process of selecting higher education institutions and courses is increasingly difficult due to the existence of various public and private institutions of higher learning that offer a wide selection of quality and competitive programs [3]. This has created intense challenges and competitions among public and private institutions of higher learning to attract the public in choosing the courses offered. As such, there is a need to understand the tendency of the public and private employees to pursue higher education and to identify the factors influencing the intention to continue their studies. In addition, the absence of studies conducted to determine the community's interest in pursuing education-related courses at UiTM Kelantan makes this study very significant. Therefore, this study is conducted to provide an accurate and recent perspectives of their intention to choose educational courses offered at UiTM Kelantan.

2. Literature Review

2.1. Intention to Enrol in Course

In social science, intention is an important indicator of an individual's readiness to accept a behaviour [4,5]. It is an aspect of motivation that affects one's behaviour and an indication of the willingness to perform an action or plan the effort required to perform an action [6]. In addition, a person's tendency to perform an action increases with the strength of one's intention [7].

This study involved the variables of attitude, subjective norm and perceived behavioural control as factors that determine one's intention to perform an act or behaviour. The choice of these three factors is based on the Theory of Planned Behaviour. The theory postulates that the more important one's intention to do something is, the higher the probability of an action being taken. With the support of previous studies, it is expected that attitude, subjective norm and perceived behavioural control will influence the intention to choose an education course at UiTM Kelantan.

2.2. Attitude

Attitude is an internal factor of an individual which is defined as the positive or negative level of a person's feelings towards a particular desire or behaviour [4]. According to [8], the students' perceptions of a domain often influence their choice of a particular subject. This means that students with negative attitudes toward a particular subject will avoid choosing it while positive students will have a positive commitment towards the subject. In relation to this study, attitude is required in the decision-making process so as to continuously and consistently choose the course offered.

Previous studies have proven the influence of positive attitude on the choice of various courses such as language [9,10], accounting [11] and entrepreneurship [12–14]. The analysis of these studies found that attitude towards certain subjects has a significant relationship with the formation of behavioural intention. Considering these findings, it is expected that a positive attitude towards an education course will also reflect the same effect on the intention to choose an education course at UiTM Kelantan. As such, the first hypothesis is presented as:

H1. *Attitude positively influences the intention of choosing an education course at UiTM Kelantan.*

2.3. Subjective Norm

Subjective norm is a part of external motivation that plays a role in persuading a person to do an important action. It also refers to a person's perception of social pressure placed on them from a particular parent, friend or community so as to perform or not to perform the desired behaviour [15]. In the context of this study, this represents a person who has the impression that individuals influence him or her, such as family, employers, friends and community, and may encourage him or her to make a choice and be motivated to do the action. In other words, subjective norm determines an individual's action.

The results of previous studies support subjective norm as one of the important factors in choosing a college or university course. Social factors which consist of family and peers are seen to be the major factors influencing the selection of a course [16,17]. A study by [17] on student selection at a university in Indonesia found that the family factor is one of the five important factors besides funding fees, reputation, distance factor and career prospects. Therefore, individuals are expected to choose the course offered at UiTM Kelantan if they are encouraged by the important individuals around them. This leads to the second hypothesis which is:

H2. *Subjective norm positively influences the intention of choosing an education course at UiTM Kelantan.*

2.4. Perceived Behavioural Control

Perceived behavioural control is an individual's perception according to which performing a behaviour is within his or her control and is often assessed by the ease or difficulty of performing the behaviour [18]. It is an internal motivation that influences one's intention and action which then enables a person to set goals for action [18]. In this study, a person with high expectations for his or her behaviour will be more likely to choose a course of action despite facing various problems such as lack of interest, course information or confidence in his or her own ability and limited economic resources.

Findings of previous studies on the choice of accounting and entrepreneurship courses showed the influence of perceived behavioural control on the intention to choose such courses [12,14]. The study of [14] showed that a student's self-confidence is one of the major sources of motivation to choose an accounting course and pursue a career in that field. As such, one is expected to choose an education course offered at UiTM Kelantan if he or she has the confidence to face various difficulties. In relation to that, the third hypothesis is written as:

H3. *Perceived behavioural control positively influences the intention of choosing an education course at UiTM Kelantan.*

3. Research Framework

The selection of Theory of Planned Behaviour (TPB) [18] as the model of this study is based on the ability of this theory to explain the intention of choosing an action or behaviour. Several previous studies using the Theory of Reasoned Action (TRA) have provided strong empirical support on the two variables of attitude and subjective norm [15]. Nevertheless, this initial theory (TRA) was criticized by scholars for its lack of factors that could explain one's self-control in performing an action [18]. Furthermore, the variable perceived behavioural control is included in the model which describes the person's actual intention and behaviour. The rationale for choosing this variable is due to the fact that a person's behaviour is usually influenced by one's confidence in the ability to perform an action. The higher the expectation for behavioural control, the higher the intention to do an action which leads to high performance [18]. In particular, attitude, subjective norm and perceived behavioural control are expected to be positively related to the intention to choose an education course offered at UiTM Kelantan.

4. Methodology

This study used a quantitative research approach through simple random probability sampling technique. The population of the study was employees working at public and private sectors in Kelantan and Terengganu. The number of the study sample was accurately described using the G*power software [19] (Version 3.1.9.6, University Kiel, Germany) which resulted into a minimum of 77 samples required for this study.

Data were obtained through questionnaire distribution using a 7-point Likert scale. A preliminary test was performed to confirm the item reliability of the study. The analysis found that the values of Cronbach's alpha for all variables exceeded 0.6, which were beyond the level suggested by [20]. Data collection was completed with a total of 355 questionnaires obtained from 400 questionnaires distributed to the respondents in the states of Kelantan and Terengganu. However, after the data cleaning process, a total of 347 data elements were usable for analysis in this study. The strength of the analysis was further enhanced by using two stages of Partial Least Square Structural Equation Modelling (PLS-SEM) analysis, namely measurement model and structural model.

5. Findings

The demographic analysis of the respondents included 248 (71.5%) women and 99 (28.5%) men who are currently working in various public and private sectors in Kelantan and Terengganu. The majority of the respondents (84.7%) were from Kelantan while only 15.3% of them were from Terengganu. A total of 46 respondents (13.3%) were in the age group of 29 years, 106 (30.5%) were aged 30 to 39 years, 111 (32.0%) were aged 40 to 49 years and 84 respondents (24.2%) were over 50 years old. Most of the respondents earned over Ringgit Malaysia RM 6000 per month (43.8%), followed by an income between RM 5000 and RM 6000 (16.7%), RM 4001 and RM 5000 (17.3%), RM 3001 and RM 4000 (7.2%), RM 2001 and RM 3000 (5.8%) and less than RM 2000 (9.2%). The mode of study that the respondents chose was part-time for 209 respondents (60.2) followed by full-time for 138 (39.8%) respondents. The respondent's choice of education course at UiTM Kelantan was driven by the following reasons: being near to home (72.1%), low costs (62.9), appropriate environment (27.4%), family support (24.4%) and others (8.1%).

5.1. Measurement Model Analysis

The evaluation of the measurement model involves convergent validity and reliability. Convergent validity refers to the level of agreement of several items or indicators in measuring the same concept or construct [21]. The results in Table 1 show that all items exceeded the 0.6 value for item reliability (factor loading) [22], the 0.7 value for rhoA path coefficient [23], the 0.5 value for AVE (Average Variance Extracted) and the 0.7 value for CR (Construct Reliability) [24]. All the values obtained exceeded the minimum requirement of convergent validity [25].

Table 1. Convergent validity analysis.

Construct	Item	Loading	Cronbach	rho_A	CR	AVE
ATT	D1	0.907	0.971	0.971	0.971	0.847
	D2	0.930				
	D3	0.916				
	D4	0.918				
	D5	0.942				
	D6	0.909				
SN	E1	0.954	0.956	0.956	0.956	0.843
	E2	0.903				
	E3	0.896				
	E4	0.918				
PBC	F1	0.871	0.885	0.886	0.885	0.794
	F2	0.911				
INT	G1	0.955	0.957	0.962	0.959	0.824
	G2	0.945				
	G3	0.927				
	G4	0.912				
	G5	0.791				

ATT: Attitude, INT: Intention, PBC: Perceived Behavioural Control, SN: Subjective Norm, rhoA: Reliability indicator, AVE: Average Variance Extracted, CR: Construct Reliability.

In addition, a discriminant validity test was performed using the Heterotrait-Monotrait (HTMT) criterion. The analysis was performed to detect the presence of a collinearity issue or a cross-loading conflict in the research items. According to [25], HTMT is the latest criterion in SmartPLS software (SmartPLS GmbH, Version 3.3.9, Boenningstedt, Germany) that can accurately determine discriminant validity. Through this test, a construct value greater than 0.85 indicates discriminant validity has been fulfilled [20]. The analysis for this study found that all constructs had values below the HTMT value of 0.85, which is the maximum value for discriminant validity. As such, discriminant validity of all constructs was achieved as presented in Table 2 below.

Table 2. Discriminant validity analysis.

	ATT	INT	PBC	SN
ATT				
INT	0.775			
PBC	0.726	0.781		
SN	0.835	0.778	0.808	

ATT: Attitude, INT: Intention, PBC: Perceived Behavioural Control, SN: Subjective Norm.

Figure 1 shows the SmartPLS measurement model results. The R^2 value of business success was 0.704, suggesting that 70.4 percent of the course selection intention can be explained by the three independent variables. Once the evaluation of the measurement model was completed, a further analysis was carried out which included structural model analysis for hypothesis testing.

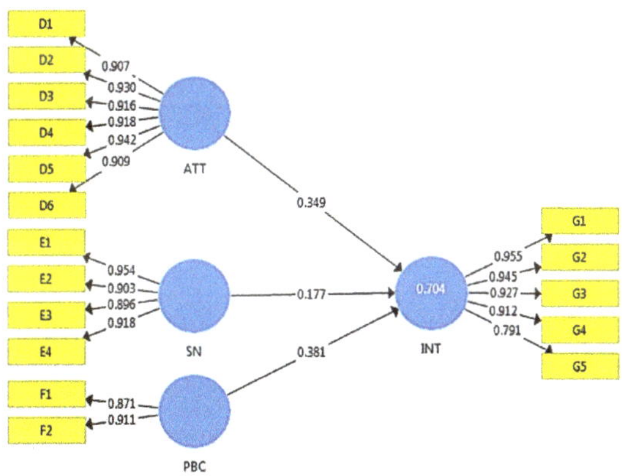

Figure 1. Measurement model.

5.2. *Structural Model Analysis*

Prior to the structural model analysis, a lateral collinearity test was performed. Even though discriminant validity and HTMT ratio showed no collinearity issue for all research constructs, conducting lateral collinearity testing is necessary. This is because the existence of this type of collinearity may impede the research findings, since it has a tendency to disrupt the causal relationship between the predictors and the dependent variables of the research model. This occurs when two variables that are fundamentally related to each other are found to measure the same construct [26].

Table 3 depicts the results for the lateral collinearity test. The Variance Inflation Factors (VIF) values for all independent variables (ATT, SN and PBC) were lesser than five [27], which indicated that lateral collinearity was not an issue in this study.

Table 3. Lateral collinearity analysis.

Construct	Intention (VIF Value)
ATT	2.93
PBC	2.32
SN	3.59

VIF $\leq$ 5.0 [27].

This study developed three hypotheses in determining the direct relationship between the constructs of the study. According to [24], a research hypothesis can be determined through bootstrapping analysis using 5000 sampling methods in SmartPLS software. The findings show that all three research hypotheses were supported by t values greater than or equal to 1.645 with attitude (ATT) (ß = 0.349, $p < 0.01$), subjective norm (SN) (ß = 0.177, $p < 0.05$) and perceived behavioural control (PBC) (ß = 0.381, $p < 0.01$) positively influenced course selection intention while the number of variance explained by these three variables accounted for 70.4%. Therefore, H1, H2 and H3 were fully supported. The R^2 value of 0.704 exceeded the value of 0.26 as proposed by [28] which showed that ATT, SN and PBC influence the intention to choose an education course at UiTM Kelantan. The structural model of hypothesis analysis is presented in Figure 2.

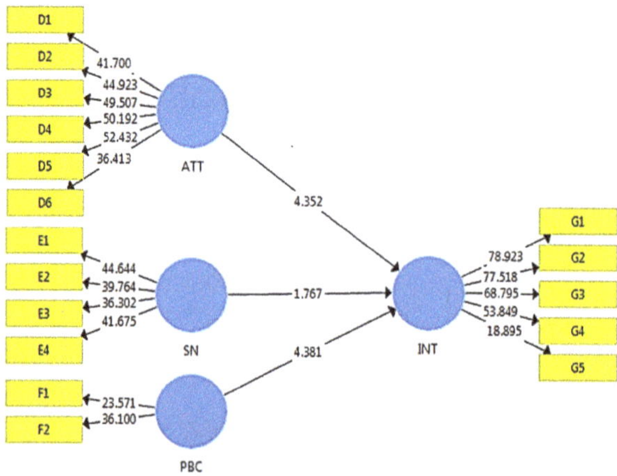

Figure 2. Structural model.

Table 4 shows the results of hypothesis analysis for this research model. The value $R^2 = 0.704$ indicates that 70.4% of the variance in the intention to choose an education course was explained by the three study variables of ATT, SN and PBC. According to [29] on the level of R^2 (0.67 = high; 0.33 = medium; 0.19 = low), the value for this study was considered high. In addition, as stated by [28] on the guideline of effect size (f^2) (0.02 = small; 0.15 = medium; 0.35 = large), this study had small and medium effect sizes since the values were between 0.023 and 0.166. Perceived behavioural control had a modest effect on the intention to choose an education course while attitude and subjective norm had little effect on the intention to choose a course. Prediction relevance value (Q^2) beyond the value of zero suggests that the variable has a predictive ability based on the intention to choose an education course [24]. The acceptable data-model fit is determined through the goodness-of-fit measures (*GoFs*) [29]. If the average square root value of AVE multiplied by R^2 exceeds the value proposed by [30] (*GoF*small = 0.01; *GoF*medium = 0.25; *GoF*large = 0.36), the model fit of the study is achieved. The analysis found that the obtained value was high at 0.694, which exceeded the proposed value level.

Table 4. Hypothesis analysis.

Hypothesis	Relationship	Beta	SE	T Value	Result	R^2	f^2	Q^2	*GoF*
H_1	ATT → INT	0.349 **	0.08	4.352	Supported	0.704	0.122	0.526	0.694
H_2	SN → INT	0.177 *	0.100	1.767	Supported		0.023		
H_3	PBC → INT	0.381 **	0.087	4.381	Supported		0.166		

ATT: Attitude, INT: Intention, PBC: Perceived Behavioural Control, SN: Subjective Norm ** $p < 0.01$, t value > 2.33; * $p < 0.05$, t value > 1.645.

6. Discussion

The objective of this study is to investigate the influence of attitude, subjective norm and perceived behavioural control on the intention to choose an education course at UiTM Kelantan. The findings support all the hypotheses of the study which demonstrates that attitude, subjective norm and perceived behavioural control influence the intention to choose an education course. The main perceived contributor is behavioural control, followed by attitude and subjective norm.

It can be seen in this study that all factors collectively influence a person's intention to perform an action. All factors accounted for 70.4% of the variance in the intention to choose an education course at UiTM Kelantan. However, the strength of the relationship is different. Perceived behavioural control was found to significantly contribute to the intention of choosing education course as compared to the other two factors with a moderate effect size ($f^2 = 0.166$). This result is similar to previous studies that examined the relationship of perceived behavioural control with the intention to select language, accounting and entrepreneurship courses [9–14]. In other words, employees involved in this study tend to choose an education course offered at UiTM Kelantan despite facing challenges such as limited economic resources and lack of adequate information on the offered course. However, they are still unsure of their interest and self-confidence in choosing an education course since there is no relationship analysis of these two items with perceived behavioural control.

Positive attitude towards an education course is the second important factor that influences the selection of an education course at UiTM Kelantan. This finding is also parallel to previous studies which showed that attitude is an important factor in influencing respondents' choice of university courses [9,10,12–14]. Based on this result, the offering of an education course at UiTM Kelantan is very much anticipated, and especially at the undergraduate level since such course is not offered at any UiTM campuses, with the exception of UiTM's main campus in Shah Alam.

The last factor influencing the respondents' selection of an education course at UiTM Kelantan is subjective norm. The finding of this study corresponds with previous studies that viewed the family factor as a source of motivation for choosing such course [14,16,17]. In this study, the family is seen to be the main motivator for the respondents to choose the course, followed by the community, employers and friends. Family and community are key references in the selection of a course, which may be due to the experience of family members or successful community members in the education field.

7. Conclusions

In conclusion, internal factors such as perceived behavioural control and positive attitude are significant factors that influence the selection of an education course in UiTM Kelantan as compared to external factors such as subjective norm. Despite the obstacles and challenges, the majority of the employees are positive towards choosing this course. Although subjective norm has the least influence within the research model, its influence is still significant since the existence of family members, friends and community is still relevant in the selection of an education course at UiTM Kelantan. Therefore, these three factors need to be taken into account as they greatly influence the intention to choose an education course offered at UiTM Kelantan.

Author Contributions: Conceptualization, Z.M. and N.A.M.; methodology, M.S.A.Y.; validation, Z.M. and N.A.M.; formal analysis, M.S.A.Y.; investigation, M.S.A.Y., W.N.W.M., Z.M. and N.A.M.; data curation, M.S.A.Y.; writing—original draft preparation, M.S.A.Y. and W.N.W.M.; writing—review and editing, W.N.W.M. and N.A.M.; visualization, M.S.A.Y. and W.N.W.M. All authors have read and agreed to the published version of the manuscript.

Funding: This research received no external funding.

Institutional Review Board Statement: Not applicable.

Informed Consent Statement: Informed consent was obtained from all subjects involved in the study.

Data Availability Statement: Not applicable.

Acknowledgments: The authors wish to thank UiTM Cawangan Kelantan for supporting this study.

Conflicts of Interest: The authors declare no conflict of interest.

References

1. Majlis Pelancaran Pelan Strategi UPM 2001–2010, Etika Kerja UPM dan Anugerah Fellowship Naib Canselor. Available online: www.upm.edu.my/website/koleksi_ucapan/2002 (accessed on 12 January 2022).
2. Mansor, F.; Tengku Kasim, T.S.A. Pembangunan Modal Insan dari Perspektif Pengurusan Islam. *J. Al-Tamaddun.* **2008**, *3*, 85–109. [CrossRef]
3. Dimon, Z.; Azizan, N.I.; Ismail, N.; Bakar, S.A.; Shaharuddin, H.N. Faktor yang mempengaruhi pelajar memilih bidang pengajian Islam: Kajian di Fakulti Pengajian Peradaban Islam, KUIS. *Attarbawiy Malays. Online J. Educ.* **2017**, *1*, 50–57.
4. Ajzen, I. Perceived behavioural control, self-efficacy, locus of control, and the theory of planned behavior. *J. Appl. Soc. Psychol.* **2002**, *32*, 665–683. [CrossRef]
5. Fishbein, M.; Ajzen, I. *Belief, Attitude, Intention and Behaviour: An Introduction to Theory and Research*; Addison-Wesley: Reading, UK, 1975.
6. Ajzen, I.; Driver, B.L. Application of the theory of planned behaviour in leisure choice. *J. Leis. Res.* **1992**, *24*, 207–224. [CrossRef]
7. Ajzen, I.; Madden, T.J. Prediction of goal directed behaviour: Attitudes, intentions, and perceived behavioural control. *J. Exp. Soc. Psychol.* **1986**, *2*, 453–474. [CrossRef]
8. Curran, J.M.; Rosen, D.E. Student attitudes toward college courses: An examination of influences and intentions. *J. Mark. Educ.* **2006**, *28*, 135–148. [CrossRef]
9. Yusoff, M.S.A.; Hassan, N.S.I.C.; Nor, N.H.M.; Yusof, N. Pengaruh sikap dan nilai Islam terhadap niat untuk menghadiri program bahasa Inggeris di masjid. *e-Acad. J.* **2018**, *7*, 98–108.
10. Bosan, B. The Effects of Role-Play on Students' Attitude and Motivation to Speak and Interact in the ESL Classroom. Master's Thesis, Universiti Teknologi Malaysia, Johor, Malaysia, 2013.
11. Hatta, M.; Agustian Riduan, O.H. Niat mahasiswa akuntasi bekarir sebagai akuntan di luar negeri: Pengujian theory of planned behavior. *J. Akunt.* **2017**, *7*, 63–68.
12. Potishuk, V.; Kratzer, J. Factors affecting entrepreneurial intentions and entrepreneurial attitudes in higher education. *J. Entrep. Educ.* **2017**, *20*, 25–44.
13. Shah, N.; Soomro, B.A. Investigating entrepreneurial intention among public sector university students of Pakistan. *Educ. + Train.* **2017**, *59*, 841–855. [CrossRef]
14. dos Santos, E.A.; de Almeida, L.B. To pursue a career in accounting or not: A study based on the Theory of Planned Behavior. *Rev. Contab. Financ.* **2018**, *29*, 114–128. [CrossRef]
15. Ajzen, I.; Fishbein, M. *Understanding Attitudes and Predicting Social Behaviour*; Prentice-Hall: Hoboken, NJ, USA, 1980.
16. Jafari, P.; Aliesmaili, A. Factors influencing the selection of a university by high school students. *J. Basic Appl. Sci. Res.* **2013**, *3*, 696–703.
17. Kusumawati, A. A qualitative study of the factors influencing student choice: The case of public university in Indonesia. *J. Basic Appl. Sci. Res.* **2013**, *3*, 314–327.
18. Ajzen, I. The theory of planned behavior. *Organ. Behav. Hum. Decis. Processes* **1991**, *50*, 179–211. [CrossRef]
19. Faul, F.; Erdfelder, E.; Buchner, A.; Lang, A.G. Statistical power analyses using G*Power 3.1: Tests for correlation and regression analyses. *Behav. Res. Methods* **2009**, *41*, 1149–1160. [CrossRef] [PubMed]
20. Kline, R.B. *Principles and Practice Of Structural Equation Modeling*; Guilford Press: New York, NY, USA, 2011.
21. Malhotra, A.; Schuler, S.R.; Boender, C. *Measuring Women's Empowerment as a Variable in International Development*; The World Bank: Washington, DC, USA, 2002.
22. Byrne, B.M. *Structural Equation Modeling With AMOS: Basic Concepts, Applications, And Programming*; Routledge: London, UK, 2016.
23. Dijkstra, T.K.; Henseler, J. Consistent partial least squares path modeling. *MIS Quart.* **2015**, *39*, 297–316. [CrossRef]
24. Hair, J.F.; Hult, T.M.; Ringle, C.M.; Sarstedt, M.A. *Primer on Partial Least Square Structural Equation Modeling (PLS-SEM)*; Sage Publications: Thousand Oaks, CA, USA, 2014.
25. Henseler, J.; Ringle, C.M.; Sarstedt, M. A new criterion for assessing discriminant validity in variance-based structural equation modeling. *J. Acad. Mark. Sci.* **2015**, *43*, 115–135. [CrossRef]
26. Kock, N.; Lynn, G.S. Lateral collinearity and misleading results in variance-based SEM: An illustration and recommendations. *J. Assoc. Inf. Syst.* **2012**, *13*, 546–580. [CrossRef]
27. Hair, J.; Hollingsworth, C.L.; Randolph, A.B.; Chong, A. An updated and expanded assessment of PLS-SEM in information systems research. *Ind. Manag. Data Syst.* **2017**, *117*, 442–458. [CrossRef]
28. Cohen, J. *Statistical Power Analysis for the Behavioural Sciences*, 2nd ed.; Lawrence Erlbaum Associates: Hillsdale, NJ, USA, 1988.

29. Chin, W.W. The partial least squares approach for structural equation modeling. In *Modern Methods for Business Research*; Marcoulides., G.A., Ed.; Lawrence Erlbaum Associates: Mahwah, NJ, USA, 1998; pp. 295–358.
30. Tenenhaus, M.; Esposito Vinzi, V.; Chatelin, Y.-M.; Lauro, C. PLS path modelling. *Comput. Stat. Data Anal.* **2005**, *48*, 159–205. [CrossRef]

proceedings

Proceeding Paper

Language Proficiency among Non-Native Chinese Language Learners: A Discriminant Analysis †

Sian Hoon Teoh [1,*], Buai Chin Heng [2], Yee Feng Neo [2] and Julia Yin Yin Tan [3]

[1] Faculty of Education, Universiti Teknologi MARA, Cawangan Selangor, Bandar Puncak Alam 42300, Selangor, Malaysia
[2] Akademi Pengajian Bahasa, Universiti Teknologi MARA, Shah Alam 40450, Selangor, Malaysia
[3] Faculty of Language Studies and Human Development, Universiti Malaysia Kelantan, Bachok 16300, Kelantan, Malaysia
* Correspondence: teohsian@uitm.edu.my
† Presented at the International Academic Symposium of Social Science 2022, Kota Bharu, Malaysia, 3 July 2022.

Abstract: The Chinese Language (CL) is difficult to learn, and CL non-native learners are finding it increasingly difficult. Therefore, additional information on how to guide students is critical. As a result, the goal of this research is to find out what criteria distinguish high and low proficiency students in Chinese learning. The data was collected through a Google form questionnaire from 79 CL non-native students who had previously studied Chinese. The findings demonstrate that the only difference between the two groups is the writing strategies used. This study concludes that various teaching methods should emphasise writing skills in order to become proficient in Chinese.

Keywords: Chinese Language (CL); discriminant analysis; non-native CL learners; proficiency

check for updates

Citation: Teoh, S.H.; Heng, B.C.; Neo, Y.F.; Tan, J.Y.Y. Language Proficiency among Non-Native Chinese Language Learners: A Discriminant Analysis. *Proceedings* **2022**, *82*, 42. https://doi.org/10.3390/proceedings2022082042

Academic Editor: Mohamad Rahimi Mohamad Rosman

Published: 14 September 2022

1. Introduction

The Chinese Language (CL) seems to have drawn more people to learn it seriously. Seen in this light, it has been observed that parents pay close attention to their children's CL learning settings. Several Malaysian studies have investigated the learning situations of these non-native CL students, including research into the causes that motivate parents to send their children to CL medium schools (known as *Sekolah Jenis Kebangsaan Cina* (SJKC). These studies were also conducted on the pupils' academic achievement and learning issues [1–6].

These studies' conclusions have unintentionally highlighted a few problems faced by CL non-native learners. The problems highlighted were the non-native students' poor academic performance at SJKCs [2] and issues with the CL being the instructional medium of mathematics and science, hindering the learning of the latter as well [3]. The other much-discussed problems were the inadequate teaching methods used by teachers [1], an inconducive learning environment at home, and low expectations and support from parents [1–3].

This investigation was prompted by the public's interest in learning the CL effectively. Hence, the goal of this study was to investigate the learning elements that induced the learners' CL mastery of the 'high' and 'poor' proficiency learners in CL classrooms at SJKC schools.

2. Literature Review

2.1. Learning Chinese Language

Learning the CL is becoming increasingly popular as a result of education globalisation [7], which has resulted in an increasing interconnectedness of societies in economics, technology, politics, culture, and language [8]. Hence, language plays a vital role in maximising these interactionist relationships.

Among the spoken languages in the world, it was reported that CL ranks the most favoured [9]. Correspondingly, non-native CL learners have benefited from economic development by meeting global needs via learning this language. Although difficult, the CL has become an important subject in the Malaysian Certificate of Education, which is taken by all fifth-form secondary school pupils in Malaysia [10]. Hence, non-native learners' CL proficiency and challenges in achieving proficiency have been gaining attention.

2.2. *The Difficulties in Learning Chinese Characters*

When learning any language, voice is frequently used during the hearing and reading stages. Listening and reading require phonological awareness, which is concerned with the sounds of the spoken language. Phonological awareness, which is based on phonologically constructed speaking practises in Western (alphabetic) languages, plays a major role in a learner developing their reading. However, the CL is a unique medium where logographic letters are used to convey meaning rather than phonological speech.

Studies also highlighted that phonological speech, which is based on sound similarities of different characteristics, has caused difficulties for CL learners [11,12]. In classrooms, learners are often asked to identify the correct characters they see based on the meaning of a sentence. However, some CL characters have similar sounds, although they are different. This adds to the learners' confusion, which intensifies the failure to identify similar sounds (but having different meanings as they are different characters).

Figure 1 shows phonological speech representing different words with different meanings. For example, the phonological speech of /shàng/ is often used in reading for different characters with different meanings. Figure 1 also shows that a character will have a different meaning when it is attached to another (character). For example, the character /shàng/ is attached to /mǎ/ to form /mǎ shàng/, which means 'immediately'. On the other hand, if it is attached to /miàn/to form /shàng miàn/, it means 'above'. Also, a single character of the different meaning (e.g., /shàng/) holds a specific meaning which is 'still'. This example shows that the same phonologic speech can have different characters with different phonologic speech.

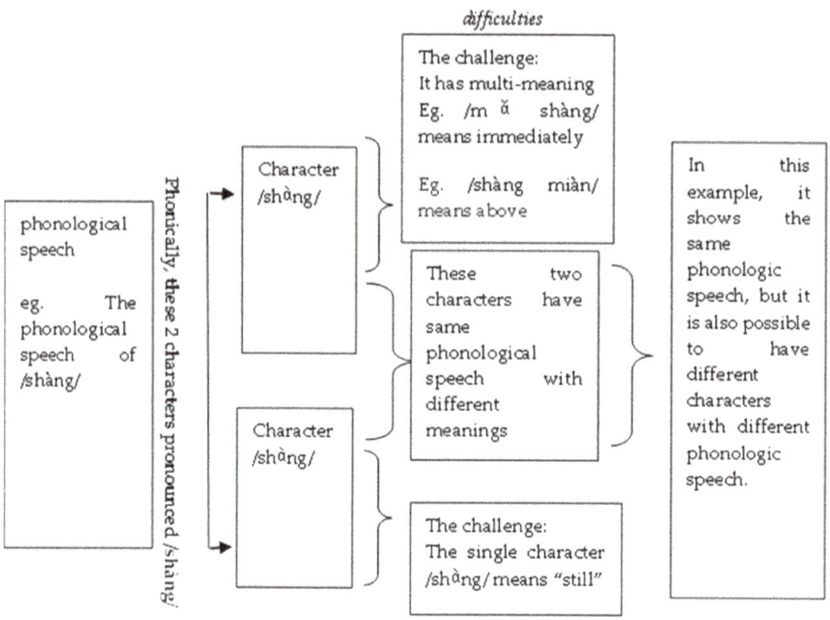

Figure 1. Complexity of Learning Chinese Language Characters: An Illustration.

Secondly, the complexity of learning this language is also challenged by the multi-meaning carried by a specific character and having different phonology for the same character with different meanings. As a result, a student's ability to read the CL is closely linked to his or her writing skills [6]. In short, writing skills are inextricably associated with CL proficiency.

2.3. Some Considerations in Learning the CL

It is noted that students' diverse origins, which may include strategies in learning, have an impact on their accomplishments. Hence in learning languages, students may be guided with specific strategies. It is important that students engage in the guided strategies when practising the four skills (reading, writing, listening, and speaking). As learning improves, learners may apply more strategies to improve acquisition and, thus, confidence. According to [13] students who apply appropriate strategies tend to be seen as good language learners. Applying appropriate strategies also requires appropriate knowledge for the learners to proceed with their ability. According to [14], carrying out appropriate strategies involves thought and behaviour. To this end, students' activities are equally vital in determining the success of language learning, which can be altered by motivation. Figure 2 shows the steps and links involved in achieving linguistic competency, showing the importance of strategy as a significant factor.

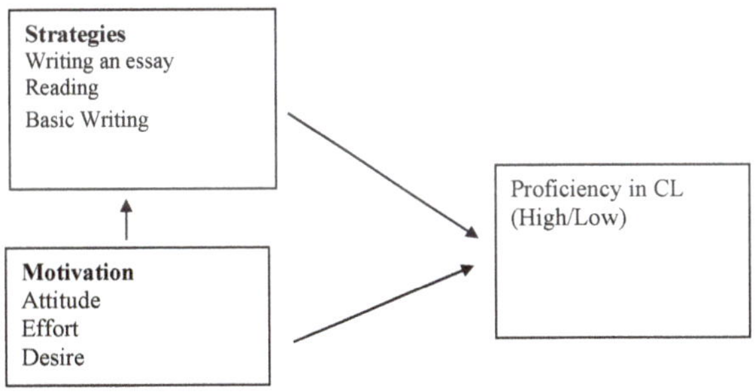

Figure 2. Proficiency in CL: Factors Involved.

3. Materials and Methods

3.1. Research Design and Sampling

This study employed a survey research design. Its sample was 79 non-native CL students who had experience in learning the CL at school. The focus of this study was on those who had shown good achievement in learning the Chinese language. Hence, the sample was former SJKC non-native CL high achievers in Malaysia. Taking note of the low percentage of high-achieving learners, which was an average of 800 students yearly, this study took 10% of 800 to respond to the questionnaires. Thus, a total of 79 students participated. The participants were then placed in two groups, namely 'low proficiency' and 'high proficiency'.

A sampling method called snowballing with volunteerism was employed in the selection of the samples. A list of samples was first identified by a few educators in Malaysia. Then, the participants' permissions were obtained to access their data. The participants who did not agree to participate were excluded. The inclusion criteria to select the samples were based on a volunteer basis, and most importantly, they consented to complete a Google form questionnaire.

3.2. Instrumentation

A questionnaire consisting of three parts was constructed following two references [15,16]. The three parts were CL proficiency, motivation, and strategies. Two experts were consulted to establish the content validity of the instrument. A pilot test was conducted on

79 participants. Its results showed a convincing and acceptable level of Cronbach's Alpha values, ranging from 0.76 to 0.85. Even though the construct 'writing an essay' showed a lower value of Cronbach's Alpha, the items contributed fittingly to the overall construct of writing, which was the combination of basic writing and essay writing (Cronbach's Alpha = 0.76).

3.3. Data Analysis

Discriminant Function Analysis (DA) was used to analyse the obtained data. The DA serves the same purpose as multiple linear regression by predicting an outcome. Hence, the DA was employed in this study since multiple linear regression is restricted to scenarios where the dependent variable is an interval variable. The regression equation also provides an estimated mean population numerical dependent variable value for specified weighted combinations of independent variable values. The 'proficiency level' was the dependent variable.

This study used two proficiency levels, 'low' and 'high', to address research question one. To address research question two, six proficiency levels were measured. The first independent variable was motivation, which consisted of attitude, effort, and desire. Another independent variable was strategy, which consisted of basic writing, essay writing, and reading. These independents were the discriminators (in regression analysis, the independent variables are predictors). This study focused on high achievers divided into two categories ('low proficiency' and 'high proficiency') in order to observe the factors that affect their proficiency.

In the DA, the independent variables are combined in weighted combinations to produce a single new composite variable, namely the discriminant score. Thus, the significant portions of the discriminant score reflect misclassifying cases into respective groups (low/high proficiency). A good DA model shows minimal misclassification, so the analysis detects the variables that primarily contribute to differentiating groups.

However, this was a simple discriminant analysis with two groups in the dependent variable. The simple discriminant analysis is provided with one set of eigenvalues: Wilks' Lambda and beta coefficients. The number of sets is always one less than the number of DV groups. Therefore, in this analysis, the data obtained were the respondents' demographic data and the answers given by them. Further, the 'proficiency level' was a nominal variable to indicate whether the learner was of high or low proficiency. The other variables were attitude, effort, desire, and writing strategies.

To reiterate, the aim of the analysis was to identify if these variables discriminate the participants' proficiency (low or high proficiency) and examine whether there were any significant differences between the 'low' and 'high' proficiency groups on each of the independent variables using group means and ANOVA.

The 79 respondents were divided into two groups based on their self-reported CL competency, which was based on the Common European Framework of Reference (CEFR) categorization. Low proficiency refers to those who rated themselves A1, A2, or B1, while high proficiency refers to the respondents rating themselves B2, C1, or C2.

4. Results

4.1. The First Finding

Research Question 1: Are the factors (motivation, namely attitude, effort, and desire; strategies, namely basic writing, writing an essay, and reading) significantly discriminate between the two groups ('low' and 'high' proficiency)?

If the Group Statistics and Tests of Equality of Group Means show that there are no significant group differences, it is not worthwhile to proceed any further with the analysis. In this study, the group statistics (group means and standard deviations) suggest that these may not be good discriminators as the separations are small. Nevertheless, to determine the significant discriminator, a Test of Equality of Group Means is used. Table 1 shows the descriptive statistics of all independent variables in the two groups (LP and HP) with overall scores of 5.20, 4.81, 5.43, 3.19, 3.23, and 3.05 for attitude, effort, desire, basic writing strategy, essay writing, and reading, respectively. Specifically, for the LP, the mean scores are 5.18 (attitude), 4.76 (effort), 5.43 (desire), 3.13 (basic writing strategy), 3.15 (writing an essay), and 3.00 (reading).

Table 1. Group Statistics.

	Level2	Mean	Std. Deviation	Valid N (Listwise)	
				Unweighted	Weighted
1.00	Attitude	5.18	0.78	62	62.00
	Effort	4.76	0.92	62	62.00
	Desire	5.43	0.63	62	62.00
	Basic Writing Strategy	3.13	0.71	62	62.00
	Writing an Essay	3.15	0.73	62	62.00
	Reading	3.00	0.75	62	62.00
2.00	Attitude	5.26	0.94	17	17.00
	Effort	4.99	0.61	17	17.00
	Desire	5.41	0.71	17	17.00
	Basic Writing Strategy	3.37	0.50	17	17.00
	Writing an Essay	3.56	0.53	17	17.00
	Reading	3.21	0.59	17	17.00
Total	Attitude	5.20	0.81	79	79.00
	Effort	4.81	0.87	79	79.00
	Desire	5.43	0.64	79	79.00
	Basic Writing Strategy	3.19	0.67	79	79.00
	Writing an Essay	3.23	0.71	79	79.00
	Reading	3.05	0.72	79	79.00

While for the HP, the mean scores are 5.26 (attitude), 4.99 (effort), 5.41 (desire), 3.37 (basic writing strategy), 3.56 (writing an essay), and 3.21 (reading). The Tests of Equality of Group Means show no significant group differences; hence, proceeding further with the analysis is not worthwhile.

Table 2 provides statistical evidence of significance in differences between the means of the two groups for all IV's, with only the 'essay writing' producing a higher F value ($F = 4.80$), with a significant value of p (=0.03) < 0.05.

Table 2. Tests of Equality of Group Means.

	Wilks' Lambda	F	df1	df2	Sig.
Attitude	1.00	0.13	1	77	0.72
Effort	0.99	0.97	1	77	0.33
Desire	1.00	0.01	1	77	0.92
Basic Writing Strategy	0.98	1.70	1	77	0.20
Writing an Essay	0.94	4.80	1	77	0.03
Reading	0.99	1.04	1	77	0.31

Table 3 shows the Box's M test to test the null hypothesis of whether there is any difference among the groups. Alternatively, it tests equal population covariance matrices. In this analysis, the Box's M is 2.30 with $F = 2.25$, which is not significant at $p > 0.05$.

Table 3. Test Results.

Box's M		2.30
F	Approx.	2.25
	df1	1
	df2	6215.96
	Sig.	0.13

For the assumption of equal variances to hold, the log determinants should also be equal. Table 4 shows that the groups have a reasonable Log Determinant with values close to each other, indicating mild variability of the groups.

Table 4. Log Determinants.

Level2	Rank	Log Determinant
1.00	1	−0.64
2.00	1	−1.28
Pooled within-groups	1	−0.74

The eigenvalues in Table 5 show the result of a function. The canonical correlation is the multiple correlations between the predictors (independent variables) and the discriminant function. The function provides an index of overall model fit, which is interpreted as being the proportion of variance explained (R2). In this study, a canonical correlation of 0.24 suggests that the model is explaining a 5.9% variation in the grouping variable.

Table 5. Eigenvalues.

Function	Eigenvalue	% of Variance	Cumulative %	Canonical Correlation
1	0.062 [a]	100.0	100.0	0.24

[a]. First 1 Canonical Discriminant Functions were used in the analysis.

Wilks' Lambda indicates the significance of the discriminant function. Table 6 shows a significant function ($p < 0.000$), providing a proportion of total variability not being explained, i.e., it is the converse of the squared canonical correlation. Therefore, a 94.1% variation is unexplained in the function.

Table 6. Wilks' Lambda.

Test of Function(s)	Wilks' Lambda	Chi-Square	df	Sig.
1	0.94	4.63	1	0.03

4.2. The Second Finding

Research Question 2: Do motivation (attitude, effort, and desire) and strategies (basic writing, writing an essay, and reading) significantly influence the learners' proficiency?

The result of Research Question 2 is further explained since the model was significant (Table 7). Table 7 shows the F-value = 15.05 with $p < 0.05$, signifying that the analysed model is significant. Nevertheless, not all the dependent variables significantly contribute to the model, as displayed in Tables 8–10.

Table 7. ANOVA [a].

Model		Sum of Squares	df	Mean Square	F	Sig.
1	Regression	14.85	1	14.85	15.05	<0.001 [b]
	Residual	76.01	77	0.99		
	Total	90.86	78			

[a]. Dependent Variable: Respondent's Self-Report CL Proficiency Level. [b]. Predictors: (Constant), Strategies.

Table 8. Model Summary.

Model	R	R Square	Adjusted R Square	Std. Error of the Estimate	Change Statistics				
					R Square Change	F Change	df1	df2	Sig. F Change
1	0.40 [a]	0.16	0.15	0.99	0.16	15.05	1	77	<0.001

[a]. Predictors: (Constant), Strategies.

Table 9. Coefficients [a].

Model		Unstandardized Coefficients		Standardized Coefficients	t	Sig.
		B	Std. Error	Beta		
1	(Constant)	0.50	0.61		0.83	0.41
	Strategies	0.74	0.19	0.40	3.88	<0.001

[a]. Dependent Variable: Respondent's Self-Report CL Proficiency Level.

Table 10. Excluded Variables [a].

Model		Beta In	t	Sig.	Partial Correlation	Collinearity Statistics
						Tolerance
1	Motivation	0.08 [b]	0.68	0.50	0.08	0.74

[a]. Dependent Variable: Respondent's Self-Report CL Proficiency Level. [b]. Predictors in the Model: (Constant), Strategies.

This analysis involved the dependent variables of six proficiency levels and the independent variables of 'motivation', measured from attitude, effort, and desire. The independent 'strategies' were measured from the items in basic writing, essay writing, and reading. A regression analysis was then employed to investigate the relationship among these three variables. The Stepwise method in the regression analysis (Table 8) shows that the model (with $p < 0.05$) is significant for the independent variable 'strategies' (Table 9) only. Hence, the variable 'motivation' (Table 10) was excluded from the analysis.

Table 9 shows that only the factor 'strategies' contributed to the model. 'Strategies' was the essay writing strategy.

Table 10 shows that the model in Table 9 will only be significant if 'motivation' is excluded from the analysis. In Table 8, the result indicates that only 16.3% (namely $R2 = 0.16$) variation in proficiency level is explained by 'strategies'. Hence, other factors might have to be considered in future research.

5. Discussion

The first finding indicated that only the 'writing an essay' strategy contributed to differentiating the two groups. It shows that just by looking into any student's essay writing skills, one can easily indicate the student's proficiency level as being 'low' or 'high'.

The second finding supported the first finding in the analysis of regression. The regression analysis indicated that 16.3% (namely, $R2 = 0.163$) variation in proficiency level was explained by the participants' writing strategies. Despite the low percentage of contributions to the proficiency level, the indicator provides some insight into the work required in essay writing to achieve proficiency by non-native CL speakers in learning the language.

These findings were in accordance with many other authors' findings that success in learning the CL requires certain writing talents [11,12]. Understanding vocabulary [11] is important, but so are the tactics and approaches to coordinate and integrate the CL characters. Students are expected to demonstrate the ability to use diverse characters in essays, such as combining the characters to bring new meanings. As a result, if a character is correctly specified, it will provide multiple inputs [17]. This situation was investigated in this study's literature review (Figure 1). The difficulties of handling characters in writing or any other form of communication have produced much uncertainty.

The acquisition of the CL characters has been the subject of investigation [18,19], which leads to the discovery that learners must be exposed to the larger workings of Chinese characters. This is particularly important in Malaysia as Malaysian tests emphasise writing rather than reading. Mindful that students thriving in the CL have a basic understanding of how to use the characters, therefore, in classrooms, the teaching and use of the characters should be enhanced. In this manner, the CL learners may be in a better position to gain a better overall mastery of the language if they concentrate on the written work.

6. Conclusions

In conclusion, knowing how to use the characters when writing essays is significant in helping non-native CL learners learn the language more effectively. In essence, the CL is naturally tough to master, which is compounded by the complexity of its characters. Achievement can only be enhanced by increasing the learners' writing ability using suitable or acceptable strategies, which, in this case, is the ability to combine and use the characters in essay writing. Although this study has a limited number of high proficiency achievers (17 learners) compared to low proficiency achievers (62 learners), it managed to increase the literature on the awareness of the role of the characters in language mastery. It was discovered that assisting CL teachers focus on improving students' abilities to express ideas in their essay writing is essential. Given these circumstances, non-native CL learners need to use relevant literature such as dictionaries and other resources to obtain acceptable knowledge and skills in handling CL characters to express their views.

On that front, discriminant analysis can only provide insights based on the limited data of high proficiency learners. Therefore, more research may be conducted on how to teach non-native CL learners to use proper characters to express their thoughts.

Author Contributions: Conceptualization, S.H.T.; methodology, S.H.T.; software, S.H.T.; validation, S.H.T., B.C.H., Y.F.N., and J.Y.Y.T.; formal analysis, Y.F.N.; investigation, B.C.H. and Y.F.N.; resources, S.H.T.; data curation, S.H.T. and J.Y.Y.T.; writing—original draft preparation, S.H.T.; writing—review and editing, B.C.H. and Y.F.N.; visualization, S.H.T.; supervision, S.H.T.; project administration, B.C.H.; funding acquisition, B.C.H. All authors have read and agreed to the published version of the manuscript.

Funding: This research was funded by Grant Inisiatif APB (GIA), Universiti Teknologi MARA, grant number 600-TNCPI 5/3/DDF (APB) (001/2021).

Institutional Review Board Statement: Not applicable.

Informed Consent Statement: Not applicable.

Data Availability Statement: The data presented in this study are available on request from the corresponding author. The data are not publicly available due to privacy restrictions.

Conflicts of Interest: The funders had no role in the design of the study; in the collection, analyses, or interpretation of data; in the writing of the manuscript; or in the decision to publish the results.

References

1. Bi, L.W.T. A study on SRJK (C) Yuk Yi (Sabah, Malaysia) Non-Chinese Language pupils' learning situation (note: A translated title from Chinese Language). Master's Thesis, Minzu University of China, Beijing, China, 2019.
2. Heng, B.C.; Neo, Y.F.; Chan, J.Y. Investigating the attributes of chinese primary school adoption among the Malay community in Malaysia. *ESTEEM J. Soc. Sci. Humanit.* **2021**, *5*, 153–166.
3. Heng, B.C.; Neo, K.S. Pendidikan Cina di Kalangan Pelajar Melayu Kelantan. In Proceedings of the 3rd Academic Conference of UiTM Kelantan, Kota Bharu, Kelantan, Malaysia, 14–15 December 2005.
4. Tan, J.Y.Y. Malay parents' perspective on admission of their children to Chinese primary school in Kelantan, Malaysia. *J. Arts Sci. Commer.* **2015**, *6*, 26–30.
5. Tan, Y.L.; Ngu, I.T.; Chew, F.P. The phenomenon of increased number of non-native students in under-enrolled Chinese Primary schools in Kuala Pilah. *J. Chin. Lit. Cult.* **2017**, *5*, 79–100.
6. Tan, L.H.; Spinks, J.A.; Eden, G.F.; Perfetti, C.A.; Siok, W.T. Reading depends on writing, in Chinese. *Proc. Natl. Acad. Sci. USA* **2005**, *102*, 8781–8785. [CrossRef] [PubMed]
7. Odinye, S.I. The Spread of Mandarin Chinese as a Global Language. Nigerian Journal Online-Published Articles. Available online: https://www.researchgate.net/publication/305639211_The_spread_of_Mandarin_Chinese_as_a_global_language (accessed on 24 March 2022).
8. Agwuele, A. Globalization, dying language and futility of saving them. *Public Culture* **2010**, *16*, 97–118.
9. Statista. The World's Most Spoken Languages. Available online: https://www.statista.com/chart/12868/the-worlds-most-spoken-languages/ (accessed on 1 March 2022).
10. Foon, W. No Easy End to SPM Chinese Saga. The Star. Available online: https://www.thestar.com.my/news/nation/2016/03/13/trend-in-weaker-spm-chinese-subject-results-is-worrying (accessed on 1 March 2022).
11. Kwarteng, M.; Acheampong, D.O. The perceived vocabulary strategies and motivation used by learners of Chinese as a foreign language. *J. World Engl. Educ. Pract.* **2021**, *3*, 22–45. [CrossRef]
12. Tahsildar, M.N. Chinese Language Complexities among International Students in China. *Educ. Q. Rev.* **2019**, *2*, 67–76. [CrossRef]
13. Hardan, A.A. Language learning strategies: A general overview. *Procedia-Soc. Behav. Sci.* **2013**, *106*, 1712–1726. [CrossRef]
14. O'Malley, J.M.; Chamot, A.U.; Stewner-Manzanares, G.; Kupper, L.; Russo, R.P. Learning strategies used by beginning and intermediate ESL students. *Lang. Learn.* **1985**, *35*, 21–46. [CrossRef]
15. Cohen, A.D.; Oxford; Chi, J.C. Language Strategy Use Inventory. Available online: https://carla.umn.edu/maxsa/documents/LanguageStrategyInventory_MAXSA_IG.pdf (accessed on 20 February 2021).
16. Gardner, R.C. *Social Psychology And Second Language Learning: The Role Of Attitudes And Motivation*; Edward Arnold: London, UK, 1985.
17. Nel, N.M.; Soezin, K.R.O.G. Factors influencing the acquisition of Mandarin Chinese as a second additional language focusing on phonetics. *Particip. Educ. Res.* **2021**, *8*, 1–27. [CrossRef]
18. Kuo, L.J.; Kim, T.J.; Yang, X.; Li, H.; Liu, Y.; Wang, H.; Park, J.H.; Li, Y. Acquisition of Chinese characters: The effects of character properties and individual differences among second language learners. *Front. Psychol.* **2015**, *6*, 986. [CrossRef] [PubMed]
19. Osborne, C.; Zhang, Q.; Zhang, G.X. Which is more effective in introducing Chinese characters? An investigative study of four methods used to teach CFL beginners. *Lang. Learn. J.* **2020**, *48*, 385–401. [CrossRef]

proceedings

MDPI

Proceeding Paper

Factors Affecting Innovative Behaviours among Students in Public Higher Learning in the Southern Region in Malaysia †

Khairunnisa Rahman [1], Mariam Setapa [2,*], Nurdia Azlin Ghazali [3], Norshahniza Sahari [1] and Nur Haslina Ramli [2]

1 Faculty of Business and Management, Universiti Teknologi MARA, Johor Branch, Segamat Campus, Jalan Universiti Off Km. 12, Jalan Muar, Segamat 85000, Johor, Malaysia
2 Faculty of Business and Management, Universiti Teknologi MARA, Kelantan Branch, Machang Campus, Bukit Ilmu, Machang 18500, Kelantan, Malaysia
3 Faculty of Computer and Mathematical Sciences, Universiti Teknologi MARA, Negeri Sembilan Branch, Seremban Campus, Persiaran Seremban Tiga/1, Seremban 3, Seremban 73000, Negeri Sembilan, Malaysia
* Correspondence: maria135@uitm.edu.my
† Presented at the International Academic Symposium of Social Science 2022, Kota Bharu, Malaysia, 3 July 2022.

Abstract: The purpose of this study is to investigate how undergraduate students perceive leadership competencies and their impact on innovative behaviour. A quantitative research approach was applied in this investigation. The study's results were validated using quantitative analysis. SPSS software was used to conduct the data analysis. Descriptive statistics were used to summarise the respondent's information. The independent t-test and analysis of variance (ANOVA) were used to compare the differences in the level of innovative behaviour based on demographic background. Besides, multiple linear regression was implemented to explore the relationship between students' leadership competency skills, cognitive development/critical analysis, interpersonal skills, and innovative behaviour. A total of 2156 students responded to an online survey; however, only 731 qualified. The respondents in this survey were undergraduate students who have held a position in any club or group on campus. The current study shows that cognitive development/critical analysis, interpersonal skills, and students' leadership competence skills all play a role in predicting student innovative behaviour. Student interpersonal skills are the most important factor influencing students' innovative behaviour. This study examined the effect of leadership skills on students' innovative behaviour. Additionally, this study examined undergraduate students who were active in any club or group on campus. This type of research has not been thoroughly studied, if at all, in academic circles.

Keywords: innovative behaviours; public higher learning students; cognitive development; interpersonal skills; leadership skill and competency

Citation: Rahman, K.; Setapa, M.; Ghazali, N.A.; Sahari, N.; Ramli, N.H. Factors Affecting Innovative Behaviours among Students in Public Higher Learning in the Southern Region in Malaysia. *Proceedings* **2022**, *82*, 43. https://doi.org/10.3390/proceedings2022082043

Academic Editor: Mohamad Rahimi Mohamad Rosman

Published: 14 September 2022

1. Introduction

As society and the workplace continually evolve, higher education institutions are under pressure to adapt to new expectations. Undergraduates must be prepared for future professions; future workers must be taught how to fulfil job duties; and, ideally, new ideas that lead to innovation must be generated. People's jobs and lifestyles have evolved as a result of the Fourth Industrial Revolution, also known as Industry 4.0. Students in higher education must set a higher standard for creativity to solve these problems because they are future employees and an organisation's major success is built on innovative people. Because technological advancements have led to substantial labour market shifts, the technology that emerges in Industrial 4.0 has the potential to replace staff functions. As a result, it is vital to encourage students to engage in innovative behaviour.

Education institutions in particular can be seen as a microcosm of the trend of global industrialisation [1]. Every educational establishment is responsible for ensuring that its students are prepared and capable enough to compete in the workforce. A business must innovate in education if it wants to produce qualified human resources [2]. Businesses must innovate if they want to maintain a competitive advantage in the modern market. Afsar et al. [3] states that businesses can employ innovation to gain a competitive advantage and succeed as an organisation. Student creativity is linked to economic performance and is viewed as a crucial ability for the twenty-first century [4].

In business, industry, entrepreneurship, and higher education, innovative behaviour has been recognised as a critical factor in long-term success [4–6]. However, in higher education, it is uncommon to hear certain students' creative behaviour. Higher education institutions must generate a workforce that is both creative and prepared to meet the challenges of the twenty-first century [7,8]. Numerous studies show that, because of the importance of education in the development of human innovative skills, higher education institutions alone will not be able to achieve these expectations [9]. According to a study, students' ability to participate in novel activities is underappreciated [10]. Particularly in higher education, educational systems have come under fire for failing to develop these professional knowledge requirements. As a result, developing new potential abilities is one of higher education's main objectives, especially for students.

Even though innovative behaviour is one of the most essential techniques to achieving academic and professional goals, research among students in educational settings is still limited [11]. According to Ailing et al. [12], universities also lack the instruments necessary to create innovative undergraduate capabilities. Innovative characteristics, leadership abilities, and competency all play a role in innovative behaviour. This gap provides the framework for this study to examine innovative behaviour among students in higher learning institutions because graduates are the university's output. There is a need to investigate innovative student behaviour in the Malaysian educational system.

The primary goal of this research is to look into the innovative behaviour of students at a public higher learning institution in Malaysia's southern region. As a result, the specific research objective of this study is to see if there are any disparities in levels of innovative behaviour based on demographics. Secondly, the objective to investigate the impact of various elements on innovative behaviour (leadership skill and competency, cognitive development/critical analysis, and interpersonal skill).

2. Literature Review and Hypotheses

This section will cover the dependent variable of innovative behaviour, and independent variables of student leadership skills and competencies, interpersonal skills, and cognitive development or critical analysis. The ability and desire of students to be creative are related to their innovative behaviour in this study. Students who can adapt to unexpected situations and ideas, accept opposing viewpoints and mistakes, experiment freely and take measured risks, and be open to innovation exhibit creativity [13]. Students can also apply what they have learnt to create fresh solutions and ideas. However, there is a dearth of studies on innovative behaviour and demographic traits. The purpose of this study is to determine whether there is a relationship between innovative behaviour and demographic traits. This statement leads to the following hypothesis.

Hypothesis 1. *There are differences in innovative behaviour based on demographic factors.*

For students to self-identify as prospective innovators in their field, Cusson [14] contends that they must possess conceptual competency in innovative behaviour. In order to better understand how students from different backgrounds make decisions and seize chances, a number of studies involving students from such backgrounds have been conducted. According to Binnawas et al. [15], students who participated in a club or organisation at school showed more confidence and drive; nevertheless, few studies on this group have looked at their innovative behaviour. The following hypothesis comes as a result of this statement.

Hypothesis 2. *There is a significant influence between students' leadership competency skills and innovative behaviour.*

Many resources are required in higher education institutions, but human resources are a crucial resource for developing creative people. One action that can be taken with the aid of coaching, instruction, and training is interpersonal skills, which, according to Hogan and Warrenfelz, are competences and behaviours that entail direct communication, such as interacting with others and forming relationships [16]. According to Mariepazh [17], the range of interpersonal skills, which include a person's ability to begin, create and maintain compassionate connections as well as fruitful ones also determines one's capacity to do so. It is divided into four categories: "disclosing oneself and trusting others, accurately communicating with one another, resolving conflict and relationship issues in a healthy manner, and supporting and valuing variety".

One of the elements that significantly influence a student's innovative conduct in pursuing the goals of higher education is their interpersonal skills. According to the findings of research by Kanthasamy [16], their analysis revealed a positive association between interpersonal skills and behaviour, and according to Mahmudi [18], interpersonal skills, group integrity, and self-efficacy have direct beneficial influences on behaviour. This statement leads to the following hypothesis.

Hypothesis 3. *There is a significant influence between students' interpersonal skill influence and innovative behaviour.*

This paper discusses critical thinking as the fundamental cognitive skill. To make the most informed decisions possible about what to believe and what to do, critical thinking is a process that activates specific cognitive skills, such as conceptualising, applying, analysing, synthesizing, and/or evaluating information that has been gathered from or generated by observation, experience, reflection, or communication [19]. By fusing its connections to adolescent development and its contributions to adolescents' learning, welfare, and positive development, Sun and Hui [20] study cognitive competence as a construct for outstanding youth growth. It demonstrates how critical thinking may be transformed into self-regulated cognitive abilities that teenagers can master and use to accomplish tasks more effectively, come up with accurate answers to issues, and arrive at the best judgments. It is thought that developing critical thinking skills in children helps them learn for life and develop holistically, as well as preparing them to lead society in the future and address societal issues and advance humankind. This statement leads to the following hypothesis.

Hypothesis 4. *There is a significant influence between students' cognitive development/critical analysis and innovative behaviour.*

The suggested conceptual framework for this inquiry is shown in Figure 1. Based on a detailed analysis of prior research, a conceptual framework is provided to recognise the linkages, as illustrated in Figure 1.

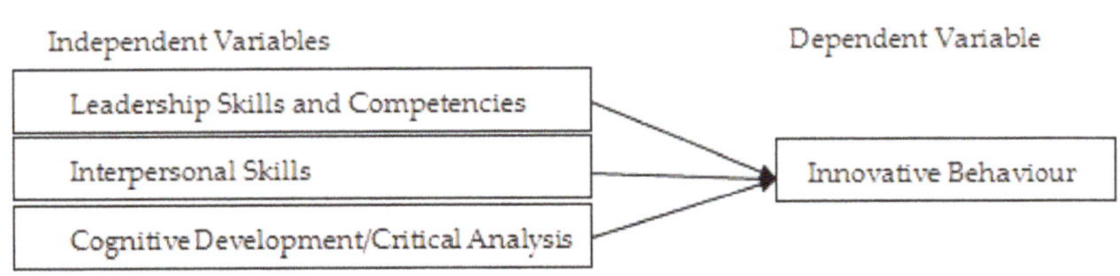

Figure 1. Research Framework.

3. Research Design

In this exploratory investigation, data and information were gathered using a questionnaire as the survey instrument. This study also used cross-sectional analysis; a type of observational study that looks at data from a population or a representative sample during a specific time period.

3.1. Sampling and Data Collection

Simple random sampling was used in this study. Each person was chosen completely at random, with each population member had an equal chance of being chosen. This study was conducted at three (3) different UiTM campuses: Campus Johor, Campus Melaka, and Campus Negeri Sembilan. The only two branches that UiTM Campus Johor has were in Segamat and Pasir Gudang. UiTM Campus Melaka has three branches: Alor Gajah, Lendu, and Bandaraya Melaka. Furthermore, there were three (3) UiTM Campus Negeri Sembilan branches situated in Seremban, Rembau, and Kuala Pilah.

As the focus of this study is on college students who held a position in any club or group on campus, the total number of students from the three campuses was 2156. The size of the sample from each campus was calculated using G-Power software [21], with a 95% confidence level and a 5% margin of error to calculate the precise number of respondents chosen. In detail, a total of 151 students from Johor Campuses answered the survey. Meanwhile, Melaka campuses stated 580 as the population and 232 as the sample; however, the total responses received were 264. The same scenario existed with Negeri Sembilan campuses, where the numbers were 1440 for the total of population and 304 for the sample; however, the actual responses were 316.

This survey had 731 individuals and the respondents were given three (3) weeks to answer the questions on the Google Form sheet. After receiving approval from the Ethics Committee, the Google Form sheet questionnaire was carefully designed, and the link to it was then shared with the responders on these three (3) campuses. A leader from each branch was chosen to oversee the responses and guarantee that the respondents provided reliable cooperation for this study.

3.2. Questionnaire Design

There were three (3) components to the questionnaire. The respondents' profiles were discussed in Part A, which included seven (7) questions regarding the respondents' campus, gender, age, education, cumulative grade point average (CGPA), race, and status in clubs or organisations. Part B, on the other hand, focused on the independent variable, which included three (3) components: interpersonal skills, cognitive growth/critical analysis, and leadership skills and competencies. Additionally, the dependent variable of innovative behaviour was the focus in Part C. The Likert scale, from 1 (strongly disagree) to 5 (strongly agree), was used in Parts B and C. The questionnaire utilised in this study was adapted by modifying the instrument to suit the study.

3.3. Data Analysis

This study used quantitative analysis to achieve the objectives. The data analysis was conducted using Statistical Package for Social Sciences (SPSS) Version 26 software (IBM Corp. Armonk, NY, USA). The descriptive statistics were analysed to summarise the information about the respondent. Next, a *t*-test and ANOVA test were carried out to compare the gender, campus, age, CGPA educational background, and position differences towards innovative behaviour. Multiple linear regression was implemented to explore the relationship between students' leadership competency skills, cognitive development/critical analysis, interpersonal skills, and innovative behaviour.

4. Results

4.1. Demographic

The results of descriptive statistics for 731 respondents are summarized in Table 1. About 316 respondents were from UiTM Negeri Sembilan (43.2%), 264 respondents were from UiTM Melaka (36.1%), and 151 respondents were from UiTM Johor (20.7%). The results show that 73.9% of the respondents were female and 26.1% were male. About 62.4% aged 18–20 years, 35.6% aged 21–23 years, and 2% were more than 23 years old. It can also be observed that 63.6% of the respondents had a diploma, while 36.5% had a degree. Looking at the CGPA, 47.3% of the respondents obtained a CGPA of 3.51 and above, and 44.5% obtained a CGPA of 3.01–3.50. For the position in the committee, 13.8% of the respondents were president, 7.5% were vice president, and 8.2% were secretary.

Table 1. Descriptive statistics of respondents' demographic background.

Variable	Description	Frequency	Percentage
Campus	UiTM Negeri Sembilan	316	43.2
	UiTM Melaka	264	36.1
	UiTM Johor	151	20.7
Gender	Female	540	73.9
	Male	191	26.1
Age	18–20	456	62.4
	21–23	260	35.6
	24–26	14	1.9
	More than 26	1	0.4
Education Level	Degree	456	63.6
	Diploma	266	36.4
CGPA	2.50 and below	6	0.8
	2.51–3.00	51	7.0
	3.01–3.50	328	44.9
	3.51–4.00	346	47.3
Position	President	101	13.8
	Vice President	55	7.5
	Secretary	60	8.2
	Assistant Secretary	19	2.6
	Treasurer	39	5.3
	Assistant Treasurer	15	2.1
	Academic Unit	38	5.2
	Multimedia Unit	79	10.8
	Protocol Unit	26	3.6
	Community Unit	22	3.0
	Activity Unit	20	2.7
	Other	257	35.2

4.2. Questionnaire Reliability

Table 2 shows the Cronbach alpha coefficients as the reliability and internal consistency of Likert scale questions results for this study. It can be seen that all of the Cronbach's alpha coefficients were more than 0.7, which suggests that the questionnaire is reliable.

Table 2. Reliability statistic for the questionnaire.

Variable	Cronbach's Alpha
leadership skill and competency	0.886
cognitive development/critical analysis	0.787
interpersonal skill	0.914

4.3. Differences in Innovative Behaviour Based on Demographic Background

Table 3 summarizes the results based on the t-test and ANOVA test, respectively. The tests were conducted to assess the first hypothesis testing, which compared the significant difference between the demographic background of students (gender, campus, age, CGPA, educational background, and position) and innovative behaviour. The results show that there were no statistically significant differences between innovative behaviour and campus, age, and CGPA, as the *p*-value was more than 0.05. It can be said that innovative behaviours of students on the three campuses are generally the same. Moreover, the innovative behaviours of students on the different groups of age and CGPA are the same.

Table 3. Testing differences in innovative behaviour on demographic background using *t*-test and ANOVA.

Variable	Description	Mean Score Innovative Behaviour	*p*-Value
Gender	Female	32.92	0.030
	Male	33.88	
Education	Degree	33.7	0.041
	Diploma	32.87	
Campus	UiTM Negeri Sembilan	32.82	0.610
	UiTM Melaka	33.32	
	UiTM Johor	33.21	
Age	18–20	32.77	0.068
	21–23	33.86	
	24–26	33.07	
	More than 26	35	
CGPA	2.50 and below	28.5	0.075
	2.51–3.00	32.92	
	3.01–3.50	32.93	
	3.51–4.00	33.51	
Position	President	35	0.000
	Vice President	34.16	
	Secretary	33.12	
	Assistant Secretary	32.71	
	Treasurer	34.64	
	Assistant Treasurer	32.71	
	Academic Unit	32.08	
	Multimedia Unit	33.56	
	Protocol Unit	30.5	
	Community Unit	33.5	
	Activity Unit	31.5	

Interestingly, there were statistically significant differences between innovative behaviour and gender (*p*-value = 0.030), as well as education level (0.041) and position (0.000). This implies that the innovative behaviours of male and female students are different. Besides that, the innovative behaviours of students are different if the students have different education levels and positions in a committee.

4.4. Relationship between Innovative Behaviour and Independent Variables

The strength of the association between creative behaviour and the four independent variables is determined by the Pearson correlation coefficient. Table 4 displays the Pearson correlation coefficient between independent variables and innovative behaviour. It is clear that there is a significant correlation between independent variables and innovative behaviour because all of the independent variables' *p*-values were less than 0.05. According to the correlation coefficient, there is a substantial association between interpersonal skill influence and innovative behaviour. One can also observe a moderate correlation between students' leadership skill and competency and cognitive factor development/critical analysis and dependent variable (innovative behaviour).

Table 4. Pearson's correlation coefficients of independents variables and innovative behaviour.

Independent Variable	Pearson's Correlation	*p*-Value	Relationship Strength
leadership skill and competency	0.643	0.000	Moderate
cognitive development/ critical analysis	0.617	0.000	Moderate
interpersonal skill influence	0.707	0.000	Strong

4.5. *Factors Influencing Innovative Behaviour*

From Table 5, 54.2% of the total variation in measuring the factors influencing innovative behaviour can be explained by students' leadership skill and competency, cognitive development/critical analysis, and interpersonal skill, while other factors explain 45.8%.

Table 5. Goodness of fit test results of the model.

R Square	Adjusted R Square
0.542	0.540

Table 6 shows the ANOVA results; the F-statistic was 286.226 and the critical value for F3, 727 at a *p*-value of 0.05 was 2.60. Thus, this suggested that the three independent variables in the model are significantly predictive of the dependent variable.

Table 6. ANOVA test results for the significance of the model.

Model	Sum of Squares	Degree of Freedom	Mean Square	F	Sig.
Regression	11,007.123	3	3669.041	286.226	0.000
Residual	9319.181	727	12.819		

4.6. *Testing the Significance Factors Influencing Innovative Behaviour*

This study examines the impact of individual independent variables (students' leadership skill and competency, cognitive development/critical analysis, and interpersonal skill) on the dependent variable (innovative behaviour) in order to test the second, third, and fourth hypotheses. Table 7 shows the results of multiple linear regression. From the results, it can be concluded that all three factors significantly influenced creative behaviour as the *p*-value was less than 0.05. The absolute value of β indicates the order of importance of the independent variable. Looking at the coefficients of each independent variable, interpersonal skill influence was the most influential factor in students' innovative behaviour.

Table 7. Factors influencing innovative behaviour.

Factors	β	t	*p*-Value
constant	5.173	5.119	0.000
leadership skills and competencies	0.275	5.926	0.000
cognitive development/critical analysis	0.313	3.334	0.001
interpersonal skill influence	0.361	10.620	0.000

The regression equation of the model for this study can be written as follows:

$$y = 5.173 + 0.275x_1 + 0.313x_2 + 0.361x_3 \quad (1)$$

where y is the value of innovative behaviour, x_1 is leadership competency skill, x_2 is cognitive development/critical analysis, and x_3 is interpersonal skill influence.

5. Discussion

The current study demonstrates that interpersonal skills, student leadership competence skills, and cognitive factor development/critical analysis strongly predict students' innovative behaviour. Students with strong interpersonal skills offer fresh perspectives on novel thoughts, opinions, and ideas to increase the potential for innovation. For students to adapt to the shifting demands of the labour market and to develop their leadership qualities, they must possess abilities such as teamwork, public speaking, problem-solving, decision-making, and other technical skills [22].

Ali [23] found that three personality traits of agreeableness, extraversion, and openness to experience are positively connected with phases of innovative behaviour in the formulation and advancement of ideas. The Big Five Personality result indicates that students with more competitive and positive interpersonal skills are welcomed in the job market. Although the majority of college students lack these skills, many businesses assert that interpersonal skills like oral communication are crucial when evaluating staff, especially potential new workers [24].

According to this study, a student's innovative behaviour differs depending on their gender, educational attainment, and position within a committee. The study's findings support the notion that a student's capacity for critical thought and factor development will have a favourable impact on their capacity for innovation. The influence of cognitive abilities in human learning activities will only be accurately portrayed in a learning environment by adding particular cognitive skills in the context of exploration, as learning activities entail a variety of unique abilities and operate together in unforeseen ways [25].

Student interpersonal relationships is the factor that has the biggest impact on students' innovative behaviour. The study's findings show that having strong interpersonal trust has a good effect on a student's innovative behaviour. More proactive and daring behaviour can be displayed by individuals (and organisations), which can encourage innovative behaviour [26,27].

6. Conclusions

This study investigated the relationship between student leadership competency and innovative behaviour through the construction and testing of a model. The undergraduate students at the centre of this study are those whose perceptions of their capacity for innovation influence those of their degree of competence. Before making generalisation about other groups, it is essential to understand and respect individual distinctions. The data were only gathered from Universiti Teknologi MARA in the southern region, because this study focused solely on undergraduate students there. To create a better generalisation, more research can be conducted to increase the sampling size across all sites. Additionally, this study advised that it is conducted in a variety of campuses, including research universities and private universities. To have a more comprehensive understanding of innovative behaviour, future studies should incorporate moderator or mediator variables. The focus of the current study was on students' innovative behaviour as it relates to leadership competencies. Finally, it is recommended that students be given access to a specific leadership development programme in order to enhance their leadership skills.

Author Contributions: Conceptualization, K.R.; methodology, N.A.G.; software, N.A.G.; validation, N.A.G.; formal analysis, N.A.G.; investigation, K.R.; resources, K.R.; data curation, M.S.; writing—original draft preparation, K.R., M.S., and N.H.R.; writing—review and editing, K.R. and M.S.; visualization, N.S.; project administration, M.S. All authors have read and agreed to the published version of the manuscript.

Funding: This research received no external funding.

Institutional Review Board Statement: Not applicable.

Informed Consent Statement: Not applicable.

Data Availability Statement: Not applicable.

Acknowledgments: Special thanks go out to the Universiti Teknologi MARA students who served as the study's sample as well as to all the other people who provided direct or indirect support for its completion.

Conflicts of Interest: The authors declare no conflict of interest.

References

1. Keung, E.K.; Rockinson-Szapkiw, A.J. The relationship between transformational leadership and cultural intelligence: A study of international school leaders. *J. Educ. Adm.* **2013**, *51*, 836–854. [CrossRef]
2. Helmy, I.; Pratama, M.P. Effect of proactive personality and psychological empowerment on innovative behavior through creative self-efficacy. *J. Pro Bisnis* **2018**, *11*, 14–21.
3. Afsar, B.; Badir, F.Y.; Saeed, B. Transformational leadership and innovative work behavior. *Ind. Manag. Data Syst.* **2014**, *114*, 1270–1300. [CrossRef]
4. Dyer, J.H.; Gregersen, H.B.; Christensen, C. Entrepreneur behaviors, opportunity recognition, and the origins of innovative ventures. *Strat. Entrep. J.* **2008**, *2*, 317–338. [CrossRef]
5. Barak, M.; Usher, M. Innovation in a MOOC: Project-Based Learning in the International Context. In *Active Learning in College Science: The Case for Evidence Based Practice*; Mintzes, J.J., Walter, E.M., Eds.; Springer Nature: Berlin, Germany, 2020; pp. 639–653.
6. Law, K.M.; Geng, S. How innovativeness and handedness affect learning performance of engineering students? *Int. J. Technol. Des. Educ.* **2018**, *29*, 897–914. [CrossRef]
7. Avvisati, F.; Jacotin, G.; Vincent-Lancrin, S. Educating higher education students for innovative economies: What international data tell us? *Tuning J. High. Educ.* **2013**, *1*, 223–240. [CrossRef]
8. Trilling, B.; Fadel, C. *21st Century Skills: Learning for Life in Our Times*; Jossey-Bass: San Francisco, CA, USA, 2009.
9. Quintana, C.D.D.; Mora, J.; Perez, P.J.; Vila, L.E. Enhancing the development of competencies: The role of UBC. *Eur. J. Educ.* **2016**, *51*, 10–24. [CrossRef]
10. Edward-Schachter, M.; Garcia-Granero, A.; Sanchez-Barrioluengo, M.; Quesada-Pineda, H.; Amara, N. Disentangling competencies: Interrelationships on creativity, innovation and entrepreneurship. *Think. Ski. Creat.* **2015**, *16*, 27–39. [CrossRef]
11. Chen, I.S.; Chen, J.K. Creativity strategy selection for the higher education system. *Qual. Quant.* **2012**, *46*, 739–750. [CrossRef]
12. Ailing, C.L.; Liping, L.; Xingsen, J.Z.; Dong, L. Study on innovation capability of college students based on extenics and theory of creativity. *Procedia Comput. Sci.* **2013**, *17*, 1194–1201. [CrossRef]
13. Roffeei, S.H.M.; Kamarulzaman, Y.; Yusop, F.D. Inculcating innovative behavior among students: Determinant of innovation culture in Malaysian higher education. *Malays. Online J. Educ. Manag.* **2017**, *5*, 1–17. [CrossRef]
14. Cusson, R.M.; Meehan, C.; Bourgault, A.; Kelley, T. Educating the next generation of nurses to be innovators and change agents. *J. Prof. Nursin* **2020**, *36*, 13–19. [CrossRef] [PubMed]
15. Binnawas, M.S.H.; Khalifa, G.S.; Bhaumik, A. Antecedents of student's behavioural intentions in higher education institutions. *Int. J. Psychosoc. Rehabil.* **2020**, *24*, 1949–1962.
16. Kanthasamy, A. *Relationship between Interpersonal Skills. Emotional Intelligence and Managerial Performance at Indah Water Consortium Sdn Bhd*; Research Report for Master Business Administration; Universiti Sains: Penang, Malaysia, 2009.
17. Mariepazh, A. Teaching efficacy. Interpersonal. Intrapersonal skills and teaching performance in the tertiary school. *IAMURE Int. J. Soc. Sci.* **2012**, *2*, 17–25.
18. Mahmudi. *Yogyakarta Public Sector Performance Management*; STIM YPKN Publishing and Printing Unit: Yogyakarta, Indonesia, 2010.
19. Flage, D.E. *The Art of Questioning: An Introduction to Critical Thinking*; Pearson: New Jersey, NJ, USA, 2004.
20. Sun, R.C.; Hui, E.K. Cognitive competence as a positive youth development construct: A conceptual review. *Sci. World J.* **2012**, *2012*, 210953. [CrossRef] [PubMed]
21. Faul, F.; Erdfelder, E.; Lang, A.-G.; Buchner, A. G*Power 3: A flexible statistical power analysis program for the social, behavioral, and biomedical sciences. *Behav. Res. Methods.* **2007**, *39*, 175–191. [CrossRef]
22. Lopes, D.C.; Gerolamo, M.C.; Del Prette, Z.A.P.; Musetti, M.A.; Del Prette, A.L.M.I.R. Social skills: A key factor for engineering students to develop interpersonal skills. *Int. J. Eng. Educ.* **2015**, *31*, 405–413.
23. Ali, I. Personality traits, individual innovativeness and satisfaction with life. *J. Innov. Knowl.* **2019**, *4*, 38–46. [CrossRef]
24. Lester, S.W.; Tomkovick, C.; Wells, T.; Flunker, L.; Kickul, J. Does service-learning add value? Examining the perspectives of multiple stakeholders. *Acad. Manag. Learn. Educ.* **2005**, *4*, 278–294. [CrossRef]
25. David, F.L. The role of nonverbal ability tests in identifying academically gifted students: An aptitude perspective. *Gift. Child Q.* **2005**, *49*, 111–138. [CrossRef]
26. Mayer, R.C.; Davis, J.H.; Schoorman, F.D. An integrative model of organizational trust. *Acad. Manag. Rev.* **1995**, *20*, 709–734. [CrossRef]
27. Rockstuhl, T.; Ng, K.Y. *The Effects of Cultural Intelligence on Interpersonal Trust in Multicultural Teams. Handbook of Cultural Intelligence*; Routledge: London, UK, 2015; pp. 224–238.

proceedings

MDPI

Proceeding Paper

Secondary Students' Mental Well-Being after the Pandemic—An Analysis According to Location and Study Levels †

Joanne Sau-Ching Yim [1,*], Puteri Rohani Megat Abdul Rahim [2] and Daljeet Singh Sedhu [2]

1 Faculty of Social Sciences and Humanities, Tunku Abdul Rahman University College, Kampar 31900, Malaysia
2 Academy of Language Studies, Universiti Teknologi MARA, Perak Branch, Seri Iskandar 32610, Malaysia
* Correspondence: yimsc@tarc.edu.my

† Presented at the International Academic Symposium of Social Science 2022, Kota Bharu, Malaysia, 3 July 2022.

Abstract: The study surveyed 1547 secondary school students' mental health when they resumed in-person classes by examining the prevalence of depressive symptoms with the abridged Beck Depression Inventory II. Chi-square test found a significant association between location and depression levels, with a higher proportion of rural students reporting a higher depression level. One-way ANOVA detected significant differences within levels of secondary classes, where post hoc analysis found higher depressive levels with final year students. As schools prepare to transit from the pandemic to endemic phase, the study highlighted the specific students who need attention to tailor specific programmes to targeted groups and enhance resources efficiency.

Keywords: secondary school students; post-COVID schooling; youth depression; rural-urban students; mental well-being; final year students' depression

Citation: Yim, J.S.-C.; Rahim, P.R.M.A.; Sedhu, D.S. Secondary Students' Mental Well-Being after the Pandemic—An Analysis According to Location and Study Levels. *Proceedings* **2022**, *82*, 44. https://doi.org/10.3390/proceedings2022082044

Academic Editor: Mohamad Rahimi Mohamad Rosman

Published: 14 September 2022

1. Introduction

The COVID-19 pandemic is a public health crisis that has disrupted many aspects of our lives. This project aims to investigate the impact of the disrupted learning system on secondary students' current mental well-being. Due to the contagious nature of the virus, the prolonged closure of in-person classes in Malaysia had inevitably transformed traditional classrooms into online classes. The closure of schools has not only disrupted the acquiring of new knowledge, but also because students have forgotten what they had learnt earlier. In 2020, it was reported that 90 percent of the world's student population was affected by the pandemic [1]. In Malaysia, the estimated learning loss ranges from 0.45 to 0.95 as indicated by learning-adjusted years of schooling (LAYS). LAYS is a standard metric that captures the quantity and quality of education, measuring the number of years of schooling a child can expect to obtain by age 18, adjusted by a country's average student achievement. In other words, students in Malaysia lost about 45% of a learning-adjusted year of schooling in the best-case scenario, whereas in the worst-case, 95% [2]. Starting from 2021, schools underwent different types of class attendance based on the severity of the COVID-19 infection rates. In general, Malaysian schools were closed for more than 200 days in 2020 and 2021 [3].

Even though the nation is moving towards resumption of in-person classes, the impact of the different modes of class attendance has already occurred and may not be able to be reversed. There were reports by local studies that found students suffered from psychological effects brought about by different study modes [4,5]. However, more investigation is needed as a majority of research had focused on students in higher educational institutions, rather than school students [5]. Moreover, not all students were affected the same and this may contribute to developmental and achievement gaps later. As the COVID-19 pandemic has entered its third year towards the endemic phase, it is important to examine the current state of students' well-being to better prepare for the recovery to normalcy. Specifically, this research aims to answer the following questions:

1. What is the current state of mental well-being in terms of depression levels among secondary school students?
2. Are there any associations between students' mental well-being in terms of depression levels and their location of residence?
3. Are there any significant differences between study levels and depression levels?

2. Literature Review

The COVID-19 pandemic has disrupted learning in a way no one could have predicted around the globe. The proliferation of online learning due to the pandemic and also the different modes of learning has sparked many studies that offer different perspectives. In India, a survey found that challenges faced by secondary level online learners include low motivation, lack of interest, and time management [6]. From Egypt, undergraduate students reported similar challenges in learning motivation, online technicalities, time management, self-discipline and regulation, as well as online communication [7]. Students reported difficulties in written communication, which is required by online learning, with only about 56% out of 450 respondents felt comfortable and competent in text communication. Such issues in online learning may be resolved once classes are conducted in-person. However, recent research has highlighted potential issues in well-being that were brought about by the transition to in-person classes. A recent Lithuanian study compared perceived differences between online and in-person classes in high schools, and found students' energy levels, sleep quality, and mental well-being to be worse than during online classes [8]. This is attributed to the fewer hours of sleep after school reopened, compared to lockdown where students had more hours of sleep.

From France, Lane [9] and her colleagues surveyed two independent samples of secondary students before and during the pandemic, and found that students reported more symptoms of generalized anxiety during the pandemic than the time before the pandemic. Evidence from the United Kingdom suggested that anxiety induced by examination significantly correlated with risk in developing emotional disorder and also school-related wellbeing [10]. There are many school factors that could affect students' well-being which they need to adapt to. One aspect of well-being that is deemed vital for students' development during the learning process is mental well-being. Chambel and Curral [11] describe mental well-being as associated with central outcomes, such as academic engagement, academic achievement, dropout, and educational aspirations. Students with positive mental health flourish in life as they have positive emotions, and they are able to function well psychologically and socially. However, students with incomplete mental health are languishing in life. These students would describe themselves as "empty" and "hollow". Subsequently, the students will be disengaged. They no longer have interest in studying. If this issue is not addressed appropriately, this will widen the learning disparity among students. Thus, it is imperative to prioritise students' well-being because it is an important platform that is necessary for post-COVID learning recovery and to enhance learning [12]. This was concluded from a case study of a school that fully opened after lockdown, where a recovery curriculum that prioritised well-being was implemented. The curriculum incorporated weekly well-being sessions, increased physical education classes, emphasised play-based education and experiential learning, as well as having competitions [12]. Such intervention improved students' work when compared against success measures.

A recent study investigated university students' well-being during the pandemic with an abridged version of Beck Depression Inventory II and measured well-being with items: sadness, pessimism, loss of pleasure, loss of interest, loss of energy, sleep, irritability, appetite, concentration, and fatigue [13]. It was found that a majority of students suffered from sadness, sleeping patterns, lack of concentration, loss of energy, and pessimism, with females having high levels of depression compared to males. However, as schools are easing back to normal operations, updated evidence is needed as this study was conducted during home confinement. In Malaysia, research has found that students' psychological well-being is undermined by conflicts that arise from fulfilling responsibilities in academic work and family/personal lives [14]. Such role conflicts were prevalent especially during lockdown and brought about elevated stress, anxiety, depression, social dysfunction, and unhappiness among university students. Another Malaysian on student well-being found issues in sleep quality, stress-induced fatigue, inactivity, and poor eating habits to be affecting secondary school students [15]. The study recommended assistance from family, schools, and media to help students adapt to new norms. Local research had also suggested that location had given rise to learning disparity between urban and rural students due to limited Internet access or poor infrastructure [16]. The abrupt transition from in-person classes to online mode had caused students to be unprepared for such learning methods, and they doubted the effectiveness of virtual teaching mode [16]. The location of schools can act as stressors due to the disparity of technological advancement, and affect students' wellbeing. It is possible that the current situation may differ as students get acquainted with different learning modes. Previous study noted the difference in mental wellbeing of rural and urban students, and updated students' perspective would be helpful to inform good practices and improvements [17,18].

Building on the above research, the present study investigated students' mental well-being by assessing their depression levels. Further examination based on the demographic factors of location (rural/urban) and also different classes levels (Form 1 to Form 6) can provide better knowledge that is beneficial for schools' preparation shift from the pandemic to endemic phase.

3. Research Methods

Participants

Respondents consisted of 1547 secondary school students from Malaysian public schools in Selangor. Permissions to conduct research were sought from the Ministry of Education (MOE), relevant authorities in state and school districts, as well as from school principals. Due to safety considerations and to prevent face-to-face collection of questionnaires, the MOE only allowed distribution of questionnaires in electronic forms. Hence, the questionnaire was converted to Google Form. A list of all secondary schools with respective email addresses was obtained from the website of the MOE, and a link to the Google Form was given in the emails sent to principals seeking for permission to conduct research. Data were collected from February 2022 to May 2022.

○ Research Instrument

To measure mental well-being, the study used an abridged version of Beck Depression Inventory II (BDI-II) that included 10 items: sadness, pessimism, loss of pleasure, loss of interest, loss of energy, sleep, irritability, appetite, concentration, and fatigue [19]. The items are measured on a 4-point scale that ranged from 0 to 3, with higher summed scores indicating greater psychopathological impairment. The levels of depression are classified as: None (0–5), mild depression (6–8), moderate depression (9–12), severe depression (13 and above) [13]. As required by MOE, the questionnaire items were translated to Malay Language as it is the National Language and thus can be better understood by the school communities. The study employed a back-translation method where three subject matter experts reviewed the scales' content suitability to ensure the items represent the intended area of investigation and match the underlying concepts [20]. A pre-test and a pilot test were conducted to ensure the face validity and reliability of the instrument and quality of data collected. The internal consistency for both pilot test and actual study were 0.75 and 0.89, respectively. Data collected were analysed with IBM SPSS Statistics Software version 26 (Armonk, NY, USA).

○ Descriptive Analysis

The present study involved 1547 respondents with 953 (61.6%) females and 594 (38.4%) males. Respondents reported a mean age of 15.20 (SD = 1.43), with 63.9% residing in urban areas while 36.1% in rural areas. They are currently enrolled in Form 1 (11.40%), Form 2 (24.30%), Form 3 (27.50%), Form 4 (19.30%), Form 5 (14.10%), and Form 6 (3.40%) at secondary school level. A majority of them (42.2%) attended school on alternate days of the week, 35.2% attended school every day, 17.2% went to school based on the severity of COVID-19 infection rates, while 5.6% studied at home. To answer research question one, Table 1 shows the descriptive statistics for depression levels. The sample reported a mean of 7.78 which is classified as mildly depressive based on the scoring method presented earlier. About 54.5% reported mild to severe depression levels.

Table 1. Descriptive statistics for Depression Level.

Depression Level	Total Sample		By Location		Mean	SD
	Frequency	Percent	Urban	Rural		
None	703	45.5%	48.10%	40.80%	7.78	6.38
Mild	271	17.5%	16.20%	19.90%		
Moderate	274	17.7%	18.20%	16.80%		
Severe	299	19.3%	17.50%	22.50%		
Total	1547	100%	100.00%	100.00%		

○ Inferential Analysis

In answering research question two, a chi-square test is employed to test association between school location and depression levels. There was a significant association found, χ^2 (3, n = 1547) = 11.98, p = 0.007, with a higher proportion of students located in rural areas having a higher depression level. Research question three looked into the depression levels among students enrolled in different levels of study. A one-way ANOVA was employed to detect any significant difference among students from Form 1 to Form 6. As shown in Table 2, there was a statistically significant difference at $p < 0.05$ level in depression levels for the six levels of study: F (5, 1541) = 5.33, p = 0.001.

Table 2. One-Way ANOVA Depression level.

	Sum of Squares	df	Mean Square	F	Sig.
Between Groups	1069.37	5	213.874	5.334	0.001
Within Groups	61,784.19	1541	40.094		
Total	62,853.56	1546			

Note: Significance at $p < 0.05$.

As the F value indicated significant differences among the mean scores of the six groups of students, a post hoc test was carried out to ascertain where the significant differences lie. Tukey's HSD test for multiple comparisons found that the mean value of depression level was significantly different between the following groups:

- Form 5 (M = 8.20) students had a higher level of depression than Form 1 (M = 5.84) and Form 2 (M = 6.42) students.
- Form 3 (M = 7.75) students had a higher level of depression than Form 1 (M = 5.84) and Form 2 (M = 6.42) students.
- Form 4 (M = 7.75) students had a higher level of depression than Form 1 students (M = 5.84).

4. Discussion

The current study surveyed the mental wellbeing of secondary school students with BDI—II. Among the items measured, "pessimism" and "loss of energy" recorded high means. Pessimism is directed at doubts about one's future, and this echoed with current literature that found students' confidence hampered by the pandemic due to interference in their academic tasks and plans [14]. "Loss of energy" refers to exertion of extra efforts in completing tasks, and this reflected existing evidence of students who returned to in-person classes reported lesser energy compared to online classes when they have better sleep quality [8]. Additionally, respondents reported mild depression on average, with 54.5% reported mild to severe depression levels. This is alarming as another study conducted in a Malaysian northern state before the pandemic only recorded 27.4% depression prevalence among secondary school students [21], whereas another study conducted during the pandemic recorded 37.3% prevalence [14]. This prevalence of poor mental health found requires attention or intervention to ensure well-being.

With regards to depression analysed according to study levels, Form 5 students had the highest level of depression. In Malaysian public schools, Form 5 is the final level of secondary school where students will sit for the Malaysian Certificate of Education to complete formal education. Thereafter, some would choose to proceed to Form 6 which is pre-university level to prepare for university entrance. A research had found final year students reported higher depression, attributed to lesser sleep, lesser outdoor activities, and lesser physical activities [8]. Similarly, it is possible the Form 5 students had devoted most of their time in preparing for their final examination and disregarded other activities. The prevalence of depression among final year secondary school students is also consistent with a longitudinal study done with an Australian sample, where stress and depression heightened towards the time for major [22]. The significant predictors of depression include anxiety, emotional self-efficacy, and connectedness with friends, and it was suggested that intervention can be directed at these areas to alleviate final year students' depression. Comparing the extent of depression of other levels, Form 1 students had lower depression than other forms. This is possible as the Form 1 students who transitioned from primary school did not undergo the Primary School Achievement Test—the major examination for students exiting primary level which was abolished in 2021 [23]. Hence, the Form 1 students did not experience this examination stressor. The present findings implied that higher form students reported higher depression, while lower form reported lower depression.

With regards to location, rural students reported a slightly higher depression prevalence than urban students. Looking at the extent of depression, both rural and urban students reported similar level categories of mild and moderate depression. However, rural students recorded a higher percentage of severe depression at 22.5%. This means at least 2 out of 10 students may be suffering from severe depression. Some research has shown that rural adolescents are more vulnerable to depression where their depressive symptoms are higher than the urban school students [17,18]. Local research had noted the challenges in implementing online learning in rural areas, in terms of poor internet connectivity and limited infrastructure, and rural students are disadvantaged in activities such as live discussion and group work [5,16]. These activities are important for interactions and the development of social skills, without which may give rise to disinterest, hamper confidence, and loss of focus—depressive symptoms measured in this study. It is possible that rural students returning for in-person classes have also experienced similar challenges.

5. Conclusions

Well-being is important for the recovery of post-COVID learning, as students' cognitive development and learning hinge on their emotional and mental well-being [12]. Schools are one of the crucial institutions that address students' mental health needs, it is important to roll-out well-being programmes for returning students. Besides receiving students who seek help from student counsellors, well-being programmes will be more inclusive to benefit the school population as a whole. As noted from success stories, physical activities can enhance students' well-being after schools' re-opening [12]. However, it must be noted that our sample reported having less energy, hence, activities may not be too strenuous. Alternatively, adjustments can be made to schooling duration, i.e., not to have too long extra-curricular activities after classes. Counselling departments in schools can also screen students to identify those in need of support, which is more inclusive than getting referrals from teachers. It can be deduced that during and after the pandemic, exploring mental well-being among students is fundamental. The findings also show that rural students experience a higher depression level compared to their counterparts in the urban area. This shows that the location of the schools and where the students reside do affect the students' level of mental health. Therefore, this indicates that more scholarly as well as policy-level and attention is needed to understand the situation and assist the students in their transition from pandemic to endemic as they return to school.

Author Contributions: Conceptualization, J.S.-C.Y.; methodology, J.S.-C.Y.; formal analysis, J.S.-C.Y.; investigation, J.S.-C.Y., P.R.M.A.R. and D.S.S.; data curation, J.S.-C.Y.; writing—original draft preparation, J.S.-C.Y.; review and editing, J.S.-C.Y., P.R.M.A.R. and D.S.S.; visualization, J.S.-C.Y., P.R.M.A.R. and D.S.S.; project administration, J.S.-C.Y. All authors have read and agreed to the published version of the manuscript.

Funding: This research received no external funding.

Institutional Review Board Statement: Not applicable.

Informed Consent Statement: Not applicable.

Data Availability Statement: Not applicable.

Acknowledgments: We thank the departments in the Malaysian Ministry of Education and the school principals for the co-operation given.

Conflicts of Interest: The authors declare no conflict of interest.

References

1. United Nations Educational, Scientific and Cultural Organization. Available online: https://unesdoc.unesco.org/ark:/48223/pf0000373992 (accessed on 21 December 2020).
2. Asian Development Bank. Available online: https://www.adb.org/sites/default/files/publication/692111/ado2021-special-topic.pdf (accessed on 11 November 2021).
3. The Star. Available online: https://www.thestar.com.my/news/nation/2022/01/12/address-pandemic-induced-learning-loss-quickly-before-it-grows-more-critical-says-expert (accessed on 12 February 2022).
4. Sundarasen, S.; Chinna, K.; Kamaludin, K.; Nurunnabi, M.; Baloch, G.M.; Khoshaim, H.B.; Hossain, S.F.A.; Sukayt, A. Psychological Impact of COVID-19 and Lockdown among University Students in Malaysia: Implications and Policy Recommendations. *Int. J. Environ. Res. Public Health* **2020**, *17*, 6206. [CrossRef] [PubMed]
5. Zaki, N.M.; Shaharuddin, M.F.; Ishak, N.A.; Hilmi, Z.S.; Azmi, N.A.; Kurnia, J.C. Psychologial impacts of COVID-19 pandemic towards Malaysia's secondary school students. *Platform J. Manage. Humanit.* **2021**, *4*, 31–44.
6. Khan, M.A.; Kamal, T.; Illiyan, A.; Asif, M. School students' perception and challenges towards online classes during COVID-19 pandemic in India: An econometric analysis. *Sustainability* **2021**, *13*, 4786. [CrossRef]
7. Kaoud, H.; El-Shihy, D.; Yousri, M. Online learning in Egyptian universities post COVID-19 pandemic: A student's perspective. *Int. J. Emerg. Technol. Learn.* **2021**, *16*, 38–52. [CrossRef]
8. Puteikis, K.; Mameniškytė, A.; Mameniškienė, R. Sleep Quality, Mental Health and Learning among High School Students after Reopening Schools during the COVID-19 Pandemic: Results of a Cross-Sectional Online Survey. *Int. J. Environ. Res. Public Health* **2022**, *19*, 2553. [CrossRef] [PubMed]
9. Lane, J.; Therriault, D.; Dupuis, A.; Gosselin, P.; Smith, J.; Ziam, S.; Dufour, M. The impact of the COVID-19 pandemic on the anxiety of adolescent in Québec. *Child Youth Care Forum* **2021**, *51*, 811–833. [CrossRef] [PubMed]
10. Putwain, D.W.; Gallard, D.; Beaumont, J.; Loderer, K.; von der Embse, N. Does test anxiety predispose poor school related wellbeing and enhanced risk of emotional disorders? *Cogn. Ther. Res.* **2021**, *45*, 1150–1162. [CrossRef]
11. Chambel, M.J.; Curral, L. Stress in academic life: Ork characteristics as predictors of student well-being and performance. *Appl. Psychol. Int. Rev.* **2005**, *54*, 135–147. [CrossRef]
12. Lynch, T. Leading school recovery from the impact of COVID-19: Two birds, one stone. *Education* **2022**, *3*, 1–18. [CrossRef]
13. Giusti, L.; Mammarella, S.; Salza, A.; Del Vecchio, S.; Ussorio, D.; Casacchia, M.; Roncone, R. Predictors of academic performance during the COVID-19 outbreak: Impact of distance education on mental health, social cognition and memory abilities in an Italian university student sample. *BMC Psychol.* **2021**, *9*, 142. [CrossRef]
14. Badri, S.K.; Yunus, W.M. The relationship between academic vs. family/personal role conflict and Malaysian students' psychological wellbeing during COVID-19 lockdown. *J. Furth. High. Educ.* **2021**, *46*, 76–88. [CrossRef]
15. Amran, M.S.; Jamaludin, K.A. The impact of unplanned school closures on adolescent behavioral health during the COVID-19 pandemic in Malaysia. *Front. Public Heal.* **2021**, *9*, 639041. [CrossRef] [PubMed]

16. Selvanathan, M.; Hussin, N.A.; Azazi, N.A. Students learning experiences during COVID-19: Work from home period in Malaysian higher learning institutions. *Teach. Public Adm.* **2021**. [CrossRef]
17. Ang, A.L.; Wahab, S.; Adb Rahman, F.N.; Hazmi, H.; Md Yusoff, R. Depressive symptoms among adolescents in Kuching, Malaysia: Prevalence and associated factors. *Pediatrics Int.* **2018**, *61*, 404–410. [CrossRef] [PubMed]
18. Zhou, M.; Zhang, G.; Rozelle, S.; Kenny, K.; Xue, H. Depressive symptoms of Chinese children: Prevalence and correlated factors among subgroups. *Int. J. Environ. Res. Public Health* **2018**, *15*, 283. [CrossRef] [PubMed]
19. Beck, A.T.; Steer, R.A.; Ball, R.; Ranieri, W. Comparison of Beck Depression Inventories -IA and -II in psychiatric outpatients. *J. Pers. Assess.* **1996**, *67*, 588–597. [CrossRef] [PubMed]
20. Brislin, R.W. Back-translation for cross-cultural research. *J. Cross Cult. Psychol.* **1970**, *1*, 185–216. [CrossRef]
21. Ishak, N.A.; Ahmad, N.S.; Omar, M.N. Issues and trend of depression among students in Malaysia. *Univers. J. Educ. Res.* **2020**, *8*, 5951–5957. [CrossRef]
22. Wuthrich, V.M.; Belcher, J.; Kilby, C.; Jagiello, T.; Lowe, C. Tracking stress, depression, and anxiety across the final year of secondary school: A longitudinal study. *J. Sch. Psychol.* **2021**, *88*, 18–30. [CrossRef]
23. Rajaendram, R. The Star. Available online: https://www.thestar.com.my/news/nation/2021/04/28/education-minister-upsr-exams-to-be-abolished-pt3-is-cancelled-this-year (accessed on 27 June 2021).

proceedings

MDPI

Proceeding Paper

Mediating Effect of Goal Acquisition on the Relationship between Personal Factor and Self-Directed Learning †

Nursyamilah Annuar [1,*], Dazmin Daud [2], Sabiroh Md Sabri [1], Nurul Labanihuda Abdull Rahman [1], Sharifah Khairol Musairah [1] and Hasyeilla Abd Mutalib [1]

1 Faculty of Business and Management, Universiti Teknologi MARA, Cawangan Perlis, Kampus Arau, Perlis 02600, Malaysia

2 Faculty of Business and Management, UCSI University, UCSI Heights 1 Jalan Puncak Menara Gading, Taman Connaught, Kuala Lumpur 56000, Malaysia

* Correspondence: nsyamilah@uitm.edu.my

† Presented at the International Academic Symposium of Social Science 2022, Kota Bharu, Malaysia, 3 July 2022.

Abstract: This study was conducted to examine the relationship between personal factors and self-directed learning, the relationship between goal acquisition and self-directed learning, and the role of goal acquisition as a mediator in the relationship between personal and self-directed learning. Empirical studies are still lacking on the role of goal acquisition as a mediator, especially between personal relationships with self-directed learning. The study was conducted on a sample of 378 students in public universities. Hypothesis testing was performed using SEM-AMOS analysis. We found that emotion, family support, and goal acquisition have a positive and significant relationship with self-directed learning, and that goal acquisition has a significant role as a mediator between personal factors and self-directed learning. The results of this study prove that goal acquisition serves as a mediator in the relationship between personal factors and self-directed learning. These findings also indicate the importance of personal elements in influencing student excellence in self-study.

Keywords: self-directed learning; human resource development; organizational behavior

Citation: Annuar, N.; Daud, D.; Sabri, S.M.; Rahman, N.L.A.; Musairah, S.K.; Mutalib, H.A. Mediating Effect of Goal Acquisition on the Relationship between Personal Factor and Self-Directed Learning. *Proceedings* **2022**, *82*, 46. https://doi.org/10.3390/proceedings2022082046

Academic Editor: Mohamad Rahimi Mohamad Rosman

Published: 14 September 2022

1. Introduction

In this study, personal factors are defined as things that are dynamic and result from the individual self, consisting of psychological and physical systems that determine one's adaptation to the environment [1]. In this study, we used the Subjective Well-Being model initiated by [2]. According to [2], individuals react differently to the same situation and evaluate the situation based on previous expectations, assessments, and experiences. Subjective well-being encompasses emotional responses, domains of satisfaction, and life satisfaction [3,4]. The domain of satisfaction covers the dimensions of work, family, health, and emotions because these dimensions are seen to have a more impact on open and distance learning students. However, in this study, we tested the dimensions of family, health, and emotions only, because these dimensions are seen to have more impact. Based on the model of Subjective Well-Being, the family dimension involves an individual's ability to gain family support. Health, on the other hand, looks at the level of fear, anxiety and anger. As for the emotional dimension, it involves pleasant and unpleasant feelings. The mediating variable in this study is goal acquisition, referring to the management of human behavior, which includes what a person thinks and feels and behaviors that lead to goal achievement, such as self-reflection [5]. As for the dependent variables, this study takes into account the maturity of students in applying the process of self-directed learning. Self-directed learning is interpreted by [6] as a process in which an individual takes the initiative, with or without the help of others, to diagnose learning needs, formulate learning goals, identify resources for learning, select and implement learning strategies, and evaluate learning outcomes. Self-directed learning is an improvement in terms of the knowledge, skills, achievement, or personal development of students who choose to use any method in any situation at any time [7–9].

Studies have found that students' self-directed learning is at a low level, especially involving students pursuing open and distance learning programs [10–13]. This is due to the burden of responsibilities that must be borne by students who work in addition to trying to achieve excellent academic performance. This lack of student maturity in this self-directed learning process invites significant losses for the student as well as the organization [14–16]. The burden that students face involves managing their time, family, and career to succeed in this self-directed learning process. From an organizational point of view, organizations suffer losses when students who are sponsored by tuition fees are unable to complete their studies within the stipulated period and are unable to contribute to organizational development [17–19]. Low self-directed learning skills will have an impact on a student's motivation [20,21], self-confidence [22], ability to control learning [23], and ability to take initiative [24] to achieve academically as well as professionally.

Studies have found that open and distance learning students have higher levels of work and study stress as compared to students of full-time learning [12]. In addition, there are also studies that have found that students experience stress not only as a result of work stress, but also due to family burdens, health problems, and an uncomfortable environment [25,26]. Students who experience high stress may fail to control their emotions [27], which in turn can involve injury as well as death [28]. There is evidence to suggest that students commit suicide as a result of academic stress [19].

Personal factors refer to a person's personality. Personality can be interpreted as behaviors that characterize a person [29]. The personal dimension is divided into five categories: work, emotions, family support, finance, and health [2]. Family and partner support, health, and emotions are the factors that determine students' behavior in self-directed learning [30–34]. Personality is also related to student emotions that influence student goal achievement [35–39]. Emotional stability helps students implement goal acquisition more effectively [21]. Students who are able to control their emotions are predicted to have high levels of goal achievement [40,41]. If students are having problems with their family or spouses, student performance in learning will be disrupted [42,43]. High family support for student learning is expected to influence student goal achievement [30–32]. Students with problems among family members have lower levels of goal achievement [42]. Similarly, good health can increase student goal achievement [30,32,33,43,44]. Physically healthy students are able to implement their goals well [16]. Conflict theory suggests that individuals with limited time and energy as well as additional roles experience stress in meeting their needs, causing even more role conflict [45,46].

Goal acquisition is considered to be a stable tendency to take personal initiative in a variety of activities and situations [47]. Goal acquisition has a positive relationship with students' self-directed learning [24,48]. Students with high goal achievement show a more self-directed attitude in their learning [49]. Students who implement goal acquisition regularly can improve their academic performance [15].

2. The Role of Goal Acquisition as a Mediator

The acquisition of goals helps individuals acquire knowledge, improve social quality, increase the onset of perseverance in performing activities, achieve better performance, and develop a sense of discipline. Goal acquisition is an incentive that forces an individual to act towards the achievement of some goal. As defined by [50], goal acquisition is interpreted as a solid target that is expected to be achieved in one's learning. Goal acquisition is a key factor in effective management of the learning process [51]. Goal acquisition also refers to the aspect of self-reflection [17].

Goal acquisition was chosen as a mediator because it is an aspect related to positive self-development outcomes and serves as a liaison for psychological aspects for a person related to external factors such as positive performance, commitment, and responsibility among students [52–54]. Studies have found that goal acquisition serves as a mediator between other variables (such as task factors) and academic excellence variables in general [55] and self-directed learning maturity in particular [52]. Based on the theory of Subjective Well-Being, students with high life satisfaction will be able to increase their appreciation for a completed task which can then positively and directly affect the level of goal achievement [56,57].

Based on the context of this study, life satisfaction is a personal factor that consists of satisfaction in terms of emotions, family support, and health, and will directly affect the level of achievement of student goals. Meanwhile, the goal acquisition relationship is able to influence the maturity of students' self-directed learning based on the self-directed learning model by [7], which emphasizes students' responsibility for learning. Research by [50] linked Goal Acquisition Theory in describing the role of goal acquisition as a mediator in the relationship between life satisfaction and positive outcomes through responsibility for learning. Responsibility for learning can promote goal acquisition and have a positive impact on work outcomes such as job performance [15,57,58]. Research by [59] also criticizes that students with goal acquisition are confident that persistent effort will lead to positive outcomes, and confidence in learning is based on this belief.

Past researchers have noted that emotional stability [60], family support [61], and health [24] can influence goal achievement. Meanwhile, the acquisition of goals has the result of responsibility for learning, such as maturity in self-directed learning. The relationship between these variables proves that goal acquisition can play a mediating role in the relationship between personal factors and self-directed learning. However, there is still no specific study that examines the role of goal acquisition as a mediator in the relationship between personal factors and self-directed learning, especially in the field of distance education. Nevertheless, there are still past studies that use goal acquisition as a mediator of the relationship between other variables. Therefore, we propose the following hypotheses:

Hypothesis 1 (H1). *Goal acquisition has a mediating effect on the relationship between personal factors and self-directed learning.*

Hypothesis 1a (H1a). *Goal acquisition has a mediating effect on the relationship between emotional dimensions and self-directed learning.*

Hypothesis 1b (H1b). *Goal acquisition has a mediating effect on the relationship between the dimensions of family support and self-directed learning.*

Hypothesis 1c (H1c). *Goal acquisition has a mediating effect on the relationship between health dimensions and self-directed learning.*

Study Framework

We attempted to integrate personal and self-directed learning factors in addition to moderating goal acquisition. We adapted different models, concepts, and theories and integrated them into the framework of the study. These include the Personal Responsibility Orientation (PRO) model, the Subjective Well-Being (SWB) model, the Theory of Multiple Perspectives, and Goal Acquisition Theory, illustrated in Figure 1. Figure 1 shows that goal acquisition mediates the relationship between personal factors and self-directed learning. Personal factors refer to psychological and physical changes due to events that occur in the environment and affect the level of maturity of students. Individual personalities influence student behavior. The personal component refers to family support, health, and emotional support. These three factors are grounded in the Theory of Multiple Perspectives and the Subjective Well-Being model. Based on Brockett and Hiemstra's PRO model, students are responsible for their own learning and also take risks on the impact of each decision made. For students who are less satisfied with their lives, they still need to be responsible and accept the consequences of decisions made in self-directed learning. High life satisfaction in personal factors helps students to achieve a high level of self-directed learning [62].

In addition, goal acquisition mediates the relationship between personal factors and self-directed learning. Goal acquisition is a strong target that is expected to be achieved in one's learning [50]. Goal setting helps students deal with challenges in learning. Increasing the level of maturity of students' skills to learn independently is the desired result that will change or improve the skills and behaviors of individuals to continue to progress and be enthusiastic to apply self-directed learning in school or their career. The framework of the study (Figure 1) shows the links between personal factors (family support, health, and emotional) and the maturity level of students' self-directed learning skills. Goal acquisition mediates the relationship between the two types of variables.

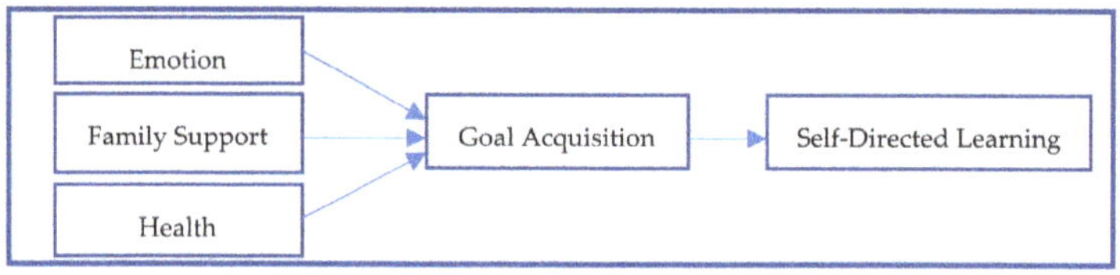

Figure 1. Conceptual framework.

3. Research Methodology

This study is a descriptive and quantitative deductive study, the conceptual framework of which is based on the conclusions made from the literature review. The sampling method used in this study is the Non-Probability Sampling method, which applies the purposive sampling technique, or judgmental sampling. Purposive sampling is a procedure in which a group of subjects with certain characteristics are selected as study respondents [63]. The respondents of this study consisted of students enrolled in a bachelor's degree program through online distance learning platforms in public universities in Malaysia.

Power analysis categorizes the ability of a study to obtain a meaningful effect to identify the sample size required in order to provide the necessary power for an effect on scientific interest [64]. Among the software built is STATISTICA, which is a comprehensive statistics package that offers the option to calculate sample size based on power analysis for the structural equation model. Therefore, this software was used to calculate the required sample size by emphasizing the values of RMSEA = 0.08, df = 27, power goal = 0.80, and error probability $\alpha = 0.05$. As a result, the proposed minimum sample size is 275.

Instruments for self-directed learning, goal acquisition, and personal factors use a Likert scale with five answer choices scaled from 1 to 5, with 1 representing "strongly disagree" and 5 representing "strongly agree". As for the demographics of the study, the measurement item contained seven questions related to gender, age, race, academic qualification and marital status. Emotion and family support were measured using the Emotional Quotient Inventory, and health was measured using the Patient Health Questionnaire. The measurement instruments evaluated emotions (11 questions), family (9 questions), and health (8 questions). We also used the Self-Directed Learning Readiness Skills questionnaire (25 questions). The reliability of the coefficients was as follows: emotion, $\alpha = 0.876$; family, $\alpha = 0.742$; health, $\alpha = 0.848$; goal acquisition, $\alpha = 0.866$; self-directed learning, $\alpha = 0.836$. Cronbach's alpha coefficient of 0.7 and above is considered a reliable measurement [65]. Therefore, the CR values for all instruments are at a suitable level.

4. Findings

The study respondents consisted of 378 people. Males (36%) represented 136 people and females (64%) represented 242 respondents. All study data for the three variables, namely, personal factors, goal acquisition, and self-directed learning, were analyzed by structural equation modeling (SEM) analysis using IBM SPSS AMOS version 2.1 software (Armonk, NY, USA). The measurement model was run first before implementing the structural equation model [65]. Table 1 shows the results of the measurement model regarding the reliability and validity of the study instruments. The recommended level is

>0.5 [65]. A total of 14 items were dropped due to a load value of less than 0.5. However, the reliability value is at the good category level where the composite reliability value is in the range of 0.742 to 0.876.

Table 1. Cronbach's alpha values.

Construct	Instrument	No. of Items	Cronbach's Alpha Values
Self-Directed Learning Skills	PRO-SDLRS [66]	25	0.836
Emotion	*Emotional Quotient Inventory* [67]	10	0.876
Family	*Emotional Quotient Inventory* [67]	10	0.742
Health	Patient Health Questionnaire [68]	6	0.848
Goal acquisition	Learning Practices [69]	15	0.866

The structural equation model used a bootstrapping procedure of 5000. The structural equation model is used to study the model and explain the direct relationship between personal factors and self-directed learning and the indirect relationship through goal acquisition intermediaries. The results of the study after evaluating the fit of the structural model show that the data are consistent with the model: χ^2 (677) = 2075.761, $p = 0.000$, $\chi^2/\text{df} = 3.066$, GFI = 0.786, CFI = 0.821, IFI = 0.822, TLI = 0.804, RMSEA = 0.074. The results show that the correspondence indices such as CFI, IFI, and TLI are very close to 0.9, which is the level of acceptance. GFI (0.786) and NFI (0.757) are also close to the acceptance criteria of 0.9. Chi-squared (χ^2/df) is below the value of 5, which is an indicator value of the acceptance of the match between the hypothesis model and the data that have been collected. The value of RMSEA is 0.074, which is a value close to the match. The results show that the direct relationships are positive and significant, complying with the set value with a significance level of $p < 0.01$. However, health showed an insignificant value. Therefore, only three study hypotheses were accepted and supported.

The maximum likelihood estimation technique was used to predict the model. The results of the path analysis hypothesis for the model structure are presented in Table 2. As illustrated in Table 2, the results show that personal factors have a significant and positive relationship with self-directed learning skills (emotion ($\beta = 0.306$, CR = 5.230, $p = 0.000$); family ($\beta = 0.076$, CR = 1.536, $p = 0.124$); health ($\beta = 0.098$, CR = 1.850, $p = 0.064$)). Therefore, based on the structural model, our hypothesis is supported.

Table 2. Regression weights in the direct hypothesis model.

Hypothesis Relationships	Standardized Regression Weights Beta	Unstandardized Regression Weights Estimate B	S.E.	C.R.	*p* Value
Emotion	0.306	0.221	0.042	5.230	***
Family	0.076	0.057	0.037	1.536	0.124
Health	0.098	0.053	0.029	1.850	0.064
Goal Acquisition	0.836	0.759	0.077	9.908	***

Note: If a *p*-value is less than 0.001, it is flagged with three stars (***).

The results of the study in Table 2 show a positive and significant relationship between goal acquisition and self-directed learning, with $\beta = 0.836$, C.R. = 9.908, and $p = 0.000$. Thus, the hypothesis is supported.

Next, we examined the significant and positive intermediate effect of goal acquisition on the relationship between personal factors and self-directed learning. The bootstrapping approach was used to this end. AMOS software can directly generate bootstrapped bias-corrected confidence intervals for indirect effects. Parallel to Multi-Model Analysis (AMM) to test the effect of intermediaries, the decision to test the intermediaries for each hypothesis was made by comparing the model directly opposite to the full intermediary model. The use of AMOS is also similar to Multi-Model Analysis (AMM) to test the effect of intermediaries, with structural models directly designed based on the hypothesis of a direct relationship between personal factors and self-directed learning skills. The direct structural model is consistent with the data: χ^2 (683) = 2378.405, p = 0.000, χ^2/df = 3.482, GFI = 0.771, CFI = 0.783, IFI = 0.784, TLI = 0.764, RMSEA = 0.081. The results show that all the appropriate model indices such as GFI, CFI, IFI, and TLI are close to the level of acceptance of 0.9. Moreover, the relative chi-squared value is below the value of acceptance of 5, and RMSEA is 0.081, which is close to appropriate.

The full intermediate structure model of the study is also consistent with data where χ^2 (677) = 2075.761, p = 0.000, χ^2/df = 3.066, GFI = 0.786, CFI = 0.821, IFI = 0.822, TLI = 0.804, RMSEA = 0.074. The results of the study show that the full structure model explains 80% of the self-directed learning skills, while the direct structure model explains only 65% of the self-directed learning skills. This finding indicates that the intermediate variables proposed are added aspects to the diversity of self-directed learning skills.

The results in Table 3 show that the standardized indirect effect (SIE) for personalities with self-directed learning skills through goal acquisition was significant (emotions (β = 0.236, p = 0.000), family (β = 0.073, p = 0.033), health (β = 0.052, p = 0.139)). The results also show that the standardized regression weight (β) for the interpersonal hypothesis with self-directed learning skills in the intermediary model was decreased but significant in both the direct model and the intermediate structural model. In other words, the indirect effect of personal factors on self-directed learning skills through goal acquisition was not empty through 95% emotional bias-corrected (bias-corrected C1 = 0.154 to 0.343), family bias-corrected (bias-corrected C1 = 0.005 to 0.156), health bias-corrected (bias-corrected C1 = 0.018 to 0.132) confidence intervals. The findings of this study indicate that goal acquisition partially mediates the relationship between personal factors and self-directed learning skills. Therefore, our hypotheses are supported by the data.

Table 3. Personal factors' indirect effects on self-directed learning, with goal acquisition as a mediator.

95% CI Bootstrap BC				
Hypothesis Path	**Beta**	***p***	**LB**	**UB**
Full Model				
Emotion—self-directed learning skills	0.382	0.000		
Family—self-directed learning skills	0.115	0.124		
Health—self-directed learning skills	0.091	0.064		
Intermediary Model				
Emotion—self-directed learning skills	0.064	0.000		
Family—self-directed learning skills	−0.001	0.034		
Health—self-directed learning skills	0.048	0.095		
Standardized Indirect Effect (SIE) Emotion	0.236	0.000	0.154	0.343
Standardized Indirect Effect (SIE) Family	0.073	0.033	0.005	0.156
Standardized Indirect Effect (SIE) Health	0.052	0.139	−0.018	0.132

Note: BC = Bias-corrected confidence interval; 5000 bootstrap samples have been requested.

5. Discussion

From the results of the study, we found that all our hypotheses were supported, except for the relationship between health and self-directed learning through the mediation of goal acquisition. The findings of the study show that the two personal dimensions of emotions and family support have a positive and significant direct relationship with goal achievement. These results are in line with findings from studies [21,33,35,36] that found that emotional stability helps improve goal acquisition among ODL students. The findings suggest that emotions can influence students' reflection on learning through goal acquisition. Therefore, the working student organization and the university should give focus and attention to the emotions of ODL students. In addition, the results of this study are also in line with the findings of studies [10,30,41] that family support is a catalyst to goal achievement among ODL students. This means that family support has an impact on students' achievement of goals. The student's family must pay attention to the student's self-development by providing full support to the student achieve goals and subsequently succeed academically and professional. However, the relationship between health and goal achievement shows an insignificant relationship despite past studies from [30,32,33] that showed significant findings. This may be due to cultural differences and the context of where this study was conducted in Malaysia, which consists of various races, and in the context of distance education.

As for the findings of the study on the relationship between goal acquisition and self-directed learning, the results are consistent with the studies by [15,57,59] which found that goal acquisition can influence the level of maturity of self-directed learning of ODL students. This indicates that when ODL students have a high level of goal achievement towards learning, it directly affects the increase in the level of maturity of learning independently. Thus, the relationships between personal aspects, goal acquisition, and self-directed learning are seen to be interrelated. These findings are also in line with the meaning of self-directed learning, which is closely related to self-development from emotional and physical aspects in order to help reduce stress due to the heavy burden of responsibility on students [26,30].

In addition, we also found an indirect relationship when the role of goal acquisition is mediated by the relationship between the three personal dimensions and self-directed learning. These findings support Goal Acquisition Theory, where goal acquisition exerts a mediating effect on the relationship between individual self-development and positive outcomes on learning [15,48,49]. Theoretically, this study shows the relationship between personal resources, namely, emotions, family support, and health, with goal acquisition and self-directed learning. The framework of this study is based on the theory of Subjective Well-Being through Goal Acquisition Theory by [50], which has been successfully tested and validated based on the findings obtained, although there are health dimensions that show an insignificant relationship due to contextual differences compared to previous studies. This indicates that family emotions and support with goal acquisition ultimately have a positive effect on the maturity of self-directed learning. This suggests that individual aspects such as emotions and seeking family support should not be set aside and become a necessity for work organizations and universities to ensure that stress due to commitment towards work and family and lack of maturity in self-directed learning among ODL students can be addressed. This is because work and study stress were found to have a relationship with the level of maturity of students' self-directed learning [19,20]. This study also proves the importance of the role of direct goal acquisition in improving self-directed learning as well as the mediating role of goal acquisition in the indirect relationship between emotional factors and family support with self-directed learning.

6. Conclusions

We found that two personal dimensions of emotion and family support have positive direct and significant relationships with self-directed learning, that goal acquisition has a positive and significant direct relationship with self-directed learning, and that goal acquisition has a significant role as a mediator in the relationship between the two dimensions of personal factors and self-directed learning. The results of this study prove that goal acquisition serves as a mediator in the relationship between personal factors and self-directed learning. These findings also indicate the importance of personal elements in influencing student excellence and maturity in self-study.

Author Contributions: Conceptualization, N.A. and D.D.; methodology, S.K.M. and N.L.A.R.; validation, H.A.M. and S.M.S.; writing—original draft preparation, N.A. and S.M.S.; writing—review and editing, N.A. and D.D.; funding acquisition, N.A. All authors have read and agreed to the published version of the manuscript.

Funding: This research was funded by the Research Management Centre, Universiti Teknologi MARA, Cawangan Perlis Kampus Arau, grant number 600-UiTMPs (PJIM&A/PI-DPPD 13). RMI File No 600-TNCPI 5/3/DDN (09) (002/2020).

Institutional Review Board Statement: Not applicable.

Informed Consent Statement: Not applicable.

Data Availability Statement: Not applicable.

Acknowledgments: The authors thank the Research Management Centre, Universiti Teknologi MARA, Cawangan Perlis Kampus Arau. This work was supported by an Internal Research Acculturation Grant awarded by Universiti Teknologi MARA, Cawangan Perlis, Kampus Arau (ref. no. 600-UiTMPs (PJIM&A/PI-DPPD 13)). RMI File No 600-TNCPI 5/3/DDN (09) (002/2020).

Conflicts of Interest: The authors declare no conflict of interest.

References

1. Werner, T.P.; Rogers, K.S. Scholar-Craftsmanship: Question-Type, epistemology, culture of inquiry, and personality type in dissertation research design. *Adult Learn.* **2013**, *24*, 159–166. [CrossRef]
2. Diener, E.; Suh, E.M. Measuring quality of life: Economic, social, and subjective indicators. *Soc. Indic. Res.* **1997**, *40*, 189–216. [CrossRef]
3. Hiemstra, R.; Brockett, R.G. From mentor to partner: Lessons from a personal journey. *New Dir. Adult Contin. Educ.* **1998**, 43–51. [CrossRef]
4. Krapu, T.M.; Meinke, L.; Kramer, L.; Friedman, R.; Voda, J. The Life Satisfaction Survey (LSS): Development of the instrument. In Proceedings of the Fourth International Coach Federation Coaching Research Symposium, St. Louis, MO, USA, 1 November 2006.
5. Higgins, E.T.; Kruglanski, A. *Motivational Science: Social and Personality Perspectives*; Psychology Press: Philadelphia, PA, USA, 2000.
6. Knowles, M.S. *Self-Directed Learning: A Guide for Learners and Teachers*; Association Press: New York, NY, USA, 1975.
7. Brockett, R.G.; Hiemstra, R. *Self-Direction in Adult Learning: Perspectives on Theory, Research, and Practice*; Routledge: New York, NY, USA, 1991.
8. Gibbons, M. *The Self-Directed Learning Handbook: Challenging Adolescent Students to Excel*; Jassey-Bass: San Francisco, CA, USA, 2002.
9. Schweder, S.; Raufelder, D. Adolescents' expectancy value profiles in school context: The impact of self-directed learning intervals. *J. Adolesc.* **2022**, *94*, 569–586. [CrossRef]
10. Abdullah, M.M.; Muniapan, S.F.; Rathakrishnan, B. Adult participation in self-directed learning programs. *Int. Educ. Stud.* **2008**, *1*, 66–72. [CrossRef]
11. Annuar, N.; Sabri, M.D.; Mutalib, H.A.; Rahman, N.L.A.; Mutalib, S.K.M.S.A.; Daud, D.; Johan, M.R.M. Empowering self-directed learning through technology, organization and personal support in open and distance learning: A conceptual review. *Turk. J. Physiother. Rehabil.* **2021**, *32*, 23303–23313.
12. Howland, J.L.; Moore, J.L. Students' perceptions as distance learners in internet-based courses. *Distance Educ.* **2002**, *23*, 183–195. [CrossRef]
13. Mead, M.S. The Effect of Self-Directed Learning Readiness and Online Course Quality Ratings on Student Satisfaction and Academic Performance in Undergraduate E-Learning. Ph.D. Thesis, University of Missouri, Kansas City, MO, USA, 2011.
14. Oldake, A.O. Measurement of Self-Directed Learning in Online Learners. Ph.D. Thesis, Capella University, Minneapolis, MN, USA, 2006.
15. Raemdonck, I.; Van Der Leeden, R.; Valcke, M.; Mien, J.T. Predictors of self-directed learning for low-qualified employees: A multi-level analysis. *Eur. J. Train. Dev.* **2012**, *36*, 572–591. [CrossRef]
16. Yuan, H.B.; Williams, B.A.; Fang, J.B.; Pang, D. Chinese baccalaureate nursing students' readiness for self-directed learning. *Nurse Educ. Today* **2012**, *32*, 427–431. [CrossRef]
17. Bostwick, D.T. The Relationship between Background Demographics and Motivational Orientation for Learning of Adult Education Students and Their Career Choices and Educational Aspirations. Ph.D. Thesis, TUI University, Cypress, CA, USA, 2010.
18. Chou, P.N. Effect of students' self-directed learning abilities on online learning outcomes; Two exploratory experiments in electronic engineering. *Int. J. Humanit. Soc. Sci.* **2012**, *2*, 172–179.
19. Douglas, E. A Study of Barriers to Adult Self-Directed Learning. Ph.D. Thesis, University of Phoenix, Phoenix, AZ, USA, 2010.
20. Lynch, D.J. Motivational factors, learning strategies and resource management as predictors of course grades. *Coll. Stud. J.* **2006**, *40*, 423–428.
21. Regan, J.A. Motivating students towards self-directed learning. *Nurse Educ. Today* **2003**, *23*, 593–599. [CrossRef]
22. Seo, E.H. Self-efficacy as a mediator in the relationship between self-oriented perfectionism and academic procrastination. *Soc. Behav. Pers. Int. J.* **2008**, *36*, 753–764. [CrossRef]
23. Dynan, L.; Cate, T.; Rhee, K. The impact of learning structure on students' readiness for self-directed learning. *J. Educ. Bus.* **2008**, *84*, 96–100. [CrossRef]
24. Li, S.T.; Tancredi, D.J.; Co, J.P.T.; West, D.C. Factors associated with successful self-directed learning using individualized learning plans during pediatric residency. *Acad. Paediatr.* **2010**, *10*, 124–130. [CrossRef] [PubMed]
25. Annuar, N.; Shaari, R. The Antecedents toward Self-Directed Learning among Distance Learner in Malaysian Public Universities. In Proceedings of the Global Summit on Education GSE, Kuala Lumpur, Malaysia, 4–5 March 2014.
26. Kim, M.; Park, S.Y. Factors affecting the self-directed learning of students at clinical practice course for advanced practice nurse. *Asian Nurs. Res.* **2011**, *5*, 48–59. [CrossRef]

27. Jang, J.; George, R.T. The Relationship Of Emotional Intelligence To Job Stress, Affective Commitment, and Turnover Intention Among Restaurant Employees. In Proceedings of the 16th Graduate Student Research Conference in Hospitality and Tourism, Houston, TX, USA, 7 January 2011.
28. De la Fuente, S.E.I.; Extremera, R.A.; Pecina, C.V.; Canadas de la Fuente, G.R. Prevalence and risk factors of burnout syndrome among spanish police officers. *Psicothema* **2013**, *25*, 488–493.
29. Diener, E.; Oishi, S.; Lucas, R.E. Personality, culture, and subjective well-being: Emotional and cognitive evaluations of life. *Annu. Rev. Psychol.* **2003**, *54*, 403–425. [CrossRef]
30. Benozzo, A.; Colley, H. Emotion and learning in the workplace: Critical perspectives. *J. Work. Learn.* **2012**, *24*, 304–316. [CrossRef]
31. Bradley, P.; Oterholt, C.; Herrin, J.; Nordheim, L.; Bjorndal, A. Comparison of directed and self-directed learning in evidence-based medicine: A randomised controlled trial. *Med. Educ.* **2005**, *39*, 1027–1035. [CrossRef]
32. Illeris, K. Workplace learning and learning theory. *J. Work. Learn.* **2003**, *15*, 167–178. [CrossRef]
33. Kim, S.; McLean, G.N. The impact of national culture on informal learning in the workplace. *Adult Educ. Q.* **2013**, *64*, 39–59. [CrossRef]
34. Lechuga, V.M.; Lechuga, D.C. Faculty motivation and scholarly work: Self-determination and self- regulation perspectives. *J. Profr.* **2012**, *6*, 59–97.
35. Idros, S.N.S.; Mohamed, A.R.; Esa, N.; Samsudin, M.A.; Daud, K.A.M. Enhancing self-directed learning skills through e-SOLMS for Malaysian learners. *Procedia–Soc. Behav. Sci.* **2010**, *2*, 698–706. [CrossRef]
36. Mayhew, J.A. Adult Learners' Perceptions of Their Employers' Leadership Behaviors and Their Own Readiness for Self-Directed Learning. Ph.D. Thesis, Indiana Wesleyan University, Marion, IN, USA, 2008.
37. Merriam, S.B. Andragogy and self-directed learning: Pillars of adult learning theory. *New Dir. Adult Contin. Educ.* **2001**, *89*, 3–13. [CrossRef]
38. Pintrich, P.R.; De Groot, E.V. Motivational and self- regulated learning components of classroom academic performance. *J. Educ. Psychol.* **1990**, *82*, 33–40. [CrossRef]
39. Zimmerman, B.J.; Tsikalis, K.E. Can computer-based learning environments (CBLEs) be used as self-regulatory tools to enhance learning? *Educ. Psychol.* **2005**, *40*, 267–271. [CrossRef]
40. Boyer, S.L.; Artis, A.B.; Solomon, P.J.; Fleming, D.E. Improving sales performance with self-directed learning. *Mark. Manag. J.* **2012**, *22*, 61–75.
41. Mitchell, F. Self-Directed support and disabled young people in transition (Part 2). *J. Integr. Care* **2012**, *20*, 223–230. [CrossRef]
42. Cho, D.; Kwon, D.B. Self-Directed learning readiness as an antecedent of organizational commitment: A Korean study. *Int. J. Train. Dev.* **2008**, *9*, 140–152. [CrossRef]
43. Miller, A.L. Reconsidering achievement goal orientation for undergraduate honors students. *J. Adv. Acad.* **2022**, *33*, 364–399. [CrossRef]
44. Poston, B. Maslow's hierarchy of needs. *Surg. Technol.* **2009**, *41*, 347–353.
45. Eng, W. Self-Directed learning with web-based sites: How well do students perceptions and thinking match with their teachers? *Teach. Sci.* **2008**, *54*, 24–31.
46. Nugraha, S.P. Self-Directed Learning among Wives of International Students at the Florida State University. Ph.D. Thesis, The Florida State University, Tallahassee, FL, USA, 2005.
47. Seibert, S.E.; Kraimer, M.L.; Crant, J.M. What do proactive people do? A longitudinal model linking proactive personality and career success. *Pers. Psychol.* **2001**, *54*, 845–874. [CrossRef]
48. Gilany, A.H.; Abusaad, F.E.S. Self-Directed learning readiness and learning styles among Saudi undergraduate nursing students. *Nurse Educ. Today* **2012**, *33*, 1040–1044. [CrossRef]
49. Hamlett, C. Self-Directed Learning and Achievement in a Blackboard-Based College Algebra Course. Ph.D. Thesis, Walden University, Minneapolis, MN, USA, 2006.
50. Austin, J.; Vancouver, J. Goal constructs in psychology: Structure, process, and content. *Psychol. Bull.* **1996**, *120*, 338–375. [CrossRef]
51. Aggarwal, I. *Web-Based Learning and Teaching Technologies: Opportunities and Challenges*; Idea Group Publishing: Hershey, PA, USA, 2004.
52. Abar, B.; Loken, E. Self-Regulated learning and self-directed study in a pre-college sample. *Learn. Individ. Differ.* **2010**, *10*, 25–29. [CrossRef]
53. Eccles, J.S.; Wigfield, A. Motivational belief, values, and goals. *Annu. Rev. Psychol.* **2002**, *53*, 109–132. [CrossRef]
54. Osman, M. The effects of self set or externally set goals on learning in an uncertain environment. *Learn. Individ. Differ.* **2012**, *22*, 575–584. [CrossRef]
55. Park, J.H.; Kwon, D.B. Employees' perceived work environment and self-directed learning readiness in Korean companies. *Hum. Resour. Dev. Int.* **2004**, *7*, 333–350. [CrossRef]
56. Friedman, D. *The Thought Exchange: Overcoming Our Resistance to Living a Sensational Life*; Library Tales Publishing: New York, NY, USA, 2011.
57. Lent, R.W.; Brown, S.D. Integrating person and situation perspectives on work satisfaction: A social-cognitive view. *J. Vocat. Behav.* **2006**, *69*, 236–247. [CrossRef]

58. Toscano, F.; Zappala, S. Overall job performance, remote work engagement, living with children, and remote work productivity during the COVID-19 pandemic: A mediated moderation model. *Eur. J. Psychol. Open* **2021**, *80*, 133–142. [CrossRef]
59. Ocak, G. Examination of the relationships between fifth graders' self-regulated learning strategies, motivational beliefs, attitudes, and achievement. *Educ. Sci. Theory Pract.* **2013**, *13*, 380–388.
60. Siamak, K.; Pole, N. Cognitive behaviour therapy and worry reduction in an outpatient with generalized anxiety disorder. *Clin. Case Stud.* **2010**, *9*, 53–62. [CrossRef]
61. Avdal, E.U. The effect of self-directed learning abilities of student nurses on success in turkey. *Nurse Educ. Today* **2013**, *38*, 838–841. [CrossRef]
62. Harwood, A. Lifelong Learning: The Integration of Experiential Learning, Quality of Life Work in Communities, and Higher Education. Ph.D. Thesis, The University of Montana, Missoula, MT, USA, 2007.
63. Chua, Y.P. *Kaedah Penyelidikan: Buku 1*, 2nd ed.; Mc Graw Hill (Malaysia) Sdn Bhd: Kuala Lumpur, Malaysia, 2011.
64. Maxwell, S.E.; Kelly, K.; Rausch, J.R. Sample size planning for statistical power and accuracy in parameter estimation. *Annu. Rev. Psychol.* **2008**, *59*, 537–563. [CrossRef]
65. Hair, J.F.; Tatham, R.L.; Anderson, R.E.; Black, W. *Multivariate Data Analysis*, 6th ed.; Pearson Prentice Hall: Upper Saddle River, NJ, USA, 2006.
66. Stockdale, S.L. Development of an Instrument to Measure Self-Directedness. Ph.D. Thesis, University of Tennessee, Knoxville, TN, USA, 2003.
67. Bar-on, R.; Brown, J.M.; Kirkcoldy, B.D.; Thome, J.P. Emotional Expression and Implications for Occupational Stress; an Application of The Emotional Quotient Inventory. *Personal. Individ. Differ.* **2000**, *28*, 1107–1118. [CrossRef]
68. Kroenke, K.; Spitzer, R.L.; Williams, J.B.W. The PHQ 9: Validity of a brief depression severity measure. *J. Gen. Intern. Med.* **2001**, *16*, 606–613. [CrossRef]
69. Ishak, R.B.; Ghani, M.F.B.A.; Siraj, S. Amalan Pembelajaran Dalam Kalangan Guru Sekolah Berprestasi Tinggi. *JuKu J. Kurikulum Pengajaran Asia Pasifik* **2013**, *1*, 52–60.

proceedings

MDPI

Proceeding Paper

The Use of Learning Media among Pre-University Students in Dungun District, Terengganu, Malaysia †

Nasir Nayan [1,*], Fatihah Hassim [1], Hanifah Mahat [1], Mohmadisa Hashim [1], Yazid Saleh [1], Edi Kurniawan [2] and Nurul Khotimah [3]

1 Faculty of Human Sciences, Universiti Pendidikan Sultan Idris, Tanjung Malim 35900, Malaysia
2 Fakultas Ilmu Sosial, Universitas Negeri Semarang, Semarang 50229, Indonesia
3 Fakultas Ilmu Sosial, Universitas Negeri Yogyakarta, Yogyakarta 55281, Indonesia
* Correspondence: nasir@fsk.upsi.edu.my; Tel.: +60-54506487
† Presented at the International Academic Symposium of Social Science 2022, Kota Bharu, Malaysia, 3 July 2022.

Abstract: This study investigates the level of learning media among form-six students. A quantitative approach using a questionnaire instrument was employed. The sample size includes 201 form-six students using a simple random sampling technique where the analysis involved is a mean score scale. The findings showed that utilization levels were moderate. Pre-university students were selected due to the importance of learning media in furthering studies at the university level. A lack of learning media knowledge will result in the students being unable to compete with others at the university level. Students have less practical experience in using learning media while in class.

Keywords: learning media; educational technology; 21st-century learning

Citation: Nayan, N.; Hassim, F.; Mahat, H.; Hashim, M.; Saleh, Y.; Kurniawan, E.; Khotimah, N. The Use of Learning Media among Pre-University Students in Dungun District, Terengganu, Malaysia. *Proceedings* **2022**, *82*, 45. https://doi.org/10.3390/proceedings2022082045

Academic Editor: Mohamad Rahimi Mohamad Rosman

Published: 14 September 2022

1. Introduction

Information and communication technology (ICT) is essential. This technology has currently dominated the world. The advancement of technology has made it much easier to pursue knowledge. Diverse applications and services are introduced every day in the search of knowledge. Certain facilities introduced have made the task of teachers more manageable and efficient in delivering a lesson during the teaching process and facilitation (TPF). This technology has also introduced various forms of teaching and, at the same time, involves students in 21st-century learning (PAK21) methods. PAK21 is a student-centered learning process based on elements of communication, collaboration, critical thinking, creativity, and the application of values and ethics [1].

Education development is highly emphasized in the era of globalization challenges and the boom of ICT technology such that students and teachers can master this information technology during the teaching and learning process (T&L) [2]. This is because learning media lean more towards using technology, resulting in the need for teachers and students to master this technology. Learning media does not discard conventional learning; it makes the TPF process easier. The education system in schools can also create smart schools when educators and students use ICT. However, teachers should be more knowledgeable and aware of these media so that the TPF process can run smoothly. When a teacher makes a difference during the TPF process, students will be more interested, and curiosity increases. With this, the teacher will automatically achieve learning objectives, and the teacher will be happy to see the diligence and attention given by his students. The existence of technology in education can also facilitate educators or students in obtaining information from around the world by running searches on the internet. This situation also shows that schoolwork or projects can be completed and facilitate their work. This can also be seen when teachers and students learn to accord to their abilities and produce a computer-literate generation. Thus, it is not surprising that there is an increasing proliferation of scholars [3–6]. In conclusion, the education system is essential in forming a TPF session. The TPF process will not occur smoothly and efficiently without an organized and systematic education system. The education system has been changed to keep pace with the times and to avoid being left behind. Based on this study, the problems studied include the type of use of learning media, the level of use, and the effectiveness of learning media among form-six students in Dungun, Terengganu.

2. Learning Media in Education

Ref. [7] attempted to identify the media technology medium used in R&D that practically impacts students and educators. This study examines the importance and necessity of the current use of media technology in learning and teaching. The findings of this study help educators and students identify the main medium of use with respect to technological media related to the exploration, mastery, and deepening of knowledge in a more systematic manner. Technology media used in R&D include Facebook, E-learning, online web, video streaming, M-learning, and YouTube. This has shown a difference with this study in which the findings show that students think that the existence of learning media is associated with the quality of their assignments and not quantity. This permitted student engagement in activities such as plagiarism, namely, copying and pasting.

The authors of [8] even attempted to explore the use of mobile applications on logical thinking and technological advances in education. The technology in mobile learning (m-learning) discusses web-based applications and the advantages of using web-based applications. Logical thinking relates to human life, which always requires decision-making skills, regardless of whether a decision is right or wrong. Nevertheless, an assumption is a decision or opinion of a person that is considered valid, even though it may not necessarily be true. The results show that exploring mobile applications in education facilitates human beings in making decisions. At the same time, the present study is different in that most respondents say that learning media such as mobile applications cause their performance to decline. The authors of [9] examined a practical example in a mobile comic that can be used as an alternative medium to stimulate sensations of young children. This has included pedagogical instruments. In addition, discussing the use of mobile comics can help teachers improve literature for young people and further hone students' understanding of literature by making the experience more enjoyable and allowing them to gain more valuable experience when using mobile comics. Previous studies have shown that mobile comics can be an alternative medium. However, other sources such as magazines, newspaper clippings, and journals have affected students who use them for current studies. The study of [10], which aims to explain various social media, can help students obtain various helpful information in order to engage in teaching and learning activities more actively. In addition, the authors described the advantages and disadvantages of social media in the teaching and learning process in higher education.

According to this study, this study's findings are related to social media and higher education. Social media is said to disseminate information that is easy to convey and can be accessed. This has been supported in daily activities, especially in the education sector. Through educators, various applications have been applied in learning and teaching processes for students in higher education. Learning and teaching, when applied using social media through mobile phones, have also been made more exciting and effective. Initially, social media was used for communication purposes; however, it has evolved, resulting in various functions. For example, students can communicate regardless of distance, making communication more easily accessible to everyone. It is easy to receive information, files, and send photos and videos, etc. Moreover, social media is a medium of teaching and learning. This is evident in the use of media, as the flow of information is now quicker without hindrance. Thus, social media is very suitable for applications in higher education as a tool to increase the level of teaching and learning. The authors of [11] attempted to investigate the effectiveness of using various teaching media based on constructivism in science's teaching and learning process to improve students' basic Science process skills (KPS). The findings also show significant differences in KPS skills between the groups that follow ICT and environmental strategies with the conventional group. This study can conclude that R&D by using PMP integrated with ICT and the environment has had a positive impact on the development of skills related to the scientific process among students.

3. Materials and Methods

3.1. Sampling and Study Instruments

The research method used is a descriptive quantitative study to examine, analyze, and identify learning media phenomena among form-six students more depth and systematic manner. This study uses a Special Package for Social Statistics (IBM SPSS V 23.0, Armonk, NY, USA) for Windows version 23.0. The study area is in the district of Dungun, Terengganu. Dungun is in the eastern part of Terengganu's state, located at coordinates 4°44′ N 103°25′ E, and it covers an area of 273,519.35 hectares. In Terengganu, there are fifty-two form-six school centers. Only five schools reside in the Dungun district. Schools with form six only are SMK Tengku Intan Zaharah (SMK TIZ), SMK Sultan Omar (SMKSO), SMK Paka (SMKP), SMK Durian Mas (SMKDM), and SMK Ketengah Jaya (SMKKJ). A total of 201 form-six students were sampled for a simple randomly selected study from 426 form-six students. The population for this study is 426 students in these five schools. Of the total, there are 152 students in SMKTIZ, 84 students in SMKSO, 80 students in SMKP, 53 students in SMKDM, and 57 in SMKKJ (Table 1). According to [12], the sample size requires 201 respondents only. This is because this study's population was 426 respondents—the breakdown of the total sample is in Table 1. The distribution of questionnaires was performed using a Google form. Teachers only need to interact and instruct students to answer the questionnaire. Table 2 shows the construct, the number of items involved, and questionnaire source. The Likert scale used is shown in Table 3.

Table 1. Number of study population and sample.

School Name	Population	SAMPLING SIZE	Sampling Percentage (%)
SMK Tengku Intan Zaharah	152	31	15.3
SMK Sultan Omar	84	52	25.7
SMK Paka	80	72	35.6
SMK Durian Mas	53	21	10.4
SMK Ketengah Jaya	57	25	12.4
TOTAL	426	201	100

Table 2. Questionnaire Information.

Part	Description	Item Number	Number of Items	Items Sources
A	Respondent background	1–11	11	Self-built according to the needs of the study
B	Level of Use of Learning Media	34–48	10	Built with modifications and referring to the study of [13]

Table 3. Likert scales used.

Scale Values	Scale	Description
1	Never (N)	Never used the learning media
2	Sometimes (S)	Sometimes uses the learning media
3	Once in a while (Occasionally) (O)	Once in a while, uses the learning media
4	Frequent (F)	Frequently uses the learning media
5	Very often (VO)	High frequency uses the learning media

3.2. Validity and Reliability of the Questionnaire

A total of three validation experts were lecturers at the Department of Geography and Environment, UPSI, and assisted in developing the questionnaires. This pilot study was conducted on thirty students studying for a Bachelor of Geography degree at Universiti Pendidikan Sultan Idris (UPSI). The data obtained will be analyzed using a Special Package for Social Statistics (SPSS) for Windows version 23.0. This SPSS is used to determine the Cronbach's alpha value as the reliability coefficient. Each item obtained an alpha value at a good level in this pilot study. As a result of the pilot study, the alpha value for the learning media usage-level construct is 0.810. This value is accepted by [14], in which alpha values between 0.60 and 0.80 were acceptable, while alpha values above 0.80 were considered good.

3.3. Data Collection

Approval from the Education Policy Planning and Research Division (BPPDP) was obtained because the Ministry of Education Malaysia's (MOE) regulations require research related to Peruvian schools to obtain approval from the ministry before it can be conducted. BPPDP is responsible for issuing permission letters to conduct studies involving schools, vocational colleges, matriculation colleges, teacher education institutes, district education offices, state education departments, and divisions under the MOE. Applications were made using the eRAS 2.0 System (online system with 2nd version, Malaysia). This system can be accessed through http://eras.moe.gov.my (accessed on 18 November 2020). However, this division still accepts applications using BPPDP form 1.2 and issues letters manually until 28 February 2018 [15]. The letter was used to obtain permission from the school principal to obtain a sample of the study in the school. Meetings with schoolteachers were held to obtain their consent to select respondents and distribute the questionnaire online.

3.4. Data Analysis

In this research study, the level analysis of each variable will be described descriptively, namely, the percentage value (%), mean (M), and standard deviation (SD). Level values are based on the cutoff point setting and [16] (Table 4). The level has been categorized at the calculation level between a higher mean score and a lower score $(5 - 1) = 4$ and divided into three categories $(4 \div 3) = 1.33$. The lowest level is between 1.00 and 2.33, which is the sum of 1.00 with 1.33. Meanwhile, the moderate level is between 2.34 and 3.67 (2.34 + 1.33), and the highest level is between 3.67 and 5.00 (3.67 + 1.33).

Table 4. Cutoff Point Levels of Each Variable.

Assessment Level	Mean Score Scale
Low	1.00–2.33
Moderate	2.34–3.66
High	3.67–5.00

Source: [16].

4. Results

A total of 201 forms six students were sampled with a simple random technique from 426 form-six students. The number of respondents according to the schools involved is thirty-one people for SMKTIZ (15.4%), fifty-two people for SMKSO (25.9%), seventy-two people for SMKP (35.8%), twenty-one people for SMKDM (10.4%), and twenty-five people for SMKKJ (12.4%), as shown in Table 1. A total of 101 (50.2%) form-six students were semester-one students, while for semester three, the total was 100 students (49.8%). Only semester-one and -three students were involved because the second semester does not exist in the third term. Second-semester students will only exist from January to June. Data collection was performed in November, when only semester-one and -three students were physically available.

Therefore, the analysis was only conducted for the construct of the level of use of learning media among form-six students. Only the level of use of learning media has sub-constructs such as interactive boards and drills, social networking tools, and public-information storage tools. To facilitate the interpretation of construct levels, the contruct has been divided to low level, medium level, and high level, using the cutoff point recommended by [16] (Table 5). Table 6 shows the percentage of learning media usage.

Table 5. Cutoff Point Level of Use of Learning Media.

Score Scale	Level
1.00–2.33	Low
2.34–3.66	Moderate
3.67–5.00	High

Source: [16].

Table 6. Percentage of Application Use in Learning.

Item	Scale				
	N	S	O	F	VO
I use the WhatsApp application to form class groups	3 (1.5%)	2 (1.0%)	17 (8.5%)	67 (33.3%)	112 (55.7%)
I use the WhatsApp app to upload and download documents or information	0 (0%)	3 (1.5%)	13 (6.5%)	41 (20.4%)	144 (71.6%)
I use the Facebook app to post information, ideas, and files	2 (1.0%)	6 (3.0%)	17 (8.5%)	51 (25.4%)	125 (62.2%)

Table 6 describes the level interpretations for each construct. Questions related to the level of use of learning media are related to students' knowledge and experience while using this learning media. Based on the data, most students are accustomed to using learning media for their learning. The evidence can be observed in the questions posed to the respondents about the level of use of learning media. For the first question, 55.7% (112 students) chose very often while three chose never with a percentage of 1.5%. For the second question, 144 students used the WhatsApp application to upload and download documents or information by choosing very often, with a percentage of 71.6%. In contrast, none of the students chose the answer never. Next, the third question showed that 62.2% voted very often, with 125 people representing the highest number of respondents, and only 2, equivalent to 1.0%, had never voted.

These results had a connection to the jigsaw-type cooperative-learning model [17,18], which is a cooperative-learning model. This is where students learn in small groups and work together to obtain the maximum learning experience and valuable experience through individuals or groups. This model is also related to teaching media through gadget applications. This is where students who engage themselves in seeking this knowledge and type of learning are also more student-centered than teacher-centered. This is evident when the respondents are more independent and use learning media through social networking tools very well. Students can shoulder their responsibilities and acquire knowledge even when absent from the classroom. Therefore, the existence of learning media using such social networking tools can facilitate and attract students in their learning experience.

Table 7 describes the results of the study. The sub-construct levels were divided into interactive boards and drills, social networking tools, and public-information storage tools, and the first sub-construct is an interactive board and drill. In an analysis of the results for item 1, the use of display media, which was a blackboard during group learning, showed that it is at a moderate level (M = 3.03 and SD = 1.06). Many students chose to use occasionally, which includes ninety-one people and is equivalent to 45.3 percent, while only eighteen people, equivalent to 9.0 percent, used it regularly. This result indicates that the blackboard has less use when students study in groups.

Table 7. Levels of Use of Learning Media Among Form Six.

Item Num.	Construct	N		S		O		F		VO		Mean	SD	Mean Score Level
		N	%	N	%	N	%	N	%	N	%			
	Interactive Boards and Drills													
C1		21	10.4	29	14.4	91	45.3	42	20.9	18	9.0	3.03	1.06	Moderate
C2		23	11.4	53	26.4	89	44.3	24	11.9	12	6.0	2.74	1.01	Moderate
C3		2	1.0	3	1.5	30	14.9	63	31.3	103	51.2	4.30	0.84	High
	Social Networking Tools													
C4		3	1.5	2	1.0	17	8.5	67	33.3	112	55.7	4.40	0.80	High
C5		0	0	3	1.5	13	6.5	41	20.4	144	71.6	4.62	0.67	High
C6		2	1.0	6	3.0	17	8.5	51	25.4	125	62.2	4.44	0.84	High
C7		71	35.3	27	13.4	73	36.3	14	7.0	16	8.0	2.38	1.25	Moderate
C8		2	1.0	11	5.5	20	10.0	52	25.9	116	57.7	4.33	0.93	High
C9		73	36.3	32	15.9	68	33.8	13	6.5	15	7.5	2.32	1.23	Low
C10		127	63.2	24	11.9	36	17.9	9	4.5	5	2.5	1.71	1.06	Low
	Cloud Information Storage Tools													
C11		51	25.4	39	19.4	66	32.8	32	15.9	13	6.5	2.58	1.20	Moderate
C12		23	11.4	29	14.4	46	22.9	55	27.4	48	23.9	3.37	1.30	Moderate
C13		20	10.0	19	9.5	40	19.9	51	25.4	71	35.3	3.66	1.31	Moderate
C14		1	0.5	3	1.5	19	9.5	36	17.9	142	70.6	4.56	0.76	High
C15		22	10.9	21	10.4	61	30.3	48	23.9	49	24.4	3.40	1.26	Moderate

Legend: N = never; S = sometimes; O = occasionally; F = frequent; VO = very often.

The focus of analysis for item 2 is the use of slideshows in the classroom teaching process and during facilitation (TPF). This statement is at a moderate level (M = 2.74 and SD = 1.01). Eighty-nine people, equivalent to 44.3 percent of students, voted occasionally, and only twelve people, equivalent to 6.0 percent, used slideshows often. This result shows that most teachers do not use slideshows frequently, and this may be because students also will not be directly involved in its application. Therefore, it is appropriate that these slideshows are at a moderate level of use. Next, the analysis of item 3 shows that teachers use various multimedia materials (video presenter, LCD projector, and many others), and this statement is high (M = 4.30 and SD = 0.84). A total of 103 people, equivalent to 51.2 percent, voted very often, while only 2 people, equivalent to 1.0 percent, voted never. By examining a selection of answers, teachers often use these multimedia materials during TPF sessions and because students always want use it.

Moreover, the next sub-construct is a social networking tool. In item 4, the use of the WhatsApp application for forming class groups showed that it was at a high level at 112 people, equivalent to 55.7 percent, and they all voted very often with respect to the application's utilization. There were only two people, equivalent to 1.0 percent, that occasionally used it. From the selection of these answers, it can be concluded that the students are familiar with the WhatsApp application, which causes them to use it very often to connect with their friends and teachers by communicating through existing groups. Item 5 uses the WhatsApp application to upload and download documents or information. This statement is at a high level (M = 4.62 and SD = 0.67). The number of students who voted for this statement reached 144 people, equivalent to 71.6 percent, while no student voted for never. As is well known, WhatsApp is important in one's life, especially as a student. Students can share information easily and quickly through this application. It is not surprising that this usage is at a high level.

The statement of item 6 uses the Facebook application to post information, ideas, and files. This is the highest level obtained (M = 2.38 and SD = 1.25). One hundred twenty-five people, equivalent to 62.2 percent, voted very often, while only two people, equivalent to 1.0 percent, voted never. This result shows that Facebook is also the student's choice in presenting and finding information about their learning. The analysis for item 7 involved using the Telegram application to communicate with friends or teachers. This application is only at a moderate level (M = 2.38 and SD = 1.25). The choice of answers for occasionally is the highest, with seventy-three people, equivalent to 36.3 percent, having voted. At the same time, only fourteen people, equivalent to 7.0 percent, are said to use it regularly. By a selection of answers, it is not surprising that this application is only at a moderate level because students do not use this Telegram application.

The analysis of item 8 used the Facebook application to follow the relevant groups of the subjects taken. The average level for this statement is high (M = 4.33 and SD = 0.93). Students numbering 116 selected this statement, equivalent to 57.7 percent, while only 2 people, equivalent to 1.0 percent, had never used it. It is not surprising that this Facebook platform is prevalent among website users, and students use it to find information about their learning processes. Item 9 comprises the use of Instagram to communicate with teachers, and it is at a low level. Many students have never used it, amounting to seventy-three people and is equivalent to 36.3 percent, compared to those that use it often, which includes thirteen people, equivalent to 6.5 percent. This value is because Instagram is unsuitable for communicating with teachers. After all, this application requires high internet speeds. Therefore, most students never use it.

The analysis of item 10 refers using e-mail to send homework to teachers. The results of this analysis are low (M = 1.71 and SD = 1.06). The majority of students also voted never, which includes 127 people, equivalent to 63.2 percent, while only 5 people, equivalent to 2.5 percent, voted very often. The results of this answer selection prove that students are more motivated to submit homework in person than online using e-mails. The last sub-construct is a public-information storage tool. In item 11, the use of laptops to keep brief notes is at a moderate level (M = 2.58 and SD = 1.20). The number of students selecting this sub-construct comprised sixty-six people, equivalent to 32.8 percent, compared to those who selected very often, which included thirteen people, equivalent to 6.5 percent. Through this statement, students do not always use laptops frequently. Students may prefer traditional methods such as making brief notes in a notebook or paper.

Item 12 refers to using laptops to download files or documents related to a subject, and this item is also at a moderate level (M = 3.37 and SD = 1.30). The majority of students also chose once in a while, which included 55 people, equivalent to 27.4 percent, compared to never, which included 23 people, equivalent to 11.4 percent. This result clearly shows that students are also not interested and less interested in using laptops. The reason may be because the cost of owning laptops is high. The analysis of item 13 includes using a smartphone to download files or documents related to a subject. The mean level for this statement was moderate (M = 3.66 and SD = 1.31). The choice of the answer very often is the highest, including seventy-one people equivalent to 35.3 percent, compared to sometimes, which included nineteen people equivalent to 9.5 percent. Although a smartphone is an essential tool in one's daily routine, students do not like to download files or documents related to a lesson using a smartphone because the space required is insufficient, and there is not enough memory size in the phone.

The analysis of item 14 refers to using Google Drive to store all files in one place, and the average level is high (M = 4.56 and SD = 0.76). The majority of students chose very often, which includes 142 people, equivalent to 70.46 percent, compared to never, which included one person equivalent to 0.5 percent. Many students choose to use it because Google Drive does not require much space and memory. The app is also easy to store and can be easily searched for. It is not surprising that students love to use it. The last analysis is item 15, which involves using Google Photos as unlimited photo storage, and it was chosen at a moderate level. A total of sixty-one people, equivalent to 30.3 percent, voted occasionally, and only twenty-one people, equivalent to 10.4 percent, chose to use it occasionally. This application is less attractive to students because students are unaware of benefits of this application and have yet to explore its benefits. This item is at a modest level.

5. Discussion

This section discusses the findings for the second objective, which is to identify the use of learning media among form-six students. Findings examined with mean scores, standard deviations, and levels for the variables and components involved are at a moderate level (M = 3.46). This result means that the level of use of learning media for form-six students is moderate and good. This study also shows that the use of the WhatsApp application to form (group) classes is at the highest level, with a mean score of 4.62 and a standard deviation of 0.67. Students choose this application because they are used to it and use it on a daily basis. The app also serves as a liaison between students, peers, and teachers in terms of communications about learning or personal matters. Therefore, it is not surprising that this application is at the highest mean score.

However, it is different from e-mailing to send homework to teachers. The mean score obtained is 1.71, and the standard deviation is 1.06. This study showed that students did not use this method because they preferred to submit their homework face to face. Their teachers also do not use this method and leave students vulnerable to the use of e-mail. The study results also found that most students did not use e-mails in learning as they chose the disagree strongly answer. This learning media level shows that students do not use them fully. They prefer to use traditional methods, such as reference books and guided teaching, rather than teachers alone. Therefore, the objective results of this study are only at a moderate level based on the conducted study.

6. Conclusions

Students' focus and responsible attitudes during teaching and learning sessions or outside the classroom should always be considered. This is because a student's success depends on one's efforts. Students should use the sophisticated technology available today and utilize it as much as possible. This is because such a learning medium benefits students in their lessons. This learning media can have various positive effects if students use it better. This study can also be seen where most students use learning media in their learning. Therefore, it is hoped that this study can help all parties directly or indirectly guide educational organizations, especially the school. This is because this learning media is vital as a learning aid in the teaching and facilitation process in school and outside the classroom. Not to be outdone, in developing the country's education system, the ministry must also provide technology and infrastructure that can meet the needs of the field of education in Malaysia. More generations are IT literate and creative, and innovative in the future.

Author Contributions: Conceptualization, N.N., F.H., and H.M.; methodology, N.N., F.H., and H.M.; software, N.N., F.H., and H.M.; validation, N.N., F.H., H.M., M.H., and Y.S.; formal analysis, N.N., F.H., H.M., M.H., Y.S., E.K., and N.K.; investigation, N.N., F.H., and H.M.; resources, N.N., F.H., and H.M.; data curation, N.N., F.H., and H.M.; writing—original draft preparation, N.N., F.H., and H.M.; writing—review and editing, N.N., F.H., H.M., M.H., Y.S., E.K., and N.K.; visualization, N.N.; supervision, N.N. and F.H.; project administration, N.N. and F.H.; funding acquisition, N.N. and F.H. All authors have read and agreed to the published version of the manuscript.

Funding: This research received no external funding.

Institutional Review Board Statement: Not applicable.

Informed Consent Statement: Not applicable.

Data Availability Statement: Not applicable.

Conflicts of Interest: The authors declare no conflict of interest.

References

1. JPN Kedah. Pembelajaran Abad ke 21 (PAK21). Available online: http://jpnkedah.moe.gov.my/images/files/Surat_Hebahan/Pembelajaran_abad_21.pdf (accessed on 24 April 2022).
2. Sidin, R.; Mohamad, N.S. ICT dalam pendidikan: Prospek dan cabaran dalam pembaharuan pedagogi. *J. Pendidik.* **2007**, *32*, 139–152.
3. Mahat, H.; Hashim, M.; Saleh, Y.; Nayan, N.; Norkhaidi, S.B. Environmental Sustainability Knowledge, Attitude and Practices among Pre-school Students. *IOP Conf. Ser. Earth Environ. Sci.* **2019**, *286*, 012003. [CrossRef]

4. Mahat, H.; Hashim, M.; Saleh, Y.; Nayan, N.; Norkhaidi, S.B. A confirmatory factor analysis of Malaysian primary school students' energy saving practices. *J. Turk. Sci. Educ.* **2018**, *15*, 51–63. [CrossRef]
5. Mahat, H.; Hashim, M.; Norkhaidi, S.B.; Nayan, N.; Saleh, Y.; Hamid, N.; Baharudin, N.H.; Mohd Faudzi, N.A. The readiness of geography teacher trainees in gamification approach. *Rev. Int. Geogr. Educ. Online* **2021**, *11*, 720–734.
6. Nayan, N.; Mahat, H.; Hashim, M.; Saleh, Y.; Norkhaidi, S.B. Climate literacy awareness among preservice teachers in Malaysia. *Cakrawala Pendidik.* **2020**, *39*, 89–101. [CrossRef]
7. Mohd Yusof, M.N.; Tahir, Z. Kepentingan penggunaan media sosial teknologi maklumat dalam pendidikan IPTA. *J. Soc. Sci. Humanit.* **2017**, *12*, 1–10.
8. Hassan, A.; Abd Rahman, F.; Yew, S.K. Meneroka pemikiran logik melalui penggunaan aplikasi mudah alih. *Int. J. Educ. Train.* **2015**, *1*, 1–7.
9. Setyowati, Y. Mobile comics' as a media to enhance literature's sense to young learners. In Proceedings of the 61st TEFLIN International Conference: English Language Curriculum Development: Implications for Innovations in Language Policy and Planning, Pedagogical Practices, and Teacher Professional Development, Solo, Indonesia, 7–9 October 2014; pp. 1024–1026.
10. Manap, F.F.; Karim, A.A. Pengaplikasian media sosial sebagai alternatif pengajaran dan pembelajaran dalam kalangan pelajar pengajian tinggi. In Proceedings of the 2nd International Conference on Learning Innovation and Quality Education, Solo, Indonesia, 28 October 2017; pp. 1–11.
11. Vebrianto, R.; Osman, K. Keberkesanan penggunaan pelbagai media pengajaran dalam meningkatkan kemahiran proses sains dalam kalangan pelajar. *J. Pendidik. Malays.* **2012**, *37*, 1–11.
12. Krejcie, R.V.; Morgan, D.W. Determining sample size for research activities. *Educ. Psychol. Meas.* **1970**, *30*, 607–610. [CrossRef]
13. Anuar, N.I. *Kesedaran Penggunaan Aplikasi Web 2.0 Dalam Kalangan Pelajar Geografi di Universiti Pendidikan Sultan Idris*; Final Year Project; Faculty of Human Sciences, Universiti Pendidikan Sultan Idris: Perak, Malaysia, 2019.
14. Tasir, Z.; Abu, M.S. *Analisis Data Komputer SPSS 11.5 for Windows*; Venton Publishing: Kuala Lumpur, Malaysia, 2003; ISBN 978-98-3266-1214.
15. Khalid, F. Permohonan Menjalankan Kajian di Institusi di Bawah KPM. Available online: https://www.farizakhalid.com/researchsupport (accessed on 24 April 2022).
16. Landell, K.H.; Withrow, D. *Management by Menu*, 4th ed.; Wiley and Sons Inc.: London, UK, 1997; ISBN 978-04-714-7577-4.
17. Johnson, D.W.; Johnson, R. *Learning Mathematics and Cooperative Learning: Lesson Plans for Teachers*; Interaction Book Company: Edina, MN, USA, 1991; ISBN 978-09-3960-3138.
18. Johnson, D.W.; Johnson, R. *Cooperative Learning Lesson Structures*; Interaction Book Company: Edina, MN, USA, 1991; ISBN 978-09-3960-3169.

proceedings

Proceeding Paper

Analysis of Speech Acts in COVID-19-Related Facebook Comments †

Nor Syamimi Iliani Che Hassan * and Nor Hairunnisa Mohammad Nor

Akademi Pengajian Bahasa, Universiti Teknologi MARA Kelantan Branch, Machang 18500, Malaysia
* Correspondence: syami@uitm.edu.my

† Presented at the International Academic Symposium of Social Science 2022, Kota Bharu, Malaysia, 3 July 2022.

Abstract: As one of the most popular social networking sites, people use Facebook to get the latest information related to the COVID-19 pandemic. Driven by this trend and the motivation to address the gap in studies on speech acts on Facebook, we sought to examine Facebook users' comments about COVID-19 by identifying the speech acts used to communicate users' thoughts in their comments. These were direct comments posted in response to COVID-19-related posts uploaded by the Malaysian National Security Council on their official Facebook account. Our analysis revealed that five speech acts were used to communicate specific purposes intended for two groups of recipients: the government and citizens.

Keywords: pragmatic; speech act; Facebook

check for updates

Citation: Che Hassan, N.S.I.; Mohammad Nor, N.H. Analysis of Speech Acts in COVID-19-Related Facebook Comments. *Proceedings* **2022**, *82*, 50. https://doi.org/10.3390/proceedings2022082050

Academic Editor: Mohamad Rahimi Mohamad Rosman

Published: 15 September 2022

1. Introduction

Starting at the end of 2019, the world was inundated with the sudden wave of the COVID-19 virus, which negatively affected people in many aspects of life. Despite that, the Internet use was increasing as technology facilitated online teaching and learning, work-from-home practices, online shopping and online communication [1–4]. Apart from these activities, the pandemic has contributed to the increase in the use of social media among Internet users such as Facebook, Instagram, Line, WhatsApp and YouTube for at least half an hour to three hours daily [5]. Indeed, social media has become an essential medium for obtaining and sharing COVID-19 information [5,6]. However, this has led to a more serious issue when social media have propagated misinformation, rumors and theories of conspiracy about the virus in the virtual world [7,8]. Although the validity of such information cannot be substantiated, there are individuals who continue to circulate the information despite not trusting its truth [9]. Looking at the adverse impact of unverified information being spread across social media, the present study was conducted to examine the content of online posts uploaded by Internet users in response to COVID-19 information. Specifically, we aimed to address these two research questions: (1) What were said by Internet users in their comments about COVID-19? (2) Who were these comments addressed to? These research questions were addressed from the speech act perspective.

2. Types of Speech Acts

A typical interaction usually involves a speaker, a hearer and an utterance that is associated with various types of acts. These acts are referred to as speech acts, the minimal units of linguistic communication intended to convey the specific functions or intentions of the speaker [10,11]. Ref. [12] distinguishes speech acts as consisting of the trichotomy of locutionary, illocutionary and perlocutionary acts. Locutionary act refers to the production of an utterance, illocutionary act is the intended act done through an utterance which can be a promise, request, suggestion and many more, and perlocutionary act is associated with the causal relation between an utterance and its causal effect on the hearer. According to [13], a speaker who speaks a language is already performing a speech act; thus, all linguistic communication is hypothesized to involve linguistic acts. He challenged Austin's illocutionary act by classifying it as:

1. Assertives–An act that makes a hearer form or attend to the belief that the speaker is committed to a certain belief (e.g., prediction, notification, confession, denial).
2. Directives–An act that gets a hearer to do things through requests for action and requests for information (e.g., command, suggestion, question).
3. Commissives–An act that commits a speaker to a future action (e.g., promise).
4. Expressives–An act that expresses feelings toward a hearer.
5. Declaratives–An act that brings about effects of immediate changes in an institutional state of affairs. It usually carries elaborate extra-linguistic institutions.

A further distinction of speech acts is offered by [14] who distinguishes declarative speech acts as effectives and verdictives.

6. Effectives–An act that is meant to change an institutional state of affairs.
7. Verdictives–An act that determines what is to be the case within the institution.

3. Speech Acts on Social Networking Sites

Internet users communicate virtually using numerous social networking sites and applications that involve various speech acts that are interesting to be researched. Studies on text messages posted by Facebook users in their personal accounts or group accounts show that they were composed using various speech acts. Most of these studies sourced their data from Facebook users' status updates [15–20] and only a couple of studies analyzed speech acts in users' comments [15,21]. Of these seven studies, four selected status updates and comments that were narrowly focused, namely, university-related topics [18], apostasy [19], politics [15] and expressives [16]. Except for two studies, analyses of the five studies were based on Searle's speech acts. For the studies based on [13] typology of speech acts, they commonly found that expressive, assertive, directive speech acts were more frequently employed than others. Commissive speech acts were similarly found to be the least frequent in Facebook users' posts.

These studies have shown that texts posted by Internet users in or through their Facebook accounts reflect the speech acts commonly found in face-to-face interactions. However, most of the studies collected their data from Facebook users' status updates, although Facebook users can also write their posts in other ways (e.g., write comments in response to other Facebook users' posts and comments). Our summary of past studies also indicated that the online posts analyzed in the studies were random and not written about specific topics or issues. Hence, we aimed to examine Facebook users' online posts, particularly their comments toward posts about COVID-19 uploaded to a government agency's official Facebook page.

4. Materials and Methods

4.1. Data Sampling

The study corpus consisted of 555 Facebook comments collected using the purposive sampling method. They were posted in response to daily reports of COVID-19 cases on the Facebook page of the Malaysian National Security Council in July and August 2021. The Malaysian National Security Council is a government agency responsible for the management and coordination of security-related policies, such as safety issues related to COVID-19 [22]. The agency fully utilizes its Facebook page by posting current updates about the pandemic on its social network. The agency's Facebook page was selected as the basis for our corpus, as it played an active role in sharing essential information with the public on a daily basis and the information shared received direct feedback from its 1.1 million followers. The comments recruited as our research sample were those written in response to the Malaysian National Security Council's Facebook posts in July and August 2021. We selected these two months because SOP relaxation was administered by the

Malaysian government after a series of movement control orders were enforced in the country. Many feared that the ease of SOP could exacerbate the spread of COVID-19. Data were selected based on the main inclusion criterion–each comment was sent as a direct response to the Malaysian National Security Council's Facebook posts. Comments were otherwise excluded if they were sent to another comment (referred to as indirect comments in this study) and contained statements that were irrelevant to the daily reports. Every comment was collected using the copy-paste method in a .doc file. There were originally 570 comments recruited into the corpus; however, 15 were irrelevant to the Facebook posts. All in all, there were 555 direct comments that were written in English and Malay languages, 32 and 523 comments, respectively.

4.2. Data Codification

We coded the data in several rounds by addressing one question at a time using ATLAS.ti version 22 as our main coding software. This software which was developed by Thomas Muhr and assisted by Jörg Hecker, can be directly downloaded from the website: https://atlasti.com/. The main advantage of using ATLAS.ti as a tool for computer-assisted qualitative data analysis (CAQDAS) is that it eases data management and data retrieval compared to performing manual data analysis. Nonetheless, our primary role in identifying the meanings and determining and applying the codes to the data was not substituted by the software because it only functioned as a tool to support our analysis process [23]. Both researchers were involved in the coding process, primarily to identify the types of speech acts in the comments, guided by the analytical framework developed for this study, adapted from [24]. The analytical framework used in this study is shown in Table 1.

Table 1. Framework of speech act analysis.

Speech Act	Speech Act Properties
Assertive	Statements of fact, getting the viewer to form or attend a belief
Directive	The sender uses this to get the receiver to do something
Commissive	The sender commits himself to do something
Expressive	Sender expresses feeling toward (though not necessarily about) the receiver
Effective	To change an institutional state of affairs
Verdictive	To determine what is the case in an institution
Quotation	The message is not originally produced by the sender

5. Results

5.1. Frequency of Speech Acts in Facebook Comments

The following chart delineates the frequency of five types of speech acts that occurred in the Facebook comments to the daily statistics about COVID-19 cases in Malaysia, as reported on the official Facebook page of the Malaysian National Security Council. Assertives, expressives and directives were the most frequent speech acts employed by the Facebook users, while the other two speech acts—commissives and quotations—were found to account for only one per cent each in the comments. Figure 1 shows the speech act distribution found in the study.

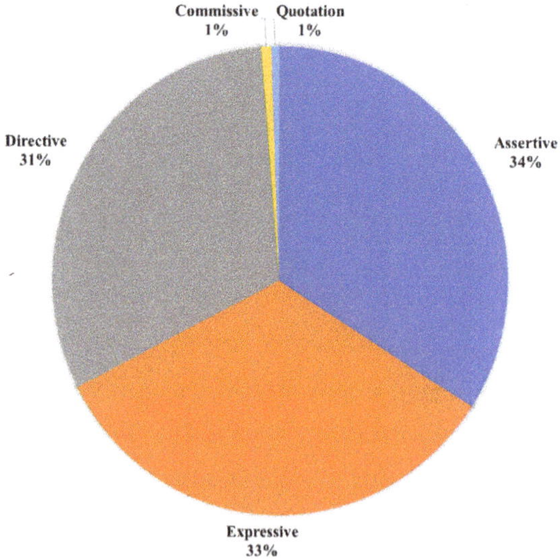

Figure 1. Frequency of speech acts in Facebook comments.

5.2. Assertive Speech Act

Assertive speech acts (34%) were the most frequently employed speech acts by Facebook users to write comments. Based on our findings, there were four ways in which this speech act was used in the comments. The four functions of assertive speech acts are summarized as follows:

1. To make statements about the National COVID-19 Immunisation Programme. 16 juta daripada 60+-juta jumlah dos. (English translation: 16 million out of 60+-million number of doses.)
2. To share information obtained from other sources. Semalam saya ada mendengar penerangan yang sangat jelas dan padu daripada Pakar Virologi yang berpengalaman lebih 30 tahun, Dr [name] mengenai COVID-19 dan kaedah menanganinya. (English translation: Yesterday I listened to a very clear and comprehensive explanation from an expert in virology with more than 30 years of experience, Dr [name], related to COVID-19 and ways to combat it.)
3. To make statements about new cases related to the COVID-19. Mti msih lbih dr 200 ye (English translation: There are still more than 200 deaths) To assert controversial remarks about the National COVID-19 Immunisation Programme saya rasa tembak air, takkan makin banyak vaksin makin banyak case (English translation: I think (they) injected plain water, how come with more (people) vaccinated there are more cases).

5.3. Expressive Speech Act

The second most frequent speech act in our corpus was the expressive speech act (33%). Because this speech act is performed by speakers to express their emotions, we found that the Facebook users in our corpus used the expressive speech act to convey a range of emotions, from positive to negative, which were triggered by the alarming rate of COVID-19 infection. Thus, the expressive speech acts in the Facebook comments were classified according to the following five functions:

1. To express shock.
   OMG: the death rate
2. To express worry and sadness.

Dua bulan ku duk umah dah,makin meningkat kes harian ... aaahhhh ku sedih (English translation: I stayed at home for two months, daily cases are increasing ... aaahhhh I am sad)

3. To express frustration and anger.
Begini kawal keselamatan negara! Mengecewakan! (English translation: Is this how to control the country's safety! Frustrating!)
4. To express sarcasm and skepticism.
Terbaik..makin ramai di vaksin makin naik kes.pelik bin ajaib. (English translation: Excellent..more are vaccinated there are more cases, utterly weird.)
5. To express hope.
Rabbi yassir wala tu'assir rabbi tammim bil khair Ya Allah permudahkanlah jangan dipersulitkan dan Kau akhirkanlah dengan kebaikan. (English translation: Oh Allah, make this task easy and do not make it difficult. Oh, Allah! Make it end well.)

5.4. Directive Speech Act

Of the speech acts found in our corpus, 31% were composed of directive speech acts. For this speech act, we identified two target recipients of the directives in our corpus, i.e., the government and the public (fellow Facebook users and fellow Malaysians). When the speech acts were directed at the government, their functions were to advise the government about dissemination of COVID-19-related information, to advise the government about The National COVID-19 Immunisation Programme, to advise the government to adapt COVID-19 management of other countries, to advise the government to enforce a stricter law on the movement control order and to request relaxation on the movement control order. Through this speech act and its specific functions, Facebook users appeared to reflect a negative attitude toward the government's efforts to curb the proliferation of the virus across the country. Therefore, their directive speech acts were most probably motivated by their intention to protect society for collective well-being. Another function of the directive speech act was to give advice to their fellow Malaysians. In contrast to the directive speech acts intended for the government, which were realized in their comments due to their dissatisfaction with the government's efforts, Facebook users who employed directive speech acts in their messages for fellow Malaysians did so because they were concerned about each other's well-being.

1. To advise the government about the dissemination of COVID-19-related information.
Pls publish these info daily instead of the usual one from MOH.
2. To advise the government about the National COVID-19 Immunisation Programme.
Suntik je lah vaksin Dari umah ke rumah ... mcm banci penduduk ... jd kurangla risiko jangkitan semasa mengambil vaksin (English translation: Just administer home to home vaccination ... like population census ... so, lesser risk of infection while getting vaccinated)
3. To advise the government to adapt COVID-19 management to other countries.
apakata mkn hantar mana2 menteri pergi wuhan n belajar handle covid case. (English translation: Why not mkn deploy any ministers to wuhan and learn how to handle covid cases.)
4. To advise the government to enforce a stricter law on the movement control order.
minta kerajaan serta MKN lockdown semua sektor untuk 2 minggu supaya kes turun serta vaksin dipercepatkan ... Takkan nak tunggu makin teruk ... (English translation: hope that the government and MKN lockdown all sectors for 2 weeks so that new cases will decrease and vaccination can be sped up ... we should not wait until it gets worse...)
5. To request the government for relaxation of the movement control order.
MKN boleh dah kot buka sektor bekerja sendiri,sebagai contoh tukang persendirian,dah dekat 3 bulan projek rumah org tak siap..kesian dekat tuan rumah bersama tukang persendirian. (English translation: Perhaps MKN could now re-open self-employed

sector like builder, nearly 3 months housing project was left incomplete pity the house owners and the builders.)

6. To advise fellow Malaysian citizens about their personal well-being.
   Terus jaga SOP, keluar bila perlu shj. #stayathome (English translation: Keep abiding by the SOP, go out only when necessary. #stayathome)

5.5. Commissive and Quotation Speech Acts

Besides the three very frequently occurring speech acts, our corpus also had instances of commissive and quotation speech acts, which occurred at a much lower frequency than those three. The commissive speech act was often used to warn the government out of their disappointment with the increasing number of COVID-19 cases.

7. Aku tak maw mngundi da smpai bila2 . . . (English translation: I'm not going to vote ever. . .)

As for the quotation speech act, the quotations were adopted from the widely used tagline derived from the Malaysian government's campaign to increase awareness among Malaysians about the importance of safeguarding themselves and their loved ones against the COVID-19.

8. #stayathome

5.6. Target Recipients of the Speech Acts

Interestingly, we found that the comments posted by the users were intended for two groups of recipients. One of the target recipients was the government, which formed the large majority of the recipients, while the other was fellow citizens. When the speech acts were directed at the government, it was obvious that the users blamed the government for the perceived failure of COVID-19 management. Through the speech acts, blaming was expressed overtly by stating the recipient as the government, "kerajaan" in general or as a specific organization, "Majlis Keselamatan Negara" (English equivalent: National Security Council). In some cases, the organization was mentioned in an abbreviated form, such as "KKM," which stands for Kementerian Kesihatan Malaysia (English equivalent: Malaysian Ministry of Health). There were also cases in which the government or its related organizations were not mentioned, but through a closer scrutiny of the speech acts, we could understand that the users were trying to blame the government for the seemingly weak government enforcement of the movement control order.

When the comments were targeted at fellow citizens, the terms of address used were 'guys', 'korang' (English equivalent: you), 'kita' (English equivalent: we), 'sesiapa' (English equivalent: anyone) and 'semua' (English equivalent: all of you), which were meant for any public in general. The use of 'kita' (English equivalent: we) indicates solidarity among Facebook users, especially in coping with the pandemic. There was an instance that addressed the recipients very specifically, "ibu mengandung," to address expecting or pregnant mothers. Other than these, the target recipients were embedded in the speech acts. In other words, they were not clearly addressed but could be identified through a close examination of the comments.

Although a large part of the blaming was directed at the government, there was still a small number of users who blamed their fellow citizens and attributed the reason for the spike in COVID-19 cases to their faults. These people were specifically addressed in the comments as "Pembawa virus senyap" (English translation: Silent virus carriers), "Cluster COVID Perhimpunan Baju Hitam" (English translation: COVID Cluster of the Black Shirt Assembly) and "kluster raya" (English translation: the raya cluster)–raya refers to the Eid al-Fitr, a festival celebrated by Muslims after the fasting month of Ramadhan.

6. Discussion

The study sought to examine what Facebook users do through their comments on the COVID-19 report posted on the official Facebook page of the Malaysian National Security

Council. Our analysis revealed that the users communicated their intent in various ways, which can be organized in different types of speech acts. The three most frequent types of speech acts through which the users communicated their intent in a textual form were assertive, expressive and directive. When applied in online communication, specifically one that transpires via social media such as Facebook, an assertive speech act is employed by a user to state a belief regardless of its truth [17]. Through our findings, this speech act was intended to state facts and opinions regarding the COVID-19 specifically about the National COVID-19 Immunisation Programme and statistics of new COVID-19 cases, vaccinated individuals and deaths. When these statements were posted on Facebook, especially those that lacked validity, it suggested that Facebook is also utilized as a medium to communicate one's personal assumptions and rumors with others in a virtual form. Social media provide an open platform for Internet users to share news, stories, personal experiences and viewpoints [25]; therefore, people can simply post unverified statements and information that can be easily circulated by others using the same or different platforms. Expressive speech acts are realized by Facebook users to express their psychological states on the online platform [17]. Our study showed that users expressed a variety of positive and negative emotions through expressive speech acts. The platform was used to convey their emotions, which were affected by the COVID-19. Furthermore, Facebook is utilized as a means to express one's positive and negative emotions, as it is considered a more appropriate medium to do so compared to other social networking sites such as Twitter and Instagram [26]. As for the directive speech act, our study found that the advice, commands, suggestions and requests were addressed to the Malaysian government and Malaysian citizens. By employing directive speech acts in their comments, the users seemed to exhibit a certain level of personal knowledge about COVID-19 and proper or practical ways to handle the issue to the two addressees. Because everyone had dealt and lived with COVID-19 for more than a year, through the speech acts, these users showed that they were experienced and well informed about COVID-19 management implemented in the country and other foreign countries. Furthermore, Malaysians were already taking precautions on their own, even before the movement control order was enforced nationwide [27]. Addressing the second research question, our analysis revealed that the comments were intended for two groups of recipients, i.e., the Malaysian government and Malaysian citizens. In their comments, the Facebook users also appeared to blame these two addressees, rather than themselves, if there was yet another alarming COVID-19 outbreak in the community due to the loopholes in the movement control order. The pandemic, which had continued for many years, triggered fear among people. Thus, it is a natural reaction to first ascribe the fault to others instead of to oneself. In other cases, people even blamed citizens of other countries and international travelers for COVID-19 upsurges [28,29]. Users might perceive themselves as self-disciplined and law-abiding, but blaming others as the causes of any COVID-19 pandemic reflects collectivity and solidarity with the Malaysian community to fight against selfless, individualistic people [29].

7. Conclusions

The present study has attempted to scrutinize Facebook users' comments sent in response to daily reports of COVID-19 cases in Malaysia uploaded by the Malaysian National Security Council on its official Facebook page, while the country was undergoing one of the phases of movement control order. Through the study, Facebook users were seen to employ at least five types of speech acts–assertive, expressive, directive, commissive and quotation. We learned that the Facebook users had a specific target audience to address their comments to and these were stated either in a covert or overt form. Although our study did attempt to fill the gap based on past studies conducted in a similar context, there are several limitations that should be addressed by future studies. First, the study collected comments that were mostly written in the Malay language and only a small number of comments posted in English. We suggest that a comparative study be conducted on online speech acts realized in these two languages. Second, the study only focused on the functions of each speech act without examining the linguistic realization of the speech acts, which should be considered by future studies. Third, the study was mostly reported in a qualitative form, with only a little report of the quantitative findings. Hence, future studies are recommended to include quantitative findings, such as statistically significant differences in the data. Online communication has become even more frequent since the development of more advanced technology. Considering this trend, more studies on speech acts in online communication should be conducted, as it is a blend of written and spoken communication that could produce new themes or sub-themes under the existing speech acts.

Author Contributions: Conceptualization, N.S.I.C.H. and N.H.M.N.; writing—literature review, N.H.M.N.; data collection and analysis, N.S.I.C.H.; proofreading and editing, N.H.M.N. All authors have read and agreed to the published version of the manuscript.

Funding: This research received no external funding.

Institutional Review Board Statement: Not applicable.

Informed Consent Statement: Not applicable.

Data Availability Statement: Not applicable.

Acknowledgments: We would like to thank the inter-raters who helped to validate the findings.

Conflicts of Interest: The authors declare no conflict of interest.

References

1. Dhawan, S. Online learning: A panacea in the time of COVID-19 crisis. *J. Educ. Technol. Syst.* **2020**, *49*, 5–22. [CrossRef]
2. Ambikapathy, M.; Ali, A. Impact and challenges towards employees work from home during COVID-19 (MCO) period. *Int. J. Soc. Sci. Res.* **2020**, *2*, 97–107.
3. Chan, X.Y.; Rahman, M.K.; Al Mamun, A.A.; Salameh, W.M.; Hussain, H.W.; Alam, S.S. Predicting the Intention and Adoption of Mobile Shopping during the COVID-19 Lockdown in Malaysia. *Sage Open* **2022**, *12*, 21582440221095012. [CrossRef]
4. Nguyen, M.H.; Gruber, J.; Fuchs, J.; Marler, W.; Hunsaker, A.; Hargittai, E. Changes in digital communication during the COVID-19 global pandemic: Implications for digital inequality and future research. *Soc. Media Soc.* **2020**, *6*, 2056305120948255. [CrossRef]
5. Saud, M.; Mashud, M.; Ida, R. Usage of social media during the pandemic: Seeking support and awareness about COVID-19 through social media platforms. *J. Public Aff.* **2020**, *20*, e02417. [CrossRef]
6. Han, X.; Wang, J.; Zhang, M.; Wang, X. Using social media to mine and analyze public opinion related to COVID-19 in China. *Int. J. Environ. Res. Public Health* **2020**, *17*, 2788. [CrossRef]
7. Radwan, E.; Radwan, A. The spread of the pandemic of social media panic during the COVID-19 outbreak. *Eur. J. Environ. Public Health* **2020**, *4*, em0044. [CrossRef]
8. Tasnim, S.S.; Hossain, M.; Mazumder, H. Impact of rumors and misinformation on COVID-19 in social media. *J. Prev. Med. Public Health* **2020**, *53*, 171–174. [CrossRef]
9. Long, V.J.E.; Koh, W.S.; Saw, Y.E.; Liu, J.C.J. Vulnerability to rumours during the COVID-19 pandemic in Singapore. *Ann. Acad. Med. Singap.* **2021**, *50*, 232–240. [CrossRef]
10. Olshtain, E.; Cohen, A. The learning of complex speech act behaviour. *TESL Can. J.* **1990**, *7*, 45–65. [CrossRef]
11. Searle, J.R. What is a speech act. In *Perspectives in the Philosophy of Language: A Concise Anthology*; Stainton, R.J., Ed.; Broadview Press: Peterborough, ON, Canada, 1965; pp. 253–268.
12. Austin, J. *How to Do Things with Words*; Oxford University Press: Oxford, UK, 1962.
13. Searle, J.R. *Speech Acts: An Essay in The Philosophy of Language*; Oxford University Press: Oxford, UK, 1969.
14. Clark, H. *Using Language*; Cambridge University Press: Cambridge, UK, 1996.
15. Tsoumou, J.M. Analysing speech acts in politically related Facebook communication communication. *J. Pragmat.* **2020**, *167*, 80–97. [CrossRef]
16. Syafitri, W. Expressive speech acts in the statuses of the Facebook. *PARADIG J. Lang. Lit. Stud.* **2020**, *3*, 1–10. [CrossRef]
17. Banikalef, A.A.A. The impact of culture and gender on the production of online speech acts among Jordanian Facebook users. *Int. J. Arab. Stud.* **2019**, *19*, 395–410. [CrossRef]
18. Nartey, M. A speech act analysis of status updates on Facebook: The case of Ghanaian university students. *Lang. India* **2013**, *13*, 114–141.
19. Mohamad, A.; Ab Rashid, R.; Yunus, K.; Rahman, S.B.A.; Darus, S.; Musa, R.; Teh, K.S.M. Speech acts in the Facebook status updates posted by an apostate. *Int. J. Engl. Linguist.* **2018**, *8*, 226–231. [CrossRef]
20. Illyas, S.; Khushi, Q. Facebook status updates: A speech act analysis. *Acad. Res. Int.* **2012**, *3*, 500–507.

21. Nisa, K.; Manaf, N.A. Analysis of illocutionary speech acts on student's social media post and comments. *Seloka J. Pendidik. Bhs. Sastra Indones.* **2021**, *10*, 62–67.
22. Majlis Keselamatan Negara. SOP Pelan Pemulihan Negara (PPN). Available online: https://www.mkn.gov.my/web/ms/sop-perintah-kawalan-pergerakan/ (accessed on 31 May 2022).
23. Paulus, T.; Woods, M.; Atkins, D.P.; Macklin, R. The discourse of QDAS: Reporting practices of ATLAS.ti and NVivo users with implications for best practices. *Int. J. Soc. Res. Methodol.* **2017**, *20*, 35–47. [CrossRef]
24. Carr, C.T.; Schrock, D.B.; Dauterman, P. Speech acts within Facebook status messages. *J. Lang. Soc. Psychol.* **2012**, *31*, 176–196. [CrossRef]
25. Zubiaga, A.; Liakata, M.; Procter, R.; Hoi, G.W.S.; Tolmie, P. Analysing how people orient to and spread rumours in social media. *PLoS ONE* **2016**, *11*, e0150989. [CrossRef]
26. Waterloo, S.F.; Baumgartner, S.E.; Peter, J.; Valkenburg, P.M. Norms of online expressions of emotion: Comparing Facebook, Twitter, Instagram, and WhatsApp. *New Media Soc.* **2018**, *20*, 1813–1831. [CrossRef]
27. Azlan, A.A.; Hamzah, M.R.; Sern, T.J.; Ayub, S.H.; Mohamad, E. Public knowledge, attitudes and practices towards COVID-19: A cross-sectional study in Malaysia. *PLoS ONE* **2020**, *15*, e0233668. [CrossRef] [PubMed]
28. Hardy, L.J.; Mana, A.; Mundell, L.; Neuman, M.; Benheim, S.; Otenyo, E. Who is to blame for COVID-19? Examining politicized fear and health behavior through a mixed methods study in the United States. *PLoS ONE* **2021**, *16*, e0256136. [CrossRef] [PubMed]
29. Labbé, F.; Pelletier, C.; Bettinger, J.A.; Curran, J.; Graham, J.E.; Greyson, D.; MacDonald, N.E.; Meyer, S.B.; Steenbeek, A.; Xu, W.; et al. Stigma and blame related to COVID-19 pandemic. *Soc. Sci. Med.* **2022**, *296*, 114803. [CrossRef] [PubMed]

proceedings

Proceeding Paper

Student Satisfaction on Teaching Strategies and Multimedia Usage in Online Arabic Language Course †

Nurul Asma Mazlan [1,*], Ijlal Saja [1], Nur Aqilah Norwahi [1], Nik Mahfuzah Nik Mat [2] and Mohd Faizulamri Mohd Saad [3]

1 Academy of Language Studies, Universiti Teknologi MARA, Jasin Campus, Merlimau 77300, Malaysia
2 Academy of Language Studies, Universiti Teknologi MARA, Negeri Sembilan Campus, Seremban 70300, Malaysia
3 Faculty of Islamic Studies, Universiti Kebangsaan Malaysia, Bangi 43600, Malaysia
* Correspondence: asmamazlan@uitm.edu.my
† Presented at the International Academic Symposium of Social Science 2022, Kota Bharu, Malaysia, 3 July 2022.

Abstract: The objective of this study is to examine student satisfaction based on online teaching strategies as well as multimedia usage in an Arabic language course. Based on a convenient sampling technique, the survey was distributed online to 213 students of UiTM Malacca branch in order to identify the factors contributing to student satisfaction. This descriptive and analytical study used quantitative methods using SPSS and a PLS-SEM procedure to analyze the relationship between variables. The results indicate a significant and positive relationship between student satisfaction and multimedia usage during Arabic language ODL sessions. Interestingly, the data analysis reported a significant relationship between teaching strategies and student satisfaction.

Keywords: online teaching strategies; multimedia usage; Arabic language; open and distance learning; student satisfaction

Citation: Mazlan, N.A.; Saja, I.; Norwahi, N.A.; Mat, N.M.N.; Saad, M.F.M. Student Satisfaction on Teaching Strategies and Multimedia Usage in Online Arabic Language Course. *Proceedings* **2022**, *82*, 51. https://doi.org/10.3390/proceedings2022082051

Academic Editor: Mohamad Rahimi Mohamad Rosman

Published: 15 September 2022

1. Introduction

Open and distance learning (ODL) has long been introduced and has been implemented in most Malaysian learning institutions, from preschool to tertiary education, since as early as the 1950s, and it was first offered by Stamford College [1]. Currently, this is an effective and much-needed approach to prevent coronavirus from spreading as it is a life-threatening pandemic [2]. According to UNESCO (2002) [3], ODL is a teaching and learning session that takes place beyond rigid time and space between instructors and students. The diversity and breadth of the scope covers broad aspects of access, curriculum as well as elements [1].

Therefore, in order to achieve student satisfaction in teaching and learning, various measures and strategies have been used by educators today [4–6]. Among the strategies adopted is to increase the use of integrated online multimedia [7]. Thus, many learning courses are conducted online including Arabic language courses at Universiti Teknologi MARA (UiTM). In short, Arabic language subjects are offered at UiTM as elective subjects that emphasize the mastery of four skills, namely reading, listening, writing and speaking [8,9].

To achieve integrated online student learning satisfaction is not as easy as expected due to many lecturers who are less skilled. Among the reasons students do not achieve a high level of satisfaction is because of the difficulty in understanding Arabic grammar. The Arabic grammar structure reportedly has a negative impact on student academic achievement [10], which is distinctively different from most students' first language, Malay language. In addition, students also face difficulties in adapting to online learning because it takes a long time and energy to be familiar with the various elements of multimedia. Moreover, different courses and lecturers may use different multimedia platforms and teaching strategies, causing students' focus to be interrupted. This then leads to an increase in student workload and decreases student satisfaction with their learning sessions. Undeniably, it is necessary to improve multimedia skills since it is also in line with the development of the current era of globalization. In addition, lecturers also need to devise various strategies so that students can master Arabic language skills well after the completion of the courses [11]. Other than that, lecturers also need to devise various strategies so that students can master Arabic language skills well. Therefore, the objectives of this study are to examine the influence of Arabic language teaching strategies on student satisfaction and also to review the influence of multimedia use on student satisfaction. This is imperative to determine a more effective and efficient learning environment, especially for the students to enhance their Arabic language skills.

2. Literature Review

2.1. Teaching Strategies

A variety of teaching strategies provide an optimal effect on Arabic language learning. Certainly, learning sessions at the university level differ from primary and secondary school levels. This is due to the lecturer as a facilitator to the students [11,12] and as a strategy to master the Arabic language, although in essence, the teaching approach is not much different for all levels of educational institutions. Nevertheless, it can be noted that the lecturer's strategy as a facilitator clearly encourages good moral growth to continue learning Arabic [12]. It also helps to continuously increase the students' learning efforts as it can foster independence and cooperation among students when the study sessions are conducted by lecturers [13].

In addition, the UiTM curriculum and syllabus prioritize basic language skills in order to facilitate students in learning the Arabic language from basic to advanced levels. This is related to the stages of human thoughts, which has been much debated by psychologists and philosophers. Humans have a unique mind with the ability to think from the basic level of knowledge up to evaluation. Previous studies have examined how student innovation development improves students' high-level thinking skills. Multimedia and innovation do not only allow student engagement but also create active interaction in the learning environment. It even generates pleasant and enjoyable learning sessions, especially during ODL [11,14].

Therefore, cooperative teaching is implemented in an orderly and structured manner through a combination of lectures and student activities. A more effective strategy practiced by lecturers is incorporating interactive games that help improve students' Arabic language skills [15]. Studies in the past have created space and opportunities for many lecturers and students to build games based on multimedia technology. This strategy has proved to have a more positive effect in improving Arabic language skills, where students are also able to acquire a quick response based on the game score and evaluation system that is automatically displayed to students. Thus, students can immediately acknowledge their level in Arabic language learning [7,16,17].

It is also crucial for lecturers to encourage the students to refer to other sources to improve their understanding on a topic in Arabic language. This is also reported by Haroon (2000) [18], who studied the strategies of mastering Arabic language through the reading of various sources other than relying solely on the modules provided to the students at the University of Malaya. It was found that students' assessment results improved significantly compared to students' results that referred to only one particular source.

Other than that, student–lecturer active discussion also helps to improve students' thinking skills. Through a study conducted, students would undergo discussion activities based on the materials prepared by the lecturer. Next, in small groups, students would discuss and present the results based on the discussion. This tremendously helped students to have a better understanding of the topics because the lecturers created an active learning session.

To ensure student satisfaction, methods such as simulation, role play and drills are also practiced in order to enhance basic Arabic language skills [19,20]. Previous studies have shown that learning Arabic at UiTM gave students the potential to learn Arabic better because it emphasized the elements of simulation, role-playing and drills. As a result, it was found that there was a significant relationship between activities during learning sessions and student satisfaction [19]. It was also reported that UiTM students were able to speak Arabic more confidently and fluently through simulation and role play activities during (ODL) [20]. Hence, simulation and role play methods positively contribute to the aspect of speaking skills while drills further strengthen students' mastery in learning Arabic language.

In general, lecturers need to be resourceful in strategizing and diversifying activities when integrating ODL and Arabic language learning. Student satisfaction evidently can be achieved if the activities incorporated are appropriate and effective to help students understand and master the Arabic language. Thus, a hypothesis for this study is that teaching strategies positively and significantly influence student satisfaction in learning Arabic.

2.2. Multimedia Usage

The subject syllabus content cannot be presented well without the use of teaching aids by lecturers. In a learning environment such as ODL, teaching aids based on multimedia technology greatly help student satisfaction in learning Arabic language. Since then, UiTM has intensified the efforts to facilitate lecturers and students to conduct teaching and learning more smoothly by developing the U-Future platform [21]. Previous studies have clarified that the U-Future platform is appropriate and up to date to assist ODL sessions. Reference materials can be directly accessed without the need to be downloaded as it will overwhelm laptop storage capacity. In addition, lecturers can also compile lecture plans according to the subject syllabus via this platform. Various types of audio, video, documents and question sets can be developed to facilitate students even beyond ODL sessions [21,22]. However, other than U-Future, other online platforms such as Google Classroom, Moodle, Edmodo and Microsoft Teams are also used to aid the teaching and learning sessions [22]. Thus, student satisfaction increases whenever ODL sessions are implemented in an orderly and systematic manner.

In addition, there are also other multimedia-based teaching aids for students to access continuously, such as the teaching videos developed by lecturers. Students with internet connection issues have the opportunity to access the video materials provided at any time with ease. Furthermore, video development accelerates the process of student comprehension as students have an array of different learning methods to suit different student learning styles. This has been proven in many past studies revealing that video is very suitable for language learning because it includes interesting pictures with sound as well as graphic features. As a result, students' interest in learning is also increased [16,23].

Moreover, due to the pandemic, Google Meet, Zoom, GoToMeeting and Cisco Webex platforms are also actively employed in almost all walks of life, including by lecturers and students. This is especially vital in language learning as these platforms can optimize two-way interaction between lecturers and students. It has been found that integrated online learning through platforms such as Google Meet, Zoom, and Cisco Webex are suitable for teaching and learning sessions involving discussion and question-and-answer activities. Through these platforms, lecturers can also demonstrate various interactive materials such as engaging Microsoft PowerPoint slides, audio and infographics [6,11,13,22,24].

Question-and-answer and interview sessions can also be conducted using applications such as WhatsApp, Telegram, Signal, WeChat, Line or Bip. Students who are embarrassed to ask during Google Meet, Zoom, GoToMeeting and Cisco Webex online meetings can also inquire through the app instead. This makes it easier for both lecturers and students to interact with ease. Other than that, it has been shown in past studies that WhatsApp and Telegram apps are most widely used during ODL sessions. This evidently shows that learning still happens continuously without limitations by using these apps as they can be easily accessed using smartphones [15,22,24].

On the other hand, the "less serious" online platforms such as Facebook, Twitter, Instagram and TikTok provide more widespread social media exposure, including in language learning. For instance, learning Arabic language should go beyond the university scope and students are able to explore Arabic holistically. Social media platforms provide students a chance to gain extensive knowledge, which benefits student's experience in learning Arabic language. It is clearly stated in a study that a social media platform, Facebook, is suitable to be applied in learning sessions as it is verily up to date [24].

Overall, teaching aids based on multimedia technology provide students with a higher level of satisfaction compared to only a rigid and traditional teaching and learning session. Contemporary and relevant teaching aids also assist ODL sessions in being a pleasant experience for both students and lecturers. Hence, another hypothesis of this study is the positive and significant influence of multimedia usage over student satisfaction in learning Arabic language.

3. Materials and Methods

The primary data for this descriptive and analytical study were obtained through questionnaires administered to the undergraduate degree students from three UiTM campuses in Melaka, which are Kampus Alor Gajah, Kampus Bandaraya as well as Kampus Jasin. A convenient sampling technique was used in this study and the survey was conducted online using Google Form to collect the responses. The entire population for this study was 460 students that came from five different faculties in UiTM Melaka, which include Faculty of Art and Design, Faculty of Business and Management, Faculty of Computer Science and Mathematics, Faculty of Hotel and Management and Faculty of Accountancy. The minimum sample size for this study was 129 students that have been identified primarily by referring to the GPower software (Aichach, Germany). Luckily, 213 students responded to the survey for this study. A seven-point Likert-based scale was adopted to represent the respondents' preference. The description of the Likert scale used included 1—Strongly Disagree to 7—Strongly Agree. The data were next analyzed using SmartPLS 3.0 software (Hamburg, Germany) to gather the responses [25].

Before data analysis was initiated, a data cleaning process involving several tests such as incomplete data detection and isolated data was performed. Of the 215 questionnaires distributed, 213 questionnaires were retained for the purpose of further analysis due to incomplete responses from the remaining two respondents. Next, the partial least square structured modeling (PLS-SEM) procedure was performed, which involved two stages of analysis, namely, the item measurement model as well as the structure measurement model. Through this software, data would be re-sampled to as many as 5000 samples to produce expected standard error and t-value [26,27]. SmartPLS 3.0 is able to assimilate sampling errors that can influence the value of correlation coefficients between the variables while improving the accuracy of theoretical tests [28]. This software will also ensure the convergent validity and discriminant validity of the data.

4. Results

4.1. Measurement Model

Under this section, the testing of the validity and reliability of the data will be performed. Three criteria are required in this section, namely, convergent validity, discriminant validity and internal reliability of the research items [29]. Convergent validity refers to the level to which several items can measure the same concept. This validity will be achieved when all AVE (average variance extracted) values exceed 0.50 and CR (composite reliability) exceeds the minimum level of 0.6 in social science research (Hair et al. 2012). The construct reliability or factors studied were tested using Cronbach's alpha and rhoA values. The minimum value required is 0.7 [30]. The findings of this study have indicated that the minimum requirements for convergent validity, discriminant validity and reliability of the questionnaire items have been met. The results of the analysis are summarized in Table 1 (Figure 1).

Table 1. Convergent Reliability Test.

Construct	Item	Loading	Cronbach	rhoA	CR	AVE
TS	SP1	0.876	0.921	0.932	0.946	0.746
	SP2	0.891				
	SP3	0.886				
	SP7	0.825				
	SP9	0.849				
	SP10	0.855				
MU	PM3	0.682	0.889	0.902	0.911	0.562
	PM4	0.709				
	PM6	0.693				
	PM8	0.747				
	PM10	0.809				
	PM11	0.736				
	PM13	0.791				
	PM14	0.817				
SS	KP1	0.784	0.932	0.950	0.940	0.757
	KP2	0.865				
	KP3	0.917				
	KP4	0.900				
	KP5	0.878				

TS: Teaching Strategies, MU: Multimedia Usage, SS: Student Satisfaction, rhoA and CR: Composite Reliability, AVE: Average Variance Extracted.

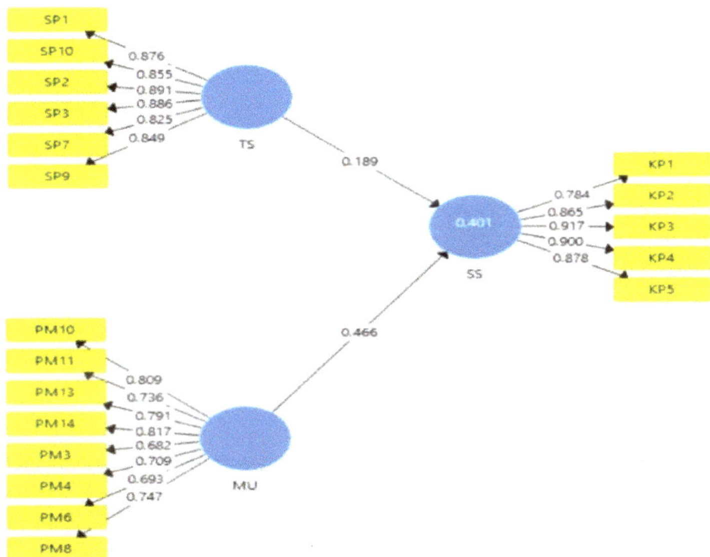

Figure 1. Reflective Measurement Model. (TS: Teaching Strategies; MU: Multimedia Usage; SS: Student Satisfaction).

4.2. Structural Model

This study was conducted to investigate the UiTM Melaka students' satisfaction in learning Arabic language online. To meet this purpose, two hypotheses were constructed among the study variables. The SmartPLS 3.0 bootstrapping function [25] was employed to test the significance level and t-value for all path coefficients in the study model. The analysis proved that a path coefficient of multimedia usage was found to affect student satisfaction significantly at the level of 0.05 with a value of $t \geq 1.645$, except for the teaching strategies factor, which was found not to affect student satisfaction.

Next, the quality of the research model was determined through the values of effect size (f^2), R^2 and Q^2 [28]. The analysis shows that the effect size (f^2) has diverse values, from a small effect size (0.038) to a large effect size (0.057). In addition, the value of R^2 is large at 0.406 while the value of Q^2 which exceeds 0 (0.387) indicates that the study model has sufficient predictive relevance [28]. Table 2 represents the analysis of the research hypothesis and quality models (Figure 2).

Table 2. Path Coefficient Test Results.

Hypothesis	Correlation	Std. Beta	Std. Error	t-Value	Result	R^2	f^2	Q^2
H1	TS → SS	0.189	0.097	1.951 *	Supported	0.406	0.038	0.387
H2	MU → SS	0.476	0.096	4.878 *	Supported		0.057	

* $p < 0.05$, t-value is larger than 1.645.

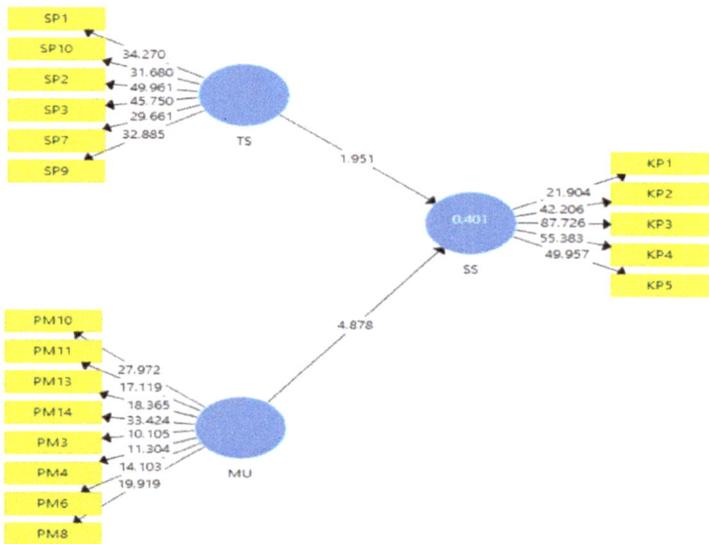

Figure 2. Structural Model. (TS: Teaching Strategies; MU: Multimedia Usage; SS: Student Satisfaction).

4.3. *Importance–Performance Matrix Analysis*

In order to obtain the diagnostic value of the model, IPMA analysis was performed [31]. This evaluation is based on a comparison between the average value of student satisfaction (SS) with PLS expectation, which calibrates the importance of each construct in the research model. In other words, through the analysis of IPMA, the significance and achievement of each factor that affects student satisfaction will be identified.

Table 3 clearly reports that the multimedia usage factor (MU) is the most important factor with importance value of 0.466 and performance value of 80.672, compared to strategic teaching (TS) with importance and performance values of 0.189 and 74.973, respectively. By considering this IPMA analysis, lecturers need to focus on these two aspects, especially multimedia usage, which has been found to be the most important factor in influencing a student to achieve learning satisfaction.

Table 3. IPMA Analysis.

Construct	Importance (Total Effect)	Performance (Index Value)
Student satisfaction	-	70.770
Teaching satisfaction	0.189	74.973
Multimedia usage	0.466	80.672

5. Discussion

Language learning, especially as third or foreign language, requires a tremendous amount of hard work as students are expected to master the skills within the duration of the course. As tertiary level students, they are usually more independent compared to younger learners, however, the lecturers' or instructors' roles as facilitators greatly help them in engaging with the lessons, especially now that ODL is being implemented due to the pandemic. Lecturers' teaching strategies are today, more than ever, highly significant as students may find it challenging to adapt to ODL. Multiple studies have proven that teaching strategies positively influence student satisfaction in traditional classrooms, and through Smart PLS analysis, this study has also revealed similar findings where teaching strategies are reported to be significant. Activities such as role plays and drills allow students to fluently demonstrate their Arabic vocabulary and construct sentences confidently, even through online platforms such as Google Meet, Cisco Webex and Zoom, among others. In addition, it is also shown in the analysis that students highly appreciate lecturers that incorporate cooperative teaching by creating activities that involve both lecturers and students.

In relation to teaching strategies, the Smart PLS analysis has also revealed a positive and significant relationship between the use of multimedia and student satisfaction at UiTM Melaka. These findings directly proved that the use of multimedia in learning Arabic language is a major key to student satisfaction. Multimedia is an interactive software or application that integrates text, color, images, graphics, sound, animated audio, and full motion video in one application. This multimedia learning system offers a better method to improve students' understanding in various languages. Studies conducted showed that the use of multimedia gathered a positive response and greatly helped students in learning language more easily and effectively [32,33]. This approach also increases effectiveness and students' interest in their tasks that are related to language learning. Therefore, it is not surprising that this learning desire will provide motivation and satisfaction to students to accept and continue studying via ODL sessions [34].

These findings evidently indicate that teaching strategies and multimedia usage are important factors in influencing students to achieve satisfaction in language learning. According to Buasuwan (2018) [35], the application of technological innovation improves social life and daily lifestyle tremendously as well as making routines easier than ever before. Undeniably, the use of technology also has a positive impact on lifestyle and education. Looking at current developments, the education sector is also highly influenced by the implementation of ODL, which is supported by technology applications. Although educators around the world would still use conventional methods in their teaching, technology-based teaching is often used effectively in the teaching and learning session.

6. Conclusions

In conclusion, lecturers are required to have technological competencies and skills in line with the Industrial Revolution 4.0 objectives. Lecturers certainly play an important role in generating students' thinking skills to ensure they have a high level of imagination and creativity in the learning process. In addition, everyone in the education sector must cooperate and commit to their roles in achieving a transformation in education. The development of technology that is ever so dynamic and multi-functional should be the transformation catalyst for an innovative education system. This means that the lecturer's role in creating an interesting learning environment is eminently critical, so that students' learning development is not neglected.

Author Contributions: Conceptualization, N.A.M. and I.S.; methodology, N.A.M.; software, N.A.M.; validation, N.A.M., I.S., N.A.N. and M.F.M.S.; formal analysis, N.A.M.; investigation, N.A.M. and I.S.; resources, N.A.N. and N.M.N.M.; data curation, N.A.M. and I.S.; writing—original draft preparation, N.A.M., I.S. and M.F.M.S.; writing—review and editing, N.A.N. and N.M.N.M.; visualization, N.A.M.; supervision, M.F.M.S.; project administration, N.A.M., I.S. and N.A.N.; funding acquisition, N.A.M., I.S., N.A.N., N.M.N.M. and M.F.M.S. All authors have read and agreed to the published version of the manuscript.

Funding: This research received no external funding.

Institutional Review Board Statement: Not applicable.

Informed Consent Statement: Not applicable.

Data Availability Statement: Not applicable.

Acknowledgments: Paper presented during a panel session at The International Academic Symposium of Social Science 2022, organized by The Faculty of Information Management, Universiti Teknologi MARA, Kelantan Branch.

Conflicts of Interest: The authors declare no conflict of interest.

References

1. Agensi Kelayakan Malaysia. Available online: https://www2.mqa.gov.my (accessed on 5 July 2021).
2. Sheth, J. Impact of COVID-19 on consumer behavior: Will the old habits return or die? *J. Bus. Res.* **2020**, *117*, 280–283. [CrossRef] [PubMed]
3. UNESCO. Available online: https://unesdoc.unesco.org/ark:/48223/pf0000128463 (accessed on 5 July 2021).
4. Greenchow, C.; Chapman, A. Social distancing meet social media: Digital tools for connecting students, teachers and citizens in an emergency. *Inf. Learn. Sci.* **2020**, *121*, 331–342.
5. Jha, A.K.; Arora, A. The neuropsychological impact of e-learning on children. *Asian J. Psychiatr.* **2020**, *54*, 102306. [CrossRef] [PubMed]
6. Upadhyay, U.D.; Lipkovich, H. Using online technologies to improve diversity and inclusion in cognitive interviews with young people. *BMC Med. Res. Methodol.* **2020**, *20*, 159. [CrossRef]
7. Prambayun, A.; Suyanto, M.; Sunyoto, A. Model gamifikasi untuk sistem manajemen pembelajaran. In Proceedings of the Seminar Nasional Teknologi Informasi Dan Multimedia, Yogyakarta, Indonesia, 6–7 February 2016.
8. Sopian, A. Kemahiran menulis bahasa arab: Analisa kelemahan pelajar UiTM Melaka. *Int. J. Humanit. Philos. Lang.* **2019**, *2*, 58–65. [CrossRef]
9. Ahmad, Z.; Abdullah, I. Metode pengajaran dan pembelajaran bahasa arab berasaskan empat kemahiran. In Proceedings of the Seminar Pengajaran dan Pembelajaran Bahasa Arab, Terengganu, Malaysia, 14 September 2013.
10. Che Mat, A.; Soon, G.Y. Cabaran–cabaran dalam pengajaran Bahasa Arab sebagai bahasa asing pilihan di IPT: Kajian kes. *AJTLHE ASEAN J. Teach. Learn. High. Educ.* **2011**, *2*, 9–21.
11. Taib, S.H.; Ismail, M.A.; Abdin Lubis, M.A.L. Inovasi kesepaduan dan strategi pengajaran dan pembelajaran di era Revolusi Industri 4.0. *ACER-J* **2019**, *3*, 38–54.
12. Saharuddin, N.; Abdullah, N.S.M. Elemen dalam program penyediaan guru sebagai pemudah cara pemikiran moral guru pelatih pendidikan moral. *JuKu Jurnal Kurikulum Pengajaran Asia Pasifik* **2015**, *3*, 1–13.
13. Mazlan, N.A.; Zolkapli, N.M.; Wan Ismail, W.M.; Norwahi, N.A.; Mohd Saad, M.F. The impact of COVID-19 lockdown on UiTM Melaka Students' learning session. *J. Lang. Linguist. Stud.* **2021**, *17*, 1045–1057. [CrossRef]
14. Abdin Lubis, M.A.L.; Taib, S.H.; Ismail, M.A. Inovasi sistem pendidikan dan strategi pengajaran bahasa arab di era milenial 4.0. *Asean Comp. Educ. Res. J. Islam Civiliz. (ACER-J)* **2019**, *3*, 9–20.
15. Mazlan, N.A.; Zailani, S.; Arifin, Z.; Norwahi, N.A.; Saad, M. Arabic language acquisition in open and distance learning: The relationship between lecturers' teaching strategies and student understanding. *Rev. Int. Geogr. Educ. Online* **2021**, *11*, 127–134.
16. Sahrir, M.S.; Alias, N.A. A study on Malaysian language learners' perception towards learning Arabic via online games. *GEMA Online J. Lang. Stud.* **2011**, *11*, 129–145.
17. Jasni, S.R.; Zailani, S.; Zainal, H. Pendekatan gamifikasi dalam pembelajaran bahasa arab. *J. Fatwa Manag. Res.* **2019**, *13*, 358–367. [CrossRef]
18. Haroon, I. *Kajian penggunaan bahan rujukan oleh pelajar UM untuk Quality Assurance Data;* Universiti Malaya: Kuala Lumpur, Malaysia, 2000.
19. Abdullah, I.; Ammar, A.; Zakaria, M.Z.; Arsyad, M.; Majid, A. Pengaruh pendekatan mass lecture dan motivasi terhadap kepuasan pembelajaran bahasa arab komunikasi. *Malays. J. Soc. Sci. Humanit.* **2021**, *6*, 90–107. [CrossRef]

20. Saja, I.; Ramli, S.M.; Sopian, A.; Abdul Razak, A.Q. The use of youtube in Arabic speaking skills during open and distance learning (ODL). *J. Sains Sos. Dan Kemanus.* **2021**, *18*, 188–202.
21. Amir, M.A.Z.; Mat Luwi, A.H.; Ab Rahman, F.A. A study of students' perception of online learning classes based on MARS Model in UiTM Seremban 3 campus. *Voice Acad.* **2022**, *18*, 67–80.
22. Ismail, U.S.; Chulan, M.; Rosaidi, N.A.; Mohamad Nor, N.A.; Ismail, N.; Abdul Pisal, N. Factors influencing students' motivation in online learning for arabic language courses at UiTM Perlis branch. *e-Acad. J.* **2021**, *10*, 154–165. [CrossRef]
23. Sitorus, M.B. Studi Literatur Mengenai Gamifikasi Untuk Menarik Dan Memotivasi: Penggunaan Gamifikasi Saat Ini Dan Kedepan. In Proceedings of the Seminar 1 Pascasarjana Teknik Elektro, Malang, Indonesia, 9 August 2016; pp. 1–10.
24. Ritonga, A.W.; Ritonga, M.; Nurdianto, T.; Kustati, M.; Rehani, R.; Lahmi, A.; Yasmadi, Y.; Pahri, P. E-learning process of maharah qira'ah in higher education during the COVID-19 pandemic. *Int. J. High. Educ.* **2020**, *9*, 227–235. [CrossRef]
25. Ringle, C.; Da Silva, D.; Bido, D. Structural equation modeling with the SmartPLS. *Braz. J. Mark.* **2015**, *13*, 56–73. [CrossRef]
26. Hair, J.; Ringle, C.; Sarstedt, M. PLS-SEM: Indeed a silver bullet. *J. Mark. Theory Pract.* **2011**, *2*, 139–152. [CrossRef]
27. Chin, W. The partial least squares approach to structural equation modeling. *Mod. Methods Bus. Res.* **1998**, *2*, 295–336.
28. Hair, J.; Hollingsworth, C.; Randolph, A.; Chong, A. An updated and expanded assessment of PLS-SEM in information systems research. *Ind. Manag. Data Syst.* **2017**, *117*, 442–458. [CrossRef]
29. Ramayah, T.; Cheah, J.; Chuah, F.; Ting, H.; Memon, M.A. *Partial Least Squares Structural Equation Modeling (PLS-SEM) Using Smartpls 3.0*; Pearson: Kuala Lumpur, Malaysia, 2018.
30. Cronbach, L. Coefficient alpha and the internal structure of tests. *Psychometrika* **1951**, *16*, 297–334. [CrossRef]
31. Martilla, J.; James, J. Importance-performance analysis. *J. Mark.* **1977**, *41*, 77–79. [CrossRef]
32. Md Salleh, S.; Tasir, Z. Pembinaan sistem pembelajaran berasaskan simulasi interaktif menerusi web bagi kursus telekomunikasi dan rangkaian. *J. Teknol. Pendidik. Malays.* **2011**, *1*, 49–61.
33. Yahaya, M.; Sahrir, M.; Nasir, M. Pembangunan laman web EZ-Arabic sebagai alternatif pembelajaran maya bahasa Arab bagi pelajar sekolah rendah Malaysia. *J. Teknol.* **2013**, *61*, 11–18. [CrossRef]
34. Abdullah, N.; Mat Lazim, N.R.J.; Ahmad Zain, R. *Teknologi Dalam Pengajaran dan Pembelajaran*; Penerbitan Multimedia Sdn. Bhd.: Puchong, Malaysia, 2009.
35. Buasuwan, P. Rethinking Thai higher education for Thailand 4.0. *Asian Educ. Dev. Stud.* **2018**, *7*, 157–173. [CrossRef]

proceedings

MDPI

Proceeding Paper

A Study on Difficulties Encountered and Perception by English as Second Language (ESL) Learners in Malaysian University Examination Test (MUET) †

Seeni Mehraj Begam Bt V. K. S. Vyzul Karnine [1,2,*], Abdul Shakour Duncan Preece [1,*], Ismail Bin Sheikh Ahmad [1,*] and Siti Salmiah Binti Muhammad [1,3,*]

1 Kulliyyah of Education, International Islamic University Malaysia, Kuala Lumpur 53100, Selangor, Malaysia
2 General Studies Department, Sultan Salahuddin Abdul Aziz Shah Polytechnic, Persiaran Usahawan, Seksyen U1, Shah Alam 40150, Selangor, Malaysia
3 Faculty of Major Language Studies, Islamic Science University of Malaysia, Bandar Baru Nilai, Nilai 71800, Negeri Sembilan, Malaysia
* Correspondence: mehrajraf@hotmail.com (S.M.B.B.V.K.S.V.K.); shakour@iium.edu.my (A.S.D.P.); drismail@iium.edu.my (I.B.S.A.); msalmiah@usim.edu.my (S.S.B.M.)
† Presented at the International Academic Symposium of Social Science 2022, Kota Bharu, Malaysia, 3 July 2022.

Abstract: Malaysian University Examination Test (MUET) is one of the English proficiency tests required for students who want to pursue a degree at Malaysian public universities. This test consists of four skill tests, namely listening, reading, speaking, and writing. However, over the past several years, many students who sit for the exam achieved an incredibly low band, and the reasons are not clear. This study aims to investigate the perceptions and difficulties encountered by learners in English as a Second Language (ESL) students in the Malaysian University Examination Test (MUET). Three Malaysian Polytechnic degree-level students who had to re-sit for MUET to complete their degree programme were chosen as the informants. The study applied the qualitative research design, and data were gathered using semi-structured interviews. It was discovered that the students faced the challenges such as preparation, lack of knowledge in vocabulary, grammar, and time constraints in answering the question in MUET. The paper concludes by suggesting that the students are taking steps to overcome MUET-related challenges. Future research is required to investigate the type of teaching methodology that can assist these students prior to MUET.

Keywords: Malaysian University English Test; proficiency test; skills; English as a Second Language

Citation: Karnine, S.M.B.B.V.K.S.V.; Preece, A.S.D.; Ahmad, I.B.S.; Muhammad, S.S.B. A Study on Difficulties Encountered and Perception by English as Second Language (ESL) Learners in Malaysian University Examination Test (MUET). *Proceedings* **2022**, *82*, 53. https://doi.org/10.3390/proceedings2022082053

Academic Editor: Mohamad Rahimi Mohamad Rosman

Published: 15 September 2022

1. Introduction

In Malaysia, the English language is considered by many as a second language after the national language, Bahasa Melayu. English is widely used in trade at domestic and international levels. Hence, fluency and mastering English is an added value for any professional or highly skilled job. However, students still struggle to learn English as a second language (ESL) in Malaysia. An average student spends about 14 years learning the English language from primary school, secondary school, and up to the Polytechnic level. In these 14 years, many of them fail to become fluent. Hunt and Beglar [1] found that English learners are continuously weak in vocabulary acquisition. This situation is supported in a newspaper, where the Malaysian central bank reported that about 15.3% of youths are unemployed [2].

According to Marlyna et al. [3], there are incidences of mistakes in 'subject-verb agreement' (SVA) and copula 'be' in learning English. In addition, graduates in Malaysia, in general, are still not fluent in English [4].

1.1. Malaysian University English Test

The Malaysian University English Test, introduced in 1999, is one of the English proficiency tests for potential students planning to pursue their studies at any local universities in Malaysia and Singapore. The Malaysian Examination Council (MEC) was the appointed council to administer the test [5]. MUET is one of the holistic methods to develop the students' abilities to master English and the pre-degree students need to take the test [6]. It measures the candidates' level of English mastery and evaluates their performance. Achievement in MUET is indicated using bands ranging from Band 1 to Band 6. MUET covers all four skills of speaking, reading, writing, and listening, which are essential skills in the process of language learning.

1.2. Perception and Difficulties Encountered by Students as the Second Language Learner

Perception is one of the elements of thinking that is subjective. On the other hand, difficulties can be referred to as challenges and obstacles that can be tangible and intangible. Students as second language learners have been struggling to master English in their academic life, as well as professionals in the workplace. There are also differences in Eastern and Western cultures that affect language acquisition [7]. Three problems identified as the result of these differences are mistakes in grammar, organisation drawbacks, and unsuitable presentation of ideas in English [8].

What are the solutions to problems faced by these learners? How can scaffolding be used in the learning process? There could also be other factors that were not discovered in their process of learning ESL.

The learners face problems from the primary level towards the secondary until the tertiary level. The entrance to university, which requires these learners to obtain a specific band, resulting in MUET, has been one of the major demotivating factors and a major obstacle in their studies at the higher education level.

2. Problem Statement

Students' learning behaviours can change with the preparation for a high-stakes test and can be seen in their determination to learn [9–11]. Learner autonomy can also be increased by determination [12,13]. The context of this paper is centred on learners who are preparing themselves for MUET, an entry test to public universities in Malaysia and Singapore. They face difficulties obtaining a higher band to enter public universities. Most of them obtain Band 2, which is a poor result, and therefore, entry to the university is not possible. Previous studies suggest that the learners are overburdened with the risk of the test, where they think that they will not be able to achieve a good result even if they put a lot of effort into the preparation [14,15]. Thus, the sample of this study involved the MUET repeaters, in which they received poor results. These informants are chosen because it is strongly believed that the difficulties and challenges they faced prior to or during the test can be investigated in-depth.

3. Purpose and Objectives of the Study

This study's goal was to investigate the perceptions and difficulties encountered by learners of ESL students in the Malaysian University Examination Test (MUET).

4. Research Questions

The central research question of this study is to investigate the difficulties encountered by ESL students in MUET.

The following are the sub-research questions of the study:

1. What are the challenges faced by ESL students in MUET?
    - 1.1 Challenges in terms of readiness;
    - 1.2 Challenges in terms of knowledge in varied vocabularies;
    - 1.3 Challenges in terms of time constraints in answering questions.
2. What are the perceptions of ESL learners toward MUET?

5. Significance of the Study

This paper is significant in terms of investigating the difficulties encountered by ESL students in MUET. It is hoped that this paper can help the students to improve their band in MUET and thus further their studies at any public university in Malaysia or Singapore. This research is also could improve the teaching methods used by the instructors, teachers, or lecturers of MUET candidates.

6. Literature Review

6.1. The Use of MUET for University Entry and Graduation Purpose

English is often the medium of instruction at the tertiary level [16]. Therefore, students' English language proficiency needs to be determined before university admission. In order to ensure the students are proficient in English, the Malaysian University English Test is used as the yardstick, and the pre-degree students as mentioned earlier need to take the test. According to Kaur and Nordin [17]. MUET is an indicator of their proficiency level in the English language, which allows the pre-degree students to enrol on the pre-degree students' desired course. In addition, MUET is administered by the Malaysian Examination Council, which is one of the well-trusted exam councils, and it is acknowledged by Malaysia and Singapore [18]. After the students have successfully entered the desired university in a specific course or program, they are expected to be able to pursue their studies effectively as they are more well versed in the English language. In fact, after completing their degree, the acceptable band in MUET is the requirement for their graduation purpose.

6.2. Research Related to the Scope of Study

Moreover, for ESL teacher education, MUET is the indicator of academic achievement [18]. Unfortunately, the students in Malaysia are facing challenges of the interference of the first language, which affects their self-confidence [19–21]. The low proficiency level among Malaysian students was investigated in a study by Che Musa et al. [20]. Other studies found that students' level of confidence is affected because they feel comfortable speaking in their native language rather than English [20–22]. Arshad et al. [23] discovered that confidence in the students' abilities plays a significant role in their academic performance. In conducting a given task, there are different reactions from students because they have various levels of confidence and abilities [24]. Rethinasamy and Chuah [25] found a correlation between students' performance in English Preparatory class and the students' MUET marks. Hence, it is significant to investigate whether there could be any improvements for the students' preparatory classes. This study can fill the gap in which an in-depth investigation should be carried out to discover improvements in the preparatory class for better marks in MUET. A study carried out by Yunus and Chien [26] discovered that mind mapping is the best strategy in MUET preparatory classes for writing skills. However, other strategies for other skills are yet to be discovered. Hence, this highlights another gap to be filled in this study. An in-depth investigation is required for further information on the specific difficulties in other skills opens an opportunity for this study to be carried out.

7. Methods

This study chose a qualitative research design due to the researchers' experience and interest in teaching preparation classes for MUET among ESL students. Qualitative research gives importance to meaning and does not produce generalised hypotheses [27]. The method of gathering data was a semi-structured interview. The objective of this method was to gain insights from experience shared in the interview about the difficulties encountered with MUET. The researchers also wanted to investigate their perception of MUET as well as their thoughts and experiences. The source of the interview questions is based on the nature of the research questions. Adequate interview questions were raised to find information to explain the research questions. All interview questions were piloted and reviewed by quali experts.

7.1. Population and Sample

This study used purposive sampling to understand the phenomenon that was investigated [28]. Three degree-level students who had to re-sit MUET to graduate from their degree programme at a Malaysian Polytechnic were purposely chosen as the sample. They were selected based on the ability to provide rich data on the challenges of taking MUET for the first time and their perception of the result obtained. The students' perspectives are significant because they had to re-sit MUET due to the low band obtained in the first attempt of MUET, which was Band 2. These students could not successfully graduate until they obtain Band 3 as the graduation requirement of their degree programme, which is relevant to the research objectives [29]. The information of the informants was kept confidential. They were addressed as Informants 1, 2, and 3.

7.2. Instrument

The three informants were interviewed using seven semi-structured questions. The seven questions asked in the interview were: (1) How do you learn to prepare yourself for MUET? (2) What are the challenges you faced prior to taking the MUET exam? (3) What is your hope/expectation/s from MUET? (4) How do you express your feeling toward your previous MUET result? (5) Why do you feel that way when you get your result? (6) What do you do to cope with that feeling? (7) What is your suggestion to improve your English? Question numbers 1 and 2 were raised to answer research question 1, and questions numbers 3 to 7 were raised to answer research question number 2. The informants who were chosen for this study were the MUET repeaters who were in their final year of studies at one of the Malaysian Polytechnic. The questions are based on the research questions. In order to ensure the quality of the interview data, an interview protocol refinement framework was used [30]. The phases are: (1) ensuring the alignment of the interview questions with research questions, (2) constructing an inquiry-based conversation, (3) receiving feedback on interview protocols, and (4) piloting the interview protocol is piloted.

7.3. Pilot Interview and Saturation of Study

A pilot interview was conducted with two informants with the same background of studies and problems. According to Hassan [31], a pilot study is a small-scale trial, which is a technique for preparing a larger study. There were comments on adding the questions from six to seven. The question that was added was about suggestions for improving English mastery. Other than that, the word expectation was not understood by the informant in the pilot interview. In the final set of interview questions, the word 'expectation' was changed to 'hope.' The interview questions were validated before conducting the actual interview with the three informants identified. Even though this study only involved three informants, Charmaz [32] claims the saturation of the study can be achieved but with modest claims.

7.4. Data Collection Procedure

Three informants, degree students, pursuing their studies in a tertiary institution, were selected from the population. Semi-structured interviews were used to gather their insights and in-depth data. According to Seidman [33], it is worth knowing the individual's stories by conducting interview-based research. Researchers can explore more details from clarifications requested during the interview. The interview protocol was prepared before the interview session to gather responses from the interviewee [34]. The interview was held using the Zoom online meeting application. This platform was chosen due to the 'Conditional Movement Control Order' implemented by the government of Malaysia during the COVID-19 pandemic. The interview was held by texting the interviewees and planning the interview time and date with them. The interview lasted between 15 and 40 min and was recorded, saved as a video, and later converted to an MP3 audio file.

7.5. Data Analysis Procedure (DCR) Steps

Data analysis is an essential part of every study. The data were extracted from the informants effectively in order to reach the saturation level. The data were collected using a semi-structured interview, and a thematic analysis was conducted to analyse it. Either the DCR or the 3-step formula by Bazeley [35] was used as the data analysis procedure.

Firstly, 'D' stands for description. It is used to provide more data about the sources of data. In this study, the sample was the fourth-semester students undergoing a degree programme in one of the tertiary level institutions in Malaysia. They were taking MUET for the first time, obtained Band 2, and were going to re-sit for the test in September 2020.

The second step is 'C' or compare. More or less, the informants suggested remarkably similar themes. The themes were generated by choosing the keywords from the interview. WH questions were used, and the informants provided many data in response.

The third step is 'R,' which stands for Relate. Generated coding was used to know and explore further the conditions from which the theme arose and form relationships between the themes. Each informant's answer was coded, and themes were generated separately. Similarities of keywords helped the researchers generate a theme that applies to all types of answers given by the informants.

7.6. Credibility and Trustworthiness

In order to ensure the generated themes are credible and trustworthy, two inter-raters were chosen from the same workplace as the researchers. They are the experts in their field of study and knowledgeable about the problem of the study. This process helped the researchers in checking the themes. There were a few comments given by the inter-raters, and the researchers' interpretation of the themes was acceptable and reliable at an average of 88% inter-raters agreement.

8. Result and Discussion

Table 1 above shows a clear picture of the major theme generated from the two main research questions and as well as the interview questions.

Table 1. Inter-raters agreement.

(Total Number of Agreements/Total Number of Responses) × 100%	
Inter-rater 1:	(8/9) × 100 = 88%
Inter-rater 2:	(8/9) × 100 = 88%
Average:	(16/18) × 100 = 88%

Table 2 shows the themes derived from the answers generated from the interview. It provide the reports on the difficulties encountered and perception by ESL learners in MUET. The given responses were based on the research questions mentioned earlier in this paper.

- Research question 1: What are the challenges faced by ESL students in MUET?

In relation to this question, the informants answered the questions by sharing the learning strategies used to prepare themselves for MUET. They stated that they studied by attending class for guidance as well as applying independent study in their learning strategies. This first generated theme is supported by the following evidence (No. 5: I#3).

> *Attending class and taking notes from the internet and YouTube (some channels) that provide teaching MUET for all the skills, so I just see the format and how to prepare and do the exercises.*

Table 2. Major themes derived from the research questions.

Research Questions	Interview Questions	Major Themes
Research Question: 1: What Are The Challenges Faced By The ESL Students in MUET? a. Challenges in Terms of Readiness b. Challenges in Terms of Knowledge in Varied Vocabularies c. Challenges in Terms of Time Constraint in Answering Question	1. How do you learn to prepare yourself for MUET? 2. What are the challenges you faced prior to taking MUET exam?	1. Dependent and Independent study 2. Lack of preparation 3. Lack of knowledge
Research Question 2: What Are the Perceptions by The Esl Students Toward Muet?	1. What is your hope/expectation/s from MUET? 2. How do you express your feeling toward your previous MUET result? 3. Why do you feel that way when you get your result? 4. What do you do to cope with that feeling? 5. What is your suggestion for improving English?	4. Improvement toward the higher band 5. Disappointment 6. Awareness in future exam 7. Sharing feelings 8. Increase effort in reading and other skills

This finding corresponds with a study carried out by Mahmud [36], in which the independent study was implemented by the informants by performing the exercises on their own as part of the preparation for the test. This finding indicates that even though the informants are facing challenges in their learning process, they still have the motivation to push themselves to work hard in order to gain a better result. However, it is not known whether, in the independent study approach, they are using the right source for the exercises and examples. Thus in future studies, this area can be explored more.

Furthermore, the study explored what challenges the students faced before taking MUET. This question was divided into three types of challenges to obtain in-depth data from the interviewees. The first challenge was in terms of readiness. In this context, the informants agreed that they were not ready at all, and the second theme is the lack of preparation generated from this answer (No. 8: I#1).

In terms of readiness, I think I am not ready I can't prepare all the things and push myself to do three papers listening writing and reading.

The study indicates that the informants were not ready for the test due to a lack of preparation in three skills, namely listening, writing, and reading. However, another study carried out by Kuen and Embi [37] showed that listening skill is highly prepared compared to the other three skills, namely reading, speaking, and writing, with the least preparation. The reason for various levels of readiness is yet to be known; hence, this leads to future studies in the area.

The third theme, which is a lack of knowledge of vocabulary and grammar, was generated from the three informants' responses which sounded similar. Hence, it was interesting to learn that they were facing the same problem, and this shows that they are very weak (No. 15: I#2).

For me I think . . . aaaa I have enough knowledge, but I think I don't have quite enough knowledge, but the grammar and vocabularies don't think to have enough knowledge because if take reading and writing I always forget the meaning of the words and also, I forget the vocabulary and grammar.

This is another key finding that exhibits the students' encountering lack of knowledge in terms of vocabulary and grammar. Hence, this result indicates that over the years, the same problem has been faced by the MUET candidates, and further studies need to be looking into how to overcome the same problem.

The fourth theme is the lack of time to answer the question. This theme was generated based on the information from the informants. It shows that they were struggling with their incapability to think and answer the questions in MUET. One excerpt from Informant 2 indicated that (No. 17: I#2),

Not enough time for answering the question for reading and listening.

This particular informant showed that these specific skills require more time to answer. Interestingly, it was supported by the first informant, pointing to the listening test (No. 14: I#1),

I can't real the apa ... aaaa the audio what the audio said sometimes it's fast for me so I can't catch the audio.

The informant added more information on the reading test (No. 16: I#1),

Reading is long passage sometimes words more too hard I don't the passage is talking and the topic about not in Malaysia the words lah difficult to understand.

In addition, Informant 2 gave another insight into the time constraint during the speaking skills test (No. 18: I#1).

Speaking sometimes we get two times, and the duration is 2 min to get the point and to elaborate and when we lack vocabulary, we don't know how to describe and know what to talk but we can't talk in front of the examiner.

Interestingly theme number four is related to theme number 3, which is a lack of knowledge in vocabulary and grammar. This is a piece of significant evidence that shows cause and effect. There is a relationship between the knowledge of vocabulary and grammar when the informants want to produce their answers through reading, writing, and speaking.

- Research question 2: What are the perceptions of ESL students toward MUET?

Research question two is a very subjective measurement that resulted in all sorts of responses from the informants. However, the answers are vital because this is key to improving MUET and the teaching methodology in MUET preparatory classes. The question that is relevant to this research question is what the students hope from MUET. The fifth theme, which is to improve the band in MUET to a higher band, was generated from this question. All three informants indicated that they want to achieve at least Band 3 and above. Band 3 was mentioned by all three of them. Evidence of these statements can be seen in the following excerpt from the informants (No. 28: I#1):

We can learn more skills and we can use them for the next life in working I hope MUET can improve my English. the expectation in terms of band I hope I can get Band 3 or four to finish my studies and my future studies.

This was supported by Informant 2 with this statement (No. 21: I#2):

I hope I can get a minimum of Band 3 and do well in listening because I have very bad listening skills. In the hall, I can't hear the voice clearly and it's too quick to catch up on what they want to say in the audio.

Informant 3 added (No. 18 I#3):

To get Band 3 and above.

All these feedbacks show the crucial needs of the informants. They want to score Band 3. Therefore, the researchers wanted to know more about why there was a mutual answer from all three informants. Hence, more questions were asked to explore the details.

The next question was about the expression of feelings when they received the previous MUET result. All three informants received Band 2 in their previous MUET. They mentioned feeling sad about the result obtained, for instance, in this statement (No. 33 I#1),

Because to enter university I need Band 3 so I am sad because I think I can't further study in degree.

The informants had to repeat the test because of their Band 2 result and achieve at least Band 3 in the next test in order to graduate.

The next theme is theme number six, which was generated from their expressions of the previous MUET result. The themes of disappointment and awareness were raised regarding improving the MUET re-sit result (No. 23: I#2).

When I see my result...yes. I feel disappointed.

Informant 3 reflected (No. 22: I#3),

Because I know I am not prepared for the exam and do not have enough time and mistakes is from me and I accept it and I know I can take it again.

Theme number seven was generated from the question regarding the reason the informants obtained their previous results. The seventh theme is lack of preparation. The excerpt from Informant 3 indicated (No. 22: I#3),

Because I know I am not prepared for the exam and do not have enough time and mistakes is from me and I accept it and I know I can take it again.

This informant realised their own mistake and had incredibly positive insights that indicated improvements were made in preparation for the next test.

Theme number 8 is managing feelings by sharing them with family members and friends. This shows that family members are one of the most important sources of motivation for these informants. The following statement shows the sharing of feelings (No. 32: I#2),

The reason for sharing because between parents because later they want to know the result because they know I am taking the result. I share it with my parents and friends and the friend's result is better than mine, so I feel disappointed.

Share with my friends because they asked if they don't ask, I will keep it to myself.

The source of Informant 3's motivation can be seen from the following answer (No. 24: I#3).

Discuss with my mom and share with mom why fail and listen to their advice and not get stressed. Other than that, I tried to correct my mistakes to overcome so I promise myself.

The last theme, or the ninth theme, is to increase reading materials and audio sources exposure to improve English. This theme was generated from the question seeking the informants' suggestions on how to improve their level of English. The following is a statement from Informant 2 (No. 36: I#2):

Mmm more reading to get a new word or new, new ... more reading to get new word ... more exercises to know the sentence the meaning of the sentence and more speak to others.

Anything else? (interviewer)

Mmm more hear a lot of English songs (No. 38: I#2)

Informant no 1 indicated a concern about improvement in knowledge about Malaysia and overseas. It also suggested that the informant had ambitious English language skill goals (No. 40: I#1).

Read articles and journals to improve my knowledge and get to know the situation in Malaysia or oversea. We can watch movies to gain knowledge from that we can hear and read subtitles so want to improve in speaking and reading especially.

All these themes suggest that the informants, as MUET candidates, need more guidance, motivation, and strategies in their preparation. Their hope of obtaining Band 3 and above is possible if effective strategies are taught in the module used in the preparatory classes along with the support from the instructors. Teaching quality can increase meaningful learning [38].

The preliminary model based on the themes generated from the findings is shown in Figure 1.

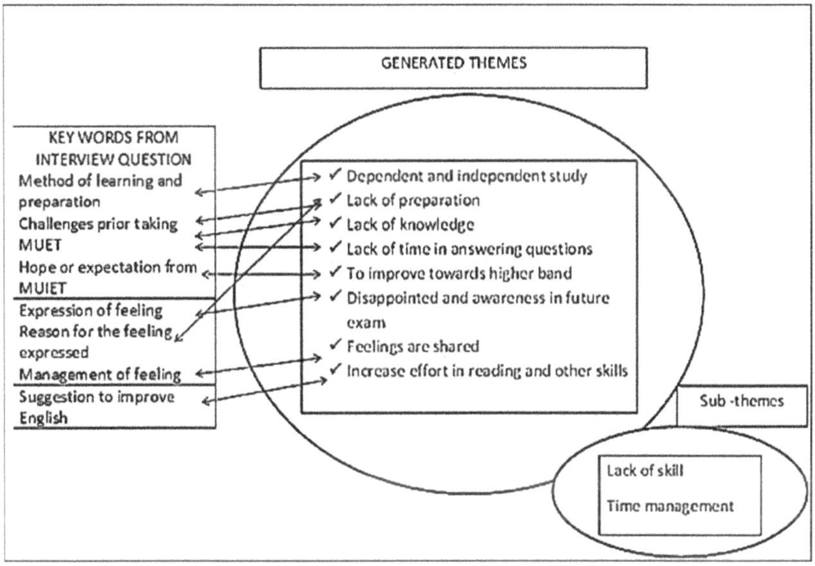

Figure 1. D3LTDLFI Model.

9. Conclusions

This paper suggests that there are difficulties faced by ESL Students during MUET. The difficulties encountered by the students were supported by their perceptions of MUET. This answered two research questions posed by this study, namely, the challenges faced by the ESL students in MUET and the ESL students' perception of MUET. Challenges described through difficulties in preparation, lack of knowledge in vocabulary, grammar, and time constraints in answering the question show that students are aware of their weaknesses. When they expressed their perceptions, fortunately, they were still motivated to improve themselves by performing an independent study in order to achieve a higher band.

Interestingly, the findings include positive actions such as the sharing of feelings with their family members and friends. The students also responded excellently when they indicated that they wanted to involve themselves with more reading materials and audio sources to improve their English level.

Some suggestions should be taken into consideration in improving the teaching methodology for the preparatory class as well as special workshops for the weak candidates. These measures can help increase their motivation and confidence level to ensure their score band will be higher, allowing them to enter the university, graduate successfully, and have a great career in the future.

Author Contributions: Conceptualisation, S.M.B.B.V.K.S.V.K., A.S.D.P., S.S.B.M. and I.B.S.A.; methodology, S.M.B.B.V.K.S.V.K., A.S.D.P., S.S.B.M. and I.B.S.A.; validation, S.M.B.B.V.K.S.V.K., A.S.D.P., S.S.B.M. and I.B.S.A.; formal analysis, S.M.B.B.V.K.S.V.K., A.S.D.P., S.S.B.M. and I.B.S.A.; investigation, S.M.B.B.V.K.S.V.K., A.S.D.P., S.S.B.M. and I.B.S.A.; resources, S.M.B.B.V.K.S.V.K., A.S.D.P., S.S.B.M. and I.B.S.A.; data curation S.M.B.B.V.K.S.V.K., A.S.D.P., S.S.B.M. and I.B.S.A.; writing—original draft preparation, S.M.B.B.V.K.S.V.K., A.S.D.P., S.S.B.M. and I.B.S.A.; writing—review and editing, S.M.B.B.V.K.S.V.K., A.S.D.P., S.S.B.M. and I.B.S.A.; visualisation, S.M.B.B.V.K.S.V.K., A.S.D.P., S.S.B.M. and I.B.S.A.; supervision, S.M.B.B.V.K.S.V.K., A.S.D.P., S.S.B.M. and I.B.S.A.; project administration, S.M.B.B.V.K.S.V.K., A.S.D.P., S.S.B.M. and I.B.S.A. All authors have read and agreed to the published version of the manuscript.

Funding: This research received no external funding.

Institutional Review Board Statement: Not applicable.

Informed Consent Statement: Not applicable.

Data Availability Statement: Not applicable.

Acknowledgments: Paper presented in International Academic Symposium of Social Science (IASSC) 2022.

Conflicts of Interest: The authors declare no conflict of interest.

References

1. Hunt, A.; Beglar, D.A. Framework for developing EFL reading vocabulary. *Read. For. Lang.* **2005**, *17*, 23–59.
2. The Star Online. Young and Jobless. Available online: https://www.thestar.com.my/business/business-news/2017/03/27/young-and-jobless/ (accessed on 8 January 2022).
3. Maros, M.; Salehuddin, K.; Hua, T.K. Everyone People Must Have A Best Friend. Interference of Malay Structures in English Written Discourse. In Proceedings of the TLEiA Conference. Teaching and Learning of English: Towards an Asian Perspective, Penang, Malaysia, 8–20 November 2009.
4. Jalaluddin, N.H.; Norsimah, M.A.; Kesumawati, A.B. The mastery of english language among lower secondary school students in Malaysia: A linguistic analysis. *Euro. J. Soc. Sci.* **2008**, *7*, 106–119.
5. Chan, S.H.; Wong, B.E. Assessing Oral Skills of Pre-tertiary Students: The Nature of the Communicative Act. In Proceedings of the International Conference on English Instruction and Assessment, Chiayi, Taiwan, 24–25 April 2004; pp. 33–48.
6. Malaysian Examinations Council, Regulations, Test Specifications, Test Format and Sample Questions. Available online: https://www.mpm.edu.my/en/muet/regulations-test-specifications-test-format-and-sample-questions (accessed on 22 May 2022).
7. Wu, Y.W.; Chen, M.C. Writing Strategies and Writing Difficulties among College Students of Differing English Proficiency. In Proceedings of the 24th ROC-TEFL, Taipei, Taiwan, 5–6 May 2007; pp. 176–189.
8. Cai, G. Beyond "Bad Writing": Teaching English Composition to Chinese ESL Students. In Proceeding of the 44th Annual Meeting of the Conference on College Composition and Communication, San Diego, CA, USA, 31 March–3 April 1993.
9. Cho, D. Use of standardized tests as university graduation requirement. *Engl. Teach. Anseonggun* **2004**, *59*, 251–266.
10. Pan, Y.C.; Newfields, T. Tertiary EFL proficiency graduation requirements in Taiwan: A study of washback on learning. *Electron. J. Foreign Lang. Teach.* **2012**, *9*, 108–122.
11. Thomas, R.M. *High-Stakes Testing: Coping with Collateral Damage*; Routledge: New York, UK, 2005.
12. Pan, Y.C. Learner washback variability in standardized exit tests. *TESL-EJ* **2014**, *18*, n2.
13. Stecher, B.M. Consequences of Large-scale, High Stakes Testing on School and Classroom Practice. In *Making Sense of Test-Based Account*; Hamilton, L.S., Stecher, B.M., Klein, S.P., Eds.; Rand Corporation: Santa Monica, CA, USA, 2002; pp. 79–100.
14. Chu, H.Y. Stakes, Needs and Washback: An Investigation of the English Benchmark Policy for Graduation and EFL Education at Two Technological Universities in Taiwan; Unpublished. Unpublished Ph.D. Dissertation, National Taiwan Normal University, Taipei, Taiwan, 2009.
15. Watanabe, Y. Does the university entrance examination motivate learners? A case study of learner interviews. *Akita Engl. Stud.* **2001**, *3*, 100–110.
16. Gill, S.K. Language policy in Malaysia: Reversing direction. *Lang. Pol.* **2005**, *4*, 241–260. [CrossRef]
17. Kaur, N.; Nordin, R. A case for reconstruction of the pedagogy of the Malaysian University English Test (MUET) through thematic units instruction. *J. Inst. Res. South East Asia* **2006**, *4*, 5–16.
18. Othman, J.; Nordin, A.B. MUET as a predictor of academic achievement in ESL teacher education. *GEMA Online J. Lang. Stud.* **2013**, *13*, 99–111.
19. Archvadze, E. *The Problems of First Language Interference in the Process of Teaching Second Languages*; Tsereteli State University: Kutaisi, GA, USA, 2005.
20. Che Musa, N.; Koo, L.Y.; Azman, H. Exploring English language learning and teaching in Malaysia. *GEMA Online J. Lang. Stud.* **2012**, *12*, 35–51.
21. Derakhshan, A.; Karimi, E. The interference of first language and second language acquisition. *Theory Pract. Lang. Stud.* **2015**, *5*, 2112–2117. [CrossRef]
22. Lekova, B. Language interference and methods of its overcoming in foreign language teaching. *Trakia J. Scie.* **2010**, *8*, 320–324.
23. Arshad, M.; Zaidi, S.M.; Mahmood, K. Self-Esteem & academic performance among university students. *J. Educ. Prac.* **2015**, *6*, 156–162.
24. Park, H.; Lee, A.R. L2 learners' anxiety, self-confidence and oral performance. In Proceedings of the 10th Conference of Pan-Pacific Association of Applied Linguistics, Edinburgh, UK, 4–5 August 2005; pp. 107–208.
25. Rethinasamy, S.; Chuah, M.K. The Malaysian University English Test (MUET) and its use for placement purposes: A predictive validity study. *Electron. J. Foreign Lang. Teach.* **2011**, *8*, 234–245. [CrossRef]
26. Yunus, M.; Chien, H.C. The use of mind mapping strategy in Malaysian University English Test (MUET) Writing. *Creat. Educ.* **2016**, *7*, 619–626. [CrossRef]
27. Crouch, M.; McKenzie, H. The logic of small samples in interview-based qualitative research. *Soc. Sci. Inform.* **2006**, *45*, 483–499. [CrossRef]

28. Creswell, J.W. *Educational Research: Planning, Conducting and Evaluating Quantitative and Qualitative Research*; Pearson: Boston, MA, USA, 2012.
29. Patton, M.Q. *Qualitative Research and Evaluation Methods*; Sage Publications: Thousand Oaks, CA, USA, 2002.
30. Castillo-Montoya, M. Preparing for interview research: The interview protocol refinement framework. *Qual. Rep.* **2016**, *21*, 811–831. [CrossRef]
31. Hassan, Z.A.; Schattner, P.; Mazza, D. Doing a pilot study: Why is it essential? *Malays. Fam. Physician* **2006**, *1*, 70–73. [PubMed]
32. Charmaz, K. *Constructing Grounded Theory: A Practical Guide through Qualitative Analysis*; Sage: Thousand Oaks, CA, USA, 2006.
33. Seidman, I. *Interviewing as Qualitative Research: A Guide Researchers in Education and the Social Sciences*, 4th ed.; Teachers College Press: New York, NY, USA, 2013.
34. Merriam, S.B.; Tisdell, E.J. *Qualitative Research: A Guide to Design and Implementation*, 4th ed.; Jossey-Bass: San Francisco, CA, USA, 2016.
35. Bazeley, P. Analysing qualitative data: More than 'identifying themes'. *Malay. J. Qual. Res.* **2009**, *2*, 6–22.
36. Mahmud, N. Investigating the Washback Effect of the MUET as a University Entry Test on Students in Malaysia. Doctoral Dissertation, University of York, York, UK, 2018.
37. Kuen, Y.L.; Embi, M.A. MUET preparation language learning strategies. *Adv. Lang. Lit. Stud.* **2012**, *3*, 84–93. [CrossRef]
38. Abdul Rahim, H.; Hayazi, M.Y. *Penggunaan Alat Bantu Mengajar (ABM) Di Kalangan Guru- Guru Teknikal Di Sekolah Menengah Teknik Daerah Johor Bahru, Johor*; Universiti Teknologi Malaysia: Johor Bahru, Malaysia, 2010; pp. 1–8.

proceedings

Proceeding Paper

A Study on Needs Analysis in English Teaching and Learning Skills for Sultan Salahuddin Abdul Aziz Shah Polytechnic Engineering Students Pursuing Degree †

Seeni Mehraj Begam Bt V. K. S. Vyzul Karnine [1,2,*], Abdul Shakour Duncan Preece [1,*] and Siti Salmiah Binti Muhammad [1,3,*]

1 Kulliyyah of Education, International Islamic University Malaysia, Kuala Lumpur 53100, Selangor, Malaysia
2 General Studies Department, Sultan Salahuddin Abdul Aziz Shah Polytechnic, Persiaran Usahawan, Seksyen U1, Shah Alam 40150, Selangor, Malaysia
3 Faculty of Major Language Studies, Islamic Science University of Malaysia, Bandar Baru Nilai, Nilai 71800, Negeri Sembilan, Malaysia
* Correspondence: mehrajraf@hotmail.com (S.M.B.B.V.K.S.V.K.); shakour@iium.edu.my (A.S.D.P.); msalmiah@usim.edu.my (S.S.B.M.)

† Presented at the International Academic Symposium of Social Science 2022, Kota Bharu, Malaysia, 3 July 2022.

Abstract: It is known that one of the most vital indicators in scaffolding teaching and learning is carrying out a needs analysis which is considered part of continuous improvement of the curriculum design. In order for the students to gain knowledge and to achieve success in their future goals, the needs are met through a needs analysis. The aim of this study is to investigate appropriate categories of skills in English language learning required for degree-level students in engineering. A total of 53 students from Electrical and Civil Engineering departments from Sultan Salahuddin Abdul Aziz Shah Polytechnic were involved in this research. A survey questionnaire was used in this investigation. Descriptive statistics were used for the analysis, and it was discovered that the prospective engineering students perceived learning English as highly important. The results also revealed that the highest skills in needs and order were writing, speaking, listening and reading. The students encountered challenges and difficulties pertaining to grammar, pronunciation, idioms, and vocabulary even though the students' ability level is Intermediate Upper in English. The needs analysis suggests directions for future research on improving the related skills.

Keywords: English language learning skills; needs analysis; importance; purpose; level of ability

Citation: Karnine, S.M.B.B.V.K.S.V.; Preece, A.S.D.; Muhammad, S.S.B. A Study on Needs Analysis in English Teaching and Learning Skills for Sultan Salahuddin Abdul Aziz Shah Polytechnic Engineering Students Pursuing Degree. *Proceedings* **2022**, *82*, 63. https://doi.org/10.3390/proceedings2022082063

Academic Editor: Mohamad Rahimi Mohamad Rosman

Published: 19 September 2022

1. Introduction

Graduates' unemployment is a long-standing issue that has recently became a matter of growing distressing concern amidst the COVID-19 pandemic outbreak and global economic downturn [1]. Engineering graduates are not spared. In fact, similarly to many other graduates, engineering students have to possess good English language proficiency skills in order to secure positions in this highly competitive and demanding work environment, both in public and private sectors. In addition, Mohamad et al. [2] reported that soft skills in graduates are considered vital in contrast with the accomplishment achieved from their academic performance. It is known that the industries are demanding very highly skilled employees to be recruited. The needs of the industries need to be fulfilled by applying appropriate pedagogical implementations in Malaysia higher education institution which in the end is expected to produce an expert labor force. This problem statement is clearly highlighting an indication of demand from the industries, the first step that needs to be implemented is to carry out a continuous improvement in the course delivered for the students in higher education institutions. Continuous improvement involves knowing the respondents' needs and wants through a needs analysis. The keystone for English for

Specific Purposes (ESP) is a needs analysis in order to carry out an attentive course [3]. Thus, a needs analysis will be an effective step in preparation for a language course to evolve. The analysis will ensure that high-standard pedagogy materials can be embedded in the teaching and learning of the course. Hence, it will lead to an effective teaching and learning. According to Chostelidou [4] an effective and efficient curriculum, syllabus and course evolution and layout can be achieved from a unique structure in the ESP scope which requires a needs analysis as the foundation. The studies in the research on ESP have been globally carried out, namely in Israel [5], the United States [6,7], Iran [8,9], Taiwan [10] and Turkey [11,12]. The previous studies which have been carried out in different contexts indicate that there is a need for a study in Malaysian context. Hence, the needs from the students pursuing their studies in tertiary level will be a vital information for this study. Therefore, the students in one of the Malaysian institutes of higher learning have been selected as a focal point of this study.

2. Methodology

This study aims to investigate the needs of degree-level students at Sultan Salahuddin Abdul Aziz Shah Polytechnic from the perspective of English language learning skills as well as the mode of teaching and learning of English course. This study sampled two different groups of students from two different departments, Facilities Management students of the Civil Engineering Department and Medical Electronics students from the Electrical Engineering Department.

The goal of this study is to reveal the main aspect of the needs; hence, this study is exploratory in nature. The preliminary stage of the study attempted to discover the needs of respondents in the process of learning the courses in their programme. The scaling method was used to gather the data in this research. According to Taherdoost [13], there are two main categories in scaling methods: namely, open-ended questions and close-ended questions in which an instrument can be constructed using scaling as one of the mediums for a measurement. In this quantitative study, closed questions were used to measure the disagreement and the degree of agreement using Likert Scale. In addition, one of the commonly used scaling methods is adapted from Davis and Cosenza [14] known as attitude scale. The sub-scale used under attitude scale is the Likert scale which is used to obtain the attitude score pertaining to the variety of statements on the respondents' choices of answer which will display their attitude.

A total of 35 items were used in the survey questionnaire. The items for this study have been adopted and adapted from a survey questionnaire by Kayaogly and Akbas [15] in their study. Some minor changes have been applied to the questionnaire. The survey questionnaire involves eight diverse themes using a five-point Likert scale format The themes are level of ability in English (5 components), the significance in the enhancement of speaking skills (4 components), the significance in the enhancement of listening skills (4 components), the significance in the enhancement of writing skills (5 components), the significance in the enhancement of reading skills (5 components), the learning context preference in an English course (3 components), the significance of learning English (3 components) and the intention of learning English (6 components).

The researchers obtained the consent from the Civil Engineering Department and Electrical Engineering Department to carry out the study. The questionnaires were given to fifty-three students who volunteered to be the respondents. The respondents took about fifteen to twenty minutes to complete the questionnaire given.

3. Validity, Reliability, and Ethical Consideration in Research

Recent literature has been examined to ensure the study's validity. This study had identified another study with a similar purpose [15] that is suitable to be adopted and adapted. The adaptation was undertaken because the current study is more applicable to degree-level students who are pursuing their engineering programs at Sultan Salahuddin Abdul Aziz Shah Polytechnic.

The components have a Cronbach's Alpha higher than 0.70, which means the questionnaire has an acceptable level of reliability [16–19]. Therefore, the questionnaire has used reliable and valid items. As part of the requirements of the ethical consideration, the information of the respondents has been kept private and confidential. The protection of the respondents' confidential information and the broad accountability of the researcher of the society must be balanced [20]. The respondents were informed about the goal of this study before the data collection. The questionnaires were completed voluntarily by the respondents. It is a violation of privacy if there is no consent obtained from the respondents [21].

4. Procedure in Analyzing the Data

The best strategy in the process of quantifying the analysis and data collection is using quantitative research [22] because the sample used in quantitative research can reflect a larger population [23]. Quantitative analyses can be performed faster by using Statistical Package for the Social Sciences (SPSS) [24]. This study used SPSS statistics for Windows (Version 25.0) which was released in 2017 and manufactured by IBM Corp. Armonk, New York. The frequency of each data was revealed using descriptive statistical analysis in SPSS, which suggested a common pattern.

5. Result and Conclusions

The table presented below demonstrates the findings from this research. The items used in the questionnaire are presented in Table 1.

Table 1. Engineering Students' Intention Learning English.

What Is Your Intention in Learning English?	$\bar{x}$	σ
To enhance communication with individuals	4.58	0.53
To carry out tasks relevant to my program	4.43	0.67
To write a paper/report	4.41	0.66
To facilitate academic gatherings (seminar, meeting with experts etc.)	4.37	0.81
To communicate with English speaking experts	4.33	0.65
To accomplish an ample grade in exams	4.22	0.64
Number of Sample (n = fifty-three)		

The intention of the students in learning English at Polytechnic is shown in the respondents' opinions (see Table 1). It is discovered that the central purpose of the respondents in acquiring knowledge in English was to enhance communication or interaction with individuals as this intention score the highest means of 4.58. This is similar to the findings in Kim's research [25], which indicated that daily conversational English is the most significant in engineering context. The respondents also indicated that the objective in learning English is to carry out tasks relevant to their program and this was shown as the second highest in the mean score. The study also discovered that the students had other intentions to learn English such as writing papers or reports, participating in any academic purpose gathering for example seminars (4.37), meeting with experts, etc., communicating with English-speaking experts in their field (4.33). Interestingly, though means score is still high, the intention to accomplish an ample grade in exam involving language under the scope of the study score the lowest means which is 4.22.

The questionnaire includes the significance of learning English as the second theme which is indicated in Table 2. It was shown that the respondents gave serious attention to their future as the future engineers. Thus, for them, the significance of learning English as a person and as a student with a degree is not important compared to their future as an engineer. This is indicated in very high means of 5.00 for the significance of learning English as a future engineer as compared to means 4.70 and 4.64 for the first two significance of learning English namely 'how learning English is important for you as a person' and 'as a student with a degree' respectively

Table 2. The Significance of Learning English.

How Learning English Is Important for You?	x̄	σ
As a person	4.70	0.50
As a student with a degree	4.64	0.62
As a future engineer	5.00	0.57
Number of Sample (n = fifty-three)		

In Table 3, the learning context preference in an English course which is the third theme also shows high means scores. It is quite intriguing and unexpected that the respondents prefer a conventional classroom compared to open and distance learning classroom and hybrid learning context as the means score for both are 4.42 and 4.11, respectively. Open and distance learning classroom score lower in the students' preferences, with means 3.30 only. It is not known why the students are not interested to learn in open and distance learning platforms; this could be investigated in the future studies.

Table 3. The Learning Context Preference in an English Course.

Which Learning Context Do You Prefer in Learning English Course?	x̄	σ
Conventional classroom	4.42	0.84
Hybrid learning	4.11	0.93
Open and Distance learning style of classroom	3.30	1.08
Number of sample (n = fifty-three)		

Table 4 indicates that the respondents considered that the most significant reading sub-skills were reading technical papers and reading handbooks/manuals in their engineering scope of study. This was in comparison to reading laboratory procedures, course notes and books related to course. All of the sub skills core high means with the highest means of 4.51 is for reading skills in technical paper, while the lowest means core of 4.19 is for reading skills involving other course related books.

Table 4. The Significance in the Enhancement of Reading Skills.

Sub-Skills: Reading	x̄	σ
Technical paper in engineering scope	4.51	0.67
Handbook/Manuals in engineering scope	4.45	0.70
Laboratory procedures	4.28	0.74
Course notes	4.26	0.68
Course related books	4.19	0.76
Number of Sample (n = fifty-three)		

Writing is vital for human communication and has been perceived with a very high percentage of importance in learning English [26]. The respondents indicated that they want to improve their writing skills when they write and prepare lab reports and this can be seen in Table 5. Writing projects have been given the second highest importance. Their field of study in polytechnic requires them to have these two writing sub-skills. In addition, the respondents are also involved with many tasks which are related with their course where they need to prepare lab reports and documents relevant to their projects in their respective departments. However, less importance was given to the other sub-skills in writing, which included making notes from written sources/origin and writing tasks for assignments and responding to exam questions. The results showed that improvements might be required in terms of their writing ability which includes producing a piece of work in writing and gaining feedback from a facilitator or an instructor, reviewing comments and subsequently, the frequent engagement in the entire process of writing [27].

Table 5. The Significance in the Enhancement of Writing Skills.

Sub-Skills: Writing	$\bar{x}$	σ
Lab reports/record	4.66	0.52
Projects description	4.58	0.60
Making notes from written source/origin	4.57	0.64
Tasks for assignments	4.47	0.67
Responding to exam questions	4.40	0.69
Number of Sample (n = fifty-three)		

The next theme, which is the significance of the enhancement of listening skills, is displayed in Table 6. It is discovered that the respondents faced problems in concentrating or listening to the lecture as well as standing in a class performing an oral presentation as all the three show significantly high means scores with listening to lecture score the highest means of 4.53 and listening to procedures score the lowest means of 4.28. Therefore, comparing audio and video sources related to courses as well as listening to procedures, it is emphasized that the respondents desire to accomplish an improvement in both sub-skills on listening.

Table 6. The Significance in the Enhancement of Listening Skills.

Sub-Skills: Listening	$\bar{x}$	σ
Lecture	4.53	0.75
Oral presentations	4.41	0.75
Audio and video sources related course	4.41	0.72
Procedures	4.28	0.82
Number of Sample (n = fifty-three)		

The questionnaire also laid out the seventh theme which is the significance in the enhancement of speaking skills, as shown in Table 7. Asking questions in the class was the highest speaking sub-skill that had been indicated for improvement by the respondents. The respondents also identified another speaking sub-skill that they need to improve, which was performing an oral presentation in class. This is a very significant finding because most of the tasks in their field of study need to be understood before being carried out. Hence failing to ask questions in class will prevent them from receiving appropriate feedback from lecturers. Consequently, the students will struggle more during their presentations in class for two major reasons: lack of understanding and low confidence level.

Table 7. The Significance in the Enhancement of Speaking Skills.

Sub-Skills: Speaking	$\bar{x}$	σ
Making enquiry in class	4.60	0.63
Performing oral presentation in class	4.39	0.69
Carrying out discussion with classmates	4.32	0.64
Project presentation to audience	4.26	0.65
Number of Sample (n = fifty-three)		

Table 8 shows half of the respondents (50.9%) have indicated that they encounter difficulties with vocabulary, idioms, grammar and pronunciation even though they have pointed out that intermediate upper is their current level of English. This finding is consistent with the studies by Alastal and Shuib [28] and Ulum [29], which indicated that majority of students claimed that they have a weakness in using grammar correctly. Significantly, many other studies have also highlighted the same findings [30–32]. This intermediate level enables them to converse in a comfortable way in a common or everyday situations and issues. However, about 41.5% of the respondents' are in advanced level. Thus, they can converse naturally with fluency on most of the topics with few difficulties in the acquisition of vocabulary, understanding idioms, proper utilization of grammar and using a correct pronunciation. Another 7.50% of the respondents indicated that intermediate lower is their present level of ability in English. Hence, this shows their communication can be carried out with fluency but with different topics and some difficulties in vocabulary acquisition, idioms, grammar, and pronunciation. The results revealed that the students' capability in English is not entirely weak. On the other hand, the struggle shows that even though they are in a higher education institution which is considered to be the tertiary level of education, poor knowledge of grammar can be challenging because it is highly needed for interactions and conversations in real life [33,34].

Table 8. Level of Ability in English.

Please State Your Current Level of Ability in English (Choose One Only)	(f)	Percentage
Mother tongue influences the vocabulary, grammar and pronunciation heavily: Basic Lower	None	None
Survival level of knowledge of vocabulary, grammar and idiom with wrong pronunciation, short conversation on small number of anticipated topic: Basic Upper	None	None
Some difficulty with vocabulary, idiom, grammar with English fluency in communication: Intermediate lower:	4.0	7.50
Some difficulty with vocabulary, idioms, grammar and pronunciation, manageable in familiar situation comfortably with familiar issues: Intermediate upper	27.0	50.99
Little difficulty with vocabulary idioms grammar and pronunciation but able to converse or communicate fluently and naturally on most issues: Advanced	22.0	41.50
Number of sample (n = fifty-three) (f) = frequency		

6. Conclusions

The data which is required from the degree-level students have successfully demonstrated the needs analysis by adapting and adopting the questionnaire created by Kayaoglu and Akbas [15]. The study has fulfilled its purpose of discovering the respondents' needs. The next step by the researchers is to plan an effective course for the degree-level students. This course is expected to fulfil the needs of the respondents in this study. This needs analysis will enable the course designer to determine the scope that is suitable for the students' level of proficiency locate the scope of the language which need to be suitable with the students' level of proficiency [35]. In addition, a needs analysis would open an opportunity for the development of a material or program by providing information on learners' knowledge, perceptions of their learning, information on their learning styles as well as their interest [36,37]. In relation to this, the main skill which the respondents are facing difficulty is writing skill. Hence, the course developed should be catering to the needs and problem areas faced by the respondents which will be the goal to be achieved in a course. This is supported by Basturkmen [38], who stated that in designing a course, a 'narrow angled' course should be produced towards an objective determined. Hence, the course provided will match the needs and wants from the engineering students.

Furthermore, teaching material can also be produced to scaffold the teaching and learning process for continuous improvement. The needs analysis has some directions for future course planning. The scope which needs improvement is listening, reading and speaking skills. A well-planned course such as English for Specific Courses should be suitable and effective for higher education teaching and learning [39]. Therefore, when the institutions recognize their students' needs, the students' motivation in learning can increase producing a positive outcome and career prospects [40,41]. This needs analysis is significant because it offers the opportunity to fulfill the needs and wants of the degree-level students which will enable the formation of a teaching and learning environment which is conducive to degree-level students. An effective course design can be constructed, which will ensure the students at the tertiary level can develop the skills necessary for them to become successful engineers with worldwide employability potential.

Author Contributions: Conceptualization, S.M.B.B.V.K.S.V.K., A.S.D.P. and S.S.B.M.; methodology, S.M.B.B.V.K.S.V.K., A.S.D.P. and S.S.B.M.; validation, S.M.B.B.V.K.S.V.K., A.S.D.P. and S.S.B.M.; formal analysis, S.M.B.B.V.K.S.V.K., A.S.D.P. and S.S.B.M.; investigation, S.M.B.B.V.K.S.V.K., A.S.D.P. and S.S.B.M.; resources, S.M.B.B.V.K.S.V.K., A.S.D.P. and S.S.B.M.; data curation S.M.B.B.V.K.S.V.K., A.S.D.P. and S.S.B.M.; writing—original draft preparation, S.M.B.B.V.K.S.V.K., A.S.D.P. and S.S.B.M.; writing—review and editing, S.M.B.B.V.K.S.V.K., A.S.D.P. and S.S.B.M.; visualization, S.M.B.B.V.K.S.V.K., A.S.D.P. and S.S.B.M.; supervision, S.M.B.B.V.K.S.V.K., A.S.D.P. and S.S.B.M.; project administration, S.M.B.B.V.K.S.V.K., A.S.D.P. and S.S.B.M. All authors have read and agreed to the published version of the manuscript.

Funding: This research received no external funding.

Institutional Review Board Statement: Not applicable.

Informed Consent Statement: Not applicable.

Data Availability Statement: Not applicable.

Acknowledgments: This study would like to express the acknowledgement to the Research and Innovation Unit, Academic Development Centre under Sultan Salahuddin Abdul Aziz Shah Polytechnic.Paper presented in International Academic Symposium of Social Science (IASSC) 2022.

Conflicts of Interest: The authors declare no conflict of interest.

References

1. D'Silva, V. More and More Graduates Are Facing Unemployment in Malaysia. Available online: https://www.nst.com.my/news/nation/2020/02/562309/more-and-more-graduates-are-facing-unemployment-malaysia (accessed on 2 March 2020).
2. Mohamad, N.H.; Selamat, A.; Ibrahim, B.; Salleh, B.M. Exploration of spiritual elements in holistic-entrepreneur (Holistic-e) among TVET graduate students. *J. Tech. Educ. Train* **2019**, *11*, 72–83. [CrossRef]
3. Rahman, M. English for specific purposes (ESP): A holistic review. *Univ. J. Educ. Res.* **2015**, *3*, 24–31. [CrossRef]
4. Chostelidou, D. A needs analysis approach to ESP syllabus design in Greek tertiary education: A Descriptive account of students' needs. *Procedia Soc. Behav. Sci.* **2010**, *2*, 4507–4512. [CrossRef]
5. Deutch, Y. Needs analysis for academic legal english courses in Israel: A model of setting priorities. *J. Engl. Acad. Purp.* **2003**, *2*, 125–146. [CrossRef]
6. Bosher, S.; Smalkoski, K. From needs analysis to curriculum development: Designing a course in health-care communication for immigrant students in the USA. *Engl. Specif. Purp.* **2002**, *21*, 59–79. [CrossRef]
7. Hoekje, B.J. Medical discourse and ESP courses for International Medical Graduates (IMGs). *Engl. Specif. Purp.* **2007**, *26*, 327–343. [CrossRef]
8. Mazdayasna, G.; Tahririan, M.H. Developing a profile of the ESP needs of iranian students: The case of students of nursing and midwifery. *J. Engl. Acad. Purp.* **2008**, *7*, 277–289. [CrossRef]
9. Dehnad, A.; Bagherzadeh, R.; Bigdeli, S.; Hatami, K.; Hosseini, F. Syllabus revision: A needs analysis study. *Procedia Soc. Behav. Sci.* **2010**, *9*, 1307–1312. [CrossRef]
10. Liu, J.Y.; Chang, Y.J.; Yang, F.Y.; Sun, Y.C. Is what I need what I want? Reconceptualising college students' needs in English courses for general and specific/academic purposes. *J. Engl. Acad. Purp.* **2011**, *10*, 271–280. [CrossRef]
11. Ulucay, S.; Demirel, Ö. Perceptions of Professionals, Academicians and Current and Graduate Students on Designing an ESP Curriculum for Logistics Department. *Procedia Soc. Behav. Sci.* **2011**, *15*, 794–800. [CrossRef]
12. Akyel, A.S.; Ozek, Y. A language needs analysis research at an english medium university in Turkey. *Procedia Soc. Behav. Sci.* **2010**, *2*, 969–975. [CrossRef]
13. Taherdoost, H. Measurement and scaling techniques in research methodology: Survey/questionnaire development. *Int. J. Acad. Res. Manag.* **2016**, *6*, 1–5.
14. Davis, D.; Cosenza, R.M. *Business Research for Decision Making*, 3rd ed.; Wadsworth Publishing: Belmont, CA, USA, 1993; ISBN1 0534932495. ISBN2 9780534932497.
15. Kayaoğlu, M.N.; Akbaş, R.D. An Investigation into medical students' english language needs. *Particip. Educ. Res.* **2016**, *3*, 63–71. [CrossRef]
16. Durdu, L.; Özden, M.Y.; Delialioğlu, Ö. Design, development and implementation of a web-based support material for English as second language. *J. Teach. Educ.* **2012**, *1*, 84–110.
17. Ganapathy, M.; Shuib, M.; Azizan, S.N. Malaysian ESL students' perceptions on the usability of a mobile application for grammar test: A case study of ESL Undergraduates in Universiti Sains Malaysia. *3L: Lang. Linguist. Lit.* **2016**, *22*, 127–140. [CrossRef]

18. Ali Mohsen, M. The use of computer-based simulation to aid comprehension and incidental vocabulary learning. *J. Educ. Comput. Res.* **2016**, *54*, 863–884. [CrossRef]
19. Abdelhalim, S.M. Developing EFL students' reading comprehension and reading engagement: Effects of a proposed instructional strategy. *Theory Pract. Lang. Stud.* **2017**, *7*, 37. [CrossRef]
20. ILTA. Codes of Ethics. 2016. Available online: http://www.iltaonline.com/images/pdfs/ilta_code.pdf (accessed on 2 May 2019).
21. Loizos, P. Video, Film, and Photographs. In *Qualitative Researching with Text Image and Sound: A Practical Handbook*; Bauer, M.W., Gaskell, G., Eds.; SAGE: Thousands Oak, CA, USA, 2000; pp. 93–107.
22. Becker, S.; Bryman, A.; Ferguson, H. *Understanding Research for Social Policy and Social Work: Themes, Methods and Approaches*, 2nd ed.; Policy Press: Bristol, UK, 2012; ISBN 978-1847428158.
23. Powers, D.E.; Powers, A. The incremental contribution of TOEIC® listening, reading, speaking, and writing tests to predicting performance on real-life English language tasks. *Lang. Test* **2015**, *32*, 151–167. [CrossRef]
24. Connolly, P. *Quantitative Data Analysis in Education: A Critical Introduction Using SPSS*; Routledge: London, UK, 2007; ISBN 9780415372985.
25. Kim, H.H. Needs analysis for english for specific purpose course development for engineering students in Korea. *Int. J. Multimed. Ubiquitous Eng.* **2013**, *8*, 279–288. [CrossRef]
26. Tang, Z.; Cheng, L.; Chen, C.C. *Improving Engineering Students' Writing Skills in a Computer-Based Course Using an Automated Writing Evaluation System*; The Hong Kong Polytechnic University Hung Hom: Kowloon, Hong Kong, 2016.
27. Burstein, J.; Chodorow, M.; Leacock, C. Automated essay evaluation: The criterion online writing service. *AI Mag.* **2004**, *25*, 27. [CrossRef]
28. Alastal, A.A.; Shuib, M. Investigating the academic English language target needs of undergraduates at the faculty of applied science at Al-Aqsa University: Students' perceptions. *Asian ESP J.* **2012**, *8*, 5–26.
29. Ulum, Ö.G. A Needs analysis study for preparatory class ELT Students. *Onl. Sub.* **2015**, *1*, 14–29. [CrossRef]
30. Alqunayeer, H.S.; Zamir, S. Needs analysis of Saudi EFL female students: A case study of Qassim University. *J. Curric. Stud.* **2016**, *5*, 87–104. [CrossRef]
31. Moslemi, F.; Moinzadeh, A.; Dabaghi, A. ESP needs analysis of Iranian MA students: A case study of the University of Isfahan. *Engl. Lang. Teach* **2011**, *4*, 121–129. [CrossRef]
32. Prachanant, N. Needs analysis on English language use in tourism industry. *Procedia Soc. Behav. Sci.* **2012**, *66*, 117–125. [CrossRef]
33. Budiarta, I.K.; Krismayani, N.W. Improving speaking skill and developing character of the students through collaboration of think-pair-share and the concept of Tri Kaya Parisudha. *J. Santiaji Pend.* **2014**, *4*, 73–80. [CrossRef]
34. Syakur, A. Using cooperative learning strategies to improve the student's reading comprehension. *J. Edu. Kaj. Ilmu Pend.* **2018**, *4*, 97–110.
35. Tzotzou, M. Designing and administering a needs analysis survey to primary school learners about EFL learning: A case study. *Presch. Prim. Educ.* **2014**, *2*, 59–82. [CrossRef]
36. Huang, W. *6th International Education, Economics, Social Science, Arts, Sports and Management Engineering Conference (IEESASM 2018)*; Atlantis Press: Amsterdam, The Netherlands, 2019; pp. 249–256.
37. Zhang, J. College English curriculum group: The construction based on needs analysis. *Engl. Lang. Teach* **2018**, *11*, 80–85. [CrossRef]
38. Basturkmen, H. *Developing Courses in English for Specific Purposes*; Palgrave Macmillan: New York, NY, USA, 2010; ISBN 978-0-230-29051-8.
39. Chostelidou, D. Needs-based course design: The impact of general English knowledge on the effectiveness of an ESP teaching intervention. *Procedia Soc. Behav. Sci.* **2011**, *15*, 403–409. [CrossRef]
40. Aysan, E. Learning science and science education in a new era. *Ann. Med. Surg.* **2015**, *4*, 158–161. [CrossRef] [PubMed]
41. Agussani, A. The Approach to Indonesian Language and Literature Curriculum in the Industrial Revolution Era 4.0 and Society 5.01. Available online: https://publication.umsu.ac.id/index.php/ht/article/view/244 (accessed on 8 February 2021).

proceedings

MDPI

Proceeding Paper

Gender and Faculty Relationship: Oral Presentation Apprehension Factors during Online Distance Learning †

Masliza Mat Zali *, Razifa Mohd Razlan, Raja Mariam Raja Baniamin and Roszainora Setia

Academy of Language Studies, Universiti Teknologi MARA (UiTM) Cawangan Terengganu, Dungun 23000, Terengganu, Malaysia
* Correspondence: masli154@uitm.edu.my

† Presented at the International Academic Symposium of Social Science 2022, Kota Bharu, Malaysia, 3 July 2022.

Abstract: Speaking is rated as the most fearful task among language learners. Since oral presentation is one of the most vital tasks for second language learners, most learners experience oral presentation apprehension during their learning life, regardless of whether this is encountered face-to-face or through online distance learning (ODL). The study aims to investigate whether there are any significant relationships concerning the state and traits of apprehension regarding oral presentations across genders and faculty. The purposive sampling method was utilized for this study, which consisted of 92 undergraduates taking an oral presentation skills course. The data collected were analyzed using correlation analysis (IBM SPSS Statistics for Windows, IBM Corp. Version 24.0. Armonk, NY, USA). The analysis revealed that gender shows a significant relationship with the state of apprehension; meanwhile, faculty has no relationship with apprehension. For future research, this study provides insights for educators and learners on how to deliver an engaging oral presentation successfully.

Keywords: oral presentations; apprehension; online distance learning; gender; faculties

check for updates

Citation: Zali, M.M.; Razlan, R.M.; Baniamin, R.M.R.; Setia, R. Gender and Faculty Relationship: Oral Presentation Apprehension Factors during Online Distance Learning. *Proceedings* **2022**, *82*, 70. https://doi.org/10.3390/proceedings2022082070

Academic Editor: Mohamad Rahimi Mohamad Rosman

Published: 20 September 2022

1. Introduction

One of the objectives of each university is to produce globally competitive and marketable graduates who can go on to develop the nation, as well as to help graduates obtain good jobs once they graduate. On the other hand, according to previous studies [1,2], fresh Malaysian graduates lack employability skills, including a poor understanding of the English language and weak communication skills. Their poor command of English includes poor speaking and oral presentation skills. This is an alarming issue as it will tarnish the recognition of the education system in Malaysia, as well as impede Malaysia's vision of becoming a developed economic country [3]. The need to produce a well-rounded graduate is strongly emphasized in the Malaysia Education Blueprint 2012–2025 and the Malaysia Education Blueprint 2015–2025.

Although English is a second language in Malaysia, its use within the country has become more crucial than ever before as society, in general, has started to see its significance in today's competitive digital age. At the university level, English is offered as a core subject for all undergraduate students. When comparing the four language skills, speaking is the most fearful among English language learners [4] because it requires a greater abundance of vocabulary, challenges regarding enunciation, articulation, and content, as well as a high confidence level [5]. Many English as a second language (ESL) learners find that oral presentations are unsatisfying, intimidating, tough, worrying [6], and represent the most problematic oral communication skill [7]. Bhati [8] found that oral assessment has a higher apprehension level compared to written assessment; therefore, we would like to look further at students' apprehension in oral assessment during online distance learning (ODL). Furthermore, Chen [9] studied the relationship between ESL students' speaking-in-class anxiety and their presentation performance and also looked into the factors that caused oral anxiety during presentations and the strategies to regulate L2 students' speaking anxiety in presentations. The study found that the impact of external factors is greater than the impact of internal factors. Chen also [9] suggested that future research on oral presentations should investigate the relationship of gender on oral presentation apprehension.

Objectives of the Study

The main objective of this study is to investigate whether there are any significant relationships between traits and states of apprehension regarding oral comprehension across genders and faculties during ODL. This research is based on the following research questions:

1. Are there any significant relationships between the trait of apprehension and gender regarding oral presentations given over ODL?
2. Are there any significant relationships between the state of apprehension and gender regarding oral presentations given over ODL?
3. Are there any significant relationships between the trait of apprehension and faculty regarding oral presentations given over ODL?
4. Are there any significant relationships between the state of apprehension and faculty regarding oral presentations given over ODL?

2. Literature Review

2.1. Apprehension in Oral Presentation

According to Hotwitz et al. [4], communication apprehension can be defined as "a type of shyness characterized by fear of or anxiety about communicating with people". They argue that a learner who has difficulty in listening to or learning via spoken messages, or has difficulty in speaking in public or in a group, is experiencing communication apprehension. However, the most widely accepted definition is the one provided by McCroskey [10], who defined communication apprehension as "an individual's level of fear or anxiety associated with either real or anticipated communication with another person or persons". Although the definitions may vary, it is noticeable that they all involve ideas of fear, distress, anxiety, shyness, and discomfort when being in touch with others. Meanwhile, Noor Hanim et al. [11] also revealed that fear of giving an oral presentation is caused by traits and states of apprehension in the learners. Students' apprehension of an oral presentation can be caused by two main factors. Also, according to Santrock [12], apprehension can be divided into traits and states of apprehension. These two factors are widely used in research as factors of apprehension in oral communication.

2.2. Theoretical Framework

By adapting the questionnaires and theoretical frameworks from Santrock [12] and Noor Hanim et al. [11], Figure 1 reveals the conceptual framework of the study. This study explores the factors of trait and state of apprehension regarding oral presentations given over ODL among students. The trait of apprehension is an internal factor related to personal characteristics, whereas the state of apprehension is an external factor from the environment.

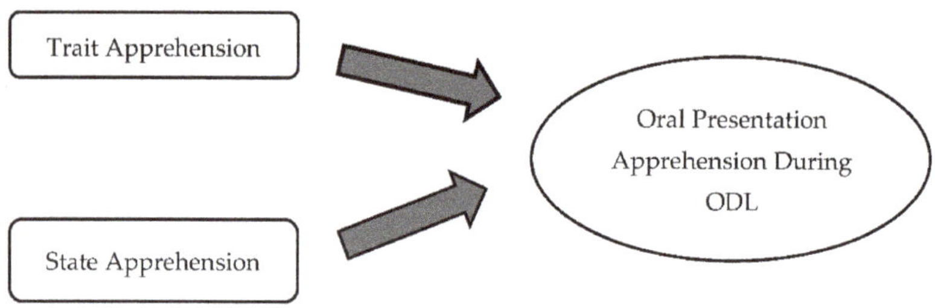

Figure 1. Theoretical Framework of Study.

The first category of apprehension is trait. This type of anxiety grows in response to a perceived threat. People with this kind of anxiety cannot work well with oral presentations due to their personal traits. A sign of concern about the trait of apprehension is that tensions from this are on-going in their expression. Second, the speakers may be too worried about their lack of preparation, comparing themselves to others or worrying that they may forget their speech. Also, he or she cannot rely on internal abilities and external appearances. Speakers can also be afraid of the audience.

The second type of apprehension is "state". This type of fear represents an unpleasant emotional experience when faced with a particular situation, request, or object or event. The state of anxiety occurs when a speaker makes a psychological assessment of some type of threat. Speakers are afraid of either their own physical representation, the environment, or even the decision-making process. Students may be afraid of the grades they get in their presentations. They may also be afraid of the size and composition of the audience or even the location and time of the presentation. They may also recognize previous performance as a threat (lower or higher). Second, students may be afraid of the negative ratings they may receive from the presentation. They are worried about the impression the examiners have on them. They are also afraid of the examiners' opinions. They are afraid to say the wrong thing when presenting themselves.

Figure 2 has shown the conceptual framework of the study. This study explores whether there are any significant relationship of trait and state apprehension for oral presentation across genders and faculties. The two main constructs are trait and state apprehension; while the variables are gender and faculties.

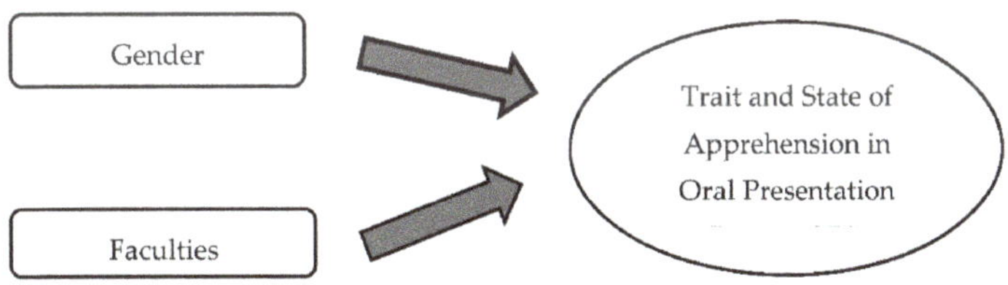

Figure 2. Conceptual Framework of Study.

3. Results and Discussion

3.1. Correlation Analysis

Pearson correlation was used to describe the strength and direction of the relationship between two variables. In this study, the relationship between the apprehension factors (state and trait) across genders and faculty were investigated using this analysis. A positive correlation indicates that as one variable increases, so does the other variable. A negative correlation indicates that as one variable increases the other decreases. A perfect correlation of 1 or −1 indicates that the value of one variable can be determined exactly by knowing the value of the other variable. On the other hand, a correlation of 0 indicates no relationship between the two variables. The Bivariate correlation procedure was subjected to a two-tailed test of significance using two different levels: highly significant ($p < 0.01$) and significant ($p < 0.05$).

Table 1 reveals a summary of the results from the correlation analysis. As shown in the table, there is a negative correlation between the factors of apprehension (state and trait) and gender. The highest correlation is between Trait Apprehension and Gender ($r = -0.296/p < 0.01$). The lowest correlation is between State Apprehension and Gender ($r = -0.269/p < 0.01$). Based on Cohen's interpretation of values (pp. 79–81, [13]), there are three guidelines that are suggested: small (r = 0.01 to 0.29), medium (r = 0.30 to 0.49), and large (r = 0.50 to 1.0). These guidelines apply whether or not there is a negative sign for the r value. Positive and negative r values have the same strength of correlation. Therefore, these correlation values indicate a small correlation between the variables, suggesting quite a weak relationship between State Apprehension and Gender, as well as

Trait Apprehension and Gender. However, State and Trait Apprehension do not correlate with Faculty (r = −0.202, r = 0.066/p > 0.05), indicating no relationship.

Table 1. Correlation Analysis.

No.			1	2	3	4
1	State Apprehension	Pearson Correlation Sig. (two-tailed) N				
2	Trait Apprehension	Pearson Correlation Sig. (two-tailed) N	0.389 ** 0.000 92			
3	Gender	Pearson Correlation Sig. (two-tailed) N	−0.269 ** 0.010 92	−0.296 ** 0.004 92		
4	Faculty	Pearson Correlation Sig. (two-tailed) N	−0.202 0.053 92	−0.066 0.530 92	0.134 0.204 92	

** Correlation is significant at the 0.01 level (two-tailed).

3.2. *Trait Apprehension across Gender*

Research question 1: Are there any significant relationships between the trait of apprehension and gender regarding oral presentations given over ODL?

Table 2 shows the results of the Pearson Correlation: −0.296. This shows the significant relationship between Trait Apprehension and Gender for oral presentations (r = −0.296/p < 0.01). At the same time, the results show a small correlation between the genders, suggesting quite a weak relationship between Trait Apprehension and Gender. In other words, Gender has a slight influence on Trait Apprehension, which represents personal traits, such as feeling nervous during presentation, fear of the audience, distress at their lack of preparation, comparing themselves to others, or worrying that they may forget their speech. He or she also could not rely on his or her own abilities and appearance. This study agrees with a study conducted by the authors of [11], which found that there is a relationship between gender and traits of apprehension regarding oral presentations. On the other hand, this study did not support a previous study [14], in which it was found that gender is not a significant factor in communication apprehension.

Table 2. Results for the relationship between Trait Apprehension and Gender for oral presentations.

		Trait Apprehension
Gender	Pearson Correlation Sig. (two-tailed) N	−0.296 ** 0.004 92

** Correlation is significant at the 0.01 level (two-tailed).

3.3. *State Apprehension across Gender*

Research Question 2: Are there any significant relationships between the state of apprehension and gender regarding oral presentations given over ODL?

Table 3 shows the results of the Pearson Correlation: −0.269. This shows a significant relationship between Trait Apprehension and Gender regarding oral presentations (r = −0.269/p < 0.01). Similarly, to State Apprehension result, the results also show a small correlation to Gender, showing a weak relationship between State Apprehension and Gender. In other words, gender has a slight influence on Trait Apprehension, which includes external factors like speakers being afraid of either their own physical representation, the environment, or even their decision-making process. Students may be afraid of the grades they receive for their presentations, the size and composition of the audience, or even the location and time of the presentation, as well as the negative ratings they may receive from the presentation. This result agrees with another previous study [11], which found that there is a correlation between gender and the state of apprehension for oral presentations; meanwhile, the results differ from those from [14], which found that gender is not a significant factor in communication apprehension.

Table 3. Results for the relationship between State Apprehension and Gender for oral presentations.

	State Apprehension	
Gender	Pearson Correlation	−0.269 **
	Sig. (two-tailed)	0.010
	N	92

** Correlation is significant at the 0.01 level (two-tailed).

3.4. *Trait Apprehension across Faculties*

Research Question 3: Are there any significant relationships between the trait of apprehension and faculty regarding oral presentations given over ODL?

Table 4 shows the result of the Pearson Correlation: −0.066. This shows that there is no significant relationship between Trait Apprehension and Gender regarding oral presentations ($r = -0.066/p > 0.05$) within the two faculties: the Faculty of Business Management and the Faculty of Hotel Management and Tourism. This study supports the findings in [11], which showed that there is no correlation between faculty and the state of apprehension in oral presentations.

Table 4. Results for the relationship between State Apprehension and Faculty for oral presentations.

	Trait Apprehension	
Faculties	Pearson Correlation	−0.066
	Sig. (two-tailed)	0.530
	N	92

3.5. *State Apprehension across Faculties*

Research Question 4: Are there any significant relationships between the state of apprehension and faculty regarding oral presentations given over ODL?

Table 5 shows the result of the Pearson Correlation: −0.066. This shows no significant relationship between Trait Apprehension and Gender for oral presentations ($r = -0.202/p > 0.05$) within the two faculties: the Faculty of Business Management and the Faculty of Hotel Management and Tourism. This result is similar with those found in [11], which found that there is no correlation between faculty and the state of apprehension regarding oral presentations.

Table 5. Results for the relationship between State Apprehension and Faculty for oral presentations.

	Trait Apprehension	
Faculties	Pearson Correlation	−0.202
	Sig. (two-tailed)	0.053
	N	92

4. Conclusions

To summarize, the results of this study reveal that gender has a significant relationship with both traits and states of apprehension regarding oral presentations given over ODL. Even it is a small correlation, the influence of gender on traits and states of apprehension regarding oral presentations is slightly evident. However, the faculties from which the students are from bears no relationship with traits and states of apprehension concerning oral presentations given over ODL. Students, regardless of faculty, exhibited traits and states of apprehension during their presentation.

Author Contributions: Conceptualization, R.S. and R.M.R.; methodology, R.M.R.; software, R.M.R.B.; validation, M.M.Z.; formal analysis, M.M.Z.; investigation, R.M.R. and R.S.; resources, R.M.R. and R.M.R.B.; data curation, R.M.R.B.; writing—original draft preparation, M.M.Z. and R.S.; writing—review and editing, M.M.Z.; visualization, R.M.R.B.; supervision, M.M.Z.; project administration, M.M.Z.; funding acquisition, M.M.Z. and R.M.R.B. All authors have read and agreed to the published version of the manuscript.

Funding: This research received no external funding.

Institutional Review Board Statement: Not applicable.

Informed Consent Statement: Not applicable.

Data Availability Statement: Not applicable.

Acknowledgments: We would like to express our gratitude to those who have been involved in and supported this study directly and indirectly, especially the 92 students involved in the study, as well as the Writing and Publication Unit (PJI) UiTM Cawangan Terengganu, under the project "Inspiring June 2022", which inspired this study.

Conflicts of Interest: The authors declare no conflict of interest.

References

1. Zahiid, S.J. Poor English Eroding Malaysian Graduates' Self-Belief. Malaymail Online. Available online: http://www.themalaymailonline.com/malaysia/article/pmpoor-englisheroding-malaysian-graduates-self-be (accessed on 10 January 2020).
2. Hossain, M.I.; Yagamaran, K.S.A.; Afrin, T.; Limon, N.; Nasiruzzaman, M.; Karim, A.M. Factors influencing unemployment among fresh graduates: A case study in Klang Valley, Malaysia. *Int. J. Acad. Res. Bus. Soc. Sci.* **2018**, *8*, 1494–1507. [CrossRef]
3. Shamsuddin, A.; Isa, K.H.M.; Aziz, M.N.; Mahfol, N.Z.N.M.; Alagari, T. Graduate unemployment: The awareness and perception of graduates towards Government's initiatives. *Int. J. Bus. Econ. Law* **2013**, *3*, 15–24.
4. Horwitz, E.K. Even teachers get the blues: Recognizing and alleviating language teachers' feeling of foreign language anxiety. *Foreign Lang. Ann.* **1996**, *29*, 365–372. [CrossRef]
5. Rahmat, N.H. Cycle of Fear in Learning: The Case for Three Language Skills. *Am. J. Soc. Sci. Humanit.* **2019**, *4*, 151–162. [CrossRef]
6. King, J. Preparing EFL learners for oral presentations. *Internet TESL J.* **2002**, *4*, 401–418.
7. Soomro, M.A.; Siming, I.A.; Shah, S.H.R.; Rajper, M.A.; Naz, S.; Channa, M.A. An investigation of anxiety factors during English oral presentation skills of engineering undergraduates in Pakistan. *Int. J. Engl. Linguist.* **2009**, *9*, 203–210. [CrossRef]
8. Bhati, S.S. The Effectiveness of oral presentation assessment in a finance subject: An empirical examination. *J. Univ. Teach. Learn. Pract.* **2012**, *9*, 6. [CrossRef]
9. Chen, Y. ESL Students' Language Anxiety In In-Class Oral Presentations. Master's Dissertation, Marshall University, Huntington, WV, USA, 2015. Available online: http://mds.marshall.edu/cgi/viewcontent.cgi?article=1967&context=etd (accessed on 10 January 2020).
10. McCroskey, J.C. Oral communication apprehension: A summary of recent theory and research. *Hum. Commun. Res.* **1977**, *4*, 78–96. [CrossRef]
11. Rahmat, N.H.; Othman, N.A.; Hasan, H.R.; Arepin, M. An Analysis of ESL Oral Presentation Using Trait and State Apprehension. *European J. of Engl. Lang. Teaching.* **2018**, *4*, 65–77.
12. Santrock, J.W. *Educational Psychology*; McGraw-Hill: Boston, MA, USA, 2009.
13. Cohen, J. *Statistical Power Analysis for the Behavioral Sciences, 2nd ed*; Lawrence Erlbaum Associates, Publishers: Hillsdale, NJ, USA, 1988.
14. Oladipo, S.E.; Agbajeola, R.O.; Adenaike, F.A. Assessing communication apprehension among in-service teachers in a university of education. *Res. Humanit. Soc. Sci.* **2012**, *2*, 27–31.

proceedings

MDPI

Proceeding Paper

Printed Modular Approach on Selected High School Learners in Infanta Pangasinan for Successful Learning Delivery Amidst Disruption †

Catherine N. Lumanta *, Narciso A. Martin, Jr., Randy Joy M. Ventayen, Ian D. Evangelista, Eddelaine B. Monte, Dhaly D. Boctuanon and Rofer-Jay S. Ferrer

PSU Laboratory High School Department, Pangasinan State University, Infanta Campus, Infanta 2412, Philippines
* Correspondence: clumanta.infanta@psu.edu.ph

† Presented at the International Academic Symposium of Social Science 2022, Kota Bharu, Malaysia, 3 July 2022.

Abstract: The coronavirus disease altered human history in terms of the normal course of life, day-to-day business transactions, and education setup. It could be observed that learning institutions across the globe have been affected much by the raging effect of the pandemic. This study was conceptualized based on the learning insertion made in basic education to continue to live up to the expectations in the academic community emanating from the grassroots. The printed modular approach became the instrument of stakeholders inserted into the actual learning process as the study described its statuses, effects, and challenges on selected High School learners.

Keywords: academic performance; grade point average; printed modular approach; self-learning module

Citation: Lumanta, C.N.; Martin, N.A., Jr.; Ventayen, R.J.M.; Evangelista, I.D.; Monte, E.B.; Boctuanon, D.D.; Ferrer, R.-J.S. Printed Modular Approach on Selected High School Learners in Infanta Pangasinan for Successful Learning Delivery Amidst Disruption. *Proceedings* **2022**, *82*, 88. https://doi.org/10.3390/proceedings2022082088

Academic Editors: Mohamad Rahimi Mohamad Rosman

Published: 29 September 2022

1. Introduction

The entrance of COVID-19 in the first quarter of the year 2020 marked a lasting imprint on everyone's heart and mind, accentuated by different emotions across different walks of life.

Therefore, the academic community was inescapable to its impact, globally. Thus, academic leaders must come up with an appropriate plan to succeed in the path of the teaching and learning processes.

In the Philippine setting [1], the coronavirus disease forced schools in the country to stop face-to-face learning activities and abruptly shift to a modular approach. In a practical sense, the modular approach situates Filipino learners to learn in the comfort of their homes. In this picture, the parents or guardians are the ones who aid in the entire homeschooling process.Research extant confirmed that the objectives of successful learning delivery even in the prevailing health crisis are still achievable. Many researchers [2] confirmed the promising benefits of modules regularly provided by the Secondary Public Schools as alternatives to attain successful learning every day. Moreso, one study [3] supported flexible learning in the learning process amidst disruption.

In the same context, it was argued that the Modular Distance Learning Modality (MDLM) warranted many benefits for teachers, learners, and parents. The study [4] concluded the many difficulties of modular implementation that could affect learners' individuality. The study recommended creating a holistic teaching approach through synchronous and blended experiential learning, increasing partnership with stakeholders, and ensuring accessibility of Learning Resources (LRs) and Instructional Materials (IMs). Nonetheless, in their study, researchers determined the leisure activities conducted by students at home during the pandemic. The concepts of the source of stress and coping mechanisms of students during the lockdown and extreme community quarantine due to the health crisis reminded us not to overlook teenagers, such as students, in helping them [5].

The opening of classes, specifically at the basic education level, was encouraged even at the heed of the "academic freeze". Professionals, educators, parents, and guardians prioritized safety and well-being. However, education remained the top priority, such as for the Department of Education, Philippines. Academic measures were issued to steadfast the set education objectives despite the current challenges and threats posed by the learning environment.

These safety precautionary measures somehow provided little confidence on the part of the members of the academic community. Schools initially strategized their workforce to combat the raging effect of the pandemic without compromising the health and safety of teachers, parents, and learners.

Among the many school strategies employed is the modular approach. This is one way of addressing the need for continuous education while little confidence is being challenged by the danger brought over time by the environment, in which we once believed we were safe and secure. Through this platform, there is the future of learners getting promoted to another grade level every year, while learning tasks could still warrant successful learning delivery.

The concept of the modular approach has been a practice among educational institutions. Through this approach, schools prepared learning references such as printed learning activities and Instructional Materials (IMs). In achieving the many objectives of the modular approach, learning institutions design their individual plan to continue educating the young ones while the pandemic is still happening.

For example, the Department of Education issued a memorandum to ensure continuous education. The Basic Education Learning Continuity Plan (BELCP) established the roadmap for handling and delivering the modular approach. It guided how to deliver education in this time of crisis. The key components were multiple learning modalities with blended and distanced learning, and preparing school leaders and teachers for multiple learning delivery modalities.

Consequently, The Pangasinan State University Laboratory High School Infanta Campus adopted the beneficial effect that the modular approach could give to its stakeholders. The school situates in a 3rd class Municipality where accessibility becomes one of the major concerns. Given the challenges surrounding this learning institution, such as provisions of basic education essentials, access to farm and market roads, student support, strong financial assistance and program, and school administration, the Campus keeps on moving forward to attain its academic goals. With the available resources concerning facilities and funding requirements, the School Administrators capacitate to continue the torch of delivering excellent and quality basic education, especially to far-flung and mountainous areas.

The Pangasinan State University Laboratory High School Infanta Campus prompted the research idea on the prevailing statuses, effects, and challenges given by the printed modular approach. Through this initiative, the School Administrators are sincere to its mission and vision for the University. Educating the youth and the community where it serves could mold them to become active agents towards driving the changes and needs of their society.

This study hoped to be of great help in describing loopholes in the actual learning process through the "insertion" of a printed modular approach enjoined by the stakeholders. Researchers envisioned continuously honing today's achievers to become leaders of tomorrow, which is the goal of the Pangasinan State University Laboratory High School Infanta Campus.

2. Materials and Methods

This study used the descriptive survey research design. The nominated school was the Pangasinan State University Laboratory High School that has offered Junior and Senior High School programs since 2016. The study used convenient sampling for the Academic Year 2020–2021. The study surveyed the statuses, effects, and challenges on the conduct of the printed modular approach as an intervention in the actual learning process of the High School Learners.

This study was conducted in the Pangasinan State University Laboratory High School Infanta Campus located at Bamban, Infanta, Pangasinan where the respondents were officially enrolled in the Junior and High School programs. The total samplings comprised one hundred twenty (120).

The study adopted an instrument [6]. The study structured one (1) set of questionnaires consisting of four (4) parts.

The first part asked the demographic profile of Grade 7 to 12 learners as to the grade level, sex, age, academic performance or grade, and honors received.

The second part revealed the statuses of the modular approach on the academic performance using a Likert scale where 1.00 (lowest) was interpreted as Least Agree (LeasA) and 5.00 (highest) was interpreted as Very Highly Agree (VHA). It comprised eight (8) benchmark statements.

The third part revealed the effects of the modular approach on academic performance using a Likert scale where 1.00 (lowest) was interpreted as Least Effective (LeasE) and 5.00 (highest) was interpreted as Very Highly Effective (VHE). It comprised fourteen (14) benchmark statements.

The fourth part revealed the effects of the modular approach on academic performance using a Likert scale where 1.00 (lowest) was interpreted as Least Serious (LeasS) and 5.00 (highest) was interpreted as Very Highly Serious (VHS). It comprised nineteen (19) benchmark statements.

The instrument had a total of forty-six (46) items of questions and statements and underwent minor modifications of wording to suit its purpose.

The instrument was originally structured using the medium of instruction and was translated by a Filipino Language Expert to ensure that the respondents fully understood the context of each question.

The instrument was piloted with eighteen (18) respondents to warrant integrity with reference to the Cronbach's alpha value [7].

This study surveyed The Pangasinan State University Laboratory High School as it also advocated quality education at the Secondary level. This study used convenient samplings where the computed number of respondents was based on expert panel judgment. The total sampling comprised one hundred twenty (120) learners. The pertinent records of High School learners, such as the total number of enrollees for the period and Quarterly Assessment Grade were requested from the office of the School Administrator. The study requested the six (6) Advisers (Grade 7 to 12) to join in the Focus Group Discussion (FGD), respectively. The researchers were the ones who personally distributed and retrieved the questionnaires because of the prevailing health crises with the consent and approval of the School Administrators and the concerned parents or guardians. The Junior and Senior High School learners answered the set of questionnaires. The questionnaires asked the respondents about the statuses, effects, and challenges through the printed modular approaches inserted into the actual learning process of The Pangasinan State University Laboratory High School Infanta Campus.

The retrieval ran from August 2020 to June 2021.

The study surveyed the demographic profile of the respondents as to their grade level, sex, age, academic performance or grade, and honors received, and presented them in tabular or graphical forms along with the given references [8–13].

3. Results and Discussion

For the profile of the respondents, data showed that only sex had a significant relationship to their Grade Point Average (GPA) with a computed t-value of 5.418 and p-value of 0.000 which were less than 0.05. Further, the grade level and age had no significant relationship to their GPA, since the p-value was 0.759 and 0.397, respectively, which were greater than 0.05. Females performed better than males in the printed modular approach, which was reflected in their GPA and with the greatest number of learners who obtained "With Honors".

For the status of the printed modular approach, the computed Pearson r was 0.375 and a p-value of 0.000 which was less than 0.05 at a 0.01 level of significance. This meant that there was a significant relationship between the status of the printed modular approach and the GPA of the learners.

For the effects of the printed modular approach, the computed Pearson r was 0.257 and a p-value of 0.005 which was less than 0.05 at a 0.01 level of significance. This meant that there was a significant relationship between the effects of the printed modular approach and the GPA of the learners.

For the challenges of the printed modular approach, the computed Pearson r was0.034 and a p-value of 0.712 which was greater than 0.05 at a 0.01 level of significance. This meant that there was no significant relationship between the challenges of the printed modular and the GPA of the learners.

For the significant difference of dependent and independent variables, ANOVA Single-Factor showed that the computed F-value was 12.08 which was greater than the F-critical value of 3.02, indicating that there was a significant difference among the three (3) variables, i.e., Status, Effects, and Challenges.

Further, the post-hoc test revealed that: (1) The computed t-value was 1.35 and the t-critical value for one-tailed was 1.66 and 1.98 for two-tailed. Since the computed t-value was less than the t-critical value, it meant that there was no significant difference between status and effects. (2) The computed t-value was 5.26 and the t-critical value for one-tailed was 1.66 and 1.98 for two-tailed. Since the computed t-value was greater than the t-critical value, it meant that there was a significant difference between status and challenges. (3) The computed t-value was 3.85 and the t-critical value for one-tailed was 1.66 and 1.98 for two-tailed. Since the computed t-value was greater than the t-critical value, it meant that there was a significant difference between effects and challenges.

4. Conclusions

Despite the risk exposures on the health and safety of the members of the academic community at the basic education level, the School Administrators propelled with the institution's mandate in providing excellent service delivery and quality education. The printed modular approach benefits the teachers, parents or guardians, and learners. However, it could not warrant the stakeholders' confidence level to health and emotional well-being for long-term perspectives.

The Local Government Unit and Barangay Officials play a significant role in the overall process of the printed modular approach when they are properly involved, well-informed, and encouraged for higher participation.

This study generated the following hypotheses: H_1—the printed modular has a difficulty index per grade level, age, and sex of the learners; H_2—the printed modular approach has provided more benefits to learners; H_3—the printed modular approach becomes more effective than the traditional approach; H_4—the printed modular approach has moderate challenges to learners; H_5—the printed modular approach warrants "authentic assessment" into the actual learning process; H_6—the printed modular approach provides an accurate basis for quarterly assessment, i.e., Grade Point Average (GPA); H_7—the printed modular approach promotes learners to the next grade level; H_8—the printed modular approach helps learners obtain "With Honors", "With High Honors", and "With Highest Honors"; and H_9—the printed modular approach helps female (sex) learners obtain honors.

Author Contributions: Conceptualization, C.N.L. and N.A.M.J.; methodology, N.A.M.J.; validation, I.D.E. and R.J.M.V.; formal analysis, C.N.L., N.A.M.J., I.D.E. and R.J.M.V.; investigation, E.B.M., D.D.B. and R.-J.S.F.; resources, C.N.L., N.A.M.J., I.D.E., R.J.M.V., E.B.M., D.D.B. and R.-J.S.F.; data curation, N.A.M.J.; writing—original draft preparation, C.N.L. and N.A.M.J.; writing—review and editing, C.N.L., N.A.M.J., I.D.E. and R.J.M.V.; visualization, I.D.E. and R.J.M.V. All authors have read and agreed to the published version of the manuscript.

Funding: This research received no external funding.

Institutional Review Board Statement: Not applicable.

Informed Consent Statement: Not applicable.

Data Availability Statement: Not applicable.

Acknowledgments: The authors wish to extend their appreciation to The Pangasinan State University Laboratory High School Infanta Campus, School Administrators, Concerned Advisers, Concerned Parents, Concerned Guardians, and Concerned High School Learners.

Conflicts of Interest: The authors declare no conflict of interest.

References

1. Marohombsar, A.M. View of Modular Approach: Its Effect on Learners' Reading Attitude and Proficiency in the New Normal. Available online: https://al-kindipublisher.com/index.php/ijlss/article/view/1932/1608 (accessed on 23 June 2022).
2. Martin Jr, N.A.; Ventayen, R.J.M.; Patacsil1, D.H.; Patacsil1, E.H. The development of game estrategia in the senior high school learning competencies: Basis for digital intelligence (phase one). *J. Phys. Conf. Ser.* **2021**, *1860*, 012021. [CrossRef]
3. Ventayen, R.J.M. Teachers' Readiness in Online Teaching Environment: A Case of Department of Education Teachers. *PSU J. Educ. Manage. Soc. Sci.* **2018**, *2*, 94–106.
4. Martin, N.A., Jr.; Ventayen, R.J.M.; Morales, M.M. Reap What You Sow: Lived Experiences of Modular Learners in The New Norm. 2020. Available online: https://papers.ssrn.com/abstract=3909822 (accessed on 23 June 2022).
5. Hero, C. Based on De Guzman Pastor et al 2020 Most of the Students Are Stressed about the. Available online: https://www.coursehero.com/file/p796dt50/Based-on-De-Guzman-Pastor-et-al-2020-most-of-the-students-are-stressed-about-the/ (accessed on 23 June 2022).
6. Aksan, J.A. Effect of Modular Distance Learning Approach to Academic Performance in Mathematics of Students in Mindanao State University-Sulu Senior High School Amidst COVID-19 Pandemic. *Open Access Indones. J Soc Sci.* **2021**, *4*, 445–467. [CrossRef]
7. Konting, M.M.; Kamaruddin, N.; Man, N.A. Quality Assurance In Higher Education Institutions: Exist Survey Among Universiti Putra Malaysia Graduating Students. *Int. Educ. Stud.* **2009**, *2*, 25–31. [CrossRef]
8. Department of Education Division Memorandum, No. 265 s. 2020. Module key Dates _1_. Available online: https://www.deped.gov.ph/2020/07/02/deped-prepares-self-learning-modules-for-educations-new-normal/ (accessed on 23 June 2022).
9. Department of Education Order, No. 001 s. 2021. Guidelines on the Evaluation of Self-Learning Modules for Quarters 3 and 4 for School Year 2020–2021. Available online: https://www.deped.gov.ph/wp-content/uploads/2021/01/DO_s2021_001-.pdf (accessed on 23 June 2022).
10. Department of Education Order, No. 12 s. 2020. Adoption-of-the-LCP-BE-LCP-for-SY-2020-2021.pdf - Google Drive. Available online: https://drive.google.com/file/d/1IPjAHDTIOmtsyPXY_3FunDzIxRsGwL07/view (accessed on 23 August 2021).
11. Department of Education Order, No. 18 s. 2020. Policy-Guidelines-for-the-Provision-of-Learning-Resources-in-the-Implementation-of-the-Basic-Education-Learning-Continuity-Plan.pdf. Available online: https://www.deped.gov.ph/wp-content/uploads/2020/08/DO_s2020_018.pdf (accessed on 23 June 2022).
12. DepEd Memo OUCI-2020-307. MEMORANDUM OUCI-2020-307 Department of Education Undersecretary For Curriculum And Instruction. 2020. Available online: https://www.deped.gov.ph/wp-content/uploads/2020/11/OUCI-2020-307_ACADEMIC-EASE-DURING-COVID-19-final-version-01Nov2020.pdf (accessed on 23 June 2022).
13. Department of Education Order, No. 36 s. 2016. Policy Guidelines On Awards And Recognition For The K To 12 Basic Education Program I. Rationale The Policy Guidelines on Awards and Recognition for the K to 12 Basic Education. Available online: https://www.deped.gov.ph/wp-content/uploads/2016/06/DO_s2016_036.pdf (accessed on 23 June 2022).

proceedings

Proceeding Paper

Practicality of Ad Hoc Online Assessments for Teaching ESP in Online Flipped Classrooms during COVID-19 †

Muhd Khudri Johari [1,*] and Nur Zaimah Jamil [2]

1 Academy of Language Studies, Universiti Teknologi MARA Melaka, Alor Gajah Campus, Alor Gajah 78000, Malaysia
2 Faculty of Modern Languages and Communication, Universiti Putra Malaysia, Serdang 43400, Malaysia
* Correspondence: muhdkhudri@uitm.edu.my
† Presented at the International Academic Symposium of Social Science 2022, Kota Bharu, Malaysia, 3 July 2022.

Abstract: The COVID-19 pandemic has severely affected the livelihood of many ever since its first detection nearly a year ago. Up until now, almost every domain of societal norms has been drastically impacted—particularly education. Many people immediately opted for Internet-based apps and programs within their capabilities and financial means to cope with the learning progress for themselves and the people around them. Learning institutions and teaching staffs worldwide quickly adopted these technologies and later adapted everything within online contexts. From there, extensive changes had been made to cope with the teaching/learning issues caused by the global pandemic. By utilizing an adapted questionnaire, this study discussed and evaluated the practicality of online assessments that had been implemented posthaste to assess an Aviation English course for aircraft maintenance undergraduates in flipped classrooms. The discussion findings indicated that while there were certainly a number of issues that can be quickly improvised in the near future, the whole implementation was definitely a success—especially with the ongoing limitations experienced by both educators and students.

Keywords: problem-based learning; Aviation English; paired sign test

Citation: Johari, M.K.; Jamil, N.Z. Practicality of Ad Hoc Online Assessments for Teaching ESP in Online Flipped Classrooms during COVID-19. *Proceedings* **2022**, *82*, 91. https://doi.org/10.3390/proceedings2022082091

Academic Editor: Mohamad Rahimi Mohamad Rosman

Published: 1 October 2022

1. Introduction

Online assessments are not something new, and it can be considered as something common amongst many learners and educators. However, most learning institutions still utilized physical assessments up until the beginning of the COVID-19 pandemic in early 2020. No one would have expected all the ongoing physical assessments to be converted into online assessments. It has been reported that an estimation of 1.6 billion learners worldwide have been affected by the pandemic. The number closely reflects more than 90% of the world's learner population from over 190 countries across all the continents. To make things worse, prior global issues concerning learners such as educational disparities and learning opportunities for all children regardless of socio-economic statuses have yet to be solved in many parts of the world [1].

Despite these hardships, educators and administrators worldwide have kept on experimenting and implementing numerous ways to overcome the problems caused by the lockdowns. One of the most common solutions is conducting online assessments [2] for all, which has always been an issue of resources and connectivity [3]. However, the emergency situation for many countries had hindered governments and policymakers from making drastic changes within a short period of time; hence, the ultimate lack of options has rendered online assessments to be the most convenient and practical method to continue the education flow [4]. In Malaysia, it all started in mid-March 2020; some of the first higher learning institutions (HLIs) that had gone online before others did the same were Universiti Malaya, Universiti Putra Malaysia, Universiti Sains Malaysia, Universiti

Kebangsaan Malaysia, Xiamen University Malaysia, Universiti Malaysia Sabah [5] and Universiti Kuala Lumpur. In the midst of rush, confusion and panic among many people during these times, there were also rational, logical and calm decision-making processes involved. With careful planning and quick adaptation skills, educators began to initiate migrating contents into digital clouds and teaching them in online assessments—including the researcher who was lecturing Aviation English courses at the time.

As a subdomain under English for Specific Purposes (ESP), it was unbeknownst to the researcher regarding any previous documentations or publications produced to specifically highlight on matters concerning teaching Aviation English courses in flipped classrooms with a full online mode. The emergency situation entailed a number of possible risks and consequences; hence, the researcher has decided to evaluate if the online assessments carried out were practical or not—from the students' perspectives.

2. Literature Review

To reiterate the previous statement, assessments carried out in online classes are definitely not something novel; even for teaching ESP, it is actually quite common and has been practiced by many for quite some time now. Quite recently, a group of researchers had even outlined extensive information on integrating ICT for teaching ESP [6]. One of the critical issues discussed was how can educators incorporate ICT elements with their highly-trained pedagogical methodology concurrently and effectively. It would be inaccurate to say that using PowerPoint slides to showcase ESP information is already sufficient to be considered as *"integrating ICT elements with teaching ESP"*. There are at least several conditions that need to be fulfilled to justify the notion.

First of all, both educators and students must have satisfactory background knowledge with technological components that may be used throughout the entire course. It would include basic system setup, possible applications, and accumulated experience of having utilized the components in previous occasions. Next, teaching ESP is ultimately still teaching how to use the English language appropriately within the intended domain [7]. Therefore, sound knowledge of assessing the acquired linguistic knowledge within the technological components' parameters is vital. In addition, this could only be carried out if the contents of the assessments were taught using an effective collaboration of pedagogical skills and technological contexts.

For instance, a group of educators from several universities located in different countries had proven that a flipped classroom can be taught online effectively [8]. Although their participants were undergraduates taking the online assessments for engineering-based courses, it would still be relevant to be adapted and systematically applied in courses from other domains. The main focus still reflects both online delivery methods and online assessments performed. As stakeholders, the students themselves would occasionally know and realize the fact that this distinctive transition from physical assessments to online assessments could possibly have short-lasting or long-lasting effects (or both) on their academic performance and psychological conditions. With the freedom to evaluate what is given to them, it was hoped that they would make the most out of this rare opportunity.

3. Methods

The objective of this study was to evaluate the practicality of online assessments implemented in an Aviation English course from the perspectives of undergraduates learning in flipped classrooms.

Prior to answering the questionnaire, the students had managed to undergo flipped classrooms (with their respective class groups) once a week; (i) in physical classes from Week 2 to Week 5, and (ii) in online mode from Week 6 to Week 12. Supposedly, there would be two written tests scheduled (Test A in Week 6 and Test B in Week 9. However, due to the major conversion (from physical class to online mode) that happened in between Week 5 and Week 6, the researcher had promptly initiated alternative media for both written tests before Week 6 started. When the time came, the lecturer asked for a clear image of their answer script after Test A questions had been displayed in students' Learning Management System (LMS) university portal. The test was conducted within 60 min, just like how it would normally be if the test were to be given in a normal physical class.

However, for Test B in Week 9, the researcher not only utilized a different medium (Google Form, for both multiple-choice and open-ended questions), but it was also carried out with extended test time. The supposed 60 min was extended to 180 min as per requested by many students right after Test A was finished. The request was deliberated with administrators, and permission to extend the test time was granted because of the extenuating circumstances faced by students. Due to the societal and financial impacts of COVID-19, a number of students had limitations when it came to technological and Internet networking capabilities.

3.1. Participants

This study utilized a specifically tailored questionnaire adapted from Chaisuriya and Shin (2019) [9] via Google Form to collect data from undergraduates enrolled in Aircraft Maintenance Technology programs at Universiti Kuala Lumpur—Malaysian Institute of Aviation Technology (UniKL MIAT), Malaysia. During the point when this study was initiated, all of the nearly 400 students who took the Aviation English course had already begun continuing their flipped classrooms in online mode by utilizing Microsoft Teams. They were divided into three different program majors; each major had at least three different groups with maximum of 28 students per group. Hence, cluster random sampling was used to ensure representatives from all three program majors were selected. For that, the Google Form link of the questionnaire was posted online—on their respective groups' Microsoft Teams announcement walls.

The data was collected from Week 15 to Week 18 of Semester January–June 2020. Every step of the data collection process had been performed virtually following the strict COVID-19 protocols imposed by the university. The virtual questionnaire was optional and not related to any of their course assessments. Hence, the number of participants was expected to be on the lower side. With the cluster random sampling technique, it was recorded that only 184 undergraduates answered the questionnaire.

3.2. Instruments

The questionnaire had four sections; two close-ended and two open-ended sections. The demographic section, which was the first section, was brief to prevent bias elements and ease students to do it online without any reservations: it had *age, current location, name of program*, and *name of major*.

The second section had 10 statements reflecting on their perceptions toward the online assessments implemented for the Aviation English course. The statements considered several criteria that can be associated with practicality such as format, convenience, awareness, availability, and frequency [9]. The third section had five statements pertaining to the assessments' contents and mode of delivery. Both the second and third sections utilized a 4-point Likert scale (**1**-*Strongly disagree*; **2**-*Disagree*; **3**-*Agree*; **4**-*Strongly Agree*) to questionnaire the students' agreement/disagreement with the statements listed. The researcher thought that it was imperative to omit any '*neutral*' standpoint in this matter (such as *neutral, not sure, don't know*, etc.) as this would concern prompt real-life implementations and therefore could possibly have implications in the near future. The values for interpreting the findings are as in Table 1 below:

Table 1. Percentages for each proficiency level.

Mean (M)	Interpretation
1.00 < M < 1.50	Strong disagreement
1.51 < M < 2.50	Disagreement
2.51 < M < 3.50	Agreement
3.51 < M < 4.00	Strong Agreement

The data collected was quantified with SPSS software to calculate the mean and standard deviation; the numbers were then be interpreted as final conclusions of the questionnaire. The fourth section was an open-ended question that gave the opportunity to all participants to write their suggestions and/or personal feedback regarding their abnormal semester, which saw a drastic change from physical assessments to online assessments in the middle of the semester.

4. Results

The results in Tables 2 and 3 show that majority of the 184 participants agreed on the practicality criteria asked in the questionnaire:

Table 2. Mean and standard deviation for second section.

Online Assessments' Practicality	M	SD	Meaning
1. The format of the test seems familiar to me.	3.50	0.789	Agreed
2. I can only attempt this test once.	3.46	0.816	Agreed
3. It is convenient for me to attempt this online test.	3.55	0.759	Strongly agreed
4. It is convenient for me to submit this online test.	3.42	0.852	Agreed
5. I am fully aware that this test exists virtually.	3.50	0.843	Strongly agreed
6. I do not need additional guidance on how to view this test online.	3.48	0.843	Agreed
7. Good Internet connection is readily available in my current location.	3.33	0.814	Agreed
8. Nowadays, I can communicate with lecturers and classmates anytime.	3.48	0.843	Agreed
9. I have no problem taking online tests for more than once this semester.	3.38	0.752	Agreed
10. Having more online assessments is better than physical assessments.	3.52	0.789	Strongly agreed

Table 3. Normality of distribution for both groups' pre-test scores.

Assessments Contents and Mode of Delivery	M	SD	Meaning
1. The number of questions in the online assessments are sufficient.	3.55	0.730	Strongly agreed
2. The time limit for the online assessments is adequate.	3.50	0.803	Agreed
3. The speaking assessment was done without individual presentation because of time constraint and extensive mobile data usage.	3.55	0.715	Strongly agreed
4. The audio files for listening assessment were clear enough.	3.46	0.788	Agreed
5. It is better to answer the open-ended question(s) in Google Form rather than writing the answers with pen and paper.	3.45	0.815	Agreed

Overall, the numbers indicated that almost all students agreed with the statements above. The highest mean (3.55) was recorded for second section's Question 3 and third section's Questions 1 and 3. The lowest mean (3.33) was recorded for second section's Question 7.

The small difference between the highest and lowest means proved that most students found the online assessments to be practical enough for their predicament of having to almost instantaneously adapt to online mode. One possibly major factor is students' demographic background. Having born in the years post-2000, which had been dubbed as the starting point of modern and digital millennium, they were already exposed to extensive usage of Internet, mobile phones, computers and other technological hardware. They could cope with the sudden change by participating in the online assessments and attempting the online assessments. Regardless of their respective assessments' marks and achievements, their perspectives on the online assessments were clear—they were practical; at the very least, they were practical enough to be carried out for typical university students.

However, even though the numbers for both sections look positive in general, the minority did express their thoughts with the Likert scale and their respective feedback. Figure 1 below shows some of the students' feedback as to why there were (at most) six students who marked "1-*Strongly disagree*" and 14 students who marked "2-*Disagree*" for certain statements.

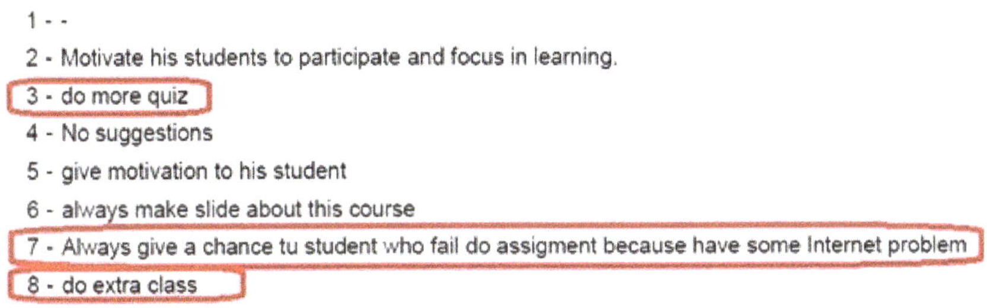
1 - -
2 - Motivate his students to participate and focus in learning.
3 - do more quiz
4 - No suggestions
5 - give motivation to his student
6 - always make slide about this course
7 - Always give a chance tu student who fail do assigment because have some Internet problem
8 - do extra class

Figure 1. Screenshot (1) of highlighted students' comments.

In this case, it can be acknowledged that a few students may have felt that the online assessments held did not meet their expectations in terms of quantity, as they requested for more assessments and also additional classes.

In addition to that, the suggestion, "*Always give a chance tu student who fail do assigment because have some Internet problem*", may have involved a handful of them who had been affected by the subpar Internet connectivity in their respective housing areas. Whilst Malaysian government and several telco operators did launch a nationwide package of free Internet data for every local citizen (both provided 1 Gb of data for every registered phone number), the coverage limitation was still an issue.

In Malaysia alone, there have been many news reports that showcased pictures, videos and social media posts that described citizens' hardship in accessing good Internet connectivity [10]. Most of them were located in rural and/or hilly areas without any telecommunications tower nearby. In addition, the free data was only made available starting from 1 April 2020, while telco operators made their free data package available starting from 10 June 2020—which was quite late for UniKL MIAT students' semester that had begun in January 2020. Nevertheless, it is also important to note that the Malaysian government did extend this free 1 Gb free data assistance until 31 December 2020; from initially within a 10 h period daily (8am–6pm) to an updated availability of 24 h period every day [11]. It was not much, per se, but it did help a lot of students to download scaled-down videos of lectures and soft copies of notes and assessments (such as *.pdf*, *.doc*, and *.ppt* files). It was also crucial for students to be able to log into their student portal and email for various purposes.

Figure 2 shows a screenshot from another class, with the additions of more suggestions. The suggestion, "*The lecturer should talk more loud and clear*", may indicate another technological problem—hardware and possibly software as well. This reflects the third section, which may be related to the audio files or the researcher's voice whenever he was lecturing. For the record, the researcher had been using Fantech HG11 headset and Realtek HD Audio Manager software preinstalled with desktop computer's Gigabyte motherboard. There have been instances when some default audio–video settings needed adjustments; for example, (i) when everyone logs into Microsoft Teams, users have the options to choose preferred audio and camera settings, and (ii) when multiple users had their microphone setting to be "*unmute*" and noises can be heard.

For some educators, they may perceive suggestions such as, "*Do physical class*", (during COVID-19) and, "*Give a lot of example in our daily life*", (while teaching Aviation English, which included real-life aviation contents and contexts) as less useful in a serious reflective teaching discussion. However, the researcher strongly believes that there were suggestions such as these because some students may need, or at the very least prefer, for their assessments to be more of an experiential learning experience rather than a plain, virtual class. Moreover, someone actually suggested, "*Give more individual assignment than group activities*", which would be an indicator of intrapersonal learner qualities.

1 - The lecturer can ask their students what have they learn on that day.
2 - okay
3 - just good
4 - -
5 - support his student
6 - No suggestions
7 - just do as usual
8 - The lecturer should talk more loud and clear.
9 - always make slide about this course
10 - Nothing
11 - He's good already
12 - Do physical class.
13 - good enough
14 - Give more individual assignment than group activities
15 - Give a lot of example in our daily life
16 - do extra class
17 - nothing
18 - -
19 - Always give a chance to students who have problem
20 - Need to smile a lot
21 - nothing
22 - he is super cool by his own

Figure 2. Screenshot (2) of highlighted students' comments.

Another interesting aspect that can be seen from the suggestions is nonverbal communication, which is clearly reflected from, "*Need to smile a lot*" (while teaching). In this case, the researcher agrees with the notion of body languages such as facial expressions and postures do play a significant role in effective teaching. For online mode, the visual limitations are to be expected as the students can only see what their educators allow them to see. The external cameras' screen resolution can never replace normal human vision. Both educators and students are bound to a limited frame, and it significantly limits the capabilities and effects of body languages.

5. Recommendations

Based on the discussions, the researcher has compiled specific recommendations for future researchers and practitioners to solve and prevent the problems faced by UniKL MIAT students not only in terms of online assessments but also some other things that had been suggested.

5.1. Number of Assessments

As per the respective syllabus requirements and teaching/learning standards set by learning institutions worldwide, educators do not have the flexibility or freedom to change the number of formal assessments for students; it can only be done under special circumstances with administrator's permissions and systematic moderation.

However, that is not the case with informal assessments that do not affect students' coursework marks. Hence, proper planning and effective communication strategies are essential, be it with colleagues who are teaching the same courses or with students. Usual pre-semester preparation will have to include additional online setups such as separate online groups and channels, various online templates for exercises and assessments, group email lists and online calendar items such as reminders and future tasks. Setting these up will indirectly ease assessment distributions later. Hence, educators can timely brief students on the possible number of assessments (for both formal and informal) from the get-go; the recommended frequency for this particular briefing is weekly or bi-weekly. Depending on the nature of courses taught, it can also be done monthly. This way, students can also prepare physically, mentally and technologically as they undergo the online learning process. It will also be advantageous should there be any sudden changes or events that can disrupt the teaching/learning flow; as everything has been planned out within specific periods, it will be easier to cross-check possible dates or sessions for replacement/retest purposes.

5.2. Internet Connectivity

In the event of weak Internet connection problem, several recommendations have been outlined based on personal experiences, directives from administrators, guidelines from the government and feedback from colleagues.

For UniKL, students had been advised to return to campus and stay indoors at all times as early as 5 April 2020. With 12 campuses across the Malaysian peninsula, students were also given options to mitigate their home-to-campus distance problem by staying at the nearest UniKL campus to their hometown. This was to avoid students from spending additional costs to pay for their travel fares. With strict enforcement from hostel fellows and security personnel, students could ensure good internal Internet connection whilst staying at a proper learning environment. This has also been adapted as standard practices amongst many public and private higher learning institutions in Malaysia as per the SOP guidelines announced by Ministry of Higher Education Malaysia [12].

Other viable options mainly include SIM cards; either specific perks and benefits at discounted prices offered by telco operators [13] with the extended awareness campaigns by university administrators; or free unlimited broadband data packages offered by universities and/or other organizations. UniKL offered this package to every eligible student who had legally registered as a person from a low-household-income family (also known as "*B40*") starting from 12 June 2020—which was roughly two weeks before the new semester started in July 2020. Other higher learning institutions had already begun these similar efforts; some even had collaborations with external organizations to boost their funding, and they were able to extend their number of recipients significantly [14].

5.3. Learner Types

As experienced educators, it would normally be easier to identify learner types of students in a physical class rather than in an online class. For example, close observations on how students speak, interact, move, and react face-to-face would normally be adequate to have some basic ideas in determining whether they are interpersonal or intrapersonal learners. However, this is difficult for online assessments especially in situations when students had to mute their audio and close their camera for various reasons.

The silver lining would be the responsibility of the educators to have both audio and camera switched on from beginning until the end of their assessments. From this perspective, it would seem as if the class is already teacher-centered. This notion is not entirely wrong because the students would naturally have minimal participation rate. Thus, the previous recommendations for online assessments must be customized to include in-class activities as well. Students must be made aware of the fact that they would always have the chance to not only participate but to do so actively and interactively with both their classmates and educators alike. Occasionally, educators would be able to pinpoint relevant learner types in the class and carry out appropriate learning activities and practices for the students.

6. Future Research

More experimental action research [15] is needed to further evaluate the practicality issues of online assessments for other courses and domains, which include (but are not limited to) aviation [16–18], engineering [19–21], health [22–24] and management [25]. The multitude of educational programs offered by various levels of learning institutions have been mostly affected by the pandemic. As highlighted, the practicality of online assessments is paramount as the education world is still adapting to the new social norms, and educators need to ensure their assessments are still applicable and effective in measuring the levels of acquired knowledge gained.

Author Contributions: Conceptualization, M.K.J. and N.Z.J.; methodology, M.K.J.; software, M.K.J.; validation, M.K.J. and N.Z.J.; formal analysis, M.K.J. and N.Z.J.; investigation, M.K.J.; resources, M.K.J.; data curation, M.K.J. and N.Z.J.; writing—original draft preparation, M.K.J.; writing—review and editing, M.K.J. and N.Z.J.; visualization, M.K.J.; supervision, M.K.J.; project administration, M.K.J.; funding acquisition, M.K.J. All authors have read and agreed to the published version of the manuscript.

Funding: The APC was funded by Universiti Teknologi MARA Melaka (*Geran Dalaman TEJA* 2022; grant number: GDT2022/1-11).

Institutional Review Board Statement: Not applicable.

Informed Consent Statement: Not applicable.

Data Availability Statement: Not applicable.

Acknowledgments: Several UniKL MIAT academicians provided guidance regarding the scoring systems (proficiency levels).

Conflicts of Interest: The authors declare no conflict of interest.

References

1. Policy Brief: Education during COVID-19 and Beyond. Available online: https://www.un.org/development/desa/dspd/wp-content/uploads/sites/22/2020/08/sg_policy_brief_covid-19_and_education_august_2020.pdf (accessed on 20 September 2020).
2. Khan, N.; Kuddus, K. Integrating ICT in English Language Teaching in Bangladesh: Teachers' Perceptions and Challenges. *Rupkatha J. Interdiscip. Stud. Humanit.* **2020**, *12*, 1–10. [CrossRef]
3. Akhter, T. Problems and Challenges Faced by EFL Students of Saudi Arabia during COVID-19 Pandemic. *Rupkatha J. Interdiscip. Stud. Humanit.* **2020**, *12*, 1–7. [CrossRef]
4. Strengthening Online Learning When Schools Are Closed: The Role of Families and Teachers in Supporting Students during the COVID-19 Crisis. Available online: http://www.oecd.org/coronavirus/policy-responses/strengthening-online-learning-when-schools-are-closed-the-role-of-families-and-teachers-in-supporting-students-during-the-covid-19-crisis-c4ecba6c/ (accessed on 26 September 2020).
5. It's Virtual Classes for Now. Available online: https://www.thestar.com.my/news/nation/2020/03/18/its-virtual-classes-for-now (accessed on 21 September 2020).
6. Muñoz-Luna, R.; Taillefer, L. *Integrating Information*; Springer International Publishing AG: Cham, Switzerland, 2018.
7. Chetia, B.; Bhatt, D. The Film Life of Pi as a Multimedia Tool in English Language Classrooms of Engineering Colleges in Gujarat—An ESP Approach. *Rupkatha J. Interdiscip. Stud. Humanit.* **2020**, *12*, 1–19. [CrossRef]
8. Tao, T.; Abuhmaid, A.; Olaimat, M.; Oudat, D.; Aldhaeebi, M.; Bamanger, E. Efficiency of flipped classroom with online-based teaching under COVID-19. *Interact. Learn. Environ.* **2020**, *28*, 1–19.
9. Chaisuriya, A.; Shin, S.Y. Examining English Test Practicality among Different Stakeholders in Thailand. *J. Asia TEFL* **2019**, *16*, 1–19. [CrossRef]
10. Malaysian Student Sits Exams in a Tree to Ensure Good Wifi. Available online: https://www.bbc.com/news/blogs-news-from-elsewhere-53079907 (accessed on 20 September 2020).
11. Free Additional 1GB Internet Data Daily until Dec 31. Available online: https://www.thestar.com.my/tech/tech-news/2020/06/05/pm-free-additional-1gb-internet-data-daily-until-dec-31#:~{}:text=economic%20recovery%20plan.%20%E2%80%94-,Malaysians%20will%20continue%20to%20receive%20free%20additional%201GB%20of%20Internet,Tan%20Sri%2 (accessed on 20 September 2020).
12. Students Returning to Hometowns Can Download SOP from Higher Education Ministry's Website. Available online: https://www.thestar.com.my/news/nation/2020/04/28/students-returning-to-hometowns-can-download-sop-from-higher-education-ministrys-website-386 (accessed on 21 September 2020).
13. Free Data: Here's What All Malaysian Telcos Are Offering during MCO. Available online: https://www.soyacincau.com/2020/03/26/malaysia-telco-covid-19-mco-free-data-access-services-comparison/ (accessed on 21 September 2020).
14. Free SIM Cards, Lessons from YTL Foundation's Collaboration with UKM. Available online: https://www.theedgemarkets.com/article/free-sim-cards-lessons-ytl-foundations-collaboration-ukm (accessed on 20 September 2020).

15. Yusof, M.A.; Ya'acob, A.M.; Zaki, M.A.M.; Rahman, Z.A.; Abidin, N.H.Z.; Padil, I.F.; Johari, M.K.; Bakar, I.A.; Hashim, H.F.M. Developing a virtual reality (VR) app for theory of flight & control as a teaching & learning aid. *Int. J. Innov. Technol. Explor. Eng.* **2019**, *28*, 86–90.
16. Abdul Samad, A.G.; Johari, M.K.; Omar, S. Preventing human error at an approved training organization using Dirty Dozen. *Int. J. Eng. Technol. (UAE)* **2018**, *7*, 71–73. [CrossRef]
17. Ariffin, M.W.Z.; Johari, M.K.; Ibrahim, H. The needs of aircraft avionics' radio line replaceable unit repair center at UniKL MIAT. *Int. J. Eng. Technol. (UAE)* **2018**, *7*, 86–88. [CrossRef]
18. Ya'acob, A.M.; Razali, D.; Anwar, U.A.; Radhi, A.H.; Ishak, A.A.; Minhat, M.; Mohd Aris, K.D.; Johari, M.K.; Casey, T. Preliminary Study on GF/Carbon/Epoxy Composite Permeability in Designing Close Compartment Processing. In Proceedings of the 1st International Conference on Aerospace and Mechanical Engineering (AeroMech 2017), Penang, Malaysia, 21–22 November 2017; pp. 1–9.
19. Azizan, M.A.; Izwan, A.; Johari, M.K. Aerodynamic analysis of aircraft wing performance realized in different wing aspect ratio. *Int. J. Eng. Technol.* **2019**, *28*, 100–103.
20. Azizan, M.A.; Hazim, M.; Johari, M.K. Analysis of energy separation acting in vortex tube. *Int. J. Eng. Technol.* **2019**, *28*, 91–95.
21. Azizan, M.A.; Hamzah, A.; Johari, M.K. The implementation of QR codes for aircraft disassemble part and NDT equipment inventory system for documentation purpose. *Int. J. Adv. Sci. Technol.* **2019**, *28*, 86–90.
22. Abdul-Samad, A.G.; Azizan, M.A.; Khairuddin, M.H.; Johari, M.K. A Review on the Mental Workload and Physical Workload for Aircraft Maintenance Personnel. In *Human-Centered Technology for a Better Tomorrow*; Lecture Notes in Mechanical Engineering; Springer: Singapore, 2022; pp. 627–635.
23. Abdul-Samad, A.G.; Azizan, M.A.; Khairuddin, M.H.; Johari, M.K. Effect of Mental Workload on Heart Rate Variability and Reaction Time of Aircraft Maintenance Personnel. In *Human-Centered Technology for a Better Tomorrow*; Lecture Notes in Mechanical Engineering; Springer: Singapore, 2022; pp. 613–625.
24. Abdul-Samad, A.G.; Azizan, M.A.; Khairuddin, M.H.; Johari, M.K. Significance of Aircraft Maintenance Personnel's Reaction Time during Physical Workload and Mental Workload. In *Human-Centered Technology for a Better Tomorrow*; Lecture Notes in Mechanical Engineering; Springer: Singapore, 2022; pp. 637–643.
25. Bardai, A.M.; Er, A.Z.; Johari, M.K.; Noor, A.A.M. A review of Kuala Lumpur International Airport (KLIA) as a competitive South-East Asia hub. In Proceedings of the 1st AEROS Conference 2017, IOP Conference Series: Materials Science and Engineering, Putrajaya, Malaysia, 12 December 2017; pp. 1–10.

proceedings

MDPI

Proceeding Paper

Problem-Based Learning (PBL) during Online Teaching †

Muhd Khudri Johari [1,*] and Nur Zaimah Jamil [2]

1 Academy of Language Studies, Universiti Teknologi MARA Melaka, Alor Gajah Campus, Alor Gajah 78000, Malaysia
2 Faculty of Modern Languages and Communication, Universiti Putra Malaysia, Serdang 43400, Malaysia
* Correspondence: muhdkhudri@uitm.edu.my

† Presented at the International Academic Symposium of Social Science 2022, Kota Bharu, Malaysia, 3 July 2022.

Abstract: The purpose of this study is to examine how problem-based learning (PBL) is adapted in an online teaching medium using Microsoft Teams and then utilized in Aviation English classrooms at a Malaysian aviation-based technical institution with the purpose of testing for any discernible effect. Using Microsoft Teams, lecturers guided final-year Aircraft Maintenance undergraduates into self-directed study and peer instruction for five consecutive weeks. A study was performed to determine the influence of online PBL instruction on learners' exam results. The paired sign test was used to analyze the non-normal data in SPSS—the mixed results showed that additional studies should be carried out for additional areas of study/curricula when pursuing PBL-Institute prestige in order to enhance the online PBL implementation processes and choose the most appropriate online PBL framework for each area of study.

Keywords: problem-based learning; Aviation English; paired sign test

check for updates

Citation: Johari, M.K.; Jamil, N.Z. Problem-Based Learning (PBL) during Online Teaching. *Proceedings* **2022**, *82*, 92. https://doi.org/10.3390/proceedings2022082092

Academic Editor: Mohamad Rahimi Mohamad Rosman

Published: 1 October 2022

1. Introduction

With several effective implementations and designs, online problem-based learning (PBL) has been customized in a variety of ways to accommodate a wide range of subjects, curricula, and programs. Developed as a result of the medical school curriculum [1], it is frequently used as an educational strategy [2] and a method of evaluation for students [3] at various levels—areas of study covered include engineering education [4].

The primary aim of this study was to determine the efficacy of online PBL in Aircraft Maintenance courses as a starting point for achieving PBL-Institute certification within several years. In some ways, this project may also be viewed as trial research with the aim of attempting to fully introduce online PBL at the university level. Online PBL was used to teach one of three Aviation English disciplines available in the Bachelor of Aircraft Engineering Technology degrees for this project; both Mechanical and Avionics students were chosen randomly. However, the number of Mechanical students examined was much greater than the total population of Avionics students, as just one Avionics group was offered the chance to be the chosen group during that term (final-year students).

2. Literature Review

Diverse Aviation English courses are presently taught at a variety of levels in higher education, ranging from certifications and A-level courses up to postgraduate programs. Certain educators continue to favor traditional instructional practices (e.g., huge lectures and public visualization, as well as experiential activities) as their approach; this is most likely due to personal opinions and perspectives [5]. Despite this, an increasing number of academics are developing and experimenting with different methods of approaching these courses with their students—notably online PBL [6].

The researchers chose to adapt an online PBL simulation as intervention procedures, and they not only tested for statistical importance but also used a pre–post-test experimental method [7]. Several noticeable variations were the overall number of students and their individual disciplines, the measuring instruments used, and the substance of the Aviation English topic. Researchers began by guiding the Experimental group students on interactive learning and peer instruction during the first week of sessions before the post-test. The second week marked the beginning of a totally learner-centered atmosphere in class; students were educated briefly on online PBL immediately following the set introduction, and they were required to answer the issues presented within 60 min. However, during the third and fourth weeks, students received their in-class tasks immediately following a predetermined induction, and researchers guided both sessions.

3. Methods

This sample was restricted to all (89) final-year Bachelor of Aircraft Engineering Technology students registered at Universiti Kuala Lumpur—Malaysian Institute of Aviation Technology (UniKL MIAT) and specializing in either a Mechanical or Avionics major. The course materials from the chapter "Writing an Aircraft Maintenance Report" were taught online via Microsoft Teams and tested over the five consecutive weeks. Participants were randomly assigned to one control group and three experimental groups from four classes (three Mechanical classes and one Avionics class). Both the pre- and post-test questions were open-ended and mirrored the complexities of real-world scenarios. Each exam required students to respond to all ten open-ended questions within 60 min. Two marks were awarded for completing all justifications and elaborations with the right answer to each question. Following consultation with the university's corresponding program coordinators and subject matter experts regarding the scoring systems, it was accepted that three proficiency levels would be used, particularly for the analysis of test scores (out of 20) in this investigation—'*Low*' (0–14), '*Moderate*' (15–17), and '*High*' (18–20).

4. Results

The overall subject proficiency level for all classes was ***moderate*** (mean = 17.629). Table 1 shows the percentages for each level.

Table 1. Percentages for each proficiency level.

Low (0–14 Marks)	Moderate (15–17 Marks)	High (18–20 Marks)
3.47% (3 out of 89 students)	75.28% (67 out of 89 students)	21.35% (19 out of 89 students)

Table 2 shows the normality of distribution for both groups' pre-test scores below.

Table 2. Normality of distribution for both groups' pre-test scores.

Group	Normality of Distribution
Experimental groups	The skewness value was −1.498, while the Kurtosis value was 5.245. Although the skewness value was still within the range between −2 and +2, the Kurtosis value was significantly outside the similar range. Even the Shapiro–Wilk's value (0.00) was lower than alpha. Hence, the data were considered to be non-normal.
Control groups	The skewness value was −2.29, while the Kurtosis value was 4.783. Both values were outside the range between −1 and +1. In addition, the Shapiro–Wilk's values (0.00 for Mechanical 1; 0.012 for Mechanical 2; and 0.032 for Avionics) were also lower than 0.05. Hence, the data were considered to be non-normal as well.

5. Analysis

5.1. Null Hypothesis

Overall, the Control group (M = 17.49, SD = 2.46) scored slightly lower than the Experimental group (M = 17.7, SD = 1.06). The significant value was 0.644, which was more than alpha, based on the findings of the one-way ANOVA. As a result, we were unable to reject the null hypothesis *'There is no significant difference between the students' pre-test scores'*. This is very reasonable given that no interventions were initiated at this point. Post-test results, on the other hand, were predicted to be altered following the initiation and implementation of the interventions.

5.2. Testing Mean Scores

Since the second hypothesis required evaluating the Experimental groups' pre- and post-test mean scores, they were all examined using the paired sample t-test. However, as previously stated, homogeneity tests revealed that the values from both the Experimental and Control groups were non-normal. As a result, the researchers chose a nonparametric test for this case—the paired sign test [8]. Prior to performing the analyses, the paired data (signs of different mean scores) were sorted as follows.

It was discovered that there were considerably more adverse differences (25 out of 45) between the values than there were favorable differences (14 out of 25) or ties (6 out of 25). The paired sign test revealed a Z-value of −1.601, which was less than alpha and, hence, rejected the null hypothesis. It may be stated that the online PBL intervention had a considerable effect on the Experimental group pupils. The mean for the Experimental groups' post-test scores declined from 17.70 (pre-test) to 16.90. Although these mean scores indicated the Experimental group students' overall results, the researchers aimed to ensure that they demonstrated identical tendencies for all students enrolled in their specific courses.

5.3. Testing Individual Scores

To assess the efficiency of online PBL in these classrooms precisely, the disparities between each student's pre- and post-test results were thoroughly evaluated. If the online PBL intervention had a genuinely measurable impact, as previously determined, the results from these specific studies would correspond to the prior pattern.

The p-value for the paired sign test in Mechanical 1 was specified as "Exact Sig. (2-tailed)"; it was 0.00, which was less than 0.05. This result corroborated the prior finding—the researchers rejected the null hypothesis and established that the online PBL therapy had a statistically significant influence on students' test results. There were twenty negative changes as opposed to a few positive differences (3) and ties (2), which could only suggest that the intervention was eventually lowering the Aviation English competency of Mechanical 1 students. It may also be supported by the fact that their post-test mean scores (14.96) were much lower than their pre-test mean scores (17.52), indicating that they fell into the Moderate category rather than achieving the High category.

For the second Mechanical group, the paired sign test findings indicated that the p-value was "Exact Sig. (2-tailed)"; it was 0.023, which was less than 0.05. Additionally, this number indicated that the researchers rejected the null hypothesis and established that the online PBL intervention had a substantial influence on the students' test results. There were 18 positive differences compared to much fewer negative differences (6) and ties (4), indicating that the intervention was effective for the Aviation English competency of the Mechanical 2 students. Although the mean scores for their post-test did not appear to be substantial due to the little increase (from 17.93 to 18.25), they were noteworthy since they crossed the barrier and achieved the higher level of High rather than retaining their competency at the very same level.

The p-value for the Avionics class was specified as "Exact Sig. (2-tailed)"; it was 1.000, which was more than alpha. This number indicated that the researchers did not reject the null hypothesis and established that the online PBL intervention had no discernible influence on the students' test results. There were five positive differences and ties as opposed to four negative differences, which could only mean that the researchers lacked compelling evidence at the 0.05 level to demonstrate a disparity in the Aviation English test scores of Avionics students following online PBL intervention compared to the prior Mechanical classes' findings—in addition, their mean scores did not change much either and remained in the Moderate range (17.57 increased to 17.64).

5.4. Effect on Post-Test Scores

The researchers failed to reject the hypothesis since the Z-value was −0.359, and the Asymp. Sig. (two-tailed) value was considerably greater than alpha at 0.720. This established that, on average, online PBL has a positive influence on students' post-test results.

6. Conclusions and Recommendations

The researchers concluded that the trial produced conflicting findings—a good impact for Mechanical 2 students, an adverse influence for Mechanical 1 students, and no noticeable impact for Avionics students in terms of their Aviation English competency. Although there were comparable positive-effect results for one class in many other experiments [9,10], the researchers also reached adverse and non-significant-effect outcomes—which had been expected. The limits are unquestionably critical in identifying the most accurate and credible results within the purview of the institution. As a result, researchers will continue to explore with students from various disciplines, learning experiences, and areas of study in order to amass as much data and information prior to attaining the ideal PBL-Institute standard.

Several suggested online PBL conditions, such as low-to-moderate class sizes, student-centered techniques, and collaborative lessons, have already been established or outlined in UniKL MIAT. Additional technological disciplines [11–16] do not present a difficulty in terms of classroom sizes, since UniKL MIAT complies to the Civil Aviation Authority of Malaysia's (CAAM) regulation of allowing a limit of 28 participants in a group at any given time [17]. Non-technological disciplines [18–21], on the other hand, may set up a bigger student population in a classroom and would need a customized online PBL approach to accommodate the course content and teaching/learning methods.

Author Contributions: Conceptualization, M.K.J. and N.Z.J.; methodology, M.K.J.; software, M.K.J.; validation, M.K.J. and N.Z.J.; formal analysis, M.K.J. and N.Z.J.; investigation, M.K.J.; resources, M.K.J.; data curation, M.K.J. and N.Z.J.; writing—original draft preparation, M.K.J.; writing—review and editing, M.K.J. and N.Z.J.; visualization, M.K.J.; supervision, M.K.J.; project administration, M.K.J.; funding acquisition, M.K.J. All authors have read and agreed to the published version of the manuscript.

Funding: The APC was funded by Universiti Teknologi MARA Melaka (*Geran Dalaman TEJA* 2022; grant number: GDT2022/1-11).

Institutional Review Board Statement: Not applicable.

Informed Consent Statement: Not applicable.

Data Availability Statement: Not applicable.

Acknowledgments: Several UniKL MIAT academicians provided guidance regarding the scoring systems (proficiency levels).

Conflicts of Interest: The authors declare no conflict of interest.

References

1. Noordegraaf-Eelens, L.; Kloeg, J.; Noordzij, G. PBL and sustainable education: Addressing the problem of isolation. *Adv. Health Sci. Educ.* **2019**, *24*, 971–979. [CrossRef] [PubMed]
2. Saqr, N.M.; Alamro, A. The role of social network analysis as a learning analytics tool in online problem-based learning. *BMC Medic. Ed.* **2019**, *19*, 1–11. [CrossRef] [PubMed]

3. Rosmilasari, D.; Adoe, D. Design and Implementation of Online Problem Based Learning (PBL) Assisted by Innovative Media to Improve Elementary School Student Learning Outcomes. *J. Educ. Tech.* **2021**, *4*, 456–464. [CrossRef]
4. Peramunugamage, A.; Usoof, H.; Hapuarachchi, J. Moodle Mobile Plugin for Problem-Based Learning (PBL) in Engineering Education. In Proceedings of the IEEE Global Engineering Education Conference, Dubai, United Arab Emirates, 8–11 April 2019; pp. 859–867.
5. Braßler, M. The Role of Interdisciplinarity in Bringing PBL to traditional Universities: Opportunities and Challenges on the Organizational, Team and Individual Level. *Interdiscip. J. Probl. Learn.* **2020**, *14*, 1–14. [CrossRef]
6. Tadjer, H.; Lafifi, Y.; Seridi-Bouchelaghem, H. A New Approach for Assessing Learners in an Online Problem Based Learning Environment. In *Learning and Performance Assessment: Concepts, Methodologies, Tools, and Applications*, 3rd ed.; IGI Global: Hershey, PA, USA, 2020.
7. Sari, R.; Perdana, R.; Riwayani; Jumadi; Wilujeng, I.; Kuswanto, H. The Implementation of Problem-based Learning Model with Online Simulation to Enhance the Student's Analytical Thinking Skill. *J. Phys. Conf. Ser.* **2019**, *1233*, 1–8. [CrossRef]
8. Sarty, G. *Introduction to Applied Statistics for Psychology Students*; University of Saskatchewan: Saskatoon, SK, Canada, 2020.
9. Nawi, A.; Zakaria, G. The Impact of Applying Religious Values through Online Problem-Based Learning among Undergraduate Students: A Quasi-Experimental Study. *Int. J. Learn. Teach. Educ. Res.* **2019**, *18*, 192–213. [CrossRef]
10. Edy, D.; Widiyanti; Basuki. Revisiting The Impact of Project-Based Learning on Online Learning in Vocational Education: Analysis of Learning in Pandemic COVID-19. In Proceedings of the 4th International Conference on Vocational Education and Training (ICOVET), Malang, Indonesia, 19 September 2020; pp. 264–267.
11. Azizan, M.A.; Hamzah, A.; Johari, M.K. The implementation of QR codes for aircraft disassemble part and NDT equipment inventory system for documentation purpose. *Int. J. Adv. Sci. Tech.* **2019**, *28*, 86–90.
12. Ariffin, M.W.Z.; Johari, M.K.; Ibrahim, H. The needs of aircraft avionics' radio line replaceable unit repair center at UniKL MIAT. *Int. J. Eng. Tech. (UAE)* **2018**, *7*, 86–88. [CrossRef]
13. Azizan, M.A.; Hazim, M.; Johari, M.K. Analysis of energy separation acting in vortex tube. *Int. J. Eng. Tech.* **2019**, *28*, 91–95.
14. Ya'acob, A.M.; Razali, D.; Anwar, U.A.; Radhi, A.H.; Ishak, A.A.; Minhat, M.; Mohd Aris, K.D.; Johari, M.K.; Casey, T. Preliminary Study on GF/Carbon/Epoxy Composite Permeability in Designing Close Compartment Processing. In Proceedings of the 1st International Conference on Aerospace and Mechanical Engineering (AeroMech 2017), Penang, Malaysia, 21–22 November 2017; pp. 1–9.
15. Azizan, M.A.; Izwan, A.; Johari, M.K. Aerodynamic analysis of aircraft wing performance realized in different wing aspect ratio. *Int. J. Eng. Tech.* **2019**, *28*, 100–103.
16. Yusof, M.A.; Ya'acob, A.M.; Zaki, M.A.M.; Rahman, Z.A.; Abidin, N.H.Z.; Padil, I.F.; Johari, M.K.; Bakar, I.A.; Hashim, H.F.M. Developing a virtual reality (VR) app for theory of flight & control as a teaching & learning aid. *Int. J. Innov. Tech. Expl. Eng.* **2019**, *28*, 86–90.
17. Approved Maintenance Training Organization (CAAM Part 147). Available online: https://www.caam.gov.my/wp-content/uploads/2021/03/an-1201-i02-MTO.pdf (accessed on 2 September 2021).
18. Abdul-Samad, A.G.; Azizan, M.A.; Khairuddin, M.H.; Johari, M.K. A Review on the Mental Workload and Physical Workload for Aircraft Maintenance Personnel. In *Human-Centered Technology for a Better Tomorrow*; Lecture Notes in Mechanical Engineering; Springer: Singapore, 2022; pp. 627–635.
19. Bardai, A.M.; Er, A.Z.; Johari, M.K.; Noor, A.A.M. A review of Kuala Lumpur International Airport (KLIA) as a competitive South-East Asia hub. In Proceedings of the 1st AEROS Conference 2017, IOP Conference Series: Materials Science and Engineering, Putrajaya, Malaysia, 12 December 2017; pp. 1–10.
20. Abdul-Samad, A.G.; Azizan, M.A.; Khairuddin, M.H.; Johari, M.K. Effect of Mental Workload on Heart Rate Variability and Reaction Time of Aircraft Maintenance Personnel. In *Human-Centered Technology for a Better Tomorrow*; Lecture Notes in Mechanical Engineering; Springer: Singapore, 2022; pp. 613–625.
21. Abdul-Samad, A.G.; Azizan, M.A.; Khairuddin, M.H.; Johari, M.K. Significance of Aircraft Maintenance Personnel's Reaction Time During Physical Workload and Mental Workload. In *Human-Centered Technology for a Better Tomorrow*; Lecture Notes in Mechanical Engineering; Springer: Singapore, 2022; pp. 637–643.

proceedings

MDPI

Proceeding Paper

Types of Resources for Blended Learning Approach to Study High School Chemistry †

Theresa Stanley

SMK Bandar Tun Hussein Onn 2, Jalan Suadamai, Bandar Tun Hussein Onn, Batu 9, Cheras 43200, Malaysia; treskanny22@gmail.com

† Presented at the International Academic Symposium of Social Science 2022, Kota Bharu, Malaysia, 3 July 2022.

Abstract: Blended learning is one of the 21st-century teaching and learning approaches used in classrooms. Many scholars have positively affected students' performances or motivation in their research on blended learning in school. The application of blended learning in the school requires an educator to prepare suitable resources to suit the subject content and the ambiguity of learners. The best practice is to utilize the resources proven to be effectively applied in blended learning, considering the subject content. Currently, no studies are conducted to review the resources an educator uses to teach high school chemistry. For these reasons, this study will address the types of resources used for blended learning in high school chemistry, its characteristics deemed suitable, and the effectiveness of blended learning in using these resources reported by the researchers.

Keywords: blended learning; chemistry; high school

Citation: Stanley, T. Types of Resources for Blended Learning Approach to Study High School Chemistry. *Proceedings* **2022**, *82*, 94. https://doi.org/10.3390/proceedings2022082094

Academic Editor: Mohamad Rahimi Mohamad Rosman

Published: 7 October 2022

1. Introduction

Blended learning is a 21st-century learning approach that is a combination of online and physical classes. The students learn by using technology, where more students control the pace of studying. In a traditional classroom, students try to understand what the teacher is saying, and students may need time to understand the subject being taught. When the students take longer to understand what is being taught, they might be left behind, as the first concept has not been understood. While in blended classroom learning, the teachers will try to detect errors in thinking, writing, reading, and listening. At the same time, they cooperate with the social constructivist theory by Vygotsky. The students learn from each other, which can promote active learning and help each other. The benefits of blended learning are that it saves time and resources, such as videos being reusable, so the teacher does not have to create resources repeatedly. Minor updates can be conducted in stages after the recording is done, and the students are in control of their learning processes. This benefits a myriad of learners, as not all students study at the same pace. Students can replay the lessons according to their studying methods. Blended learning allows students to clear misconceptions in class while studying at home.

Teachers can build a better student and teacher relationship; teachers are more focused on the students at the same time while fostering critical thinking skills. Since the students will watch the recorded class lesson as homework, the teacher may focus on individual students. Many scholars suggest that the positive outcome of a blended learning approach is aiding students' difficulty in learning chemistry. The abstractness of chemical concepts, teacher-imposed learning models, and lack of teacher support are why students find studying chemistry problematic [1,2]. The challenging factor of blended learning is combining the online and physical lessons run by teachers in classrooms [3].

For this reason, a guided approach toward implementing blended learning will be fruitful. Blended learning combines the right technologies in lessons and develops a feasible teaching and learning environment. Well-prepared teachers can create a successful blended learning classroom [4]. By understanding blended learning, teachers will know how to choose suitable materials for the school [5]. Blended learning gives flexibility. Some variations include personalized learning, social interaction, and direct content learning in the school. In this research, reviewing the myriad of resources reported by different scholars in the past ten years will help teachers choose the best resource for blended learning in high school teaching and learning chemistry. This study investigates nine papers published in the *Scopus Journal* after filtering 28 articles from 2018 to 2021.

2. Methodology

2.1. Resource

The purpose of this research is to investigate the published papers obtained from the Scopus database from 2008 to 2021 by implementing a filtration search using keywords "blended" and "learning", "chemistry", and "high school". Thirty-seven papers in Scopus were suitable for this study. Twenty-two articles were non-related documents unrelated to chemistry and did not involve high school students, six were non-available pieces, and only eight were used in this research. Figure 1 displays the fundamental steps of searching using the Scopus database searching steps.

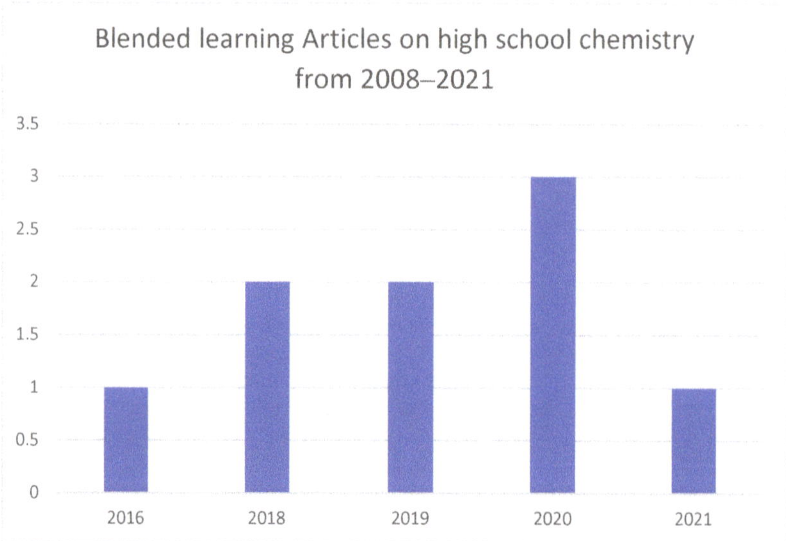

Figure 1. Blended learning articles on high school chemistry from 2018 to 2021.

2.2. Data Distribution

Based on the analysis of the search results, this research paper on blended learning resources in high school chemistry from 2008 to 2021 was analyzed based on the topic, resources, method of teaching, samples of findings, and country of origin. Five articles were published every two years, from 2016 to 2021. Before 2016, there was no research contribution to the literature on blended learning resources in chemistry. Since 2016, researchers have been keen to investigate the research topic of blended learning, with two papers published every two years. There were more papers published at the beginning of 2020. In addition, there was one piece published in 2016. In 2018 and 2019, two papers were published and peaked to reach the topmost publication number of three articles in 2020. After peaking at three articles, the Figure 2 dropped to one document in 2021. The number of publications' dramatic growth was in 2020, with three pieces.

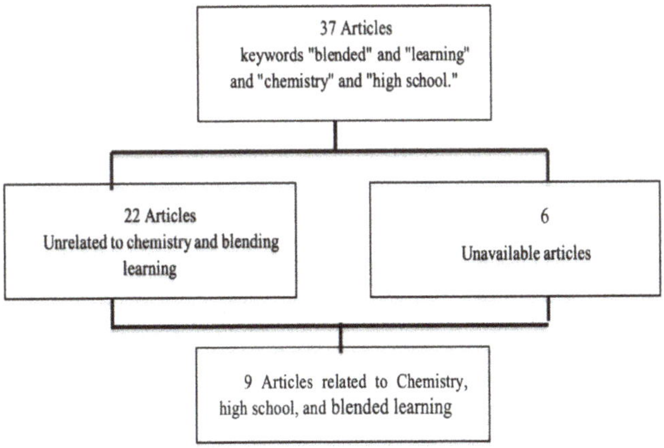

Figure 2. The fundamental steps of searching using Scopus.

3. Results

From 2008 to 2013, the topic of interest was oxygen preparation experiments, formation of hydrogen bonds and titration, redox reaction, and thermochemistry. Apart from that, there were hydrocarbon, salt, mole, and learning of organic chemistry. All these topics were the topics taught in high school chemistry. The methods of teaching are divided into experimental teaching, inquiry-based learning, self-developed four-dimensional model, modules, gamification, offline and traditional, followed by online learning.

3.1. Experimental Teaching

Based on the research from China and the United States, using experimental learning suggests using virtual and real experiments, as both approaches complement each other. Both types of research indicate the effectiveness of combining learning activities that can improve students' conceptual ideas [6,7].

3.2. Gamification

Research shows that using gamification in the teaching and learning of hydrocarbon is only effective when combined with android-based games and blended learning, compared to android games alone [8]. Based on the research, educational games with hands-on laboratory experiments will coherently explain scientific phenomena together with authentic scientific inquiry [9].

3.3. Learning Management System (LMS)

The use of modules by scholars in blended learning to teach thermochemistry using LMS [10]. The use of LMS attracted the student's attention to chemistry and improved their metacognition skills. LMS provides the resources for education to prepare more creative material for blended learning. In Malaysia, further support for LMS and its composite learning resources has improved students' understanding [11].

3.4. Video, Self-Developed Device, Assigned Task

Research from the United States found that using video for blended learning could improve students' understanding of studying chemistry [12]. The self-developed device from Indonesia suggested that blended learning promotes critical thinking [13]. A study from Vietnam presented the assigned task used in project-based learning and further proposed the need to investigate the other learning conditions appropriate before incorporating blended learning [14].

4. Limitation

This research is only limited from 2008 to 2013, and the articles searched were limited to chemistry in high school and blended learning. The article searches were limited to only using the *Scopus Index* journal. There should be further research on more resources related to blended learning, as it is a 21st-century classroom approach.

5. Conclusions

Blended learning resources are similarly served by or available through a system, especially a computer or telecommunications system (such as the internet). The resources used in high school chemistry include video games, modules, learning materials in LMS, and virtual experiments. Two studies state that the use of virtual experiments as resources for blended learning involves experiments accompanied by actual experiments. Scholars have noted that blended learning resources will be more effective if not limited to one type of resource and can always be a string of resources or materials, including assessment.

Blended learning has increased students' understanding, achievements, motivation, metacognitive, and critical thinking skills. Blended learning is more prevalent in the current decade because it contributes to 21st-century learning skills. Chemistry is one of the STEM subjects that require the understanding of many abstract concepts and needs an educator, particularly those teaching high school chemistry, to be creative in their teaching methods to foster interest in learning among the students before moving to the more complex part of the subjects in higher education.

Funding: This research received no external funding.

Institutional Review Board Statement: Not applicable.

Informed Consent Statement: Not applicable.

Data Availability Statement: Not applicable.

Acknowledgments: Paper presented during a panel—organized by IASSC 2022 Organizing Committee—in the context of the advances in online learning, communication, innovation, and creativity conference hosted by Assoc. Filmor J. Murillo, MDC, during the International Academic Symposium of Social Science (IASSC) 2022.

Conflicts of Interest: The authors declare no conflict of interest.

References

1. Sirhan, G. Learning Difficulties in Chemistry: An Overview. *J. Turk. Sci. Educ.* **2007**, *4*, 2–20.
2. Woldeamanuel, M.M.; Atagana, H.; Engida, T. What Makes Chemistry Difficult? *Afr. J. Chem. Educ.* **2014**, *4*, 31–43.
3. Garrison, D.R.; Heather, K. Blended Learning: Uncovering Its Transformative Potential in Higher Education. *Internet High Educ.* **2004**, *7*, 95–105. [CrossRef]
4. Alqarni, A. Blended Learning and Flipped Classroom Approaches. *ARJHSS* **2018**, *4*, 1–6.
5. Picciano, A.G. *Introduction to Blended Learning: Research Perspectives*; Routledge: London, UK, 2013.
6. Shen, Q. Blended Learning Activities in A Chemistry Experiment. *World Trans. Eng. Technol. Educ.* **2016**, *14*, 70–76.
7. Jihad, T.; Edward, K.; Patrick, G.; Chappel, S.; MacRae, M.; Yongjun, L.; Jin, K.M. Perspectives on Blended Learning through the On-Line Platform, Lab Lessons, For Chemistry. *J. Sci. Educ. Technol.* **2018**, *7*, 34–44. [CrossRef]
8. Fitriyana, N.; Antuni, W.; Jaslin, I.; Kristian, H.S. Android-Based-Game and Blended Learning in Chemistry: Effect on Students' Self-Efficacy and Achievement. *Cakrawala Pendidik.* **2020**, *39*, 507–521. [CrossRef]
9. Hodges, G.W.; Lu, W.; Juyeon, L.; Allan, C.; Yoonsun, J. An Exploratory Study of Blending the Virtual World and the Laboratory Experience in Secondary Chemistry Classrooms. *Comput. Educ.* **2018**, *122*, 179–193. [CrossRef]
10. Nababan, K.; Budi, H.; Nurma, Y.I. Blended Learning in High School Chemistry to Enhance Students' Metacognitive Skills and Attitudes towards Chemistry: A Need Analysis. *AIP Conf. Proc.* **2019**, *2194*, 020068.
11. Ayob, N.S.; Abd Halim, N.D.; Zulkifli, N.N.; Zaid, N.M.; Mokhtar, M. Overview of Blended Learning: The Effect of Station Rotation Model on Students' Achievement. *Crit. Rev.* **2020**, *7*, 320–326.
12. Ruedas-Caletrio, J.; Martín-García, A.V. Blended Learning: Convergence between Technology and Pedagogy. *REEC* **2022**, *40*, 271–273.

13. Hadisaputra, S.; Ihsan, M.S.; Ramdani, A. The Development of Chemistry Learning Devices Based Blended Learning Model to Promote Students' Critical Thinking Skills. *J. Phys. Conf. Ser.* **2020**, *1521*, 042083. [CrossRef]
14. Dai, N.V.; Vu, Q.T.; Chu, V.T.; Kieu, P.H.; Dao, T.V.A. Project-Based Teaching in Organic Chemistry through Blended Learning Model to Develop Self-Study Capacity of High School Students in Vietnam. *Sci. Educ.* **2021**, *11*, 346. [CrossRef]

proceedings

MDPI

Proceeding Paper

Reliability Analysis of the Effect of Digital Literacy on Performance among Secondary School Students in Malaysia †

Mohd Arif Fahmi Bidin, Shamila Mohamed Shuhidan * and Noor Zaidi Sahid

Faculty of Information Management, Universiti Teknologi MARA, Puncak Perdana 40150, Selangor, Malaysia
* Correspondence: shamila@uitm.edu.my

† Presented at the International Academic Symposium of Social Science 2022, Kota Bharu, Malaysia, 3 July 2022.

Abstract: This paper intends to classify the factors relating to digital literacy and their effect on student performance. School teachers have been chosen in preliminary studies and support from past literature and pilot tests via survey have been carried out with 30 secondary school students. The influence of three dimensions on student performance have been studied, which include individual factors, the learning environment (physical), and socio-cultural aspects. From the pilot test, the results show that self-efficacy has the lowest value of 0.584, and the highest value is educational materials with 0.875 and is considered as one of the main factors of digital literacy.

Keywords: digital literacy; individual factors; learning factors; socio-cultural factors; student performance

Citation: Bidin, M.A.F.; Shuhidan, S.M.; Sahid, N.Z. Reliability Analysis of the Effect of Digital Literacy on Performance among Secondary School Students in Malaysia. *Proceedings* **2022**, *82*, 102. https://doi.org/10.3390/proceedings2022082102

Academic Editor: Mohamad Rahimi Mohamad Rosman

Published: 19 October 2022

1. Introduction

A holistic approach to education and learning in the 21st century syndicates a distinct focus on student performance and computational thinking, which comprise of multidimensional abilities [1]. They will learn and comprehend their relationship with the world around them, utilise technology to conduct research and communicate with others, feel at ease working in teams, and acquire the strength and abilities necessary to accept leadership roles [1,2]. Digital literacy, ICT, and other contemporary technology-based abilities are crucial for the education of learners in the twenty-first century. Many current elementary and secondary school curricula have begun to include components of informatics education, training pupils to be familiar with the fundamental tools of the digital world. Other studies discuss informatics education in several nations, especially western countries and Japan. The science behind IT is variably referred to as "Computer Science" (the phrase most often used in the United States), "Computing Science," and "Informatics". Therefore, students in the 21st century must acquire new skills, knowledge, and ways of learning that are linked with the demands of living and working in a digital economy with a complex information environment [3]. This suggests that students should be exposed to digital literacy to help them improve their capacity to learn and to prepare them for a lifetime of learning and developing future employment skills that will push the digital economy forward [4]. The digital economy is anticipated to contribute to 18.2% of Malaysia's GDP by 2020 (Economic Planning Unit, Office of the Prime Minister of Malaysia), and 90% of all future employment will need digital capabilities.

2. Literature Review

According to [5], in a digital learning setting, digital literacy is defined as the capacity and awareness to utilise developing digital tools to fulfil tasks while displaying the proper attitude. Meanwhile, [6] defined digital literacy as offline or online cognitive, technological, and socioemotional learning. A cognitive element includes technology selection, information search, appraisal, and selection employing critical thinking abilities, and so on. Technical abilities are an important part of digital literacy, and one of its characteristics is the capacity to use digital learning systems. The socio-emotional dimension relates to an individual's behaviour when using digital technologies. Students nowadays are being taught to use applications and utilize the internet for academic purposes. Nonetheless, teachers often presume that the majority of students nowadays are digital natives [7], owing to their continual exposure to digital technology, such as computers, mobile phones, video games, and the internet, but not every student is a digital native and has the skills to properly utilize technology to obtain information for academic work [8]. Most students lack self-efficacy to learn something new that will help them excel academically. Few studies have been conducted to determine students' levels of self-efficacy and reading literacy [9–11]. On the other hand, the availability of pictures and graphics encourages students to explore more of a digital content site [8,12]. Students who do not have a positive correlation between their self-efficacy beliefs and their performance are regarded as poorly calibrated. Self-efficacy is vital in education, according to [13]. Students who overestimate their abilities may sometimes fail, which might reduce their drive. Students who underestimate their abilities, on the other hand, may be hesitant to make an attempt, resulting in a reduction in skill gain. According to [14], self-efficacy beliefs might also be connected to gender and culture.

Students are the fastest increasing demographic of smart phone users, with usage generally centred on internet access, based on the study by [15]. Despite their ubiquitous usage, little is known about the variables that contribute to their usage. Research has emphasised the relevance of the home environment in fostering and sustaining both safe and risky online behaviour development and emphasised the need for addressing parental usage and family behaviours on the use of technology in the home [15,16]. Parents sometimes underestimate the significant influence their child's usage of technology has on them [17]. Other than in the home environment, effective 21st century learning environments in the classroom improve the teaching–learning process [15–18]. Furthermore, the absence of, or poor quality of, infrastructure and connection in schools threatens to exacerbate students' disadvantages [16,19,20]. Infrastructure and technology utilisation in schools are both critical enablers for digital technology-based education, as well as increasing teaching and learning. Furthermore, the issue stems from a lack of teacher qualifications and a poor teaching approach in the classroom. When teachers lack digital literacy, students witness it firsthand, and this may create a divide which impedes the growth of digital culture [20–22]. Similar to the findings from [23,24] which demonstrated a positive correlation between digital literacy and student performance, the results of the current research are compatible with the findings of [25–27].

3. Development of Research Framework

Preliminary studies found that the students who took computer science lacked information on the topic and had mediocre digital literacy abilities. This is due to the self-efficacy of the student itself doing programming exercise and practice. The school has inadequate facilities for conducting lessons, and the instructors need to be updated. In Malaysia, there is limited information and literature pertaining to the implementation of digital literacy (DL) in school, even though they already exercised the computer science subject. On the study method in the literature framework, a few of the components of the frameworks were analysed with regards to the relevant subject of this research. The proposed theoretical framework is shown in Figure 1.

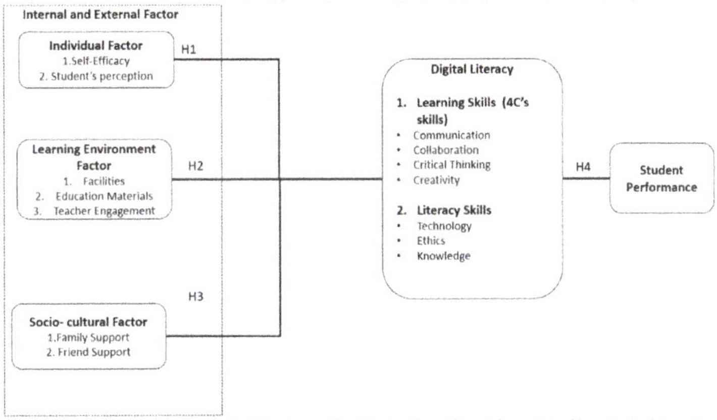

Figure 1. Research Framework.

In the 21st century, students must be highly competent and technologically literate learners [1,2,27]. They must be adaptable and receptive to new ideas and circumstances [27,28]. This should include the capacity to appreciate the underlying nature of technology phenomena and the ethical and societal consequences of using online technologies and information. These concerns are accompanied by digital literacy characteristics that demonstrate their effects on the link between 21st-century learning rudiments and student performance and their incorporation into this relationship. Students must possess the four C's: critical thinking, creativity, cooperation, and communication skills, in order to succeed in school. Digital literacy enhances engagement and participation processes and enables pupils to become active as opposed to passive in interpersonal circumstances [29]. From the Figure 1, there are three independent variables found which are Individual facto, Learning Environment Factor and Socio-cultural factor that have significant relationship with digital literacy. Furthermore, this study also extends to measure on the effect of digital literacy towards student performance. Generally, there are four hypotheses have been observed in this study. Therefore, the hypothesis for this study is as in Table 1:

Table 1. Hypothesis Statement.

	Hypothesis
H1	Individual factor has a significant relationship towards digital literacy.
H2	Learning environment has a significant relationship towards digital literacy
H3	Socio-cultural factor has a significant relationship towards digital literacy.
H4	Digital literacy has a significant relationship with student performance.

4. Pre-Testing, Validity, and Reliability of Research Instrument

The pre-tested and pilot test provided a meaningful finding that helped the researcher to identify the appropriate variables for further exploration. Thus, the result of this preliminary study supports the rationality and dependability of the study. Preliminary research was directed to confirm the usefulness of the initial questionnaire to draw the information and items needed. To this effect, two schoolteachers were found eligible to be interviewed and granted permission from their school. Two computer science teachers were chosen as they are the ones who are actively involved with digital literacy programs and initiatives within school environments. According to [30], digital literacy should be part of the teaching plan for every subject. However, due to the COVID-19 outbreak, the researcher did not manage to conduct personal visits to the respective teachers and school. Structured interviews were conducted online and via calls. The aims of these interviews were to gather as much as possible on the experience and reality of their personal involvement in digital literacy in school. The topics covered in these interviews focused on the management support from school committees, including students, teachers, parents, and management. Various aspects of teaching and learning within the individual factor, environment factor, and socio-cultural factor have also been discussed with the teachers.

In developing the survey questionnaire, five professionals were chosen to review the instrument. The survey was verified by two senior lecturers at the Faculty of Education from Public Universities. These academicians are highly experienced and qualified in the field of education, digital literacy, and research methodologies for social sciences. Besides, three teachers from secondary schools were also selected for this expert reviews. These teachers were selected as they are the ones who engage actively with education, students' digital literacy, and the environment of education in Malaysia. In order to approach the evaluators, a formal letter endorsed by the Faculty of Information Management was sent to seek permission to conduct the questionnaire pre-testing. Once the permission was granted via email, the researcher made a follow-up call and arranged the date to see the respective experts. The researcher sent the questionnaire and pre-testing checklist via email before meeting them for face-to-face interviews. One week to two weeks were given for the evaluators to read and understand the content of the questionnaire. A few concerns were highlighted especially regarding the usage of some jargon and terms to match with the understanding of the school students' level. All of these reviews were highly useful in helping the researchers finalize the questionnaire set with the appropriate modifications.

The survey questions are classified into seven sections with closed-ended types and open-ended types of questions. The structure of the survey questionnaire began with Section 1 concerning demographic content, consisting of 11 questions. The classification of the data will be easier, since the information from all of the respondents can be categorized based on gender, age, type of school, and ICT skills. In Sections 2–6, the questionnaire consisting of 45 questions focused on the independent variables and dependent variables of the study. All sections use the five Likert scale ranging from "strongly agree" (coded 5) to "strongly disagree" (coded 1). Additionally, two open-ended questions were added in which allowed respondents to respond subjectively.

A survey method is used to collect data from a predefined group of respondents to gain information and insights into various topics of interest. In this study, the data will be collected from secondary school students through a personally administered structured questionnaire. The respondents will consist of the students from Klang Valley. All of the participants are students from form one, two, and four. The schools chosen for this research are based on the Smart School Qualification Standard (SSQS) 2018 with rankings of four stars and above. The main advantage of this technique is time management, since the respondents only spend around 10 min answering the questions, compared to the interview session. In this survey, only one method of survey will be conducted, which is a written survey.

5. Reliability Analysis Result

As revealed in Table 2, Cronbach's alpha coefficients determine the reliability and internal consistency of the scales exercised in the study. All factors had Cronbach's alpha values greater than 0.6, suggesting that the reliability level is appropriate and acceptable, except for in the dimension of self-efficacy where the value of Cronbach's alpha are below 0.6, which is unsatisfactory. According to the SPSS analysis results, the overall consistency, or Cronbach's alpha values, of all 56 items for each dimension contained in the instrument was between 0.584 and 0.875. Educational materials and tools show the highest value of 0.875, while self-efficacy shows the lowest value of 0.584; this means that there is room for refining the instruments in this study. One factor that might impact the value of Cronbach's alpha is the fact that the pilot test was conducted during the COVID-19 outbreak and the students had to attend online classes instead of normal classroom teaching and learning. It is also suggested that assessing students' self-efficacy could be advantageous for teachers to design appropriate teaching approaches during the COVID-19 crisis, as is reported by [31]. According to [32,33], a value of Cronbach's alpha of less than 0.6 will affect the validity of the data. Two possibilities happen in this scenario: either the number of items in the self-efficacy dimension are not enough or the items are partly correlated with each other, or there is diversity in the construction of the instrument. Therefore, the items need to be revised or removed. One of the approaches used in refining the value of Cronbach's alpha is adding more related items to test the same concept. This result implies that, particularly at this point in the investigation, the overall index of the scale's internal consistency within the instrument is reliable with no unexpected abnormalities found in the data. The pilot test has been carried out with 30 students and the result of the pilot test is summarised in Table 2.

Table 2. Cronbach's Alpha Analysis Result.

Variables		Number of Items	Cronbach's Alpha
Individual factor	Self-efficacy	5	0.584
	Student's perception	5	0.664
Learning environment factor	Facilities	5	0.706
	Educational materials	5	0.875
	Teacher engagement	5	0.748
Socio-cultural factor	Family support	5	0.819
	Friend support	5	0.848
Digital literacy		5	0.873
Student performance		5	0.765
Overall		**45**	**0.946**

6. Discussion and Conclusions

The resolution of this article is to present the summary of an ongoing research project on the effect of digital literacy in school on student performance. Literature reviews and preliminary studies have been conducted with computer science teachers in secondary school and found that three dimensions, namely individual/personal factors, learning environment influences, and socio-cultural aspects of digital literacy, impact student performance. Preceding the actual research, the instrument underwent pre-testing with three subject-matter experts to ensure the reliability of each question and understandability of the items. A pilot study was also carried out and was participated in by 30 respondents among secondary school students in Klang Valley. The analysis results of the collected data suggest that the overall index of the scale's internal consistency within the instrument is reliable, as stated in this paper. Consequently, it is essential to conduct research on digital literacy, particularly on the relationship of the individual factor, learning environment factor, socio-cultural factor on digital literacy, and the correlation between digital literacy towards student performance. According to studies, digital literacy shows a significant relationship to the individual factor, learning environment factor, socio-cultural factor. Besides, digital literacy also made a major contribution towards student performance. Therefore, this study was able to provide some empirical evidence on the subject. Extensive research methods and systematic analysis were used to verify the empirical evidence in this study. Instrument development for this study was guided by the results of instrument validation and hypothesis testing. The instrument needs to be validated in order for the results to be accepted and published in a reputable journal and presented at academic conferences. Despite the fact that secondary school students were the primary focus of this study, the instrument can be used for any level in other types of industries. Furthermore, the questionnaire items can be used or adapted in a separate study to identify the specific factors involved in the selected dimensions. This is due to the fact that the developed instrument may reduce the potential problems associated with the individual factor, learning environment factor, and socio-cultural factors and digital literacy. The findings of this study can be used to improve students' performance in terms of academic performance. The findings provide a small amount of support for policymakers and practitioners, particularly the Ministry of Education and school administrators to reconsider their policies development frameworks and align our education structure of the Malaysia Education Blueprint for 2013–2025 with the digital literacy set skills to meet future job demands. In practice, this research aids principals and school administrators to develop an improved thoughtfulness for digital literacy and its importance in 21st century learning. With the findings from this study, school administrators were able to see the level of digital literacy for each construct. From there, it can be improved by conducting related programs and activities. Given the potential influence of the individual factor, learning environment factor, and socio-cultural factor, schools must develop policies that allow for the integration of digital literacy aspects in the classroom. Digital literacy is crucial in the 21st century learning method. This implementation and these resources can help to boost digital literacy in school and improve the existing curriculum standards. This will also be able to increase students' interest in learning. Through the existence of this policy, various innovative ideas can be developed. Following that, school will become the centre of innovation and digital literacy in society. Furthermore, technological tools can aid in the learning process.

Author Contributions: Conceptualization, M.A.F.B. and S.M.S.; methodology, M.A.F.B. and S.M.S.; software, M.A.F.B. and S.M.S.; validation, S.M.S.; formal analysis, S.M.S.; investigation, M.A.F.B.; resources, M.A.F.B.; data curation, M.A.F.B. and S.M.S.; writing—original draft preparation, M.A.F.B.; writing—review and editing, S.M.S.; visualization, M.A.F.B.; supervision, S.M.S.; project administration, M.A.F.B., S.M.S. and N.Z.S. All authors have read and agreed to the published version of the manuscript.

Funding: This research received no external funding.

Institutional Review Board Statement: Not applicable.

Informed Consent Statement: Not applicable.

Data Availability Statement: Not applicable.

Acknowledgments: The researchers would like to extend their gratitude to Universiti Teknologi MARA (UiTM)–Faculty of Information Management Puncak Perdana for providing the necessary funding for publication. The researchers also would like to thank schools, teachers, and respondents who provided feedback and participated in this study.

Conflicts of Interest: The authors declare no conflict of interest.

References

1. Abbas, Q.; Hussain, S.; Rasool, S. Digital Literacy Effect on the Academic Performance of Students at Higher Education Level in Pakistan. *Glob. Soc. Sci. Rev.* **2019**, *IV*, 108–116. [CrossRef]
2. Ado, T. Influence of Learning Environment on Students ' Academic Achievement in Mathematics: A Case Study of Some Selected Secondary Schools in Yobe State–Nigeria. *J. Educ. Pract.* **2015**, *6*, 40–44.
3. Afshari, M.; Ghavifekr, S.; Siraj, S.; Jing, D. Students' Attitudes towards Computer-assisted Language Learning. *Procedia-Soc. Behav. Sci.* **2015**, *103*, 852–859. [CrossRef]
4. Appel, G.; Grewal, L.; Hadi, R.; Stephen, A.T. The future of social media in marketing. *J. Acad. Mark. Sci.* **2020**, *48*, 79–95. [CrossRef]
5. Shariman, T.P.N.T.; Razak, N.A.; Noor, N.F.M. Digital Literacy Competence for Academic Needs: An Analysis of Malaysian Students in Three Universities. *Procedia-Soc. Behav. Sci.* **2012**, *69*, 1489–1496. [CrossRef]
6. Barrett, P.; Treves, A.; Shmis, T.; Ambasz, D.; Ustinova, M. *The Impact of School Infrastructure on Learning*; The World Bank: Washington, DC, USA, 2019; pp. 1–71.
7. Chu, S.K.W.; Reynolds, R.B.; Tavares, N.J.; Notari, M.; Lee, C.W.Y. Twenty-First Century Skills and Global Education Roadmaps. In *21st Century Skills Development Through Inquiry-Based Learning*; Springer: Singapore, 2017; pp. 17–32.
8. Danet, M. Parental Concerns about their School-aged Children's Use of Digital Devices. *J. Child Fam. Stud.* **2020**, *29*, 2890–2904. [CrossRef]
9. Elfil, M.; Negida, A. Sampling methods in Clinical Research: An Educational Review. *Emergency* **2017**, *5*, 52.

10. Ghavifekr, S.; Rosdy, W.A.W. Teaching and learning with technology: Effectiveness of ICT integration in schools. *Int. J. Res. Educ. Sci.* **2015**, *1*, 175–191. [CrossRef]
11. Gilavand, A. Investigating the Impact of Environmental Factors on Learning and Academic Achievement of Elementary Students: Review The Comparison of the Dental Students' Incentives in Choosing Their Field of Study View Project Faculty Members' Rank Promotion View Project. Available online: www.ijmrhs.com (accessed on 4 August 2022).
12. Gutiérrez-Núñez, S.E.; Cordero-Hidalgo, A.; Tarango, J. Implications of Computational Thinking Knowledge Transfer for Developing Educational Interventions, Contemp. *Educ. Technol.* **2022**, *14*, ep367.
13. Harshbarger, R. Learning in the 21st Century: A Study Addressing Educational Trends and Implications. Available online: http://stars.library.ucf.edu/cgi/viewcontent.cgi?article=6043&context=etd (accessed on 4 August 2022).
14. Jain, S.; Angural, V. Use of Cronbach's alpha in Dental Research. *Med. Res. Chron.* **2017**, *4*, 285–291.
15. Joynes, C.; Serena, R.; Amonoo-Kuofi, E.F. *21st Century Skills: Evidence of Issues in Definition, Demand and Delivery for Development Contexts*; UK Government's Department for International Development: London, UK, 2019; pp. 1–75.
16. Kim, S.; Raza, M.; Seidman, E. Improving 21st-century teaching skills: The key to effective 21st-century learners. *Res. Comp. Int. Educ.* **2019**, *14*, 99–117. [CrossRef]
17. Konca, A.S. Digital Technology Usage of Young Children: Screen Time and Families. *Early Child. Educ. J.* **2022**, *50*, 1097–1108. [CrossRef]
18. Koşar, G.; Akbana, Y.E.; Yakar, L. Development and Validation of a Reading Self-Efficacy Scale. *Int. J. Assess. Tools Educ.* **2022**, *9*, 203–219. [CrossRef]
19. Lai, P. The Literature Review of Technology Adoption Models and Theories for the Novelty Technology. *J. Inf. Syst. Technol. Manag.* **2017**, *14*, 21–38. [CrossRef]
20. Li, D. A Review of Self-efficacy of Learners through Online Learning. *J. Humanit. Educ. Dev.* **2020**, *2*, 526–533. [CrossRef]
21. Lin, G.Y. Self-efficacy beliefs and their sources in undergraduate computing disciplines: An examination of gender and persistence. *J. Educ. Comput. Res.* **2016**, *53*, 540–561. [CrossRef]
22. Malik, R.S. Educational Challenges in 21St Century and Sustainable Development. *J. Sustain. Dev. Educ. Res.* **2018**, *2*, 9. [CrossRef]
23. Merkouris, A.; Chorianopoulos, K.; Kameas, A. Teaching Programming in Secondary Education Through Embodied Computing Platforms: Robotics and Wearables, ACM Trans. *Comput. Educ.* **2017**, *17*, 1–22. [CrossRef]
24. Meyer, K. Talents, abilities and educational justice. *Educ. Philos. Theory* **2021**, *53*, 799–809. [CrossRef]
25. Mutta, M.; Pelttari, S.; Salmi, L.; Chevalier, A. Digital Literacy in Academic Language Learning Contexts: Developing Information-Seeking Competence. Available online: https://www.researchgate.net/publication/279840004 (accessed on 4 August 2022).
26. Ng, W. Can We Teach Digital Natives Digital Literacy? *Comput. Educ.* **2012**, *59*, 1065–1078. [CrossRef]
27. Passey, D. Computer Science (CS) in The Compulsory Education Curriculum: Implications for Future Research. *Educ. Inf. Technol.* **2017**, *22*, 421–443. [CrossRef]
28. Aguilera-Hermida, A.P. College Students' Use and Acceptance of Emergency Online Learning Due to COVID-19. *Int. J. Educ. Res. Open* **2020**, *1*, 100011. [CrossRef] [PubMed]
29. Md-Saadat, H.; Sadaquat, A.K. Self-Efficacy Of Teachers: A Review Of The Literature: A Review. *Jamshedpur Res. Rev.* **2022**, *1*, 110–116.
30. Perera, M.U.; Gardner, L. Analysing the relationships between digital literacy and self-regulated learning of undergraduates—A preliminary investigation. In *Advances in Information Systems Development*; Springer: Cham, Switzerland, 2018; Volume 26, pp. 1–19.
31. Raman, R.; Venkatasubramanian, S.; Achuthan, K.; Nedungadi, P. Computer science (CS) education in Indian schools: Situation analysis using darmstadt model. *ACM Trans. Comput. Educ.* **2015**, *15*, 1–36. [CrossRef]
32. Ramli, N.F.; Talib, O. Can Education Institution Implement STEM? From Malaysian Teachers' View. *Int. J. Acad. Res. Bus. Soc. Sci.* **2017**, *7*, 2222–6990.
33. Roche, T. Assessing the role of digital literacy in English for Academic Purposes university pathway programs. *J. Acad. Lang. Learn.* **2017**, *11*, 71–87.

proceedings

MDPI

Proceeding Paper

Exploring the Effectiveness of Interactive Simulation as Blended Learning Approach in Secondary School Physics †

Mohd Nizar Mohd Najib *, Aizan Yaacob and Ruzlan Md-Ali

School of Education, Universiti Utara Malaysia, Sintok 08000, Malaysia
* Correspondence: m.nizar_m.najib@ahsgs.uum.edu.my
† Presented at the International Academic Symposium of Social Science 2022, Kota Bharu, Malaysia, 3 July 2022.

Abstract: The mastery of physics and the ability to apply this knowledge may contribute to the development of the country. However, the skills of secondary school students concerning physics are less impressive. This exploratory research aims to examine the effectiveness of interactive simulation by PhET as blended learning in physics. The results showed that there was a significant difference in the pre- and post-test mean scores for the experimental group by providing a positive impact on students' interest, motivation, the pleasure of studying in a group or self-learning, and help to study for exams. This proved that simulation as an aspect of blended learning could improve students' achievement in physics.

Keywords: interactive simulation; blended learning; PhET simulation

check for updates

Citation: Najib, M.N.M.; Yaacob, A.; Md-Ali, R. Exploring the Effectiveness of Interactive Simulation as Blended Learning Approach in Secondary School Physics. *Proceedings* **2022**, *82*, 103. https://doi.org/10.3390/proceedings2022082103

Academic Editor: Mohamad Rahimi Mohamad Rosman

Published: 20 October 2022

1. Introduction

The field of science that explains various phenomena and situations found in the universe is physics. Thus, all the events that occur around us in everyday life is related to physics [1]. Physics is one of the important areas to explore in this modern era, which can be applied in technology and engineering, which is certainly beneficial for developing countries [2]. However, fewer students studying physics compared to other subjects at higher education levels because this subject is considered as difficult, boring, less well-liked, and irrelevant [3]. These problems arise when students have difficulty connecting the physics concept to their scientific reasoning skills to explain a phenomenon due to the difficulty of the subject, inefficient teaching, and unclear direction, respectively [2–5]. Previous studies on physics in Malaysia has shown that a lack of understanding about physics concept has become a serious phenomenon in upper secondary school [5]. Students found that they were difficult to understand the basic concepts of physics, and tend to focus more on numerical operations [3,6]. As a result, the number of students who avoid taking physics-related courses at high level institutions is very high [3]. Also, it is found that students' motivation to learn physics has decreased below acceptable levels [3,7,8]. There are two main contexts in learning, namely active learning and passive learning [9,10]. Passive learning is the passive acceptance of information, while active learning involves the interaction between the learners and the materials during the process of analyzing, comparing, inferencing, or critically evaluating [11]. Teaching that triggers active learning is a new format of teaching, which includes the use of simulations, lectures, and experiments directly in the curriculum [12]. The use of teaching technologies that enable active learning improves teachers' self-efficacy and the quality of the teaching and learning process [12]. Researchers and educators have suggested that active learning is an approach where the learning process through the students' construction of knowledge and understanding that occurs while they are in school [13]. In Malaysia, all science textbooks incorporate special features with an emphasis on science, technology, engineering and mathematics (STEM), thinking skills, scientific skills, and computational thinking (CT). The objectives are to equip learners with 21st century skills, and encourage scientifically minded individuals and active learning [14]. However, most secondary schools still use conventional methods with less technology in the learning processes, especially in physics. Hence, the Ministry of Education's objective will not be achieved unless schools extensively integrate technology into their teaching and learning processes.

The term 'blended learning' is frequently used in the context of education today, but there is ambiguity about what it actually means [15]. There is agreement that the main content in blended learning is teaching and learning that takes place involving face-to-face meetings as well as online [16–18]. Studies involving blended learning can provide benefits in improving the academic performance of learners. Thus, it can be said that studies related to blended learning contribute to the field of teaching and learning, including learning physics [18]. For example, a study using an experimental design conducted by Hrastinski [18] related to the effect of blended learning on student performance proved that blended learning was more effective than face-to-face instruction.

1.1. Problem Statement

Students at school tend to stay away from physics, whether related to the subject, either directly or indirectly, because the students find this subject to be dreadful and difficult to study [3,19]. This problem started in the classrooms, which still practice conventional teaching techniques and lack exposure to the use of modern teaching and learning in line with the 21st century [19]. Learning physics using conventional learning invites a lot of negative feelings, such as the students finding the subject to not be interesting, difficult to understand, and irrelevant to real-life situations [20,21]. Students found difficulty understanding the basic concepts of physics and totally rely upon numerical operations [6]. To overcome this problem, changes are needed, from using teacher-centered to student-centered methods. This is in line with the education evolution in the 21st century, where STEM is being implemented [19]. Teaching and learning using simulations have been conducted by previous studies and proven to show positive results [5,22–24]. It is easier for students to understand a concept or theory and attracts students to study the subject [21]. This is because simulated learning is clearer and more efficient than conventional learning, which does not have any animation. Simulated learning proves that students are actively involved in learning [25]. However, the studies on simulations at the secondary and tertiary levels conducted in Malaysia are minimal compared to other countries [23]. As a result, the application of simulation is not fully maximized in schools, although most schools have computer labs for students to use. Previous studies on simulations are more focused on biological and chemical subjects [23,26]. Therefore, this study attempted to examine the effectiveness of learning physics through interactive simulation as blended learning and the comparison with conventional learning through the views of students at the secondary school level.

1.2. Research Questions

The research questions for this study are as follows:

1. Is there a significant difference in student achievement between pre and post-tests after using simulation as blended learning in the experimental group (EG)?
2. Is there a significant difference in student achievement between pre and post-tests after using conventional learning in the control group (CG)?
3. What are the secondary school students' thoughts on their usual classroom learning of physics?
4. What are the secondary school students' thoughts on the learning physics subject using simulation as blended learning?

1.3. Hypotheses

Two hypotheses were formulated based on the first and second research questions:

H1: *There is no significant difference between pre and post-test results in the EG, which used simulation as blended learning for physics subjects.*

H2: *There is no significant difference between pre and post-test results in the CG, which used conventional learning for physics subjects.*

2. Materials and Methods

2.1. Research Design

This study employed an exploratory research design by conducting quantitative and qualitative methods on 60 Form Four Pure Science students (age 16) in one of the districts in Kedah, Northern Peninsular Malaysia. The samples were randomly selected from two schools and were later divided into EG and CG, as shown in Figure 1. The pre and post-test were conducted in a quantitative method, followed by interviewing two of the participants from each group in a qualitative method [27].

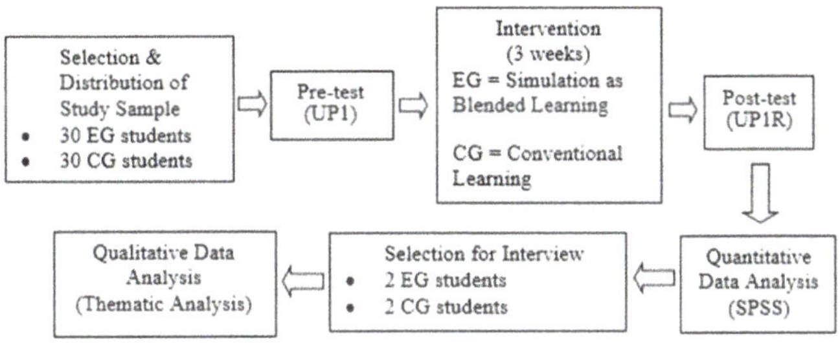

Figure 1. This is a figure of researchers' exploratory research design in this study.

2.2. Research Instruments

In this study, the EG, who underwent blended learning in physics, used the Physics Interactive Simulation Learning Module (MoPSIF) as a guide. MoPSIF was specifically designed to suit the use of the web-based interactive simulation known as Physics Education Technology (PhET). PhET was developed by experts from the University of Colorado Boulder to aid students in learning physics and other STEM subjects through simulated learning. The Simulated Daily Lesson Plan (SDLP) has also been provided to physics teachers who use the PhET interactive simulation in the teaching process. For the quantitative data collection, two sets of multiple choice type questions, pre-test (UP1) and post-test (UP1R), were used for the EG and CG. For the qualitative data collection, an interview protocol was prepared by the researchers and evaluated by a group of experts. The examples of questions in the protocol are as follows:

1. Can you tell me the learning instructions in physics that usually happen in your classroom?
2. How do you usually feel during physics lesson?
3. To what extent the textbook and the reference books helped you in learning physics?
4. Do you agree that learning in small groups in class can help you understand physics better? Why?
5. Do you always revise the physics topics at home?
6. Which do you prefer between learning physics in school or online learning at home?
7. What can you share about learning the topic of *Forces and Motion: Forces in Equilibrium* using the PhET simulation and MoPSIF module?

All of the research instruments were evaluated by experts in the field of education with at least five years of experience.

3. Results

3.1. Validity and Suitability Tests

Well-designed modules, known as MoPSIF and SDLP, were designed by researchers to aid PhET simulation via blended learning in physics. The topic covers Forces and Motion: Forces in Equilibrium. Both the module and SDLP were validated by three expert lecturers in the field of education from three different institutions. Each aspect of the validity category with the mean score value is shown in Table 1. The average mean score for all of the aspects was 3.57. Based on the validity result, the module and SDLP were categorized as very good in terms of their suitability.

Table 1. This is a table of module and SDLP evaluation by experts based on the aspects of suitability and validity.

Validity	Panels			Average	Category
	P1	P2	P3		
Introduction	2.00	2.75	2.50	2.42	Good
Learning outcomes	4.00	4.00	4.00	4.00	Very Good
Materials and Technology Usage	4.00	3.75	4.00	3.92	Very Good
Teaching Process and Procedure	4.00	4.00	4.00	4.00	Very Good
Assessment Activities	4.00	4.00	4.00	4.00	Very Good
Conclusion	3.00	4.00	4.00	3.67	Very Good
Professional Writing	3.00	3.00	3.00	3.00	Very Good

3.2. Reliability Test

A survey was constructed in addition to the module validity from the experts. To determine the appropriateness of the survey, a reliability test was conducted on five (5) dimensions of the survey questionnaire, which consisted of a total of 40 items. According to Nunnaly [28] and Nunnaly and Bernstein [29], alpha reliability values above 0.70 were consistent for each dimension in this study. The reliability test results for each dimension of the survey are shown in Table 2.

Table 2. This is a table of reliability test of survey questionnaire for each dimension of the study.

Dimensions	Number of Items	Cronbach's Alpha Value
Students' perception of simulated learning using module	10	0.77
Students' attitude of simulated learning using module	10	0.70
Teacher guidance	7	0.71
Teacher's knowledge of learning by simulation & module	7	0.70
Group work support by teacher in the classroom	6	0.71

3.3. Analysis of the Effectiveness of Simulationas Blended Learning

Based on Table 3, the paired sample t-test was significant (t (29) = −11.47, $p < 0.05$). A p-value of <0.05 indicates that there is a significant difference between the pre-test and post-test results [27]. The result of the study successfully rejected H1. These results proved that there was a significant difference between the pre-test and post-test results in the experimental group. The mean score (13.37) after the use of interactive simulation as blended learning was higher than the mean score (9.90) before it was conducted.

Table 3. This is a table of results using blended learning with simulation in physics of the experimental group.

Tests	N	Mean	Standard Deviation	df	t	p
Pre	30	9.90	2.20	29	−11.47	0.00 *
Post	30	13.37	1.65			

* $p < 0.05$.

As indicated in Table 4, the paired sample t-test was insignificant (t (29) = 0.44, $p > 0.05$). A p-value of >0.05 indicates that there is no significant difference between the pre-test and post-test results [27]. The results of the study failed to reject H2. They also proved that there was no significant difference between the pre-test and post-test results in the experimental group. The mean score (9.33) after undergoing conventional learning was almost the same as the mean score (9.47) before undergoing conventional learning in physics.

Table 4. This is a table of results using conventional learning in physics of the control group.

Tests	N	Mean	Standard Deviation	df	t	p
Pre	30	9.47	2.43	29	0.44	0.67 *
Post	30	9.33	2.73			

* $p > 0.05$.

3.4. Thematic Analysis

The thematic analysis of the interview data found that the learning experiences in classrooms were related to internal aspects, which are the assessment of self-achievement in physics and the external aspects involving the physics teacher. These students' thoughts have provided insights into the real situation in the classroom, which should be considered as the basis for improving and enhancing the learning process of physics.

3.4.1. Students' Thoughts on Conventional Learning in Physics

Schools preferred to use alternative references rather than using textbooks in conventional teaching and learning approaches. This finding is in line with Yap [30], who states that teachers should focus more on the content of textbooks and notes in conventional teaching. However, respondent S1 found that it was difficult to imagine a phenomenon or process related to the content of the lesson through reading. S1 states that the images in the reading materials, reference books, and textbooks were stationary. According to S1,

> *"Teachers like to use various reference books in class ... They seldom use the textbook." (S1).*

> *"The pictures or images in the books are all static. Not moving ... like frozen. It is often too difficult for me to imagine how things work." (S1).*

According to S1, physics lessons in the classroom are boring. This is because of a failure to understand during the teaching and learning process, to the extent that S1 stopped paying attention to what the teacher was explaining.

> *"I felt bored .... I was also confused with the teacher's explanation ... Sometimes my mind just shut off. That was why I became sleepy in class. When it felt hard to grasp what the teacher was trying to say, I just stopped listening ... " (S1).*

S1 preferred to study on his own than in a group when asked about group learning in physics via conventional learning because the focus was always interrupted. According to S1, group activities via conventional learning were less helpful. Participation in group work was only limited to presentations and conducting experiments during physics lessons [3,19]. According to S2, the lack of diversity in group activities was also a factor in students becoming bored and less interested in physics.

> *"When the teacher gave us activities to be in done in groups, usually I would not focus because there would be a lot of chatting and laughing, and less work was done. We learnt only a little from the activities ... We only sat in groups when there was a presentation ... I did not really contribute much ..... it was more like I did not help or contribute anything at all" (S1).*

> *"Other than presentation, there was no other group work activity ... for example discussion. It was always group presentation and doing experiments in the lab. That was all" (S2).*

Self-learning physics at home was difficult if the reference resources were limited because they must only rely on books and notes. Explanations of any incomprehensible concept would be asked of the teacher the next day because no one can help at home. As stated by S1 and S2,

"At home, I have only the textbook and notes to refer to. I did only a bit of revision and homework ... If I did not understand a certain point, I would ask the teacher the next day. But, sometimes I forgot to do so ... There was nobody at home who could help me. I only relied on the text book, notes and reference books ... I would say these resources have not helped me that much" (S1).

"At home, I did the learning alone. There was nobody whom I could ask for help. I could not have a study group outside school hours because all my friends live far away from each other. I would say that it was difficult to even organise a study group once a month outside school hours" (S2).

3.4.2. Students' Thoughts on Interactive Simulation via Blended Learning in Physics

S3 stated that the topic taught was easy to understand and not boring because the process or phenomenon could be easily described. The simulation images moved, and the method of calculation could be performed easily. According to S3, the simulation activity carried out was a repetitive simulation similar to the characteristics of simulated learning stated by Joyce et al. [29].

"This new method made it is for me to understand ... I have problems in this sub-topic ... resolution of forces ... ... ! I am still not clear about its calculation. It is hard for me to imagine it when the teacher is explaining. When I look at the picture or the diagram, I cannot understand because it doesn't move ... Through simulation, I can understand it easily. I can also calculate easily because I can understand the concept. The pictures move ... and we can repeat the process ... It is not boring ... This is the best way to learn!" (S3).

Learning experiences with PhET simulation via blended learning were more meaningful than conventional learning. S3 felt that it was quite difficult to digest the information provided in writing in the textbook. The use of simulation via blended learning helped the students imagine the process that occurred when the values were changed, and it was easy for the students to acquire the concept of Forces and Motion. This learning style also helped students summarize lesson content after understanding the topic. S4 repeatedly tried and observed the changes that occurred when different values were inserted.

"Through simulation learning, we can actually see how the process works compared to pictures in books. It is the same with a certain phenomenon. Explanation in words does not help us imagine or understand the process" (S3).

"It is difficult to imagine a certain phenomenon or concept. I am afraid that I might have a different imagination from others. Sometimes I could not imagine at all a certain concept ... So, this simulation has helped me a lot" (S3).

"We could see the actual motion ... If the horizontal plane was tilted, we could actually see its influence on the force ... And that made me understand things easily. My friends and I could repeat the process just by changing the values and see what the results are. It was really easy to understand" (S4).

Learning using PhET Simulation as blended learning was made easier with MoPSIF in Forces and Motion: Forces in Equilibrium. The initiative to 'play' with the simulation even after completing all the activities in MoPSIF was demonstrated by S3 and classmates. This proved that the simulation activities encouraged self-exploration, either in a group or self-learning. MoPSIF, built by the researchers, has also supported blended learning. The students would first ask their peers, followed by referring to the MoPSIF module, before asking the teacher if the problem still persisted. According to S3,

"First, we did the experiment alone. Then, we repeated using other values ... we also tried many different things which were not in the module just to see the result ... If there was anything that we did not understand we referred to the module. If we still could not understand, we asked the teacher" (S3).

"I find it difficult to do it at home at first, because if I don't understand or I'm stuck there will be no one to help. But, the module was very helpful. There was a guide on how to do the experiment ... it was easy to follow and it was easy to understand ... The tasks in the module also helped me how to go about doing the simulation" (S3).

The concern about learning physics at home was overcome by online support learning such as PhET web-based simulation learning, and ran smoothly with MoPSIF. This could overcome the learning problems during the pandemic because, indirectly, the effect was better compared to utilizing only one type of learning style, such as the conventional method.

"It is a very suitable learning method ... especially during the pandemic. Simulation learning is the best way, in my opinion. The link to the website was already there. We just needed the internet. And there was also the module. The module really helped us how to do the task. It was really helpful especially when we could not attend school" (S3).

"The simulation was easy because there was the module as guide, with the pandemic; we have to observe social distancing. Our house is the only safe place. So, when teacher gave us the simulation and the module, I felt it was the suitable way of learning. It was more efficient than the previous method. This new method was also more interesting. I could easily understand the concept. I am now interested to learn more in Physics subject. It is hightime we use the new method of teaching and learning" (S4).

Learning physics using the simulation became more interesting because the students felt that it was similar to playing a game. This proved that there was an element of gamification that attracted their focus and distracted them from boredom when studying the topic [20]. These findings are in line with the research from the University of Colorado, Boulder, where their objective was to produce a game-like simulation to encourage exploration and discovery. This supports past research findings, which state that the involvement of students in learning through teaching technology, such as simulation, is active and can improve the quality of learning [12,25].

"When it comes to the part where we have to do a lot of reading, I get bored easily. However, with the simulated learning I could interact with it and understand easily the function and process of certain things. I enjoyed best when I could replace the values just to see what happens. I didn't feel bored at all. And if I made a mistake somewhere, there was always the reset button. Just click and it would go back to normal. It was just like playing the computer games" (S3).

4. Discussion

From the analysis of the results above, students who learn Forces and Motion: Forces in Equilibrium in physics via simulation as blended learning have shown great achievement mean scores compared with the conventional learning method. The thematic analysis results indicated that the use of the conventional learning method in the CG was boring, dreadful, not interesting, and limited by references, such as textbooks and reference books, when self-learning. According to students in the CG, these factors arise when distraction in group learning occurrs, the fear of being scolded by the teacher when asking questions, and a lack of diversity in teaching and learning processes, except for practical (experimental) activities in the physics laboratory. The EG students stated that the use of the simulation learning method as blended learning was interesting, easy to understand, increased their imagination, encouraged self-exploration, and had the elements of gamification. The students in the EG also stated that they were able to summarize learning content and aid their learning while revising, whether in a group or self-learning.

5. Conclusions

In conclusion, blended learning using simulation can improve students' achievement in physics compared to conventional methods. Creativity in diversifying the teaching and learning approach of physics topics by teachers is needed to boost students' positive attitude towards the subject of physics. The use of simulation as blended learning, together with well-designed MoPSIF and SDLP, is a new approach to learning physics topics in secondary schools. Simulation learning as blended learning has the potential to attract students' interest in physics and support meaningful active learning as opposed to conventional learning. Blended learning in physics with good interactive simulation by PhET, accompanied by simulation activities and guidance, such as MoPSIF and SDLP, are relevant in learning nowadays. It has been proven that misconceptions in physics can be eliminated by using PhET interactive simulations via blended learning. The repetition of PhET simulations can be performed as many times as they please until an understanding of the topic is achieved. Simulation via blended learning helps students to learn physics subjects with or without the aid of a teacher, and this encourages students to learn physics either in groups or independently. As such, future research needs to examine the impact on teachers' motivation to teach students using simulation methods as blended learning in physics.

Author Contributions: Conceptualization, M.N.M.N. and R.M.-A.; methodology, M.N.M.N.; software, M.N.M.N.; validation, M.N.M.N., A.Y. and R.M.-A.; formal analysis, M.N.M.N.; investigation, M.N.M.N.; resources, M.N.M.N.; data curation, M.N.M.N.; writing—original draft preparation, M.N.M.N.; writing—review and editing, R.M.-A.; supervision, R.M.-A. and A.Y. All authors have read and agreed to the published version of the manuscript.

Funding: This research received no external funding.

Institutional Review Board Statement: The study was conducted in accordance with the Educational Research Application System (ERAS), and approved by the MINISTRY OF EDUCATION (MOE) MALAYSIA.

Informed Consent Statement: Informed consent was obtained from all subjects involved in the study.

Data Availability Statement: Not applicable.

Acknowledgments: I would like to thank to PhET team from the University of Colorado Boulder for allowing us to use the interactive simulation freely.

Conflicts of Interest: The authors declare no conflict of interest.

References

1. Aykutlu, I.; Bezen, S.; Bayrak, C. Teachers opinions about conceptual challenges experienced in teaching physics curriculum topics. *Procedia Soc. Behav. Sci.* **2015**, *174*, 390–405. [CrossRef]
2. Gunta, S.; Ousman, K. Problems in the teaching and learning of Physics in the secondary and preparatory schools, the cases of Wolaita and Dwuro zones. *Glob. J. Hum. Soc. Sci.* **2015**, *15*, 1–5.
3. Saleh, S. Malaysian students' motivation towards physics learning. *Eur. J. Sci. Math. Ed.* **2014**, *2*, 223–232. [CrossRef]
4. Srisawadi, N.; Kroothkeaw, S. Supporting students' conceptual development of light refraction by simulation-based open inquiry with dual-situated learning model. *J. Comput. Educ.* **2014**, *1*, 49–79. [CrossRef]
5. Abdullah, S.; Shariff, A. The effects of inquiry-based computer simulation with cooperative learning on scientific thinking and conceptual understanding of Gas Laws. *EJMSTE* **2008**, *4*, 387–398. [CrossRef]
6. Kolcak, D.Y.; Mogol, S.; Unsal, Y. A Comparison of The Effects of Laboratory Method and Computer Simulations to Avoid Misconceptions in Physics Education. *Educ. Sci.* **2014**, *175*, 154–171.
7. Utusan Malaysia. Dasar 40:60 Pelajar Sastera Sains Belum Tercapai. Available online: http://www.utusan.com.my/utusan/info.asp?y=2009&dt=0327&pub=Utusan_Malaysia&sec=Dalam_Negeri&pg=dn_07.htm (accessed on 19 March 2017).
8. Sidin, R. Socialization of science and technology: A standard proposal. *J. Pendidik. UKM* **2004**, 47–63.
9. Ferguson, K.J. *Facilitating Student Learning, in an Introduction to Medical Learning*; Jeffries, W.B., Hugget, K.N., Eds.; Springer: New York, NY, USA, 2010; pp. 1–10.
10. Ferguson, C.; Hickey, R.; Ferns, S.; Savelle, R. International collaborations: Inspiring active learning in higher education. *Int. J. Cross-Discip. Educ.* **2019**, *10*, 4120–4126. [CrossRef]
11. Nalevska, G.P.; Kuzmanovska, M. Teaching methods as a factor of students' learning motivation. *J. Educ. Res.* **2020**, *2*, 40–50.

12. Tien, E.C.; Hamid, H. Use of technology in active learning teaching practices to enhance lecturers' self-efficacy in technical university environment. *IJET* **2020**, *9*, 436–443.
13. Cambridge Assessment International Education. Active Learning. Available online: https://www.cambridgeinternational.org/Images/271174-active-learning.pdf (accessed on 4 August 2022).
14. Ministry of Education. Malaysia Education for All: End Decade Review Report 2000–2015. Available online: www.moe.gov.my (accessed on 4 August 2022).
15. Oliver, M.; Trigwell, K. Can 'blended learning' be redeemed? *E-Learn. Digit. Media* **2005**, *2*, 17–26.
16. Cross, J. *Blended Learning Systems. The Handbook of Blended Learning: Global Perspectives, Local Designs*; Bonk, C.J., Graham, C.R., Eds.; Pfeiffer: San Francisco, CA, USA, 2006; pp. 3–23.
17. Garrison, D.R.; Kanuka, H. Blended learning: Uncovering its transformative potential in higher education. *Internet High. Educ.* **2004**, *7*, 95–105. [CrossRef]
18. Hrastinski, S. What Do We Mean by Blended Learning? *TechTrends* **2019**, *63*, 564–569. [CrossRef]
19. Bunyamin, M.A.H.; Finley, F. STEM Education in Malaysia: Reviewing the Current Physics Curriculum. In Proceedings of the International Conference of Association for Science Teacher Education, Reno, NV, USA, 7 January 2016; pp. 1–16.
20. Yunzal, A.N., Jr.; Casinillo, L.F. Effect of Physics education technology (PhET) simulations: Evidence from STEM students' performance. *J. Educ. Res. Eval.* **2020**, *4*, 221–226.
21. Liu, M.; Horton, L.; Omalson, J.; Toprac, P. A study of learning and motivation in a new media enriched environment for middle school science. *Educ. Technol. Res. Dev.* **2011**, *59*, 249–265. [CrossRef]
22. Chen, Y.-L.; Pan, P.-R.; Sung, Y.-T.; Chang, K.-E. Correcting misconceptions on electronics: Effects of simulation-based learning environment backed by a conceptual change model. *Educ. Technol. Soc.* **2013**, *16*, 212–227.
23. Elangovan, T.; Ismail, Z. The effect of realistic simulation and non-realistic simulation on biology students' achievement. *Sci. Acad. Publ.* **2013**, *3*, 231–241.
24. Ulen, S.S.; Cagran, B.; Slavinec, M.; Gerlic, I. Designing and evaluating the effectiveness of Physlet-based learning materials in supporting conceptual learning in secondary school physics. *J. Sci. Educ. Technol.* **2014**, *23*, 658–667. [CrossRef]
25. Ajredini, F.; Zajkov, O.; Mahmudi, N. Case study on simulations and real experiments on higher order skills. *Maced. Phys. Teach.* **2012**, *48*, 29–34.
26. Al-Mashaqbeh, I. Computer simulation instruction: Carrying out chemical experiments. *IJMECS* **2014**, *5*, 1–7. [CrossRef]
27. Creswell, J. *Educational Research: Planning, Conducting and Evaluating Quantitative and Qualitative Research*, 4th ed.; Pearson: Essex, UK, 2015.
28. Nunnaly, J.C. *Psychometric Theory*, 2nd ed.; McGraw-Hill: New York, NY, USA, 1978.
29. Nunnaly, J.C.; Bernstein, I.H. *Psychometric Theory*, 3rd ed.; McGraw-Hill: New York, NY, USA, 1994.
30. Yap, W.L. Transforming conventional teaching classroom to learner-centred teaching classroom using multimedia-mediated learning module. *IJIET* **2016**, *6*, 105–112.

proceedings

MDPI

Proceeding Paper

Virtual Professional Communication Project Presentation: Examining Students' Speaking Anxiety in a Malaysian Public University †

Sheik Badrul Hisham Jamil Azhar [1], Ameiruel Azwan Ab Aziz [2,*], Amirah Mohd Juned [2], Nazarul Azali Razali [2], Nor Atifah Mohamad [2] and Nuramirah Zaini [1]

1 Akademi Pengajian Bahasa, Universiti Teknologi MARA Cawangan Melaka, Jasin Campus, Jasin, Melaka 77000, Malaysia
2 Akademi Pengajian Bahasa, Universiti Teknologi MARA Cawangan Melaka, Alor Gajah Campus, Melaka 78000, Malaysia
* Correspondence: ameirul@uitm.edu.my
† Presented at the International Academic Symposium of Social Science 2022, Kota Bharu, Malaysia, 3 July 2022.

Abstract: ELS304 Professional Communication Exercise is a final semester course that the LG120 Diploma in English for Professional Communication students must complete before graduating and was offered for the first time in September 2020 via Open Distance Learning (ODL). For this, students are required to undertake a project that is related to their field of study and present it to a panel of examiners consisting of faculty lecturers and industry panel members. As carrying out the presentation in English was already stressful enough for the students, the presence of these external examiners could lead to a higher level of stress and anxiety. With this in mind, this study intends to examine the students' perceptions of communication apprehension, fear of negative evaluation and anxiety during the final course presentation and whether the presence of these two additional factors, the ODL method and the inclusion of external examiners could cause communication apprehension on the part of the students that could lead to additional emotional pressure, fear and anxiety. The purposive sampling method was used to obtain responses from 63 final semester LG120 Diploma in English for Professional Communication students from Universiti Teknologi MARA (UiTM) Cawangan Melaka and Johor who had recently completed the ELS304 Professional Communication Exercise course and had gone through the online presentation involving external examiners. The instrument for this study was administered using Google Forms, which utilized a 5-point rating scale for three components categorized as Communication Apprehension, Fear of Negative Feedback and Test Anxiety. The findings revealed that this group of LG120 students who conducted the presentation of their final project via ODL and in the presence of industry panel members generally experienced a moderate to high level of anxiety. Several other conclusions were made based on the findings. Overall, this study has shown that, although the mode of this assessment was changed to ODL, similar anxiety factors in public speaking that affect students remained a constant presence.

Keywords: communication apprehension; emotional pressure; speaking anxiety; test anxiety; public speaking; open and distance learning

check for updates

Citation: Azhar, S.B.H.J.; Aziz, A.A.A.; Juned, A.M.; Razali, N.A.; Mohamad, N.A.; Zaini, N. Virtual Professional Communication Project Presentation: Examining Students' Speaking Anxiety in a Malaysian Public University. *Proceedings* **2022**, *82*, 104. https://doi.org/10.3390/proceedings2022082104

Academic Editor: Mohamad Rahimi Mohamad Rosman

Published: 20 October 2022

1. Introduction

Speaking is one of the four skills that students are required to master when learning any language. However, out of the four skills—reading, writing, listening and speaking—learning to speak is regarded as the most difficult and often causes anxiety among students, especially undergraduates [1,2]. It can also be seen that when learning English as a second language, most students would face anxiety that could negatively affect their speech performance, whether inside or outside of the classroom [3,4].

To make matters worse, ever since the emergence of the COVID-19 pandemic in late 2019 and the Movement Control Order (MCO) announcement made by the Government of Malaysia which started on 18 March 2020, the method of teaching and learning has shifted toward Open Distance Learning (ODL). This situation has presented new challenges for academicians and students as all forms of teaching, learning, and assessments must now be completed online.

Concerning the course of Professional Communication Exercise (ELS304), which is a final semester course for the LG120 English for Professional Communication diploma program at UiTM Melaka and Johor branches, the students are required to present their project not only to assigned examiners amongst faculty lecturers but also to external examiners from the industry in assessing their final project presentation. The external panel members are chosen based on the general areas covered by the LG120 program, such as Translation, Proofreading, Creative Writing and Desktop Publishing.

The presence of these external examiners could have added more unwelcomed stress on the students on top of the already present anxiety. This situation raises a question as to whether the students are able to cope with the anxiety of presenting their project both via the ODL method and with the presence of external examiners watching and evaluating the presentations. This issue needs to be studied as the presence of these two additional factors, the ODL method and the inclusion of external examiners from the industry, could cause communication apprehension on the part of the students that could lead to additional emotional pressure, fear and anxiety.

This research aims to understand the ELS304 Professional Communication Exercise students' perceptions of the methods employed to carry out the final course presentation and whether they have any anxiety or fear for carrying out their presentations online and in the presence of external examiners. For these reasons, the study examines the factors contributing to students' anxiety during their online final project presentation. The findings would allow researchers to understand better the factors leading to anxiety amongst students so that a more student-friendly environment can be created to ease the tensions.

2. Literature Review

2.1. Foreign Language Classroom and Anxiety

Anxiety among language learners can greatly impact on language learning, that could lead to avoidance of participation in the classroom [5]. In understanding how language learners are affected by anxiety, language teachers need to be sensitive to learners' communication apprehension, test anxiety, and fear of negative evaluation [6]. For this purpose, The Foreign Language Classroom Anxiety Scale (FLCAS) was developed to facilitate teachers and educators in identifying students who are experiencing negative anxiety that affects their language learning experience [5,6].

Anxiety is considered as a form of disorder associated with feelings of worry, tension and fright. This disorder is a prevalent form of mental disorder [7]. This happens when one tends to overthink what will happen in the future. This feeling can start to manifest when a person starts having mood disturbances or over-thinking and when there is a need to undergo changes in behavior or physiological activity [7]. As the world begins to move faster, this so-called disability is becoming a major health concern worldwide and becoming increasingly prevalent [8].

Another important factor concerning anxiety is that although it is known that this condition usually starts to develop during early adulthood, this group of people rarely get the support they need to overcome the issue [9]. It was projected that the number of college students who meet at least one of the diagnostic criteria of anxiety could be around 50% [10]. This projection puts college students in the high-risk category of experiencing anxiety. The stress and anxiety faced by students would put them at risk, as the psychological distress would have an unwanted effect on their academic performance [11].

The stress and anxiety could be attributed to many reasons. The common factors that lead to anxiety are biological factors, such as age and gender [12]. Further investigation also found that being a female would increase the possibility of a person suffering from anxiety [12]. Other anxiety-related reasons could be traced to financial burdens [13]. However, the best explanation for causes of anxiety amongst college students could very well be the pressure from examinations and workload [14]. Other factors include lack of leisure time for oneself, competition with other students and fear of not meeting the parents' expectations [14].

2.2. *Fear of Public Speaking*

Anxiety is one of the most common problems observed concerning language learning [15]. Many factors can contribute to this anxiety, lack of confidence or the fear of performing the public speaking itself. Language anxiety originates from the worry and negative emotional reaction aroused when learning or using a second language [16]. The fear of public speaking, or glossophobia, is quite common among those learning a second language, and some researchers found that as many as 77% of the world's population would face some form of anxiety when performing public speaking [17]. However, other researchers discovered that the situation existed even before the students participated in public speaking [16]. This anxiety could appear even days before the speaking task, leading some students to skip the task altogether [16]. These findings align with the claim that language anxiety is related to situational anxiety experienced in students' second language classrooms or the classroom setting [18]. This means that situations in the background are also contributing to the anxiety, aside from having to present in the second language itself.

Other studies relating to foreign language learning found three anxiety categories that contributed the second language speaking-anxiety which are: the learner (the learner's personal and interpersonal anxiety and the learner's beliefs about language learning), the teacher (the instructor's beliefs about language teaching and the instructor–learner interactions), and the instructional practice (classroom procedures and language testing) [19]. This further proves that the speaking anxiety faced by students goes beyond speaking in the target language alone and that the situation and environment in which the speech production is performed leads to the feeling of anxiety. This is in line with the theory that anxiety faced by second language learners could be attributed to communication apprehension, fear of negative evaluation and test anxiety.

It cannot be denied that other factors could contribute to speaking anxiety by second language learners, as pointed out by many researchers. A comparative study on anxiety-coping strategies between Malaysian and Japanese second language students found that second language learners were also concerned when they felt that the audience might judge the way they dressed, as well as their stage presence [20]. Furthermore, the fear of looking stupid in front of an audience would also create fear and anxiety since they became the center of attention [21]. On the other hand, most students become a silent crowd when any speaking activity is performed, while some would just stutter their responses [22]. While this could merely be due to a lack of confidence, other reasons could also be the students' lack of interest and unwillingness to participate in the conversation [22]. Whatever the reasons, it cannot be denied that some forms of anxiety are happening in the classroom environments which would directly affect the speech production of the students.

3. Methodology

3.1. *Sample and Research Procedures*

The purposive sampling method was used to obtain responses from 63 final semester LG120 Diploma in English for Professional Communication students from the Universiti Teknologi MARA (UiTM) Cawangan Melaka and Johor. These respondents were explicitly selected as they had recently completed the ELS304 course and had gone through the online presentation involving external panel members from the industry.

The data were collected virtually where respondents were asked to answer an online survey conducted via Google Forms in the final week of their study after they had completed the ELS304 presentation. The respondents were given one week to complete and submit the Google Forms for analysis. This was considered the best time to conduct the survey as the core aspect of this study is to understand the fear and anxiety that the students experienced while carrying out their presentation online and in front of external panel members from the industry.

The data obtained through the survey were analyzed using descriptive statistical analysis and interpreted through means and standard deviations.

3.2. Instrument

The instrument for this study was developed using Google Forms. The instrument for this study was adapted from the Students Foreign Language Classroom Anxiety Scale [5], which utilized the 5-point rating scales from Strongly Disagree to Strongly Agree with the lowest score being for Strongly Disagree. The survey consisted of three components categorized as *Communication Apprehension, Fear of Negative Feedback* and *Test Anxiety.*

3.3. Data Analysis

Upon conducting fundamental data analysis, Cronbach's Alpha was performed to ensure the reliability of the data. Table 1 shows that the Cronbach's Alpha value for Section One: Communication Apprehension was 0.725, Section Two: Fear of Negative Evaluation was 0.751 and Section Three: Test Anxiety was 0.740. These indicate that the items are reasonably reliable for the respondents of the study.

Table 1. Cronbach's Alpha for Student's Speaking Performance Components.

Components	Cronbach's Alpha Value	No of Items
Communication Apprehension	0.725	13
Fear of Negative Evaluation	0.751	4
Test Anxiety	0.740	7

4. Results and Discussion

The study results obtained are presented based on the research questions through mean values and standard deviation. Descriptive statistical analysis of means and standard deviations were applied to measure and interpret the responses obtained from the 63 final semester LG120 students.

4.1. Students' Perception of Communication Apprehension Experienced While Carrying out Their ELS304 Final Presentation

Mean values and standard deviation of the students' perception of communication apprehension attributes/sub-attributes experience during the ELS304 Final Presentation are illustrated in Table 2. The highest mean score of 3.59 was recorded by *'feel anxious even if I am well-prepared to speak'*. High scoring means of 3.52 and 3.41 were also recorded for *'would feel comfortable to be around students with good oral proficiency'*, and *'feel at ease preparing myself to speak well'*, respectively. The next highest was for *'am panic to speak without preparation'* at 3.29. Seven of the 13 items listed under this category scored below 3.20. The lowest mean score recorded was for *'find myself thinking of things unrelated to the presentation'* at 2.40.

Table 2. Communication Apprehension.

Components	Mean	SD
Communication Apprehension		
"never feel quite sure of my speaking performance"	2.95	1.20
"tremble when I am asked to present my project"	2.73	1.21
"get so nervous that I forget what to say"	2.76	1.24
"get frightened/nervous when I don't understand what the panel says"	2.75	1.32
"find myself thinking of things unrelated to the presentation"	2.40	1.28
"feel more tense and nervous than my other presentations in other subjects"	3.21	1.17
"get nervous and confused when I am speaking"	2.68	1.13
"feel very sure and relaxed before the presentation starts"	2.94	1.06
"feel at ease preparing myself to speak well"	3.41	0.94
"am panic to speak without preparation"	3.29	1.34
"feel anxious even if I am well-prepared to speak"	3.59	1.03
"would not be nervous speaking to audience with good oral proficiency"	3.38	1.10
"would feel comfortable to be around students with good oral proficiency"	3.52	1.03

4.2. Students' Perception of the Fear of Negative Evaluation by External Panels While Carrying out Their ELS304 Final Presentation

The Fear of Negative Evaluation component's findings from the survey is shown in Table 3. All four items listed under this category show scores below 3.50, with the highest being 3.4 for *'get nervous to answer questions without preparation'*. This is followed by *'keep thinking that other students are better in English than I am'* at 3.38, and *'feel conscious speaking in front of audiences'* at 3.33. The lowest mean recorded was for the descriptor *'am afraid others will laugh at me when I am speaking English'* with a score of 2.54.

Table 3. Fear of Negative Evaluation.

Components	Mean	SD
Fear of Negative Evaluation		
"feel conscious speaking in front of audiences"	3.33	1.03
"keep thinking that other students are better in English than I am"	3.38	1.31
"am afraid others will laugh at me when I am speaking English"	2.54	1.32
"get nervous to answer questions without preparation"	3.40	1.20

4.3. Students' Perception of the Test Anxiety They Experienced While Carrying out Their ELS304 Final Presentation

Table 4 below illustrates the findings for the *Test Anxiety* component. Out of seven items, the highest score obtained here was for *'am worry about the consequences of failing the assessments'* at 4.08. This was followed by *'am afraid the panels will correct all my speaking errors'* with a score of 3.00. The remaining five items all scored below 3.00, with *'am usually at ease during the assessments'* at 2.97 and *'get upset for not understanding the panels' questions or comments'* with a score of 2.94. The lowest score obtained for this category is 2.30 for *'often feel like not presenting because of fear of failing the assessment'*.

Table 4. Test Anxiety.

Components	Mean	SD
Test Anxiety		
"am usually at ease during the assessments"	2.97	1.06
"am worry about the consequences of failing the assessments"	4.08	1.07
"get upset for not understanding the panels' questions or comments"	2.94	1.19
"am afraid the panels will correct all my speaking errors"	3.00	1.31
"get more confused when I study more for the assessments"	2.73	1.14
"feel overwhelmed by the number of rules to speak English"	2.70	1.12
"often feel like not presenting because of fear of failing the assessment"	2.30	1.16

5. Conclusions

The findings of this study suggested that this group of LG120 students who conducted the presentation of their final project via ODL and to industry panel members generally experienced moderate to high levels of anxiety. Based on the data analysis, while they are mainly driven by the anxiety to perform well and not fail the assessment, the findings indicated constant concerns when examined across communication comprehension, fear of negative evaluation and test anxiety.

Firstly, the students were nervous even when they believed they were well-prepared for the presentation. Despite indicating that they feel at ease with the audience and that they believe they have a good command of the language, they still feel anxious and nervous about the idea and action of presenting in front of others. The existing literature noted that students felt uncomfortable speaking in front of many people [23]. They do not feel confident and are under pressure as they worry about making mistakes during the presentation [23]. Their confidence in their ability to present and use the language was occasionally affected by the sense that they will make language or speaking errors and that these errors will impact on their presentation, grading and self-esteem.

Secondly, the respondents indicated a high anxiety level when faced with an unprepared situation. Students' anxiety levels increased when students were asked about an unfamiliar topic [24]. This will make the students feel uneasy and stressed about the question from the industry panel members as they do not have prior knowledge about the topic. Students are becoming panicked about speaking without preparation [25]. This area could be related to situations where they must address questions by the examiners and industry panel members. These queries about the students' projects often vary according to examiners and industry panel members, making it quite difficult for students to prepare the answers, thus affecting their anxiety level.

Several conclusions can be made based on the current study. Firstly, this group of students believes that they have a good grasp of the language and are confident in their presentation skills. However, their anxiety levels are heightened in situations in which they have no control and for fear of making language mistakes while presenting. Therefore, the students are encouraged to participate in collaborative work, and this will help the students to gain confidence in speaking [26].

Further research should be conducted to examine the factors that affect these types of students in handling the online mode of the presentation, focusing on the concerns of balancing the assessment and the online delivery platform. Another area worth looking into will be comparing students' anxiety levels while presenting their projects to familiar faculty lecturers with industry panel members with whom they have not previously engaged. Other than that, inferential analysis such as correlation and regression analysis can examine the relationship between speaking anxiety and academic performance. This study has shown that although the mode of this assessment was changed to ODL, similar anxiety factors in public speaking that affected the students remained a constant presence.

Author Contributions: Conceptualization, S.B.H.J.A., A.A.A.A. and N.A.R.; methodology, A.A.A.A. and A.M.J.; software, A.M.J.; validation, S.B.H.J.A., A.A.A.A. and A.M.J.; formal analysis, N.A.M. and N.Z.; investigation, N.Z. and N.A.M.; resources, A.M.J. and N.A.M.; data curation, N.Z.; writing—original draft preparation, S.B.H.J.A. and N.A.R.; writing—review and editing, S.B.H.J.A. and A.M.J.; visualization, A.A.A.A.; supervision, A.A.A.A. and N.A.R.; project administration, A.M.J.; funding acquisition, A.A.A.A. All authors have read and agreed to the published version of the manuscript.

Funding: This research was funded by the Research and Industrial Linkages Department of Universiti Teknologi MARA Melaka under the Skim Geran Dalaman TEJA (GDT2022/1-2).

Institutional Review Board Statement: Not applicable.

Informed Consent Statement: Not applicable.

Data Availability Statement: Not applicable.

Acknowledgments: The researchers would like to thank the Research and Industrial Linkages Department of Universiti Teknologi MARA Melaka, which provided funding under the Skim Geran Dalaman TEJA (GDT2022/1-2).

Conflicts of Interest: The authors declare no conflict of interest.

References

1. Evans, J.F.; Clinton, E.; Cookson, G.; Brown, S.; Woods, D. The Stress of Public Speaking Increases Cortisol Levels in Undergraduates: Is increased preparation really the best remedy? *In Vivo* **2017**, *38*, 75.
2. Long, K.L.Y.; Yih, Y.J.; Lin, S.S.P. Undergraduates' Speaking Anxiety in English as Second Language (ESL) Classrooms. *IJSMSust* **2020**, *4*, 25–40.
3. Ahmed, N.; Pathan, Z.H.; Khan, F.S. Exploring the Causes of English Language Speaking Anxiety among Postgraduate Students of University of Balochistan, Pakistan. *Int. J. Engl.* **2017**, *7*, 99–105. [CrossRef]
4. Batiha, J.M.; Noor, N.M.; Mustaffa, R. Speaking Anxiety Among English as a Foreign Language Learner in Jordan: Quantitative Research. *Int. J. Educ. Res.* **2016**, *4*, 63–82.
5. Horwitz, E.K.; Horwitz, M.B.; Cope, J. Foreign Language Classroom Anxiety. *Mod. Lang.* **1986**, *70*, 125–132. [CrossRef]
6. Horwitz, E. Preliminary Evidence for the Reliability and Validity of a Foreign Language Anxiety Scale. *TESOL Q.* **1986**, *20*, 559–562. [CrossRef]
7. Munir, S.; Takov, V. Generalized Anxiety Disorder. In *StatPearls*; StatPearls Publishing: Treasure Island, FL, USA, 2022. [PubMed]
8. Wainberg, M.; Scorza, P.; Shultz, J.; Helpman, L.; Mootz, J.; Johnson, K.; Neria, Y.; Bradford, J.; Oquendo, M.; Arbuckle, M. Challenges and Opportunities in Global Mental Health: A Research-to-Practice Perspective. *Curr. Psychiatry Rep.* **2017**, *19*, 1–16. [CrossRef] [PubMed]
9. Sarokhani, D.; Delpisheh, A.; Veisani, Y.; Sarokhani, M.; Manesh, R.; Sayehmiri, K. Prevalence of Depression among University Students: A Systematic Review and Meta-Analysis Study. *Depress Res Treat.* **2013**, *2013*, 373857. [CrossRef] [PubMed]
10. Bruffaerts, R.; Mortier, P.; Kiekens, G.; Auerbach, R.; Cuijpers, P.; Demyttenaere, K.; Green, J.; Nock, M.; Kessler, R. Mental Health Problems in College Freshmen: Prevalence and Academic Functioning. *J. Affect. Disord.* **2018**, *225*, 97–103. [CrossRef]
11. Tosevski, D.; Milovancevic, M.; Gajic, S. Personality and Psychopathology of University Students. *Curr. Opin. Psychiatry* **2010**, *23*, 48–52. [CrossRef] [PubMed]
12. Bangasser, D.; Curtis, A.; Reyes, B.; Bethea, T.; Parastatidis, I.; Ischiropoulos, H.; Van Bockstaele, E.; Valentino, R. Sex Differences in Corticotropin-Releasing Factor Receptor Signaling and Trafficking: Potential Role in Female Vulnerability to Stress-Related Psychopathology. *Mol. Psychiatry* **2010**, *15*, 896–904. [CrossRef] [PubMed]
13. Kruisselbrink-Flatt, A. A Suffering Generation: Six Factors Contributing to the Mental Health Crisis in North American Higher Education. *College Q.* **2013**, *16*, 1–17.
14. Kumaraswamy, N. Academic Stress, Anxiety and Depression Among College Students: A Brief Review. *Int. Rev. Soc. Sci. Humanit.* **2013**, *5*, 135–143.
15. Çağatay, S. Examining EFL Students' Foreign Language Speaking Anxiety: The Case at a Turkish State University. *Procedia Soc.* **2015**, *199*, 648–656. [CrossRef]
16. Heeren, A.; Ceschi, G.; Valentiner, D.P.; Dethier, V.; Philippot, P. Assessing Public Speaking Fear with the Short Form of the Personal Report of Confidence as a Speaker Scale: Confirmatory Factor Analyses Among a French-Speaking Community Sample. *Neuropsychiatr. Dis. Treat.* **2013**, *9*, 609–618. [CrossRef]
17. Khusnia, A.N. Students' Speaking Anxiety in English Foreign Language. *J. Educ. Train.* **2017**, *11*, 83–90.
18. MacIntyre, P.; Clement, R.; Dornyei, Z.; Noels, K. Conceptualising Willingness to Communicate in a L2: A Situational Model of L2 Confidence and Affiliation. *Mod. Lang.* **1998**, *82*, 545–562. [CrossRef]
19. Young, D. Creating a Low-Anxiety Classroom Environment: What Does Language Anxiety Research Suggest? *Mod. Lang.* **1991**, *75*, 426–437. [CrossRef]
20. Hammad-Rafada, S.; Ahmad Madini, A. Major Causes of Saudi Learners' Speaking Anxiety in EFL Classrooms. *J. Engl. Lang. Educ.* **2017**, *5*, 54–71. [CrossRef]
21. Ab Aziz, A.; Yatim, A.; Nordin, N.; Norwahi, N. Stress and Coping Strategies in Public Speaking: Comparative Case Studies of Japanese and Malaysian Undergraduates. *Int. J. Acad. Res. Bus. Soc. Sci.* **2021**, *11*, 1581–1591. [CrossRef]
22. Akkakoson, S. Speaking Anxiety in English Conversation Classrooms Among Thai Students. *Malays. J. Learn. Instr.* **2016**, *13*, 63–82. [CrossRef]
23. Daud, A.; Ras, F.; Novitri, N.; Audia, C.P. Factors Contributing to Speaking Anxiety: A Case Study of Pre-Service English Teachers. *J. Educ. Sci.* **2019**, *3*, 412–422. [CrossRef]
24. Hasibuan, S.H.; Manurung, I.D.; Ekayati, R. Investigating Public Speaking Anxiety Factors Among EFL University Students Through Online Learning. *AREL* **2022**, *11*, 21–38.
25. Damayanti, M.; Listyani, L. An Analysis of Students' Speaking Anxiety in Academic Speaking Class. *J. Lang. Teach. Res.* **2020**, *4*, 152–170. [CrossRef]
26. Nik-Fauzi, S.F.B.; Ambi, S.H.; Abang Madaud, A.F.; Unin, N. Students' Perceptions of Online Learning Experiences for ESL Speaking Activities. *J. Cogn. Dev.* **2022**, *8*, 186–201. [CrossRef]

proceedings

MDPI

Proceeding Paper

Association between Lifestyle-Related Behaviors and Academic Performance among Students †

Siti Nurhafizah Mohd Shafie [1,*], Mardhiah Shamsul Shahri [1], Nur Najwa Najihah Irwan Izuddi [1], Nursyazana Mohamad Shukri [1], Nasuhar Ab Aziz [1], Azzah Amran [2] and Mohd Noor Azam Nafi [1]

1 Faculty of Computer & Mathematical Statistics, Universiti Teknologi MARA Kelantan, Kota Bharu Campus, Kota Bharu 15150, Malaysia
2 Faculty of Business and Management, Universiti Teknologi MARA Kelantan, Kota Bharu Campus, Kota Bharu 15150, Malaysia
* Correspondence: sitinurhafizah@uitm.edu.my
† Presented at the International Academic Symposium of Social Science 2022, Kota Bharu, Malaysia, 3 July 2022.

Abstract: A healthy lifestyle is one factor that minimizes the risk of becoming seriously ill or dying early. A student with a healthy mind and body would be a better learner and thus would achieve a better academic performance. However, the lifestyle of university students was disrupted due to online learning introduced by higher education institutions in response to the outbreak of the coronavirus. Therefore, this study aimed to identify the association between students' lifestyle-related behavior (physical activity, poor diet, healthy diet, sleep, screen time and body mass index) and the academic performance of all full-time students from the October 2021 to February 2022 session in UiTMKB excluding part 1 students, practical students and part-time students. A cross-sectional study was carried out and proportionate stratified random sampling technique was applied to select a sample from the Faculty of Computer and Mathematical Sciences and Faculty of Business and Management. A total of 281 students (24.20% students from the Faculty of Computer and Mathematical Sciences and 75.80% students from the Faculty of Business and Management) were surveyed through questionnaires using Google forms. Binary logistic regression was used to assess the relationship between the lifestyle-related behaviour and the academic performance of the UiTMKB students. The results showed poor diet, healthy diet and sleep have statistically significant influence on the academic performance of the UiTMKB students; meanwhile, the other variables (physical activity, screen time and body mass index) have no significant influence on the academic performance. Thus, this study will significantly assist the students to improve academic performances in terms of the factor of the student's lifestyle-related behavior.

Keywords: academic performance; lifestyle behavior; logistic regression

check for updates

Citation: Shafie, S.N.M.; Shahri, M.S.; Izuddi, N.N.N.I.; Shukri, N.M.; Aziz, N.A.; Amran, A.; Nafi, M.N.A. Association between Lifestyle-Related Behaviors and Academic Performance among Students. *Proceedings* **2022**, *82*, 105. https://doi.org/10.3390/proceedings2022082105

Academic Editor: Mohamad Rahimi Mohamad Rosman

Published: 20 October 2022

1. Introduction

A healthy lifestyle is a manner of life that reduces the chances of being seriously ill or dying prematurely [1]. However, maintaining health has become a challenge during the pandemic because evidence from previous outbreaks has demonstrated that when a pandemic progress, it has a significant influence on lifestyle-related behaviors [2]. For instance, Malaysia had introduced Movement Control Order (MCO) that started on March 18th, 2020, in response to the respiratory illness caused by the SARS-CoV-2 coronavirus that had been declared a "pandemic" by the World Health Organization (WHO) as of March 2020 [3]. Consequently, educational institutions were highly affected where students had to proceed to learning online, also known as open and distance learning (ODL).

Higher education institutions (HEIs) were severely impacted at their core due to the effect of the COVID-19 pandemic, as some studies conducted during the pandemic showed unfavorable results with regard to academic performance [4]. This has led to isolation among the younger generation in public and private higher education institutions (HEIs). Moreover, a study by [5] revealed that more than one-third of undergraduate nursing students in the nursing school in the Philippines were reported to have low academic achievement. In addition, the results of a study in a university in Italy showed that approximately 60% of all students did not meet their curriculum objectives [6]. University students aim for a high cumulative grade point average (CGPA) as it is one of the qualifications that employers in the professional field seek when hiring employees. This is a big concern because many studies suggest that a healthier lifestyle has a positive impact on academic performance [7,8]. Studies during the pre-pandemic phase showed that students' academic performance are related to lifestyle-related behavior. The implementation of online learning however has indirectly altered the lifestyle of university students specifically in terms of physical activity, dietary habits, sleep, screen time and body weight status.

Describing this scenario, ref. [9] emphasized that numerous studies throughout the world show a significant reduction in students' physical activity, ranging from 48 to 61 percent, when compared to the years prior to the advent of the coronavirus (COVID-19). These affect the students academically because some studies identified that academic performance has been proven to have a significant correlation with physical activity [10,11]. In addition, during the period of university, students are exposed to a variety of unhealthy dietary options, including meals high in saturated fat and refined sugar, which may impair cognitive function [12]. Valladares et al.12 have proven that students who practice a healthy diet are more likely to perform well academically, therefore, consuming a healthy diet is to be encouraged among students. Moreover, a study of sleep quality among students and administration staff in a university in Italy by [13] during the pandemic found that there is a greater impact of sleeping habits on the students rather than the staff in terms of bedtime and wake up time. Thus, sleep matters for students. The authors of [14] highlighted that most studies found an association between better academic achievement and optimum sleep duration. Furthermore, screen time is related to academic performance as shown in a study by [15]. Their study found an independent inverse relationship between screen time and academic performance. Lastly, ref. [16] discovered that body weight was reported to increase amid the pandemic due to negative eating habits and more sedentary time.

This is a big concern because many studies suggest that a healthier lifestyle has a positive impact on academic performance [8,17]. Therefore, this study was conducted to analyze the factors associated with the academic performance in terms of lifestyle-related behaviors among UiTMKB students during the pandemic.

- Academic performance: the word "academic performance" has typically been used to refer to any means of expressing a student's scholastic status. Frequently, grade-point average is used as a measure of academic accomplishment, particularly at the college level [18]. Academic performance has been a prominent issue among university students [19]. As [20] mentioned, students' academic excellence will lead to more employment. Therefore, any university's primary goal is to improve the academic performance of its students. Živčić-Bećirević et al. [21] highlighted that the performance of university students is typically represented in terms of a grade point average (GPA). College GPA is commonly thought of as a measure of academic success.
- Physical activity: numerous studies [11,15,22] found that engaging in physical exercise enhances academic performance. A study by [15] was examined the relationship of weight status, physical activity and screen time with academic achievement in Chilean adolescents. The result showed that adolescents classified with medium–low physical activity are less likely to obtain high academic achievement compared to those with high levels of physical activity. Moreover, refs. [11,22] identified that academic performance has been reported to have a significant correlation with physical activity.
- Poor diet: Kim et al. [23] investigated how the frequency of different diet items and mealtimes affects school performance. The study discovered that consuming soft

drink, instant noodle, fast food, and confectionery more than seven times per week had a detrimental impact on school performance.

- Healthy diet (fruits and vegetables): Rehman et al. [24] aimed to see if there is an association between university students' self-reported academic achievement and various socio-demographic characteristics, health behaviours and mental health among students in Bahria University in Karachi. In the study, they uncovered that more than half of the students said they ate breakfast every day, and this was linked to higher grades. In adddition, ref. [12] has discovered that students who practice a healthy diet are more likely to perform well academically.
- Sleep: Adelantado-Renau et al. [7] analyzed the association of sleep patterns with academic and cognitive performance in adolescents. The results showed that sleep quality was associated with academic performance and analysis of covariance revealed that higher grades among adolescents were associated with better sleep quality. Meanwhile, ref. [25] emphasised that sleep habits and sleep duration are affected by prolonged screen time as the blue light emitted from the screens of devices before bedtime is associated with sleep inefficiency and shortened sleep duration. In addition to this, sleep matters for students because [7] highlighted that higher grades were associated with better sleep quality.
- Screen time: Yan et al. [26] identified the amount of time spent by adolescents in Wuhan, China on screen-based behaviours and its associations with academic performance. The results indicated that more than four hours spent on social networking sites on school days and non-school days, watching television from two to four hours and watching videos less than one hour on school days or not every day were all negatively associated with academic performance. Meanwhile, ref. [15] has proven that screen time is one of the aspects of lifestyle that is related to academic performance. Their study found an independent inverse relationship between screen time and academic performance.
- Body mass index: Wehigaldeniya et al. [27] conducted a cross-sectional study at the University of Kelaniya in Sri Lanka to evaluate the association between academic achievement and body mass index (BMI) of undergraduate students. Their study found that there was a positive association between BMI and university students' academic performance. A study by [28] among youth in Canadian schools has shown that body weight status is related to academic performance as their study found that obesity was linked to a lower academic performance.

The main purpose of this research generally is to determine the association between lifestyle-related behaviours and academic performance among Universiti Teknologi MARA Kota Bharu (UiTMKB) students during the COVID-19 pandemic. Specifically, there are two major objectives. The first objective is to determine the associations between individual indicators of lifestyle-related behaviours (physical activity, poor diet, healthy diet, sleep, screen time and body mass index) and academic performance among UiTMKB students. Secondly, to identify factors of lifestyle-related behaviors (physical activity, poor diet, healthy diet, sleep, screen time and body mass index) that influence UiTMKB students' academic performance.

2. Methodology

2.1. Data Collection

A cross-sectional study was carried out in this study to determine the association between individual indicators of lifestyle-related behaviour (physical activity, poor diet, healthy diet, sleep, screen time, body mass index) and academic performance among UiTMKB students. The population is 1037 of all full-time students (251 students from Faculty of Computer and Mathematical Sciences and 786 students from Faculty of Business and Management) from October 2021 to February 2022 session in UiTMKB, excluding part one students, practical students and part-time students. The questionnaires in Google forms were distributed online by personally administering to 281 students (68 students from the Faculty of Computer and

Mathematical Sciences and another 213 students from the Faculty of Business and Management) that were randomly selected using proportionate stratified random sampling. Table 1 shows the research instruments were used in this study.

Table 1. Research Instruments.

Section	Item	Scale of Measurement
Demographic	Gender/Faculty Program/Current semester Age	Nominal Ordinal Interval
Academic performance [29]	An open-ended question on academic performance required students to self-report their latest GPA in a ratio scale.	Scale of measurement: • poor (0.00–1.99); • average (2.00–2.99); • good (3.00–4.00).
Physical activity [30]	Physical activity = 150 min divided by 7 days (22 min/day) Physical activity was assessed by taking 150 minutes divided by 7 days to obtain an average least duration in minutes recommended on a day	Scale of measurement: • 0–2 days; • 3–5 days; • 6–7 days.
Dietary habits [22]	Dietary habit of 'poor diet' and 'healthy diet' • 'Poor diet': seven questions • 'healthy diet': five questions	Scale of measurement: • Almost never = 1; • Less than once a week = 2; • Every week = 3; • Once a day = 4; • More than once a day = 5.
Sleep [31]	Scale with a range from 0 to 40	Scale of measurement: • Never = 1; • Less than once a week = 2; • Once or twice a week = 3; • 3 to 5 nights/days = 4; • Almost every day/night = 5.
Screen time [22]	The questions of typical usage of a variety of screens on typical weekdays and weekend days were used to calculate average daily screen time. The average of these two totals was calculated to represent average daily screen time during a typical week.	Scale of Measurement: • less than 2 h = 1; • 2 to less than 4 h = 2; • 4 to 7 h = 3; • more than 7 h = 4.
Body mass index [32]	Body mass index (BMI) required students to report their height in metre (m) and weight in kilogram (kg). These values used to calculate BMI (kg/m^2)	Scale of measurement: • 15.0 to 19.9 = underweight; • 20.0 to 24.9 = normal weight; • 25.0 to 29.9 = overweight; • 30.0 or greater = obese.

2.2. Data Analysis

Descriptive analysis was used to describe demography, lifestyle-related behaviors and academic performance of the respondents. Frequency and percentage were applied to describe and summarise the data obtained from categorical variables, which were physical activity, screen time and body mass index. Meanwhile, mean and standard deviation were used to examine continuous data such as poor diet, healthy diet and sleep. Binary logistic regression was applied as a main inferential analysis in order to identify the factors of students' lifestyle-related behavior that influenced academic performance. The model was assessed by analysing the omnibus test, Cox and Snell R Square and Nagelkerke R Square, goodness-of-fit and classification table.

The target variable with two values was academic performance which are the success event ($Y = 1$) for good, whereas ($Y = 0$) for other than good. In estimating the logistic regression model, the probability was expressed in terms of odds $\left(\frac{p}{1-p}\right)$ as shown in Equation (1).

$$logit(p) = ln(odds) = ln\left(\frac{p}{1-p}\right) = \beta_0 + \beta_1 X_1 + \beta_2 X_2 + \beta_0 X_1 + \ldots + \beta_k X_k \quad (1)$$

where

β_0 = the intercept (the value of Y when all $X_i = 0$)
β_1 = the parameter of the model
p = the probability of interest ($Y = 1$)

3. Results and Discussion

Table 2 shows the descriptive statistics for this study. Firstly, female respondents had the highest percentage of 77.22% (217 respondents) followed by male respondents with the percentage of 22.78% (64 respondents). Next, most of the respondents achieved a good academic performance with a percentage of 71.89% (202 respondents), compared to the respondents that did not achieve a good academic performance with a percentage of 28.11% (79 respondents). Then, based on physical activity, most of the respondents achieved at least 22 min of physical activity for 0–2 days with a percentage of 53.4% (150 respondents) followed by 3–5 days with a percentage of 35.9% (101 respondents). Whereas respondents who achieved at least 22 min of physical activity for 6–7 days had the lowest percentage, of 10.7% (30 respondents). Meanwhile, in the variable screen time, the respondents who spent more than 7 h using screens had the highest percentage of 63.3% (178 respondents) followed by respondents who spent 4–7 h using screens, which is 14.2% (40 respondents). The second lowest wa the respondents that spent 2–4 h using the screen with a percentage of 11.4% (32 respondents), while the lowest was respondents that spent less than 2 h using the screen with a percentage of 11% (31 respondents). Next, in the variable body mass index, most of the respondents appeared to be normal weight with a percentage of 40.9% (115 respondents) followed by respondents who are underweight with a percentage of 29.5% (83 respondents). The next two lowest percentage of respondents' body mass index were overweight and obese with a percentage of 19.9% (56 respondents) and 9.6% (27 respondents) respectively.

Table 2. Descriptive Statistics of Categorical Variables.

Variables	Frequency of Respondents	Percentage (%)
Gender		
Male	64	22.78
Female	217	77.22
Academic performance		
Good	202	71.89
Other than good	79	28.11
Physical activity (days/week active for at least 22 min)		
0–2 days	150	53.4
3–5 days	101	35.9
6–7 days	30	10.7
Screen time		
Less than 2 h	31	11.0
2 to less than 4 h	32	11.4
4 to 7 h	40	14.2
More than 7 h	178	63.3
Body mass index		
Underweight	83	29.5
Normal weight	115	40.9
Overweight	56	19.9
Obese	27	9.6

Table 3 shows the basic summary statistics of three continuous variables which are poor diet, healthy diet and sleep. The total scores of poor diet ranges from 7 to 31, with a mean of 16.22 and standard deviation of 4.965. Next, the total scores of healthy diet ranges from 8 to 25, with a mean of 18.07 and standard deviation of 4.046. Lastly, total scores of sleep ranges from 8 to 40, with a mean of 22.73 and standard deviation of 6.566.

Table 3. Descriptive Statistics of Continuous Variables.

	Minimum	Maximum	Mean	Standard Deviation
Poor diet	7	31	16.22	4.965
Healthy diet	8	25	18.07	4.046
Sleep	8	40	22.73	6.566

Table 4 indicates several statistical tests for assessing criteria for multiple logistic regression model. The significant value (*p*-value) of the omnibus tests is less than 0.05 with chi-square value is 251.373, meaning that the model is statistically significant. Then, the value 0.850 of Cox and Snell R Square and Nagelkerke R Square shows that the independent variable was able to explain the variability of the dependent variable by 85% while the rest is explained by other variables not included in the model. The Hosmer and Lemeshow test to recognise the results of the goodness-of-fit test indicate that the model fits the data well, as the value of the *p*-value is larger than 0.05. Moreover, the classification accuracy was 95.7%, which indicates that the model is 95.7% accurate in classifying the students of "good" and "other than good" academic performance. Since the classification accuracy is greater than 80%, the model has a good prediction. The classification error rate means that a fraction of predictions was incorrect is 4.3%.

Table 4. Assessing Criteria for Multiple Logistic Regression.

Model Assessing Criteria	Value (*p*-Value)
Results of omnibus tests of model coefficient (Chi-square)	251.373 (<0.05)
Cox and Snell R Square	0.591
Nagelkerke R Square	0.850
Hosmer and Lemeshow test	10.33 (0.243)
Results of classification table (classification accuracy)	95.7%

Table 5 tabulates the results of Wald chi-squared test, indicating that poor diet, healthy diet and sleep were significant to the model developed in this study, since all the *p*-values are less than significance level (0.05).

Table 5. Multiple Logistic Regression Results (Coefficient).

	B	Exp (B)	Wald	*p*-Value
Constant	−7.070	0.00085	9.503	0.002
Poor diet	−0.351	0.704	34.397	0.000
Healthy diet	0.395	1.484	7.701	0.006
Sleep	0.368	1.445	11.671	0.001

Therefore, the model that can be derived from the results of the multiple logistic regression is in Equation (2).

$$logit(p) = ln\left(\frac{p}{1-p}\right) = -0.7070 - 0.351poordiet + 0.395healthdiet + 0.368sleep \quad (2)$$

Based on forward selection method, three variables were significant, which were poor diet, healthy diet and sleep, and included in the final model as shown in Table 5.

The odds ratio from the results of the multiple logistic regression of the variable poor diet is 0.704 which means that one unit increase in poor diet will decrease the odds of achieving a good academic performance by 29.6%, holding all other predictors constant. Secondly, the odds ratio of variable healthy diet is 1.484 which shows that one unit increase in healthy diet will increase the odds of achieving a good academic performance by 48.4% holding all other predictors constant. Lastly, the odds ratio of variable sleep is 1.445 which indicates that one unit increase in sleep will increase the odds of achieving a good academic performance by 44.5%, holding all other predictors constant.

4. Conclusions

This study investigated the association between lifestyle-related behaviour and academic performance among UiTMKB students. From this study, variable poor diet has a statistically significant influence on academic performance. This study coincided with a study by [23]. Their studies found that an unhealthy diet has a detrimental impact on academic performance. The current study also reveals that a healthy diet is associated with good academic performance, and the results of this study matched with the study conducted by [24]. The study concluded that consumption of healthy diet among students has a positive impact on academic performance. Similarly, the current study also found that sleep has a significant relationship on academic performance. The result is supported by the previous literature that found a significant relationship between sleep and academic performance among students [7].

However, the current study identified that physical activity does not have significant influence on academic performance among the UiTMKB students. The finding is in contrast with the study by [15] because their study discovered that academic performance is associated with physical activity. Moreover, the current study also showed that there is no significant correlation between screen time and academic performance which is inconsistent with the previous studies investigated by [26]. The study proved that there is an inverse relationship between screen time and students' academic performance. Lastly, this study found that body mass index has no association with academic performance which is in contrast with the previous studies that showed a significant association between BMI and academic performance among students [27].

Author Contributions: Conceptualization, S.N.M.S., N.A.A., A.A. and M.N.A.N.; methodology, S.N.M.S., N.A.A., A.A. and M.N.A.N.; validation, M.S.S., N.N.N.I.I. and N.M.S.; formal analysis, M.S.S., N.N.N.I.I. and N.M.S.; writing—original draft preparation, M.S.S., N.N.N.I.I. and N.M.S.; writing review and editing, S.N.M.S., N.A.A., A.A. and M.N.A.N. All authors have read and agreed to the published version of the manuscript.

Funding: The registration fee is funded by Pembiayaan Yuran Prosiding Berindeks (PYPB), Tabung Dana Kecemerlangan Pendidikan (DKP), Universiti Teknologi MARA (UiTM), Malaysia.

Institutional Review Board Statement: The study was conducted in accordance with the ICH Good Clinical Practice Guidelines, Malaysian Good Clinical Practice Guidelines and Declaration of Helsinki, and approved by the Branch Research Ethics Committee (BREC) of Unit of Research Management and Innovation (URMI) Universiti Teknologi MARA (protocol code BERC/01/2022 (UG/MR/20) and date of approval 27 January 2022).

Informed Consent Statement: Informed consent was obtained from all subjects involved in the study.

Data Availability Statement: The datasets used in the study are publicly available to everyone and can be accessed at https://drive.google.com/drive/folders/1KwaFZ_-3dWzxfNeXAx8TXH2g7mqUUpA-?usp=sharing.

Acknowledgments: The authors would like to thank the reviewers for their constructive suggestions.

Conflicts of Interest: The authors declare that there are no conflict of interest regarding the publication of this paper.

References

1. World Health Organization. Healthy Living. In *Nutrition Policy, Infant Feeding and Food Security*; WHO: Geneva, Switzerland, 1999; p. 24.
2. Chopra, S.; Ranjan, P.; Malhotra, A.; Sahu, A.; Dwivedi, S.N.; Baitha, U.; Goel, A.; Kumar, A. Development and Validation of a Questionnaire to Evaluate the Impact of COVID-19 on Lifestyle-Related Behaviours: Eating Habits, Activity And Sleep Behaviour. *Public Health Nutr.* **2021**, *24*, 1275–1290. [CrossRef] [PubMed]
3. Mucci, F.; Mucci, N.; Diolaiuti, F. Lockdown and Isolation: Psychological Aspects of COVID-19 Pandemic in the General Population. *Clin. Neuropsychiatry* **2020**, *17*, 63. [PubMed]
4. Aristovnik, A.; Keržič, D.; Ravšelj, D.; Tomaževič, N.; Umek, L. Impacts of the COVID-19 pandemic on life of higher education students: A global perspective. *Sustainability* **2020**, *12*, 8438. [CrossRef]
5. Oducado, R.M.; Estoque, H. Online Learning in Nursing Education During the COVID-19 Pandemic: Stress, Satisfaction, and Academic Performance. *J. Nurs. Pract.* **2021**, *4*, 143–153. [CrossRef]
6. Giusti, L.; Mammarella, S.; Salza, A.; Del, V.S.; Ussorio, D.; Casacchia, M.; Roncone, R. Predictors of Academic Performance During the COVID-19 Outbreak: Impact of Distance Education on Mental Health, Social Cognition and Memory Abilities in an Italian University Student Sample. *BMC Psychol.* **2021**, *9*, 1–17. [CrossRef]
7. Adelantado-Renau, M.; Diez-Fernandez, A.; Beltran-Valls, M.R.; Soriano-Maldonado, A.; Moliner-Urdiales, D. The Effect of Sleep Quality on Academic Performance is Mediated by Internet Use Time: DADOS Study. *J. Pediatr.* **2019**, *95*, 410–418. [CrossRef]
8. Rajendran, S.R.; Chamundeswari, S. Understanding the impact of lifestyle on the academic performance of middle-and high-school students. *J. Sociol. Res.* **2019**, *10*, 1–20. [CrossRef]
9. Talapko, J.; Perić, I.; Vulić, P.; Pustijanac, E.; Jukić, M.; Bekić, S.; Meštrović, T.; Škrlec, I. Mental Health and Physical Activity in Health-Related University Students During the COVID-19 Pandemic. *Healthcare* **2021**, *9*, 801. [CrossRef]
10. Faught, E.L.; Gleddie, D.; Storey, K.E.; Davison, C.M.; Veugelers, P.J. Healthy Lifestyle Behaviours are Positively and Independently Associated with Academic Achievement: An Analysis of Self-Reported Data from a Nationally Representative Sample of Canadian Early Adolescents. *PLoS ONE* **2017**, *12*, e0181938. [CrossRef]
11. La, C.C.; Maniaci, G.; Palummo, A.; Saia, G.F.; Pinetti, G.; Zarbo, M.; Ferraro, L.; La, B.D. Healthy Lifestyles and Academic Success in a Sample of Italian University Students. *Curr. Psychol.* **2021**, *40*, 5115–5123.
12. Valladares, M.; Durán, E.; Matheus, A.; Durán-Agüero, S.; Obregón, A.M.; Ramírez-Tagle, R. Association Between Eating Behavior and Academic Performance in University Students. *J. Am. Coll. Nutr.* **2016**, *35*, 699–703. [CrossRef] [PubMed]
13. Marelli, S.; Castelnuovo, A.; Somma, A.; Castronovo, V.; Mombelli, S.; Bottoni, D.; Leitner, C.; Fossati, A.; Ferini-Strambi, L. Impact of COVID-19 Lockdown on Sleep Quality in University Students and Administration Staff. *J. Neurol.* **2021**, *268*, 8–15. [CrossRef] [PubMed]
14. Lemma, S.; Berhane, Y.; Worku, A.; Gelaye, B.; Williams, M.A. Good Quality Sleep is Associated with Better Academic Performance Among University Students in Ethiopia. *Sleep Breath.* **2014**, *18*, 257–263. [CrossRef] [PubMed]
15. García-Hermoso, A.; Marina, R. Relationship of Weight Status, Physical Activity and Screen Time with Academic Achievement in Adolescents. *Obes. Res. Clin. Pract.* **2017**, *11*, 44–50. [CrossRef] [PubMed]
16. Reyes-Olavarría, D.; Latorre-Román, P.Á.; Guzmán-Guzmán, I.P.; Jerez-Mayorga, D.; Caamaño-Navarrete, F.; Delgado-Floody, P. Positive and Negative Changes in Food Habits, Physical Activity Patterns, and Weight Status During COVID-19 Confinement: Associated Factors in the Chilean Population. *Int. J. Environ. Res. Public Health* **2020**, *17*, 5431. [CrossRef] [PubMed]
17. Adelantado-Renau, M.; Jiménez-Pavón, D.; Beltran-Valls, M.R.; Moliner-Urdiales, D. Independent and combined influence of healthy lifestyle factors on academic performance in adolescents: DADOS Study. *Pediatric Res.* **2019**, *85*, 456–462. [CrossRef]
18. Zhang, W.; Wang, Y.; Wang, S. Predicting Academic Performance using Tree-Based Machine Learning Models: A Case Study of Bachelor Students in an Engineering Department in China. *Educ. Inf. Technol.* **2022**, 1–16. [CrossRef]
19. Bakouei, F.; Omidvar, S.; Seyediandi, S.J.; Bakouei, S. Are Healthy Lifestyle Behaviors Positively Associated with the Academic Achievement of the University Students? *J. Adv. Med. Educ. Prof.* **2019**, *7*, 224–229.
20. Razak, W.M.; Baharom, S.A.; Abdullah, Z.; Hamdan, H.; Aziz, N.U.A.; Anuar, A.I. Academic Performance of University Students: A Case in a Higher Learning Institution. *KnE Soc. Sci.* **2019**, *31*, 1294–1304.
21. Živčić-Bećirević, I.; Smojver-Ažić, S.; Dorčić, T.M. Predictors of University Students' Academic Achievement: A Prospective Study. *Drus. Istraz* **2017**, *26*, 457–476.
22. Faught, E.L.; Ekwaru, J.P.; Gleddie, D.; Storey, K.E.; Asbridge, M.; Veugelers, P.J. The Combined Impact of Diet, Physical Activity, Sleep and Screen Time on Academic Achievement: A Prospective Study of Elementary School Students in Nova Scotia, Canada. *Int. J. Behav. Nutr. Phys. Act.* **2017**, *14*, 1–13. [CrossRef] [PubMed]
23. Kim, S.Y.; Sim, S.; Park, B.; Kong, I.G.; Kim, J.H.; Choi, H.G. Dietary Habits are Associated with School Performance in Adolescents. *Medicine* **2016**, *95*, e3096. [CrossRef]
24. Rehman, R.; Zafar, A.; Mohib, A.; Hussain, M.; Ali, R. Self-Reported Academic Performance in Relation to Health Behaviours Among Bahria University Students. *J. Pak. Med. Assoc.* **2018**, *68*, 195–199. [PubMed]
25. Majumdar, P.; Biswas, A.; Sahu, S. COVID-19 Pandemic and Lockdown: Cause of Sleep Disruption, Depression, Somatic Pain, and Increased Screen Exposure of Office Workers and Students of India. *Chronobiol. Int.* **2020**, *37*, 1191–1200. [CrossRef] [PubMed]

26. Yan, H.; Zhang, R.; Oniffrey, T.M.; Chen, G.; Wang, Y.; Wu, Y.; Zhang, X.; Wang, Q.; Ma, L.; Li, R.; et al. Associations Among Screen Time and Unhealthy Behaviors, Academic Performance, and Well-Being in Chinese Adolescents. *Int. J. Environ. Res. Public Health* **2017**, *14*, 596. [CrossRef] [PubMed]
27. Wehigaldeniya, W.G.; Oshani, P.A.; Kumara, I.M. Height, Weight, Body Mass Index (BMI) and Academic Performance (AP) of University Students in Sri Lanka: With Special Reference to the University of Kelaniya. *Int. J. Sci. Res. Publ.* **2017**, *7*, 217–219.
28. Livermore, M.; Duncan, M.J.; Leatherdale, S.T.; Patte, K.A. Are Weight Status and Weight Perception Associated with Academic Performance Among Youth? *J. Eat. Disord.* **2020**, *8*, 1–10. [CrossRef] [PubMed]
29. Aziz, A.A.; Ismail, N.H.; Ahmad, F.; Hassan, H. A Framework for Students Academic Performance Analysis using Naive Bayes Classifier. *J. Teknol.* **2015**, *75*, 13–19.
30. Bull, F.C.; Al-Ansari, S.S.; Biddle, S.; Borodulin, K.; Buman, M.P.; Cardon, G.; Carty, C.; Chaput, J.P.; Chastin, S.; Chou, R.; et al. World Health Organization 2020 Guidelines on Physical Activity and Sedentary Behaviour. *Br. J. Sports Med.* **2020**, *54*, 1451–1462. [CrossRef]
31. Veldi, M.; Aluoja, A.; Vasar, V. Sleep Quality and More Common Sleep-Related Problems in Medical Students. *Sleep Med.* **2005**, *6*, 269–275. [CrossRef] [PubMed]
32. Nuttall, F.Q. Body Mass Index: Obesity, BMI, and Health: A Critical Review. *Nutr. Today* **2015**, *50*, 117–128. [CrossRef] [PubMed]

proceedings

MDPI

Proceeding Paper

Teachers' Continuous Intention to Use the Virtual Learning Environment (VLE) Platform: Do Gender Differences Still Matter †

Hapini Awang [1,*], Nur Suhaili Mansor [1], Ramlan Mustapha [2] and Isyaku Uba Haruna [3]

1 Institute for Advanced and Smart Digital Opportunities (IASDO), School of Computing, Universiti Utara Malaysia, Sintok 06010, Malaysia
2 Akademi Pengajian Islam Kontemporari, Universiti Teknologi MARA, Pahang Raub Campus, Raub 27600, Malaysia
3 School of Computing, Federal University Dutse, Dutse 720101, Nigeria
* Correspondence: hapini.awang@uum.edu.my

† Presented at the International Academic Symposium of Social Science 2022, Kota Bharu, Malaysia, 3 July 2022.

Abstract: The lacking and obsolete empirical evidence on the manner in which gender influences VLE adoption indicates the need to conduct new research regarding this matter. Presently, the usage trend of ICT gadgets, equipment, or any sort of technology including VLE has dramatically changed, which perhaps created the possibility of gender effect abolishment, although its existence was previously proven in certain studies conducted in previous decades. Therefore, this study investigates the difference between male and female teachers' continuous intention to use the VLE platform. The data were gathered from primary and secondary school teachers across the northern region of Malaysia and were analyzed using *t*-test analysis. The findings confirmed that there is no difference in terms of gender when it comes to the extent of teachers' intention to continue using VLE in their pedagogical routines. This implies that increased familiarity with ICT tools has eliminated the gender effect in VLE adoption among teachers.

Keywords: gender; Information Systems; Virtual Learning Environment; Google Classroom; VLE adoption; intention to use

Citation: Awang, H.; Mansor, N.S.; Mustapha, R.; Haruna, I.U. Teachers' Continuous Intention to Use the Virtual Learning Environment (VLE) Platform: Do Gender Differences Still Matter. *Proceedings* **2022**, *82*, 107. https://doi.org/10.3390/proceedings2022082107

Academic Editor: Mohamad Rahimi Mohamad Rosman

Published: 21 October 2022

1. Introduction

The past twenty years have seen rapid advances in the field of school education, congruent with the increasing awareness of Information and Communication Technology (ICT) that has occurred globally. As a result, educational authorities continue to invest heavily in the acquisition of educational technology, including implementing a Virtual Learning Environment (VLE) in schools. There are various justifications behind this investment, but one of the expected outcomes should be an increase in teachers' productivity, particularly in terms of effectiveness, efficiency, and quality [1]. This is mainly because VLE offers unlimited opportunities for teachers and other related parties such as students, school administrators, and parents. Nevertheless, the provision of VLE does not guarantee its sustainable usage, especially when considering that it is voluntarily used. As noted by [2], the usage patterns of any Information System (IS) may vary significantly when the nature of its use is optional. Moreover, this notion is ratified to be true as the resistance toward VLE among teachers has recently been recognized as a serious, worldwide educational technology concern [3–6].

Thus far, the existing literature on VLE has unveiled various factors that contribute to its usage, which also provides a fundamental understanding of the reasons behind teachers' resistance to this platform [7]. Despite this, the evidence of the gender effect on continuous VLE intention is inconclusive. Presently, there is a general lack of empirical research conducted to compare the extent of male and female teachers' intention to continuously use

VLE in their teaching routines. In fact, the available literature on gender roles in technology adoption has mostly been conducted within recent decades [8–10]. However, with the progress of time, caution must be applied, as the findings might not be relevant anymore. Today's society is familiar with technology, regardless of personal traits, where some doubt may exist on the possibility of gender effect abolishment. This notion makes sense because increased familiarity with technology usage will enhance the user's knowledge structure, which will help in the learning process [8,11]. As a result, there is a possibility of alleviation in the resistance to a particular technology, in this case, VLE, that will lead to further enhancement of their attitude (intention to use). However, thus far, the empirical evidence on this issue is still scarce. Therefore, the aim of the paper is to clarify the existence of the gender effect in VLE adoption among Malaysian teachers. This information can be used to develop targeted interventions aimed at establishing sustainable usage of VLE, which should also raise the standards of the Malaysian school education system.

1.1. VLE in Malaysian Schools' Education

VLE is known as a type of E-Learning platform that is widely used in higher education institutions and schools, with the ability to support both teaching and learning. as well as education management [12]. It is also universally identified as an Internet-based platform that underpins different educational undertakings, including online courses, quizzes, and tutorials [13]. Although VLE is occasionally referred to as Learning Management Systems (LMS), there is actually a slight difference between these E-Learning platforms. In light of this, [14] claims that VLE is an appropriate term to be used for E-Learning platforms that are used for educational purposes, while LMS should be referred to as an E-Learning platform specially developed for training. Despite the variety of features and abilities of VLE, it is best known for its capability to support both asynchronous and synchronous teaching and learning [15]. In addition, VLE also enhances the essence of conventional learning by promoting flexibility in six aspects, namely, time, place, space, technology, interaction, and control [16].

The history of VLE implementation in Malaysian school education began in 2012, with the introduction of Frog VLE, a platform that was adopted from the United Kingdom (UK). Frog VLE has an astonishing success record in its origin country and is currently implemented in many other countries around the world [17]. Therefore, with the embedment of Frog VLE into the school education system, the Ministry of Education Malaysia (MOE) aimed to take advantage of digital education to eliminate the proximity effects between rural and urban schools. Furthermore, it is hoped that this initiative would accelerate the quality of Malaysian education to the same level as educationally advanced countries such as the UK, Finland, and the United States (US). Viewed as an investment for the long haul, Frog VLE execution is predicted to be used for at least 13 years, and MOE believed that it would change the Malaysian education landscape by advancing the enduring utilization of ICT in aspects of both pedagogy and education management [18,19].

Unfortunately, despite this ambitious and costly investment to make Frog VLE available in Malaysian schools, it has nonetheless produced unexpected outcomes. During implementation, statistics have demonstrated very low usage of this platform, particularly among its main users—teachers [20,21]. From day one, this phenomenon continued to be an untreated issue until it reached its peak, when the MOE had to declare the termination of the Frog VLE service in mid-2019, and it was replaced by the Google Classroom (GC) platform [22]. Similar to Frog VLE, the GC platform offers almost the same basic VLE features, such as assignment, communication, collaboration, and information sharing. However, two main characteristics make GC dominant over Frog VLE. First, it is free of charge [23], and therefore, it is cost-effective for MOE, especially when implementing VLE on a nationwide scale. Second, GC is relatively easier to use, particularly for users of Google applications, for example, Gmail, YouTube, Google Drive, Google Form, and so on [24]. Undeniably, the Malaysian generation today is familiar with these Google applications. As GC has the ability to link the platform with other Google applications, this will provide advantages by eliminating the reliance on training, attracting users, especially teachers, and further leading to sustainable utilization of the VLE platform.

During the Frog VLE implementation period, the resistance and low usage of the platform have caused panic among stakeholders, which can be seen in situations where teachers are forced to use the platform up to a certain target of usage [25]. The local educational authorities likely did this in order to justify MOE's mega-investment in the provision of Frog VLE. Perhaps this action could increase the usage statistics of VLE among teachers. This, however, in turn, has unethically shifted the nature of the VLE platform from voluntary to mandatory. Although the success of VLE implementation is indicated by its usage [26,27], this concept involved interdependent relationships between other success factors such as teacher satisfaction and net benefits [28]. By changing the usage nature of VLE from voluntary to mandatory, the platform was exposed to a dramatic decrement in teacher satisfaction and net benefits. Therefore, this practice should be banned to avoid (another) failure of VLE implementation. Over the past decades, a number of researchers have sought to determine the factors that contribute to this resistance phenomenon. As a result, several factors such as teachers' workload, quality dimensions, and teacher readiness have been recognized as influential factors in VLE adoption [25,29–31]. Nevertheless, much uncertainty still exists about this, especially in terms of gender roles in VLE implementation during the digital era.

1.2. Gender Differences in Technology Adoption

In IS adoption, gender is found to be influential, especially to determine the strength of usage [8,32]. To elaborate, men are task-oriented and usually require a better quality of information, especially in terms of perceived usefulness when performing certain tasks [10,32]. On the other hand, women are instead discovered to be more sensitive and detail-oriented, especially in making decisions [33]. They occasionally digest information in an organized way, the opposite of men, who usually discard pertinent details in order to process the information from a broader perspective [34]. Ref. [8] took note of this and put forth the theory that women would respond more quickly to variations in the environment, which would further influence their intentions. A few empirical IS studies that uncovered the greater impact of perceived ease of use (one of the measurements for system quality) among women have reinforced this suggestion [10,32]. This evidence indicates that women anticipate a good-quality system that is easy to use [32], and consequently, if they perceive that the specific system is convoluted in nature, they will most likely demand improved service quality. To summarize, the preceding discourse on gender roles has shown that the intention to use VLE among teachers tends to be affected by gender dissimilarities. Male teachers want improved information quality, while female teachers tend to consider the system and service quality. A probable consequence is that teachers' gender could affect their perception and dependency on the information, system, and service quality, and thus, could possibly influence their determination of VLE continuous usage. Therefore, considering the preceding discussion, it is noteworthy to take into account the male and female differences when investigating VLE acceptance or VLE success among teachers.

2. Research Methodology

This study applied a descriptive quantitative approach to investigate and compare the level of intention to use VLE among male and female Malaysian teachers. Based on the simple random sampling procedure, data collection was performed using a cross-sectional survey among primary and secondary teachers across the northern region of Peninsular Malaysia. To ensure the accuracy of the findings, the survey questionnaire was systematically developed by considering several aspects of reliability, content validity, and face validity. This procedure involves the participation of six experts in the field of languages, IS, E-Learning, and statistics. In addition, 16 and 150 respondents were involved in the pre-testing and pilot test of the instrument, respectively. As a result, four valid and reliable items were produced to examine teachers' intention to continue using the VLE platform. The questionnaire also gathered information regarding respondents' gender in order to investigate the possibility of gender differences in the extent of intention to continue using the VLE platform. Later, 850 sets of questionnaires were randomly distributed. As a result (after four months), 643 were returned and usable, giving a valid response rate of approximately 75.6%.

Analysis and Findings

The analysis starts with the procedures of data cleaning and preparation, which freed the data from missing values and outliers. The normality test based on skewness and kurtosis values also indicated that the data were normally distributed [35], as shown in Table 1. This allows parametric analyses, such as an independent sample *t*-test, to be conducted on the usable cases (N = 643), comprising 380 female and 263 male teachers.

Table 1. Normality test based on skewness and kurtosis values.

Variable	Skewness	Kurtosis
Gender	0.37	−1.87
Intention to Use	−0.06	−0.01

Note. Cut-Off Value = ±2.

Next, the main analysis investigating the differences in the intention to continue using the VLE platform between male and female teachers was performed using an independent sample *t*-test. Based on the preceding discussion (in the previous section), the following null hypothesis was derived.

H_0. *There is no difference in teachers' intention to continue using the VLE platform between male and female teachers.*

There was homogeneity of variance as assessed by Levene's Test for Equality of Variances, and therefore, an independent sample *t*-test was run on the data with a 95% confidence interval (CI) for the mean difference. The analysis indicates that there was not a significant difference in the level of intention to use the VLE platform between female (M = 4.18, SD = 1.24) and male (M = 4.18, SD = 1.24) teachers, $t(641) = -0.70$, $p = 0.48$. Therefore, the null hypothesis is accepted. This result suggests that gender differences do not have any effect on teachers' attitudes towards VLE platform adoption. Specifically, it proves that teachers' perseverance in continuously using the VLE platform is not determined by their gender, either male or female.

3. Discussion and Conclusions

First, it is vital to be clear on the term 'intention to use' before the discussion on the study's findings can proceed. The controversy about the definition of 'intention to use' certain technology or IS has raged unabated for over a decade [28,36,37]. Since then, only a small number of studies have taken part in this debate [37–39], and still, the differences between the intention to use in pre-adoption (potential) and continuous users are still not clear. Moreover, an analysis of past literature demonstrated uncertainties and misuse of the term 'intention to use' in prior studies [40,41] that need to be clarified here. The 'intention to use' is usually related to the attitude of two types of IS users, namely, potential users (those who have never had an experience using the system but intend to use it in the future) and continuous users (those who already use the system and intend to reuse it the future). However, the term 'intention to use' is more appropriate for continuous users, while 'intention to adopt' is more appropriate for potential users [36]. Accordingly, this was made clear by the body of literature, which shows that the majority of researchers tend to measure the intention for continuous use, especially in investigating the relationship between attitude (intention to use) and behavior (actual usage) [42–44]. As in the context of VLE implementation in Malaysia, the 'intention to use' should be accredited to continuous users because the platform has been around since 2012, and the majority of teachers have initial experience using it. Indeed, during the Frog VLE period, all teachers were required to create a VLE account, and they were provided with personal IDs by 1BestariNet through the VLE administrator in schools [45]. This evidenced that Malaysian teachers are continuous users of the VLE platform.

In this study, no difference in continuous intention to use the VLE platform was found between male and female teachers. VLE platforms, either Frog VLE or Google Classroom, are part of the technology bombardment that occurs in the education sector nowadays. Thus, it is almost certain that this unexpected result may be due to the current digital lifestyle of teachers. As ICT utilization has become routine in one's lifestyle, the same has occurred in the school education system. ICT gadgets such as phones and tablets have become a necessity for human beings, regardless of gender, and the Internet is widely used for various purposes. Therefore, it can be assumed that familiarity with ICT tools has eliminated the effects of teachers' gender in the context of continuous intention to use the VLE platform. The implication of this is that it has contradicted the opinions of prominent IS scholars [8,10] and several local E-Learning researchers [25,46] who advocated that personal attributes, including gender, have a meaningful influence on regulating teachers' extent of intention to continuously use the VLE platform. However, when investigating only the level of teachers' intention, caution must be applied, as the findings might not be accurate and/or incomplete. This is mainly because teachers' attitudes towards VLE, including the continuous intention to use it, are most likely attributed to certain antecedent factors such as information quality, system quality, and service quality offered by a particular VLE platform [28]. In light of this, it would be appealing if future research could elaborate more by examining the role of gender as a moderator in these relationships. By doing so, the relevancy of the gender effect in VLE adoption could be validated.

Author Contributions: Conceptualization, H.A. and N.S.M.; methodology, H.A.; software, R.M.; validation, I.U.H., H.A. and R.M.; formal analysis, N.S.M.; investigation, H.A.; resources, I.U.H.; data curation, H.A.; writing—original draft preparation, H.A.; writing—review and editing, N.S.M.; visualization, N.S.M.; supervision, H.A.; project administration, N.S.M.; funding acquisition, H.A. All authors have read and agreed to the published version of the manuscript.

Funding: This research was funded by UNIVERSITI UTARA MALAYSIA College Grant, Grant Number: UUM/RIMC/P-30/10 Jld.4, Research ID: 10008988, S/O Code: 14826.

Informed Consent Statement: Informed consent was obtained from all subjects involved in the study.

Data Availability Statement: Not applicable.

Acknowledgments: This research is supported by the Universiti Utara Malaysia College Grant: S/O Code: 14826.

Conflicts of Interest: The authors declare no conflict of interest. The funders had no role in the design of the study; in the collection, analyses, or interpretation of data; in the writing of the manuscript, or in the decision to publish the results.

References

1. Thompson, R.L.; Higgins, C.A.; Howell, J.M. Influence of Experience on Personal Computer Utilization: Testing a Conceptual Model. *J. Manag. Inf. Syst.* **1994**, *11*, 167–187. [CrossRef]
2. Lee, D.M.S. Usage Patterns and Sources of Assistance for Personal Computer Users. *MIS Q.* **1986**, *10*, 293–304. [CrossRef]
3. Ghavifekr, S.; Razak, A.Z.A.; Ghani, M.F.A.; Ran, N.Y.; Meixi, Y.; Tengyue, Z. ICT Integration In Education: Incorporation for Teaching & Learning Improvement. *Malays. Online J. Educ. Technol.* **2014**, *2*, 24–54.
4. Hinostroza, J.E.; Labbé, C.; Brun, M.; Matamala, C.; Enrique Hinostroza, J.; Labbé, C.; Brun, M.; Matamala, C.; Hinostroza, J.E.; Labbé, C.; et al. Teaching and Learning Activities in Chilean Classrooms: Is ICT Making a Difference? *Comput. Educ.* **2011**, *57*, 1358–1367. [CrossRef]
5. Peeraer, J.; Van Petegem, P. ICT in Teacher Education in an Emerging Developing Country: Vietnam's Baseline Situation at the Start of "The Year of ICT." *Comput. Educ.* **2011**, *56*, 974–982. [CrossRef]
6. Tezci, E. Factors That Influence Pre-Service Teachers' ICT Usage in Education. *Eur. J. Teach. Educ.* **2011**, *34*, 483–499. [CrossRef]

7. Haruna, I.U.; Nadzir, M.M.; Awang, H.; Mohamed, L. A Conceptual Model of E-Taxation Satisfaction: How Can Taxpayers Be Tickled Pink with the Smart Web-Based Taxation Application? In Proceedings of the Asian Conference on Intelligent Computing and Data Sciences (ACIDS), Perlis, Malaysia, 24–25 May 2021; pp. 1–8.
8. Venkatesh, V.; Thong, J.Y.L.; Xu, X. Consumer Acceptance and Use of Information Technology: Extending the Unified Theory of Acceptance and Use of Technology. *MIS Q.* **2012**, *36*, 157–178. [CrossRef]
9. Morris, M.G.; Venkatesh, V.; Ackerman, P.L. Gender and Age Differences in Employee Decisions About New Technology: An Extension to the Theory of Planned Behavior. *IEEE Trans. Eng. Manag.* **2005**, *52*, 69–84. [CrossRef]
10. Venkatesh, V.; Morris, M.G. Why Don't Men Ever Stop to Ask for Directions? Gender, Social Influence, and Their Role in Technology Acceptance and Usage Behavior. *MIS Q.* **2000**, *24*, 115–139. [CrossRef]
11. Alba, J.W.; Hutchinson, J.W. Dimensions of Consumer Expertise. *J. Consum. Res.* **1987**, *13*, 411. [CrossRef]
12. Mueller, D.; Strohmeier, S. Design Characteristics of Virtual Learning Environments: State of Research. *Comput. Educ.* **2011**, *57*, 2505–2516. [CrossRef]
13. Abdelhag, M.E.; Osman, S.E.F. SOA for Effective Data Integration of Virtual Learning Environment Systems. *Int. J. Adv. Res. Comput. Sci. Softw. Eng.* **2014**, *4*, 680–685.
14. Pinner, R. What Is The Difference between an LMS and a VLE? Available online: https://elearningindustry.com/difference-between-lms-and-vle (accessed on 30 January 2022).
15. Halonen, R.; Thomander, H.; Laukkanen, E. DeLone & McLean IS Success Model in Evaluating Knowledge Transfer in a Virtual Learning Environment. *Int. J. Inf. Syst. Soc. Chang.* **2010**, *1*, 36–48.
16. Piccoli, G.; Ahmad, R.; Ives, B. Web-Based Virtual Learning Environments: A Research Framework and a Preliminary Assessment of Effectiveness in Basic IT Skills Training. *MIS Q.* **2001**, *25*, 401–426. [CrossRef]
17. Nor-Azlah, M.J.; Fariza, K. Keberkesanan Kemahiran Komunikasi Di Kalangan Guru Dalam Penggunaan Persekitaran Pembelajaran Maya (Frog VLE). *Pengajaran Sumber Dan Teknol. Mklm. Impaknya Ke Atas Penyelid. Dalam Pendidik* **2014**, *11*, 63–69.
18. Nor-Fadzleen, S.; Halina, M.D. Knowledge Management Enhancement in Virtual Learning Environment (VLE) in Malaysian Schools. In Proceedings of the International Conference on Virtual Learning Environment (ICVLE), Subang Jaya, Malaysia, 27–29 November 2013; pp. 1–9.
19. Cheok, M.L.; Wong, S.L. Predictors of E-Learning Satisfaction Among the Malaysian Secondary School Teachers. In Proceedings of the 22nd International Conference on Computers in Education: Asia-Pacific Society for Computers in Education, Nara, Japan, 30 November–4 December 2014; pp. 33–36.
20. Bahagian Teknologi Pendidikan. *Dokumentasi Kajian Dan Laporan Pemantauan*; Bahagian Teknologi Pendidikan: Putrajaya, Malaysia, 2017.
21. Kementerian Kewangan Malaysia. *Maklum Balas Ke Atas Laporan Ketua Audit Negara 2013 Siri 3*; Kementerian Kewangan Malaysia: Putrajaya, Malaysia, 2014.
22. Kementerian Pendidikan Malaysia. Kenyataan Media: Perkhidmatan Talian Internet Di Sekolah. Available online: https://www.moe.gov.my/index.php/pemberitahuan/kenyataan-media/5484-kenyataan-media-perkhidmatan-talian-internet-di-sekolah (accessed on 30 June 2019).
23. Beaumont, K. Google Classroom: An Online Learning Environment to Support Blended Learning. *Compass J. Learn. Teach.* **2018**, *11*, 1–6. [CrossRef]
24. Bondarenko, O.V.; Mantulenko, S.V.; Pikilnyak, A.V. Google Classroom as a Tool of Support of Blended Learning for Geography Students. In Proceedings of the CEUR Workshop Proceedings, Hotel Plejsy, Slovakia, 21–25 September 2018; Volume 2257, pp. 182–191.
25. Cheok, M.L.; Wong, S.L. Frog Virtual Learning Environment for Malaysian Schools: Exploring Teachers' Experience. In *ICT in Education in Global Context*; Huang, R., Kinshuk, P.J.K., Eds.; Springer Science+Business Media: Singapore, 2016; pp. 201–209.
26. Awang, H.; Zahurin, M.A.; Wan Rozaini, S.O.; Ishak, M.S. Examining Virtual Learning Environment Success Using DeLone and McLean IS Success Model. In Proceedings of the PACIS 2018 Proceedings, Yokohama, Japan, 26–30 June 2018; p. 50.
27. Awang, H.; Wan Rozaini, S.O.; Zahurin, M.A. Model to Evaluate Virtual Learning Environment among Malaysian Teachers. *Int. J. Innov. Technol. Explor. Eng.* **2019**, *8*, 59–63.
28. DeLone, W.H.; McLean, E.R. The DeLone and McLean Model of Information Systems Success: A Ten-Year Update. *J. Manag. Inf. Syst.* **2003**, *19*, 9–30.
29. Awang, H.; Zahurin, M.A.; Wan Rozaini, S.O. The Moderating Effect of Workload in Determining the Continuous Usage of Virtual Learning Environment amongst School Teachers. In Proceedings of the PACIS 2018 Proceedings, Yokohama, Japan, 26–30 June 2018; p. 51.
30. Awang, H.; Wan Rozaini, S.O.; Zahurin, M.A. A Conceptual Model to Evaluate Virtual Learning Environment among Malaysian Teachers. *J. Telecommun. Electron. Comput. Eng.* **2018**, *10*, 59–63.
31. Copriady, J. Self-Motivation as a Mediator for Teachers' Readiness in Applying ICT in Teaching and Learning. *Procedia-Soc. Behav. Sci.* **2015**, *176*, 699–708. [CrossRef]
32. Venkatesh, V.; Morris, M.G.; Davis, G.B.; Davis, F.D. User Acceptance of Information Technology: Toward a Unified View. *MIS Q.* **2003**, *27*, 425–478. [CrossRef]
33. Meyers-Levy, J.; Tybout, A.M. Schema Congruity as a Basis for Product Evaluation. *J. Consum. Res.* **1989**, *16*, 39. [CrossRef]

34. Meyers-levy, J.; Maheswaran, D. Exploring Differences in Males' and Females' Processing Strategies. *J. Consum. Res.* **1991**, *18*, 63–70. [CrossRef]
35. Garson, G.D. *Testing Statistical Assumption*; Statistical Associate Publishing: Asheboro, NC, USA, 2012; pp. 1–52.
36. Karahanna, E.; Straub, D.W.; Chervany, N.L. Information Technology Adoption Across Time: A Cross-Sectional Comparison of Pre-Adoption and Post-Adoption Beliefs. *MIS Q.* **1999**, *23*, 183. [CrossRef]
37. Agarwal, R.; Prasad, J. The Role of Innovation Characteristics and Perceived Voluntariness in the Acceptance of Information Technologies. *Decis. Sci.* **1997**, *28*, 557–582. [CrossRef]
38. Kongarchapatara, B. Factors Affecting Adoption versus Behavioral Intention to Use QR Code Payment Application Factors Affecting Adoption versus Behavioral Intention to Use QR Code Payment Application Boonying Kongarchapatara* and Chalida Rodjanatara College of Management. In Proceedings of the 2018 International Conference on e-Commerce, e-Administration, e-Society, e-Education, and e-Technology, Osaka, Japan, 1–3 April 2018; pp. 1–22.
39. Kupfer, A.; Schöb, S.; Ableitner, L.; Tiefenbeck, V. Technology Adoption vs. Continuous Usage Intention: Do Decision Criteria Change When Using a Technology? In Proceedings of the AMCIS 2016: Surfing the IT Innovation Wave-22nd Americas Conference on Information Systems, San Diego, CA, USA, 11–14 August 2016; Volume 2, pp. 1–10.
40. Laumer, S.; Maier, C.; Eckhardt, A. Why Do They Resist?–An Empirical Analysis of an Individual's Personality Trait Resistance Regarding the Adoption of New Information Systems. In Proceedings of the 18th European Conference on Information Systems (ECIS); Association for Information Systems, Pretoria, Africa, 7–9 June 2010; pp. 1–14.
41. Hsu, M.W. An Analysis of Intention to Use in Innovative Product Development Model through TAM Model. *Eurasia J. Math. Sci. Technol. Educ.* **2016**, *12*, 487–501. [CrossRef]
42. Zhu, D.S.; Lee, R.Z.C.; Kuo, M.J.; Lin, T.S. A Study on the Continuous Using Intention of Travelling Website. In Proceedings of the 2013 IEEE/ACIS 12th International Conference on Computer and Information Science, ICIS 2013-Proceedings, Niigata, Japan, 16–20 June 2013; pp. 55–261.
43. Ramayah, T.; Ahmad, N.H.; Lo, M.C. The Role of Quality Factors in Intention to Continue Using an E-Learning System in Malaysia. *Procedia-Soc. Behav. Sci.* **2010**, *2*, 5422–5426. [CrossRef]
44. Khader, A.A.M. A Hybrid Proposed Framework Based on Quality Factors (QF) and Technology Acceptance Model (TAM) for Mobile Learning Process: Higher Education Students in Jordanian Universities. *Int. J. Inf. Bus. Manag.* **2015**, *7*, 200–213.
45. 1BestariNet Bagaimana Log Masuk Dan Cari Alamat VLE Sekolah. Available online: http://1bestarinet.net/?page_id=21 (accessed on 19 November 2015).
46. Raman, K.; Yamat, H. Barriers Teachers Face in Integrating ICT During English Lessons: A Case Study. *Malays. Online J. Educ. Technol.* **2014**, *2*, 11–19.

proceedings

MDPI

Proceeding Paper

Student Behavior, Sociocultural, Learning Environment and Information Literacy: A Proposed Framework in Managing Indigenous Knowledge †

Nur'Ainaa Abdul Rahim, Shamila Mohamed Shuhidan * and Haslinda Husaini

Faculty of Information Management, Universiti Teknologi MARA, Shah Alam 401501, Malaysia
* Correspondence: shamila@uitm.edu.my

† Presented at the International Academic Symposium of Social Science 2022, Kota Bharu, Malaysia, 3 July 2022.

Abstract: Educational achievement is an essential component of the development indicator of a country. The phrase 'education for all' covers all levels of life, including indigenous students (ISs). Commonly, education among indigenous people (IP) is not very encouraging. Many previous studies have highlighted the factors involved in ISs' refusal to attend school and how to solve this problem. However, there are only a few studies on the relationship between the importance of obtaining literacy information in schools and managing knowledge for a guaranteed future. Previous studies found that a lack of information literacy skills, low motivation, cultural background, and learning environment contributed to supporting teaching and learning among ISs. Therefore, this paper proposes a conceptual framework to examine the relationship between student behavior, the sociocultural learning environment, and information literacy in managing IK among ISs in Malaysia. This future research will be carried out among indigenous primary school students aged 10 to 12 years. This proposed research framework will later adopt a quantitative research approach and use an appropriate questionnaire based on the level of education, age, and background after being reviewed by an expert. The expected contribution is the proposed new framework, since literature about ISs is lacking in Malaysia. It is expected that instruments will be developed and used for research, perhaps for indigenous secondary students in Malaysia.

Keywords: information literacy; indigenous knowledge (IK); indigenous student (IS); student behavior; sociocultural; learning environment

check for updates

Citation: Rahim, N.A.; Shuhidan, S.M.; Husaini, H. Student Behavior, Sociocultural, Learning Environment and Information Literacy: A Proposed Framework in Managing Indigenous Knowledge. *Proceedings* **2022**, *82*, 108. https://doi.org/10.3390/proceedings2022082108

Academic Editor: Mohamad Rahimi Mohamad Rosman

Published: 21 October 2022

1. Introduction

Regarding learning in the 21st century learning, the transformation of education around the world has emphasized national education. Its foundation is to build a knowledge base society. The Sustainable Development Goal 4 of UNESCO highlighted the quality of education, which aims to ensure inclusive and equitable quality education and promote lifelong learning opportunities for all, including indigenous students (ISs) among indigenous people (IP) [1]. IP create, manage, and disseminate their indigenous knowledge (IK) from one generation to another to survive in their lifestyle context. They need to manage their knowledge for sustainable development goals. This is due to the knowledge they use for their entire life and improve their next-generation lifestyle to obtain a better lifestyle for IP. Managing and maintaining IK will aid in poverty reduction, equity, environmental protection, and sustainable development, and improve local engagement in development [2]. However, ISs have lower access to and a worse level of education than other populations. Their education frequently lacks curriculum and teaching approaches that acknowledge their communities' cultures, pedagogies, and traditional knowledge. ISs frequently do not have access to education in their native languages or teaching techniques are not adequately included [3]. Thus, ISs are reluctant to obtain proper education from schools, leading to a lack of literacy knowledge. Literacy was one of the important elements for the Orang Asli (one of the indigenous people in Malaysia) in formal education and schooling [4].

2. Literature Review

IPs are cultural groups that have shared ancestral links to the lands and natural resources they live on. Their identities, livelihoods, and physical and spiritual well-being are all intricately tied to the land and natural resources they rely on. Most IP continue to speak the languages of the nation or region they live in [5]. In Malaysia, 13.8 percent of the 31,660,700 million Malaysian population are IP [6]. The three largest ethnic groups of IP are Negrito, Senoi, and Proto Malays. In peninsular Malaysia, the states with the highest IP number are Pahang and Perak. Pahang recorded 67,504 IP while Perak recorded 53,299 IP. The Senoi is Malaysia's largest indigenous ethnic group [7]. In Sarawak, their IP, which is Dayak and/or Orang Ulu, is estimated have 15 groups, which is 70.5% of the Sarawak population. IP in Sabah or Anak Negeri have 39 ethnic groups, which is 58.6% of the Sabah population.

As part of Malaysia's population, IP also follow the same education systems that support Sustainable Development Goal 4: quality education by UNESCO (2030). However, studies found that the majority of ISs only have basic school education. According to a past study, parental influence contributes to ISs' poor academic performance [8]. Most IP, particularly parents, are unaware of the value of education in improving their and their children's lives. This is because the vast majority have never received a formal education, and a tiny percentage continue their studies until elementary school but do not complete it [9]. This group of minorities is still far behind. It has a high education rate at all stages of schooling, from preschool to institutions of higher learning [10]. According to studies, nearly half of ISs drop out after Year 6 [11], 39.1 percent of ISs finish sixth grade but do not register for grade one [12], and only 30 percent of ISs complete secondary school, which is less than half of the national average [13]. In this situation, the government proactively supports and encourages any possibilities by creating awareness of the importance of education among ISs, including the sustainability of social, environmental, and knowledge development among IP. Indigenous knowledge (IK) refers to the understandings, skills, and philosophies developed by societies with long histories of interaction with their natural surroundings. For IP, IK informs decision-making about fundamental aspects of day-to-day life [14]. IK is information that is unique to a particular culture or community. This is known as tribal knowledge, men's awareness, or traditional research. A previous study has defined IK as a collection of local people's knowledge, talents, and skills gathered over many years of experience, learning, development, and transmission [15]. To manage this IK, IP require information literacy, which is part of the management of education systems [16]. Literacy means the ability to read and write.

Nevertheless, there are various types of literacy, such as audio-visual literacy, print literacy, computer literacy, media literacy, web literacy, technical literacy, functional literacy, library literacy, and information literacy. Nominal and active literacy focuses on making people aware of reading and writing in their day-to-day activities [17]. Others define a low literacy rate as an important challenge among IP in knowing their own cultural identity. To recognize their communities, they require the development of a certain set of skills, which is why they need literacy to manage and understand their IK [18].

3. Problem Statement

The educational development of ISs is a critical challenge in achieving the aim of education for all. Yet, despite the government of Malaysia's many measures, there is still a significant incidence of attrition among ISs in schools. This is because school instructional programs and educational activities are geared toward mainstream education and tend to overlook the IP culture's conventional integration. This might be one reason why ISs reject the traditional education system [19]. The issue that has been highlighted among ISs is that the lack of information literacy skills of ISs results in many problems [20]. In sharp contrast to the national literacy rate of 93.1 percent in 2010, the Orang Asli literacy rate was only 51 percent in 2008 [4]. They do have tremendous IK but not the set skills of formal literacy to manage it. Studies also found that indigenous students have low motivation and behavior in terms of learning, including their cultural background and learning environment [20]. This is due to their feelings about living with their community and social ecology. Some ISs have low motivation to go to school and do not prioritize their education [21]. Lastly, undocumented IK leads to the extinction of all precious local history and heritage because of the illiterate and inadequate management of IK [22].

4. Proposed Framework

The framework developed for this research is discussed in this section. The interaction of variables was derived from the literature in this research. It was then conceptualized to make an original contribution to knowledge, and to explore the relationships between student behavior, sociocultural, learning environment, and information literacy in managing indigenous knowledge. The independent variables (IVs) developed for this research are student behavior, sociocultural, and learning environment. The dependent variable (DV) that is attached is information literacy in managing indigenous knowledge. The framework is shown in Figure 1 below:

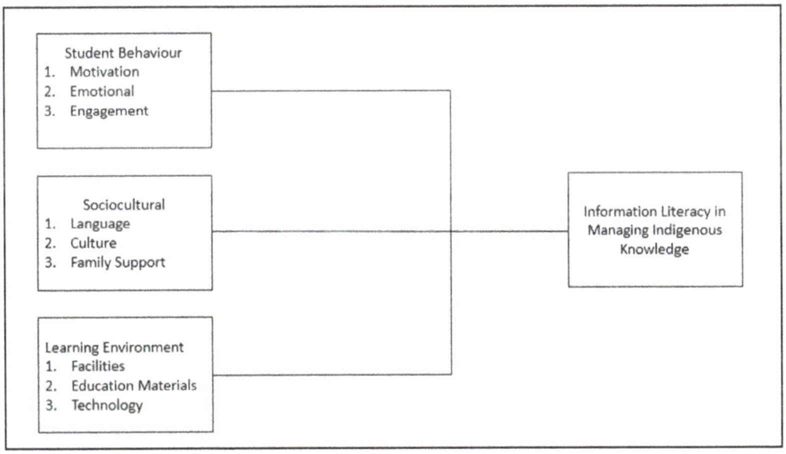

Figure 1. Research framework of the study.

Figure 1 shows the variables that have been measured in this study have been discussed below.

4.1. Student Behavior

Students are taught new behaviors by observing the behavior of others, which is known as social learning or modeling: they see how other individuals behave and copy them.

4.1.1. Motivation

Students have an essential role in determining academic success. Although numerous adjustments and reforms have been undertaken, educational progress will not be made unless the student is willing to adapt to succeed. Student motivation elements have a significant impact on student accomplishment. They believe that learning is not as crucial to their life as they think. Even if their parents are not highly educated, they can still live 'well' according to the average [23]. ISs' highly motivated attitudes towards education will positively impact ISs because they are eager to acquire knowledge in school. This is because they know that obtaining a formal education in schools with teachers is mandatory to ensure that they are not left behind from obtaining knowledge [21]. Self-motivation also consists of a sense of confidence to study in school and work hard even if it is difficult for ISs [24]. Children with a poor quality of life may feel inadequate, have low self-esteem, or have other self-related issues. This leads to personality flaws in children, such as low self-esteem, passive talking to teachers, a lack of cognitive abilities in the context of learning, and classroom actions that can be classified as learned helpless behaviors [24].

4.1.2. Emotional

Emotions are complicated and intricately linked to intellect and motivation [25]. Emotional experiences are attached to situations that are important to the individual. Emotions are likely to be elicited when a scenario, event, or circumstance is relevant to us or when we are affected by anything. Learners will feel delighted, frustrated, worried, proud, or satisfied only if the learning topic or method is meaningful to them [26–28]. A previous study focused on the passionate side of incorporating the emotions of IP that some educators might overlook [29]. If teachers do not level their leadership gaps in schools with a majority of ISs, where teachers are conducting a teaching and learning process between ISs and non-ISs, these ISs will be left behind if they do not understand something because the influence of culture results in them remaining silent and holding back without a reaction even if they feel offended by the teacher's words, especially in school. Parents of ISs are more likely to trust their children's concerns and parents of ISs easily believe in their children's complaints, which contributes to the school dropout factor and will reduce their interest in learning.

4.1.3. Engagement

Student engagement is described as "the willingness, need, desire, and compulsion of students to participate in and succeed in the learning process" [30]. Student engagement has also been defined as students' degree of interest, how they interact with others in the course, and their drive to learn about subjects [31]. Students are more likely to be engaged in their education when they are driven to achieve well in their classes, involved or invested in their desire to learn, and willing to expend the effort demanded by their professors [32].

4.2. Sociocultural

Sociocultural variables heavily influence individual growth and functioning. Because sociocultural support, stressors, and other variables frequently have major facilitative or debilitative impacts on learning outcomes, they often have a significant role in the learning process. As a result, these elements are commonly considered in most approaches to learning ability.

4.2.1. Culture

The strong cultural influence in the lives of IP is not a foreign thing in their daily lives, including aspects of education. The influence of culture in education for ISs can have positive and negative effects. A previous study stated that the teachers involved in teaching in schools had incorporated elements such as IK as one of their subjects [16]. This can encourage ISs to have an interest in learning. A rural school geography teacher said he adds topics such as forest conservation to his teaching so that "children may learn from rural people. The pupils can ask the community elders how to maintain the forest". Teachers from an urban elementary school and a teacher of Lao language and literature discussed the necessity of instilling traditional moral values in children. Learning about traditional culture, for example, "may help pupils conduct appropriately and respect things" according to one of the elementary school teachers, and "Teachers may learn to appreciate students, and students can learn to respect teachers." Learning about Lao culture, according to the other main instructor, "allows kids to learn" and learn about morals.

4.2.2. Language

In the context of education for indigenous students (ISs), the language issue is an important issue because the language used by IP is not the same as the language used by teachers at school. The literacy practices and expectations in school are different for indigenous children and children from other marginalized populations [33–35]. As a result of this detachment from their world, children do not do well in school, which adds to indigenous children encountering greater academic obstacles than children from other groups [36,37]. The Orang Asli children's academic performance has been hindered by the lack of awareness of the relevance of their culture in the school curriculum [38]. Because the community's literacy experiences were not properly recognized, Malaysian language education did not effectively allow ISs to engage in meaningful learning in schools [4]. Many past studies have found that language is important in IS education development. Use of the correct language can increase students' desire to continue to strive to gain knowledge and make the learning meaningful to them [39].

4.2.3. Family Support

The family institution significantly impacts children's development since the parent is the agent of strong socialization and the amount of time children spend at home is greater than the time spent at school [40]. The factor that affects individual development is parents' academic level, being a low, medium, or high degree. ISs do not master literacy as this relates to their parents' academic background [41]. Past studies have proven that environmental factors influence the formation and development of students' personalities covering family circumstances and intellectual-level parents. Parents with low educational levels do not serve as role models for their children. For example, they do not read in front of their children. As a result, their children's enthusiasm for reading decreases, which influences their literacy skills. Apart from this, parents provide their children with less assistance and direction [41]. Indigenous people know the importance of a good education, especially for their children. Good mastery of literacy can impact ISs more significantly [21]. One impact is that they can manage their indigenous knowledge in order to improve the standard of living and the family economy [42,43].

4.3. Learning Environment

The learning environment is one of the important factors in promoting a positive environment for students. In addition, factors such as good facilities at school with perfect classrooms encourage students to go to school.

4.3.1. Facilities

A good learning environment is an important aspect for ISs. A bad environment will result in ISs not mastering literacy and even reacting negatively to learning. Disruptions in the IS environment prevent them from obtaining literacy such as a lack of transportation facilities and discomfort in the space used for education. For example, it has been identified that the space in the homes of ISs is used for sleeping, eating, and studying simultaneously [41]. It can be said that the provision of good facilities in schools equipped with electricity and water sources will provide a good learning experience for ISs [21].

4.3.2. Educational Materials

Providing adequate teaching and learning materials for ISs is compulsory. The materials should also be appropriate to the level of education for the IS group. This is because not all ISs can easily accept this with the curriculum learning syllabus provided by the Ministry of Education [44]. The Malaysian Ministry of Education has developed a specific curriculum syllabus tailored to the cognitive level and abilities of Orang Asli children, which incorporates cultural components within the framework of the Orang Asli community [45]. The Orang Asli and Penan Curriculum (KAP), for example, is an early childhood learning curriculum program aimed at improving ISs' 3M literacy skills (reading, writing, and counting) [46].

4.3.3. Technology

The educational process, both formal and informal, includes the use of tools and devices to socialize or cultivate an individual to develop in a balanced or commensurate manner in cognitive, affective, physical, and social aspects in line with the current developmental demand [47,48]. All students, including ISs, should be exposed to every change that occurs so that they are ready and always open to gaining knowledge to face the challenges ahead. The role of ICT literacy culture is channeled through teaching and learning processes such as ICT literacy programs. The application of ICT interests, knowledge, skills, and readiness in the early stages of education is a priority among ISs to prevent them being left behind by other mainstream institutions [49]. Nowadays, the use of the internet is not just to browse entertainment sites; it is also an important requirement for managing a business. Unfortunately, the internet network in some areas is not as good, for example, in the IP hinterlands [50].

Nevertheless, studies on internet use among ISs have shown a positive effect. Access to a computer and the internet enables pupils' ICT proficiency to increase. With suitable infrastructure, such as computers and the internet, students may gain experience using them, which indirectly helps to increase their ICT competency. A previous study revealed that pupils with ICT access at home have a higher ICT experience [51].

4.4. *Information Literacy in Managing Indigenous Knowledge*

The importance of information literacy among IP, according to past studies, is to understand how to integrate relevant and appropriate information with what they already know to create new construct knowledge that increases their capacity to improve their daily lives or resolve demands relating to specific situations that have arisen [16]. The application of information literacy elements in ISs' education will increase the knowledge that ISs use to manage their IK. The benefits will become apparent: students' self-confidence will grow and lead to independent thinking. This positive effect will help to increase the skills of ISs used to innovatively manage their IK. A high-literacy individual may play an important part in conserving cultural traditions and introducing contemporary innovations into the community, and helping to amplify indigenous perspectives over time and places [52]. This is evident in research on digital literacy, which has a favorable influence on student learning. Some indigenous communities have embraced digital technology literacy to manage their IK. To overcome IP's inadequate language and literacy abilities, attempts have been made to establish teaching and learning methodologies using digital technology literacy [53]. IP frequently interact via text messaging and social media, and they utilize photography and video to generate content to self-represent and define their identity and achieve visibility within and outside their communities [54]. Furthermore, according to research, the utilization of digital technology has increased their exposure to and practice of basic interpersonal communication skills in their native languages and English [55]. ISs remarked that being in touch with their relatives via phone and email helped them to preserve their original tongue [56].

5. Discussion and Conclusions

This study is significant in its support of the Malaysia Education Blueprint 2013–2025 and Sustainable Development Goals of UNESCO (2030), which is to provide a quality education for all with no one left behind. The importance of conducting this study is that researchers can identify the problems faced by IP in preserving IK, which may become extinct if is no effort is made to manage it. Furthermore, this can help them to improve individual literacy rates to find information on how to manage IK so they can improve their quality of life and economic status and reduce the gap with the outside community. Authors have found several factors that influence learning desire among ISs. First, there is student behavior, which divided into the several aspects of student motivation, emotion, and engagement with school, even though the constraints faced by ISs are well known. The sociocultural aspect contains the aspects of language, family support, and culture. This biggest aspect is most frequently found in previous studies on IS education. Lastly, the learning environment contains the aspects of facilities, education materials, and technology. These three factors are independent variables in the research development while the dependent variable is the information literacy in managing indigenous knowledge, which is mainly used to test IS literacy rates. Information literacy is a major prerequisite in sustaining IK as cultural heritage as individuals rely on information to survive in society. It is hoped that the incorporation of information literacy in managing IK as cultural heritage is vital for IP to sustain their socio-economy and knowledge development and improve their lifestyles. This also supports the Sustainable Development Goals (SDGs) 1 (No Poverty), 4 (Quality of Education), and 11 (Sustainability Cities and Communities).

Author Contributions: Conceptualization—N.A.R. and S.M.S.; methodology, N.A.R. and S.M.S.; writing—original draft preparation, N.A.R.; writing—review and editing, N.A.R., S.M.S., and H.H.; supervision, S.M.S. and H.H.; project administration, N.A.R., S.M.S., and H.H.; resources, N.A.R. and S.M.S. All authors have read and agreed to the published version of the manuscript.

Funding: This research was funded by Fundamental Research Grant Scheme grant number FRGS/1/2021/SS0/UITM/02/16.

Institutional Review Board Statement: Not applicable.

Informed Consent Statement: Not applicable.

Data Availability Statement: Not applicable.

Acknowledgments: The authors would like to thank everyone who participated and willing to contributed ideas and thought in the research process and writing.

Conflicts of Interest: The authors declare no conflict of interest.

References

1. The SDGs in Action. Available online: https://www.undp.org/sustainable-development-goals?utm_source=EN&utm_medium=GSR&utm_content=US_UNDP_PaidSearch_Brand_English&utm_campaign=CENTRAL&c_src=CENTRAL&c_src2=GSR&gclid=Cj0KCQjw_7KXBhCoARIsAPdPTfgzNWjlBu3p427JfBB4jCedTG0NfT7ZLOkC36PjkDq1z6i18Ey1GucaAg8hEALw_wcB (accessed on 4 August 2022).
2. Yunnus, F. *Preservation of Indigenous Knowledge (IK) by Public Libraries in Westcliff, Chatsworth, Durban*; Master-University of Western Cape: City of Cape Town, South Africa, 2017.
3. Education, Language, and Indigenous Peoples. Available online: https://blogs.worldbank.org/education/education-language-and-indigenous-peoples (accessed on 7 August 2019).
4. Renganathan, S. Literacy and Development for the Orang Asli in Malaysia: What Matters? *Prospects* **2016**, *46*, 479–490. [CrossRef]
5. Indigenous Peoples. Available online: https://www.worldbank.org/en/topic/indigenouspeoples (accessed on 4 August 2022).
6. The Indigenous World 2022: Malaysia. Available online: https://www.iwgia.org/en/malaysia.html (accessed on 4 August 2022).
7. Orang Asli. Available online: https://www.jakoa.gov.my/orang-asli/ (accessed on 24 March 2022).
8. Deli, M.M.; Ruhizan, M.Y. Community-Based Learning Center of Renewable Energy Sources for Indigenous Education. *Int. J. Acad. Res.* **2017**, *7*, 828–837. [CrossRef]
9. Sawalludin, A.F. The Struggle of Orang Asli in Education: Quality of Education. *MJSSH* **2020**, *5*, 46–51. [CrossRef]
10. Mohamad, J.S.; Norul, K.I.; Nur, A.A.A.; Nurul, H.Y.; Siti, A.H. Kajian Terhadap Kesedaran Pendidikan Di Kalangan Masyarakat Orang Asli, In Proceedings of Persidangan Kebangsaan Pendidikan Luar Bandar, Kota Kinabalu, Sabah, Malaysia 3–5 February 2009.
11. Mohamad, H.A.K.; Muammar, G.H. Perubahan budaya masyarakat Orang Asli Suku Temiar di Kampung Bukit Cermin, Kuala Kangsar, Perak. *Geografia: Malays. J. Soc. Space* **2021**, *17*, 298–312.
12. Rosnon, M.R. Indigenous Education Policy In Malaysia: A Discussion of Normalisation in Schooling. *J. Edu. Soc. Sci.* **2016**, *4*, 25–32.
13. Dasar Pengoperasian Kemasukan Awal Kanak-Kanak Di Bawah Umur Ke Tahun 1. Available online: https://www.moe.gov.my/en/primary-school/dasar-pengoperasian-kemasukan-awal-kanak-kanak-di-bawah-umur-ke-tahun-1 (accessed on 25 July 2022).
14. Local, Indigenous and Scientific Knowledge. Available online: https://www.unesco.org/en/biodiversity/knowledge (accessed on 4 August 2022).

15. Grenier, L. *Working with Indigenous Knowledge: A Guide for Researchers*; International Development Research Centre: Ottawa, ON, Canada, 1998; pp. 1–130.
16. Doner, D.; Gorman, G.E. Indigenous Knowledge and The Role of Information Literacy Education. In Proceedings of the World Library and Information Congress: 74th Ifla General Conference and Council, Quebec, QC, Canada, 10–14 August 2008; pp. 1–16.
17. What Are the 13 Types of Literacy? Available online: https://www.theedadvocate.org/what-are-the-13-types-of-literacy/ (accessed on 4 August 2022).
18. Abdullah, R. Penilaian Pelaksanaan Program LINUS Dalam Kalangan Murid Orang Asli Di Perak. Ph.D. Thesis, Universiti Utara Malaysia, Changlun, Malaysia, 2018. Available online: https://etd.uum.edu.my/6897/ (accessed on 11 April 2022).
19. Kamaruddin, K.; Osman, J. Educational Policy and Opportunities of Orang Asli: A Study on Indigenious People in Malaysia. *J. Hum. Resour. Manag.* **2008**, *4*, 86–97.
20. Mohd, M.M.N.; Mohamad, F.S.; Mohd, N.M.N. Kesedaran pendidikan dan minat kerja pelajar Orang Asli suku kaum Jakun di Rompin, Pahang. Geografia Malays. *J. Soc. Space* **2018**, *14*, 72–86.
21. Awang, M.M. Ketidakhadiran Ke Sekolah: Faktor Penyebab, Aktiviti Ketika Ponteng dan Aspirasi Budaya Belajar Pelajar. *Orang. Asli. Malays. J. Learn. Instr.* **2022**, *19*, 277–308. [CrossRef]
22. Public Libraries as Conduits for Indigenous Knowledge in South Africa. Available online: https://www.researchgate.net/profile/Patrick-Ngulube/publication/324078990_Public_Libraries_as_Conduits_for_Indigenous_Knowledge_in_South_Africa/links/5ac386c5a6fdcccda660f57d/Public-Libraries-as-Conduits-for-Indigenous-Knowledge-in-South-Africa.pdf?origin=publication_detail (accessed on 4 August 2022).
23. Wan, A.W.H.; Shaharuddin, A.; Noraziah, A. Faktor Budaya Dan Persekitaran Dalam Prestasi Pendidikan Anak Orang Asli Malaysia: Kajian Kes Di Kelantan, Geografia Malays. *J. Soc. Space* **2014**, *10*, 107–122.
24. Yew, W.C. Persepsi Ibu Papa Orang Asli Terhadap Pengalaman Pendidikan Anak-anak Orang Asli di Sekolah. *MJSSH* **2021**, *6*, 24–39. [CrossRef]
25. Hascher, T. Learning and Emotion: Perspectives for Theory and Research. *Eur. J. Educ. Res.* **2010**, *9*, 13–28. [CrossRef]
26. Peter, S.; Axel, M. *Emotion Eliciting Events in the Workplace: An Intercultural Comparison*; Springer: Berlin/Heidelberg, Germany, 2007; pp. 236–248.
27. Eynde, P.; Jeannine, E.T. Focusing on the Complexity of Emotion Issues in Academic Learning: A Dynamical Component Systems Approach. *Educ. Psychol. Rev.* **2006**, *18*, 361–376. [CrossRef]
28. Veronika, B.; Jürgen, H.O. *Handbuch der Allgemeinen Psychologie: Motivation und Emotion*; Hogrefe Verlag: Stockholm, Sweden, 2009; pp. 1–13.
29. Budaya Pemikiran Pelajar Orang Asli Membantut Proses Pengajaran Dan Pembelajaran Mata Pelajaran Pendidikan Seni Visual Dismk Tapah. Cadangan Mengatasi. Available online: http://Sitihayatimohdyusoff.Blogspot.Com/2009/08/Budaya-Pemikiran-Pelajar-Orang-Asli.Html (accessed on 4 August 2022).
30. Bomia, L.; Beluzo, L.; Demeester, D.; Elander, K.; Johnson, M.; Sheldon, B. *The Impact of Teaching Strategies on Intrinsic Motivation*; ERIC: Champaign, IL, USA, 1996; pp. 1–29.
31. Educational Measurement. Available online: https://onlinelibrary.wiley.com/doi/10.1111/emip.12423 (accessed on 4 August 2022).
32. Mandernach, B.J. Assessment of Student Engagement in Higher Education: A Synthesis of Literature and Assessment Tools. *Int. J. Learn. Teach. Educ. Res.* **2015**, *12*, 1–14.
33. Arnold, D.H.; Greta, L.D. The early education of socioeconomically disadvantaged children. *Annu. Rev. Psychol.* **2003**, *54*, 517–545. [CrossRef] [PubMed]
34. Macnaughton, S. *Considering Culture in Research Based Interventions to Support Early Literacy*; The Guilford Press: New York, NY, USA, 2006; Volume 2, pp. 229–240.
35. Neuman, S.B. *The Knowledge Gap: Implications for Early Education*; The Guilford Press: New York, NY, USA, 2006; pp. 29–40.
36. Valdes, G. *Bridging the Distances between Culturally Diverse Families and Schools: An Ethnographic Portrait*; Teachers College Press: New York, NY, USA, 1995; pp. 1–239.
37. Anderson, I. Indigenous and Tribal Peoples' Health (The Lancet-Lowitja Institute Global Collaboration): A Population Study. *Lancet* **2016**, *388*, 131–157. [CrossRef]
38. Errico, S. *The Rights of Indigenous Peoples in Asia*; ILO: Bangkok, Thailand, 2017; pp. 1–192.
39. Letchamanan, H.; Nur, S.M.A.; Kamal, S.F. Language Education for Orang Asli Children in Malaysia, Pertanika. *J. Soc. Sci. Hum.* **2021**, *29* (Suppl. S3), 443–458.
40. Talib, J.; Nazri, M. Bagaimana Kanak-Kanak Orang Asli Gagal Di Sekolah? *J. Pengaj. Umum* **1970**, *8*, 51–76.
41. Sani, N. Merapatkan Jurang Literasi Murid-Murid Orang Asli: Apa Cabarannya? *J. Pers. Pelajar* **2014**, *17*, 19–30.
42. Al-Roubaie, A. Building Indigenous Knowledge Capacity for Development. *World J. Sci. Technol. Sustain. Dev.* **2010**, *7*, 113–129. [CrossRef]
43. Chin, Y.W.; Sia, M.K. The Socioeconomic Condition of Peninsular Malaysia's Indigenous Households Involved In Small-Scale Oil Palm Farming. *Int. J. Educ. Hum. Soc. Sci.* **2021**, 189–193.
44. Lasan, C.; Zamri, M. Pembelajaran Timbal Balik Bahasa Melayu Dalam Kalangan Murid Peribumi Bidayuh Di Sarawak. *J. Pendidik. Malays.* **2018**, *43*, 1–15.

45. Wahab, N.A.; Ramlee, M. Reflections on Pedagogical and Curriculum Implementation at Orang Asli Schools in Pahang. *Procedia Soc.* **2015**, *172*, 442–448. [CrossRef]
46. Hamidah, A.; Saemah, R. Persepsi Murid Orang Asli Terhadap Kurikulum Dan Relevansi Dalam Kehidupan. *PASAK3* **2018**, *3*, 105–124.
47. Sidin, R.; Nor, S.M. ICT Dalam Pendidikan: Prospek Dan Cabaran Dalam Pembaharuan Pedagogi. *J. Pendidik.* **2007**, *32*, 139–152.
48. Tella, A.; Mutula, S.M. Gender Differences in Computer Literacy Among Undergraduate Students at The University Of Botswana: Implications For Library Use". *Malays. J. Libr. Inf. Sci.* **2008**, *13*, 59–76.
49. Celik, V.; Etem, Y. Attitudes to Technology, Perceived Computer Self-Efficacy and Computer Anxiety as Predictors of Computer Supported Education. *Comput. Educ.* **2013**, *60*, 148–158. [CrossRef]
50. Manaf, S.Z.A.; Rosseni, D. Kesahan Dan Kebolehpercayaan Soal Selidik Kebolehgunaan Modul Pembelajaran Kendiri Menggunakan Model Pengukuran Rasch. *J. Adv. Res.* **2015**, *8*, 1–11.
51. Luu, K.; John, G.F. An Analysis of the Relationship Between Information and Communication Technology (ICT) and Scientific Literacy in Canada and Australia. *Comput. Educ.* **2011**, *56*, 1072–1082. [CrossRef]
52. Singleton, G. Youth Empowerment and Information and Communication Technologies: A Case Study of a Remote Australian Aboriginal Community. *GeoJournal* **2009**, *74*, 403–413. [CrossRef]
53. Li, J.; Amareen, B.; Novera, R. The Use of Digital Technology to Enhance Language and Literacy Skills for Indigenous People: A Systematic Literature Review. *CAEO* **2021**, *2*, 1–57. [CrossRef]
54. Brady, F.; Laurel, E.D.; Tina, A. Indigenous Adoption of Mobile Phones and Oral Culture. In *Proceedings of the Cultural Attitutdes Towards Communication and Technology*; Murdoch University: Perth, Australia, 2008; pp. 384–398.
55. Cummins, J. *Language, Power and Pedagogy: Bilingual Children in The Crossfire*; Multilingual Matters Ltd: Bristol, UK, 2001; Volume 25, pp. 405–412.
56. Bussert-Webb, K.; Díaz, M. Digital literacy, language, and latinos: L1.4Word. *J. Lit. Res.* **2013**, *14*, 2–50.

proceedings

MDPI

Proceeding Paper

Design and Development of Attendance and Temperature Recording System: A Smart Companion for the Current VLE Implementation in Malaysian Schools [†]

Nur Suhaili Mansor [1], Hapini Awang [1,*], Ahmed B. Ndanusa [2,3], Rizatulazri Idris [4] and Abdulrazak F. Shahatha Al-Mashhadani [5]

1 Institute for Advanced and Smart Digital Opportunities (IASDO), School of Computing, Universiti Utara Malaysia, Sintok 06010, Kedah, Malaysia
2 Department of Computer Science, University of Abuja, Abuja 900105, Nigeria
3 Department of Engineering and Space Systems, National Space Research and Development Agency, Abuja 900105, Nigeria
4 Department of Operation, MAP2U SDN BHD, Nilai 71800, Negeri Sembilan, Malaysia
5 Faculty of Business, Sohar University, Sohar 311, Oman
* Correspondence: hapini.awang@uum.edu.my

† Presented at the International Academic Symposium of Social Science 2022, Kota Bharu, Malaysia, 3 July 2022.

Abstract: With the prominence of COVID-19 in our lives and with the looming threat of other pandemics raising in the future, keeping track of everyone's temperature as a clear symptom of the disease is imperative for society moving forward. This applies to schools and other educational institutions too. This study aims to give a comprehensive solution that would be adaptable to multiple institutions. A web application that can run on most consumer devices would let schools keep track of their students' temperature on top of easing the process of tracking class attendance by correlating both. At the moment, Malaysian schools have already implemented several ICT initiatives, including for data management (APDM, SAPS, etc.) as well as for teaching and learning (Virtual Learning Environment). Therefore, the proposed online system for attendance and temperature recording would be a good complement to support the digitalization of Malaysian schools.

Keywords: Virtual Learning Environment (VLE); ICT in education; attendance recording system; web-based computing

Citation: Mansor, N.S.; Awang, H.; Ndanusa, A.B.; Idris, R.; Shahatha Al-Mashhadani, A.F. Design and Development of Attendance and Temperature Recording System: A Smart Companion for the Current VLE Implementation in Malaysian Schools. *Proceedings* **2022**, *82*, 111. https://doi.org/10.3390/proceedings2022082111

Academic Editor: Mohamad Rahimi Mohamad Rosman

Published: 26 October 2022

1. Introduction

Schools throughout the world are adapting to the pandemic situation brought about by the COVID-19 virus. Educational institutions need to make sure that their students are safe in the school environment. One way of doing this, which is comprehensive all over the world and is seen everywhere, is the use of temperature scanners. Temperature is a good indicator that a person is currently infected with a disease. As such, being able to track everyone's temperature is an important aspect of the future of society and every department and institution should not only collaborate with the rest of the country's institutions, but they should also have the right and the tools to manage and track people's temperature to create safer environments. Attendance tracking is almost a natural correlation to the issue of tracking temperature in a school setting [1]. Since most students are registered in the internal school system, and their attendance must be tracked (whether manually or otherwise) in order to receive a passing grade or mark, it is clear that tracking temperature in addition to attendance is the preferred way of doing it. Manually tracking temperatures would be time-consuming and mentally taxing for most educators. Manually calling out a name and marking next to the individual's name is simple enough. However, the addition of manually scanning temperatures and correctly recording them is more complicated. Like many other educational institutions, SK Tanah Rata is facing these issues and they reached out to create an application that would suit their needs.

Attendance systems in schools have gradually progressed over the years, due to the students' data being stored in a school database, the wide adoption of mobile devices, and the ease of using an automated system for attendance compared to doing it manually [2]. However, not all schools implement automated attendance systems, especially in institutions for younger students. This is mainly because despite the adoption of mobile devices being great in most areas, many parents do not allow younger children to use mobile devices until they grow older. Temperature, being a key indicator of human illness, plays an essential role in maintaining the safety of all students, educators, and staff members. Manually keeping track of it is a risk of close contact infections. An automated system to keep track of both attendance and temperature would ease many of these problems.

2. Methodology

The study was conducted following the Agile application development. Agile software development refers to a group of software development methodologies based on iterative development, where requirements and solutions evolve through collaboration between self-organizing cross-functional teams. Agile methods or Agile processes generally promote a disciplined project management process that encourages frequent inspection and adaptation, a leadership philosophy that encourages teamwork, self-organization, and accountability, a set of engineering best practices intended to allow for rapid delivery of high-quality software, and a business approach that aligns development with customer needs and company goals. Agile development refers to any development process that is aligned with the concepts of the Agile Manifesto. The manifesto was developed by a group of fourteen leading figures in the software industry and reflects their experience of which approaches do and do not work for software development. Figure 1 illustrates the Agile phases.

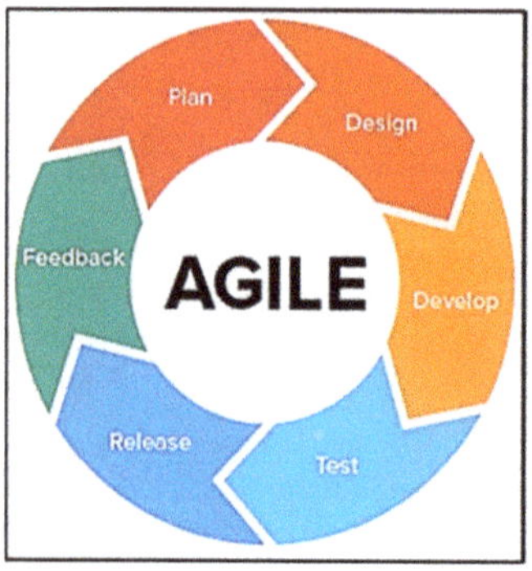

Figure 1. Agile Phases.

2.1. Plan

As the Agile application development indicates, the first step was to plan the development of the application, this was done by gathering the requirements (Table 1).

Table 1. List of Requirements for Creating and Managing Attendance and Temperature.

Functional Requirements			
Num.	**Req. ID**	**Requirements Description**	**Priority**
1.	**TSS_01**	**Login**	
	TSS_01_01	Users shall login to the system by inputting their username and password for authentication.	M
	TSS_01_02	The system can authenticate usernames and passwords.	D
	TSS_01_03	The system can store authentication information in the browser cookies for easy access.	D
	TSS_01_04	Users can request a new password via email in case they forgot it.	O
2.	**TSS_02**	**Manage Users**	
	TSS_02_01	Admin shall add users to access the system.	M
	TSS_02_02	Admin can also delete a specific user.	O
	TSS_02_03	Admin can also edit the details of a specific user.	O
	TSS_02_04	The system can verify if the username and password of a new user are acceptable to be added.	D
3.	**TSS_03**	**Manage Students**	
	TSS_03_01	Users shall add students to the system.	M
	TSS_03_02	Users can also delete a specific student.	O
	TSS_03_03	Users can also edit the details of a specific student.	O
	TSS_03_04	The system can verify the completion of the details of a new student.	D
4.	**TSS_04**	**Manage Classes**	
	TSS_04_01	Admin shall add a new class to the system.	M
	TSS_04_02	Admin can also delete a specific class.	O
	TSS_04_03	Admin can also edit the contents of a class.	O
5.	**TSS_05**	**Manage Attendance**	
	TSS_05_01	Admin shall add a student from the system to a class.	M
	TSS_05_02	Admin can also delete a specific student from a class.	O
6.	**TSS_06**	**Manage Temperature**	
	TSS_06_01	Users shall append a new value for the temperature to a student of a class.	M
	TSS_06_02	Users can also delete a specific temperature from a student in a class.	O
	TSS_06_03	Users can also edit the temperature of a student in a class.	O
7.	**TSS_07**	**Add Classes**	
	TSS_07_01	Teachers shall add a new class to the system.	M
8.	**TSS_08**	**Add Attendance**	
	TSS_08_01	Admin shall add a student from the system to a class.	M
9.	**TSS_09**	**Logout**	
	TSS_09_01	Users shall log out of the system.	M
	TSS_09_02	The system can force a logout if a certain amount of time passes.	D

2.2. Design

Before Prototype: The overall design of the application was made using pen and paper, and the main elements, buttons, and links for the users to interact with UML diagrams were created to visualize the elements and features (Figures 2 and 3).

During Prototyping: When the prototyping phase started, the design was made using Figma (Figures 4 and 5), an interactive tool that allows the creation of prototypes that are very functional and helps show the connection between functions and parts of the application with one another.

During Development: When developing the application, the design elements were created using HTML and CSS and using Bootstrap to create elements such as buttons and clean up the look of the application.

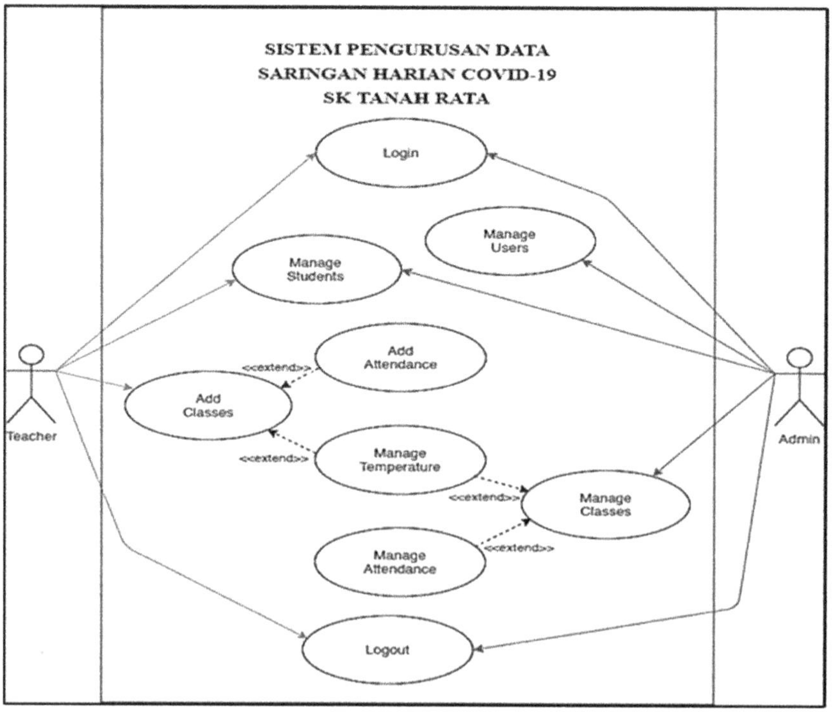

Figure 2. Use Case Diagram.

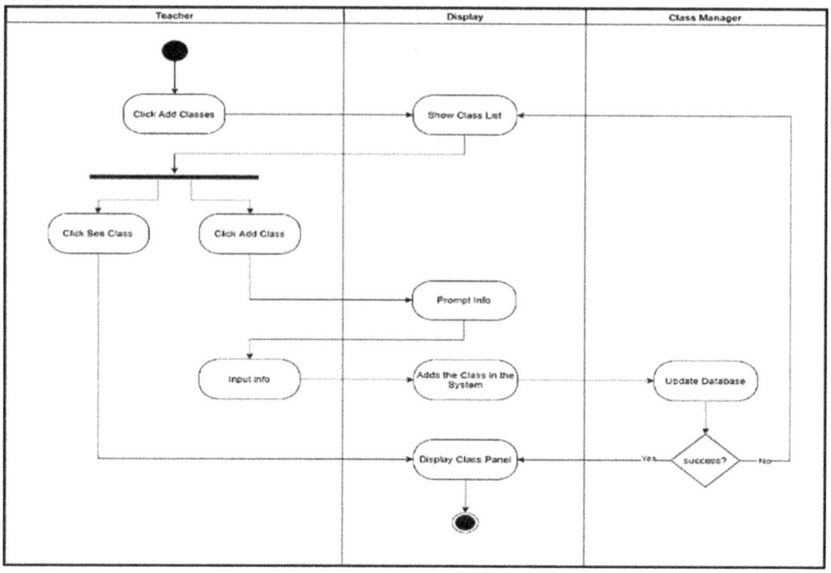

Figure 3. Activity Diagram for Add Classes Use Case.

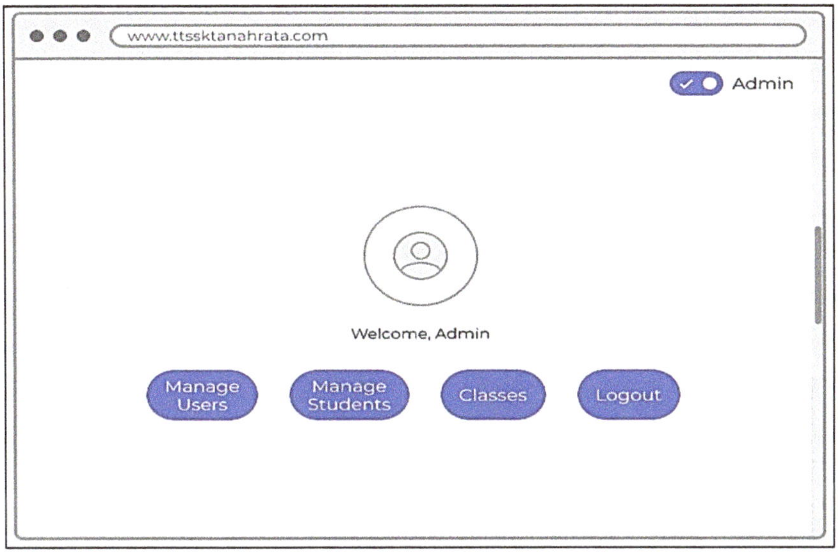

Figure 4. Landing Page Using Figma.

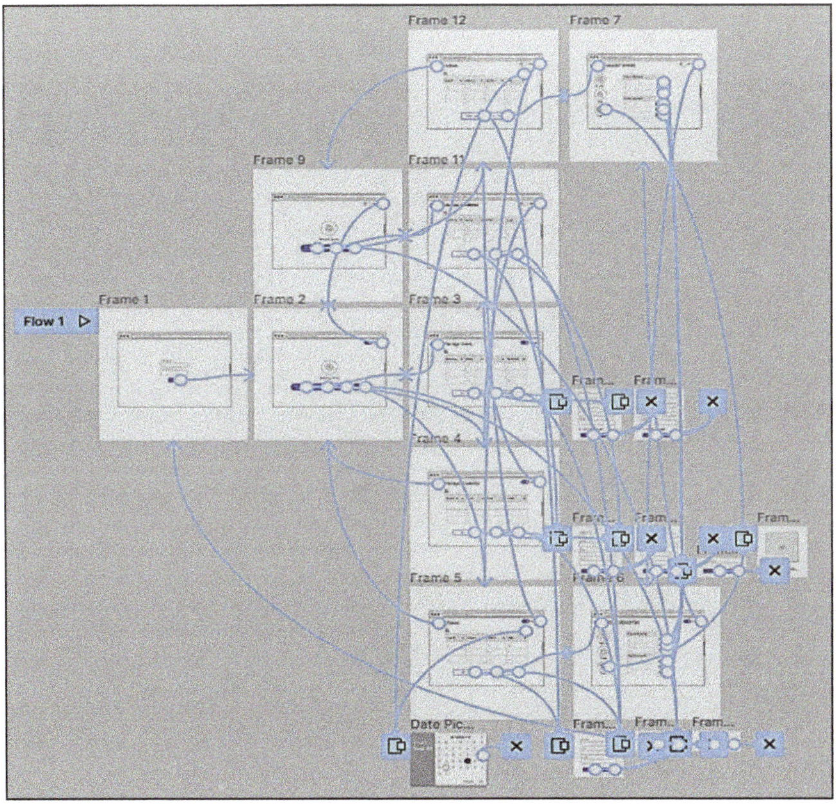

Figure 5. Prototype Interactions in Figma.

2.3. Development

The application was developed using the Ruby language, and the Ruby on Rails application framework written in Ruby [3–5]. Ruby on Rails is a server-side model–view–controller (MVC) framework, that provides default structures for a database, a web service, and web pages (Figure 6). It encourages and facilitates the use of web standards such as JSON or XML for data transfer and HTML, CSS, and JavaScript for user interfacing. In addition to MVC, Rails emphasizes the use of other well-known software engineering patterns and paradigms, including convention over configuration (CoC), do not repeat yourself (DRY), and the active record pattern [6,7]. Active record is a key component in the development of this application.

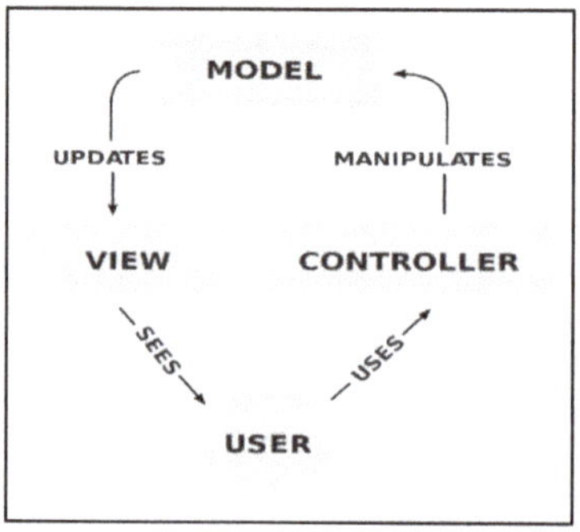

Figure 6. MVC Model.

All of the development processes can be tracked on GitHub, where the open-source code of the application resides, and all the comments with information regarding the improvements can be seen. The first step was to install Ruby, and all the dependencies needed by Rails in the main computer, this included packages such as yarn and node.js. Once Ruby was installed, gems (Ruby development packages) could be installed, followed by Rails. Once Rails was installed in the system, we generated a repository using Rails, which created all the directories and files needed to run a barebones application. Most of the code was written using Visual Studio Code, an open-source editor from Microsoft. The code was run using Microsoft PowerShell. Version control was implemented using Git, and a remote repository was created on GitHub to mirror all the changes made to the application public for interested parties to see.

We utilized the Ruby Gem 'devise' for user authentication. We created routes, models, controllers, and views with all the CRUD actions for each of the elements we wanted to integrate: users, students, classes, and attendance. The database was handled by sqlite3 in the development environment, however, we planned on using PostgreSQL for the production environment. To create the models and migrate them into the database, Rails implemented what is known as migration files, these files were created to generate the tables, columns, and relationships of the different elements of the application. These migration files were meant to be run only once and any updates to a table had to be done by creating a new migration file. Once the migrations were complete, we started to write the code for every CRUD action in the system. It was an iterative process in which we tested each feature, identified the errors, attempted to fix them, made appropriate changes, uploaded the code, tested the system, and so on. There are still things to be improved. The Entity-Relationship Diagram (ERD) of the different resources of the system is presented in Figure 7.

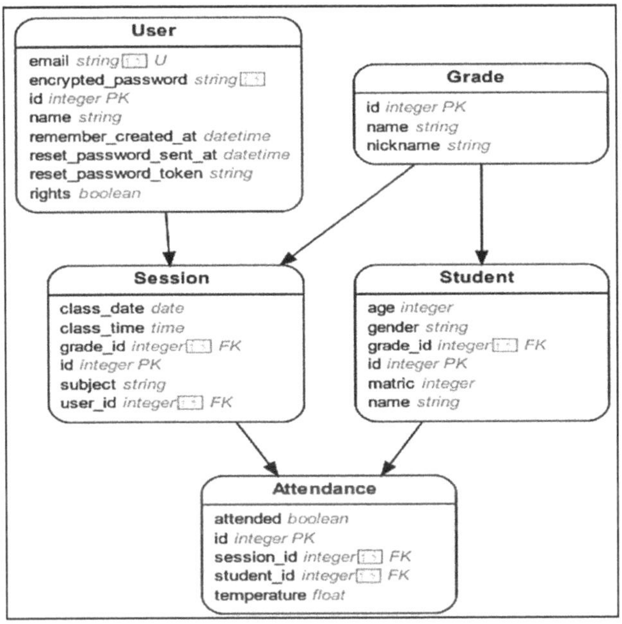

Figure 7. Entity Relationship Diagram of the Different Resources.

2.4. Testing

Testing was done in Google Chrome, using localhost:3000 as the root address of the application. We used Google Chrome to test features such as JavaScript, HTML, and CSS. The testing processes are presented in Figure 8.

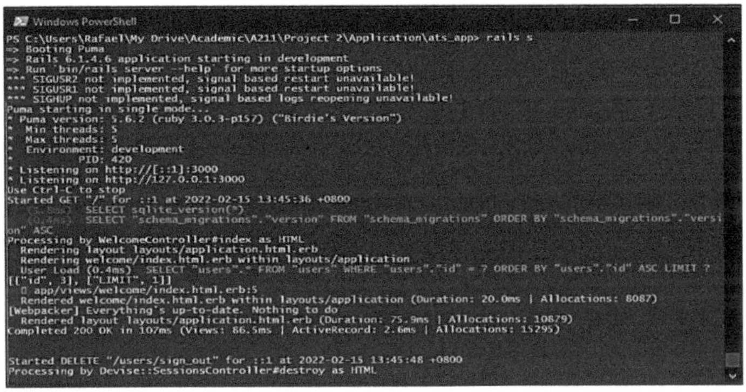

Figure 8. Server running on Windows via Microsoft PowerShell.

2.5. Deployment

Deployment of the application was done on Heroku, a subsidiary of Salesforce. Heroku is a cloud platform as a service (PaaS) supporting several programming languages [8]. Applications that are run on Heroku typically have a unique domain used to route HTTP requests to the correct application container or dyno. Each of the dynos is spread across a "dyno grid" which consists of several servers. Heroku's Git server handles application repository pushes from the permitted user. All Heroku services are hosted on Amazon's EC2 cloud-computing platform. A Heroku account was created, a Heroku application was initiated, and the code was pushed to the Heroku git, which then ran the code on their Ruby server. Migration files were run and any updates to the software were carried over Git push. The application is currently online and can be found on https://rafael-ats.herokuapp.com/ (accessed on 25 March 2021) An example of the system's interfaces is shown in Figure 9.

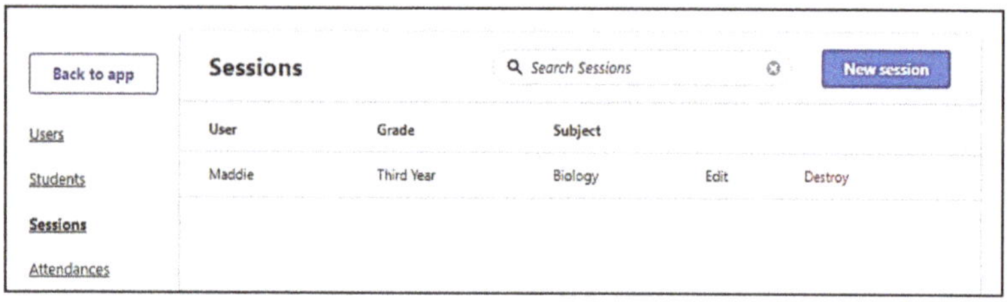

Figure 9. Application Dashboard Running on Heroku.

3. Conclusions and Future Work

In line with the Malaysian vision to digitalize its education, the proposed online system for attendance and temperature recording would support the existing implementation of VLE [9–11]

Towards a future of limited close contact, constant health checks, and close dependency on technology, we must adjust to the new paradigms in order to continue living fulfilling lives and producing products and services on par or better than what we had before. Applications such as the one described in this paper will aid in the improvement of different institutions and sectors in the country and help provide better services and education. There are still many ways in which to improve applications such as this one. For example, the use of face recognition, geolocation and long-range temperature devices, would remove the middleman and provide a truly automated attendance system with a temperature infrastructure.

Author Contributions: Conceptualization, N.S.M. and H.A.; methodology, A.B.N. and R.I.; testing, A.F.S.A.-M. and N.S.M.; formal analysis, H.A.; investigation, N.S.M.; resources, R.I.; writing—original draft preparation, N.S.M. and H.A.; writing—review and editing, A.B.N. and A.F.S.A.-M. All authors have read and agreed to the published version of the manuscript.

Funding: This research received no external funding.

Institutional Review Board Statement: Not applicable.

Informed Consent Statement: Not applicable.

Data Availability Statement: Not applicable.

Acknowledgments: This research is supported by the Universiti Utara Malaysia College Grant: S/O Code: 14826.

Conflicts of Interest: The authors declare no conflict of interest.

References

1. Patel, A.; Joseph, A.; Survase, S.; Nair, R. Smart Student Attendance System Using QR Code. In Proceedings of the 2nd International Conference on Advances in Science & Technology (ICAST), Mumbai, India, 8–9 April 2019; pp. 1–4.
2. Wei, X.; Manori, A.; Devnath, N.; Pasi, N.; Kumar, V. QR Code Based Smart Attendance System. *Int. J. Smart Bus. Technol.* **2017**, *5*, 1–10. [CrossRef]
3. Hartl, M. *Ruby on Rails Tutorial: Learn Web Development with Rails*, 3rd ed.; Addison-Wesley: Boston, MA, USA, 2015.
4. Tate, B.A.; Hibbs, C. *Ruby on Rails: Up and Running*; O'Reilly Media: Farnham, UK, 2008.
5. Tate, B.A.; Hibbs, C. *Ruby on Rails: Up and Running: Lightning-Fast Web Development*, 1st ed.; O'Reilly Media: Sebastopol, CA, USA, 2006.
6. Bächle, M.; Kirchberg, P. Ruby on Rails. *IEEE Softw.* **2007**, *24*, 105–108. [CrossRef]
7. Ruby, S.; Copeland, D.B.; Thomas, D. Agile Web Development with Rails 6. In *Pragmatic Programmers*; Pragmatic Bookshelf: Raleigh, NC, USA, 2020.
8. Richardson, L.; Ruby, S. *RESTful Web Services*; O'Reilly Media: Sebastopol, CA, USA, 2008.

9. Awang, H.; Zahurin, M.A.; Wan, R.S.O. Measuring Virtual Learning Environment Success from the Teacher's Perspective: Scale Development and Validation. In Proceedings of the 3rd International Conference on Applied Science and Technology, Penang, Malaysia, 10–12 April 2018; pp. 1–6.
10. Awang, H.; Zahurin, M.A.; Wan, R.S.O. The Moderating Effect of Workload in Determining the Continuous Usage of Virtual Learning Environment amongst School Teachers. In Proceedings of the Pacific Asia Conference on Information Systems, Yokohama, Japan, 26–30 June 2018; p. 51.
11. Awang, H.; Wan, R.S.O.; Zahurin, M.A. A Conceptual Model to Evaluate Virtual Learning Environment among Malaysian Teachers. *J. Telecommun. Electron. Comput. Eng.* **2018**, *10*, 59–63.

proceedings

MDPI

Proceeding Paper

Major Trends in Ageing Population Research: A Bibliometric Analysis from 2001 to 2021 †

Sabiroh Md Sabri [1,*], Nursyamilah Annuar [1], Nurul Labanihuda Abdull Rahman [1], Sharifah Khairol Musairah [1], Hasyeilla Abd Mutalib [1] and Iwan Kurniawan Subagja [2]

1 Faculty of Business and Management, Universiti Teknologi MARA Cawangan Perlis, Arau 02600, Malaysia
2 Faculty of Economics, Universitas Krisnadwipayana Jakarta, Jakarta 13077, Indonesia
* Correspondence: sabir707@uitm.edu.my; Tel.: +60-122-992-725
† Presented at the International Academic Symposium of Social Science 2022, Kota Bharu, Malaysia, 3 July 2022.

Abstract: People aged 65 years and above currently outnumber children below five years old. Globally, there were 727 million persons aged 65 years or over in the world in 2020. This increase in the ageing population has called for more research to be conducted concerning this group. This study presents a bibliometric analysis of the publications on ageing population research retrieved from the Scopus database between 2001 and 2021. The findings show that there has been a continuous growth in publications in research on the ageing population for 20 years since 2001. The results obtained from this research will provide the researcher with an overview of the trends in research related to the ageing population and allow them to strategize further research in different areas in the future.

Keywords: ageing; ageing population; bibliometric analysis; elderly; older people

check for updates

Citation: Sabri, S.M.; Annuar, N.; Rahman, N.L.A.; Musairah, S.K.; Mutalib, H.A.; Subagja, I.K. Major Trends in Ageing Population Research: A Bibliometric Analysis from 2001 to 2021. *Proceedings* **2022**, *82*, 19. https://doi.org/10.3390/proceedings2022082019

Academic Editor: Mohamad Rahimi Mohamad Rosman

Published: 9 September 2022

1. Introduction

According to reports, a significant change in demographics known as the "age quake" is currently taking place across the world. Gerontologists refer to the sudden shift in the world's population as the "age quake", which they believe began in the early 21st century. This phenomenon occurs when a growing percentage of the population is anticipated to live into elderly age [1–3]. History reveals that young children have always outnumbered older individuals since the beginning of human history [4]. However, things are starting to change now. People over the age of 65 will outnumber youngsters under the age of five for the first time in history. This tendency is not only happening in one place, it is happening everywhere around the world. According to [3,5], the world population's rising mortality rate and declining fertility rate are consistent with this prognosis. Additionally, studies conducted by the National Institute on Aging and the National Institute of Health in the United States of America revealed that 8% of the world's population—approximately 500 million people—are over the age of 65 today [3]. The percentage of people over 60 climbed globally from 9 percent in 1994 to 12 percent in 2014 and was projected to reach 21 percent by 2050. The World Population Organization (WPO) of the United Nations (UN) states that a population is considered to be ageing when the proportion of people 60 or older in a country's total population reaches at least 10% of the total population. As a result, this nation or area ages. Figures 1 and 2 below show the proportions of young children, older adults, and persons 60 years of age or older in significant regions in the years 1994, 2014, and 2050, respectively. These figures make it abundantly evident that the elderly is becoming a larger portion of the worldwide population.

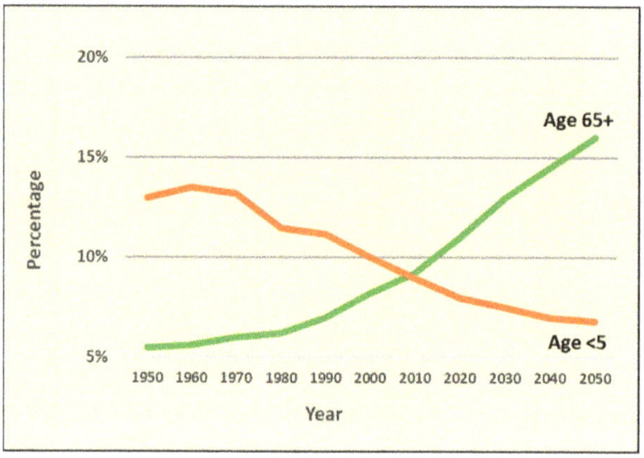

Figure 1. Young children and older people as a percentage of the global population [3].

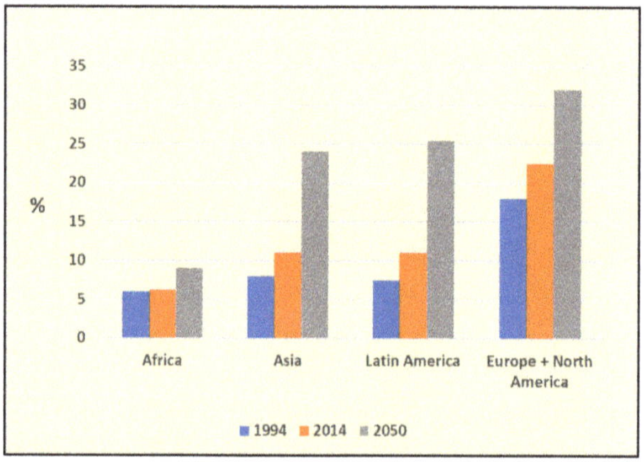

Figure 2. Percentage of population aged 60 years or older for major areas for the years 1994, 2014 and 2050 [3].

By the year 2050, the population of elderly people will grow quickly and surpass that of children. It is clear that by the year 2050, the world's population will consist primarily of older people because the number of older individuals has increased beyond the percentage that the UN uses to define an elderly population. People with various points of view analysing the effects of an ageing population on nations have responded in a variety of ways to ageing population challenges. According to reports from the World Economic Forum 2011, the phenomenon of an ageing population has significant effects on a country's economy and society due to the increase in elderly persons and changes in working-age populations. Along with this, a variety of research projects have been carried out to assist authorities in addressing the challenges and opportunities presented by an ageing population.

This study is being conducted to better understand the direction of current ageing population-related research by analysing the trends and advancements in the field. This study will help researchers comprehend the whole perspective of ageing population research and therefore plan for future research.

2. Methodology

In this study, publications related to the ageing population were retrieved from the Scopus database. Scopus is a bibliographic database that covers nearly 36,377 titles from roughly 11,678 publishers, of which 34,346 are peer-reviewed journals in the top-tier subject fields of life sciences, social sciences, physical sciences, and health sciences. Based on this, Scopus was chosen as it is the largest electronic database available. The search was made on all types of documents published between 2001 and 2021, thereby allowing the search engine to identify the broad spectrum of research in the literature. This review adopted PRISMA (Preferred Reporting Items for Systematic Reviews and Meta-Analyses) [6,7] guidelines for conducting systematic reviews of research. The search string "ageing population" was used to retrieve documents from the Scopus database. This initial search yielded 3954 documents which were then used in the bibliometric analysis. The 3954 Scopus-indexed documents' bibliographic details, which describe their characteristics, made up the data studied for this analysis. These "meta-data" include the names, titles, publication dates, author affiliations, and Scopus citation details of the documents. Microsoft Excel software was used to conduct the frequency analysis; for citation metrics and analysis, Harzing's Publish or Perish was used, and VOSviewer was used for data visualisation. The publications' trends, collaboration, authorship patterns, productive authors, countries' contributions, most active institutions, favourite journals, and most-cited articles were some of the common bibliometric variables used in this study to summarise the findings.

3. Results and Discussion

This section presents the results obtained from this study; it includes document profiles, research trends analysis, the geographical distribution of publications analysis, authorship and affiliation analysis, source title analysis, citation analysis, and keyword analysis.

3.1. Document Profiles

Document profiles refer to the recording and analysis of the characteristics of a document. In this study, a total of 3954 documents were retrieved—the majority of which (2543; 64.31%) were research articles. The second most common type of document was review articles (354; 8.95%). Other types of the document were book chapters, conference papers, editorials, notes, letters, books, errata, and short surveys. Details about the types of retrieved documents are shown in Table 1.

Table 1. Document type.

Document Type	Total Publications (TP)	Percentage (%)
Article	2543	64.31
Review	354	8.95
Book Chapter	302	7.64
Conference Paper	259	6.55
Editorial	181	4.58
Note	118	2.98
Letter	56	1.42
Book	51	1.29
Erratum	50	1.26
Short Survey	40	1.01
Total	**3954**	**100.00**

The majority of retrieved documents were published in English (3671; 91.11%). Other commonly encountered languages included French (69; 1.71%), Spanish (59; 1.46%), Portuguese (38; 0.94%), and German (31; 0.77%). Most of the published documents were research works related to the field of medicine (2011; 50.86%). Other subject areas included social sciences (1012; 25.59%), biochemistry, genetics, and molecular Biology (454; 11.48%), economics, econometrics, and finance (386; 9.76%), and nursing (350; 8.85%). Details of the top ten subject areas are presented in Table 2 below.

Table 2. Top ten subject areas.

Subject Area	Total Publications (TP)	Percentage (%)
Medicine	2011	50.86
Social Sciences	1012	25.59
Biochemistry, Genetics, and Molecular Biology	454	11.48
Economics, Econometrics, and Finance	386	9.76
Nursing	350	8.85
Engineering	214	5.41
Psychology	186	4.70
Environmental Science	181	4.58
Business, Management and Accounting	178	4.50
Computer Science	170	4.30

3.2. Research Trend Analysis

Examining the documents according to the year of publication enables the researcher to track the development and soaring interest in the research topic through time [8]. With a total of 364 documents produced, 2021 saw the most output—while 2002 saw the lowest, with just 60 publications. The number of documents published throughout the study period showed an increasing trend within the last 20 years, as seen in Figure 3. However, the number of citations illustrated by the line showed a decreasing trend. Due to the short period that had passed since their publication, the number of citations per publication was highest for documents released in 2005 (43.04 citations per publication) and lowest for those published in 2020 (1.42 citations per publication).

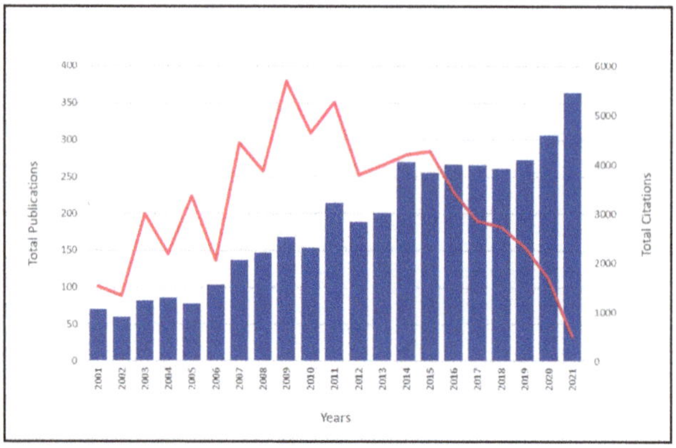

Figure 3. Research trends 2001–2021.

3.3. Geographical Distribution of Publications Analysis

The release of the collected materials involved researchers from 114 different nations. With a total of 1044 documents (26.44%), the United States of America came out on top—followed by the United Kingdom (470; 11.88%), China (317; 8.01%), Australia (244; 6.17%), and Canada (236; 5.97%). In Figure 4, the map displays 114 nations arranged into five distinct groups—each of which is coloured differently.

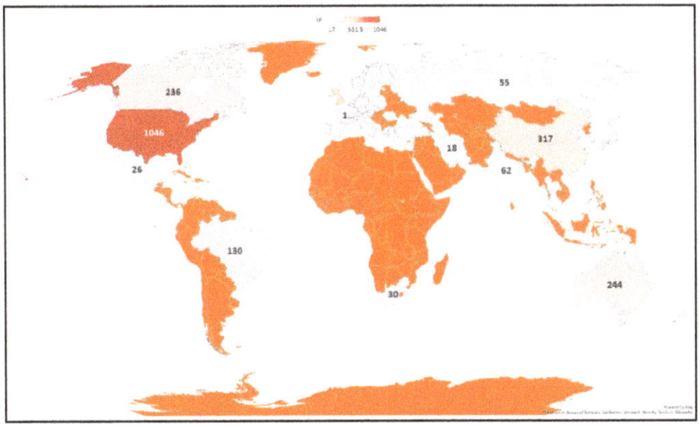

Figure 4. Geographical distribution of publications.

3.4. Authorships and Affiliations Analysis

Table 3 provides a summary of the top institutions with at least five publications. With a total of 48 publications, the University of Alberta in Canada is the most productive institution in this discipline (1.21%). Following this were University College London (42; 1.06%), King's College London (38; 0.98%), University of Cambridge (35; 0.88%) and Inserm (33; 0.83%).

Table 3. Most influential institutions with a minimum of five publications.

Affiliation	TP	NCP	TC	C/P	C/CP	*h*	*g*
University of Alberta	48	31	317	6.60	19.00	10	16
University College London	42	38	1477	35.17	21.00	19	38
King's College London	38	34	1564	41.16	20.00	17	38
University of Cambridge	35	34	2149	61.40	19.00	18	35
Inserm	33	30	1156	35.03	19.00	17	33
Fudan University	33	31	361	10.94	17.00	12	17
National Institute on Aging	33	30	1845	55.91	21.00	18	33
Pontifícia Universidade Católica do Paraná	33	22	167	5.06	9.00	8	12
Lunds Universitet	33	31	831	25.18	18.00	15	28
Karolinska Institutet	30	27	892	29.73	18.00	13	29

Notes: TP (total number of publications); NCP (number of cited publications); TC (total citations); C/P (average citations per publication); C/CP (average citations per cited publication); *h* (h-index); and *g* (g-index).

As shown in Table 4, the most productive author was Tonelli, M. (University of Calgary, Canada), with a total of 36 publications (0.91%). He was also the author with the highest number of cited publications. Other productive authors in the area included Riella, M. (32: 0.81%). Lima-Costa, M.F. (22; 0.56%), Brayne, C. (18; 0.46%), and Elmståhl, S. (16;0.40%).

Table 4. Most productive authors.

Author's Name	TP	NCP	TC	C/P	C/CP	*h*	*g*
Tonelli, M.	36	24	169	4.69	7.04	8	12
Riella, M.	32	20	124	3.88	6.20	6	10
Lima-Costa, M.F.	22	21	383	17.41	18.24	11	19
Brayne, C.	18	17	1122	62.33	66.00	12	18
Elmståhl, S.	16	14	465	27.35	33.21	7	17
Firmo, J.O.A.	15	16	644	40.25	40.25	13	16
Mason, A.	15	13	243	16.20	18.69	10	15
Scherbov, S.	13	13	1151	88.54	88.54	8	13
Guest, R.	11	10	52	4.73	5.20	5	6
Lee, S.H.	10	9	114	11.40	12.67	3	10

Notes: TP (total number of publications); NCP (number of cited publications); TC (total citations); C/P (average citations per publication); C/CP (average citations per cited publication); *h* (h-index); and *g* (g-index).

3.5. Source Title Analysis

Table 5 lists the top ten journals that have published research related to ageing populations. PLOS One ranked first with 45 documents (1.14%), followed by the Journals of Gerontology - Series A Biological Sciences and Medical Sciences (32; 0.181%), Archives of Gerontology and Geriatrics (29; 0.73%), International Journal of Environmental Research and Public Health (27; 0.68%), and Lecture Notes in Computer Science (Including Subseries Lecture Notes in Artificial Intelligence and Lecture Notes in Bioinformatics; 26; 0.66%).

Table 5. The most active source titles.

Source Title	TP	TC	Cite Score	SJR 2020	SNIP 2020
PLOS One	45	877	5.3	0.99	1.349
Journals of Gerontology—Series A Biological Sciences and Medical Sciences	32	934	9.1	2.134	1.771
Archives of Gerontology and Geriatrics	29	499	4.3	0.985	1.361
International Journal of Environmental Research and Public Health	27	277	3.4	0.747	1.356
Lecture Notes in Computer Science (Including Subseries Lecture Notes in Artificial Intelligence and Lecture Notes in Bioinformatics)	26	112	1.8	0.249	0.628
Sustainability Switzerland	25	151	3.9	0.612	1.242
Age and Ageing	20	999	9.6	2.014	2.753
Population Ageing in Central and Eastern Europe Societal and Policy Implications	20	40	N/A	N/A	N/A
Cadernos De Saude Publica	19	203	2.3	0.633	1.267
BMC Geriatrics	18	289	4.5	1.414	1.758

Notes: TP (total number of publications); TC (total citations).

A network visualisation map for the co-citation analysis of journals with a minimum of 50 citations is shown in Figure 5 below. PLOS One was frequently co-cited by other journals, showing the maximum number of linking lines from other journals. Additionally, this publication had the largest circle size, indicating that it had the most citations for research on ageing populations.

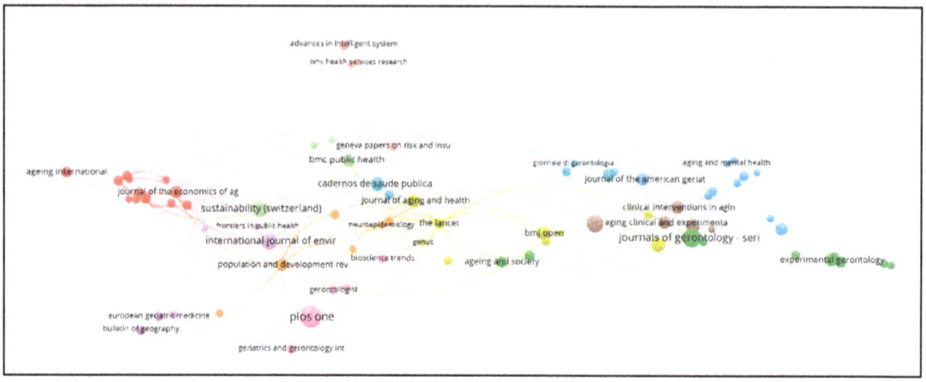

Figure 5. Co-citation analysis of source titles network visualization map.

3.6. Citation Analysis

Citation analysis is a method of determining the relative importance or impact of an author, an article, or a publication by calculating the number of times that an author, article, or publication has been cited by other works. This paper examined the impact of the documents retrieved from the Scopus database that are related to ageing population research. The citation metrics are summarized in Table 6 below. A total of 3954 documents with articles titled "ageing population" published from 2001 to 2021 were retrieved from the database. A total number of 67,140 citations were recorded for all these publications.

Table 6. Citation Metrics.

Metrics	Data
Papers	3954
Citations	67,140
Years	20
Cites_Year	3197.14
Cites_Paper	16.98
Cites_Author	23,670.96
Papers_Author	1865.4
Authors_Paper	3.44
h_index	105
g_index	170

The top ten cited articles related to ageing populations are shown in Table 7. The article titled "Ageing populations: the challenges ahead" authored by Christensen, Doblhammer, Rau, and Vaupel was published in The Lancet—a medical journal—in 2009 and received the highest number of citations. It received a total of 2207 citations and was the most impactful article based on citations per year (169.77 citations/y). This was followed by the article "The coming acceleration of global population ageing" authored by Lutz, Sanderson, and Scherbov (847; 60.5 citations/y); "The Aging Population and Its Impact on the Surgery Workforce", authored by Etzioni, Liu, Maggard, and Ko (578; 30.42 citations/y); "Inflammatory markers in population studies of aging", authored by Singh and Newman (518; 47.09 citations/y); and "Immunosenescence: Emerging challenges for an ageing population", authored by Aw, Silva, and Palmer (578; 30.42 citations/y).

Table 7. Most highly cited articles.

Authors	Title	Cites	Cites per Year
Christensen, Doblhammer, Rau, and Vaupel (2009)	Ageing populations: the challenges ahead	2207	169.77
Lutz, Sanderson, and Scherbov (2008)	The coming acceleration of global population ageing	847	60.5
Etzioni, Liu, Maggard, and Ko (2003)	The Aging Population and Its Impact on the Surgery Workforce	578	30.42
Singh and Newman (2011)	Inflammatory markers in population studies of aging	518	47.09
Aw, Silva, and Palmer (2007)	Immunosenescence: Emerging challenges for an ageing population	511	34.07
Bell, Tsai, Yang, Pidsley, Nisbet, Glass, Mangino, Zhai, Zhang, Valdes, Shin, Dempster, Murray, Grundberg, Hedman, Nica, Small, Dermitzakis, McCarthy, Mill, Spector, and Deloukas (2012)	Epigenome-wide scans identify differentially methylated regions for age and age-related phenotypes in a healthy ageing population	483	48.3
Smit, Brinkman, Geerlings, Smit, Thyagarajan, van Sighem, de Wolf, and Hallett (2015)	Future challenges for clinical care of an ageing population infected with HIV: A modelling study	417	59.57
Fox, Richardson, Maidment, Savva, Matthews, Smithard, Coulton, Katona, Boustani, and Brayne (2011)	Anticholinergic medication use and cognitive impairment in the older population: The medical research council cognitive function and ageing study	408	37.09
Pollack (2005)	Intelligent technology for an aging population: The use of AI to assist elders with cognitive impairment	391	23
Fried, Carlson, Freedman, Frick, Glass, Hill, McGill, Rebok, Seeman, Tielsch, Wasik, and Zeger (2004)	A Social Model for Health Promotion for an Aging Population: Initial Evidence on the Experience Corps Model	369	20.5

3.7. *Keywords Analysis*

A keyword analysis is a process of evaluating and analysing keywords and search phrases available in a document to ascertain which ones are used most in an article. Using VOSviewer software, author keywords with minimum occurrences of 25 were mapped. After the core search, query-related keywords were eliminated, and the results revealed that the terms "humans", "ageing", "aged", "male", "female", "middle-aged adult", "aged", "80 and beyond", "major clinical study", "extremely elderly", "risk factor", and "age" were the most often-found author terms.

Figure 6 above depicts a network visualization map of the authors' keywords. Circles of the same colour group imply that the publications share a common subject; each represents a subfield of the field of ageing population research. The 111 total items were divided into seven clusters. The green cluster was the largest, originated by the keyword "ageing", and had 656 occurrences, followed by population ageing (380 occurrences), elderly (157 occurrences), epidemiology (85 occurrences), and dementia (59 occurrences).

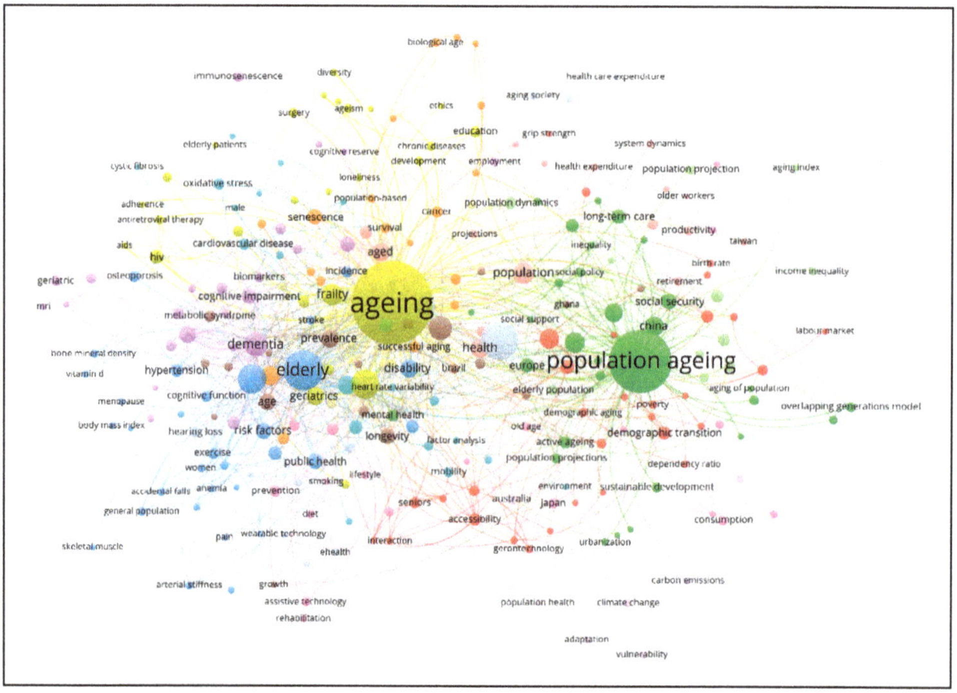

Figure 6. Network visualization map of author keywords.

4. Conclusions

The impact of demographic changes on ageing populations in many countries has made it important that governments and other stakeholders return to the subject which is the ageing population and find ways to enhance contributions toward this group through research. In this paper, we analysed 3954 documents retrieved from the Scopus database using bibliometric analysis. The majority of the documents were research articles published in the English language related to the field of medicine. The research trends analysis results showed that there was an increasing trend in publications related to ageing populations—starting from the year 2001 with a total of 70 publications and increasing to 366 publications in the year 2021. The results also showed the United States of America as the largest contributor to research related to ageing populations, followed by the United Kingdom,

China, Australia, and Canada. Malaysia recorded only 43 publications in the past 20 years, which is considered very low, and ranked number 25 in the Scopus database in terms of the number of publications. The most productive institution in this research area was the University of Alberta, located in Canada, with a total of 48 publications—while the most productive author was Tonelli, M. from the University of Calgary, Canada. PLOS One remained the main source of titles related to ageing population research. In conclusion, even though it has been observed that there is an increasing number of research works on the ageing population area, the number is still considered low relative to the duration of 20 years of observation. Many research works have focused on the medicine domain and research is still lacking in other domains such as the social sciences, information technology, business and management, and so on. This study can provide comprehensive information for future researchers to better understand the development trends of research related to ageing populations and, subsequently, to strategize for more research studies to be carried out concerning ageing populations, for the betterment of the nation.

Author Contributions: Conceptualization, S.M.S. and S.K.M.; methodology, S.M.S. and N.A.; data analysis, S.M.S., N.L.A.R. and H.A.M.; results and discussion, S.M.S. and I.K.S.; writing—original draft preparation, S.M.S. and N.A.; writing—review and editing, S.K.M. and I.K.S.; visualization, H.A.M. and N.L.A.R. All authors have read and agreed to the published version of the manuscript.

Funding: This research was funded by an Internal Research Acculturation Grant awarded by Universiti Teknologi MARA, Cawangan Perlis (Ref No. 600-TNCPI 5/3/DDN (09) (020/2020)).

Institutional Review Board Statement: Not applicable.

Informed Consent Statement: Not applicable.

Data Availability Statement: Not applicable.

Acknowledgments: The authors wish to thank the Research, Industry Network, Community and Alumni Division, Universiti Teknologi MARA, Cawangan Perlis.

Conflicts of Interest: The authors declare no conflict of interest.

References

1. Hooyman, N.; Kiyak, H.A. Aging in other countries and across cultures in the United States. In *Social Gerontology: A Multidisciplinary Perspective*, 9th ed.; Pearson: London, UK, 2011; pp. 43–68.
2. Miller, S. Get ready-the age-quake is heading our way. Need for major changes seen to care for aging population. *The Baltimore Sun*, 5 December 1991.
3. United Nations, Department of Economic and Social Affairs, Population Division. *World Population Ageing 2015*; United Nations: New York, NY, USA, 2015; ST/ESA/SER.A/390.
4. Dobriansky, P.J.; Suzman, R.M.; Hodes, R.J. *Why Population Aging Matters—A Global Perspective*; Department of State: Cambridge, UK, 2007.
5. Kochhar, R.; Oates, R. *Attitudes about Aging: A Global Perspective*; Pew Research Center: Washington, DC, USA, 2014.
6. Moher, D.; Liberati, A.; Tetzlaff, J.; Altman, D.G. Preferred reporting items for systematic reviews and meta-analyses: The PRISMA statement. *BMJ* **2009**, *339*, 332–336. [CrossRef] [PubMed]
7. Zakaria, R.; Ahmi, A.; Ahmad, A.H.; Othman, Z. Worldwide melatonin research: A bibliometric analysis of the published literature between 2015 and 2019. *Chronobiol. Int.* **2021**, *38*, 27–37. [CrossRef] [PubMed]
8. Ahmi, A.; Mohamad, R. Bibliometric Analysis of Global Scientific Literature on Web Accessibility. *Int. J. Recent Technol. Eng.* **2019**, *7*, 250–258. [CrossRef]

proceedings

Proceeding Paper

Information Literacy and Local Knowledge Transfer: A Pilot Study among the Minangkabau Community in the Agam District, Indonesia †

Elva Rahmah [1,2], Shamila Mohamed Shuhidan [1,*] and Wan Aida Wan Yahaya [3]

1 Faculty of Information Management, Universiti Teknologi MARA Cawangan Selangor, Shah Alam 40150, Selangor, Malaysia
2 Faculty of Language and Art, Universitas Negeri Padang, Padang 25171, Indonesia
3 College of Creative Arts, Universiti Teknologi MARA, Shah Alam 40450, Selangor, Malaysia
* Correspondence: shamila@uitm.edu.my
† Presented at the International Academic Symposium of Social Science 2022, Kota Bharu, Malaysia, 3 July 2022.

Abstract: This article aims to examine sample data on the level of information literacy and local knowledge. Based on the literature review, a 40-item questionnaire was constructed. To ensure the validity, reliability, and normality of the data, a data collection instrument was devised and then reviewed by experts. A pilot study was also conducted and participated in by 50 respondents. The data were processed using SPSS. The findings reveal the instrument was reliable, and there were no abnormalities in the data with the highest value of 0.848 and the lowest value of 0.760. It means that the questionnaire can be used for data collection.

Keywords: information literacy; local knowledge; socio-economy; knowledge management; Minangkabau Community

Citation: Rahmah, E.; Shuhidan, S.M.; Yahaya, W.A.W. Information Literacy and Local Knowledge Transfer: A Pilot Study among the Minangkabau Community in the Agam District, Indonesia. *Proceedings* **2022**, *82*, 31. https://doi.org/10.3390/proceedings2022082031

Academic Editor: Mohamad Rahimi Mohamad Rosman

Published: 13 September 2022

1. Introduction

Literacy is a mirror of society, and a fundamental consideration is its literary. Hence, literacy programs can be used as a means of transmitting culture. Cultural literacy is a type of literacy that can be found in addition to reading and writing. The younger generation must recognize and retain their own regional culture in order to reinforce and conserve Indonesian culture. Traditional values are upheld by maintaining national standards. Cultural preservation strives to support global cultural values. The lifelong learning process of information literacy (IL) allows people from all walks of life to efficiently access, analyze, and produce information in order to fulfil their personal, social, occupational, and educational goals [1].

Local knowledge (LK) is a body of knowledge that exists within or is acquired by local people over time through the accumulation of experiences, relationships between society and nature, community practices, and institutions [2,3]. Thus, LK appears as a portrait of identity and the worth of life in a culture. Personal experience and traditional cultural rules and norms are two subcategories of LK [4]. The knowledge of local society is characterized by unstructured tacit knowledge and is stored in the memory of the knowledge owner offering four models of knowledge transfer formation, or what is also known as the SECI model by Nonaka (1994) and Nonaka & Takeuchi (1995), which explain how knowledge is transferred into an organization or in daily life activities that can be derived either from tacit or explicit knowledge. More so, explicit and tacit knowledge interact with each other in a continuous process. Four methods of information conversion are provided, such as social socialization (tacit to tacit), externalization (tacit to explicit), mixing (explicit to explicit), and internalization (explicit to tacit) [5]. Conceptual and technological socialization enables a model of implicit awareness. Through observation, imitation, and practice, tacit information can be acquired. Externalization is the form of explicit concepts in the form of metaphors, analogies, theories, or models and is the process of articulating implicit information. In merging various knowledge experiences, the combination is the process of structural principles into a knowledge structure. Via media such as records, explicit information is transmitted. This knowledge categorization gives rise to new knowledge. The process of translating explicit knowledge into implicit knowledge is internalization and is similar to the notion of experience since it can be called learning by doing. The four processes illustrate that the transfer of information relies on an agreement between the knowledge owner and the knowledge user. The process takes a long time to identify how local knowledge transfer (LKT) occurs in personalization mechanisms, hybrid approaches, and codification mechanisms.

LK takes distinct forms in different places and across time, and it may conflict with scientific and technological notions of universal knowledge and development. The use of LK implies a transition that comes from local communities and represents the confidence of the local community in LK as a socio-economic advancement tool. LK is more in the form of hidden knowledge (tacit knowledge), namely, knowledge inherent in the attitudes of certain individuals or communities, opinions, practices, or experiences, making it difficult to codify and organize. However, in this context, there should be an effort to transform LK as tacit knowledge into explicit knowledge so that it can be accessed, studied, and utilized. Another challenge faced in the effort to revitalize LK is that we have to compete with time because LK tends to disappear or become extinct along with the swift currents of modernization or globalization. Meanwhile, the generations of people who are supposed to inherit this knowledge are often careless about conservation efforts, let alone patenting it. LK is very important in planning for community development. With the approval of the Sustainable Development Agenda, the international community has committed to resolving a slew of issues, many of which directly harm the lives of indigenous people.

2. Problem Statement

This study investigates the status of IL and L KT in the MC in the Agam district, Indonesia. IL is crucial in our daily life which includes the lives of the Minangkabau Community (MC). Studies by Seenu (2016) show that knowing one's cultural identity and recognizing one's communities demands a certain set of skills; however, the low literacy rate among local groups is a significant problem to the community [6]. Local people must include IL into the preservation of LK as cultural legacy in order to maintain their socioeconomic development, advance their knowledge, and enhance their quality of life. IL can be investigated through six steps consisting of: defining the problem, information seeking strategies, finding and accessing information, using information, synthesizing, and evaluating. In this context, LK is more in the form of hidden knowledge (tacit knowledge); however, there should be an effort to transform local knowledge as tacit knowledge into explicit knowledge so that it can be accessed, studied, and utilized. Using a four-factor approach, Sommer et al. (2021) analyzed the information literacy self-efficacy scale (ILSES). The first three steps are: developing a search strategy; understanding, interpreting and synthesizing the information; and evaluating the process and product [7]. According to Bharun et al. (2022), curriculum adaptation, literacy monitoring, improvement of human resources, knowledge exchange, and evaluation are used to establish a literacy culture based on local knowledge [8]. Therefore, it is an interest of this research to examine the context of the IL level and the status of LKT. As a result of the initial search and review of the literature to be discussed below, two factors were identified: IL and LKT.

3. Literature Review

We acquired, analyzed and reviewed the literature pertaining to IL and LKT. Particular emphasis was paid to LK in the community.

3.1. Information Literacy

According to Patah (2014), IL skills lead to skill-based literacy, which encompasses the ability to responsibly seek, categorize, use, and present information [9]. Critical information assessment as a component of media and IL must be understood in terms of how it is carried out in the acquisition of knowledge and the development of ignorance and doubt, as well as how it is utilized to evaluate the dependability of information [10]. IL is an integrated set of skills, knowledge, practices, and dispositions that prepare people to responsibly discover, interpret, and generate information while learning how information systems interact to produce and distribute news, information, and knowledge [11].

3.2. Local Knowledge

LK is information produced through time by a group of people or groups who have lived in a particular place for a long time, and it is usually based on observation, experience, and analysis among local populations [12]. LK, which has a long history of interacting with the environment, preserves and produces a complex and cumulative body of information, know-how, practices, and representations that is referred to as LK. It is also dynamic and adaptive, not static, altering in line with social, economic, cultural, etc., developments in society. It is collectively owned by the community and belong to it [13]. LK refers to the indigenous, traditional, and innate knowledge related to the way of life of local rural people [14]. Knowledge, communication processes, process interpretation, and knowledge interpretation are all steps of the phenomenon of LK transfer [15].

3.3. Sustainable Development Goals (SDGs)

A road map for peace and prosperity for people and the planet in the present and the future is envisaged by the 2030 Agenda for Sustainable Development, which has the support of all UN member states. The major objective of the agenda is for developed and developing nations to collaborate in order to achieve the 17 sustainable development goals (SDGs) [16]. The principles of sustainable development can be summarized as: development must be environmentally sustainable, socially just, economically efficient, and culturally consistent with society [17]. Global development goals, such as the SDGs, emphasize the importance of local solutions for global sustainable development, particularly those rooted in traditional ecological and local community knowledge [18]. Although the relationships between the SDGs and local values are not always obvious, many of them are linked. The SDGs should benefit from local knowledge that promotes integration, thorough understanding, and practice in terms of clarity, meaning, purpose, and related activities. According to Castellanos et al., knowledge has evolved into consideration when making decisions about social, economic, and environmental issues, as well as means of subsistence for rural families [19].

4. Methodology

A 40-item of questionnaire with a five (5) Likert scale was constructed as a result of a review of the literature to elicit the perspectives of the MC, the level of IL, and the status of LK. Feedback and suggestions from five academic experts in IL, knowledge management and preservation, and language and culture sectors were successfully acquired for pre-testing purposes. A number of points were raised, including the need for more appropriate words and verbs to improve question clarity and the inclusion of examples to aid the understanding of the questions by respondents. Based on their insightful feedback, necessary amendments and additions were implemented. Later, a pilot test was conducted to verify the understanding of the questionnaire items by respondents and that no questions were ambiguous. The data from this study, which comprised 50 respondents, were analyzed using SPSS version 26. Finally, Cronbach's alpha was used to assess the scale reliability of the questionnaire.

The research design was established using the conceptual framework, and the interaction of factors was obtained from the literature. The re-conceptualization provided an original contribution to knowledge and explored LI and LKT. Through the research process on the literature framework using the Big 6 model (which consists of the definition of information needs, information-seeking strategies, location and access to information, information usage, synthesis—information organization, and evaluation), the results and processes of a few of the frameworks were analyzed for the relevant subject of this research. This model was developed by Michael Eisenberg and Robert E. Berkowitz and is known as the 6 steps of IL. The six steps consist of defining the problem, information-seeking strategies, finding and accessing information, using information, synthesizing, and evaluation. Defining problems refers to defining tasks and information problems. Strategies in information seeking involve the listing of all sources and selecting the best source. Seeking and accessing information is performed by determining the location of the source and accessing information from within the information source. Using information is by reading, listening, gathering information from the source, and extracting information from the source. Synthesis is performed by combining information from various sources and creating an information product. Finally, evaluate the product. The concept of tasks is the first stage defined in the Big 6 model and includes being able to define the problem of information and identify the appropriate information. This implies knowing what one is supposed to do [20]. Since this model is more versatile than other data literacy models, the Big 6 model can be extended to virtually all human decision-making problems that use knowledge as a basis for decision-making [21].

LK thus appears as a portrait of identity and the value of life in a culture. LK is characterized as unstructured tacit knowledge and is stored in the memory of the knowledge owner offering four models of knowledge transfer formation or what is known as the SECI model. In the field of knowledge management, the SECI model has been proven to be one of the most stable and continues to be implemented in a variety of settings. One of its main strengths is its simplicity, both in terms of recognizing the basic tenets of models and in terms of being able to easily internalize knowledge management models and implement them. One of its key drawbacks is that it does not seem to be adequate to clarify all the stages involved in the management of information by validity. The model of Nonaka and Takeuchi focuses on the transition of knowledge between implicit and explicit knowledge, but the model does not address substantial questions about how both forms of knowledge are used to make decisions.

The review of literature was conducted and has been helpful in the process of developing the conceptual framework. The questionnaire collected the demographic profile of the participants in Section A. Section B required respondents to determine the source of information. Section C required respondents to rank the usage of IL. Section D required respondents to rank the usage of the status of LK. Section E required respondents to answer challenges and recommendations in LKT among the MC. The instrument developed for this study employed the five Likert scale for all the constructed items.

Pre-Testing and Validity and Reliability of Research Instrument

The questionnaire was distributed to five experts. The panel of expertise was selected based on experiences, academic qualifications, and their involvement in LK. The expert panel comprised (n = 2) academic experts in IL, (n = 1) academic experts in knowledge management and preservation, and (n = 2) academic experts in language and culture in order to review and identify potential issues with the questions. The field experts comprised three lecturers in library and information science from the Universitas Indonesia and Universitas Diponegoro and two lecturers in language and culture from the Universitas Negeri Padang. Four academics were Ph. D holders, and one was a professor. Several concerns were highlighted, including the need for more precise and specific alternatives when giving options to the respondents, the use of more suitable words and verbs to increase the clarity of questions, and the inclusion of examples to help the understanding of the questions by respondents. Their reviews were highly valuable since they allowed for the incorporation of all pertinent comments and suggestions into the current questionnaire, with appropriate modifications and additions.

The evidence that the instrument, approach, or procedure employed to assess a notion actually measures the intended idea is known as validity. The questionnaire was then validated in a pilot study, which allowed for the investigative questions to be answered, and the verification of collected data could be performed accordingly [22]. The pilot study was conducted to ensure that respondents understood the questionnaire items and that there were no ambiguous questions [23]. This initial study included 50 participants, and the data were analyzed using SPSS version 26.

The quality of reliability is concerned with the question of whether the researcher is receiving reliable data. They are considered to be reliable if they can be reused. Reliability is the extent to which a measure is devoid of random error and hence yields consistent results [24]. Reliability and validity of measurement play important roles in generalizing. Cronbach's alpha is one of the most popular measures of scale reliability [25]. Finally, Cronbach's alpha was used to determine the scale reliability of the questionnaire and how closely related a group of questions was. Four parts were included in the questionnaire. The first section had ten items pertaining to demographic data. The second section had 24 items that covered six topics: the definition of information needs, information-seeking strategies, location and access to information, information usage, synthesis—information organization, and evaluation. Following that, three sections emphasized the status of LK and sixteen more items. As a result, 4 items were eventually included in the instrument prior to the actual execution of the study. Additionally, two open-ended questions were added to the fourth section of the questionnaire to allow respondents to subjectively respond.

5. Results of the Reliability Test

The reliability and internal consistency of the scales used in the study were assessed using Cronbach's alpha coefficients, as shown in Table 1. Cronbach's alpha values for each factor were greater than 0.6, indicating that the dependability level was suitable and acceptable. According to the SPSS analysis results, the overall consistency, or Cronbach's Alpha values, of all 40 items for each dimension contained in the instrument was between 0.760 and 0.848. This result implies that, particularly at this point of the investigation, the overall index of the internal consistency of the scale within the instrument was reliable with no unexpected abnormalities found in the data. Information-seeking strategies showed the highest value of 0.848, while information usage showed the lowest with 0.760. This questionnaire went through a validation process with four experts and a pilot study with 50 MC in the Agam Regency. The results from the pilot study showed all instruments are reliable and valid to investigate the level of IL and status of LKT.

Table 1. Reliability analysis result.

Variables		Number of Items	Cronbach's Alpha
Information Literacy (The Big 6 Model)	Definition of information needs	4	0.780
	Information-seeking strategies	4	0.848
	Location and access to information	4	0.807
	Information usage	4	0.760
	Synthesis: information organization	4	0.788
	Evaluation	4	0.821

Table 1. *Cont.*

Variables		Number of Items	Cronbach's Alpha
Status Local Knowledge (SECI Model)	Tacit knowledge: socialization	4	0.808
	Tacit knowledge: externalization	4	0.838
	Explicit knowledge: combination	4	0.829
	Explicit knowledge: internalization	4	0.825
Overall		40	0.810

6. Discussion and Conclusions

The results suggest that information-seeking strategies showed the highest value of 0.848, while information usage showed the lowest with 0.760 which indicates that the MC in the Agam Regency has a significant level of information literacy and is related to LK. The aim of this article is to provide the outline of an ongoing research project in the MC on the level of IL and the status of LKT. Prior to the actual investigation, the instrument was initially pre-tested to identify any possible issues as well as to determine the degree of understandability of its items. Five subject-matter experts have reviewed the instrument and provided meaningful feedback. Additionally, 50 responders from the MC participated in a pilot survey. The analysis result of the collected data suggests that the overall index of the internal consistency of the scale within the instrument is reliable, with no unexpected abnormalities discovered in the data. The validity and reliability of the questionnaire are extensively discussed in this work. This is because the validity and reliability of the measuring questionnaire have a significant impact on the credibility of the research findings. In order to be valid, a questionnaire must be trustworthy. Validity is predicated on reliability. A valid questionnaire, on the other hand, may not be reliable. A valid questionnaire, on the other hand, must be trustworthy. As a result, it is critical that the questionnaire is both valid and reliable.

According to Mannan (2019), the results of study show that people in Jombang, Indonesia, have a high level of literacy, particularly when it comes to: defining and articulating information needs; knowing where to find and how to access the information they need; and evaluating, organizing and using the information they find: they have discovered, saved, communicated, and used it ethically [26]. Ong Choon et al. (2019) investigated the connection between the lifelong learning outcomes and goal orientation of adult learners, and IL self-efficacy. According to the conclusions of the study, mastery goal orientation and lifelong learning are significantly correlated [27]. Irwan (2019) discusses the challenges of transferring LK to empowering communities as not all teaching initiatives for transferring local knowledge for the development of basic literacy are successful [28].

Finally, this research makes significant theoretical, methodological, and practical contributions to the study of IL and other fields of study that are related to the value of LK to the larger community. Outcomes from this research provide valuable insights for the Government and policymakers into the protection of LK in the policies of Indonesia relating to LK.

Author Contributions: Conceptualization, E.R., S.M.S., and W.A.W.Y.; methodology, E.R. and S.M.S.; software, E.R. and S.M.S.; validation, S.M.S. and W.A.W.Y.; formal analysis, S.M.S.; investigation, E.R.; resources, E.R.; data curation, E.R.; writing—original draft preparation, E.R.; writing—review and editing, S.M.S. and W.A.W.Y.; visualization, E.R.; supervision, S.M.S. and W.A.W.Y.; project administration, E.R.; funding acquisition, E.R. All authors have read and agreed to the published version of the manuscript.

Funding: This research received no external funding.

Institutional Review Board Statement: Not applicable.

Informed Consent Statement: Not applicable.

Data Availability Statement: Not applicable.

Acknowledgments: The authors would like to thank everyone who participated in the survey and contributed their ideas and thoughts to this research, especially the fifty persons who volunteered their time to answer the questionnaire. Our appreciation and gratitude are extended to the Faculty of Information Management, Universiti Teknologi MARA, Malaysia, and the Universitas Negeri Padang, Padang, Sumatera Barat, Indonesia.

Conflicts of Interest: The authors declare no conflict of interest.

References

1. Manimekalai, S. Relevance of information literacy in digital environment. *J. Inf. Technol. Rev.* **2019**, *10*, 48–51. [CrossRef]
2. Setten, G.; Lein, H. We draw on what we know anyway: The meaning and role of local knowledge in natural hazard management. *Int. J. Disaster Risk Reduct.* **2019**, *38*, 101184. [CrossRef]
3. Nugroho, K.; Carden, F.; Antlov, H. *Local Knowledge Matters: Power, Context and Policymaking in Indonesia*; Policy Press: Bristol, UK, 2018.
4. De Lucia, C.; Pazienza, P.; Balena, P.; Caporale, D. Exploring local knowledge and socio-economic factors for touristic attractiveness and sustainability. *Int. J. Tour Res.* **2020**, *22*, 81–99. [CrossRef]
5. Vasarhelyi, M. Dynamic Theory Knowledge of Organizational Creation. *Organ. Sci.* **2017**, *5*, 14–37. Available online: https://www.svilendobrev.com/1/Nonaka_1994-Dynamic_theory_of_organiz_knowledge_creation.pdf (accessed on 22 January 2021).
6. Seenu, S.K. Information literacy and indigenous librarianship, challenges before information professionals. In Proceedings of the Information literacy and embedded librarianship challenging role of college libraries in digital era, Bijapura, India, 18–19 November 2016.
7. Sommer, M.; Kohnen, A.M.; Ritzhaupt, A.D.; Hampton, J. Investigation of the validity evidence of the information literacy self-efficacy scale (Ilses) among undergraduate students. *Commun. Inf. Lit.* **2021**, *15*, 1–23. [CrossRef]
8. Baharun, H.; Wahid, A.H.; Qodriyah, K.; Wahyuni, S.; Fatmawati, F. Building a literacy culture based on local wisdom through women's leadership. *J. Obs. J. Pendidik. Anak Usia Dini* **2022**, *6*, 2533–2543. [CrossRef]
9. Mulyono, D.; Ansori, A. Literasi informasi dalam kerangka pengembangan pendidikan masyarakat. *Comm-Edu Community Educ. J.* **2020**, *3*, 1. [CrossRef]
10. Haider, J.; Sundin, O. Information literacy challenges in digital culture: Conflicting engagements of trust and doubt. *Inf. Commun. Soc.* **2020**, *25*, 1176–1191. [CrossRef]
11. Head, A.J.; Fister, B.; MacMillan, M. Information literacy in the age of algorithms-student experiences with news and information, and the need for change. *Inf. Lit. Age Algorithms* **2019**, 1–55.
12. Cahyono, S.A.; Wuryanta, A.; Lastiantoro, C.Y. The local knowledge to mitigate the landslide disaster in Beruk village, Jatiyoso sub-district, Karanganyar regency. *IOP Conf. Ser. Earth Environ. Sci.* **2021**, *874*, 012015. [CrossRef]
13. Haseena, V. ISSN NO: 2347-6648 Volume IX, Issue III, March/2020 Page No: 1710 ISSN NO: 2347-6648 Page No: 1711. *Parishodh J.* **2020**, *IX*, 1710–1717. Available online: http://www.parishodhpu.com/gallery/561-p-march-1163.pdf (accessed on 20 June 2022).
14. Sufo Kankeu, R.; Tsayem Demaze, M.; Krott, M.; Sonwa, D.J.; Ongolo, S. Reprint of governing knowledge transfer for deforestation monitoring: Insights from REDD+ projects in the Congo Basin region. *For. Policy Econ.* **2020**, *114*, 105105. [CrossRef]
15. Andesfi, A.; Prasetyawan, Y.Y. Pemindahan pengetahuan lokal komunitas nelayan tradisional Desa Kedungmalang. *Anuva J. Kaji Budaya* **2019**, *3*, 257–271. Available online: https://ejournal2.undip.ac.id/index.php/anuva/article/view/6469 (accessed on 20 June 2022). [CrossRef]
16. Negi, V.S.; Pathak, R.; Thakur, S.; Joshi, R.K.; Bhatt, I.D.; Rawal, R.S. Scoping the need of mainstreaming indigenous knowledge for sustainable use of bioresources in the Indian Himalayan Region. *Environ. Manag.* **2021**, *2*, 1–12. [CrossRef] [PubMed]
17. Hosseinnia, E.; Shoja, B. The Role of handicrafts in the sustainable development of rural tourism with an emphasis on indigenous knowledge. *Int. J. Geogr. Geol.* **2017**, *6*, 1. [CrossRef]
18. Brown, L.D.; Atapattu, S.; Stull, V.J.; Calderón, C.; Huambachano, M.; Houénou, M.; Snider, A.; Monzón, A. From a three-legged stool to a three-dimensional world: Integrating rights, gender and indigenous knowledge into sustainability practice and law. *Sustainability* **2020**, *12*, 9521. [CrossRef]
19. Gutiérrez García, G.A.; Gutiérrez-Montes, I.; Hernández Núñez, H.E.; Suárez Salazar, J.C.; Casanoves, F. Relevance of local knowledge in decision-making and rural innovation: A methodological proposal for leveraging participation of Colombian cocoa producers. *J. Rural Stud.* **2020**, *75*, 119124. [CrossRef]
20. Dipetso, C.M.; Moahi, K.H. Information Literacy skills of high school students in Botswana. *Sch. Libr. Worldw.* **2019**, *25*, 99–115. Available online: http://content.ebscohost.com/ContentServer.asp?T=P&P=AN&K=135811454&S=R&D=lih&EbscoContent=dGJyMMvl7ESeqLc4yOvqOLCmr1GeqLBSr6y4TbOWxWXS&ContentCustomer=dGJyMPGut1Gxp7VNuePfgeyx4YHs1%2BaE (accessed on 10 January 2021). [CrossRef]

21. Marlini, R.E. Information literacy level of students of Universitas Negeri Padang Using the Big 6 Model. In Proceedings of the 1st Progress in Social Science, Humanities and Education Research Symposium (PSSHERS 2019), Padang, Indonesia, 23–24 September 2019. Available online: https://doi.org/10.2991/assehr.k.200824.035 (accessed on 10 January 2021). [CrossRef]
22. Aithal, A.; Aithal, P.S. Development and validation of survey questionnaire & experimental data—A systematical review-based statistical approach. *Int. J. Manag. Technol. Soc. Sci.* **2020**, *5*, 233–251. [CrossRef]
23. Fraser, J.; Fahlman, D.; Arscott, J.; Guillot, I. Pilot testing for feasibility in a study of student retention and attrition in online undergraduate programs. *Int. Rev. Res. Open Distance Learn.* **2018**, *19*, 260–278. [CrossRef]
24. Parsons, S.; Kruijt, A.-W.; Fox, E. Psychological Science needs a standard practice of reporting the reliability of cognitive-behavioral measurements. *Adv. Methods Pract. Psychol. Sci.* **2019**, *2*, 378–395. [CrossRef]
25. Hayes, A.F.; Coutts, J.J. Use Omega Rather than Cronbach's Alpha for Estimating Reliability. *But … Commun. Methods Meas.* **2020**, *14*, 1. [CrossRef]
26. Mannan, E.F. The role of village libraries to improve information literacy in rural communities. *Libr. Philos. Pract.* **2019**, *8*, 616–623.
27. Hee, O.C.; Ping, L.L.; Rizal, A.M.; Kowang, T.O.; Fei, G.C. Exploring lifelong learning outcomes among adult learners via goal orientation and information literacy self-efficacy. *Int. J. Eval. Res. Educ.* **2019**, *8*, 616–623. [CrossRef]
28. Irwan, I. Basic literacy problems in digital age: Problems in Transferring Local Knowledge for Empowering Community. *Proceeding Iain Batusangkar* **2019**, *3*, 173–178.

proceedings

MDPI

Proceeding Paper

Measuring the Reliability Analysis of Heutagogy Learning, Student Motivation and Digital Literacy among Indonesian Students †

Endang Fitriyah Mannan [1,2,*], Shamila Mohamed Shuhidan [1] and Mohamad Noorman Masrek [1]

1 Faculty of Information Management, Universiti Teknologi Mara Selangor Branch, Shah Alam 40150, Malaysia
2 Faculty of Vocational Studies, Universitas Airlangga, Surabaya 60286, Indonesia
* Correspondence: endang.mannan@vokasi.unair.ac.id

† Presented at the International Academic Symposium of Social Science 2022, Kota Bharu, Malaysia, 3 July 2022.

Abstract: This paper's aim was to measure the reliability of heutagogy learning, student motivation, and digital literacy among Indonesian students. This study used a quantitative approach. Respondents are secondary school students in Indonesia. Data were collected using a questionnaire. The instruments that were compiled were also reviewed by experts. A pilot study was also conducted and followed by 54 respondents. The data were processed using SPSS. The findings revealed the instrument was reliable and there were no abnormalities in the data with the highest value of 0.917 and the lowest value of 0.720. It means that the questionnaire can be used for data collection

Keywords: heutagogy learning; student motivation; digital literacy; Indonesian student

check for updates

Citation: Mannan, E.F.; Shuhidan, S.M.; Masrek, M.N. Measuring the Reliability Analysis of Heutagogy Learning, Student Motivation and Digital Literacy among Indonesian Students. *Proceedings* **2022**, *82*, 33. https://doi.org/10.3390/proceedings2022082033

Academic Editor: Mohamad Rahimi Mohamad Rosman

Published: 13 September 2022

1. Introduction

The Information and communication technology (ICT) has changed the world, including in the education method [1]. Technology advancement and expansion translate into increased educational access via devices used to deliver distance and online education. The ICT has also shifted traditional methods of teaching and learning to a new method and experience. This means that the new method allows teachers and students to explore more resources in a more flexible manner. However, not all teachers and students can use all the information available on ICT. In addition, to adopt ICT as an innovative teaching and learning practice, the education system must be transformed. The current form of education is heutagogy, which is focused on self-determined learning [2]. Heutagogy requires students to be more independent learners, and the role of the teachers is only as a guidance and the students have the freedom to set their own learning experience [3]. In addition, [4] stated heutagogy is relevant to the needs of learners in the 21st-century, particularly in the development of individual capability. Therefore, to cope with the current situation, the student needs to equip with digital literacy skills.

Studies have found that there is a relationship between digital literacy and student motivation [5–7]. Student motivation is critical in fostering a desire to begin engaging in and pursuing educational goals [8]. Moreover, [9] as cited by [3] found that students will have motivation if their teachers allowed them to find their way for study. Students must be free to choose their course or topic that is related to their interests. For example, students can utilize technology such as the internet as a facilitator for studying. The freedom method of study allows a student to expand not only their experience but also their knowledge in certain areas and fields of interest. Then, students show their highest confidence and motivation to search the topics in which they are interested.

2. Problem Statement

The heutagogy approach is still new and not all schools in Indonesia have implemented it. A prior study argued that heutagogy promotes students to reflect on daily unstructured experiences of learning [10]. Furthermore, in the heutagogy method, students and teachers have the opportunity and freedom to choose, utilize, and gain from many resources of information regarding problems in the school [2,11]. In the digital age, students must have adequate literacy skills. Literacy in Indonesia remains low in comparison to other countries [12]. As a result, digital literacy is essential as part of 21st-century education [13]. Finally, the implementation of digital literacy necessitates the collaboration of various parties, including the government, schools, parents, and society [14]. Digital literacy education for the millennial generation is necessary. According to [4] who stated that "the internet usage of primary school children found that the experiences, knowledge, familiarity, motivation and proffer guidance in literacy skill will help them to improve their learning experience". Therefore, an understanding of digital literacy skills is required in the heutagogy method to improve student motivation.

3. Literature Review

Previous studies [10,11,15–18] have shown that the heutagogy learning approach is to prepare the students to be self-determined and that it requires a new skillset specially to deal with complex information age. This section will discuss the literature review on heutagogy, digital literacy, and student motivation. The way of teaching and learning is continually experiencing changes simultaneous with the development of the country's vision of creating younger generations that are occupied with facing new challenges in the era of globalization. Heutagogy methods addressing the self-directed learning needs of the autonomous professional learner include distance learning and the use of technologies. The Pedagogy-Andragogy-Heutagogy (PAH) continuum was developed to help understand the learning processes that may increase learner agency as the learner develops new learning skills.

Today the topic of digital literacy is widely discussed in this era. Students and teachers are raised in social knowledge. Being advanced digital literacy is an existing skill that needs computerized education abilities in each part of their life. To be educated in the modern sense, it is essential to be digitally literate [15,16]. According to the previous study, digital literacy is the use of digital devices to establish meaning and communicate effectively with some accessories such as digital texts, navigate non-linear digital documents, and evaluate digital information [15,19,20]. In addition, [21] defined digital literacy as the use of computers as accessible and productive devices to collect, build, transform, and securely use information. Furthermore, Information Communication and Technology (ICT) and Digital Literacy (DL) are pre-mandatory of the digital world. Consequently, it can be concluded that technological devices can enhance student motivation and attitude and the DL of students.

Motivation can affect how students approach school in general. Student learning motivation must be a serious concern in developing 21st-century skills. The learning environment needs to be developed in such a way as to make it more interesting so that learning does not merely transmit knowledge ('standard transmission'). Increased involvement in and control over the learning process by the learner; self-initiated learning; the opportunity to develop and share patterns; and its significance [10,22] argued that "being digitally literate today involves the knowledge, attitudes, and skills needed for operating technologies, using the internet, understanding the media, and managing information. However, the convergence of literacies into the digital is, however, more than the sum of its single elements". According to the study findings, a student's technological skill is a significant predictor of attitudes toward ICT. This attitude has the potential to influence student's perceived and actual DL and ICT skills. Better ICT skills are thought to improve student's motivation and attitude. Thus, a better DL improves positive attitude towards technology.

From the underpinning theories and models of Heutagogy Learning, Digital Literacy and Student Motivation are derived the research framework as shown in Figure 1. The proposed framework aims to investigate the relationship between heutagogy learning and student motivation. The independent variable in this study is Heutagogy learning which has four (4) dimensions, namely linear agency, capability, reflection, and non-linear design which will be considered to have an influence on student learning motivation which includes dimensions of attention, confidence, and satisfaction. An important addition to the framework is digital literacy skills which are expected to be strongly correlated with the use of electronic resources among students. Thus, Digital literacy skills have been identified as a moderating variable, with the introduction of which the effect of the four (4) dimensions on the independent variable is expected to be strengthened. It is predicted that the dependent variable will have a relationship with the independent and moderating variables mentioned above.

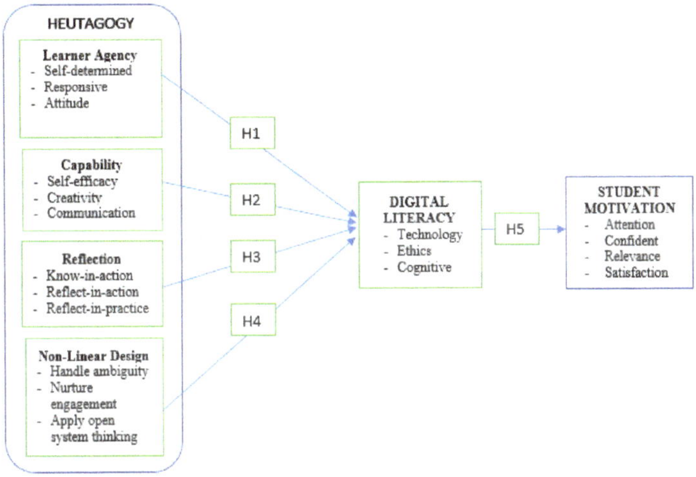

Figure 1. Research Framework.

4. Methodology

The questionnaire was constructed by adapting and adopting various literature to obtain the perspectives of students in senior high schools in East Java, Indonesia, on the research topic. A total of 87 question items were successfully constructed which were previously reviewed by experts consisting of academics and practitioners. Feedback from the experts was used to improve the questionnaire in the hope that students as respondents can understand the questions. Several points were raised, including the need for more precise words and verbs to increase the clarity of the questions and the inclusion of examples to help respondents understand the questions. After that, a trial was conducted to verify that the respondents understood the questionnaire items and that there were no ambiguous questions. Data from this preliminary study, consisting of 54 respondents, were analyzed using SPSS version 26.

5. Pre-Testing Validity and Reliability of Research Instrument

A pilot test will provide meaningful findings that help researchers identify appropriate variables for further exploration. Thus, the results of this preliminary study help researchers determine the research instrument's validity and reliability. The sampling method for this research was simple random sampling. In this research, the method used was purposive random sampling because the elements in the population had several known possibilities of being selected as sample subjects. The survey items were distributed to five experts in digital literacy in education in Indonesia: lecturers in the Library and Information Science, school supervisors and teachers. All these reviews are constructive for the researcher to complete the questionnaires with appropriate modifications.

The survey questionnaire structure begins with part 1: Demographic profile consisting of nine queries. The survey questions are classified into four sections with closed question types. In Sections 2–4, a questionnaire consisting of 87 questions focused on the independent and dependent variables of the study. All units used a five Likert scale ranging from "strongly agree" (coded 5) to "strongly disagree" (coded 1). The schools selected for this research have Semester Credit System (SKS) services, where the characteristics of the school are like the study.

6. Reliability Test Results

Table 1 shows that Cronbach's alpha was used to determine the scale's reliability and internal consistency in this study. Cronbach's alpha values for all factors were greater than 0.6, indicating that the level of reliability was appropriate and acceptable. According to the results of the SPSS analysis, the overall consistency, or Cronbach's Alpha value, of all 87 items for each dimension contained in the instrument was between 0.764 and 0.940. In the technology dimension, the Solution Common Technical section shows the highest value of 0.940, while the Problems Apply open system thinking measurement offers the lowest value with 0.764. The lowest value was still above 0.6, so all questionnaires can be used. These results imply that, particularly at this point of the investigation, the overall index of internal consistency of the scale in the instrument is reliable without any unexpected abnormalities found in the data.

Table 1. Hypotheses.

	Hypothesis
H1	Learner agency has relationship with digital literacy.
H2	Capability has relationship with digital literacy
H3	Self-reflection has relationship with digital literacy
H4	Non-linear learning design has relationship with digital literacy
H5	Digital Literacy has relationship with student motivation
H6	Digital literacy has mediated the relationship between heutagogy learning and student motivation

According to [23] a Cronbach's alpha value of less than 0.6 will have an impact on the data validity. Therefore, the items need to be revised or omitted as items. One approach to improving Cronbach's alpha value is adding more related articles to test the same concept. The Table 2 shows the reliability of the test results. It has been carried out with 30 students, and the results of the tests are as follows:

Table 2. Reliability of the Test Results.

Variables		Number of Items	Cronbach's Alpha
Learner Agency	Self-determined	5	0.896
	Responsive	4	0.787
	Attitude	5	0.917
Capability	Self-efficacy	4	0.720
	Creativity	4	0.814
	Communicate	5	0.835

Table 2. *Cont.*

Variables		Number of Items	Cronbach's Alpha
Reflection	Know-in-action	4	0.747
	Reflect-in-action	4	0.783
	Reflect-in-practice	4	0.834
Non-linear Design	Ability to Handle ambiguity	4	0.799
	Nurture engagement	5	0.813
	Apply open system thinking	4	0.775
Student Motivation	Attention	5	0.856
	Relevance	6	0.787
	Confident	4	0.803
	Satisfaction	5	0.793
Digital Literacy	Technology	4	0.829
	Ethics	6	0.813
	Cognitive	5	0.839
Overall		87	0.813

7. Discussion & Conclusions

The Research framework was expected to provide useful and insightful results regarding the relationship between heutagogy learning and student motivation. Furthermore, digital literacy skills are a moderating variable that was expected to influence both the dependent and the independent variable's relationships. All these variables will be measured in a study with 54 students in the pilot test. Before the actual investigation, the instrument was initially pre-tested to identify possible problems and determine the level of understanding of the items. Five experts reviewed the instrument and provided meaningful feedback. The results of the collected data indicate that the overall index of internal consistency of the scale in the instrument is reliable, with no unexpected abnormalities found in the data. The pilot results show that the overall consistency, or Cronbach's Alpha value, of all 87 items for each dimension contained in the instrument was between 0.720 and 0.917. In the attitude dimension on the variable learner agency shows the highest value of 0.917, while the self-efficacy dimension on the variable capability measurement offers the lowest value with 0.764. All factors have a value greater than 0.6, indicating that all questions in the questionnaire can be used for data collection [24]. The instruments will be used for the next phase of the study and hopefully, they will contribute to the improvement and implementation of heutagogy learning emphasis on a digital literacy case study in Indonesia. It can assist teachers in preparing digital literacy skill in heutagogy learning to enhance student motivation.

Author Contributions: Conceptualization, E.F.M. and S.M.S.; methodology, E.F.M. and S.M.S.; software, M.N.M.; validation, S.M.S. and M.N.M.; formal analysis, S.M.S.; investigation, E.F.M.; resources, E.F.M.; data curation, E.F.M.; writing—original draft preparation, E.F.M.; writing—review and editing, S.M.S.; visualization, E.F.M.; supervision, M.N.M.; project administration, E.F.M.; funding acquisition, E.F.M. All authors have read and agreed to the published version of the manuscript.

Funding: This research was funded by Universitas Airlangga, grant number 573/UN3.15/PT/2021.

Institutional Review Board Statement: Not applicable.

Informed Consent Statement: Not applicable.

Data Availability Statement: Not applicable.

Acknowledgments: The authors would like to thank everyone who participated in the survey and contributed their ideas and thoughts to this research, especially the students who volunteered their time to answer the questionnaire. Thanks also to the Faculty of Information Management, Universiti Teknologi MARA, Malaysia, and the Universitas Airlangga in Surabaya, Jawa Timur, Indonesia.

Conflicts of Interest: The authors declare no conflict of interest.

References

1. Shoikova, E. Models of technology and change in higher education. *Proc. Int. Spring Semin. Electron. Technol.* **2004**, *3*, 500–505. [CrossRef]
2. Halsall, J.P.; Powell, J.L.; Snowden, M. Determined learning approach: Implications of heutagogy society based learning. *Cogent Soc. Sci.* **2016**, *2*, 1–11. [CrossRef]
3. Shuhidan, S.M. Information-Seeking Processes among Primary School Children in Australia and Malaysia. Ph.D. Thesis, RMIT University, Melbourne, Australia, 2013.
4. Hase, S.; Kenyon, C. (Eds.) *Self-Determined Learning: Heutagogy in Action*; A&C Black: London, UK, 2013.
5. Chan, B.S.K.; Churchill, D.; Chiu, T.K.F. Digital literacy learning in higher education through digital storytelling approach. *J. Int. Educ. Res.* **2017**, *13*, 1–16. [CrossRef]
6. Kane, S. *Literacy And Learning In The Content Areas*; Routledge: London, UK, 2017.
7. Niemi, H.; Multisilta, J. Digital storytelling promoting twenty-first century skills and student engagement. *Technol. Pedagog. Educ.* **2016**, *25*, 451–468. [CrossRef]
8. Reeve, J. In Building autonomous learners. In *Autonomy-Supportive Teaching: What It Is, How to Do It*; Springer: Singapore, 2016; pp. 129–152.
9. Bilal, D. Web search engines for children: A comparative study and performance evaluation of "Yahooligans!," "Ask Jeeves for kids," and "super snooper". In Proceedings of the 62nd ASIS Annual Meeting, Washington, DC, USA, 31 October–4 November 1999.
10. Hase, S.; Kenyon, C. *From Andragogy To Heutagogy*; Ulti-BASE In-Site: Melbourne, Australia, 2000.
11. Blaschke, L.M. Strategies for Implementing Self-Determined Learning (Heutagogy) within Education: A Comparison of Three Institutions (Australia, South Africa, and Israel). Master's Thesis, Oldenburg, Germany, 2016.
12. Hewi, L.; Shaleh, M. Refleksi Hasil PISA (The Programme For International Student Assesment): Upaya Perbaikan Bertumpu Pada Pendidikan Anak Usia Dini. *J. Golden Age.* **2020**, *4*, 30–41. [CrossRef]
13. Gündüzalp, S. 21 st century skills for sustainable education: Prediction level of teachers' information literacy skills on their digital literacy skills. *Discourse Commun. Sustain. Educ.* **2021**, *12*, 85–101. [CrossRef]
14. Dianimdri, S.D.; Yuliani, W.D. Digital age literacy for Indonesian elementary school student. *Soc. Humanit. Educ. Stud.* **2018**, *1*. [CrossRef]
15. Eshet-Alkalai, Y. Digital literacy: A conceptual framework for survival skills in the digital era. *J. Educ. Multimed. Hypermedia* **2004**, *13*, 93–106.
16. Hargittai, E. Digital Na(t)ives? Variation in internet skills and uses among members of the "net Generation". *Sociol. Inq.* **2010**, *80*, 92–113. [CrossRef]
17. Rhema, A.; Miliszewska, I. Analysis of student attitudes towards e-learning: The Case of engineering students in Libya. *Issues Inf. Sci. Inf. Technol.* **2014**, *11*, 169–190. [CrossRef]
18. Abdullah, Z.D. Students' attitudes towards information technology and the relationship with their academic achievement. *Contemp. Educ. Technol.* **2015**, *6*, 338–354. [CrossRef]
19. Bulger, M.E.; Mayer, R.E.; Metzger, M.J. Knowledge and processes that predict proficiency in digital literacy. *Read. Writ.* **2014**, *27*, 1567–1583. [CrossRef]
20. Ng, W. Can we teach digital natives digital literacy? *Comput. Educ.* **2012**, *59*, 1065–1078. [CrossRef]
21. Fraillon, J.; Ainley, J.; Schulz, W.; Friedman, T.; Gebhardt, E. *Preparing for Life in a Digital Age: The IEA International Computer and Information Literacy Study*; International Report; Springer Nature: Berlin, Germany, 2014; p. 308.
22. Janssen, J.; Stoyanov, S.; Ferrari, A.; Punie, Y.; Pannekeet, K.; Sloep, P. Experts' views on digital competence: Commonalities and differences. *Comput. Educ.* **2013**, *68*, 473–481. [CrossRef]
23. Jain, S.; Angural, V. Use of Cronbach's alpha in dental research. *Jain S Angural V Med. Res. Chron.* **2017**, *4*, 285–291.
24. Hair, J.F.; Sarstedt, M.; Matthews, L.M.; Ringle, C.M. Identifying and treating unobserved heterogeneity with FIMIX-PLS: Part I—Method. *Eur. Bus. Rev.* **2016**, *28*, 63–76. [CrossRef]

proceedings

MDPI

Proceeding Paper

The Moderating Role of Information Security Behaviour (ISB) on the Relationship between Digital Literacy (DL) and Information Security Culture (ISC): A Proposed Research Framework †

Mohd Sharulnizam Kamarulzaman [1], Shamila Mohamed Shuhidan [2,*] and Khalid Abdul Wahid [3]

1 Cybersecurity Malaysia, Level 4, Tower 1, Menara Cyber Axis, Jalan Impact, Cyberjaya 63000, Malaysia
2 Faculty of Information Management, Universiti Teknologi Mara Selangor Branch, Shah Alam 40150, Malaysia
3 Faculty of Information Management, Universiti Teknologi Mara Kelantan Branch, Machang 18500, Malaysia
* Correspondence: shamila@uitm.edu.my; Tel.: +60-136-922-881
† Presented at the International Academic Symposium of Social Science 2022, Kota Bharu, Malaysia, 3 July 2022.

Abstract: Information Security Culture (ISC) has been suggested as a means of strengthening employees' information security in the workplace. However, the role of employees and their digital skills in the process has been largely overlooked. Establishing an information security culture that strives to secure information by having employees who are digitally literate, as well as exerting influence on the employees' behaviour regarding security, is one of the measures that may be utilised to mitigate risks provided by humans. Therefore, the purpose of this paper is to establish a research framework that moderates the effect of information security behaviour (ISB) on the relationship between digital literacy (DL) and information security culture (ISC). The study will improve DL and ISCB among employees who implement the country's development plans, such as enhancing administrative functions, social infrastructure, and economic growth performance in accordance with MyDigital initiatives, as well as the government's ISC.

Keywords: information security culture; information security behaviour; digital literacy

Citation: Kamarulzaman, M.S.; Shuhidan, S.M.; Wahid, K.A. The Moderating Role of Information Security Behaviour (ISB) on the Relationship between Digital Literacy (DL) and Information Security Culture (ISC): A Proposed Research Framework. *Proceedings* **2022**, *82*, 35. https://doi.org/10.3390/proceedings2022082035

Academic Editor: Mohamad Rahimi Mohamad Rosman

Published: 13 September 2022

1. Introduction

Information security is vital nowadays. Our professional and personal lives depend on information. Many organizations could not exist without their information assets; thus, protecting them is a key responsibility, according to Alhogail [1], citing Van Niekerk and Von Solms [2]. Previous studies show that employee irresponsibility causes most data breaches [3,4]. 30% of security concerns were triggered by employees, 27% by former employees, and 23% by unknown hackers, according to PriceWaterhouseCoopers' 2018 study [5]. A security breach may compromise sensitive data from an organization, which thus this illustrates the necessity to build a culture of information security among Malaysian organizations. Information security culture is "the combination of views, beliefs, attitudes, assumptions, knowledge and skills that regulate human involvement with information assets within an organization" [6]. Masrek et al. [7] defined ISC as personnel having the essential awareness, abilities, and understanding of information security processes and procedures. Excellent ISCs offer employees the appropriate training, expertise, and awareness. For an effective ISC, examining each employee's DL skill and knowledge and understanding their security behaviour is key to a healthy information security culture. Humans' everyday contacts with technology impact everything they do, at work and at home. This is in line with the aspiration of the Malaysia Digital Economy Blueprint, MyDIGITAL, to improve the digital literacy of employees, as required for Thrust 1: to drive digital transformation in the public sector with the goal of educating all levels of government employees on digital literacy; and Thrust 6: to build trusted, secure, ethical digital environments with the goal of raising cyber security awareness [8].

2. Literature Review

An internet connection is vital for Malaysian organizations, services, and people involved in Industry 4.0 and My Digital. Businesses rely extensively on democratised technology, including social, mobile, IoT, Big Data, cloud computing, and AI, all reliant on connectivity [9]. Employees must be digitally competent as digital technology evolves, and everything connected to the Internet faces information security concerns. Information security is now becoming everyday routine. Every aspect of our lives uses information. Many businesses are unable to function without the use of information, and therefore must take great care to protect their information assets [2]. Every company needs an information security solution as a counter measure [10]. By fostering a security-conscious culture, information assets are safer [11]. Despite this risk, employees may help to reduce information asset risk by adherence to security rules and procedures that can improve information security [12]. The strongest link in an organization's architecture is a well-trained employee with adequate digital literacy abilities [13]. Employees should be digitally literate in order to comply with information security rules and legislation [12]. This facilitates the incorporation of information security into the organization's culture. Organizational culture should affect employees' security behaviours [13]. Information security academics and specialists increasingly emphasise these organizational human capital as a component of mitigating security risks and threats [14–16]. Despite modern technical security safeguards, employees (often unknowingly) contribute to security breaches by risky behaviour due to a poor information security culture [16,17]. An organization's information system security depends on their employees' online behaviour. The human factor is one of the most neglected components of information security in organizations [10]. Proper employee behaviour may greatly improve an organization's information security system and culture [11].

3. Problem Statement

A decade ago, the idea of enabling people to work remotely every day seemed inconceivable. Employers worried about productivity and security when employees worked remotely or from home, particularly with the current epidemic. When employees work remotely, organizations are more vulnerable to cyberattacks. Employees may potentially be a weakness in an organization's IT security systems, enabling cybercriminals to get access to business data, customer data, and intellectual property. In this instance, employees' digital literacy is vital, particularly security expertise, which impacts how they handle valuable data and information. Organizations must cultivate a culture of information security in order to guarantee that their employees engage in secure information handling practices. As the economy becomes more digital, most firms struggle to tackle the cyber threat. Employees' personal information, corporate data, consumer data, intellectual property, and critical infrastructure are all at risk. COVID-19's long-term impact on organization digitization is unknown, although it has boosted the process. Moreover, many employees now work remotely, which raises the risk of cyberattacks [18]. These research problems are listed below.

3.1. Vulnerability of Employee Behaviour

According to the results of the 2017/18 research by the Institute of Information Security Professionals (IISP), over 80% of security professionals cited "people" as the industry's greatest concern, as opposed to "technology and processes." Performgreen UK 2019 [19], an information security provider, argues that changing employee behaviour is the most effective way to enhance an organization's information security. Whether planned or inadvertent, insiders may be responsible for most data breaches, says a study by Verizon, 2009 [20]. Similar studies have shown that insiders represent a threat to information security [21–23] and that organizations must decrease the risk employees offer. While research emphasizes the necessity of managing employee behaviour to secure information assets [24–27] the most effective countermeasure is to include a mix of controls, not only technological ones.

3.2. Remote Working and Safety Issue

Remote working is not new, but security was not always a concern. A 2006 research study discovered significant user awareness and security gaps, leaving them unprepared for remote working [22]. Home in this research is a location where employees dwell permanently or temporarily outside of work hours. Remote employment is not the norm for a large portion of the workforce, even if it is secure. The sudden change to remote work likely took many employees by surprise. In many cases, employees are given the tools (e.g., computers and other devices) for remote work but not the training. Some of these people may have been requested to work remotely using just their IT talents [28]. Considering the attitude and psychology of remote working, it can be understood how this may be dangerous. In a more comfortable and familiar atmosphere, employees may be less obliged by workplace regulations. The question is whether organizations that have established excellent practices in an office environment can do so remotely with their employees. Digital natives and digital immigrants must be able to adopt information security best practices while working remotely in order to prevent security issues.

3.3. Insufficient Digital Literacy Skill

Information technology advancements have led to widespread Internet usage [29]. However, internet hackers may take advantage of people's ignorance or lack of knowledge. Increasing digital literacy and security knowledge among the public and employees is vital for minimizing "hacker" damage. Inclusion of digital literacy, particularly security, in today's educational institutions may have a significant influence [30,31]. Digital skills have become "important" and must be complemented with "soft skills" such as the ability to interact successfully online and offline. Large-scale digital innovations such as artificial intelligence, machine learning, and big data analytics requie the creation of new skill sets, and this in turn affects the capacity building and skill development of the digital economy. Information security and privacy measures are generally seen to be vulnerable to the actions of individuals, who are seen as the "weakest link" [12,32]. If employees do not defend their privacy, security, or copyright rights or those of others, the organization's information asset is at risk due to individual users that lack skills, knowledge, and proper behaviour, according to research [33]. To guarantee that employees are working in a safe online environment, this study examines the requirement for digital literacy and the cultivation of acceptable information security behaviour as a discipline within information security culture. In the case of pandemics, this is especially important when employees are working remotely or in circumstances where they are responsible for their own safety.

3.4. Digital Literacy Skill Knowledge Gap and Scarcity of DL Studies on Employees in the Government Sector, Security Aspects, Information Security Behaviour and Information Security Culture

To effectively identify the relationships under discussion here, there must first be enough research to address the issue. In the local context, many stakeholders, including academics, have already given much attention to digital literacy, especially in recent times, but much of the focus has been on the education sector. According to Ahsan 2021's [34] systematic literature review from 2010 to 2021, only 24.3% of the total research was oriented to people outside the education sector, such as urban and household respondents, adolescents, native people, pharmacists, and business owners and managers in Malaysia. Lack of focus on DL and cybersecurity in Malaysian organizations, especially among employees, causes a knowledge gap. Digital literacy is an essential part of everyday life, yet many overlook it. Engelbrecht found five key weaknesses in DL's coverage of cybersecurity in a 2017 paper. A preliminary literature search shows that no publication has addressed the study trends of digital literacy in relation to cybersecurity, in the context of information security behaviour and culture among Malaysians especially while they work remotely. A study by Nasir, 2020 [35] indicates a lack of research homogeneity in Malaysian ISC models, and there is no consistent set of criteria that can be used for all Malaysian organizations. These new results by Nasir [35] show that ISC is not properly addressed in Malaysian organizations and that there are no defined models or paradigms for an information security culture when employees work remotely. Not only should ISC be implemented in the workplace, but it should also be done while people are working remotely, especially given the current pandemic situation.

4. Proposed Framework

The research proposes evaluating the relationship between digital literacy, information security culture, and information security behaviour. The ISC was expected to be influenced by employees' DL. The assessment of DL and ISC is critical in order to identify the importance of DL for a successful ISC in the organizations. Adapted in Ng's 2012 [36] Digital Literacy Model, the Digital Competence Framework established by Calvani in 2008 [37] consists of three linked components: cognitive, technical, and social-emotional. This framework is optimal for evaluating DL. Cognitive, technical, and social-emotional aspects are the key indications of DL's influence on ISC. This research will investigate the overlap between the social-emotional and cognitive components as part of this research framework, with technology serving as the medium for initiating the DL advancement. This research will assess all of DL's primary components using sub-characteristics. Human nature is the weakest link in any security system, making ISB an essential moderating variable [38]. Performgreen UK 2019 [19], an information security provider, argues that changing employee behaviour is the most effective way to enhance an organization's information security. ISB helps researchers to examine if digitally literate personnel in an organization affect ISC. This framework is intended to offer a clear picture of whether ISB can assist in establishing the relationship between DL and ISC. In addition, the components of DL should be able to assess the acceptability of DL as a factor that will influence ISC for this research. In addition to bridging the DL, ISC, and ISB gaps, a full grasp of all factors is sought for.

This proposed research framework will be used to develop the research instrument by adopting a mixed-methods approach; both qualitative and quantitative data, based on the scope of the research topic. Because qualitative data is commonly available but does not have preset answers, while quantitative data is typically dependent on closed responses, such as those found on questionnaire instruments, a mix of qualitative and quantitative approaches is essential. Figure 1 shows the research framework and the derived hypotheses.

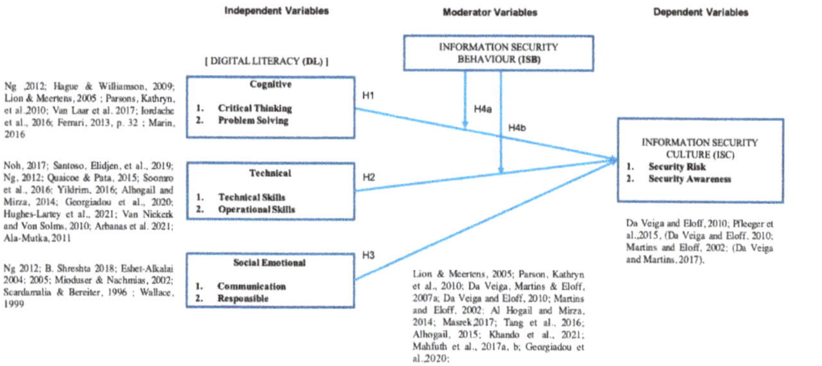

Figure 1. Proposed Research Framework.

H1: *The cognitive component has a significant relationship with information security culture.*

H2: *The technical component has a significant relationship with information security culture.*

H3: *The social-emotional component has a significant relationship with information security culture.*

H4a: *Information security behaviour will moderate the relationship between the cognitive component and information security culture.*

H4b: *Information security behaviour will moderate the relationship between the technical component and information security culture.*

4.1. Digital Literacy

Digital literacy (DL) means that information and digital technology are being utilised with self-assuredness and critical thinking to enhance personal, academic, and professional results for everyone involved. A person with DL skills can acquire and analyse data from many sources, use it in different settings, helping with problem solving and create new information in a digital environment [39]. The primary component of DL in this framework is shown by Ng's 2012 [36] Digital Literacy Model.

4.1.1. Cognitive Component

The cognitive component of Ng's [36] digital literacy paradigm is the ability to think critically throughout the search, appraisal, and creative stages of using digital information. Cognitive refers to an employee's ability to search, analyse, select, and use digital information to fulfil a work-related task while adhering to information security culture. Personality and cognitive differences may affect risk perception and tendency [40]. The study by Kathryn Parsons et al. [41] suggests that people may be categorised based on how they deal with risk, ranging from those who actively seek out danger to those who are very risk-averse. These disparities are anticipated to have an effect on how people understand the information around them, hence influencing the security behaviour that characterises the information security culture.

Table 1 shows the sub-components of the cognitive component. First, critical thinking is the ability to utilise ICT to make informed judgments and decisions about information and communication using reflective reasoning and appropriate evidence [42]. Problem solving is the second cognitive sub-component. Digital skills and competencies may be improved through problem-solving skills, which help users choose the correct digital tools to accomplish their goals and utilise digital technologies to handle conceptual and technical difficulties [43].

Table 1. Cognitive Sub-components.

No	Digital Literacy (DL)	
	Main Component	**Sub-Components**
1	Cognitive	• Critical Thinking • Problem Solving

4.1.2. Technical Component

The technical part of digital literacy is the possession of the technical and operational skills necessary to use ICT for everyday learning activities [44]. Digitally literate individuals are proficient with technology. "Technical" refers to employees' ability to upload, download, and install software for work and daily activities [36,45]. The sub-components of the technical component are shown in Table 2.

Table 2. Technical Sub-components.

No	Digital Literacy (DL)	
	Main Component	**Sub-Components**
2	Technical	• Technical Skills • Operational Skills

In the framework, technological components were classified as "technical," which contained three factors: antivirus protection, authentication and backup, and authorisation. These sub-components were chosen because they represent some of the most basic technical controls that organizations have built and address the three essential qualities of information security: confidentiality, integrity, and availability [46]. The second sub-component—operational abilities—pertains to computer and Internet software and hardware operation. In this context, it is crucial for individuals to understand how their data and information are shared, accessed, or used by governments and corporations; and, more importantly, they need the skills to protect themselves from disclosing information they may not need or want to disclose. Lack of the appropriate privacy settings and of critical skills can lead to loss of control and privacy [43,47].

4.1.3. Social Emotional

DL's third component is social emotional. The Internet and other digital communication platforms have introduced new dimensions and opportunities for collaborative learning through information-sharing and discussion groups, knowledge communities, and chat rooms, among other forms [48,49]. To take advantage of these new possibilities, users need sociological and emotional abilities that allow them to "understand the rules of the game" and overcome the obstacles in the information and communication of cyberspace. They need to understand the information security aspect of cyberspace and why organizations should adopt a proper information security culture [50].

Under the social-emotional component, there are two sub-components (Table 3). Communication involves using the Internet properly for conversing, interacting, and learning by following 'netiquette'. Internet etiquette governs the use of the Internet to communicate, socialise, and learn while adhering to the same rules of decorum as face-to-face encounters, such as using courteous language to prevent misunderstanding and misinterpretation [36]. The second dimension is responsible behaviour. Users are responsible for maintaining their own safety and privacy by preserving their privacy and keeping their personal information as secret as possible, not disclosing more personal information than required, and recognising when they are threatened and how to manage it.

Table 3. Social Emotional Sub-components.

No	Digital Literacy (DL)	
	Main Component	**Sub-Components**
3	Social Emotional	• Communications • Responsible Behaviour

4.2. Information Security Behaviour

Information security culture must include individual features, behaviours, and cognitive skills [51]. This research shows that ISB moderates cognitive and ISC implementation by organizations, supporting the idea that cognition and employee behaviour may impact risk perception (and desire to take risks) and reflect information security culture [40]. Gray and Ropeik [52] noted that an organization's culture and environment may impact how individuals think and behave. Even if employees are digitally literate and have cognitive and technical skills, understanding an organization's culture may explain why certain behaviours do or do not occur.

4.3. Information Security Culture

Information security culture is the attitudes, assumptions, beliefs, values, and knowledge which employees/stakeholders use to engage with the organization's systems and processes. Every organization wants an information security solution [10]. Despite improved technologies, companies struggle to manage information security [17]. An organization's information system security depends on employees' online behaviour. The human factor is one of the most neglected areas of IT security in organizations [10]. Focusing on staff behaviour and skills may improve an organization's information system security [11].

Information Security Culture (ISC) is recognised as an effective way to promote safe behaviour and manage security hazards in an organization. The researcher sought for framework evidence and information security culture indicators. When analysing an organization's information security culture, the security risk comes first (Table 4). To improve an organization's ISC, cutting-edge technology is given to employees (strategic component). A risk assessment (risk management component) identifies security hazards and determines how to minimise them. The second component of ISC is security awareness (Table 4), which is defined as an understanding of security threats, their negative ramifications, and the cost of security failures. "Security culture" refers to people's awareness and understanding of security issues and rules in the context of information security [53].

Table 4. Information Security Culture Components.

No	Information Security Culture (ISC)	
	Main Component	**Components**
4	Information Security Culture (ISC)	• Security Risk • Security Awareness

5. Discussion and Conclusions

It is crucial to understand employees' digital literacy and behaviour towards a successful information security culture, particularly in remote work environments. The research will give insight into employee digital literacy and how the remote work environment affects information security behaviour based on the organization's information security culture both when there are no additional security restrictions and when there are guidelines for working in a remote environment. Understanding the knowledge gap concerning the relationship between these factors, this research intends to contribute to this new research area and enhance present efforts and work in DL, ISC and ISB, applying the framework described above, and also developing guidelines. This research aims to improve the mapping of employees' digital literacy competency by combing it with an information security culture framework. It seeks to influence information security behaviours associated with working in a remote work environment, as well as assisting the government, policymakers, and organizations to remedy the knowledge gap concerning the said variables.

Author Contributions: Conceptualization, M.S.K. and S.M.S.; methodology, M.S.K. and S.M.S.; writing—original draft preparation, M.S.K.; writing—review and editing, M.S.K., S.M.S. and K.A.W.; supervision, S.M.S. and K.A.W.; project administration, M.S.K. and S.M.S.; resources, M.S.K. and S.M.S. All authors have read and agreed to the published version of the manuscript.

Funding: This research received no external funding.

Institutional Review Board Statement: Not applicable.

Informed Consent Statement: Not applicable.

Data Availability Statement: Not applicable.

Acknowledgments: Special thanks to the Faculty of Information Management, Universiti Teknologi Mara Selangor Branch, Shah Alam, particularly the supervisors, for their assistance and guidance, as well as the event secretariats for the opportunities provided. Paper presented during a panel—organized by Universiti Teknologi MARA—in the context of the International Academic Symposium of Social Science 2022 (IASSC 2022).

Conflicts of Interest: The authors declare no conflict of interest.

References

1. AlHogail, A. Design and validation of information security culture framework. *Comput. Hum. Behav.* **2015**, *49*, 567–575. [CrossRef]
2. Van Niekerk, J.F.; Von Solms, R. Information security culture: A management perspective. *Comput. Secur.* **2010**, *94*, 476–486. [CrossRef]
3. Cheng, L.; Liu, F.; Yao, D. Enterprise data breach: Causes, challenges, prevention, and future directions Wiley interdisciplinary reviews. *Data Min. Knowl. Discov.* **2017**, *7*, e1211. [CrossRef]
4. MyCert. 2017 Data Breaches Known So Far. Available online: www.mycert.org.my/data/content_files/27/831.pdf (accessed on 8 May 2020).
5. PriceWaterhouseCoopers. The Global State of Infor—Mation Security®Survey 2018. Available online: https://www.pwc.com/us/en/services/consulting/cybersecurity/library/infor-mation-security-survey.html (accessed on 2 October 2019).
6. AlHogail, A.; Mirza, A. Information security culture: A definition and a literature review. In Proceedings of the 2014 World Congress on Computer Applications and Information Systems (WCCAIS), Hammamet, Tunisia, 17–19 January 2014; IEEE: Hammamet, Tunisia, 2014.
7. Masrek, M.N. Assessing information security culture: The case of Malaysia public organization'. In Proceedings of the 4th International Conference on Information Technology, Computer, and Electrical Engineering (ICITACEE), Semarang, Indonesia, 18–19 October 2017; IEEE: Hammamet, Tunisia, 2017.
8. Malaysia Digital Economy Blueprint. Available online: https://www.epu.gov.my/sites/default/files/2021-02/malaysia-digital-economy-blueprint.pdf (accessed on 3 January 2022).
9. Aziz, K.A.; Norhashim, M.; Halim, E.M. Information security and information technology governance: A Malaysian case study. *Int. J. Manag. Pract.* **2011**, *4*, 331. [CrossRef]
10. Nel, F.; Drevin, L. Key elements of an information security culture in organisations. *Inf. Comput. Secur.* **2019**, *27*, 146–164. [CrossRef]
11. Da Veiga, A.; Eloff, J.H.P. A framework and assessment instrument for information security culture. *Comput. Secur.* **2010**, *29*, 196–207. [CrossRef]
12. Bulgurcu, B.; Cavusoglu, H.; Benbasat, I.I. Quarterly special issue information security policy compliance: An empirical study of rationality-based beliefs and information security awareness. *MIS Q.* **2010**, *34*, 523–548. [CrossRef]
13. Thomson, K.L.; Von Solms, R.; Louw, L. Cultivating an organizational information security culture. *Comput. Fraud Secur.* **2006**, *10*, 7–11. [CrossRef]
14. Gordon, L.A.; Loeb, M.P. *Managing Cybersecurity Resources: A Cost-Benefit Analysis*; McGraw-Hill: New York, NY, USA, 2005.
15. Orehek, Š.; Petrič, G. A Systematic Review of Scales for Measuring Information Security Culture. *Inf. Comput. Secur.* **2020**, *29*, 133–158. [CrossRef]
16. Tsohou, A.; Karyda, M.; Kokolakis, S. Analyzing the role of cognitive and cultural biases in the internalization of information security policies: Recommendations for information security awareness programs. *Comput. Secur.* **2015**, *52*, 128–141. [CrossRef]
17. Singh, N.; Gupta, A.M.; Ojha, A. Identifying factors of organizational information security management. *J. Enterp. Inf. Manag.* **2014**, *27*, 644–667. [CrossRef]
18. Kontakte Klaus Julisch Managing Partner. Cybercrime—The Risks of Working from Home. Deloitte Switzerland. 2020. Available online: https://www2.deloitte.com/ch/en/pages/risk/articles/covid-19-cyber-crime-working-from-home.html (accessed on 16 March 2022).
19. *Information Security Behaviour Index. Perform Green*; Perform Green Limited: Cheltenham, UK, 2019.
20. Verizon. Data Breach Investigations Report. 2009. Available online: http://www.verizonbusiness.com/resources/security/reports/2009databreachrp.pdf (accessed on 7 June 2022).
21. Andric, M. Fighting the enemy within. *IT WEB Spec. Rep.* **2007**, *95*, 54.
22. Furnell, S.M.; Jusoh, A.; Katsabas, D. The challenges of understanding and using security: A survey of end-users. *Comput. Secur.* **2006**, *25*, 27–35. [CrossRef]
23. Walton, C.B.R.; Limited, W.-M. Balancing the insider and outsider threat. *Comput. Fraud. Secur.* **2006**, *11*, 8–11. [CrossRef]
24. Albrechtsen, E. A qualitative study of users' views on information security. *Comput. Secur.* **2007**, *26*, 276–289. [CrossRef]
25. Kraemer, S.; Carayon, P. Human errors and violations in computer and information security: The viewpoint of network administrators and security specialists. *Appl. Ergon.* **2007**, *38*, 143–154. [CrossRef]
26. Stanton, J.M. Analysis of end user security behaviours. *Comput. Secur.* **2005**, *24*, 124–133. [CrossRef]
27. *COBIT Security Baseline—An Information Security Survival Kit*; IT Governance Institute: Schaumburg, IL, USA, 2004.

28. Furnell, S.; Navin Shah, J. Home working and cyber security—An outbreak of unpreparedness? *Comput. Fraud Secur.* **2020**, *2020*, 6–12. [CrossRef]
29. Mentsiev, A.U. The impact of digital technology on the study of languages and the development of digital education. *J. Physics. Conf. Ser.* **2019**, *1399*, 033085. [CrossRef]
30. Mentsiev, A.U.; Chebieva, H.S. Modern internet security threats and countermeasures (overview). *Eng. Her. Don.* **2019**, *3*, 15.
31. Pritam, S.N.; Vineeta, N.; Akhilesh, C.P. Impact of Information Technology on Learning, Teaching and Human Resource Management in Educational Sector. *Int. J. Comput. Sci. Telecommun.* **2011**, *2*, 66–72.
32. Boss, S.R. If someone is watching, I'll do what i'm asked: Mandatoriness, control, and information security. *Eur. J. Inf. Syst. Off. J. Oper. Res. Soc.* **2009**, *18*, 151–164. [CrossRef]
33. Burkell, J.A.; Fortier, A.; Di Valentino, L.; Roberts, S. Enhancing key digital literacy skills: Information privacy, information security, and copyright/Intellectual Property. *FIMS Publ.* **2015**, *35*, 67.
34. Ahsan, M.H.; Ayub, N.; Azman, N.S. Digital literacy in Malaysia: A systematic literature review on digital literacy in Malaysia: A systematic literature review on methodological approaches. *Malays. J. Qual. Res.* **2021**, *7*, 125.
35. Nasir, A. Information security culture model for Malaysian organizations: A review. *Int. J. Adv. Trends Comput. Sci. Eng.* **2020**, *9*, 117–121. [CrossRef]
36. Ng, W. Can we teach digital natives digital literacy? *Comput. Educ.* **2012**, *59*, 1065–1078. [CrossRef]
37. Calvani, A.; Cartelli, A.; Fini, A.; Ranieri, M. Models and Instruments for Assessing Digital Competence at School. *J. E-Learn. Knowl. Soc.* **2008**, *4*, 183–193.
38. Da Veiga, A.; Martins, N. Improving the information security culture through monitoring and implementation actions illustrated through a case study. *Comput. Secur.* **2015**, *49*, 162–176. [CrossRef]
39. Martin, A.; Grudziecki, J. *DigEuLit*: Concepts and tools for digital literacy development. *Innov. Teach. Learn. Inf. Comput. Sci.* **2006**, *5*, 249–267. [CrossRef]
40. Lion, R.; Meertens, R.M. Security or opportunity: The influence of risk-taking tendency on risk information preference. *J. Risk Res.* **2005**, *8*, 283–294. [CrossRef]
41. Parsons, K.; Mccormac, A.; Butavicius, M.; Ferguson, L. Human Factors and Information Security: Individual, Culture and Security Environment. Science and Technology, (DSTO-TR-2484). 2010. Available online: http://www.dtic.mil/dtic/tr/fulltext/u2/a535944.pdf (accessed on 7 June 2022).
42. van Laar, E. The relation between 21st-century skills and digital skills: A systematic literature review. *Comput. Hum. Behav.* **2017**, *72*, 577–588. [CrossRef]
43. Iordache, C.; Mariën, I.; Baelden, D. Developing digital skills and competences: A quick-scan analysis of 13 digital literacy models. *Ital. J. Sociol. Educ.* **2017**, *9*, 6–30.
44. Noh, Y. A study on the effect of digital literacy on information use behavior. *J. Librariansh. Inf. Sci.* **2017**, *49*, 26–56. [CrossRef]
45. Quaicoe, J.S.; Pata, K. The teachers' digital literacy: Determining digital divide in public basic schools in Ghana. *Commun. Comput. Inf. Sci* **2015**, *552*, 154–162. [CrossRef]
46. Arbanas, K.; Spremic, M.; Hrustek, N.Z. Holistic framework for evaluating and improving information security culture. *ASLIB J. Inf. Manag.* **2021**, *73*, 699–719. [CrossRef]
47. Ala-Mutka, K. *Mapping Digital Competence: Towards a Conceptual Understanding (Technical Note No." JRC67075-2011)*; European Commission Joint Research Centre: Seville, Spain, 2011.
48. Mioduser, D.; Nachmias, R. *WWW in Education'. Handbook on Information Technologies for Education and Training*; Springer: Berlin/Heidelberg, Germany, 2002; pp. 23–43.
49. Scardamalia, M.; Bereiter, C. Engaging students in a knowledge society. *Educ. Leadersh.* **1996**, *54*, 6–10.
50. Wallace, P. *The Psychology of the Internet*; University Press: Cambridge, UK, 1999.
51. Da Veiga, A.; Martins, N.; Eloff, J.H.P. Information security culture—Validation of an assessment instrument. *Afr. J.* **2007**, *11*, 147–166.
52. Gray, G.M.; David, P.R. Dealing with the dangers of fear: The role of risk communication. *Health Aff. (Proj. Hope)* **2002**, *21*, 106–116. [CrossRef]
53. Lawrence, P.S.; Deanna, D.C. Leveraging Behavioral Science to Mitigate Cyber Security Risk. *Comput. Secur.* **2012**, *31*, 597–611. [CrossRef]

proceedings

MDPI

Proceeding Paper

Review: Challenges and Obstacles of Trusted Elements for Mobile Health Records Management †

Fatin Nur Zulkipli [1,2,*], Nurussobah Hussin [1], Saiful Farik Mat Yatin [1] and Azman Ismail [3]

1 Faculty of Information Management, Universiti Teknologi Mara Selangor Branch, Shah Alam 40150, Malaysia
2 Faculty of Information Management, Universiti Teknologi Mara Kelantan Branch, Machang 18500, Malaysia
3 Institute of Marine Engineering Technology, Universiti Kuala Lumpur, Lumut 32200, Malaysia
* Correspondence: fatinnurzul@uitm.edu.my
† Presented at the International Academic Symposium of Social Science 2022, Kota Bharu, Malaysia, 3 July 2022.

Abstract: This study aims to identify the significant difficulties concerned with trusted elements of mobile health records management (MHRM) systems. For instance, data migration and lack of staff training have created huge challenges for organizations using such technologies. As such, this study presents potential improvement methods for electronic medical records (EMRs) by promoting the application of the latest technologies and relevant data security techniques. Secondary sources published between 2017 and 2020 were qualitatively reviewed to collect relevant information about EMR advancement. These data allowed reviewers to identify EMRs' reliability and challenges efficiently. In short, this paper identified four major challenges in records management systems: data migration, culture shock among staff, insufficient training, and data privacy. To counter these challenges, this review demonstrated that these difficulties can be effectively resolved through strategic inclusion and proper initiative of the latest innovative technologies.

Keywords: electronic health record challenges; personal health record; health information technology; medical record trustworthiness

Citation: Zulkipli, F.N.; Hussin, N.; Yatin, S.F.M.; Ismail, A. Review: Challenges and Obstacles of Trusted Elements for Mobile Health Records Management. *Proceedings* **2022**, *82*, 47. https://doi.org/10.3390/proceedings2022082047

Academic Editor: Mohamad Rahimi Mohamad Rosman

Published: 15 September 2022

1. Introduction

Advancement in the healthcare technology domain is generally known to improve patient outcomes and enhance healthcare services. The healthcare industry also helps reputable healthcare providers to offer high-quality services with great reliability to the public. Data play a significant role in ensuring that the available healthcare facilities can be improved to meet public needs and expectations. For instance, at the Ulsan University Hospital (UUH) in South Korea, a patient's medical records are transferred either via paper or phone when a patient is referred to another hospital [1]. Unfortunately, traditional paper and phone methods may lack reliability. Likewise, for patients who are sent to the emergency department, it can be difficult to know the patient's medical history, particularly in cases when the elderly cannot recall prior illnesses or the types of medication previously consumed. Therefore, the electronic medical records (EMR) management system plays a vital role in ensuring patients can be diagnosed accurately and treated without further delay. Hence, this paper aims to assess significant challenges and barriers related to EMRs using a mobile application platform.

Although hospitals were first hesitant to adopt the EMR approach in the early 21st century, the use of this approach has grown significantly since the year 2009. Published articles contributed to this trend by reporting the advantages of the EMR approach, such as big data analytics and the ability to incorporate the Internet of Things into a real-time mobile health records management (MHRM) system. As a result, adopting the EMR approach has granted healthcare practitioners easy access to their patients' data, which further enables them to quickly make decisions regarding their patients' wellbeing. For instance, a 3D print on a liver may be required for treatment planning if a cancer lesion is found at a complex location. With the help of EMR, clinical practitioners can swiftly review previous notes and any complex clinical issues previously addressed regardless of how dated or where those issues were first addressed. Additionally, instead of relying on a third-party vendor to develop the 3D print, hospitals may choose to establish an in-house 3D printing facility, which will enhance the turnaround time (TAT) and eventually improve patient management. Furthermore, implementation of a mobile approach is timely for those in a medical environment, as healthcare personnel require a system that allows them to be mobile and provide appropriate patient care. With MHRM, healthcare practitioners can access their patients' data quickly and with ease. Nevertheless, despite the benefits accrued from this platform, the adoption process does not take place without its own set of challenges and obstacles. Therefore, this review paper aims to identify the challenges and obstacles present in the use of the MHRM approach to help the healthcare industry implement the system effectively. Through an examination of the body of literature related to MHRM use, this review investigates the issues affecting the platform's implementation to answer the following questions: (i) What are the challenges of trusted elements affecting MHRM implementation? (ii) What are the obstacles of trusted elements affecting MHRM implementation?

2. Methods

In this review, the discussion will commence by first presenting the challenges of trusted elements in the MHRM system followed by the obstacles of trusted elements in the MHRM system. Then, the final bulk of this review article will focus on how the issues presented can be addressed to ease the implementation of the platform and data migration. During the research process, several keywords including medical records and challenges of MHRM were used for database search purposes; online databases (PubMed and Scopus) were used to obtain literature related to the topic. The studies reviewed were selected based on four pre-determined criteria: (i) empirical studies on medical records management; (ii) review studies analyzing medical data migration and adoption; (iii) studies on mobile health and health information technology; and (iv) studies investigating medical records' trustworthiness. A systematic search of research articles was conducted from the period of 2016 to 2020, and 35 articles were deemed relevant to the investigated topics. The review was conducted qualitatively based on issues highlighted in the selected articles. Later, the findings were categorized under two general themes: Challenges of Trusted Elements for MHRM, and Obstacles of Trusted Elements for MHRM. Tables 1 and 2 are discussed in depth in the following sections.

Table 1. Challenges of Trust Elements in the MHRM management system.

Theme	Types of Challenges	Explanation	Author(s)
Challenges	Data Migration	Migrating data to the mobile platform is costly and requires effort to ensure the migration process is successful	[2–4]
	Staff Resistance	Staff reluctance to accept or adopt the technology as part of their work routine	[5–9]
	Lack of Staff Training Sessions	Staff are not exposed to how the mobile platform works and how to troubleshoot technical disruptions	[10,11]
	Data Privacy	Concerns regarding the safety of patients' data after data are migrated to the mobile application	[12–15]

Table 2. Obstacles of Trust Elements in the MHRM management system.

Theme	Types of Obstacles	Explanation	Author(s)
Obstacles	Cost Structure	Organizations are deterred from adopting MHRM due to the high costs involved when selecting, implementing, and optimizing the system	[16]
	Technical Limitations	Absence of a reliable technology that supports the adoption process	[17–20]
	Standardization Limitations	Patients' data are not standardized and validated, which may lead to unethical use of patients' records and misdiagnoses	[21–23]
	Lack of Organizational Infrastructure	Mobile E-health record (EHR) implementation is impeded by the lack of healthcare infrastructure	[5,11,24–26]
	Poor Communication Tools	Doctors using mobile EHR think that the application is not user-friendly and limits their access to patients	[3,23]
	Changes in Workflow	Healthcare providers' workflows and efficacy may be affected, as tracking the patients' medical issues during digital consultations can be difficult, as data are captured digitally	[27–29]

3. Results and Discussion

The greatest challenge in implementing the MHRM system is trust. For the EMR management system to successfully convert into the mobile platform, a practical assessment must be conducted to identify any underlying drawbacks. This assessment is important to identify limitations of the healthcare system and function as a resource to inform the process of formulating mitigation strategies prior to large-scale integration of the system within the existing infrastructure [30]. According to [31], early identification of the issues can help reduce the outcomes of misdiagnosis that affect patients' treatment plans while safeguarding healthcare providers' reputations. The authors in [32] reported that a long-term study conducted in multiple countries including Malaysia revealed that 50% of physicians aged 30 to 45 years old display intentions to adopt the EMR system. In this context, well-known vendors offering quality EMR systems include Epic, Cerner, and Allscripts, and the availability of these databases, have been noted to encourage healthcare workers to provide their medical services to patients within a shorter period [33]. Based on the analysis, common challenges encountered by mobile EMR users are related to data migration, staff resistance, lack of staff training sessions, and data privacy. The findings were in line with [34], who posited that the underlying issues of EMRs are linked to its application using the mobile platform, which reduces the opportunities to apply appropriate security policies.

3.1. Challenges of Trusted Elements for Mobile Health Records Management

3.1.1. Data Migration

A mobile EMR interface has been proven to provide healthcare workers with numerous benefits that may also result in improved patient experiences when the system is properly optimized. Despite this, one of the biggest challenges for the implementation of this platform is clinical data migration [2]. Data migration refers to the process of selecting and transferring data from one storage device to another. Unfortunately, incorrect migration of clinical data can result in a huge impact on a patient's outcomes. Furthermore, errors that occur during data migration may lead to loss of access to the entire database and consequently will require additional time and cost to correct the migration error. When additional costs are incurred, management may find the process to be less affordable than expected. In addition, during data migration, the presence of large amounts of data within the mobile application might burden the backend servers, which eventually causes data logistics issues. Therefore, to prevent data logistics issues when migrating data from one platform to another (see Figure 1), the authors in [3] explained that the migrated data must be digitized to mirror a paper-based document. Unfortunately, digitizing data and creating online forms are time-consuming, and the authors in [4] reiterated that unintentional errors should be avoided when filling out online forms. Understandably, such carelessness can result in healthcare providers administering the wrong treatment.

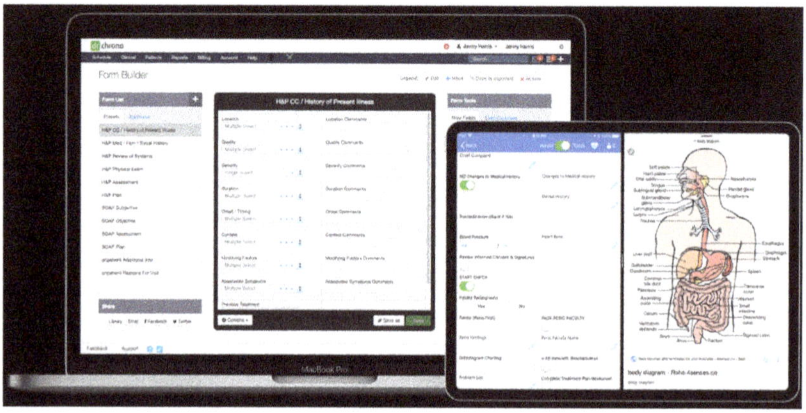

Figure 1. Electronic health record in mobile device [35].

3.1.2. Staff Resistance

Enjoying the benefits of the mobile EMR platform requires dedicated contributions from all levels of staff including clinical and non-clinical personnel. According to [5], healthcare staff are responsible for ensuring that patients' data are readily available, particularly for those who are awaiting treatment or medical assistance. However, the implementation of this platform does not always occur without resistance from healthcare staff because of the lack of knowledge or technical skills in personnel that allows them to use the mobile EMR system effectively. Notably, although the move to digitize medical records improves the healthcare industry, many healthcare workers are not ready to shift to a paperless format, as they doubt their ability to manage the technicalities of deploying the system on a mobile platform [6–8]. Furthermore, it has been noted that healthcare staff's refusal to use the EMR platform has led to incidences of miscommunication [9]. Thus, to minimize staff resistance towards the mobile EMR system, there must be effective change enacted from the management level. Additionally, practices supporting the conversion must be applied to begin platform implementation.

3.1.3. Lack of Staff Training Session

From an advancement perspective, healthcare providers are encouraged to adopt the EMR approach instead of the traditional system. However, a lack of relevant technical knowledge supporting the use of EMR is prevalent among healthcare workers and thus restricts them from being able to use the technology. Therefore, offering healthcare staff adequate and effective training sessions can improve their overall technical knowledge base. On the other hand, it must also be noted that Malaysia's healthcare industry lacks the required capacity to produce suitable training sessions for individuals, including the time, financial costs, and other resources required. As a result, the authors in [10] believe that the scarcity of available training programs in the country has discouraged the improvement of technical skills among healthcare staff that may further boost the healthcare technology domain. Considering techniques and strategies acquired from healthcare professionals' experiences may also be helpful to provide practical methods to reduce the system's underlying issues [11].

3.1.4. Data Privacy

With the implementation of the EMR mobile platform, it is possible to assume that all clinical and non-clinical personnel will be able to access patient information from anywhere via a mobile device. As such, this situation poses a challenge for data privacy. According to [12], the documentation process of patient healthcare data exposes sensitive information, and thus any mandatory divulging of patient data for treatment purposes require data confidentiality. In 2016, Tucker and colleagues reported that the application of reduction techniques has helped to reduce risks targeted at patient privacy [13]. Data privacy is particularly important, as [14] posited that the unintentional loss of confidentiality associated with patient data can negatively affect the healthcare organization's reputation in the market. Therefore, to prevent a breach in patient privacy, the authors in [15] suggested that protecting data with a suitable security protocol is essential.

3.2. Obstacles of Trusted Elements for Mobile Health Records Management

While there are no complete records management systems, there are a few features that can develop the reliability of the management system. Unfortunately, implementing such a system at a national level is not an easy task due to the following identified obstacles.

3.2.1. Cost Structure

The overall cost of implementing an EMR mobile platform is high because the implementation process usually involves several steps including hardware setting, training of staff, software cost and maintenance, and network fees. Furthermore, implementation costs may even be higher in the event of unfavorable situations occurring during the process of optimizing the system, which may subsequently affect the system's efficacy. In an average system, healthcare providers typically lean towards public company funding to implement the system. Unfortunately, a major challenge faced by the government is that it is typically unable to place a large sum of money into the electronic healthcare record system. For instance, the implementation of the EHR system offers transparency in healthcare charges. This has resultantly led to an increase in the prices of medical treatments. Thus, it can be said that the system is not helpful to average-income earners within the population to access the medical treatments that they need. These terms and conditions will undoubtedly create obstacles within the healthcare system and limit the support of the medical system due to the higher level of medical costs [16].

3.2.2. Technical Limitations

An optimized mobile EHR platform requires several specialized pieces of equipment to allow the system to function effectively within the healthcare industry. However, the authors in [17] have demonstrated that a lack of technical integration is present among the major barriers hindering the implementation of the mobile EHR system in healthcare facilities. In addition, mobile EHR requires artificial intelligence (AI) to ensure its cybersecurity, which can be extremely costly. Automation is also not included in mobile EHR, thus leading to data error and a lack of information in healthcare services. Therefore, managing the system by addressing its technical needs is imperative [18]. For example, mobile EHR depends on technological innovation for its fiber healthcare information recording system. However, the authors in [19] and [20] have proven that the lack of technical innovation makes such devices powerless in healthcare facilities, as many healthcare personnel are initially reluctant to use the mobile EHR system.

3.2.3. Poor Data Standardization

Another obstacle identified in this study is poor data standardization in the healthcare industry. Although mobile EHR follows cybersecurity and data protection protocols, the system is not inspected or authenticated by the central authority, thus potentially leading to misdiagnoses. In such circumstances, the authors in [21] stated that misdiagnosis could be life-threatening to patients. Furthermore, the data standardization process in the mobile EHR platform is complex and critical and requires good technical development knowledge [36]. However, there is a lack of competent and skillful staff and technological opportunities to carry out such a process. Due to these limitations, healthcare providers have failed to use the adopted technology correctly [22]. Medical records registered unethically have also been unable to serve as proper medical records for patients due to the risk of causing complications during medical treatments [23].

3.2.4. Lack of Organizational Infrastructure

The organizational culture in a hospital can also pose an obstacle to EHR implementation. The authors in [11] highlight that manually recording and maintaining patients' medical records can lead to poor documentation. Hence, technology is believed to be more effective to standardize information. Unfortunately, the current organizational infrastructure in hospitals is not suitable to implement the mobile recording system, as many individuals in the low-income community are not comfortable recording their health issues on the mobile EHR platform [5,24,25]. As such, healthcare providers, too, are not eager to record patients' data using mobile devices. To address this issue, the government must take proper initiatives to manage mobile EHR systems in all proposed healthcare services [26].

3.2.5. Poor Communication Tools

Many clinicians have highlighted that mobile EHR is not an effective communication tool to communicate with patients. Clinicians have found the platform to be challenging when attempting to assist patients as patients' medical histories are digitally recorded. Apart from the inconvenience caused by mobile EHR, authors in [23] reported that the mobile EHR platform has limited clinicians' ability to have physical access to patients. Records displayed on the system have also been found to be incorrect and inadequate at times. As a result, healthcare providers find it difficult to access patients' actual problems and health issues. Accessing patients' records is a process that requires highly designed communicative tools for both patients and healthcare providers [3], which may, unfortunately, pose a challenge for users.

3.2.6. Changes in Workflow

It is commonly highlighted that the mobile EHR platform can make significant workflow changes by reducing medical services' work requirements as consultation sessions are captured digitally. However, according to [27], this may also result in the possibility of losing data due to a lack of significant investment in training and operation. For instance, several doctors have reported that using a digital platform may reduce the number of patient visitations [28]. For this reason, the authors in [29] offer that this change may be challenging for healthcare providers who must identify patients' medical issues accurately. Moreover, it has also been argued that medical services become complicated when devices are used to control medical records.

4. Conclusions, Limitations, and Future Implications

The challenges faced in the implementation of the mobile EHR system are: (i) data migration; (ii) staff resistance; (iii) a lack of staff training, and (iv) data privacy. In terms of obstacles, the system is impeded by: (i) cost; (ii) technical limitations; (iii) poor data standardization; (iv) a lack of organizational infrastructure; (v) a lack of communication tools, and (vi) changes in workflow. Unfortunately, the reviewed literature does not provide enough information on the contents that should be present in an ideal mobile medical record system. There is also little mention of how these contents can influence healthcare behavior and perception. Based on this review, it is recommended that greater emphasis be placed on the importance of mobility in medical records management, so that it may be possible to understand how mobile health records can be successfully integrated into healthcare providers' medical record systems.

Author Contributions: Data curation: F.N.Z. and A.I.; Formal analysis: F.N.Z., N.H., S.F.M.Y. and A.I.; Funding acquisition: F.N.Z., N.H. and S.F.M.Y.; Investigation: F.N.Z. and A.I.; Methodology: F.N.Z. and N.H.; Project administration: F.N.Z.; Resources: F.N.Z. and A.I.; Software: F.N.Z. and A.I.; Supervision: N.H. and S.F.M.Y.; Validation: N.H. and S.F.M.Y.; Visualization: F.N.Z. and A.I.; Writing—original draft: F.N.Z.; Writing—review and editing: F.N.Z. and A.I. All authors have read and agreed to the published version of the manuscript.

Funding: This research received no external funding.

Institutional Review Board Statement: Not applicable.

Informed Consent Statement: Not applicable.

Data Availability Statement: Not applicable.

Acknowledgments: This paper was presented during the International Academic Symposium of Social Science 2022, organized by Universiti Teknologi MARA (UiTM) Kelantan, Malaysia.

Conflicts of Interest: The authors declare no conflict of interest.

References

1. Choi, W.; Park, M.; Hong, E.; Kim, S.; Ahn, R.; Hong, J.; Yeo, S. Early experiences with mobile electronic health records application in a tertiary hospital in Korea. *Healthc. Inform. Res.* **2015**, *21*, 292–298. [CrossRef] [PubMed]
2. Dornan, L.; Pinyopornpanish, K.; Jiraporncharoen, W.; Hashmi, A.; Dejkriengkraikul, N.; Angkurawaranon, C. Utilisation of electronic health records for public health in Asia: A review of success factors and potential challenges. *BioMed Res. Int.* **2019**, *2019*, 7341841. [CrossRef] [PubMed]
3. Lee, E. Taming performance variability of healthcare data service frameworks with proactive and coarse-grained memory cleaning. *Int. J. Environ. Res. Public Health* **2019**, *16*, 3096. [CrossRef] [PubMed]
4. McCracken, S.S.; Edwards, J.S. Implementing a knowledge management system within an NHS hospital: A case study exploring the roll-out of an electronic patient record (EPR). *Knowl. Manag. Res. Pract.* **2017**, *15*, 1–11. [CrossRef]
5. Ganiga, R.; Pai, M.M.M.; Sinha, R.K. Security framework for cloud-based Electronic Health Record (EHR) system. *Int. J. Electr. Comput. Eng.* **2020**, *10*, 455–466. [CrossRef]
6. Xia, Q.; Sifah, E.B.; Smahi, A.; Amofa, S.; Zhang, X. BBDS: Blockchain-based data sharing for electronic medical records in cloud environments. *Information* **2017**, *8*, 44. [CrossRef]
7. Asiri, A.; Obeidat, Q.; Alenezi, M.A. Cloud-based cross-enterprise imaging framework. *i-Manag. J. Softw. Eng.* **2017**, *12*, 6.
8. Drigas, A.; Dede, D.E.; Dedes, S. Mobile and other applications for mental imagery to improve learning disabilities and mental health. *Int. J. Comput. Sci. Issues (IJCSI)* **2020**, *17*, 18–23.
9. Liu, L.; Chen, W.; Nie, M.; Zhang, F.; Wang, Y.; He, A.; Wang, X.; Yan, G. iMAGE cloud: Medical image processing as a service for regional healthcare in a hybrid cloud environment. *Environ. Health Prev. Med.* **2016**, *21*, 563–571. [CrossRef]
10. Evans, R.S. Electronic health records: Then, now, and in the future. *Yearb. Med. Inform.* **2016**, S48–S61. [CrossRef]
11. Epstein, R.H.; Dexter, F.; Schwenk, E.S. Provider access to legacy electronic anesthesia records following implementation of an electronic health record system. *J. Med. Syst.* **2019**, *43*, 1–8. [CrossRef] [PubMed]
12. Roehrs, A.; da Costa, C.A.; da Rosa Righi, R.; da Silva, V.F.; Goldim, J.R.; Schmidt, D.C. Analysing the performance of a blockchain-based personal health record implementation. *J. Biomed. Inform.* **2019**, *92*, 103140. [CrossRef] [PubMed]
13. Helmers, R.; Doebbeling, B.N.; Kaufman, D.; Grando, A.; Poterack, K.; Furniss, S.; Burton, M.; Miksch, T. Mayo Clinic Registry of Operational Tasks (ROOT): A Paradigm Shift in Electronic Health Record Implementation Evaluation. *Mayo Clin. Proc. Innov. Qual. Outcomes* **2019**, *3*, 319–326. [CrossRef] [PubMed]
14. Choi, E.; Xiao, C.; Stewart, W.F.; Sun, J. Mime: Multilevel medical embedding of electronic health records for predictive healthcare. In Proceedings of the 32nd Conference on Neural Information Processing Systems, Montreal, QC, Canada, 3–8 December 2018; pp. 1–11.
15. Mukamba, N.; Beres, L.K.; Mwamba, C.; Law, J.W.; Topp, S.M.; Simbeza, S.; Sikombe, K.; Padian, N.; Holmes, C.B.; Geng, E.H.; et al. How might improved estimates of HIV programme outcomes influence practice? A formative study of evidence, dissemination and response. *Health Res. Policy Syst.* **2020**, *18*, 121. [CrossRef]

16. Hussien, H.M.; Yasin, S.M.; Udzir, S.N.I.; Zaidan, A.A.; Zaidan, B.B. A systematic review for enabling of developing a blockchain technology in healthcare application: Taxonomy, substantially analysis, motivations, challenges, recommendations and future direction. *J. Med. Syst.* **2019**, *43*, 320. [CrossRef] [PubMed]
17. French-Baidoo, R.; Asamoah, D.; Opoku Oppong, S. Achieving confidentiality in electronic health records using cloud systems, I.J. *Comput. Netw. Inf. Secur.* **2018**, *1*, 18–25.
18. Simpao, A.F.; Tan, J.M.; Lingappan, A.M.; Gálvez, J.A.; Morgan, S.E.; Krall, M.A. A systematic review of near real-time and point-of-care clinical decision support in anesthesia information management systems. *J. Clin. Monit. Comput.* **2017**, *31*, 885–894. [CrossRef]
19. Xu, J.; Williams-Livingston, A.; Gaglioti, A.; McAllister, C.; Rust, G. A practical risk stratification approach for implementing a primary care chronic disease management program in an underserved community. *J. Health Care Poor Underserved* **2018**, *29*, 202–213. [CrossRef]
20. Kruse, C.S.; Smith, B.; Vanderlinden, H.; Nealand, A. Security techniques for the electronic health records. *J. Med. Syst.* **2017**, *41*, 1–9. [CrossRef]
21. Barak-Corren, Y.; Castro, V.M.; Javitt, S.; Hoffnagle, A.G.; Dai, Y.; Perlis, R.H.; Nock, M.K.; Smoller, J.W.; Reis, B.Y. Predicting suicidal behavior from longitudinal electronic health records. *Am. J. Psychiatry* **2017**, *174*, 154–162. [CrossRef]
22. Cowie, M.R.; Blomster, J.I.; Curtis, L.H.; Duclaux, S.; Ford, I.; Fritz, F.; Goldman, S.; Janmohamed, S.; Kreuzer, J.; Leenay, M.; et al. Electronic health records to facilitate clinical research. *Clin. Res. Cardiol.* **2017**, *106*, 1–9. [CrossRef] [PubMed]
23. Kruse, C.S.; Stein, A.; Thomas, H.; Kaur, H. The use of electronic health records to support population health: A systematic review of the literature. *J. Med. Syst.* **2018**, *42*, 1–16. [CrossRef] [PubMed]
24. Ratwani, R.; Fairbanks, T.; Savage, E.; Adams, K.; Wittie, M.; Boone, E.; Gettinger, A. Mind the gap: A systematic review to identify usability and safety challenges and practices during electronic health record implementation. *Appl. Clin. Inform.* **2016**, *7*, 1069. [PubMed]
25. Larrison, C.R.; Xiang, X.; Gustafson, M.; Lardiere, M.R.; Jordan, N. Implementation of electronic health records among community mental health agencies. *J. Behav. Health Serv. Res.* **2018**, *45*, 133–142. [CrossRef]
26. Hassen, D.B. Mobile-aided diagnosis systems are the future of health care. *East. Mediterr. Health J.* **2020**, *26*, 135–1140. [CrossRef]
27. Mason, P.; Mayer, R.; Chien, W.W.; Monestime, J.P. Overcoming barriers to implementing electronic health records in rural primary care clinics. *Qual. Rep.* **2017**, *22*, 2943–2955. [CrossRef]
28. Parah, S.A.; Sheikh, J.A.; Ahad, F.; Loan, N.A.; Bhat, G.M. Information hiding in medical images: A robust medical image watermarking system for E-healthcare. *Multimed. Tools Appl.* **2017**, *76*, 10599–10633. [CrossRef]
29. López-Martínez, F.; Núñez-Valdez, E.R.; García-Díaz, V.; Bursac, Z. A case study for a big data and machine learning platform to improve medical decision support in population health management. *Algorithms* **2020**, *13*, 102. [CrossRef]
30. Alhajeri, M.; Shah, S.G.S. Limitations in and solutions for improving the functionality of picture archiving and communication system: An exploratory study of PACS professionals' perspectives. *J. Digit. Imaging* **2019**, *32*, 54–67. [CrossRef]
31. Choi, E.; Xu, Z.; Li, Y.; Dusenberry, M.; Flores, G.; Xue, E.; Dai, A. Learning the graphical structure of electronic health records with graph convolutional transformer. In Proceedings of the AAAI Conference on Artificial Intelligence, New York, NY, USA, 7–12 February 2020; Volume 34, pp. 606–613.
32. Aldosari, B. Causes of EHR projects stalling or failing: A study of EHR projects in Saudi Arabia. *Comput. Biol. Med.* **2017**, *91*, 372–381. [CrossRef]
33. Anastas, T.; Waddell, E.N.; Howk, S.; Remiker, M.; Horton-Dunbar, G.; Fagnan, L.J. Building behavioral health homes: Clinician and staff perspectives on creating integrated care teams. *J. Behav. Health Serv. Res.* **2019**, *46*, 475–486. [CrossRef] [PubMed]
34. Crowley, K.; Mishra, A.; Cruz-Cano, R.; Gold, R.; Kleinman, D.; Agarwal, R. Electronic health record implementation findings at a large, suburban health and human services department. *J. Public Health Manag. Pract.* **2019**, *25*, E11. [CrossRef]
35. DrChrono. The EMR of the Future. Less Screen-Time. More Face-Time. Available online: https://www.drchrono.com/electronic-health-record-ehr/ (accessed on 14 September 2022).
36. Xiao, C.; Choi, E.; Sun, J. Opportunities and challenges in developing deep learning models using electronic health records data: A systematic review. *J. Am. Med. Inform. Assoc.* **2018**, *25*, 1419–1428. [CrossRef] [PubMed]

proceedings

MDPI

Proceeding Paper

The Relationship between Library Technology, Support, Environment, and Postgraduate Students' Utilization of Web-Based Library and Information Services in Malaysian Academic Libraries †

Husain Hashim *, **Shamila Mohamed Shuhidan**, **Norizan Anwar** and **Mohd Nizam Yunus**

Faculty of Information Management, Universiti Teknologi MARA, Shah Alam 40150, Malaysia
* Correspondence: husainha@uitm.edu.my

† Presented at the International Academic Symposium of Social Science 2022, Kota Bharu, Malaysia, 3 July 2022.

Abstract: This research examines the utilization of Web-Based Library and Information Services (WBLIS) in academic libraries. Digital technology promotes the use of WBLIS, including during the COVID-19 pandemic. Few Malaysian studies have investigated utilization and its factors. Three factors of WBLIS utilization were identified: library technology, support, and the environment. WBLIS's output and outcome were emphasized. A conceptual model was developed and tested using non-probability sampling. A 38-item, five-point Likert Scale online survey was distributed to postgraduates from 20 public universities. Raosoft sampled 383 research, comprehensive, and focused universities using stratified sampling. SMARTPLS version 3 was used to test hypotheses on 527 respondents. Harmon's Single Factor test eliminated single-source bias. All measurement and model criteria were met. All hypotheses on the relationships between library technology, support, and environment on WBLIS utilization were supported. The findings will contribute to academic librarianship and related fields. Malaysian universities and the Ministry of Higher Education will benefit from improving academic libraries' impact on learning, research, and universities' institutional value. Future research may include private university, polytechnic, and community college students and academicians. Comparative studies and qualitative research can be conducted.

Keywords: web-based library and information services; academic library; utilization; output; outcome; library technology; library support; library environment

Citation: Hashim, H.; Shuhidan, S.M.; Anwar, N.; Yunus, M.N. The Relationship between Library Technology, Support, Environment, and Postgraduate Students' Utilization of Web-Based Library and Information Services in Malaysian Academic Libraries. *Proceedings* **2022**, *82*, 65. https://doi.org/10.3390/proceedings2022082065

Academic Editors: Mohamad Rahimi Mohamad Rosman

Published: 20 September 2022

1. Introduction

For centuries, academic libraries have been one of the main catalysts that play critical roles in providing university students with access to diverse information, knowledge, and research resources. Despite their long-standing establishment and tradition, academic libraries' existence was originally merely established to serve their parent institutions [1]. This is still the same aspiration behind their presence today, with even more pressure to prove their impact, strategic value, and visibility as well as to adhere to the need to ensure all library investments are justified [2–4]. In the Malaysian higher education system, academic libraries are set to uphold their responsibilities with regard to the provision of effective library services and resources that support teaching, learning, and research goals aligned with the quality expectations of all academic programs designed and offered at both the undergraduate and postgraduate levels. The Malaysian Qualifications Agency (MQA) expresses that this is one of the requirements necessary if an academic program is to obtain and maintain accreditation.

It is necessary to emphasize the trend in the Malaysian educational system. Since the entire globe is being driven by the massive rise of digital technology, universities are being challenged to develop innovative technologies to enhance their academic communities' learning and research pursuits. It has become a top priority to educate and prepare Malaysian youths to make a meaningful contribution to society's ever-changing industries and needs. Malaysia's Minister of Higher Education, Datuk Seri Noraini Ahmad, wrote in a prominent daily local newspaper that the provision of high-quality education is essential if Malaysians of all ages and backgrounds are to be empowered to gain new knowledge and improve their skills through access to a variety of online learning and educational resources [5]. Indeed, this is where Malaysian academic libraries can play their role in shaping their services to adapt to the emerging and increasing information wants, needs, and behaviors of university library users. In this regard, it can be said that academic libraries in Malaysia are well-equipped with Internet-driven and Web-based library and information services (WBLIS), giving their users better access, control, and flexibility to use the educational and research resources available and accessible online. Since this paper focuses on postgraduate students at Malaysian public universities as a specific type of library users, it seeks to examine the relationship between relevant factors that influence their utilization of WBLIS, as will be described in further detail below.

2. Problem Statement

In this study, the evaluation of the factors that influence the utilization of WBLIS in higher learning institutions reveals important gaps, including in knowing the frequency of use, as well as the learning and research outcomes achieved by students. In Malaysia, many studies of academic libraries focus on surveys to determine user satisfaction and perceived service quality. At the Universiti Malaysia Pahang (UMP), [6] students' perceptions and levels of satisfaction with service quality were investigated. [7] Similar research conducted at the Universiti Malaya (UM) library on service quality and customer satisfaction focused on academics, using SERVQUAL dimensions. Though the respondents in [7] stated that library services had a positive impact on their teaching, learning, and research, there is no further explanation given with regard to the outcome of WBLIS in terms of changes in their skills, competencies, and behaviors [8] and the extent to which WBLIS, for example, has stimulated their new thinking [9]. Furthermore, there is also a gap found in Malaysia in terms of studies that give emphasis to the frequency of use, which according to [2,3,10] has the importance of implying libraries' usage from the output perspective. Therefore, it becomes an interest of this present research to examine the context of the utilization of WBLIS from two perspectives: output (implicit) and outcome (explicit). Regarding the factors that influence the utilization of WBLIS, what is most fundamental is to align them accordingly with students' learning and research needs, and with their decisions and choices in utilizing the services. As a result of the initial search and review of the literature to be discussed below, three factors were identified: library technology, support, and environment.

3. Literature Review

The literature pertaining to the utilization of WBLIS was searched, acquired, analyzed, and reviewed. Particular emphasis was placed on the academic libraries at the institutions of higher education, with a focus on the research and learning needs of students.

3.1. Utilization of WBLIS

This study focuses on the utilization of WBLIS, which are comprised of five service points: library websites, WebOPAC, online databases, digital reference services, and institutional repositories. As highlighted previously in the problem statement section, the utilization of WBLIS examined in this study focuses on output and outcome perspectives. The output dimension is concerned with the frequency of use of WBLIS as outlined by [3] and [10] and considers the following:

- Visiting the library website as the primary gateway to information;
- Finding/retrieving books or other materials via WebOPAC;
- Finding/retrieving/downloading articles from online databases;

- Sending inquiries via digital reference services;
- Finding/retrieving/downloading a university's intellectual materials via a library's institutional repository.

For the purpose of data collection, the source of data is a survey that asks respondents to indicate their self-reported frequency of use of every WBLIS service point, as suggested by [2]. In addition, the same survey it aims to collect data from respondents in relation to the outcome dimension, with reference to [9], believing that utilization of WBLIS generates results as follows:

- Inspires new ideas/thinking;
- Helps to justify focus of study;
- Contributes to improved results of study;
- Helps to enhance knowledge and research skills.

3.2. Library Technology

According to [11], users' continuous desire to use a digital library is correlated with the perceived usefulness and the convenience of use of the WBLIS, which includes WebOPAC and electronic resources. In focus group interviews conducted to study the digital library (DL) system in the context of military education, the majority of respondents emphasized the need for providing sufficient information on each link to the library services on the DL system. Without enough description of the links to WebOPAC and electronic resources, for example, the users might not know the purpose of the services and would waste time navigating pages to access the needed content. This finding is in line with [12], revealing that if users found such services to be difficult to use and not user-friendly, they may turn their attention to finding alternatives. Scholars in [13] highlight postgraduate students' preferences for utilizing different databases and sources of information obtained in libraries or from other platforms. In fact, as per their research, postgraduate students prefer and use Google Search, Google Scholar, Wikipedia, Yahoo, and Amazon (weekly, 57.9%) above the library's electronic databases such as Emerald (weekly, 26.7%) and Ebsco (monthly, 14.3%). The reasons for this were that the students viewed Google Search, Google Scholar, and other search engines as being more accessible, less limited, and easier to access, use, explore, learn, and understand, as compared to subscribed library databases. Additionally, accessibility is a factor that may affect users' desire to utilize information resources available online or via mobile platforms [14]. Mobile internet was rated highly by 88% of 227 students because it is available from anywhere, at any time, and is more convenient to use, increasing access to library resources and services.

3.3. Library Support

As demonstrated in a survey at La Trobe University in Australia, 97% of respondents stated the academic library helped their learning and research success, and 90% claimed that being in the library connected them to the university and university life [2]. Meaningfully, the study disclosed that 69% of students with a self-reported A or B grade average utilized the online library every day or multiple times each week. In a more recent study, [15] presented findings on users' perceptions of library facilities at the University of Cape Coast. Due to their intensive research, postgraduate students were found to use the internet facility at the university library more than undergraduate students. This is also connected to the fact that their engagement in research requires considerable use of electronic resources. However, the participating postgraduate students all agreed that the academic library did not provide these resources in sufficient quantity. As a result, while the survey found that 86% of postgraduate students were aware of the library's internet and electronic resources, only 66% truly used the library's electronic services. This reveals how important library support and facilities are in shaping how students think their information needs are met and how much their positive or negative experiences would influence how much they utilize the library services. [16] found that at Sheffield University, students had misconceptions about academic librarians' functions. As the data revealed, out of 237 respondents, 146 and 135 believed librarians' work related to book shelving and reservations, while only 38 and 35 recognized librarians' functions in providing subject-specific research assistance and teaching them information and research skills. This failure to understand the functions and expertise of academic librarians has caused the library support provided to be poorly utilized.

3.4. Library Environment

Academic libraries must have both physical and virtual presences to adapt to the evolving learning and research information system defined by online access to digital and global materials. As stated by [17], the existence of academic libraries at universities should not be limited by time or geography. One of the most important responsibilities of academic librarians, according to [18], is to adapt library services to students' shifting learning patterns. In this regard, [18] also underlined the necessity of analyzing and collecting user feedback on library spaces. [19] revealed their findings that using academic libraries as an area to study alone, in groups, or with technology has proven to be beneficial to first-year students. However, as students progress to the upper level of study, they will find that information resources become increasingly important, with less emphasis on library space. Regarding postgraduate students, [20] disclosed that whether enrolled in a taught or research program, these students have a specialized preference in terms of space use, placing the silent study room above the group space and communal spaces as their top priority. From [21]'s qualitative research on academic library users' preferences for virtual reference services, it was revealed that the aspects of personalness and informality are important to users. It is concerned with the convenience and synchronicity of communication methods with the librarians; for example, it considers convenient methods such as live chat provided by the virtual platform that influences them to utilize virtual reference services.

3.5. Conceptual Research Framework

Based on the research problems and the review of the literature, a conceptual research framework is developed and assessed to examine the relationship between three factors: library technology, support, and environment (known as TSE factors), as well as the utilization of WBLIS (known as WBLISU) in academic libraries that focus on postgraduate students in Malaysia. The conceptual research framework is shown in Figure 1.

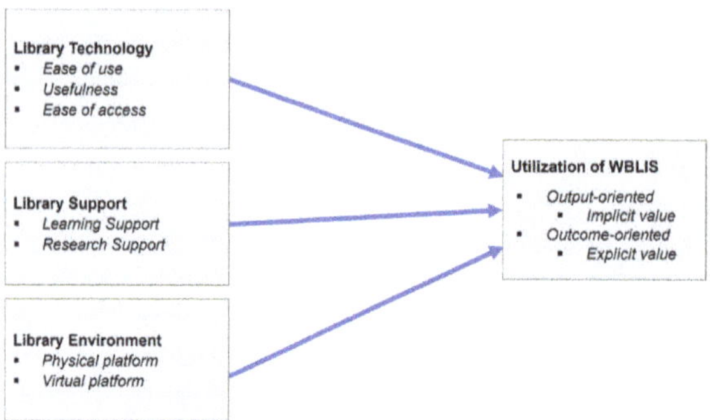

Figure 1. Conceptual research framework.

Based on the framework, the purpose of this study is to examine the relationship between TSE factors as the independent variables (IV) and the utilization of WBLIS as the dependent variable (DV). Each variable has its dimensions as depicted in Figure 1. In line with the framework, three hypotheses have been developed, as stated below:

- *H1: Library technology has a significant relationship with the utilization of WBLIS.*
- *H2: Library support has a significant relationship with the utilization of WBLIS.*
- *H3: Library environment has a significant relationship with the utilization of WBLIS.*

4. Methodology

Using the technique of non-probability sampling, a research framework was tested with postgraduate students from 20 public institutions. Prior to the actual data collection, the instruments' validity and reliability were determined by preliminary testing. Experts in the field of Library and Information Management provided reviews and feedback, which were followed by a pilot study involving 49 postgraduate students from public universities in Malaysia. The goal of the pretesting in this study is to assess the effectiveness of the questionnaire so that potential respondents can understand all items, and to offer the most direct evidence for the validity of the questionnaire data for each item. For the actual data collection, the sample size of 383 was determined by the Raosoft calculator, and stratified sampling was utilized to represent three university clusters: research, comprehensive, and targeted universities. Using a five-point Likert Scale with 38 distinct items, a web-based questionnaire was developed. To evaluate the hypotheses, data from 527 respondents were gathered and analyzed using SMARTPLS version 3 software. In this investigation, the single factor test of Harmon was employed to remove the bias produced by data collection from a single source. In this investigation, all measurement model and structural model requirements were satisfied. Finally, an examination of the structural model was conducted in order to evaluate the hypothesized relationships between variables.

5. Findings

This section describes the demographic profile of respondents who have participated in this study. Next, it discusses the reliability results, and finally, the related findings that are aligned with the research objective to examine the relationship between three factors, including library technology, support, and environment, as well as postgraduate students' utilization of WBLIS in academic libraries.

5.1. Demographic Profile

362 of the 527 respondents were female, while 165 were male, as shown by the respondents' demographic profiles. According to their composition by university clusters, 184 responses were collected from research universities, 112 questionnaires were gathered from focused universities, and 231 responses were obtained from comprehensive universities. Next, 344 respondents were enrolled in research-based programs, 141 respondents were enrolled in coursework-based programs, and 42 respondents were enrolled in mixed programs. The study received 275 responses from master's degree students. Additionally, 249 doctoral students but just 3 postgraduate diploma students participated in this investigation.

5.2. Common Method Bias

The Common Method Bias (CMB) test is required for this study to guarantee that the measurements are reliable and valid. To address concerns regarding how CMB may affect the study's results, Harman's one-factor test was used. In Harman's test, all items (measure latent variables) are consolidated into one common factor. In this study, all 38 items from all constructs under examination were analyzed and limited to a single factor. As a result, the single component amounted for only 33.327% of the overall variance, which is less than the benchmark value of 50%, indicating that common technique bias was unlikely to be a contaminant of the study.

5.3. Reliability Analysis

Cronbach's alpha coefficients were used to measure the reliability and internal consistency of the scales used in the study, as shown in Table 1. Cronbach's alpha values for all factors were more than 0.6, indicating that the reliability level was sufficient and acceptable (George and Mallery, 2003). The overall consistency, or Cronbach's Alpha values, of all 38 items for each dimension featured in the instrument ranged between 0.639 and 0.932. The usefulness dimension was discovered to be the most important, with a score of 0.932.

Research support obtained the lowest value of 0.639, and according to George and Mallery (2003), this dimension should be kept because dropping any of its elements would not significantly raise the value. This indicates that the overall index of the scale's internal consistency within the instrument is reliable, without unexpected abnormalities in the data.

Table 1. Reliability analysis results.

Variables		Number of Items	Cronbach's Alpha
Library Technology	Ease of Use	4	0.896
	Ease of Access	4	0.857
	Usefulness	4	0.932
Library Support	Learning Support	5	0.795
	Research Support	4	0.639
Library Environment	Physical Space	4	0.857
	Virtual Space	4	0.861
Utilization of WBLIS	Output	5	0.722
	Outcome	4	0.913

5.4. Findings: The Relationship between TSE Factors and the Utilization of WBLIS

The examination of structural models was conducted to evaluate the relationship between the structural models' importance and their relevance. Refs. [22,23] recommended the use of bootstrapping to identify the significance and relevance of structural model linkages. Table 2 displays the path coefficient values for the structural model. In this investigation, the T-value varies between 4.32 and 9.499. This shows a statistically significant connection (T-value greater than 1.645 and p-value below 0.05). In this study, the p-value is 0.00 for all variables, which is statistically significant. This is supported by Hair et al. (2017), who stated that a p-value of 0.05 is the minimum acceptable level for one-tailed and two-tailed tests.

Table 2. Path Coefficient of the Structural Model.

	T Values	p Values
Library Technology -> WBLISU	9.499	0.000
Library Support -> WBLISU	4.222	0.000
Library Environment -> WBLISU	4.321	0.000

Derived from the analysis results depicted in Table 2, below are the findings of the hypotheses testing shown in Table 3.

Table 3. The Findings of Hypotheses Testing.

Variables	Hypotheses	Findings
Utilization of WBLIS; Library Technology	H1: Library technology has a significant relationship with the utilization of WBLIS.	Supported (T–9.499, p–0.000)
Utilization of WBLIS; Library Support	H2: Library support has a significant relationship with the utilization of WBLIS.	Supported (T–4.222, p–0.000)
Utilization of WBLIS; Library Environment	H3: Library environment has a significant relationship with the utilization of WBLIS.	Supported (T–4.321, p–0.000)

6. Discussion and Conclusions

As this study demonstrates a significant correlation between library technology and the utilization of WBLIS among postgraduate students, this conclusion is congruent with those of previous studies. The authors of [24] determined in a study on the impact of undergraduate students' adoption and usage of mobile library applications in academic libraries at Joongbu University that perceived usefulness, interactivity, and ease of use of mobile apps have a significant impact on students' attitudes and intentions regarding adopting mobile applications. According to [25], the use of web library services by students is determined by their views on how they help to complete learning courses and tasks efficiently and on time. In addition, [12] noted that the lack of use of digital library resources to enhance teaching and learning among users was due to their perceived usefulness and usability in the context of Web resources. Regarding ease of access, it is necessary for academic libraries to take note of the various types of effort that users prefer to minimize and how it is possible to enable them to do so, with the ultimate goal of improving their information-seeking experience that will boost their desire to utilize WBLIS via online or mobile platforms [14].

The statistical results suggest that library support has a significant relationship with the utilization of WBLIS by postgraduate students. In this situation, library support involves encouraging postgraduate students to use WBLIS in their academic libraries in support of their learning and research activities. This finding is supported by previous studies. According to [26], the role and involvement of academic librarians in supporting the educational mission through their instructional efforts that result in student learning outcomes would increase the visibility and value of the academic library to the parent institution. On the contrary, as pointed out by [16], students' failure to comprehend and acknowledge the roles and skills of academic librarians is the root cause of their underutilization of library resources and services.

This study reveals that the library environment is a significant predictor of WBLIS utilization among postgraduate students. Academic library users should be able to utilize WBLIS and effectively communicate and complete their work in library environments that are regarded as including both physical and virtual spaces. According to [27], academic libraries must be always readily accessible and visible to their users in any location. In this context, to sustain their relevance in the rapidly developing learning and research information environment, which is defined by online access to digital and ubiquitous content, academic libraries must have a virtual presence. Meanwhile, the physical space remains important to students. The researchers of [20] disclosed that the academic library is perceived as primarily a place for learning and information-seeking by students, whether undergraduate or postgraduate.

Finally, this research is anticipated to offer substantial theoretical, methodological, and practical contributions to the field of academic librarianship and related fields of study. In addition, the management of Malaysian universities and the Ministry of Higher Education will be able to understand more about crucial aspects involved in enhancing the positive effects academic libraries have on students' learning and research, as well as the universities' institutional impact and value.

Author Contributions: Conceptualization, H.H., S.M.S. and N.A.; methodology, H.H., S.M.S. and N.A.; software, H.H.; validation, H.H., S.M.S., N.A. and M.N.Y.; formal analysis, H.H., S.M.S. and N.A.; investigation, H.H.; resources, H.H.; data curation, H.H.; writing—original draft preparation, H.H.; writing—review and editing, H.H., S.M.S. and N.A.; visualization, H.H. and M.N.Y.; supervision, S.M.S. and N.A.; project administration, H.H.; funding acquisition, H.H. All authors have read and agreed to the published version of the manuscript.

Funding: This paper was supported by Universiti Teknologi MARA, Malaysia (UiTM). 600-RMC/ LESTARI COVID/5/3 (015/2020).

Institutional Review Board Statement: Not applicable.

Informed Consent Statement: Not applicable.

Data Availability Statement: Not applicable.

Conflicts of Interest: The authors declare no conflict of interest.

References

1. Quinn, M.E. *Historical Dictionary of Librarianship*; Rowman & Littlefield: Playmouth, UK, 2014.
2. Salisbury, F.; Peasley, J. Measuring the academic library: Translating today's inputs and outputs into future impact and value. *Inf. Learn. Sci.* **2018**, *119*, 109–120. [CrossRef]
3. Oakleaf, M.; Whyte, A.; Lynema, E.; Brown, M. Academic libraries and institutional learning analytics: One path to integration. *J. Acad. Librariansh.* **2017**, *43*, 454–461. [CrossRef]
4. Montenegro, M.; Clasing, P.; Kelly, N.; Gonzalez, C.; Jara, M.; Alarcón, R.; Sandoval, A.; Saurina, E. library resources and students' learning outcomes: Do all the resources have the same impact on learning? *J. Acad. Librariansh.* **2016**, *42*, 551–556. [CrossRef]
5. Noraini, A. *Pushing Education for Sustainable Development to Greater Heights*; New Straits Times Press: Federal Territory of Kuala Lumpur, Malaysia, 4 January 2022.
6. Suziyana, M.D.; Mohd Yusof, T.; Nurhaizan, M.Z.; Fadzida, I. surveying users' perception of academic library services quality: A case study in Universiti Malaysia Pahang (UMP) Library. *J. Acad. Librariansh.* **2016**, *42*, 38–43.
7. Kiran, K. Service quality and customer satisfaction in academic libraries: Perspectives from a Malaysian University. *Libr. Rev.* **2010**, *59*, 261–273. [CrossRef]
8. Poll, R.; Payne, P. Impact measures for libraries and information services. *Libr. Hi Tech* **2006**, *24*, 547–562. [CrossRef]
9. Tenopir, C.; Christian, L.; Kaufman, J. Seeking, reading, and use of scholarly articles: An international study of perceptions and behavior of researchers. *Publications* **2019**, *7*, 18. [CrossRef]
10. Soria, K.M.; Fransen, J.; Nackerud, S. Stacks, serials, search engines, and students' success: First-year undergraduate students' library use, academic achievement, and retention. *J. Acad. Librariansh.* **2014**, *40*, 84–91. [CrossRef]
11. Abdul Rahman, A.R.; Mohezar, S. Ensuring continued use of a digital library: A qualitative approach. *Electron. Libr.* **2020**, *38*, 513–530. [CrossRef]
12. Matusiak, K.K. Perceptions of usability and usefulness of digital libraries. *J. Humanit. Arts Comput.* **2012**, *6*, 133–147. [CrossRef]
13. Ankrah, E.; Atuase, D. The use of electronic resources by postgraduate students of the University of Cape Coast. *Libr. Philos. Pract.* **2018**, *1632*, 1–37.
14. Chaputula, A.H.; Mutula, S. Factors impacting library-related uses of mobile phones by students in Public Universities in Malawi. *South Afr. J. Libr. Inf. Sci.* **2018**, *84*, 35–46. [CrossRef]
15. Parbie, S.K.; Phuti, R.M.S.; Barfi, K.A. Users' perception of library facilities: Evidence from the University of Cape Coast. *Libr. Philos. Pract.* **2021**, *4874*, 1–16.
16. Bickley, R.; Corrall, S. Student perceptions of staff in the information commons: A survey at the University of Sheffield. *Ref. Serv. Rev.* **2011**, *39*, 223–243. [CrossRef]
17. Beard, J.; Dale, P. library design, learning spaces and academic literacy. *New Libr. World* **2010**, *111*, 480–492. [CrossRef]
18. Montgomery, S.E. Library space assessment: User learning behaviors in the library. *J. Acad. Librariansh.* **2014**, *40*, 70–75. [CrossRef]
19. Stemmer, J.K.; Mahan, D.M. Investigating the relationship of library usage to student outcomes. *Coll. Res. Libr.* **2016**, *77*, 359–375. [CrossRef]
20. Kim, J.-A. User perception and use of the academic library: A correlation analysis. *J. Acad. Librariansh.* **2017**, *43*, 209–215. [CrossRef]
21. Mawhinney, T. User preferences related to virtual reference services in an academic library. *J. Acad. Librariansh.* **2020**, *46*, 1–18. [CrossRef]
22. Ramayah, T.; Cheah, J.; Chuah, F.; Ting, H.; Memon, M.A. *Partial Least Squares Structural Equation Modeling (PLS-SEM) Using SmartPLS 3.0: An Updated Guide and Practical Guide to Statistical Analysis*, 2nd ed.; Pearson: London, UK, 2018.
23. Hair Jr, J.F.; Hult, G.T.M.; Ringle, C.M.; Sarstedt, M. *A Primer on Partial Least Squares Structural Equation Modeling (PLS-SEM)*; Sage Publications: Thousand Oaks, CA, USA, 2014.
24. Yoon, H.Y. User acceptance of mobile library applications in academic libraries: An application of the technology acceptance model. *J. Acad. Librariansh.* **2016**, *42*, 687–693. [CrossRef]
25. Arif, M.; Ameen, K.; Rafiq, M. Factors affecting student use of web-based services: Application of UTAUT in the Pakistani Context. *Electron. Libr.* **2018**, *36*, 518–534. [CrossRef]
26. Blummer, B.; Kenton, J.M. Academic libraries and student learning outcomes. *Perform. Meas. Metr.* **2018**, *19*, 75–87. [CrossRef]
27. Maceviciute, E. Research libraries in a modern environment. *J. Doc.* **2014**, *70*, 282–302. [CrossRef]

proceedings

MDPI

Proceeding Paper

The Successful Elements on Knowledge Sharing Constructed Social Media among Academic Staff †

Nor Erlissa Abd Aziz [1,*], Siti Aishah Mokhtar [1], Noor Arina Md Arfin [1], Jashira Jamin [2], Wan Adibah Hanis Wan Aziz [3] and Nurulannisa Abdullah [1]

1 Faculty of Information Management, Universiti Teknologi MARA Kelantan Branch, Machang 18500, Malaysia
2 Faculty of Information Management, Universiti Teknologi MARA Rembau Branch, Rembau 71300, Malaysia
3 Faculty of Computer & Mathematical Sciences, Universiti Teknologi MARA Kelantan Branch, Machang 18500, Malaysia
* Correspondence: erlissa@uitm.edu.my
† Presented at the International Academic Symposium of Social Science 2022, Kota Bharu, Malaysia, 3 July 2022.

Abstract: This paper aims to explore influencing organizational elements that inspire social media based on knowledge sharing among academic staff in Malaysian Higher Education Institutions (HEIs). Nowadays, alongside the emergence of Information Communication and Technology (ICT), knowledge sharing could be considered essential in educational institutions, regarding their effectiveness and longer-term survival. Nonetheless, to certify the success of knowledge sharing within Higher Education Institutes (HEIs), there is a need to understand the role of knowledge-sharing activities in the life of these communities. Social media have emerged as essential forums for locating, connecting, collaborating, and sharing ideas among individuals. However, only limited research has addressed the role of social media in facilitating knowledge sharing between academic staff. This research adopted a qualitative approach to understanding the subject matter in depth. Data were gathered from thirty-two academic staff members from two public and two private universities. The findings resulted in the identification of seven key themes, namely: organizational structure, technology infrastructure, organizational strategy, organizational culture, management support, people and skills, which all need to be considered as influential elements for social media.

Keywords: knowledge sharing; social media; academic staff

check for updates

Citation: Aziz, N.E.A.; Mokhtar, S.A.; Arfin, N.A.M.; Jamin, J.; Aziz, W.A.H.W.; Abdullah, N. The Successful Elements on Knowledge Sharing Constructed Social Media among Academic Staff. *Proceedings* **2022**, *82*, 66. https://doi.org/10.3390/proceedings2022082066

Academic Editor: Mohamad Rahimi Mohamad Rosman

Published: 20 September 2022

1. Introduction

HEIs, by definition, are knowledge-concentrated environments, and their primary purpose is to generate and disseminate knowledge through teaching and learning. Due to the fact that knowledge is such a noteworthy asset, it is considered to be a valuable resource for organizations and individuals. Thus, knowledge could be considered a resource required by the management. Knowledge sharing is essential, as much precise work is carried out among teams comprising people who are not physically located in the same area and interact online [1].

Today, in many knowledge organizations, most work is largely unseen because it is difficult for us to obtain information about others and connections. HEIs are among those essential to supplying future generations with knowledge. HEIs are confronting several challenges and social and economic pressures which require developing strategies to respond to the current era's realities effectively. The uncertainty in knowledge work can hinder knowledge sharing, and thus may bring many adverse effects on the organization [2].

Current trends in HEIs encompass the expectation that academic staff members would be willing to move from knowledge hoarding toward knowledge sharing [3]. According to the authors, Ref. [4] stated that knowledge could be a basic asset for HEIs. The authors mentioned that knowledge sharing had become a part of many organizations, especially in HEIs where knowledge is created, obtained, and distributed. Furthermore, such sharing of knowledge between academic staff members necessarily includes the formation of interpersonal relations, as well as enhancing the education of future generations.

HEIs can apply social media to communicate and share knowledge or ideas among academic staff. One article [5] pointed out that social media could deliver a forum for learning and act as a virtual center for HEIs to encourage knowledge sharing and contribute towards research and teaching methods and a myriad of skills among academics for succeeding in educational competitiveness. Social media are considered critical aspects within the university to facilitate knowledge sharing and the performance of significant tasks to support daily communication [6]. However, knowledge-sharing difficulties while using social media in HEIs are inevitable [7].

In the Malaysian context, HEIs must adopt knowledge-sharing approaches using social media to improve knowledge management and performance and ensure long-term survival in a competitive environment. As suggested by [6–8], knowledge sharing could enhance organizational performance as a requirement for corporate competitiveness. Hence, the adoption of knowledge sharing could improve the performance of organizational competitiveness. The knowledge created by academics, embedded in their minds, is the institutional capital. Competitive advantage can be obtained if command of this knowledge can be shared with those who need it to derive benefits. Thus, Ref. [9] argues that social media and sharing of institutional knowledge by the staff will bring meaningful improvements to the institutions, with potential benefits to the broader HEIs, and hence the performance of educational institutions as a whole will be enhanced.

Yet, knowledge sharing employing using social media is still underdeveloped in HEIs in Malaysia. Most studies exploring knowledge sharing and social media are done abroad. Thus, it is important to study which strategies could be adopted to implement knowledge sharing in Malaysian HEIs. Numerous studies have explored these issues, but only in the broader context of knowledge sharing between employees of organizations. However, there is a paucity of comprehensive research on knowledge sharing between academic staff in HEIs. In this regard, Ref. [10] revealed that research to understand knowledge sharing in academic teams within HEIs is limited compared with other sectors.

Thus, this study focuses on social media and knowledge sharing, as well as investigating the influence of organizational elements that contribute to using social media for knowledge sharing. Apart from that, this research identified the knowledge and methods they shared within the knowledge-sharing between individuals, such as in an academic team or within broader institutions.

2. Literature Review

2.1. *Knowledge Sharing*

In generating new knowledge, knowledge sharing, whereby individuals share their knowledge, now plays a vital role within institutions [11]. Nevertheless, the shared understanding should be of such quality as to add value to the organization [11]. Knowledge sharing between individuals can be viewed as a form of social interaction in which individuals participate. However, if knowledge sharing effectively improves the quality of operations, informed decision-making, problem-solving, and the development of highly skilled employees must be effectively managed [8]. Thus, in recent years, knowledge has emerged as an essential dimension of strategic management, innovation, and information systems in most organizations [12].

2.2. *Organizational Elements Could Significantly Influence Knowledge Sharing*

Organizations could be viewed as systems of core, elaborating, independent and inconsistent elements and interconnections among all or part of these elements. The elements representing resources, activities, processes, networking technology and policies are essential for the viability of the organization [2]. Organizational elements could be defined as areas in which results, if they are satisfactory, will ensure successful competitive performance for the organization [12].

Previous research has led to the identification of several organizational elements considered to exert significant influences on knowledge sharing. Among these elements, there is a consensus among the authors that seven are especially important in terms of their influence on organizational culture in the context of knowledge sharing by using social media, namely: organizational culture, organizational structure, management support, technology infrastructure, skills, people and organizational strategy. Each of these is examined in the following sections. There is a general acceptance in the literature that knowledge sharing is essential to organizations.

2.3. Social Media in HEIs

Social media represent a second wave in web developments by offering tools and applications such as weblogs, wikis, Ajax, Really Simple Syndication (RSS) and tagging. However, some authors prefer to view media platforms as a dimension of online social interaction which also permits the creation and exchange of information. Other researchers approach social media from a technological perspective which can provide services such as wikis, blogs or weblogs really simple syndication (RSS), Ajax, instant messaging and podcasts. In the context of the current research, social media can be considered as authoring tools that are easy to use for people who do not have an information technology background but who can avail of its many facilities for sharing knowledge by means of creating weblogs, social networking applications and services such as Facebook and Myspace.

In addition, [5] draws attention to university websites which are mines of information and knowledge with lecturers' notes and PowerPoint presentations as well as students' contributions. These are accessible through permission from the college library, but they often have a section accessible to the wider public also. However, knowledge which has accumulated on those sites have to be managed as valuable assets, and institutions remain dubious about those sites opening up freely to the general public. Nevertheless, Rowley does envisage a widening of access, but more in the direction of explicit rather than tacit knowledge. In-house, however, such sharing may become more common, especially within departments, but again, will probably not embrace tacit knowledge. The advantage of technological approaches lies in its facility for systematizing and synchronizing information shared by academics.

3. Research Approach

This study used a qualitative approach to identify the organizational elements of knowledge sharing based on social media. Qualitative research methods have been used to understand the subject matter in depth. It is aligned epistemologically with an interpretive research approach. Interpretivist thinking considers understanding the "meaning" of social phenomena for its participants. Interpretivism also acknowledges human agency and reflexivity. The case study has been chosen which intends to identify and influence organizational elements of social media based on knowledge sharing. The results of this paper are based on interpreting the results of semi-structured interviews with academic staff in selected universities from Malaysian HEIs within the public and private sectors.

4. Findings

As is illustrated in Table 1 below, a total of 32 participants were willing to participate in the research study and, subsequently, 32 semi-structured interviews were conducted. The findings explain how knowledge sharing using social media can be applied, as conveyed in the case studies' data. In addition, the results also illustrated the influences of and how the influential organizational elements contribute to social media based on knowledge sharing in Malaysian HEIs.

Table 1. Summary of Participants and their positions selected from Case Study HEIs.

University	Number Management Team	Number of Senior Staff	Number of Junior Staff	Number of Participants
University A	2	3	4	9
University B	2	2	4	8
University C	2	3	3	8
University D	1	3	3	7
Sub Total	7	11	14	Final Total 32

Overall, seven elements were identified from the literature review data collected, specifically: organizational strategy, organizational structure, organizational infrastructure, organizational, organizational culture, management support, people, and skills. The codes are used to identify citations taken from interview transcripts. All interviews were conducted face to face. There were thirty-two participants (academic staff) involved in the interviews, and the interviewees were allocated letters according to their position in the university and ID number e.g., MT1, MT2, SS1, SS2, JS1, JS2) to maintain anonymity.

According to the interviews conducted with most of the participants, it was found that to share their knowledge with another person, a relationship must first be established between them. A participant took it one step further and pointed out that not only is the amount of knowledge sharing affected by the type of knowledge or the relationship but was also dependent upon the strength and quality of the relationship.

One of the participants' experiences advised that the communication approaches and knowledge sharing methods he used were based on the type of relationship he had with a person. In some cases, he might pick up the phone, he might send an email message, or he might get up out of his chair and physically approach the person. In some cases, he even travelled by airplane to either maintain an important relationship, repair a relationship or form a new relationship.

The participants agreed on the importance of skills and abilities that are built from the practices and experiences in the use of social media for knowledge sharing over time. In addition, social media could be a source that might lead to increased confidence. It could be used as an effective method to develop and maintain good communication for sharing knowledge. When asked about their views on why social media platforms are not popularly used by academics in MHEIs, the academic staff suggested that the use of social media platforms might increase their workload since they would have to re-design their activities to fit the use of social media, extend consultation hours beyond normal working time, cope with their own unfamiliarity with the social media tools and learn how to incorporate the social media platforms into their work.

They also saw the social media tool as an informal tool and a distraction. And, of course, they were concerned about the issue of privacy and security when social media platforms are used for academic purposes. For instance, one senior academic staff says: 'And then I just think, 'You know what? Let somebody else do it.' And I wouldn't have thought that in my academic careers because I still thought I could make a difference. It's time, I can't carry on struggling using this social media. I can't be bothered' (senior academic staff in management position).

5. Conclusions and Implications of the Study

There is increased responsiveness and identification of social media as an approach method for knowledge sharing in organizations. However, limited research has been conducted on social media applications for organizational knowledge sharing, especially in HEIs. Presently, Malaysian academics need encouragement and support to use social media that could benefit their work. Therefore, it is needed to encourage them with the influence elements to use social media. The research for this paper was conducted with academic staff, because of time constraints, but in the future, it might be worth considering conducting similar research with non-academic staff. As well, it could also be interesting to investigate the perception of social media based on knowledge sharing between academic and non-academic staff in private and public universities.

This study investigated the influence of organizational elements that contribute to social media-based knowledge sharing by Malaysian-based HEIs. It implies a need for HEIs to recognize that social media can be an effective medium for such knowledge sharing by academic staff, consequently leading to HEIs' identification of organizational elements that influence the use of social media. In seeking to determine the drivers and barriers to sustainable use, this research should interest practitioners and researchers undertaking similar projects.

Author Contributions: Thank you to all committed authors N.E.A.A., S.A.M., N.A.M.A., J.J., W.A.H.W.A. and N.A. who willingly dedicated their precious time in completing this paper. All authors have read and agreed to the published version of the manuscript.

Funding: This research received no external funding.

Institutional Review Board Statement: Not applicable.

Informed Consent Statement: Not applicable.

Data Availability Statement: Not applicable.

Acknowledgments: This research was partially supported by Universiti Teknologi MARA Kelantan Branch, Malaysia.

Conflicts of Interest: The authors declare no conflict of interest.

References

1. Ahmad, A.R.; Jameel, A.S.; Raewf, M. Impact of social networking and technology on knowledge sharing among undergraduate students. *Int. Bus. Educ. J.* **2021**, *14*, 1–16.
2. Zhao, Y.; Zhang, X.; Wang, J.; Zhang, K.; Ordóñez de Pablos, P. How do features of social media influence knowledge sharing? An ambient awareness perspective. *J. Knowl. Manag.* **2020**, *24*, 439–462. [CrossRef]
3. Fullwood, R.; Rowley, J.; Fullwood, R.; Rowley, J.; Delbridge, R. Knowledge sharing amongst academics in UK universities. *J. Knowl. Manag.* **2013**, *17*, 123–136. [CrossRef]
4. Kim, M.; Jun, M.; Han, J. The relationship between needs, motivations and information sharing behaviors on social media: Focus on the self-connection and social connection. *Asia Pac. J. Mark. Logist.* 2022, *ahead-of-print*. [CrossRef]
5. Fauzi, M.A.; Nya-Ling, C.T.; Thurasamy, R.; Ojo, A.O. An integrative model of knowledge sharing in Malaysian higher learning institute. *Kybernetes* **2018**, *47*, 1031–1052. [CrossRef]
6. Al-Rahmi, W.M.; Alias, N.; Othman, M.S.; Marin, V.I.; Tur, G. A model of factors affecting learning performance through the use of social media in Malaysian higher education. *Comput. Educ.* **2018**, *121*, 59–72. [CrossRef]
7. Ali, A.A.; Dominic, P. Organizational and individual factors impact on knowledge sharing practice: The association with cost reduction. In Proceedings of the 3rd International Conference on Computer and Information Sciences (ICCOINS), Kuala Lumpur, Malaysia, 15–17 August 2016.
8. Asunka, S. Use of social media for knowledge sharing by instructors in a higher education institution: An exploratory case study. In *Social Media for Knowledge Management Applications in Modern Organizations*; IGI Global: Hershey, PA, USA, 2018; pp. 116–143.
9. Baker, K.A. Organizational Culture. 2002. Available online: http://www.au.af.mil/au/awc/awcgate/doe/benchmark/11 (accessed on 24 August 2017).
10. Al-Kurdi, O.; El-Haddadeh, R.; Eldabi, T. Knowledge sharing in higher education institutions: A systematic review. *J. Enterp. Inf. Manag.* **2018**, *31*, 226–246. [CrossRef]
11. Nonaka, I.; Ichijo, K. Knowledge as a competitive advantage in the age of increasing globalization. In *Knowledge Creation & Management: New Challenges for Managers*; Oxford University Press: Cary, NC, USA, 2007.
12. Muda, M.N.; Yusof, Z.M. Information and communication technology (ICT) in knowledge sharing practices: Needs for the establishment of repository. *Mediterr. J. Soc. Sci.* **2015**, *6*, 69. [CrossRef]

proceedings

MDPI

Proceeding Paper

Validating the Instrument Measuring the Influence of Information Authenticity and Travel Selfies on Malaysian Online Destination Images †

Sathish Kumar Velayuthan [1,2,*] and Noor Hazarina Hashim [1]

1 Azman Hashim International Business School, Universiti Teknologi Malaysia, Johor Bahru 81310, Malaysia
2 Faculty of Hospitality, Tourism and Wellness, Universiti Malaysia Kelantan, Kota Bharu 16100, Malaysia
* Correspondence: sathishvelayuthan@gmail.com

† Presented at the International Academic Symposium of Social Science 2022, Kota Bharu, Malaysia, 3 July 2022.

Abstract: This paper aims to determine an instrument's validity in measuring the context of information authenticity that may influence online destination images through the eyes of selfie tourists. Selfie tourists have been reached via the snowball sampling technique through surveymonkey.com and the data were analysed using SmartPLS software version 3.3. The outcome favours the confirmatory factor analysis, convergent validity, discriminant validity, and the instrument's internal consistency analysis. This research is anticipated to cultivate literature on information authenticity, travel selfies, online destination images, and computer-mediated communication by providing a validated tool for future empirical research.

Keywords: validity; instrument; travel selfies; information authenticity; online destination image

Citation: Velayuthan, S.K.; Hashim, N.H. Validating the Instrument Measuring the Influence of Information Authenticity and Travel Selfies on Malaysian Online Destination Images. *Proceedings* **2022**, *82*, 81. https://doi.org/10.3390/proceedings2022082081

Academic Editor: Mohamad Rahimi Mohamad Rosman

Published: 26 September 2022

1. Introduction

Capturing travel selfies (*trafies*) and posting them on social media, more precisely, on Instagram, is a 21st-century trend amongst tourists worldwide. Tourists share their travel photos on social media as part of eWOM [1], and social media seems to be the leading platform accommodating selfies [2]. For example, #selfie, #travelselfies, and #travelphotos are popular hashtags actual travellers use when posting travel selfies on Instagram, capturing viewers' attention. Despite the fact that the popularity of travel selfies is familiarized entirely by the popular media as a medium for destination marketing, the quest to authenticate the information shared by tourists (actual tourists and paid reviewers) and destination hosts for public view on social media, and how it could impact the destination's image, remains under-researched topics. Understanding the context of information authenticity within the tourism communication perspective is indeed crucial in this digital age, to rival disinformation and misinformation spread through social media, as penned by Rubin [3]. With the assumption that information authenticity could positively impact the destination image of a particular tourism destination, the present study began to explore the influence of information authenticity on the formation of online destination images by incorporating the context of travel selfies.

The context of destination images has been growing since the 1970s. Now, after the rise of research focusing on antecedents of destination images, user-generated content, and how tourists reproduce destination images through social media posts (see, i.e., [1,4–7]), the context of destination images can be expected to evolve uninterruptedly. Scholars have mutually accepted destination images as "the sum of beliefs, ideas and impressions that a person has towards a destination" [8] (p. 19).

Accepting this definition fully, renowned scholars have introduced attributes of destination images, such as a cognitive image [8]; affective image [9]; cognitive, affective, and conative [10]; and holistic and attributive, functional and psychological, and common and unique [11]. The present study adapted three primary constructs from destination images: cognitive image, affective image, and overall image. Delving further into the context of information authenticity, few scholars have established their works within the communication realm (see, i.e., [12–16]). The present study precisely refers to information authenticity as the truthfulness of information and imageries shared by tourists via travel selfies for general view on social media. Aligned with the direction of the present study, the framework proposed by Gilpin et al. [13], which shapes authenticity in the social media context, seemed appropriate. Nevertheless, overall, a valid instrument that has proven to empirically depict the present topic remains scarce in the tourism literature. The present study strongly believes that a valid instrument is fundamental in gauging a particular study's objectives. Therefore, this paper aims to validate the questionnaire developed based on the instruments designed and used by Crompton [8], Baloglu and McCleary [17], Beerli and Martin [18], Echtner and Ritchie [19], Ekinci and Hosany [20], Rawlins [21], Gilpin et al. [13], Ponzi, Fombrun, and Gardberg, [22], Kaakinen et al. [23], Michael et al. [4], and Khan and Jan [24]. The present study considered using the constructs developed by these scholars and adapted them within the scope of the study by incorporating the context of online computer-mediated communication and travel selfies. The questionnaire designed for the present study portrays the influence of information authenticity and travel selfies on Malaysian online tourism destination images through the eyes of selfie tourists. It is anticipated to provide a comprehensive overview of computer-mediated communication that occurs in the online landscape within the Malaysian selfie tourism context.

2. Methodology

As part of the quantitative research, the present study began with questionnaire development. The items for the construct were established by adapting the scales developed by several renowned scholars within the area. The instrument was drafted and developed with eight sections: socio-demographic, information authority, author's identity, engagement, transparency, cognitive image, affective image, and overall destination image. Each item within the section depicts the objective to be measured within the scope of the study. The instrument was universally designed with several qualifying questions, open-ended questions, and exit points, as a measure to reach the targeted respondents. Table 1 illustrates the development of the instrument based on the sections, variables, and sources.

Table 1. Instrument Development.

Section	Variable	No of Items	Sources
A	Socio-Demographic	6	Self-constructed
B	Information Authority	11	[13,22,24]
C	Author's Identity	7	[13,23,24]
D	Engagement	7	[13,21]
E	Transparency	7	[13,21]
F	Cognitive Image	21	[4,8,17–20]
G	Affective Image	5	[17]
H	Overall Destination Image	6	[4,17]

Section A gathers information about the demographic profile of selfie tourists; several qualifying and disqualifying questions were also included in this section. Section B gathers information on the selfie tourist's trust towards the authors' expertise and credibility on social media. Section C depicts the selfies tourist's trust towards the authors' authentic identity. The engagement variable gathers information on the selfie tourist's trust of the interaction between the author and members in social media communication. Section E gathers information on the selfie tourist's trust towards the level of openness in social media communication. Section D, the cognitive image, gathers information about the selfie tourist's belief and factual knowledge regarding Malaysian tourism destinations' physical attributes as portraits in travel selfies on social media. Affective images gather information about the selfie tourist's inner or subjective feeling towards Malaysian tourism destinations based on travel selfies posted on social media. Finally, the overall destination image gathers information about the selfie tourist's behavioural intention based on the cognitive and affective impression made towards the tourism attraction through travel selfies. For all items in Sections B to H, a five-point Likert scale was used.

As a measure to ensure the dependability of the instrument drafted, validity tests were first conducted ahead of the pilot test. Ghauri and Gronhaug [25] declared that validity represents how accurately the instrument could engulf the objective of the study. Two types of validity test were covered to ensure the dependability of the instrument in the present study: content validation and construct validation. Content validity has been done through expert review by referring the instrument to several expert panels, varying from subject matter experts, methodology experts, and language experts, ahead of the pilot study. The experts referred to are academic and tourism industry panellists. Based on their constructive remarks, the instrument was revised and sent out for the pilot test. The pilot test was conducted via an online survey created through surveymonkey.com. A total of 100 selfie tourists were reached through the universally designed questionnaire. The snowball sampling technique, part of non-probability sampling, was used to reach the participants and obtain data with the highest point of generalization. The snowball sampling method was deemed suitable for the respondents as the target group that the present study aimed to reach was secluded. To ensure data quality, potential respondents were clearly notified about the study requirement on the main screen of the online survey. Upon completion of the data collection process for the pilot test, the construct validity test comprising convergent validity and discriminant validity was conducted and the results are presented in Tables 3 and 4. Due to the study's exploratory nature, SmartPLS software version 3.3 was used during the analysis processes, as suggested by Hair et al. [26], and we considered this software to be expedient for examining small samples and when testing both convergent and discriminant validity.

3. Findings

Table 2 presents the demographic profile of the respondents. In total, 66% of our respondents were female, with 66 responses received, followed by 34 responses from male respondents. Most of the respondents are known to be earning between RM 3001 and RM 4000, which carries a weight of 53%, followed by 31 respondents earning less than RM 3000 and 3 responses received from those with no income. The majority of our responders are between the ages of 31 and 40, with a total of 58 responses received followed by 42 responses from those aged between 18 and 30 years old. The result shows that the mainstream respondents are employed in the private sector, with 67%, or 67 respondents, followed by business personnel and government servants, with 17 responses and 10 responses, respectively. As for marital status, 61 respondents, or 61%, are known to be single, followed by 39 respondents who are married. Lastly, most of our respondents reported to have completed their tertiary education (76 respondents), followed by 24 respondents who have completed a higher degree.

Mata et al. [27] performed a PLS-SEM confirmatory factor analysis (CFA) using Smart-PLS software to establish a scale's structural validity. Correspondingly, the present study decided to employ CFA due to the nature of the instrument, which was developed based on references from previous scholars. As recommended, the result from the present study reached the minimum value of 0.5 for all the loading factors' performance. Both the t-value and p-value for the present study were accepted and significant at the $p < 0.001$ level. In addition, Hair et al. [28] suggested that two types of validity must be met in confirming the measurement model: convergent validity and discriminant validity. As per the authors, convergent validity includes the average variance extracted (AVE) and composite reliability (CR). Ngah et al. [29] highlighted that convergent validity can be established should the factor loading and AVE reach more than 0.50, and the CR more than 0.75. Table 3 below exhibits the convergent validity result, comprising the AVE, CR, and Cronbach's alpha values. The factor loading for all items (64 items) within the study passed the minimum value of 0.50. Items that failed to reach the minimum value were removed and the analysis was recommenced. AVE and CR for this study were achieved after both reached more than 0.50 for AVE and more than 0.75 for CR, as defined by Hair et al. [28].

Table 2. Demographic profile.

Variable	Frequency	Percentage (%)
Gender		
Male	34	34
Female	66	66
Income		
No Income	3	3
Less than RM 3000	31	31
RM 3001–RM 4000	53	53
RM 4001–RM 5000	6	6
RM 5001 and above	7	7
Age		
18–30	42	42
31–40	58	58
41–50	0	0
51–60	0	0
61 and above	0	0
Occupation		
General Government Servant	10	10
Private Sector Employee	67	67
Business and Self Employed	17	17
Home Maker	0	0
Student	6	6
Retiree	0	0
Unemployed	0	0
Marital Status		
Single	61	61
Married	39	39
Divorced/Widowed	0	0
Education Level		
Higher Degree—Masters/PhD	24	24
Tertiary Education—Diploma/Degree	76	76
Secondary/High School Education	0	0
Primary/Elementary Education	0	0

With the Cronbach's alpha value exceeding the minimum value of 0.70, as required by Hair et al. [30], all items were found reliable through internal consistency analysis. Hence, the instrument is deemed fit to proceed with discriminant validity.

Table 3. Convergent validity.

Items	CR	AVE	Rho_A	Cronbach's Alpha
Information Authority	0.906	0.520	0.884	0.883
Author's Identity	0.903	0.572	0.879	0.875
Engagement	0.930	0.656	0.914	0.913
Transparency	0.922	0.628	0.903	0.901
Cognitive Image	0.966	0.578	0.964	0.963
Affective Image	0.917	0.688	0.893	0.887
Overall Destination Image	0.943	0.735	0.929	0.927

Discriminant validity can be tested using three prime tests: cross loading, Fornell–Larcker criterion, and heterotrait–monotrait (HTMT). This present study approached discriminant validity through HTMT. Franke and Sarstedt [31] uttered that, HTMT values should be less than 0.85 to fulfil the validity test. Table 4 indicates the HTMT values for information authority, author's identity, engagement, transparency, affective image, cognitive image and overall destination image, which is less than 0.85. The result approves the validity test requirement.

Table 4. Discriminant Validity.

Items	1	2	3	4	5	6	7
Information Authority							
Author's Identity	0.451						
Engagement	0.395	0.791					
Transparency	0.390	0.802	0.815				
Cognitive Image	0.131	0.277	0.301	0.395			
Affective Image	0.105	0.262	0.320	0.345	0.703		
Overall Destination Image	0.138	0.281	0.346	0.396	0.613	0.740	

4. Discussion

The objective of this study was to investigate the validity of a research instrument that aims to measure the influence of information authenticity and travel selfies on Malaysian online tourism destination images. Instrument validity, encompassing CFA, content validity, and construct validity, and instrument reliability, comprising internal consistency, were tested to fulfil the aim of the study. The result suggests that the instrument established in this study is reliable, valid, and capable of producing the relevant statistical results upon application. This study has employed PLS-SEM measurements through SmartPLS version 3.3 (Joe F. Hair, Mobile, AL, USA). It provides greater statistical power to the research and can analyse complex structural equations consisting of many indicators and constructs [32]. Notably, the fundamental Cronbach's alpha value of more than 0.7 assures the reliability of the item used to measure the construct [33]; the present study thus projected that the result was satisfactory, indicating that all the seven constructs had a high level of reliability. This suggests that the instrument can be relied upon to explain the subject matter.

As part of convergent validity, the fundamental rules—these being an external loading rating value greater than 0.5 [34], a CR value greater than 0.75 [28], and an AVE value for each construct greater than 0.7 [35]—were rigorously monitored. Hence, convergent validity for the present study was established. HTMT, as part of discriminant validity, requires values less than 0.85 [36] or 0.90 [37]. Given that the HTMT values from the present study are all below 0.9, discriminant validity was established, proving that the constructs are not highly correlated. Some items were eliminated in the initial phase for failure to meet the minimum requirement regarding the factor loading rating, and so the data were retested. After accomplishing the test requirement, the overall result implies that all of the constructs and items reported in the study are valid and reliable.

5. Conclusions

The present study has employed and validated seven constructs—information authority, author's identity, engagement, transparency, affective image, cognitive image, and overall destination image—which were evaluated through the eye of selfie tourists. As part of the theoretical contribution, the validated instrument has integrated the role of information authenticity as represented by the information authority, author's identity, engagement, transparency, and destination images—represented by cognitive and affective images—through the eyes of actual selfie tourists. Exclusively, this study is anticipated to nourish the literature on information authenticity, travel selfies, destination images, and computer-mediated communication, by providing a validated instrument for future empirical research. The findings from this study will also be an eye-opener for destination hosts and prospective tourists about the rise of information authenticity within the online landscape, which could directly or indirectly impact online destination image. Forthcoming scholars interested in the subject area are welcome to use the instrument, giving great attention to sample size, mode of data collection, and the target respondents.

Author Contributions: Conceptualization, S.K.V. and N.H.H.; methodology, S.K.V. and N.H.H.; software, S.K.V.; validation, S.K.V. and N.H.H.; formal analysis, S.K.V. and N.H.H.; investigation, S.K.V.; resources, S.K.V. and N.H.H.; data curation, S.K.V. and N.H.H.; writing—original draft preparation, S.K.V. and N.H.H.; writing—review and editing, S.K.V. and N.H.H.; visualization, S.K.V. and N.H.H.; supervision, N.H.H.; project administration, S.K.V. and N.H.H.; funding acquisition, N.H.H. All authors have read and agreed to the published version of the manuscript.

Funding: This research received no external funding.

Institutional Review Board Statement: Not applicable.

Informed Consent Statement: Not applicable.

Data Availability Statement: Not applicable.

Conflicts of Interest: The authors declare no conflict of interest.

References

1. Mata, I.L.; Fossgard, K.; Haukeland, J.V. Do visitors gaze and reproduce what destination managers wish to commercialise? Perceived and projected image in the UNESCO World Heritage area ' West Norwegian Fjords. *Int. J. Digit. Cult. Electron. Tour.* **2018**, *2*, 294–321. [CrossRef]
2. Senft, T.M.; Baym, N.K. What does the selfie say? Investigating a global phenomenon. *Int. J. Commun.* **2015**, *9*, 1588–1606.
3. Rubin, V.L. Disinformation and misinformation triangle: A conceptual model for "fake news" epidemic, causal factors and interventions. *J. Doc.* **2019**, *75*, 1013–1034. [CrossRef]
4. Michael, N.; James, R.; Michael, I.; Michael, N. Australia'scognitive, affective and conative destination image: An Emirati tourist perspective. *J. Islamic Mark.* **2018**, *9*, 36–59. [CrossRef]
5. Natasha, N.; Fawzi, M.; Hussin, S.R.; Hashim, H. The on destination image formation influence of culture: A critical review. *Glob. Rev. Res. Tour. Hosp. Leis. Manag.* **2018**, 577–596.
6. De La Hoz-Correa, A.; Muñoz-Leiva, F. The role of information sources and image on the intention to visit a medical tourism destination: A cross-cultural analysis. *J. Travel Tour. Mark.* **2019**, *36*, 204–219. [CrossRef]
7. Lam, J.M.S.; Ismail, H.; Lee, S. From desktop to destination: User-generated content platforms, co-created online experiences, destination image and satisfaction. *J. Destin. Mark. Manag.* **2020**, *18*, 1–13. [CrossRef]
8. Crompton, J.L. An assessment of the image of Mexico as a vacation destination and the influence of geographical location upon that image. *J. Travel Res.* **1979**, *17*, 18–23. [CrossRef]
9. Russel, J.A.; Prat, G. A description of affective quality attributed to environment. *J. Personal. Soc. Psychol.* **1980**, *38*, 311–322. [CrossRef]
10. Gartner, W. Image formation process. *J. Travel Tour. Mark.* **1993**, *2*, 191–215. [CrossRef]
11. Echtner, C.; Ritchie, J.R.B. The measurement of destination image: An empirical assessment. *J. Travel Res.* **1993**, *31*, 3–13. [CrossRef]
12. Cohen, E. Authenticity in tourism studies: Apres ia lutte. *Tour. Recreat. Res.* **2007**, *32*, 75–82.
13. Gilpin, D.R.; Palazzolo, E.T.; Brody, N. Socially mediated authenticity. *J. Commun. Manag.* **2010**, *14*, 258–278. [CrossRef]
14. Ismail, S.; Latif, R. Authenticity Issues of Social Media: Credibility, Quality and Reality. *Proc. World Acad. Sci. Eng. Technol.* **2013**, *74*, 265–272.
15. Morhart, F.; Malär, L.; Guèvremont, A.; Girardin, F.; Grohmann, B. Brand authenticity: An integrative framework and measurement scale. *J. Consum. Psychol.* **2015**, *25*, 200–218. [CrossRef]
16. Tran, V.; Keng, C. The brand authenticity scale: Development and validation. *Contemp. Manag. Res.* **2018**, *14*, 277–291. [CrossRef]
17. Baloglu, S.; McCleary, K. A model of destination image formation. *Ann. Tour. Res.* **1999**, *26*, 808–889. [CrossRef]
18. Beerli, A.; Martín, J. Factors influencing destination image. *Ann. Tour. Res.* **1999**, *31*, 657–681. [CrossRef]
19. Echtner, C.; Ritchie, J. The meaning and measurement of destination image. *J. Tour. Stud.* **2003**, *14*, 37–48.
20. Ekinci, Y.; Hosany, S. Destination personality: An application of brand personality to tourism destinations. *J. Travel Res.* **2006**, *45*(2), 127–139. [CrossRef]
21. Rawlins, B.L. Measuring the relationship between organizational transparency and employee trust. *Public Relat. J.* **2009**, *2*, 1–21.
22. Ponzi, L.J.; Fombrun, C.J.; Gardberg, N.A. RepTrak pulse: Conceptualizing and validating a short-form measure of corporate reputation. *Corp. Reput. Rev.* **2011**, *14*, 15–35. [CrossRef]
23. Kaakinen, M.; Sirola, A.; Savolainen, I.; Oksanen, A. Shared identity and shared information in social media: Development and validation of the identity bubble reinforcement scale. *Media Psychol.* **2018**, *23*, 25–51. [CrossRef]

24. Khan, M.F.; Jan, A. A measure of social media marketing: Scale development and validation. *Jindal J. Bus. Res.* **2019**, *8*, 158–168. [CrossRef]
25. Ghauri, P.; Gronhaug, K. *Research Methods in Business Studies*; FT/Prentice Hall: Harlow, UK, 2005.
26. Hair, J.F.; Risher, J.J.; Sarstedt, M.; Ringle, C.M. When to use and how to report the results of PLS-SEM. *Eur. Bus. Rev.* **2019**, *31*, 2–24. [CrossRef]
27. Mâță, L.; Clipa, O.; Tzafilkou, K. The development and validation of a scale to measure university teachers' attitude towards ethical use of information technology for a sustainable education. *Sustainability* **2020**, *12*, 6268. [CrossRef]
28. Hair, J.F.; Ringle, C.M.; Sarstedt, M. Partial least squares structural equation modeling: Rigorous applications, better results and higher acceptance. *Long Range Plan.* **2013**, *46*, 1–12. [CrossRef]
29. Abdul Hafaz, N.; Gabarre, S.; Eneizan, E.; Asri, N. Mediated and moderated model of the willingness to pay for halal transportation. *J. Islamic Mark.* **2020**, *12*, 1425–1445.
30. Hair, J.F.; Black, W.C.; Babin, B.J.; Anderson, R.E. *Multivariate Data Analysis: A Global Perspective*; Pearson Education: New York, NY, USA, 2010.
31. Franke, G.; Sarstedt, M. Heuristics versus statistics in discriminant validity testing: A comparison of four procedures. *Internet Res.* **2018**, *29*, 430–447. [CrossRef]
32. Hair, J.F., Jr.; Hult, G.T.M.; Ringle, C.; Sarstedt, M. *A Primer On Partial Least Squares Structural Equation Modelling (PLS-SEM)*; Sage: Thousand Oak, CA, USA, 2016.
33. Nunnally, J.; Bernstein, I. *Psychometric Theory*, 3rd ed.; McGraw-Hill: New York, NY, USA, 1994.
34. Hulland, J. Use of Partial Least Squares (PLS) in strategic management research: A Review of Four Recent Studies. *Strateg. Manag. J.* **1999**, *204*, 195–204. [CrossRef]
35. Bartlett, J.E.; Kotrlik, J.W.; Hiiggiinss, C.C. Organizational Research: Determining Appropriate Sample Size in Survey Research. *Inf. Technol. Learn. Perform. J.* **2001**, *19*, 43–50.
36. Kline, R.B. *Principles and Practice of Structural Equation Modeling*, 3rd ed.; Guilford Press: New York, NY, USA, 2011.
37. Gold, A.H.; Malhotra, A.; Segars, A.H. Knowledge management: An organizational capabilities perspective. *J. Manag. Inf. Syst.* **2001**, *18*, 185–214. [CrossRef]

proceedings

MDPI

Proceeding Paper

Information Authenticity and Tourism Destination Image Formation through Computer-Mediated Communication: A Proposed Framework †

Sathish Kumar Velayuthan [1,2,*] and Noor Hazarina Hashim [1]

1 Azman Hashim International Business School, Universiti Teknologi Malaysia, Johor Bahru 81310, Malaysia
2 Faculty of Hospitality, Tourism and Wellness, Universiti Malaysia Kelantan, Kota Bharu 16100, Malaysia
* Correspondence: sathishvelayuthan@gmail.com

† Presented at the International Academic Symposium of Social Science 2022, Kota Bharu, Malaysia, 3 July 2022.

Abstract: As part of the organic image, the proper use of travel selfies could elevate the tourism industry to the next level of improved products and services that align with tourists' needs. However, due to the rise of fake reviewers and tourists who write reviews for personal incentives, the authenticity behind every post and how it could shape the destination image in the eye of prospective tourists remains sparse within the academic literature. Therefore, this study attempts to propose a comprehensive framework highlighting the role of information authenticity towards tourism destination image formation within online computer-mediated communication by incorporating travel selfies as a medium for information sharing.

Keywords: framework; computer-mediated communication; information authenticity; tourism destination image; travel selfies

Citation: Velayuthan, S.K.; Hashim, N.H. Information Authenticity and Tourism Destination Image Formation through Computer-Mediated Communication: A Proposed Framework. *Proceedings* **2022**, *82*, 82. https://doi.org/10.3390/proceedings2022082082

Academic Editor: Mohamad Rahimi Mohamad Rosman

Published: 26 September 2022

1. Introduction

Visualize the following scenario: You and your spouse have decided to finalize your honeymoon trip as a newlywed couple. Given that your spouse is a selfie enthusiast and an active Instagram user, you both decide to look for island trip options on Instagram to make your voyage unique and snap exciting travel selfies for the memories. Suddenly, a selfie post pops up on your Instagram wall. Someone who has recently visited that island's restaurant posted a selfie backed to a sight of a *'Roti Canai'* (Malaysian local flatbread) seller spitting on the dish and hashtagged the post #worsttripever. You zoom into the selfie to ensure what you see is true, and certainly that freaks you out! However, when you do a quick read of the comments section, the authors mention that it is just the style of the seller who whistles while making the *'Roti Canai'*, which is associated with the shop name *'Roti Canai Whistle'*. After seeing such a post, will you and your partner cancel your trip? Or will you ignore the post and plan ahead since someone unknown had posted it? Furthermore, regardless of whether true or not, will that specific selfie post create a bad image of the island in your mind?

In light of the above situation, social media undeniably appears to be one of the fastest platforms for disseminating information to people of various ages. With the support of computer-mediated communication (CMC) and the internet of things, the chain of information transfer is now effective, and more and more people are engaging in social networks to share their pieces of information. The present study strongly believes that travel selfies taken by actual tourists during their trips and posted on social media can be transformed into destination brands that can benefit prospective tourists and destination hosts.

Tamaki [1], based on previous studies, affirmed that image-based posts could influence tourists' decision-making and post-travel psychological judgments about tourism destinations. Hence, being part of the organic image, the proper use of travel selfies within the tourism industry could elevate the industry to the next level of improved products and services that align with tourists' needs. However, due to the rise of fake reviewers and tourists who write reviews for personal incentives [2], the authenticity behind every post and how it could shape the destination image in the eye of prospective tourists remains sparse within the academic literature. On the one hand, although false information sharing within the tourism context could tarnish the image of any tourism attraction, less empirical work has been conducted within the tourism literature highlighting the context of information authenticity within the online CMC and its relationship with the destination image. Iordanova and Stainton [3], with their research on travel blogs, have empirically proven the relationship between trust in the content shared by bloggers and cognitive and affective image formations among local tourists. On the other hand, studies in the literature have stated that the context of authenticity within the tourism realm is shapeless and greatly depends on the area of investigation. Nevertheless, a comprehensive framework combining the context of information authenticity and destination image formation incorporating travel selfies, evaluated through the eye of tourists, remains novel for exploration within the tourism literature. Therefore, the present study attempts to propose a framework highlighting the role of information authenticity towards tourism destination image formation within online computer-mediated communication by incorporating travel selfies as a medium for information sharing.

2. Literature Review

2.1. Computer-Mediated Communication

CMC has received attention from various fields and much from communication research. Eun and Soo [4] noted that CMC is a term that encompasses multiple forms of human communication through networked computers, which can be synchronous or asynchronous and involve one-to-one, one-to-many, or many-to-many exchanges of text, audio, and/or video messages. This covers the entire circle of human communication nature. To be precise, this paper refers only to CMC on social media platforms addressing the current communication trend. Social media is one of the hubs that ultimately assists the communication fostered not only in terms of the number of users but also through a significant number of platforms [5]. The growth of new communication technology has brought transformations to the communicative potential, and the dawn of social media has opened up more significant perspectives for interpersonal and organizational communication [6]. Allied with the upsurge of fake news, generally, users do not stick to any single social media platform but go through numerous platforms based on trends [7], and information users will have to rely on their verdict because it is open for anyone to recommend anything online [8].

2.2. Travel Selfies and Social Media

The Oxford dictionary first introduced the universal concept of the selfie in 2013, defining selfies as a photograph that one has taken of oneself, typically one taken with a smartphone or webcam and shared via social media. The definition establishes the strong tie between selfies and social media sharing behavior. Dinhopl and Gretzel [9] indicated that selfies are 'not confined to the single type of technology or genre of photos or videos but categorized by the desire to frame the "self" in a picture taken to be shared with online audience travel' (p. 127). Understanding the changes brought in by technology in this information edge, the view on selfies asserted by Dinhopl and Gretzel [9] seems agreeable. Paris and Pietschnig's [10] study appeared to be the first to define selfies as 'the imbrications and construction of the self within a network of actors' within the tourism literature.

Authors have associated their arguments by relating travel selfies to Larsen's [11] stand on digital photography, as it could be represented in many ways depending on how they are accumulated, made memorable, and performed in a specific circumstance. Though that stand dates long ago, it seems logical with the trend that appears as a norm in our daily routine. Establishing the link between social media and selfies, Dinholp and Gretzel [9] and Senft and Baym [12] have supported that internet-enabled smartphones primarily contribute to the popularity of selfies these days with front-facing cameras and photograph-based social media platforms. Across the globe, the selfie phenomenon is indeed becoming a trend. Lyu [13] made it clear that social media users will engage in self-presentation, while Canniford and Rokka [14] noted that this platform also serves as a space for consumer brands. Instagram seems to be one of the popular sites for selfies and travel selfies. It contains several layers of auto-generated data (geotags and timestamps), user-added data (hashtags), and user-added data (comments and likes) [15], which benefit both users and business stakeholders. Inadvertently, with the rise of social channels, tourists can now instantly share their travel selfies, experience, and travel knowledge [13]. This way, it is crystal clear that the advent of travel selfies as a communication tool is becoming a social norm, and the information shared in a particular post by tourists to be gazed at by both known and unknown peers can educate them about the good and the bad of a specific destination.

2.3. *Destination Image*

Destination image is never a new subject, as it has been well discussed since the 1970s in many mediums. Yet today, the research on destination images is still evolving and will continue to grow due to the development brought in by technology. Scholars have mutually agreed on the definition by Crompton [16] of destination image as 'the sum of beliefs, ideas, and impressions that a person has towards a destination' (p. 19). The tourist's image formation model established by Fakeye and Crampton [17] (p. 11) undoubtedly displays that the tourists' perception of or towards a particular destination can be influenced through three entries: organic image, induced image, and complex image. The formation of organic image habitually happens through information reported through newspapers, magazines, and other non-tourism information sources, while induced images form based on the tourism destination-directed information, for instance, advertisements, posters, pamphlets, and tourism-related materials. Thirdly, the image of a destination is further improved (complex image) after one has experienced the destination services and validated the existing pieces of information.

It is understood that the context of the destination image first emerges from an image that serves as a tool to impress an individual's choice. The present study witnesses the presence of travel selfies posted on social media by actual tourists at the organic image stage, which could create the impression of prospective tourists towards a destination. Distinctively, the present study puts forward its novel idea by combining the dimensions of tourist photography, information authenticity, and its relationship towards destination image in the social media context through the eye of selfie tourists. The attributes of destination image have been introduced by renowned scholars such as cognitive image [16], affective image [18], cognitive, affective, and conative [19], holistic and attributive, functional and psychological, and common and unique [20]. Three key variables from the destination image were applied for this study: cognitive image, affective image, and overall image. Beerli and Martin [21] noted that cognitive image is the individual's own knowledge and belief about a particular destination. Cognitive image for this study is referred to as selfies tourists' beliefs and factual knowledge regarding Malaysian tourism destinations' physical attributes as portrayed in travel selfies on social media. Affective image is one's feeling toward a particular destination or an emotional response toward a destination [18,22].

Affective image for this study is referred to as a selfie tourist's inner or subjective feelings towards Malaysian tourism destinations based on travel selfies posted on social media. The overall image of a destination is formed by both perceptual/cognitive and affective assessments [23].

2.4. *Information Authenticity*

A review of the literature on authenticity provides an outlook that the context of authenticity is shapeless and entirely depends on its committed context. Martin [24] stated that instead of delving into the definition of authenticity, researchers should look at how it can be used. In contemporary modern society, the status of what is and is not authentic is the result of manipulated interpretation [25], and it can be understood that facilities are '*commodified*' to fit the present market needs.

Therefore, it is worth noting that the context of 'information authenticity' within the online scape involving tourist communication and destination image formation is novel to be researched; nonetheless, significant works have been noticed in communication and computer behavior research. Introducing information authenticity as a potential antecedent for destination image in this study, Gunn [26] has affirmed since then that one should first understand the factors that influence destination image formation. This will ease the process of understanding the target market [27], as tourists are keen on information search regardless of platforms. The literature has been updated with some prominent antecedents such as information sources, past experience, culture prior to visitation, race, age, socio-demographic factors, motivation, corporate image, destination personality, destination familiarity, and many more [16,17,21,23,28–35] However, one undiscovered possible contributing aspect is the notion of 'truthfulness of information shared and the authenticity in the wake of information' conveyed explicitly in the social media that occurs freely without any screening. With the assumption that information authenticity could be an antecedent that influences destination image within the computer-mediated communication context involving travel selfies, this study managed to look further into the context of authenticity within the study scope.

Few scholars have developed their works in the area of communication, namely, authenticity in tourism [36], authenticity in social media communication [37], social media authenticity [38], brand authenticity [39], and brand authenticity scale [40] . However, Gilpin et al [37] offered a paradigm that determines authenticity in the social media context comprising four factors: authority, engagement, identity, and transparency, which seems to suit the current study's objectives.

3. Research Framework

Figure 1 refers to the multi-dimensional framework designed for the present study by incorporating online computer-mediated communication, information authenticity, destination image, and travel selfies.

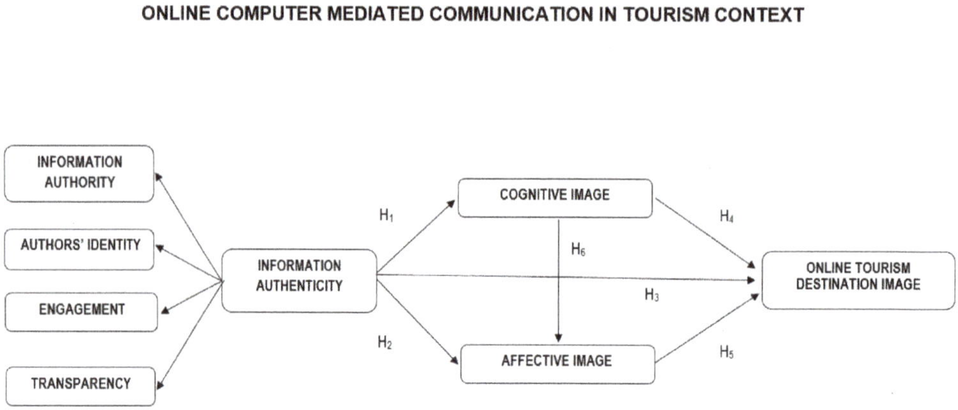

Figure 1. Research framework.

In this study, information authenticity is a reflective variable that is supported by information authority, authors' identity, engagement, and transparency. With information authenticity established, this study anticipates an impact on cognitive image development and affective image creation, which will influence overall image formation. Innovatively, this study incorporates travel selfies into the research by utilizing travel selfies as stimuli to evaluate the outcome of cognitive image and affective image development towards overall image formation.

Hypothesis Formation

As research on destination image evolves with the rise of technology, social media, and user-generated content, Rubin [41] stated that the world has begun to eye the rise of disinformation, and in the tourism realm, fake reviews are never an exception. Should the information shared on social media by some irresponsible user be forged, this will undoubtedly contribute to the erosion of the destination image of any particular tourism attraction. Roig [42] added that tourists' judgment on authenticity is often self-related. Parallel to that, this study observes the dimension of information authenticity through selfie tourists' beliefs towards the travel selfies and information posted by actual tourists on social media. Aligned with the aim of this study to explore how an image is formed in the eye of tourists after viewing travel selfies on social media based on the influence of information authenticity, the study by Iordanova and Stainton [3], which has proven the relationship between trust in the content shared by bloggers towards cognitive and affective image formation, appears noteworthy to the assumption put forward by the present study. Hence, the following hypotheses were proposed:

H1: *Information authenticity significantly influences cognitive image.*

H2: *Information authenticity significantly influences affective image.*

H3: *Information authenticity significantly influences overall destination image.*

Contrary to the previous assumptions, the relationship between cognitive image, affective image, and overall image has been vastly explored and established (see [3,21,23,29,34,43–48]). Hence, the following hypotheses were proposed based on the empirical evidence from previous research.

H4: *Cognitive image significantly influences overall destination image.*

H5: *Affective image significantly influences overall destination image.*

H6: *Cognitive image significantly influences affective image.*

4. Methodology

Figure 2 depicts the practical approach taken in this study to reach the proposed framework. Beginning with problem identification, the study's objective was set to propose a framework highlighting the role of information authenticity towards tourism destination image formation within online computer-mediated communication by incorporating travel selfies. Subsequently, an extensive literature search was conducted by first delving into the three key themes (travel selfies, information authenticity, and destination image). Relevant literature in the area was reviewed, and all particular variables were identified. The variables found were cross-checked with the needs of the present study, and the relationship between variables was prudently observed. Based on previous findings and gaps, the present study established assumptions and hypotheses to lead the study further. As a final point, the conceptual framework was developed in response to the gap within the literature considering the study's hypotheses. The study's objective was revisited to ensure that the proposed framework addresses the problem aimed to be solved.

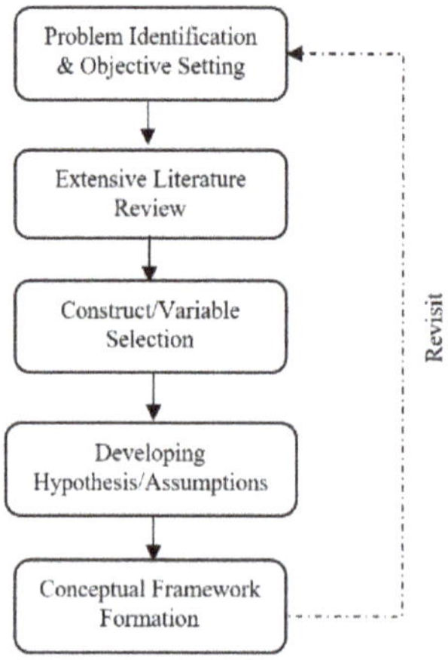

Figure 2. Research approach.

5. Conclusions

Computer-mediated communication is a growing area that sees both boon and bane. Today, with the rise of multiple social media platforms, people in general and tourists have the chance to write, capture, and share every angle of their thoughts for the public view. The receivers consume the information at large without hassle, anytime, anywhere. This is true with the dawn of visual information, travel photography, and travel selfies posted instantly on social media. Though it is convincing that destination managers are receiving free promotions, information falsification, or the state of sharing inauthentic information, is genuinely saddening. The failure to address this situation will alter the destination image and redirect the prospective tourist's intention and travel decision making.

First and foremost, the proposed comprehensive framework in this study linking the context of information authenticity and destination image formation incorporating travel selfies, evaluated through the eye of tourists, will undoubtedly be a novel contribution to the CMC and tourism communication literature.

Within the CMC literature, this paper is expected to shed light by replacing text-only-based conversation popularized in the past with the rise of image-based online conversation, mainly travel selfies. Secondly, the empirically proven context of the proposed framework is expected to verify the capability of information authenticity as a potential antecedent for tourism destination image formation. Thirdly, addressing the need for a framework by Rivera [49] to understand the rise of disinformation within the servicescape, the proposed framework is expected to be timely and can be adapted in the various areas within the service industry considering the target market, business scope, and communication scape. To be truthful, the rise of travel selfies as a potent influencing agent has been widely spoken about in popular media. Unfortunately, in the academic literature, travel selfie research, which is still at its beginning stage, is perceived as a product of narcissism and attraction shading elements (see [50,51]). Thusly, this study is expected to be an eye-opener for future researchers involving selfies for destination management. As such, the proposed framework is anticipated to craft a new path for the tourism destination management realm.

Author Contributions: Conceptualization, S.K.V. and N.H.H.; methodology, S.K.V. and N.H.H.; software, S.K.V.; validation, S.K.V. and N.H.H.; formal analysis, S.K.V. and N.H.H.; investigation, S.K.V.; resources, S.K.V. and N.H.H.; data curation, S.K.V. and N.H.H.; writing—original draft preparation, S.K.V. and N.H.H.; writing—review and editing, S.K.V. and N.H.H.; visualization, S.K.V. and N.H.H.; supervision, N.H.H.; project administration, S.K.V. and N.H.H.; funding acquisition, N.H.H. All authors have read and agreed to the published version of the manuscript.

Funding: This research received no external funding.

Institutional Review Board Statement: Not applicable.

Informed Consent Statement: Not applicable.

Data Availability Statement: Not applicable.

Conflicts of Interest: The authors declare no conflict of interest.

References

1. Tamaki, S. Likes on image posts in social networking services: Impact of travel episode. *J. Destin. Mark. Manag.* **2021**, *20*, 100615. [CrossRef]
2. Denise, M.O. Fake News and the Spread of Misinformation. 1 September 2017. Available online: https://journalistsresource.org/studies/society/internet/fake-news-conspiracy-theories-journalism-research/ (accessed on 30 May 2022).
3. Iordanova, E.; Stainton, H. Cognition, emotion and trust: A comparative analysis of Cambodia's perceived and projected online image. *Tour. Stud.* **2019**, *19*, 496–519. [CrossRef]
4. Eun, J.L.; Soo, Y.O. Computer Mediated Communication. 5 May 2017. Available online: http://www.oxfordbibliographies.com/view/document/obo-9780199756841/obo-9780199756841-0160.xml (accessed on 10 May 2022).
5. Edson, C.T., Jr.; Chen, L.; Velyn, L.H.M. Platform-swinging in a poly-social-media context: How and why users navigate multiple social media platforms. *J. Comput. Mediat. Commun.* **2019**, *24*, 21–35.
6. Lovejoy, K.; Saxton, G.D. Information, community, and action: How nonprofit organizations use social media. *J. Comput. Mediat. Commun.* **2012**, *17*, 337–353. [CrossRef]
7. Madianou, M. Smartphones as polymedia. *J. Comput. Mediat. Commun.* **2014**, *19*, 667–680. [CrossRef]
8. Westerwick, A. Effects of sponsorship, web site design, and google ranking on the credibility of online information. *J. Comput. Mediat. Commun.* **2013**, *18*, 194–211. [CrossRef]
9. Dinhopl, A.; Gretzel, U. Selfie-taking as touristic looking. *Ann. Tour. Res.* **2016**, *57*, 126–139. [CrossRef]
10. Paris, C.M.; Pietschnig, J. But First, Let me Take a Selfie, Personality Traits as Predictors of Travel Selfie Taking and Sharing Behaviors. In Proceedings of the TTRA International Conference, Portland, OR, USA, 15–17 June 2015; pp. 1–7. Available online: http://scholarworks.umass.edu/ttra%0Ahttp://scholarworks.umass.edu/ttra/ttra2015/Academic_Papers_Oral/1 (accessed on 30 May 2022).
11. Larsen, J. Practices and flows of digital photography: An ethnographic framework. *Mobilities* **2008**, *3*, 141–160. [CrossRef]
12. Senft, T.M.; Baym, N.K. What does the selfie say? Investigating a global phenomenon. *Int. J. Commun.* **2015**, *9*, 1588–1606.
13. Lyu, S.O. Travel selfies on social media as objectified self-presentation. *Tour. Manag.* **2016**, *54*, 185–195. [CrossRef]
14. Canniford, R.; Rokka, J. Heterotopian selfies : How social media destabilizes brand assemblages. *Eur. J. Mark.* **2016**, *50*, 1789–1813.
15. Tifentale, A. *The Selfie : More and Less than a Self-Portrait*; Neumüller, M., Ed.; Routledge Companion to Photography and Visual Culture London; Routledge: New York, NY, USA, 2018; pp. 44–58.
16. Crompton, J.L. An assessment of the image of Mexico as a vacation destination and the influence of geographical location upon that image. *J. Travel Res.* **1979**, *17*, 18–23. [CrossRef]
17. Fakeye, P.C.; Crompton, J.L. Image differences between prospective, first-time, and repeat visitors to the Lower Rio Grande Valley. *J. Travel Med.* **1982**, *30*, 10–16. [CrossRef]
18. Russel, J.A.; Prat, G. A description of affective quality attributed to environment. *J. Personal. Soc. Psychol.* **1980**, *38*, 311–322. [CrossRef]
19. Gartner, W. Image formation process. *J. Travel Tour. Mark.* **1993**, *2*, 191–215. [CrossRef]
20. Echtner, C.; Ritchie, J.R.B. The measurement of destination image: An empirical assessment. *J. Travel Res.* **1993**, *31*, 3–13. [CrossRef]
21. Beerli, A.; Martin, J. Factors influencing destination image. *Ann. Tour. Res.* **2004**, *31*, 657–681. [CrossRef]
22. Baloglu, S.; Brinberg, D. Affective images of tourism destinations. *J. Travel Res.* **1997**, *35*, 11–15. [CrossRef]
23. Baloglu, S.; McCleary, K. A model of destination image formation. *Ann. Tour. Res.* **1999**, *26*, 808–889. [CrossRef]
24. Martin, K. Living pasts: Contested tourism authenticities. *Ann. Tour. Res.* **2010**, *37*, 537–554. [CrossRef]
25. Mcintosh, A.; Prentice, R.C. Affirming authenticity: Consuming cultural heritage. *Ann. Tour. Res.* **1999**, *26*, 589–612. [CrossRef]
26. Gunn, C.A. *Vacationscape: Designing Tourist Regions*; Bureau of Business Research, University of Texas: Austin, TX, USA, 1972.
27. Goodall, B.; Ashworth, G. (Eds.) *How Tourist Choose Their Holidays: An Analytic Framework*; Routledge: London, UK, 1988.
28. Ahmed, Z.U. Marketing your community: Correcting a negative image. *Cornell Hotel Restaur. Adm. Q.* **1991**, *31*, 24–27. [CrossRef]
29. Baloglu, S.; Mangaloglu, M. Tourism destinations images of Turkey, Egypt, Greece, and Italy as perceived by US-based tour operators and travel agents. *Tour. Manag.* **2001**, *22*, 1–9. [CrossRef]
30. Reisinger, Y.; Turner, L. *Cross-Cultural Behaviour in Tourism: Concepts and Analysis*; Butterworth-Heinemann: Oxford, UK, 2003.
31. Hosany, S.; Ekinci, Y.; Uysal, M. Destination image and destination personality. *Int. J. Cult. Tour. Hosp. Res.* **2007**, *1*, 62–81. [CrossRef]
32. Tasci, A.; Gartner, W. Destination image and its functional relationships. *J. Travel Res.* **2007**, *45*, 413–425. [CrossRef]
33. Papadimitriou, D.; Kaplanidou, K.K.; Apostolopoulou, A. Destination Image Components and Word-of-Mouth Intentions in Urban Tourism: A Multi- Group Approach. *J. Hosp. Tour. Res.* **2015**, *42*, 503–527. [CrossRef]

34. Tapia, M.; Martínez, B.D.A.; Robles, E.M.G. Factors influencing destination image in distant culture countries: The role of corporate image. *J. Tour. Hosp. Manag.* **2017**, *5*, 95–105.
35. Michael, N.; James, R.; Michael, I.; Michael, N. Australia's cognitive, affective and conative destination image: An Emirati tourist perspective. *J. Islamic Mark.* **2018**, *9*, 36–59. [CrossRef]
36. Cohen, E. Authenticity in tourism studies: Apres ia lutte. *Tour. Recreat. Res.* **2007**, *32*, 75–82.
37. Gilpin, D.R.; Palazzolo, E.T.; Brody, N. Socially mediated authenticity. *J. Commun. Manag.* **2010**, *14*, 258–278. [CrossRef]
38. Ismail, S.; Latif, R. Authenticity Issues of Social Media: Credibility, Quality and Reality. *Proc. World Acad. Sci. Eng. Technol.* **2013**, *74*, 265–272.
39. Morhart, F.; Malär, L.; Guèvremont, A.; Girardin, F.; Grohmann, B. Brand authenticity: An integrative framework and measurement scale. *J. Consum. Psychol.* **2013**, *25*, 200–218. [CrossRef]
40. Tran, V.; Keng, C. The brand authenticity scale : Development and validation. *Contemp. Manag. Res.* **2018**, *14*, 277–291. [CrossRef]
41. Rubin, V.L. Disinformation and misinformation triangle: A conceptual model for "fake news" epidemic, causal factors and interventions. *J. Doc.* **2019**, *75*, 1013–1034. [CrossRef]
42. Roig, M.E. Identity and authenticity in destination image construction. *Anatolia* **2015**, *26*, 574–587. [CrossRef]
43. Stern, B. A revised communication model for advertising: Multiple dimensions of the source, the message, and the recipient. *J. Advert.* **1994**, *23*, 5–15. [CrossRef]
44. MacKay, K.J.; Fesenmaier, D.R. Pictorial element of destination in image formation. *Ann. Tour. Res.* **1997**, *24*, 537–565. [CrossRef]
45. Kim, S.; Yoon, Y. The hierarchical effects of affective and cognitive components on tourism destination image. *J. Travel Tour. Mark.* **2003**, *14*, 1–22. [CrossRef]
46. Konecnik, M.; Gartner, W.C. Customer-based brand equity for a destination. *Ann. Tour. Res.* **2007**, *34*, 400–421. [CrossRef]
47. Martín, H.S.; Bosque, I.A.R. Exploring the cognitive-affective nature of destination image and the role of psychological factors in its formation. *Tour. Manag.* **2008**, *29*, 263–277. [CrossRef]
48. Lam, J.M.S.; Ismail, H.; Lee, S. From desktop to destination: User-generated content platforms, co-created online experiences, destination image and satisfaction. *J. Destin. Mark. Manag.* **2020**, *18*, 100490. [CrossRef]
49. Rivera, M. Fake News and hospitality research. *Int. J. Hosp. Manag.* **2020**, *85*, 102473. [CrossRef]
50. Taylor, D.G. Putting the "self"in selfies: How narcissism, envy and self-promotion motivate sharing of travel photos through social media. *J. Travel Tour. Mark.* **2020**, *37*, 64–77. [CrossRef]
51. Christou, P.; Farmaki, A.; Saveriades, A.; Georgiou, M. Travel selfies on social networks, narcissism and the "attraction-shading effect". *J. Hosp. Tour. Manag.* **2020**, *43*, 289–293. [CrossRef]

proceedings

MDPI

Proceeding Paper

The Revisit Intention of Customers at Old Town White Coffee Outlets in Shah Alam, Malaysia during the Post-COVID-19 Pandemic Era †

Nor Saadah Che Deraman, Mohd Hairi Jalis *, Nur Farrah Yasmin Abdul Latib and Zul Hazam Mohd Piah

Department of Culinary and Gastronomy, Faculty of Hotel and Tourism Management, Universiti Teknologi MARA Cawangan Terengganu, Dungun Campus, 23000 Dungun, Malaysia
* Correspondence: mohdhairijalis@uitm.edu.my
† Presented at the International Academic Symposium of Social Science 2022, Kota Bharu, Malaysia, 3 July 2022.

Abstract: This study examined the relationship between the emotional triggers and revisit intention of customers at Old Town White Coffee outlets in Shah Alam, Malaysia during the post-COVID-19 pandemic era. Nevertheless, the motivations and attributes of customers returning to the same restaurant during the post-COVID-19 pandemic era remain unclear. By using three triggers of emotions: (i) food, (ii) environment, and (iii) staff, which were derived from the existing literature review for this study, a total of 309 customers participated, and all data were analysed using SPSS version 27. The multiple regression analysis reveals that food has the greatest effect towards the emotional trigger of customers to revisit restaurants.

Keywords: customer; theme restaurant; revisit intention; emotion

check for updates

Citation: Deraman, N.S.C.; Jalis, M.H.; Latib, N.F.Y.A.; Piah, Z.H.M. The Revisit Intention of Customers at Old Town White Coffee Outlets in Shah Alam, Malaysia during the Post-COVID-19 Pandemic Era. *Proceedings* **2022**, *82*, 2. https://doi.org/10.3390/proceedings2022082002

Academic Editor: Mohamad Rahimi Mohamad Rosman

Published: 6 September 2022

1. Introduction

Today, the foodservice business is recognised as a global industry with a broadening of customers worldwide [1–3]. This development can be seen from the array of foodservice establishments blooming in all sub-sectors including catering, hotels, food retail, food manufacturing, and restaurants. The restaurant sub-sector is experiencing an incredible expansion [1,4]. In the global marketplace, the demand for food and beverages keeps increasing, and competition is always stiff. Several factors such as the behavioural intention of customers, humanistic approach, task performance by service staff [4,5], physical environment [6–8], and positive customer emotions [9] have all influenced this positive development.

At the end of 2019, a novel coronavirus (popularly known as COVID-19) was detected and became one of the worst health disasters to hit the entire global population. Following the outbreak of COVID-19, almost every country declared some form of movement control order (MCO) to restrict human contact and minimise the chances of becoming infected. At the same time, the World Health Organisation (WHO) officially recognised COVID-19 as a pandemic. Due to continuous increase in the number of people becoming infected and the rise in the number of deaths, everyone had to stay at home, and there were strict restrictions on the performance of social, business, and trading activities. This has caused significant economic damage all over the world.

In the case of Malaysia, due to MCO enforcement, all restaurant businesses were required to halt dine-in services during the first wave of the pandemic, which began in mid-March 2020, and only takeout, drive-thru, or delivery services were allowed [10]. According to studies conducted by the National Restaurant Association in 2020, the restaurant business lost more than USD 120 billion in sales by May 2020, and eight million people were laid off or furloughed [11]. It was reported that the COVID-19 pandemic had directly increased the cost of restaurant operations to USD 240 billion by the end of 2020 [11]. Additional requirements had to be implemented by restaurant operators as a result of the social distancing initiatives imposed by the Government. In order to thrive during the COVID-19 pandemic, many restaurant businesses designed various strategies to be implemented as their survival modus operandi. Such strategies, at the same time, allowed the restaurant operators to acquire a deep understanding of the needs, wants, and perceptions of their customers, which was crucial in order to keep on attracting and retaining them [12,13].

Similar to other countries, Malaysia offers a wide range of food outlets including stalls, cafeterias, restaurants, fast food outlets, fine dining restaurants, canteens, and many more [14,15]. Such a variety of food outlets can cater to various categories of customers, from higher to lower income groups. The casual dining full-service restaurant, such as a theme restaurant, is one of the best choices for customers to go to if they wish to dine out because of its various concepts and the types of food available [16].

A theme restaurant is a dining establishment furnished with a variety of unique features that are unrelated to the act of eating [17]. As a result, a customer dining experience at a theme restaurant is an experience of consumption produced through the effects of the servicescape of the restaurant and the appraisal of real experiences. For example, a recent study on servicescape (referring to external features such as decoration, music, costumes, and service, as well as interactions between customers and personnel to share information about historical events and the culture of a theme) [17] found that favourable evaluations of these unique experiences by customers often lead to the building of attitudes, which can improve the behavioural intention of customers to return to the same theme restaurant [17].

Meanwhile, another study on understanding the degree to which an individual may recognise authenticity (i.e., psychological cue) found that a role in the construction of a favourable dining experience in a theme restaurant is crucial [18]. The involvement of customers in a restaurant is driven by their perception of authenticity, which creates pleasant emotions or attachments [17]. In addition to this, scholars believe that the concept of emotion helps to differentiate restaurant operators from their competitors and thus position their products in the marketplace [18,19]. It is a useful component to use to appeal to customers and create memorable customer experiences, particularly when they dine at special concept restaurants, such as theme restaurants [17,20].

However, the role of these emotion-evoking elements in triggering the theme restaurant revisit intention of customers remains questionable, especially with the sudden outbreak of the COVID-19 pandemic and the enforcement of MCOs. More specifically, to what extent do emotions related to the theme restaurant experience during the post-COVID-19 pandemic era trigger revisit intention? A comprehensive literature review has pointed out three main features, namely, food, service, and environment, as the main factors which influence the selection by customers of a theme restaurant to dine in. Therefore, this study examined the three emotional triggers that would lead customers to revisit their favourite theme restaurant during the post-COVID-19 pandemic era in Malaysia.

2. Materials and Methods

2.1. *Theme Restaurant*

Theme restaurants are a sub-set of the casual restaurant category, which makes up the largest share of the full-service restaurant market [21,22]. A theme restaurant is a restaurant that follows a formula in which all of its aspects, including the culinary menu, the ambience, and the décor, revolve around a single theme, which is usually related to the leisure industry [21]. The following are examples of appealing themes: (i) Hollywood and movies; (ii) sports and sporting events; (iii) reminiscence of simpler times; (iv) radio, music, and television; (v) travel and transportation; (vi) the environment; and (vii) the world around us. The themes are intended to create a memorable experience, which is bolstered by marketing methods that aim to entice diners to visit a different era, moment, or place. Aside from food and beverage consumption, the staff at a theme restaurant encourage diners to purchase souvenirs under the guise of acquiring a memento to commemorate their experiences [21,23]. The sale of such souvenirs generates a significant amount of revenue, which ranges from 20% to 55% of their sales mix, depending on the brand of the theme restaurant.

2.2. Revisit Intention

Revisit intention is defined as the judgment of an individual about using a designated service from the same company again, taking into account his or her current situation and likely circumstances [17,24]. Revisit intention is similar to visit intention except for the component of the experience. Revisit intention refers to the possibility of the customer returning to the restaurant in the future. It is further noted that revisit intention signifies the willingness of an individual to make another purchase from the same firm, based on their previous experiences. However, revisit intention is concerned with subjective viewpoints of individuals regarding their own personal behaviour in the future and may differ from actual revisit behaviour [25,26]. Studies on repeat visitation are vital for food and beverage businesses in order to gain a better understanding of food habits from an economic perspective. Food habits can be defined as a set of behaviours related to food preferences and social experiences. The level of liking for particular types of food is essential in the choice of food and the ability to directly create satisfaction through the eating experience and to build revisit intention for the restaurant [12,27]. It was also denoted that due to globalization, changes in eating habits, economic growth, and cultural influences, the hospitality industry has become highly competitive. In gaining customer loyalty and satisfaction, an understanding of customer perceptions is vital among restaurateurs. Several studies in the field have postulated that it is possible to induce actions such as repurchasing and revisiting by stimulating the memories of consumers [1,28–30]. For this reason, many companies have employed emotional elements in their advertising. A study has found that stimulating personal emotions in advertising is an effective way to attract customers [31]. Therefore, this study has attempted to employ nostalgia triggers in predicting revisit intention.

2.3. Emotional Triggers

2.3.1. Food

The key factor influencing restaurant choices is food [14,21,32,33] and its quality [1,16]. Food relates to sensory stimuli that connects with psychological perspectives perceived from such human senses [21,34]. Food consumption is the one area which is most likely to take people back into their past. It calls on several different senses, which is a valuable vector for emotion. The link between emotion and food consumption is almost obvious [35,36]. In communication strategies, one or several themes characterizing food nostalgia can be used to promote products. The use of childhood, happy family times, and intergenerational transmission are particularly suitable as communication strategies to be adopted by food brands [35,36]. In addition, food helps to create emotion by evoking memories of a particular experience or situation in the past [23,37]. For example, the chicken noodle soup that is served by the grandmother when one is sick. It is distinctive and thus memorable. This memory sets the stage for a product to be nostalgically linked to deep feelings about the grandmother. It is assumed that a person who experiences eating chicken noodle soup at a restaurant will recall such memories. Furthermore, the experience factor helps the restaurateurs in securing customer loyalty. This has been proven, especially as those experiences influence ongoing food-related behaviours and reveal the nature of current behaviours [38–41]. Additionally, the relative consequence of this attribute is that it becomes the main criteria of the customers in deciding to repeat their visit to the restaurant [32,42].

2.3.2. Environment

The environment of a restaurant is a critical factor in determining the dining experience of its customers [17,32]. There are two categories of environment in a business context: external and internal [12,15]. The external environment includes related factors present outside of the boundaries of the restaurant. In contrast, the internal environment includes physical and social factors present within the boundaries of the restaurant. On the other hand, an organization needs to study external forces that lead to changes, which the organization can prepare for, in order to secure and improve its performance in the future [43]. An earlier study undertaken by Banker in 1986 pointed out another factor, ambience, as an internal factor. Ambience refers to the atmospheric elements in a restaurant. Table setting, seat space, restaurant layout, and architecture are all physical atmospheric elements, while social factors consist of customers and the service personnel of the internal environment. When customers step into the restaurant, they start to enjoy the service provided in a comfortable environment [17,44]. Based on studies on the determinants of customer behavioural intentions in terms of environmental elements, the results showed that spatial layout, interior design, and ambiance are the key environmental elements that can influence customer emotion [9,35]. Nowadays, many people go to dine in restaurants not only to enjoy the food served but also to satisfy their need for social stimulation and entertainment [39,45–47]. Thus, restaurants need to present an attractive environment to influence the positive emotional responses of customers [23,35,41,48].

2.3.3. Staff

The employees or staff working in a restaurant play an important role in determining the service level of the restaurant and often impact the dining experiences of customers [9,49]. The engagement of employees towards their jobs is directly linked to customer service satisfaction levels. Employee characteristics such as the display of a caring attitude, responsiveness, solicited and unsolicited actions, and reliability are possible elements that can influence the casual dining experiences of customers [39–41,50]. A study found that service staff must pay attention to all the needs of the guest, which is of utmost significance in building a customer base that will regularly return to the restaurant [21,42].

2.4. *Design and Population*

The state of Selangor was selected as a demographic population background due to the availability of varying customer socio-demography. It is one of the most developed business and industrial hubs in Malaysia with many multinational industrial estates, educational institutions, rapidly growing commerce hubs, good transport services, and basic amenities. This study has been designed by selecting the population of Shah Alam. Shah Alam has a total population of 481,654, which makes up approximately 8.02% of the total population of Selangor [46]. Permission to conduct this study was obtained from the Old Town White Coffee outlets in Shah Alam. Prior to this, the researchers also approached other theme restaurants in Shah Alam; however, no feedback was received. Due to time and financial constraints, Old Town White Coffee was selected as an ideal theme restaurant model (i.e., national chain restaurant with standardised setup, service, and menu) for the purposes of the main inquiry of this study.

2.5. *Survey Instrument*

This quantitative study employed a questionnaire made up of three sections. The first section of the questionnaire required information regarding the demographic profile of respondents including gender, monthly income, race, and education level. Meanwhile, the second section contained questions regarding the emotions of respondents on revisit intention. The intention of respondents to visit theme restaurants was questioned in the third section of the questionnaire. This construct was measured using a multiple-item scale of 1–7 with the value of (where 1 = *strongly disagree* and 7 = *strongly agree*) to determine customer perception of these cues. To measure customer emotions, the study adopted constructs from [34]. A total of 300 respondents were selected, and this sample size was considered sufficient for this study [47]. The author proposed the criterion to follow when determining sample size, which is that the number of participants in a questionnaire should be larger than 30 and less than 500. The respondents were customers who dined in Old Town White Coffee outlets in Shah Alam who had different educational backgrounds and who earned different levels of income per month. This selection depended on the capability of respondents to answer and comprehend the needs of the survey. Eventually, a total of 350 questionnaires were distributed to guarantee that at least 300 usable questionnaires could be collected. Out of these 350 questionnaires, a total of 309 questionnaires were later completed and returned (88% response rate). No questionnaires were eliminated as all of them were found fit to be analysed. The following table reports the mean scores and standard deviation of emotion items obtained from the descriptive statistics.

2.6. Procedures

Figure 1 summarizes the data collection standard operating procedure (SOP) employed for this study.

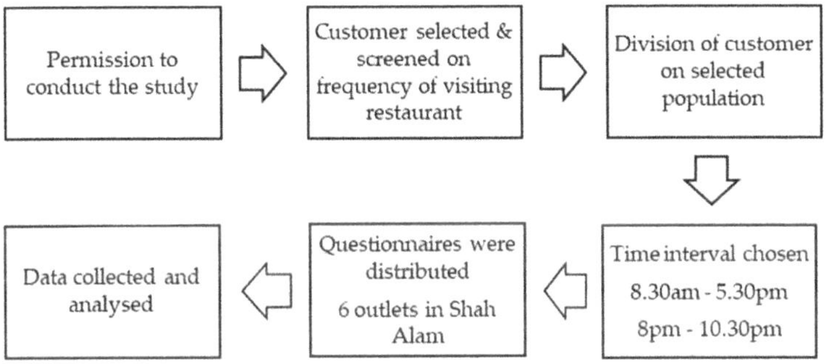

Figure 1. Summary of standard operating procedure (SOP) for data collection.

3. Results and Discussion

3.1. Profiles of Respondents

The frequency analysis reveals that more than half (52.8%) of the respondents were female, while 47.2% were male. In terms of race, 61.8% of the sample were Malays, while less of the respondents (27.2%) were Chinese, followed by Indian (9.4%), and the balance was made up of respondents from other races (1.6%). Regarding their educational level, almost half of the respondents (44.7%) had a bachelor's degree. This was followed by 28.5% who had a diploma, 15.5% who had a master's degree, and only 1.9% who had a PhD. In addition, respondents were asked about their monthly income: 66.7% of the respondents reported having a monthly income of MYR 2001 and above, 16.5% of the respondents reported earning between MYR 1500 to MYR 2000, and 13.3% of the respondents reported earning between MYR 1001 and MYR 1500. Only 3.6% of the respondents stated that they earned a monthly income of below MYR 1000. More than 75% (77.0%) of respondents claimed they frequently dined at the restaurant, while 23.0% said they did not frequently dine there. Lastly, they were required to state the frequency of their dining at the Old Town White Coffee outlet in a month. Approximately half of them (55.0%) responded that they dined at the outlet twice a month, 28.5% stated their frequency of visit as being once a month, while 16.5% claimed that they dined there every weekend.

3.2. Analysis of the Emotional Triggers of Customers

Table 1 reports the overall mean scores and standard deviation of food, environment, and service (staff) items that were obtained from descriptive statistics.

Table 1. Overall mean score of food, environment, and service (staff).

Construct	Minimum	Maximum	*M*	*SD*
Food	3.00	6.66	5.43	0.798
Environment	2.43	6.57	5.34	0.770
Service (staff)	2.57	5.85	4.46	0.625

Based on three dimensions of emotional triggers, food ($M = 5.43$, $SD = 0.798$) emerged as the most important component of how customers perceived emotion towards Old Town White Coffee in relation to other dimensions. This was followed by environment ($M = 5.34$, $SD = 0.770$) and then service (staff) ($M = 4.46$, $SD = 0.625$). This analysis proves that all three predictors spark emotions among customers, but in order to see whether these predictors relate to revisit intention, multiple regression analysis was then employed.

Multiple Regression Analysis

Multiple regression was employed to assess the significant relationship among variables in the study framework. A significant level of 0.05 was applied. In addition, it was used to determine the best predictor on revisit intention. Table 2 shows a summary of multiple regression analysis.

Table 2. Summary of multiple regression analysis.

	B	SE B	B
Constant	3.77	0.418	
Food	0.269	0.048	0.23 *
Environment	0.623	0.050	0.72 *
Service (staff)	0.364	0.062	0.37 *

Note: $R^2 = 0.07$, * $p < 0.001$.

Among the three predictors entered into the model, namely food, environment and service (staff), all three made statistically significant contributions to revisit intention with significant values of 0.002, 0.000, and 0.030 each, when $p < 0.05$. In terms of importance, environment made the largest unique contribution to the model with $\beta = 0.72$. The model was significant [F (3305) = 7.96, $p < 0.001$]. It made a significant contribution to revisit intention, and it explained 7% of the variation in revisit intention ratings. The remaining 93% is explained by other predictors. Even though the percentage was small, the ANOVA table showed that the model was highly significant and can be used to explain or predict cross-shopping behaviour among consumers.

The results indicate that food (i.e., the menu choices on offer and specific meals that bring nostalgia to a person) is the most significant element affecting the perception of emotional triggers of customers in Old Town White Coffee regarding revisit intention during the post-COVID-19 pandemic era. The best quality food choices with authentic taste bring nostalgia and meet the expectations of customers. It was found that the taste of food can bring back memories for customers but not to the extent of influencing them to return. This finding is aligned with previous studies which identified that food not only helps to create an emotional response by evoking a memory of an experience or situation in the past, but it also assists customers in choosing a restaurant based on their special food interest (SFI) [37]. This is because not all restaurants serve food in the same way as theme restaurants where customers are excited to experience again the dining moment and the feeling of enjoying the food. These one-of-a-kind experiences, if positively rated by customers, contribute to the formation of attitudes, which can increase the behavioural intention of customers to return to the restaurant.

On the other hand, the service (staff) obtained only a medium score in terms of the perception of emotions of customers in the restaurant. It shows that the staff at the Old Town White Coffee outlets in Shah Alam did not contribute towards making the restaurant memorable to customers; however, the results show that staff do influence the customers to revisit. Apart from that, it was found that environment was a key determinant in predicting the revisit intention of customers. Previous studies have found that spatial layout, interior design, and ambience are the key elements of environment that can influence customer emotion, whereby when customers step into the restaurant, they can start to enjoy the comfortable environment [9,14,44,47].

4. Conclusions and Recommendation

In conclusion, the findings of this study reveal that there are significant associations between food, environment (i.e., restaurant setup), staff (i.e., service providers) and the revisit intention of customers during the post-COVID-19 pandemic era, specifically in the case of Old Town White Coffee outlets in Shah Alam. Food and environment play very important roles in evoking the experiences and memories of customers. Therefore, efforts to maintain and sustain the existing restaurant setup and business operations should be continuously made. This allows for the restaurant to build its brand and image in the minds and eyes of every customer who visits and dines at the restaurant. Furthermore, providing appropriate food choices on the menu card which blend well with the theme restaurant itself would be able to trigger the emotions of customers. In the end, this is a competitive advantage for Old Town White Coffee outlets in Shah Alam, and it could be a great business model for other theme restaurants too. Future studies can further expand the theory of emotional triggers by breaking the theory into sub-dimensions such as social aspects, sensory inputs, and events.

Similarly, the dimensions can be divided into people, friends, food, and environment [4,31,49]. In fact, comparison between different theme restaurants would also be beneficial. Given that other selection factors (such as food price, choice, and style of service) are important in restaurants, their relative value should be checked and compared. In addition, customer behaviour models often link psychological elements (e.g., personality, imagery-related characteristics) to behavioural intention, while linking aspects of attribute (e.g., product attributes, service quality) to attitude formation [34,35] should be extended. As a result, further exploration of the relationship between psychological characteristics, such as authenticity and attitude, would be required.

Author Contributions: Conceptualization, N.S.C.D. and M.H.J.; methodology, N.S.C.D.; software, N.S.C.D.; validation, N.S.C.D., M.H.J., N.F.Y.A.L. and Z.H.M.P.; formal analysis, N.S.C.D. and M.H.J.; investigation, N.S.C.D., M.H.J., N.F.Y.A.L. and Z.H.M.P.; resources, N.S.C.D., M.H.J., N.F.Y.A.L. and Z.H.M.P.; data curation, N.S.C.D. and M.H.J.; writing—original draft preparation, N.S.C.D.; writing—review and editing, N.S.C.D. and M.H.J.; visualization, N.S.C.D. and M.H.J.; supervision, M.H.J.; project administration, N.F.Y.A.L. and Z.H.M.P.; funding acquisition, N.S.C.D., M.H.J., N.F.Y.A.L. and Z.H.M.P. All authors have read and agreed to the published version of the manuscript.

Funding: This research received no external funding.

Institutional Review Board Statement: Not applicable.

Informed Consent Statement: Not applicable.

Data Availability Statement: Not applicable.

Acknowledgments: Paper presented during a panel—organized by the Faculty of Information Management and DIGIT Special Interest Group—in the context of the International Academic Symposium of Social Science (IASSC) 2022 conference, organized and hosted by the UiTM Kelantan Branch.

Conflicts of Interest: The authors declare no conflict of interest.

References

1. Brewer, P.; Sebby, A.G. The effect of online restaurant menus on consumers' purchase intentions during the COVID-19 pandemic. *Int. J. Hosp. Manag.* **2021**, *1*, 94. [CrossRef] [PubMed]
2. Hanafiah, M.H.; Harun, M.F.; Jamaluddin, M.R. Bilateral trade and tourism demand. *World Appl. Sci. J.* **2010**, *10*, 110–114.
3. Kong, M.; Jogaratnam, G. The influence of culture on perceptions of service employee behavior. *Manag. Serv. Qual.* **2007**, *17*, 275–297. [CrossRef]
4. Hyun, S.S.; Han, H. A model of a patron's innovativeness formation toward a chain restaurant brand. *Int. J. Contemp. Hosp. Manag.* **2012**, *24*, 175–199. [CrossRef]
5. Kim, W.G.; Cha, Y. Antecedents and consequences of relationship quality in hotel industry. *Int. J. Hosp. Manag.* **2002**, *21*, 321–338. [CrossRef]
6. Ryu, K.; Han, H.; Kim, T.H. The relationships among overall quick-casual restaurant image, perceived value, customer satisfaction, and behavioral intentions. *Int. J. Hosp. Manag.* **2008**, *27*, 459–469. [CrossRef]

7. Ryu, K.; Jang, S.C.S. The Effect of Environmental Perceptions on Behavioral Intentions through Emotions: The Case of Upscale Restaurants. *J. Hosp. Tour. Res.* **2007**, *31*, 56–72. [CrossRef]
8. Ryu, K.; Jang, S. Influence of restaurant's physical environments on emotion and behavioral intention. *Serv. Ind. J.* **2008**, *28*, 1151–1165. [CrossRef]
9. Liu, Y.; Jang, S.C. The effects of dining atmospherics: An extended Mehrabian-Russell model. *Int. J. Hosp. Manag.* **2009**, *28*, 494–503. [CrossRef]
10. Wida, E.C. Which States Have Closed Restaurants and Bars Due to Coronavirus? Today. Available online: https://www.today.com/ (accessed on 30 March 2020).
11. National Restaurant Association. *Coronavirus Information and Resources*; National Restaurant Association: Washington, DC, USA, 2020.
12. Sirimongkol, T. The effects of restaurant service quality on revisit intention in pandemic conditions: An empirical study from Khonkaen, Thailand. *J. Foodserv. Bus. Res.* **2022**, *25*, 233–251. [CrossRef]
13. Gilbert, G.R.; Veloutsou, C. A cross-industry comparison of customer satisfaction. *J. Serv. Mark.* **2006**, *20*, 298–308. [CrossRef]
14. Voon, B.H. Service environment of restaurants: Findings from the youth customers. *J. ASIAN Behav. Stud.* **2017**, *2*, 67.
15. Lewis, H. The Foodservice Market Outlook to 2012—Management Briefing: The Tricky Task. 2006. Available online: http://www.just-food.com (accessed on 7 November 2021).
16. Ahmad, F.; Ghazali, H.; Othman, M. Consumers preference between fast food restaurant and casual dining restaurant: A Conceptual Paper. In Proceedings of the 3rd International Conference on Management (3rd ICM 2013), Penang, Malaysia, 10–11 June 2013; pp. 315–325.
17. Meng, B.; Choi, K. An investigation on customer revisit intention to theme restaurants: The role of servicescape and authentic perception. *Int. J. Contemp. Hosp. Manag.* **2018**, *30*, 1646–1662. [CrossRef]
18. Liu, S.Q.; Mattila, A.S. Ethnic dining: Need to belong, need to be unique, and menu offering. *Int. J. Hosp. Manag.* **2015**, *49*, 1–7. [CrossRef]
19. Kessous, A.; Roux, E. A semiotic analysis of nostalgia as a connection to the past. *Qual. Mark. Res.* **2008**, *11*, 192–212. [CrossRef]
20. Sierra, J.J.; McQuitty, S. Attitudes and emotions as determinants of nostalgia purchases: An application of social identity theory. *J. Mark. Theory Pract.* **2007**, *15*, 99–112. [CrossRef]
21. Berumen Calderón, M.F.; Estolano Cristerna, D.; Sterling Zozoaga, A.S.; Berumen Calderón, A.R. Model to assess the selection of the optimal location for a restaurant, a quantitative approach. Case study: Theme restaurants in Cancun, Mexico. *J. Foodserv. Bus. Res.* **2021**, *24*, 457–501. [CrossRef]
22. Cichy, R.; Phillip, J.; Hickey, J. *Managing Service in Food and Beverage Operations*, 4th ed.; Educational Institute of the American Motel Association: East Lansing, MI, USA, 2012.
23. Weiss, R.; Feinstein, A.H.; Dalbor, M. Customer satisfaction of theme restaurant attributes and their influence on return intent. *J. Foodserv. Bus. Res.* **2004**, *7*, 23–41. [CrossRef]
24. Hellier, P.K.; Geursen, G.M.; Carr, R.A.; Rickar, J.A. Customer repurchase intention. *Eur. J. Mark.* **2003**, *37*, 1762–1800. [CrossRef]
25. Choe, J.Y.; Kim, J.J.; Hwang, J. Perceived risks from drone food delivery services before and after COVID-19. *Int. J. Contemp. Hosp. Manag.* **2021**, *33*, 1276–1296. [CrossRef]
26. Rust, R.T.; Zahorik, A.J.; Keiningham, T.L. Return on Quality (ROQ): Making Service Quality Financially Accountable. *J. Mark.* **1995**, *59*, 58–70. [CrossRef]
27. Razzaque, M.A. Indian Restaurants and Sydneysiders: A Perceptual Study. Presented at the Australia and New Zealand Marketing Academy Conference 2009, Melbourne, Australia, 30 November–2 December 2009.
28. Braun-Latour, K.A.; Latour, M.S.; Zinkhan, G.M. Using Childhood Memories to Gain Insight into Brand Meaning. *J. Mark.* **2007**, *71*, 45–60. [CrossRef]
29. Sierra, J.J.; McQuitty, S. Service providers and customers: Social exchange theory and service loyalty. *J. Serv. Mark.* **2005**, *19*, 392–400. [CrossRef]
30. Fournier, S.; Yao, J.L. Reviving brand loyalty: A reconceptualization within the framework of consumer-brand relationships. *Int. J. Res. Mark.* **1997**, *14*, 451–472. [CrossRef]
31. Havlena, W.J.; Holak, S.L. The good old days: Observations on nostalgia and its role in consumer behavior. *Assoc. Consum. Res.* **1991**, *27*, 323–329.
32. Ponnam, A.; Balaji, M.S. Matching visitation-motives and restaurant attributes in casual dining restaurants. *Int. J. Hosp. Manag.* **2014**, *37*, 47–57. [CrossRef]
33. Jaafar, S.N.; Lumbers, M.; Eves, A. Does Food Really Matters in the 'Eating Out' Experience in Restaurants? 2006. Available online: https://openresearch.surrey.ac.uk/esploro/outputs/conferencePresentation/Does-food-really-matter-in-the/99516218802346 (accessed on 13 December 2021).
34. Berumen Calderón, M.F.; Estolano Cristerna, D.; Sterling Zozoaga, A.S.; Munoz, M.A.M.V. Qualitative Approach for the Design Stage of a Performance Measurement System to Increase Gross Profit in Restaurants. Case Study: Hard Rock Cafe Mexico Restaurants. *Eur. Sci. J.* **2022**, *18*, 58. [CrossRef]
35. Hwang, J.; Hyun, S.S. The impact of nostalgia triggers on emotional responses and revisit intentions in luxury restaurants: The moderating role of hiatus. *Int. J. Hosp. Manag.* **2013**, *33*, 250–262. [CrossRef]

36. Cha, S.-S.; Shin, M.-H. Author Assist professor F. The Effect of Delivery Food on Customer Emotional Response and Repurchase Intention. *Korean J. Food Health Converg.* **2021**, *7*, 1–10. [CrossRef]
37. Vignolles, A.; Pichon, P.E. A taste of nostalgia: Links between nostalgia and food consumption. *Qual. Mark. Res.* **2014**, *17*, 225–238. [CrossRef]
38. Renko, S.; Bucar, K. Sensing nostalgia through traditional food: An insight from Croatia. *Br. Food J.* **2014**, *116*, 1672–1691. [CrossRef]
39. Blake, C.E.; Bisogni, C.A.; Sobal, J.; Devine, C.M.; Jastran, M. Classifying foods in contexts: How adults categorize foods for different eating settings. *Appetite* **2007**, *49*, 500–510. [CrossRef] [PubMed]
40. Che Ishak, F.A.; Mazlan, N.; Lokman, N.A.; Abd Karim, M.S.; Mohamad, S.F. Exploring the Issues and Challenges in Managing Themed Restaurant in Klang Valley, Malaysia. *Int. J. Acad. Res. Bus. Soc. Sci.* **2021**, *11*, 1121–1135. [CrossRef]
41. Bojanic, D.C.; Rosen, L.D. Measuring service quality in restaurants: An application of the servqual instrument. *J. Hosp. Tour. Res.* **1994**, *18*, 3–14. [CrossRef]
42. Yüksel, A.; Yüksel, F. Market segmentation based on tourists' dining preferences. *J. Hosp. Tour. Res.* **2002**, *26*, 315–331. [CrossRef]
43. Fisher, D.P. Restaurant Staff Service Tips. 2003. Available online: https://silo.tips/download/restaurant-staff-service-tips-next (accessed on 8 December 2021).
44. Chung, Y.S. Identification of Economic Value Drivers Impacting Operational Cash Flows in the Casual Theme Restaurant Industry. Master's Thesis, Virginia Polytechnic Institute and State University, Blacksburg, VA, USA, 2005.
45. Hou, M.; Hu, Z. Patronage, only for happiness?—An analysis on coexistence of multiple consumption emotions. *Bus. Manag. Res.* **2013**, *2*, 116–127. [CrossRef]
46. Walker, J.R. *Introduction to Hospitality*, 5th ed.; Pearson Prentice Hall: Upper Saddle River, NJ, USA, 2008.
47. Department of Statistic Malaysia. The Sources of Malaysia's Official Statistics. 2020. Available online: https://www.dosm.gov.my/v1/index.php?r=column/cone&menu_id=TWNzZjZSZS9RUzhHYzdXODhFSEhLdz09 (accessed on 7 November 2021).
48. Roscoe, J.T. *Fundamental Research Statistics for the Behavioral Sciences*, 2nd ed.; Holt, Rinehart and Winston: Austin, TX, USA, 1975.
49. Kim, W.G.; Moon, Y.J. Customers' cognitive, emotional, and actionable response to the servicescape: A test of the moderating effect of the restaurant type. *Int. J. Hosp. Manag.* **2009**, *28*, 144–156. [CrossRef]
50. Holbrook, M.B. Nostalgia and Consumption Preferences: Some Emerging Patterns of Consumer Tastes. *J. Consumer Res.* **1993**, *20*, 245–256. [CrossRef]

proceedings

Proceeding Paper

The Prospective Economic Cooperation between Somalia and Some ASEAN Nations †

Abdi Omar Shuriye

School of Business and Social Sciences, Albukhary International University, Jalan Tun Abdul Razak, Alor Setar 05200, Kedah, Malaysia; abdi.shuriye@aiu.edu.my

† Presented at the International Academic Symposium of Social Science 2022, Kota Bharu, Malaysia, 3 July 2022.

Abstract: The Association of Southeast Asian Nations (ASEAN) was established to address economic and political issues for the member nations; today the economy remains the centre stage of ASEAN. The group launched its Economic Community platform in 2015. The AEC is a milestone promoting ASEAN economic integration. Somalia is a potential economic hub in the East African region, and the Somali people are vibrant entrepreneurs nicknamed "the Chinese of Africa". This paper contends that Somalia and ASEAN should enhance their economic collaboration and ASEAN should not neglect or leave this part of the world to the Chinese or western economic exploitation. Given political culture of ASEAN, Somalia can learn from the enormous experiences of ASEAN both in economics and in political stability. The ASEAN region is projected to have the fourth largest gross domestic product (GDP) by 2050, and it continues to perform far better than the world's average. Somalia already has economic connections with major ASEAN economies including Malaysia, Thailand, and Indonesia. The paper provides practical suggestions for economic collaboration between ASEAN and Somalia. The methodology employed is historical and conceptual to highlight the economic and historical values of both sides. The research problem addressed is that there is a scarcity of literature on the economic relations between Somalia and ASEAN.

Keywords: ASEAN; Somalia; economic; cooperation

Citation: Shuriye, A.O. The Prospective Economic Cooperation between Somalia and Some ASEAN Nations. *Proceedings* **2022**, *82*, 3. https://doi.org/10.3390/proceedings2022082003

Academic Editor: Mohamad Rahimi Mohamad Rosman

Published: 6 September 2022

1. Introduction

It was as early as 1955 when representatives from 29 Asian and African nations met in Indonesia to establish the non-aligned alliance between the two continents. President Sukarno initiated this meeting, entitled the Bandung Conference. These countries agreed to establish a framework for trade among themselves. Interestingly 50 years later from the first meeting, that is, in 2005, 106 nations from the two continents met again in Indonesia to establish the New Asian-African Strategic Partnership (NAASP), this time focusing on economic ties. From there on, the ASEAN-African business continue to expand.

Hence, there have been various projects and programs under the NAASP banner, from diplomatic training and technical cooperation to a business forum. Subsequently, according to the Asian Development Bank, from a base of ASEAN-Africa trade of merely US$2.8 billion in 1990, trade levels reached US$42.5 billion in 2012, equating to an annual growth rate of 14 percent. This has made Africa the second fastest-growing continent for ASEAN trade other than Asia. Africa is also becoming a destination for investment from ASEAN. Additionally, based on the UN World Investment Report in 2012, foreign direct investment (FDI) from Singapore reached $15.9 billion, making it ASEAN's largest investor in Africa. Similarly, the FDI from Malaysia and Indonesia is not far behind.

There are currently more than 200 companies from ASEAN operating in Africa, predominantly involved in agribusiness, manufacturing, oil and urban development. Currently, in 2021, among the ASEAN nations, the biggest traders with Africa are Thailand (US$11.6 billion), Indonesia (US$10.7 billion) and Singapore (US$9.5 billion), while South

Africa, Nigeria and Egypt have the largest import markets in Africa for ASEAN goods [1]. It is important to highlight that seven ASEAN nations have formed the ASEAN-Pretoria Committee to boost trade with South Africa, and Vietnam, Cambodia and Laos have seen sharp increases in trade since the Organisation Internationale de la Francophonie strengthened economic ties with French-speaking African nations.

Likewise, ASEAN and African nations are in the process of building a holistic body of knowledge on the relations between ASEAN and Africa as they attempt to develop a network of scholars from ASEAN countries with an expertise on Africa and a network of African scholars and institutions with expertise the ASEAN region. This is spearheaded by the ASEAN Studies Centre of Thailand.

2. Brief Delineation of ASEAN

The Association of Southeast Asian Nations (ASEAN) was founded on the 8th of August in 1967 in Bangkok, Thailand. The first to sign the ASEAN Declaration were Malaysia, Singapore, Indonesia, Thailand, and the Philippines. As agreed, the ASEAN Declaration obligates the signatory nations to cooperate for the purpose of social progress, cultural development, economic growth, and regional peace as well as political stability. Apart from the signatory nations, the other nations in the region started to join, as ASEAN expanded in the region. Brunei Darussalam joined the group on 7 January 1984, while Viet Nam became a signatory on 28 July 1995. Subsequently, both Lao PDR and Myanmar became members on 23 July 1997, while Cambodia decided to come aboard on 30 April 1999; making ASEAN a group of ten Member States. ASEAN aims to accelerate the economic growth, social progress, and cultural development in the region through promoting regional peace and stability through abiding respect for justice and the rule of law in the relationship among countries of the region and adherence to the principles of the United Nations Charter through joint endeavors in the spirit of equality and partnership to strengthen the foundations for a prosperous and peaceful community of Southeast Asian Nations [2].

3. Somalia and Its Economy

Somalia obtained its independence from the Italians and then, in 1960, from the British. By then Somalia was already one of the East African nations renowned for its trading with the Arabian Peninsula, Persia, India, and China. The country went through the first phase of development under an Arab sultanate from the XIIIth to the XVIth century. Ibn Battuta, the Muslim traveler, visited Somalia in 1331 and provided historians with a first-hand description of this nation. The military regime tried to build government institutions through tyranny and an authoritarian form of government; nonetheless, after the defeat in the Ogaden war against Ethiopia in 1978, the fragile political and economic situation of the nation deteriorated and eventually collapsed in 1990 [3].

From there on, the country has been maintained by the international community and its income is mainly determined by them and by the Somali diaspora. The United Nations, therefore, classified Somalia's economy as one of the least developed nations in the world; almost 85% of the nation's population depends on agriculture and livestock for their livelihood. The IMF and the World Bank are the few reliable sources supporting Somalia's economic activities. The IMF believes that the Somali economy has expanded by 5.9% since 2014. The Somali diaspora are the key investors in Somalia, in the transportation, communications, fishery equipment, airlines, hotels, telecommunications, education, health and construction sectors.

The economy of the nation, in its larger sense, consists of both traditional and modern production but gradually moves towards modern industrial techniques. According to the World Bank, the nation's economy suffers because of state failure. Based on the information available from the African Development Bank, it is almost impossible to capture Somali economic data as the nation is characterised by a severe lack of basic economic and social statistics, with the civil war and institutional collapse as the main cause of this situation. Nonetheless, according to the World Bank data, the nation's GDP has risen from $917.0 million in 1990 to an annual GDP of $6.2 billion in 2018 [4].

In this pastoral economy, agriculture accounts for about 78% of the GDP and employs 70% of Somalia's workforce, while livestock contributes about 48% of the Somali GDP and more than 57% of the export earnings of the nation. Other main exports constitute sugar, bananas, fish, corn, and meat. It is significant to note that Somali traders are challenging major meat exporters such as Australia in the Persian Gulf livestock and meat market, offering quality animals at a comparatively lower price. In 2012, Somalia exported 3 million sheep to the Middle East overtaking Australia's 2 million. Somalia has a modest industrial sector based on the processing of agricultural products, accounting for 15% of the nation's GDP. Although the security situation hinders the formation of large manufacturing entities, small and medium firms are appearing in major cities, thanks to the Somali diaspora. Many of these small-scale plants have opened or re-opened in recent years and newer ones have been created. These plants include fish-canning, meat-processing, pasta processing, mineral water, and plastic bags. The Somali diaspora developed various telecommunications companies providing the nation with basic communication systems. This basic but useful initiatives are owned and funded by local entrepreneurs and backed by expertise from China and Turkey. The firms offer affordable services, bringing Somalia ahead of many parts of Africa in this sector.

4. The ASEAN Economy

The ASEAN economy is one of the most developed economies in the world and it produces goods and services that Somalia needs most. In fact, even in challenging times such as 2021, ASEAN economies were poised for robust recovery with 6% real GDP growth. As stated by Global Data: "The six largest ASEAN nations (Indonesia, Malaysia, the Philippines, Singapore, Thailand, and Vietnam) are expected to witness positive real GDP growth rates in 2022" [5].

Malaysia has successfully diversified its economy, from an agriculture and commodity-based economy to one that is robust in manufacturing and services.The economy has propelled the country into becoming a leading exporter of electrical appliances and related components. Besides, Malaysia developed an open economy and trade relations with the world, with a trade to GDP ratio averaging over 130% since 2010. Indeed, openness to trade and investment has been instrumental in employment creation and income growth, with about 40% of jobs in Malaysia linked to export activities. With COVID-19 still a threat, Malaysia's short-term economic outlook will be dependent on government measures to help the private sector weather the shock of COVID-19 reduced export-led growth; meanwhile, the exhausted fiscal space limits public investment-led expansion. Over the longer term, nonetheless, as Malaysia converges with high-income economies, incremental growth will depend less on factor accumulation and more on raising productivity to maintain higher potential growth.

The government's ongoing structural constraints will be vital to support and sustain Malaysia's development strategy. The World Bank ranks Malaysia 55th out of 157 countries on its Human Capital Index; if Malaysia is truly interested to realize its vision of developed status, it must further progress in the key sectors, such as education and health. Social harmony is also included in the key important areas. In Southeast Asia, the country ranks fourth according to the IMF; and in the world, it stands as the 36th largest economy according to the 2020 report. The Malaysian economy has adopted high labour productivity as it maintains a high density of knowledge-based industries; it also upholds cutting-edge technology for manufacturing and digital economy compared to its regional neighbours including Thailand, Indonesia, the Philippines, or Vietnam. This makes its economy the 27th most competitive in the world. In addition, the country's economy is highly diversified with the export value of high-tech products standing at US$57.258 billion, the second highest after Singapore. It is also important to note that the country is the second largest exporter of palm oil products globally, after its neighbour, Indonesia [6].

Meanwhile Indonesia's economy is the largest in the region; and competitively one of the emerging markets in the world. As a middle-income nation and member of the G20, the country is categorized as a newly industrialized nation. Indonesia is the 15th largest economy in the world, and 7th largest in GDP. Some ten years ago, in 2012, Indonesia replaced India as the second-fastest-growing G-20 economy, behind China. COVID-19, however, has stressed the Indonesian economy, with economic growth collapsing to almost −2.07%, making it the worst growth since the 1997 Asian economic crisis [7].

Since Indonesia is the world's largest producer of palm oil, agriculture is a key sector which contributes more than 15.45% to the GDP of the nation. This sector has around 35% of land area that is uses for agricultural purposes and it employs more than 52 million workers; this is almost 45% of the total workforce in the country. Another industry classified within the agriculture sector is the production of seafood which has reached over 25.55 million metric tons, valued at around 23.35 billion US dollars in 2016 [8]. The manufacturing sector plays a key role in the economic development of the country. In fact, this sector contributes over 25% to the GDP. With these economic capacities taken into consideration, the Indonesian government is ambitiously planning to push the country up into the top ten economies in the world by 2030. The manufacturing sector could be seen as the heart of this economic determination and of the vision of 2030; it includes but is not limited to, food and beverages, textiles and garments, electronics, chemicals and automotive industries. It is this sector also that makes Indonesia the 10th-largest manufacturing country in the world. Other notable sectors are renewable energy and the automotive industry [9].

Interesting to note is the fact that over 65% of Thailand's economy depends on exports. As a newly industrialised nation with a GDP of US$508 billion, Thailand remains the 8th largest economy of Asia. The agricultural sector produces 10.6% of GDP; whereby the trade, logistics and communication sectors combined produce 14.5%. The construction and mining sectors are responsible for 4.8% of the nation's GDP, while sectors such as education, restaurants and hotels produce over 25% of the nation's GDP [10].

This second-largest economy in Southeast Asia, which also ranks fourth in Southeast Asian per capita GDP, holds US$238.6 billion in international reserves. That is the second largest in Southeast Asia. Moreover, Thailand has a surplus in the current account balance which ranks tenth of the world as it ranks also second in Southeast Asia in external trade volume. In fact, the World Bank calls it "one of the great development success stories". Thailand is a member of the WTO and AFTA and it has diverse types of free trade agreements with Australia, China, United States, and the European Union. Notably, COVID-19 has had a slighter impact on Thailand's economy comparatively; subsequently, the economy remained in good condition, having the highest growth in Southeast Asia. The three main economic sectors, namely agriculture, manufacturing, and services, contribute the most to the nation's export ability. Manufactured goods stand at over 85%, including electronics at 16%, vehicles at 15%, machinery and equipment at 8% and foodstuffs at 8%. Agricultural goods such as rice and rubber, contribute 9% GDP. Other key sectors include textiles and garments, plastics, footwear, electronics, integrated circuits, computers and components, automobiles and parts, and cement [11].

The economy of Singapore is a genuinely free market economy and the most open market in the world. It is also the 3rd least corrupt country in the world. Thus, it has an environment and government which are the most pro-business in the world; it remains the second highest GDP per capita by purchasing power, with low tax rates. This business-friendly nation has attracted international companies and hosts the headquarters of APEC. It has benefited from the inward flow of FDI from global investors and institutions due to its highly attractive investment climate and a stable political environment in recent years. It may be noted that much of Singapore economy remains in the hands of the Sovereign Wealth Fund Temasek Holdings which dominates the stakes in the nation's largest companies. Other important factors to note in the Singaporean economy are FDI, wealth management, an extended concept of intermediary trade and purchasing raw goods and refining them for re-export. Vietnam's economy on the other hand, focuses on electronics including computers and electrical products and textiles. According to the World Bank and IMF, Vietnam has over performed in building its economic sector in the last ten years.

The 34th largest economy of the world is that of the Philippines. According to the IMF report 2021, the economy of this nation is the 12th largest in Asia. As an emerging market, the Philippines is one of the three largest economies in ASEAN. As a newly industrialised nation, its economy is in transition from one based on agriculture to one based more on services and manufacturing. The nation's purchasing pawer parity is estimated to be US$1.49 trillion and the 18th in the world. The Philippines' exports include but are not limited to semiconductors and electronic products, transport equipment, garments, copper products, petroleum products, coconut oil, and fruits. Its major trading partners include China, United States, Japan, some European nations, and ASEAN. Besides, the country is one of Asia's fastest growing economies, and the economy of this nation is projected to be the 4th largest in Asia and 13th largest in the world by 2050 [12].

Cambodia, although, comparatively small economy, follows an open market system, and the nation has seen rapid economic progress in the last decade. Its US$24.58 billion GDP is rapidly increasing; tourism and textile industries are the nation's largest economic generators. Nonetheless, agriculture remains the key source of income for ordinary Cambodians, particularly those living in rural areas. However, 83% of the country's exports come from the textile industry, which represents the largest portion of Cambodia's manufacturing sector. The sector employs 336,600 workers, of which 93% are female. Agriculture is the traditional backbone of the Cambodian economy. Agriculture accounted for 96% of the nation's GDP in the 1980s and employed approximately 87% of the working citizens. Cambodia produces rice, maize, cassava, soyabeans, sweet potatoes, sesame seeds, groundnuts, as well as dry beans. Rubber and rice are, nonetheless, the principal commercial crops of the nation. Tourism, since the late 1990s, is becoming the nation's second largest industry, after textile manufacturing. The country's cultural heritage is popular, with many foreign tourists visiting the ancient Hindu temples and other attractions.

The economy of Laos is termed as that of a "developing economy"; this socialist nation, however, is rapidly growing its economy and has lower-middle income status currently. The Laos market model is a combination of high degrees of state ownership with an openness to FDI. Since the era of the planned economy the nation underwent several economic reforms as part of economic restructuring that aimed to integrating the nation to the world economy. Laos currently ranks amongst the fastest growing economies in the world, averaging 8% a year in GDP growth. The government initiated the latest round of state-owned enterprise reform in 2019; these reforms are meant to ensure the state-owned enterprises (SOEs) remain profitable ventures and sustainable resources for the nation. Agriculture, mainly rice farming, dominates the economy, employing an estimated 88% of the nation and producing 53% of the nation's GDP. Laos's agricultural products include vegetables, corn, sweet potatoes, coffee, sugarcane, tobacco, cotton, tea, peanuts, rice, buffalo, pigs, and cattle.

Laos opened itself to the world in the early 1990s, making tourism one of the fastest growing industries in the economy and it plays a vital role to this date. Brunei on the other hand, has a hybrid economic approach, with a variety of private actors combined with centralized economic government regulations. As a member of APEC and ASEAN, as well as the Trans-Pacific Partnership (TPP), the country aims for economic transformation. Its economy is entirely dominated by the export of crude oil and natural gas, as this sector accounts for over a half of the GDP and the nation's revenue. Nonetheless, the government has made remarkable progress in diversifying some basic policies of the economy away from oil and gas. The country is the third-largest oil producer in Southeast Asia [13].

Funding: This research received no external funding.

Institutional Review Board Statement: Not applicable.

Informed Consent Statement: Not applicable.

Data Availability Statement: Not applicable.

Conflicts of Interest: The author declares no conflict of interest.

References

1. ASEAN Secretariat; United Nations Conference on Trade and Development (UNCTAD). *ASEAN Investment Report 2020–2021 Investing in Industry 4.0*; ASEAN Secretariat: Jakarta, Indonesia, 2021; pp. 3–307.
2. The Association of Southeast Asian Nations (ASEAN). *ASEAN Political-Security Community Blueprint*; ASEAN Secretariat: Jakarta, Indonesia, 2007; pp. 1–16.
3. Assessment Capacities Project (ACAPS). *ACAPS Briefing Note: Somalia—Food Insecurity (16 November 2021)*; ReliefWeb: Nairobi, Kenya, 2021; pp. 1–7.
4. Khayre, A.M. Somalia: An Overview of the Historical and Current Situation. *SSRN* **2016**, 1–43. [CrossRef]
5. DataReportal. Digital 2021: Global Overview Report. 2021. Available online: https://datareportal.com/reports/digital-2021-global-overview-report (accessed on 2 July 2022).
6. Keling, M.F.; Som, H.M.; Saludin, M.N.; Shuib, M.S.; Ajis, M.N. The Development of ASEAN from Historical Approach. *Asian Soc. Sci.* **2011**, *7*, 169–189. [CrossRef]
7. International Monetary Fund. *International Monetary Fund Annual Report 2012: Working Together to Support Global Recovery*; International Monetary Fund: Yew York, NY, USA, 2012; pp. 1–72.
8. United Nations (UN). *Asian Foreign Direct Investment in Africa: Towards a New Era of Cooperation among Developing Countries*; United Nations: Nairobi, Kenya, 2007; pp. 1–190.
9. Yusof Ishak Institute (ISEAS). *The State of Southeast Asia: 2021 Survey Report*; ASEAN Secretariat: Jakarta, Indonesia, 2021; pp. 1–55.
10. The Association of Southeast Asian Nations (ASEAN). *ASEAN Development Outlook: Inclusive and Sustainable Development*; ASEAN Secretariat: Jakarta, Indonesia, 2021; pp. 16–287.
11. The World Bank. *World Development Report 2021: Data for Better Lives*; The World Bank: Washington, DC, USA, 2021; pp. 3–326.
12. ASEAN Secretariat. The ASEAN Economic Integration Brief. 2021. Available online: https://asean.org/serial/asean-economic-integration-brief-number-10-november-2021/ (accessed on 2 July 2022).
13. Kohnert, D. ASEAN and African Relations: Towards a Renewed Partnership? *SSRN* **2021**, 1–27. [CrossRef]

proceedings

MDPI

Proceeding Paper

The Employees-Organizational Performance Needs Model: The Mapping of Maslow's Hierarchy of Needs and Balanced Scorecard into the Maqasid Al-Shari'ah Model †

Noor Liza Adnan [1,*], Wan Noor Hazlina Wan Jusoh [2], Rokiah Muda [1] and Rohana Yusoff [3]

[1] Faculty of Accountancy, Universiti Teknologi MARA (UiTM) Cawangan Terengganu, Dungun 23000, Terengganu, Malaysia
[2] Academy of Contemporary Islamic Studies, Universiti Teknologi MARA (UiTM) Cawangan Terengganu, Dungun 23000, Terengganu, Malaysia
[3] Faculty of Computer and Mathematical Sciences, Universiti Teknologi MARA (UiTM) Cawangan Terengganu, Dungun 23000, Terengganu, Malaysia
* Correspondence: noorliza@uitm.edu.my
† Presented at the International Academic Symposium of Social Science 2022, Kota Bharu, Malaysia, 3 July 2022.

Abstract: Organizations employ a performance management system to ensure sustainable performance. This paper aims to develop a model of performance management that may fulfil the basic needs of an employee and simultaneously ensures organizational performance. A review of the literature concerning Maslow's Hierarchy of Needs and Balanced Scorecard was conducted. As these two models are criticized for failing to recognize the spiritual needs, they are then mapped into the Maqasid al-Shari'ah model to develop a more comprehensive performance management model, named the Employees-Organizational Performance Needs Model (EOPNM), which may contribute to the body of knowledge in the area of performance management.

Keywords: performance management; Maslow's Hierarchy of Needs; Balanced Scorecard; Maqasid al-Shariah; Employees-Organizational Performance Needs Model; spiritual needs

check for updates

Citation: Adnan, N.L.; Jusoh, W.N.H.W.; Muda, R.; Yusoff, R. The Employees-Organizational Performance Needs Model: The Mapping of Maslow's Hierarchy of Needs and Balanced Scorecard into the Maqasid Al-Shari'ah Model. *Proceedings* **2022**, *82*, 4. https://doi.org/10.3390/proceedings2022082004

Academic Editor: Mohamad Rahimi Mohamad Rosman

Published: 6 September 2022

1. Introduction

To ensure sustainability, organizations frequently strive to optimize their performance. Various methods are employed to improve the organizational performance. At an organizational level, the performance measurement system (PMS) is used to constantly monitor, evaluate, and control an organizational operation and performance by setting the standard that needs to be achieved, known as the key performance indicators (KPI) [1–4]. On the other hand, a performance appraisal system (PAS) is in place to evaluate, control, and reward employees' performance against the KPI sets [5–7], as employees form a crucial part of an organization who execute organizations' strategies and plans and ensure the targets set are achieved. As such, PMS and PAS are intertwined, each supporting the success or failure of another, in a broader performance management system employed in an organization.

PMS and PAS are being extensively used by organizations to control their employees and also to reward them once the targets are achieved. However, in the period of intense competition nowadays, companies strive to be ahead of competitors, often by driving their employees to reach a new limit that is sometimes unreasonable even at the expense of their employees' wellbeing [3,8,9]. This can be done by altering the KPI sets where measured criteria are added to include even those that are not under the control of the employees. Hence, it is not surprising when the extant literature has reported that PMS has become the key factor that causes job burnout and stress among employees [9,10]. Consequently, some literature has also reported that such PMS employed may lead to certain dysfunctional behaviors that deprive the organizations of the good performance they initially set to achieve [3,11]. In other words, such PMS and PAS are said to have failed to satisfy the human needs of the employees, making them less motivated, often lowering their self-esteem.

One most discussed PMS is known as the Balanced Scorecard (BSC) introduced by Kaplan and Norton in the 1990s. At the organizational level, BSC has been extensively researched to measure and manage organization performances [12]. Its four dimensions of customer satisfaction, internal process, learning and growth, and financial performance provide a balanced view of the organizational performance. Interestingly, when viewed from the employees' perspective, these dimensions may also represent an employee's success perspective or well-being, indicating the extent of the fulfillment of their basic needs. An employee may be viewed as successful and has their needs fulfilled when these four perspectives are achieved that enhance their job satisfaction. However, BSC has totally ignored the spiritual or humanistic aspect of an employee, which is highly vital for the happiness or self-actualization of an employee [13,14]. Hence, when looked at from an individual level, it can be concluded that many PMS and PAS employed have failed to fulfill the human needs as stipulated in Maslow's Hierarchy of Needs (MHON). MHON has been the most referred to model of human needs, which shows that humans are motivated to fulfill certain needs, like physiological, safety, social, self-esteem, and self-actualization, to survive and then to achieve a certain potential [15]. However, even MHON has been criticized by many, as it also does not consider the spiritual need of humans [16,17]. In short, it seems that either BSC or MHON is still inadequate to explain the fulfillment of needs of an employee that would collectively lead to the organizational performance.

From an Islamic perspective, such needs are outlined in the Maqasid Shari'ah (MQS), which divides human basic needs into five fundamental values, i.e., the safeguarding of faith (Al-Deen), life (Al-Hayah), mind (Al-Aql), dignity (Al-Mur'uah), and properties (Al-Mal). As MQS is aimed to preserve the wellbeing of humankind, it is believed that making it the foundation of all human systems will lead to the best outcome for everybody, be it an employee or an organization. Therefore, this paper aims to propose a performance model where both employees and organization may reap the most benefits if properly implemented. It uses the framework of MQS to map the organizational performance (using the BSC model) with the employee's needs and performance (using the Maslow's model). Even though MQS is considered as applicable to all Muslims, the application of MQS and this model can also be applied to all employees and organizations, as Islamic values are actually universal values, applicable to everyone and every organization. Not much research has tried to link these three models in order to harmonize human (employee) needs with organizational needs [18]. This paper will start with the review of MHON, followed by the BSC and MQS. Another section will then map all three models so as to create a need and performance model that will benefit both employees and organizations.

2. Human Needs from The Perspective of Maslow's Hierarchy of Needs (MHON)

MHON has been one of the most widely referred to models in many diverse areas of study, such as psychology, management, marketing, economics, sociology [17], and many more, since it was introduced in the 1950s up until now. MHON explains that humans are creatures of needs, and there are certain factors that motivate them to achieve these needs. Maslow postulated that human needs can be categorized and structured in a hierarchy of prepotency and probability of appearance, which is depicted in a pyramid of a five-stage hierarchical levels model of human needs. The model has gained worldwide acceptance due to its eye-pleasing pyramid that can be easily understood and accepted by people's common sense [17].

From the bottom of the hierarchy upwards, these needs are arranged in the following sequence: (i) physiological needs (food, shelter, and clothing); (ii) safety needs (security, law, and freedom from fear); (iii) love and belonging needs (friendship, trust, and giving affection and love); (iv) esteem needs (dignity, achievement, and prestige); and (v) self-actualization needs (realizing personal potential and personal growth). Humans strive to achieve the most potent needs, such as food and water (situated at the bottom of the pyramid), before progressing to a need at a higher stage (self-actualization at the top), indicating that needs lower down in the hierarchy must be first satisfied before individuals can advance to needs higher up. Maslow then further divided the five-stage model into two categories of deficiency and growth needs [19].

Due to various criticism of his five-stage model, Maslow expanded his model to include cognitive and aesthetic needs (to become a seven-stage model) and, later, transcendence needs (to become an eight-stage model) in the 1970s ([15] cited from Maslow (1943 and 1954)). Cognitive needs (knowledge and understanding) are arranged at the fifth level of the hierarchy, while aesthetic needs (appreciation and search for beauty) are placed at the sixth level, pushing self-actualization to the seventh level. Transcendence needs, are is described as the needs motivated by values that surpass the personal self (service to others, religious faith, etc.) is then placed at the eighth level, completing the eight-stage model.

Maslow's model of the hierarchy of needs, though being pervasive in many areas, is still being criticized on different grounds. The authors of [18] argued that Maslow made an observation on his immediate surroundings, resulting in the insufficient coverage of his study population. In addition, his study is extremely subjective and is open to bias, which reduces the validity of the data, making the operational definition of self-actualization somewhat doubtful and not generalizable [15]. Though the first two levels of physiological and safety may be applied universally, the other three levels may not be regarded the same by people in other cultures.

Maslow's assumption that lower needs must be met before a person can progress to achieve and fulfill the higher order needs has also received a lot of criticism, as things do not always follow this order [15,17]. A hungry man is still motivated to love and protect his loved ones even if his hunger is not yet satisfied, and the Prophets' Companions' motivated behaviors, which always prioritized their self-actualizing needs rather than their physiological needs [16], are examples put forward. The authors postulated that the Prophets' Companions ensured that religion always precedes their life, which should be the correct order of hierarchy of needs for all Muslims, specifically, and all humans, generally.

All Muslims should seek to achieve the pleasure of their Creator by obeying His commandments, as they will be rewarded in the Hereafter; hence, self-transcendence needs should be placed at the lowest level instead of the top. This might be the biggest flaw in Maslow's model that ignores the need for religion or spirituality in humans [16,17]. Though transcendence needs were added to the model, being placed at the top depicts them as something achievable only to certain worldly successful people instead of everybody born as human. Considering this need, [17] called for the model to be inverted, where self-transcendence needs should be placed at the bottom level as the initial stage that defines a person's purpose of life before the fulfillment of physiological needs. Based on this argument, the next sections elaborate on why and how it is possible.

3. Balanced Scorecard and Human Needs

Balanced Scorecard (BSC) is the most commonly discussed PMS, since it was introduced by Kaplan and Norton in the 1990s as a systematic tool to measure organizational performance and improvement through the use of four measurement perspectives: financial, customer, internal business process, and learning and growth [20]. BSC is built on the basic premise that a company's performance evaluation system is tailored to its well-defined mission and a strategy for fulfilling that mission [21]. Consequently, it determines the targets to be achieved by its employees, which should serve as a guideline for them to reach the company's mission and indirectly serves as their source of motivation. In short, BSC clarifies a company's needs, comprising both financial and nonfinancial measures, signifying that a company would achieve a sustainable performance if it could improve its internal business processes and enhance its employees' learning and growth as a means to satisfy its customers' satisfaction, which would eventually lead to financial success.

In summary, BSC links people to the strategy and, finally, to the performance. Though BSC clarifies a company's needs to perform and be successful, when viewed from the employees' perspectives, it may also be used to map out employees' needs to succeed in a company, specifically, and also in life, generally. Hence, it might also explain the fulfillment of human (or employee) needs as described by MHON. By being employed, employees earn their salary that they use to pay for food, clothes, shelter, etc. that fall within their physiological needs. They also abide by rules, certain employment contracts, and company rules and regulations, which promise them security that fulfill their safety needs. Simultaneously, they are surrounded by friends and colleagues that fulfill the social needs of love and to be loved, sense of belonging, togetherness, etc. Through achieving company and individual goals and then being rewarded for such achievements, employees reach their esteem needs. Self-actualization needs are then achieved when they keep striving for better, to improve themselves and, indirectly, the company. This is closely related to the learning and growth perspective, which leads to improvement in business processes. When looked at this way, it becomes obvious that, though BSC is meant to aid in a company's performance improvement, it may also explain the fulfillment of human basic needs as outlined by Maslow.

Quesado et al. [22] argued that, if BSC is used effectively, it performs as a rich source of information and knowledge, enables participants to communicate regularly with each other, allows the association of a long-term strategy with short-term actions, and produces strategic awareness among employees. With regards to this, when applied to an employees, BCS serves to fulfill their love and belonging, esteem, and self-actualization needs. Being together in a department or company creates a sense of belonging or esprit de corps that, in time, makes one feel love and be loved. Besides close family, colleagues become the people someone turns to in times of despair or times of joy. The information shared through effective regular communication does not only develop a bond among employees but also becomes a source of knowledge that forms the learning process towards their self-growth. Being evaluated by various performance measures employed in the BSC is seen as a necessary stop-and-check process to see if one is on the right track, and if unsatisfactory, the reasons for underachievement can be determined and actions can be taken to rectify it. These are necessary to reach the self-actualization needs once the esteem needs have been reached. Subsequently, the company also benefits from competent and high-esteemed employees, as business processes are much improved, which leads to high customer satisfaction.

One interesting advantage of BSC is it includes external stakeholders instead of only the internal stakeholders [12], making an organization become transparent, where external stakeholders are seen as an integrated part that play a necessary role to contribute to their overall success. In short, it exhibits the interconnectedness of various parts, which reflects that humans need to work together to create harmony and meaningful life so as to reach their self-actualization or even self-transcendence needs.

However, BSC has also been criticized for using too many indicators, causing a lack of key success factors being defined to identify the KPIs [23], as this might lead organizations to lose focus and fail to find the linkages between those indicators. In terms of human needs, it signifies that humans live in interrelatedness, and we need each other to fulfill our needs, especially to achieve our growth needs. Failure to account for this causes stress or burnout, where employees often feel overloaded with work that is sometimes not even within their control [3]. It also leads to the mismatch of practices and expectations of PMS, resulting in conflicting situations and tensions between top management and employees [9]. Somehow, this is rooted from the adoption of excessive quantitative measures while ignoring the qualitative measures, normally in the form of humanity or spiritual aspects in any PMS employees.

Stemming from this weakness, [13] introduced Islamic perspective BSC, known as the Hadhari Scorecard, with an additional measure related to humanistic and spiritual strength besides the four classical measures introduced by Kaplan and Norton. The Hadhari Scorecard proposed the balance between the nonmaterial and material aspects of a performance in managing a formulated strategy. Spirituality is later widely discussed as a potential additional dimension in enhancing the traditional BSC. Spirituality supports the organizational performance [24] and improves BSC's efficiencies and effectiveness, since it steers the employees' well-being, organizational commitment, and productivity in addition to the organization's financial performance [25]. However, integrating spirituality into a PMS is still inconclusive due to its ambiguity, which has become the quest in this paper.

4. Human Needs from The Perspective of *Maqasid Al-Shari'ah*

Islam is a comprehensive religion that treats all issues in various aspects of human life, including the administration of an organization [26]. In this regard, Shari'ah should be the foundation of any form of organization, be it Islamic or secular organizations. The ultimate goal of the Islamic organization is to achieve the Shari'ah objectives, that is, Maqasid al-Shari'ah (MQS). MQS refers to the objectives that are being sought to be realized by the Shari'ah when deciding on a ruling aimed at realizing and safeguarding benefits to the people and protecting them from evil (see [27,28]). In this view, underlining an organization's management with Shari'ah results in happy and satisfied employees who contribute more to the success of an organization while ensuring employees are treated with respect and justice.

MQS can be divided into three broad categories, namely al-dharuriyyat (the essentials), al-hajiyyat (the complementary), and al-tahsiniyyat (the embellishments) [27,29–32]. Al-dharuriyyat are the interests on which the religion and worldly affairs of the people depend upon [28,33], and its absence leads to total disruption and disorder, and it could lead to an evil ending [27,28,31–33]. The protection of these interests can be established by two means: safeguarding their existence and protecting them from being destroyed [28,33]. These must be protected, and all measures that aim at safeguarding them must be taken, either by the individual or by government authorities. Al-dharuriyyat can be further divided into the protection of the five indispensable values (al-dharuriyyat al-khamsah) that are essential to human life [29], which are religion, life, intellect, lineage or dignity, and wealth [27,28,32–37]. The framework of MQS does not have a limit, as it covers all parts of the rulings in human life, including the organizational part. Organizations must ensure that all these five objectives are espoused in their organizations if they want to sustain their competitive advantage, as this results in retaining talented employees.

Al-hajiyyat are defined as interests that seek to remove severity and hardship in cases where such severity and hardship do not pose a threat to the very survival of normal life if the interests are absent [27,28,32,36]. On the other hand, al-tahsiniyyat refers to interests that require generosity, which seeks to achieve refinement and perfection in the conduct of people at all levels [28,36]. These three categories of MQS are subjective and can be applied in any area or field. In an organizational context, the al-dharuriyyat, al-hajiyyat, and al-tahsiniyyat concepts can be used to determine the level of necessities of any actions or decisions taken by the leaders or top management. In other words, it gives certain guidelines to leaders to make decisions based on priorities.

Chapra [34] elaborated how MQS can be applied in an organizational setting. The safeguarding of faith (Al-Deen) needs to come first before other elements, because it controls and influences human personalities, such as behaviors, lifestyle, preferences, and others. This can be done by providing a comfortable space to pray or ensure work matters do not contradict with an employee's faith and organizing religious talks to the employees. The safeguarding of life (Al-Hayah) involves taking employees' lives and welfare into account, such as offering an employment injury scheme or pension scheme. The safeguarding of the mind (Al-Aql) and dignity (Al-Mur'uah) are necessary to alleviate job stress and burnout while promoting a sane mind and self-esteem. Employees with a sane mind and high self-esteem are highly motivated and make wise decisions while not feeling tempted to commit fraud. The safeguarding of properties (Al-Mal) may have a huge impact on an organization's success, as it controls against the misappropriation of assets, hence maximizing the use of resources. Thus, the principles of MQS need to be articulated and promoted to Muslims and non-Muslims alike. When the Shari'ah can be realized and clarified through universal principles and purposes (Maqasid), an organization is functioning at its best, as it is staffed with highly dedicated employees who are not only accountable to their employer but also to their colleagues and the Almighty God.

5. The Employees-Organizational Performance Needs Model: Maslow's Hierarchy of Needs and Balanced Scorecard (BSC) As Mapped by Maqasid Al-Shari'ah

Islam is Al Deen, which means a way of life. It acknowledges different human needs and drives to survive, to procreate, and to worship. The Islamic way of life does not allow these needs and drives to be suppressed but teaches its believers to channel their needs and drives rightfully through the observance of the Islamic rules. This fact is clearly portrayed in a hadith through the words of Salman telling Abu al-Darda' "Your Lord has a right over you, your body has a right over you, and your family has a right over you, so you should give the rights of all those who have a right over you". Another hadith narrated by 'Abd-Allaah ibn 'Umar said, from the Prophet (peace be upon him), "Each of you is a shepherd and each of you will be questioned (about those under his care). The leader is a shepherd and will be questioned. A man is the shepherd of his family and will be questioned. A woman is the shepherd of her husband's house and she will be questioned. A slave is the shepherd of his master's wealth and will be questioned. Each of you is a shepherd and each of you will be questioned [about those under his care]" (al-Bukhaari, 4892; Muslim, 1829).

These hadiths are among several other hadiths that clearly demand Muslims to fulfill their obligations to themselves (to maintain good health, bodily, and spiritually); to their families, society, and country; and to Allah, their Creator. Islam also lays down clear guidelines to regulate business trading, professional life, and economic activities. It encourages Muslims to work and strive for perfection and wealth but should never turn to material wealth as the main purpose in life. The Islamic approach is to maintain a balance between material (worldly needs) and spiritual needs. In all circumstances of a person's life, whether wealthy or hardship, stability or unpredictability, happiness or sorrow, he or she must comply with the Shari'ah. Hence, a model of human needs proposed by Maslow (1970) is unable to acknowledge the reality of the interaction between the material and spiritual aspects of the human being in different life situations [17].

In managing the performance of an organization, both the human needs as well as organizational needs must be taken into consideration. One of the popular measures of organizational performance is the BSC [20], but it does not acknowledge the spiritual need of a human but focuses on measuring the organizational performance to achieve higher material success. Therefore, an organization that wants to strike a balance between workers' human needs (material and spiritual) and organizational needs will have to consider all the three models: MQS, MHON, and BSC. Among the three models, MQS is a more comprehensive model that incorporates both the Maslow and BSC models.

Rosbi and Sanep [38] attempted to integrate the MHON and MQS models to map human needs. Self-actualization needs are combined with the need for religion, as both needs are necessary to achieve the pleasure of the Creator, which leads to the fulfilling of life potential. This is indispensable in reaching harmony in life and the Hereafter. The life dimension of MQS is seen as related to the safety needs of Maslow's, as humans need some rule, law, and order to feel safe and secure, which ensures their survival. The intellectual (knowledge) needs are integrated with the social needs (love and belongingness) of Maslow's in the argument that humans need knowledge and know-how in order to socialize with others. The needs of lineage complement the self-esteem needs of Maslow's hierarchy of needs. Humans are seen as social creatures who need to possess some self-esteem, which becomes essential attributes to be respected and accepted by their community. Lastly, the wealth needs are combined with the physiological needs, as humans require certain wealth to fulfil their physiological needs for food, water, shelter, etc.

Abdullah et al. [39], on the other hand, only assimilated Maslow's self-actualization needs into all the five dimensions of MQS in their paper in explaining the linkage of conscientiousness, as one of the personality traits, to job satisfaction. Doing this, unfortunately, has equated the five dimensions in MQS to only self-actualization needs, which makes MQS become too narrow while ignoring the other aspects. Therefore, a new model that integrates both elements of the material and spiritual is necessary so that employees' needs can be fulfilled, leading to improved organizational, as well as individual, employee performances. This model is exhibited in Figure 1, named the Employees-Organizational Performance Needs Model (EOPNM), which illustrates the relationship between the dimensions of MQS, MHON, and BSC.

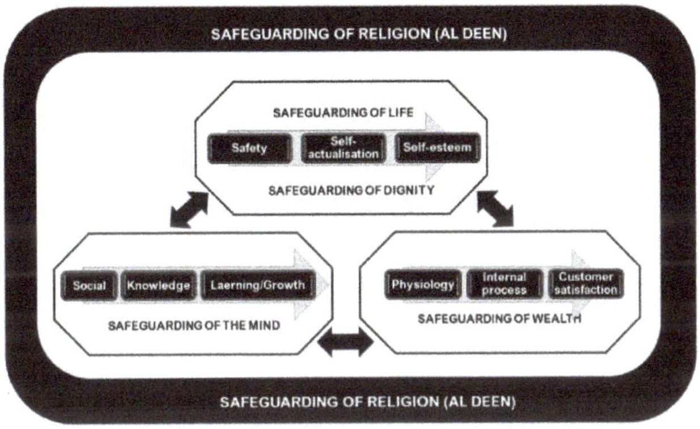

Figure 1. The Employees-Organizational Performance Needs Model (EOPNM): Maslow's Hierarchy of Needs and Balanced Scorecard as mapped by Maqasid al-Shari'ah.

Referring to EOPNM, the Safeguarding of Religion is placed at the outermost layer of the circles, because it incorporates all the other dimensions in the three models. The fundamental Islamic concepts: worshipping Allah ('ibādah), all provisions are provided by Allah (rizq), relying on Allah (tawakkul), and the belief in fate and predestination (qaḍā' and qadar) are all expressive of the fact that spiritual concepts permeate the entire life, actions, and decision-making processes of the individual and community [17]. Hence, the Safeguarding of Al-Deen is preserving the Islamic way of life that includes the Safeguarding of Life, Dignity, the Mind, and Wealth.

The Safeguarding of Life is to ensure the safety and protection of human beings; in an organizational context, it means safety and protection of workers. Workers should be able to enjoy the basic needs of life such as healthcare, food, housing, clothing, and transportation. In addition, job security and stability motivate workers to perform their best. The Safeguarding of Dignity is to protect the pride, dignity, and lineage of family and ancestors. This includes self-esteem, as well as having a reputation and respect from others in the society. Workers in an organization should be given due respect and freedom to earn self-esteem in their respective work. Muslims are required to strive their best in whatever they do (Itqan) while, at the same time, abstaining from committing any conduct against the Shari'ah. At a personal level, they will try to attain the highest reward in the Hereafter. Thus, by safeguarding his/her religion, a Muslim shall attain self-fulfillment, as well as spiritual growth. Self-Actualization is achieved through spiritual growth experienced by Muslims who lead a life that seeks Allah's blessings in whatever activities they do. Self-actualization is therefore the most prevalent at any level of human existence.

Muslims are required to safeguard their minds, because they will be accounted for in their deeds as long as they have a sound mind. To be able to contribute to society, a person must have a sound mind. Maslow's Social need, to love and to belong to a society and to have interpersonal relationships with people, requires a good heart and a sound mind. A mad person is a problem and a burden to society. Thus, to safeguard the Mind is a precondition for Maslow's Social needs in humans that requires one to also develop the mind by acquiring knowledge and skills so that they can benefit themselves, family, society, and the organization they are working for. Thus, the Knowledge dimension in BSC is incorporated into the Maqasid's dimension Safeguarding the Mind.

The Safeguarding of Wealth acknowledges the human physiological and biological needs to gain wealth for sustenance. Islam allows economic activities to generate in-come to buy nondurable goods such as food and clothing, as well as fixed assets and durable goods. Men can possess as much wealth and assets as they can, as long as the way of possession is not prohibited by Shari'ah. Hence, satisfying customers; identifying the critical process, skills, competencies, and technologies that will deliver a valuable proposition to customers, as well as to the current and future organizational success; and ensuring financial gain are part of the efforts that are applauded by Islam.

With the integration of humanity or spirituality, the performance management system is seen as fairer, as it may invoke a horizontal accountability, which is the obligation to maintain a harmonious relationship with coworkers, and vertical accountability, which is the need to keep a close relationship with the Creator. In such a case, employees are induced to give their best in whatever they do as a result of their struggle to achieve better outcomes to satisfy their worldly needs, which coincides with their spiritual needs that view work as part of an 'ibadah. In such a harmonious organization, unintended consequences of the performance measurement system may be reduced that may indirectly lead to reduced stress and burnout among employees. Since Islamic values prescribed in MQS are universally applicable, then such a model is not only applicable to Islamic organizations but also in any secular organizations

6. Conclusions

To compete and be sustained, businesses need to continuously strive to improve their operations; and employees are the key element needed to determine their success. Subsequently, organizations employ various methods to ensure their employees are performing at their peak, and performance measurements and appraisal systems are among the examples. Through the systems, measures are set and used to appraise and reward employees' performances, some to the extent that disrupts their work–life balance, which may hinder the fulfilment of their needs as a human.

Though BSC is admittedly the most complete organizational performance system, it has been criticized due to the lack of humanistic or spiritual aspects. For employees to stay motivated, not only basic human needs as stipulated in MHON have to be satisfied, but the spiritual aspects must also be emphasized. Due to that, this paper has tried to map the Maslow's needs model (representing the individual needs) and the BSC (representing the organizational needs) with the MQS, which represents the most comprehensive needs of mankind that may give rise to highly performing employees, resulting in sustainable organizations. Dissimilar to other models that ignore spirituality needs, this new model may give some guidance as to how a performance management system should be in an organization so that employees always stay motivated to give their best, as they are inclined to be horizontally and vertically accounted for. It happens when they view work as a fulfillment of their most fundamental needs, consisting of both worldly and spiritual needs. Contented employees give their best to ensure the success of their organizations. This model may be practically applied in any organization, be it an Islamic or secular organization, as it actually is in line with human nature, but it still needs to be empirically researched to determine its fundamental requirements.

Author Contributions: Conceptualization—N.L.A. and R.Y.; writing—original draft preparation—N.L.A., W.N.H.W.J., R.M. and R.Y.; writing—review and editing—N.L.A. and W.N.H.W.J.; visualization—R.M; supervision—N.L.A.; project administration—N.L.A.; and funding acquisition—N.L.A. All authors have read and agreed to the published version of the manuscript.

Funding: This research was funded by the Ministry of Higher Education, Malaysia, grant number 600-IRMI/FRGS 5/3 (139/2019).

Institutional Review Board Statement: Not applicable.

Informed Consent Statement: Not applicable.

Data Availability Statement: Not applicable.

Conflicts of Interest: The authors declare no conflict of interest.

References

1. Franco-Santos, M.; Lucianetti, L.; Bourne, M. Contemporary performance measurement systems: A review of their consequences and a framework for research. *Manag. Account. Res.* **2012**, *23*, 79–119. [CrossRef]
2. Ittner, C.D.; Larcker, D.F.; Randall, T. Performance implications of strategic performance measurement in financial service firms. *Account. Organ. Soc.* **2003**, *28*, 715–741. [CrossRef]
3. Kallio, K.M.; Kallio, T.J. Management-by-results and performance measurement in universities—implications for work motivation. *Stud. High. Educ.* **2014**, *39*, 574–589. [CrossRef]
4. Kallio, K.M.; Kallio, T.J.; Tienari, J.; Hyvönen, T. Ethos at stake: Performance management and academic work in universities. *Hum. Relat.* **2016**, *69*, 685–709. [CrossRef]
5. DeNisi, A.S.; Murphy, K. Performance appraisal and performance management: 100 years of progress? *J. Appl. Psychol.* **2017**, *102*, 421–433. [CrossRef]
6. Selvarajan, T.T.; Singh, B.; Solansky, S. Performance appraisal fairness, leader member exchange and motivation to improve performance: A study of US and Mexican employees. *J. Bus. Res.* **2018**, *85*, 142–154. [CrossRef]
7. Bayo Moriones, J.A.; Galdon-Sanchez, J.E.; Martinez-de-Morentin, S. Performance Appraisal: Dimensions and Determinants. *Int. J. Hum. Resour. Manag.* **2020**, *31*, 1984–2015. [CrossRef]
8. Adnan, N.L.; Muda, R.; Jusoh, W.N.H.W.; Yusoff, R. Is the vitality of Malaysian academics at stake? *J. Appl. Res. High. Educ.* 2021; *in press*.
9. Franco-Santos, M.; Doherty, N. Performance management and well-being: A close look at the changing nature of the UK higher education workplace. *Int. J. Hum. Resour. Manag.* **2017**, *28*, 2319–2350. [CrossRef]
10. Franco-Santos, M.; Rivera, P.; Bourne, M. *Performance Management in UK Higher Education Institutions: The Need For A Hybrid Approach*; Research and Development Series; School of Management, Cranfield University: Bedford, UK, 2013; Series 3: Publication 8.1.
11. Adnan, N.L.; Che Ku Kasim, C.K.H.; Ali, R. Ethical ideology and information manipulation: Are the relativists really the bad apples? *Int. J. Ethics Syst.* **2019**, *35*, 513–530. [CrossRef]
12. Fijalkowska, J.; Oliveira, C. Balanced scorecard in universities. *J. Intercult. Manag.* **2018**, *10*, 57–83. [CrossRef]
13. Ramli, M.I.; Janor, R.M. Hadhari Scorecard: An Islamic perspective Balanced Scorecard. Presented at Accounting, Commerce and Finance: The Islamic Perspective, International Conference, Manamah, Bahrain, 28–30 March 2006.
14. Gawankar, S.; Kamble, S.S.; Raut, R. Performance measurement using balance scorecard and its applications: A review. *J. Supply Chain Manag. Syst.* **2015**, *4*, 6–21.
15. McLeod, S. *Maslow's Hierarchy of Needs*. Simply Pschology: 2007. Available online: https://www.simplypsychology.org/maslow.html (accessed on 23 February 2022).
16. Alias, A.; Samsudin, M.Z. *Psychology of Motivation From an Islamic Perspective*; 3rd International Seminar on Learning and Motivation: Langkawi, Malaysia, 2005.
17. Bouzenita, A.I.; Boulanouar, A.W. Maslow's hierarchy of needs: An Islamic critique. *Intellect. Discourse* **2016**, *24*, 59–81.
18. Tweedie, D.; Wild, D.; Rhodes, C.; Martinov-Bennie, N. How does performance management affect workers? Beyond human resource management and its critique. *Int. J. Manag. Rev.* **2019**, *21*, 76–96. [CrossRef]
19. Wahba, M.A.; Bridwell, L.G. Maslow reconsidered: A review of research on the need hierarchy theory. *Organ. Behav. Hum. Perform.* **1976**, *15*, 212–240. [CrossRef]
20. Kaplan, R.S.; Norton, D.P. The balanced scorecard: Measures that drive performance. *Harv. Bus. Rev.* **2005**, *83*, 71–79.
21. Chen, C.C.; Jones, K. Are Employees Buying the Balanced Scorecard? *Manag. Account. Q.* **2009**, *11*, 36–44.
22. Quesado, P.; Guzmán, B.A.; Rodrigues, L.L. Advantages and contributions in the balanced scorecard implementation. *Intang. Cap.* **2018**, *14*, 186–201. [CrossRef]
23. Awadallah, E.A. A Critique of the balanced scorecard as a performance measurement tool. *Int. J. Bus. Soc. Sci.* **2015**, *6*, 91–99.
24. Karakas, F. Spirituality and performance in organizations: A literature review. *J. Bus. Ethics* **2010**, *94*, 89–106. [CrossRef]
25. Al-Hosaini, F.F.; Sofian, S. An exploration of inclusion of spirituality into the balanced scorecard (BSC) to support financial performance: A review. *Asian Soc. Sci.* **2015**, *11*, 289. [CrossRef]
26. Salleh, M.J.; Mohamad, N.A. *Islamic Principles of Administration: Implications on Practices in Organization, in Technology, Science, Social Sciences and Humanities International Conference—TeSSHI 2012*; Universiti Teknologi MARA Kedah, One Helang Hotel: Langkawi, Malaysia, 2012.
27. Zaydan, A.A.-K. *Al-Wajiz fi Usul al Fiqh*; Mu'assasah al-Risalah: Bayrut, Lebanon, 2006.

28. Al-Zuhayli, W. *Usul al-Fiqh al-Islami;* Dar al-Fikr: Dimashq, Syria, 1986; Volume 2.
29. Abu-Zahrah, M. *Usul al-Fiqh;* Dar al-Fikr al-'Arabi: Al-Qahirah, Egypt, 2003.
30. Al-Zuhayli, W. *Usul al-Fiqh al-Islami;* Dar al-Fikr: Dimashq, Syria, 1986; Volume 1.
31. Ibn 'Ashur, M.Ù. *Maqasid al-Shari'ah al-Islamiyyah;* Dar al- Nafa'is: Amman, Jordan, 2011.
32. Khallaf, A.a.-W. *Ilm Usul al-Fiqh;* Dar Ibn al-Haitham: Al-Qahirah, Egypt, 2009.
33. Al-Shatibi, A.I. *Al-Muwafaqa fi Usul al-Shari'ah,* 2nd ed.; Dar al-Kutub al-'Ilmiyyah: Beirut, Lubnan, 2009.
34. Chapra, M.U. *The Islamic Vision of Development in the Light of Maqāsid Al-Sharī'ah;* Islamic Development Bank, Islamic Research and Training Institute: Jeddah, Saudi Arabia, 2008.
35. Laldin, M.A. *Introduction to Shari'ah and Islamic Jurispudence,* 2nd ed.; CERT Publications Sdn. Bhd: Kuala Lumpur, Malaysia, 2008.
36. Kamali, M.H. *Principles of Islamic Jurisprudence,* 2nd ed.; Ilmiah Publishers Sdn. Bhd.: Subang Jaya, Selangor, Malaysia, 2009.
37. Hurayra, M.A. Achievement of maqasid-al-shariah in islamic banking: An evaluation of Islami bank bangladesh limited. *Glob. J. Comput. Sci. Technol.* **2015**, *15*, 9–15.
38. Rosbi, A.R.; Sanep, A. Keperluan asas berdasarkan Maqasid Al-Shari'ah: Ke arah pencapaian matlamat dan penilaian komprehensif agihan zakat. Zakat Transformation: From Subsistence to productive. Presented at the 4th ISDEV International Islamic Development Management Conference (IDMAC 2010), Penang, Malaysia, 21–22 December 2010.
39. Abdullah, H.; Arifin, N.M.; Mohd Salleh, A.M. Can the New Fusion of Maslow and Maqasid Al Sharia Outlooks Explain the Linkage of Big Five Personality Traits on Job Satisfaction? *Int. J. Innov. Creat. Chang.* **2019**, *7*, 17–32.

proceedings

Proceeding Paper

Testing the Validity of Purchasing Power Parity: Panel Cointegration Approaches with Big Mac Index †

Jing-Wen Wee and Hock-Ann Lee *

Labuan Faculty of International Finance, Universiti Malaysia Sabah, Jalan Sungai Pagar, Labuan 87000, Malaysia
* Correspondence: hockann@ums.edu.my

† Presented at the International Academic Symposium of Social Science 2022, Kota Bharu, Malaysia, 3 July 2022.

Abstract: Purchasing power parity (PPP) is vital for determining exchange rates. It has been evaluated in a variety of ways, but the findings have been inconsistent. Traded products have been the focus of much investigation relating to PPP. When it comes to non-traded products and services, the PPP deviation is especially noticeable. Against this background, this study examines the PPP from 1999 to 2021 across 27 economies, using the Big Mac as an example of an alternative non-traded good. The data analysis is conducted using the panel unit root and panel cointegration approaches. Both approaches confirm the validity of weak-form PPP. While the evidence for weak-form PPP is robust, it is inconclusive for strong-form PPP. The findings provide market participants in foreign exchange markets with guidance for the future movements of exchange rates when making investment decisions.

Keywords: purchasing power parity; Big Mac index; panel cointegration

check for updates

Citation: Wee, J.-W.; Lee, H.-A. Testing the Validity of Purchasing Power Parity: Panel Cointegration Approaches with Big Mac Index. *Proceedings* **2022**, *82*, 5. https://doi.org/10.3390/proceedings2022082005

Academic Editor: Mohamad Rahimi Mohamad Rosman

Published: 7 September 2022

1. Introduction

Purchasing power parity (PPP) is the oldest method of determining the value of one currency relative to another. According to PPP, the price ratio of two economies should be equivalent to the exchange rate between their currencies. Alternatively, PPP can also help produce economic data to compare different market conditions. It is used to equalize gross domestic product (GDP). Because purchasing power varies from country to country, GDP based on PPP is typically different than nominal GDP.

In the short run, there is a possibility that exchange rates deviate from PPP; however, the vast majority of economic theories suggest that PPP holds in the long run. In the 19th century, classical economists such as David Ricardo were among the first to propose the principle behind this theory. However, it was Swedish economist Gustav Cassel who popularized PPP in the 1920s.

Despite the fact that theoretical and empirical studies have produced significant information on long-run deviations in PPP over the years [1–4], the weak and strong forms of PPP remain less than fully understood [5]. Initially, researchers examined the unity's slope coefficient by regressing the exchange log on the relative prices. PPP was supported by evidence in hyperinflationary countries during the 1920s [6], but it was rejected by evidence in industrialized countries during the 1970s, according to [7]. These early studies were later called into question by the discovery of the nonstationarity of the exchange rate and prices, which invalidated early empirical approaches. Studies [8,9] failed to find that the slope coefficient equals 1 when they employed the maximum-likelihood cointegration approach developed by [10] in their analyses.

Because single-country testing methods may reduce the PPP's testing power, some studies have used panel data approaches to investigate the validity of the PPP. The results, however, were mixed. Studies Refs. [11–16] are among those who found evidence of PPP, whereas studies [17–19] are among those who rejected PPP.

During this same time period, Rogoff [1] invented the PPP conundrum, which is where the concept that the mean-reverting qualities of RERs may arise with a nonlinear process was first presented. Therefore, a considerable amount of empirical research has tested the PPP using a variety of nonlinear methodologies and nation groupings [20–25].

The PPP deviation is especially striking when it comes to non-traded goods or services, such as a hair cut or a McDonald's Big Mac. The theoretical discussion on the deviation of the PPP of non-traded goods was initiated by Refs. [26–30], all of whom contributed to expanding the understanding of the significance of non-traded goods in relation to PPP. The author of Ref. [31] was one of the pioneers who factored in non-traded goods in empirical research. The authors of Ref. [32] suggested that the presence of non-traded goods caused the failure of PPP to hold by using disaggregated consumer prices. The authors of Ref. [5] explored PPP using US and Mexican price data. Their findings validated the findings of Ref. [32]. In light of the above discussion, the aim of this paper is to make an attempt to fill a gap in the existing literature by investigating the weak and strong forms of PPP across nations by utilizing the Big Mac as an example of an alternative non-traded good.

2. Methodology

2.1. Data

This study examines the real exchange rate from the log of the nominal exchange rate and the log of the Big Mac price from 1999 to 2021 on an annual basis to assess the validity of PPP. It covers 27 economies: Australia, Britain, Canada, Czech Republic, Denmark, Euro Area, Hong Kong, Japan, New Zealand, Poland, Singapore, South Korea, Sweden, Switzerland, Taiwan, United States, Argentina, Brazil, Chile, China, Hungary, Indonesia, Malaysia, Mexico, South Africa, Russia and Thailand. The US dollar acts as the benchmark in this study.

2.2. Theoretical Model

The following is how the PPP exchange rate is defined:

$$\mathrm{EPPP} = \mathrm{P}/\mathrm{P}^* \quad (1)$$

where EPPP is PPP exchange rate, P is local price and P* is foreign price. PPP stands for purchasing power parity, which is a hypothetical exchange rate that equalizes local and international prices.

Using the above definition, the real exchange rate (RER) may be defined as:

$$\mathrm{RER} = (\mathrm{E} \times \mathrm{P}^*)/\mathrm{P} \quad (2)$$

RER quantifies the difference between the real exchange rate and the purchasing power parity (PPP). Thus, RER is equal to E/EPPP, as seen above. An inflation-adjusted value is denoted by the term "real".

2.3. Econometric Models

2.3.1. Panel Unit Root Tests

The first thing that needs to be performed is the determination of whether or not PPP holds by determining the stationarity of real exchange rate (RER) using the panel unit root test. Weak-form PPP holds when the panels are stationary. The following equation tests PPP validity:

$$lnE = \alpha lnP_i - \beta ln\beta_i^* \quad (3)$$

where

InP_i: logarithm of local Big Mac price;

$InP_i{}^*$: logarithm of the US Big Mac price.

The nominal exchange rate eliminates relative price movements, while the real rate stays unchanged over time, and time series are stationary with no unit roots. There are 5 panel unit root tests applied in this study which are the Levin–Lin–Chu (LLC), Breitung, Fisher-ADF, Hadri LM tests and Harris–Tzavalis test.

2.3.2. Panel Cointegration Tests

Kao and Pedroni panel cointegration tests are two examples of the panel cointegration tests that are analyzed in this study. These tests investigate the residuals that result from a linear combination of I (1) variables applied to 27 economies throughout the period of 1999–2021. The cointegration tests are used to investigate whether or not there is a connection between the logarithms of local Big Mac price and the US Big Mac price in local currency.

$$InP_i = \beta USP_i + \varepsilon \quad (4)$$

where

InP_i: log of local Big Mac price;

USP_i: the US Big Mac price in local currency.

Cointegration methods have been utilised by many empirical studies for the purpose of testing PPP by estimating the equation presented here:

$$y_{it} = \gamma_i + \beta x_{it} + e_{it} \quad (5)$$

where

γ_i: panel-specific fixed effects.

Kao's Cointegration Test

Study Ref. [33] assumed all panels share a common slope coefficient with the same cointegration vector in Equation (5), where $\beta_i = \beta$.

Test statistic based on ADF regression:

$$ADF_t = \frac{\frac{\hat{\rho}}{\widehat{SE}(\hat{\rho})} + \frac{6N\ \widehat{\sigma_v}}{\widehat{2\omega_v}}}{\sqrt{\frac{\widehat{\omega_v^2}}{2\sigma_v^2} + \frac{\widehat{3\sigma_v^2}}{10\omega_v^2}}} \quad (6)$$

The asymptotic distribution of all test statistics converges to N (0, 1).

Pedroni's Cointegration Test

Refs. [19,34] assumed the individual slope coefficients have an AR-parameter-based panel-specific cointegrating vector in Equation (5). The test statistics allow heterogeneity to run under the short-run dynamics and the long-run slope and intercept coefficients.

There are 3 panel-specific AR test statistics:

$$Group\ \rho = TN^{-1/2}\sum_{i=1}^{N}\left(\sum_{t=1}^{T}\hat{e}_{i,t-1}^2\right)^{-1}\sum_{t=1}^{T}\left(\hat{e}_{i,t-1}\Delta\hat{e}_{i,t} - \hat{\lambda}_i\right) \quad (7)$$

$$Group\ t = N^{-1/2}\sum_{i=1}^{N}\left(\hat{\sigma}_i^2\sum_{t=1}^{T}\hat{e}_{i,t-1}^2\right)^{-1/2}\sum_{t=1}^{T}\left(\hat{e}_{i,t-1}\Delta\hat{e}_{i,t} - \hat{\lambda}_i\right) \quad (8)$$

$$Group\ ADF = N^{-1/2}\sum_{i=1}^{N}\left(\sum_{t=1}^{T}\hat{s}_i^{*2}\hat{e}_{i,t-1}^2\right)^{-1/2}\sum_{t=1}^{T}\hat{e}_{i,t-1}\Delta\hat{e}_{i,t} \quad (9)$$

Residual:

$$\hat{\lambda}_i = \frac{1}{2}\left(\hat{\sigma}_i^2 - \hat{s}_i^2\right)$$

There are 4 same-AR test statistics:

$$Panel\ v = T^2N^{3/2}\left(\sum_{i=1}^{N}\sum_{t=1}^{T}\hat{L}_{11i}^{-2}\hat{e}_{i,t-1}^2\right)^{-1} \quad (10)$$

$$Panel\ \rho = T\sqrt{N}\left(\sum_{i=1}^{N}\sum_{t=1}^{T}\hat{L}_{11i}^{-2}\hat{e}_{i,t-1}^2\right)^{-1}\sum_{i=1}^{N}\sum_{t=1}^{T}\hat{L}_{11i}^{-2}\left(\hat{e}_{i,t-1}\Delta\hat{e}_{i,t} - \hat{\lambda}_i\right) \quad (11)$$

$$Panel\ t = \left(\widetilde{\sigma}^2_{N,T} \sum_{i=1}^{N} \sum_{t=1}^{T} \hat{L}_{11i}^{-2} \hat{e}^2_{i,t-1}\right)^{-1/2} \sum_{i=1}^{N} \sum_{t=1}^{T} \hat{L}_{11i}^{-2} (\hat{e}_{i,t-1} \Delta \hat{e}_{i,t} - \hat{\lambda}_i) \quad (12)$$

$$Panel\ ADF = \left(\widetilde{s}^{*2}_{N,T} \sum_{i=1}^{N} \sum_{t=1}^{T} \hat{L}_{11i}^{-2} \hat{e}^2_{i,t-1}\right)^{-1/2} \sum_{i=1}^{N} \sum_{t=1}^{T} \hat{L}_{11i}^{-2} \hat{e}_{i,t-1} \Delta \hat{e}_{i,t} \quad (13)$$

Residual:

$$\widetilde{\sigma}^2_{N,T} = \frac{1}{N} \sum_{i=1}^{N} \hat{L}_{11i}^{-2} \hat{\sigma}^2_i \quad (14)$$

$$\hat{L}_{11i} = \hat{\omega}^2_{u,i} - \hat{\Omega}_{u\epsilon,i} \hat{\Omega}_{\epsilon,i} \hat{\Omega}'_{u\epsilon,i} \quad (15)$$

$$\widetilde{s}^{*2}_{N,T} = \frac{1}{N} \sum_{i=1}^{N} \hat{s}^{*2}_i \quad (16)$$

The asymptotic distribution of all test statistics converges to N (0, 1).

The cointegration test hypothesis proposed by Kao and Pedroni reads as follows:

H$_0$: *No cointegration between InP_i and $USDP_i$ (weak-form PPP fails to hold);*

H$_1$: *InP_i and $USDP_i$ are cointegrated (weak-form PPP does hold).*

The fact that the test is rejected indicates the stationarity of the series. Additionally, it indicates that InP_i and USP_i are cointegrated; hence, PPP holds.

Pedroni's PDOLS:

$$y_{i,t} = \alpha_i + \beta_i x_{i,t} + \sum_{j=-P}^{P} \gamma_{i,j} \Delta x_{i,t-j} + \mu^*_{it} \quad (17)$$

where

P: number of lags;
β_i: slope coefficient;
$x_{i,t}$: explanatory variable.

$$\hat{\beta}^*_{GM} \& = \left[\frac{1}{N} \sum_{i=1}^{N} \left(\sum_{t=1}^{T} z_{i,t} z'_{i,t}\right)^{-1} \left\{\sum_{t=1}^{T} z_{i,t} \left(y_{i,t} - \underline{y}_i\right)\right\}\right] \quad (18)$$

$$t_{\hat{\beta}^*_i} \& = (\hat{\beta}^*_i - \beta_0) \left\{\hat{\sigma}_i^{-2} \sum_{t=1}^{T} (x_{i,t} - \underline{x}_i)^2\right\}^{\frac{1}{2}} \quad (19)$$

$$t_{\hat{\beta}_{GM}} \& = \frac{1}{\sqrt{N}} \sum_{i=1}^{N} t_{\hat{\beta}^*_i} \quad (20)$$

H$_0$: *the slope coefficient is equal to 1, $\beta_i = \beta_0$ (strong-form PPP does hold);*

H$_1$: *the slope coefficient is inequal to 1, $\beta_i \neq \beta_0$ (strong-form PPP does not hold).*

When the null hypothesis of slope coefficient = 1 cannot be rejected, strong-form PPP is assumed to hold. Local and foreign Big Mac prices are thus equal when translated to the same currency.

3. Empirical Results

This section reports the findings of the validity of PPP through panel unit root tests and cointegration tests.

In Table 1, the LLC, Harris–Tzavalis and Breitung test results are statistically significant, indicating that the null hypothesis that panels contain some unit roots is rejected. The Fisher-ADF and Hadri LM tests show that some panels have unit roots. As a result, weak-form PPP holds for the LLC, the Harris–Tzavalis and the Breitung tests.

Table 1. Panel unit root tests.

Variable	Levin–Lin–Chu (LLC)	Harris–Tzavalis	Breitung	Fisher-ADF	Hadri LM
Real Exchange Rate	−7.3673 ** (0.0164)	0.7221 * (0.000)	−2.6128 * (0.0045)	46.0739 (0.7699)	26.2205 * (0.0000)

Note: * and ** represent 1% and 5% at significant level, respectively. The numbers in brackets are p-values.

Table 2 present the Kao cointegration test results. The test extends the Engle–Granger two-stage cointegration framework. It compares the H_0: no cointegration with the H_1: cointegration with the assumption of a homogenous cointegrating vector across economy pairs. The results reveal that the test fails to reject the null hypothesis that InP_i and $USDP_i$ are not cointegrated, implying that none of the panels is cointegrated. As a result, PPP does not hold.

Table 2. Kao panel cointegration test.

	Statistic	p-Value
ADF	−0.3373	0.3679

Note: the test statistic is statistically insignificant

The Pedroni cointegration test results are shown in Table 3. The Engle–Granger two-stage cointegration is also extended by the test. The test, on the other hand, allows for heterogeneous cointegrating intercept and slope coefficients by relaxing the assumption of a homogenous cointegrating vector across economy pairs. While rows 2–5 present the within-dimension panel FMOLS test results, rows 6–8 report the between-dimension test results. It compares the H_0: no cointegration hypothesis to the H_1: cointegration hypothesis. The Pedroni's panel cointegration test finds the presence of relationships between the logarithms of Big Mac price and US Big Mac price in local currency, with five out of seven test statistics being significant and the null hypothesis of no cointegration being rejected, showing that weak-form PPP does exist.

Table 3. The Pedroni cointegration test.

Tests	Statistics	p-Value
Panel v-statistic	0.5224	0.3007
Panel rho-statistic	−2.343 *	0.0096
Panel t-statistic	−3.629 *	0.0001
Panel adf-statistic	−1.865 **	0.0311
Group rho-statistic	−1.478 ***	0.0697
Group t-statistic	−4.118 *	0.0000
Group adf-statistic	−0.6919	0.2445

Note: *, **, *** represent 1%, 5%, 10% at significant level, respectively.

This study now focuses on determining whether or not strong-form PPP actually exists. Examining the slope coefficients in the cointegrating vector is a necessary step for this. The calculated slope coefficients, also known as elasticity coefficients, are 0.5274 (see Section A of Table 4). The t-test does not support the elasticity parameter being equal to one. The result demonstrates that strong-form PPP is not valid. Furthermore, this result means that a 1% rise in the price of a Big Mac sold in the US will result in a 0.5 percent increase in the price of a Big Mac sold outside the US.

Table 4. Pedroni's PDOLS.

Country	Beta	*t*-Statistic
A. All	0.5274 *	−25.41
B. Individual PDOLS		
Britain	−0.4705 *	−5.096
United States	−0.01267 *	−16.43
Australia	−0.2211 *	−19.76
Canada	−0.1499 *	−8.521
Denmark	2.554 *	3.088
Hong Kong (China)	0.02194 *	−6.863
Japan	1.065	0.3414
Sweden	1.185	0.3709
Singapore	0.2934 *	−8.634
South Korea	0.4396 ***	−1.352
Brazil	0.7377	−0.5021
Hungary	−0.4107 *	−6.076
Argentina	0.9384	−0.4749
China	−0.1801 *	−21.47
Russia	0.6483 *	−3.605
Malaysia	0.6759	−1.025
Mexico	0.3681 *	−8.38
Switzerland	1.119 *	3.882
Thailand	−0.8791 *	−11.18
Chile	1.447 **	1.695
Poland	0.5408 ***	−1.382
Taiwan (China)	1.048	0.1433
Czech Republic	0.6118 **	−1.913
New Zealand	0.702 *	−7.486
South Africa	0.4262 *	−9.619
Indonesia	0.3278 *	−2.906
Euro Area	1.413	1.142

Note: *, **, *** represent 1%, 5%, 10% at significant level, respectively.

Individual *t*-tests on the elasticity coefficients produced by group DOLS estimate (see Section B of Table 4) are used to further investigate the failure of strong-form PPP. Overall, only 7 of 27 pairs of economies support the cointegration between log local Big Mac price and the US price in domestic currency, implying the existence of strong-form PPP.

4. Conclusions

Purchasing power parity, known as PPP, has been investigated through a variety of different methods, but the results have been inconclusive. The deviation in PPP is notably obvious with regard to non-traded goods and services. Against this backdrop, this study evaluated the PPP from 1999 to 2021 across 27 economies, utilizing a Big Mac as an example of an alternative non-traded good using panel cointegration techniques. This study found robust evidence of weak-form PPP across different econometric approaches. Unlike the evidence for the weak-form PPP, the evidence for the strong-form PPP is inconclusive. This is because some pairs of economies support strong-form PPP. The findings of weak- and strong-form PPP offer market participants in foreign exchange markets a direction for the future fluctuations in exchange rates, which is useful for when they are deciding whether or not to make a long-term investment. It is essential to keep in mind that applying the concept of purchasing power parity is not always the most effective trading approach for the short term. Because the strategy does not take into consideration volatility over a short period of time, it is most beneficial for trading strategies that include longer time frames. Because of this, it is essential to use PPP as one component of a comprehensive fundamental analysis approach and to make use of it in conjunction with indications derived from technical analysis.

Author Contributions: Methodology, H.-A.L. and J.-W.W.; software, J.-W.W.; formal analysis, J.-W.W.; writing—original draft preparation, J.-W.W.; writing—review and editing, H.-A.L.; supervision, H.-A.L.; funding acquisition, H.-A.L. All authors have read and agreed to the published version of the manuscript.

Funding: This research was partially funded by University Malaysia Sabah Acculturation Grant Scheme grant number SGA0022-2019.

Institutional Review Board Statement: Not applicable.

Informed Consent Statement: Not applicable.

Data Availability Statement: The data presented in this study are available on request from the corresponding author.

Conflicts of Interest: The authors declare no conflict of interest. The funders had no role in the design of the study; in the collection, analyses or interpretation of data; in the writing of the manuscript; or in the decision to publish the results.

References

1. Rogoff, K. The purchasing power parity puzzle. *J. Econ. Lit* **1996**, *34*, 647–668.
2. Froot, K.A.; Rogoff, K. Perspectives on PPP and long-run real exchange rates. In *Handbook of International Economics*; Grossman, G., Rogoff, K., Eds.; Elsevier: Amsterdam, The Netherlands, 1995; pp. 1647–1688.
3. Sarno, L.; Taylor, M.P. Purchasing power parity and the real exchange rate. *IMF Staff Pap.* **2002**, *49*, 65–105.
4. Taylor, A.M.; Taylor, M.P. The purchasing power parity debate. *J. Econ. Perspect.* **2004**, *18*, 135–158. [CrossRef]
5. Robertson, R.; Kumar, A.; Dutkowsky, D.H. Weak-form and strong-form purchasing power parity between the US and Mexico: A panel cointegration investigation. *J. Macroecon.* **2014**, *42*, 241–262. [CrossRef]
6. Frenkel, J.A. Purchasing power parity: Doctrinal perspective and evidence from the 1920s. *J. Int. Econ.* **1978**, *8*, 169–191. [CrossRef]
7. Frenkel, J.A. The collapse of purchasing power parities during the 1970's. *Eur. Econ. Rev.* **1981**, *16*, 145–165. [CrossRef]
8. Cheung, Y.W.; Lai, K.S. Long-run purchasing power parity during the recent float. *J. Int. Econ.* **1993**, *34*, 181–192. [CrossRef]
9. Pippenger, M.K. Cointegration tests of purchasing power parity: The case of Swiss exchange rates. *J. Int. Money Financ.* **1993**, *12*, 46–61. [CrossRef]
10. Johansen, S.; Juselius, K. Maximum likelihood estimation and inference on cointegration: With applications to the demand for money. *Oxf. Bull. Econ. Stat.* **1990**, *52*, 169–210. [CrossRef]
11. Frankel, J.A.; Rose, A.K. A panel project on purchasing power parity: Mean reversion within and between countries. *J. Int. Econ.* **1996**, *40*, 209–224. [CrossRef]
12. MacDonald, R. Panel unit root tests and real exchange rates. *Econ. Lett.* **1996**, *50*, 7–11. [CrossRef]
13. Oh, K.Y. Purchasing power parity and unit root tests using panel data. *J. Int. Money Financ.* **1996**, *15*, 405–418. [CrossRef]
14. Papell, D.H.; Theodoridis, H. The choice of numeraire currency in panel tests of purchasing power parity. *J. Money Credit Bank* **2001**, *33*, 790–803. [CrossRef]
15. Coakley, J.; Fuertes, A.M. New panel unit root tests of PPP. *Econ. Lett. 1997 57*, 17–22. [CrossRef]
16. Nagayasu, J. *Does the Long-Run PPP Hypothesis Hold for Africa? Evidence from Panel Co-Integration Study*; IMF Working Paper WP/98/123; International Monetary Fund, African Department: Dar es Salaam, Tanzania, 1998.
17. O'Connell, P.G. The overvaluation of purchasing power parity. *J. Int. Econ.* **1998**, *44*, 1–19. [CrossRef]
18. Wu, J.L.; Chen, S.L. Are real exchange rates stationary based on panel unit-root tests? *Int. J. Finance Econ.* **1999**, *4*, 243–252. [CrossRef]
19. Pedroni, P. Panel cointegration: Asymptotic and finite sample properties of pooled time series tests with an application to the PPP hypothesis. *Econom. Theory* **2004**, *20*, 597–625. [CrossRef]
20. Arize, A.C.; Malindretos, J.; Nam, K. Cointegration, dynamic structure, and the validity of purchasing power parity in African countries. *Int. Rev. Econ. Financ.* **2010**, *19*, 755–768. [CrossRef]
21. Narayan, P.K. New evidence on purchasing power parity from 17 OECD countries. *Appl. Econ.* **2005**, *37*, 1063–1071. [CrossRef]
22. Su, C.W.; Chang, T.Y.; Chang, H.L. Purchasing power parity for fifteen Latin American countries: Stationary test with a fourier function. *Int. Rev. Econ. Financ.* **2011**, *20*, 839–845. [CrossRef]
23. Bahmani-Oskooee, M.; Wu, T.P. PPP in the 34 OECD countries: Evidence from quantile-based unit root tests with both smooth and sharp breaks. *Appl. Econ.* **2018**, *5*, 2622–2623. [CrossRef]
24. Mike, F.; Kızılkaya, O. Testing the theory of PPP for emerging market economies that practice flexible exchange rate regimes. *Appl. Econ. Lett.* **2019**, *26*, 1411–1417. [CrossRef]
25. Doganlar, M.; Kızılkaya, O.; Mike, F. Testing the long-run PPP for Turkey: New evidence from the Fourier quantile unit root test. *Appl. Econ. Lett.* **2020**, *27*, 729–735. [CrossRef]
26. Balassa, B. The purchasing power parity doctrine. *J. Political Econ.* **1964**, *72*, 584–589. [CrossRef]
27. Samuelson, P.A. Theoretical notes on trade problems. *Rev. Econ. Stat.* **1964**, *46*, 145–154. [CrossRef]

28. Hsieh, D. The determination of the real exchange rate: A productivity approach. *J. Int. Econ.* **1982**, *12*, 355–362. [CrossRef]
29. Bergstrand, J.H. Structural determinants of real exchange rates and national price levels: Some empirical evidence. *Am. Econ. Rev.* **1991**, *81*, 325–334.
30. Obstfeld, M. *Model Trending Real Exchange Rates*; CIDER Working Papers 233187; Department of Economics, University of California-Berkeley: Berkeley, CA, USA, 1993.
31. Engel, C. Real exchange rates and relative prices: An empirical investigation. *J. Monet. Econ.* **1993**, *32*, 35–50. [CrossRef]
32. Jenkins, M.A.; Snaith, S.M. Tests of purchasing power parity via cointegration analysis of heterogeneous panels with consumer price indices. *J. Macroecon.* **2005**, *27*, 345–362. [CrossRef]
33. Kao, C. Spurious regression and residual-based tests for cointegration in panel data. *J. Econom.* **1999**, *90*, 1–44. [CrossRef]
34. Pedroni, P. Critical values for cointegration tests in heterogeneous panels with multiple regressors. *Oxf. Bull. Econ. Stat.* **1999**, *61*, 653–670. [CrossRef]

proceedings

Proceeding Paper

Introducing Tax Education in Non-Accounting Curriculum: Evidence from Academicians †

Nor Azrina Mohd Yusof [1,*], Roshidah Safeei [1] and Chai Chuen Lee [2]

1 Faculty of Accountancy, Universiti Teknologi MARA Cawangan Kedah, Merbok 08400, Kedah, Malaysia
2 Academy of Language Studies, Universiti Teknologi MARA Cawangan Kedah, Merbok 08400, Kedah, Malaysia
* Correspondence: yina1437@uitm.edu.my; Tel.: +60-44562551

† Presented at International Academic Symposium of Social Science 2022, Kota Bharu, Malaysia, 3 July 2022.

Abstract: Taxation is the key source of revenue for most developing countries in the world. Despite their reliance on taxes, most governments confront comparable revenue collection issues. For instance, Malaysia had the fourth lowest tax ratio among Asian and Pacific nations. This is quite surprising because a self-assessment system (SAS) has been in place for over two decades, yet voluntary compliance is still missing. Despite this, past research has revealed that Malaysian taxpayers appear to be tax illiterate. Thus, the purpose of the study was (i) to analyze the need for introducing tax subjects in non-accounting curriculums among academicians from non-accounting fields, (ii) to measure tax knowledge among non-accounting academicians, and (iii) to determine the preferences of non-accounting academicians on taxation topics if tax subjects are integrated into non-accounting curriculums. In September 2021, an online questionnaire was sent to all non-accounting academicians within the Universiti Teknologi MARA (UiTM) system over a three-week period. There were 358 responses in all, with 349 questionnaires usable for data analysis. According to the findings, (i) most respondents probably agreed that tax subjects should be taught in all faculties, (ii) 50.4% of respondents demonstrated tax literacy, and (iii) real property and gains tax, tax planning for small businesses, and small company tax were the three most preferred tax topics. Despite the fact that Malaysia's self-assessment system has been in place for over two decades, 41.8% of respondents had just a moderate understanding of taxes. The findings offer insight into how the Malaysian Ministry of Higher Education and the Malaysian Inland Revenue Board may work together to deliver tax education to non-accounting students at the tertiary level. As a result, because the sample was confined to all UiTM academicians from non-accounting fields, the findings should be interpreted and generalized with caution.

Keywords: tax education; non-accounting curriculum; academicians

Citation: Mohd Yusof, N.A.; Safeei, R.; Lee, C.C. Introducing Tax Education in Non-Accounting Curriculum: Evidence from Academicians. *Proceedings* **2022**, *82*, 8. https://doi.org/10.3390/proceedings2022082008

Academic Editor: Mohamad Rahimi Mohamad Rosman

Published: 7 September 2022

1. Introduction

Taxation is the key source of revenue for most developing countries in the world. Most governments rely on taxes to provide public goods and services to their residents, such as schools, hospitals, universities, and public transportation. Despite their reliance on taxes, most governments confront comparable revenue collection issues. According to the [1] study, certain nations have an extremely low tax percentage. Malaysia had the fourth lowest tax ratio among Asian and Pacific nations in the [1] report. It should be emphasized that the tax ratio relates to the ratio of taxes to gross domestic product (GDP). Surprisingly, the [1] report was based on tax revenue data from 2018, meaning that the low tax percentage has nothing to do with the present problem, the COVID-19 pandemic, which has caused serious challenges to tax systems and economies across Asia and the Pacific. In Malaysia, for example, a self-assessment system (SAS) has been in place for over two decades, yet voluntary compliance is still missing [2]. Ref. [3] contends that the influence of the SAS on a low tax ratio cannot be separated. This is because a SAS expects all taxpayers to be honest and responsible when it comes to returning their tax forms and keeping accurate records. With the SAS in hand, tax officials' responsibilities have passed to them. Ref. [4] revealed that one year after SAS was implemented for individual taxpayers in Malaysia, taxpayers had to comprehend, interpret, and obey all laws and regulations while filling their income tax return forms to avoid fines for non-compliance. To submit a valid tax return and obey tax rules and regulations, all taxpayers must have a fundamental awareness of personal taxation, allowable charges, refunds, exemptions, and relief.

Despite this, past research [5–9] has revealed that Malaysian taxpayers appear to be tax illiterate. Ref. [5] found, for example, that 30% to 50% of Malaysian taxpayers are tax ignorant. According to [9], only 10.67 percent of 995 respondents had a high degree of tax knowledge. Their findings revealed that the lack of tax knowledge could be attributed to non-accounting undergraduates who never received formal tax instruction. According to their findings, the majority of respondents wanted to learn more about taxation. Their findings are consistent with [8], who discovered that more than 90% of respondents thought tax subjects should be taught in all undergraduate faculties. Ref. [10] argued that integrating tax law and tax knowledge in social science education programs would be a step in the right direction. Meanwhile, Ref. [11] noted that the greater corporate tax compliance rate in Japan was mostly due to the efforts of the Japanese National Tax Administration (NTA), which included tax education for primary school pupils. According to the NTA, these primary school pupils are their future taxpayers; thus, teaching tax education at an early age may help them understand their role and responsibility as citizens to submit proper returns and pay their tax obligations. According to [12], the American Internal Revenue Service has dedicated a part of their website to "Understanding Taxes" for secondary school pupils and other learners. The Ministry of Education in Malaysia has recently integrated tax education into the mathematics syllabus for Form 5 students, probably because of the successful introduction of tax education to elementary school pupils in several countries, such as Japan. As pointed out by [13], because the vast majority of school-aged students are future taxpayers, implementing tax education at the school level should have the highest likelihood of success.

Furthermore, in addition to formal tax education, several tax authorities also implement informal tax education initiatives, such as the Indonesian government's annual Pajak Bertutur, which aims to raise tax knowledge among young people [14]. This annual program seeks to provide a variety of tax-related educational activities to students at all levels of education, including elementary, middle, high, and university. In Malaysia, the Inland Revenue Board of Malaysia (IRBM) has launched several informal education programs for schoolchildren, including tax camps, junior tax officers at Kidzania, speech contests, quizzes, and trips to IRBM offices [15]. It is undeniably beneficial for pupils to understand taxes at a young age, since they will be paying taxes in the future.

Previous research has shown that tax knowledge has a significant impact on tax compliance. In Malaysia, studies by [16,17] found a relationship between tax knowledge and tax compliance. Meanwhile, Ref. [3] found a significant difference in tax knowledge between Indonesian students who received tax education and those who did not. Another research study conducted by [18] in Indonesia found that tax education improves tax knowledge. However, as [16] notes, only accounting and certain business management students are exposed to tax subjects at university level in many higher learning institutions across the world, including Malaysia. Tax subjects, for example, were formerly solely offered to undergraduate students from the Faculty of Accountancy at Universiti Teknologi MARA (UiTM). Meanwhile, for the non-accounting fields, tax subjects were only offered as an elective to undergraduate students from the Faculty of Administrative Science and Policy Studies. Future taxpayers are rationally anticipated, and this is predicted to have a direct influence on tax compliance. Perhaps this is one of the elements contributing to Malaysia's low tax ratio, which is the result of low tax awareness and understanding among taxpayers. Even though tax knowledge may be learned through self-learning, formal education, and informal education, Ref. [3] asserted that few people are eager to study it. According to [19], most individuals see taxes as a burden that should be avoided. Therefore, is it still essential to teach taxation to students across all faculties? Or has digitization made it simpler for all taxpayers to access tax information, such as through the IRBM website or the MyTax app? What are academics' opinions on the importance of tax education? They are Malaysian individual taxpayers, some of whom have been taxpayers for many years. As a result, this study aimed to bridge the gap.

The purpose of this study is threefold: first, to analyze the need for introducing tax subjects in non-accounting curriculums among academicians from non-accounting fields; second, to measure the level of tax knowledge among non-accounting academicians; and third, to determine the preferences of non-accounting academicians regarding taxation topics if tax subjects are integrated into non-accounting curriculums. This study will draw attention to the importance of tax education that could have a direct impact on tax compliance in Malaysia.

2. Materials and Methods

The study adopted a quantitative approach, employing an online survey to collect the opinions and experiences of non-accounting academicians at UiTM regarding integrating tax subjects into non-accounting curriculums. At the time of the study, based on an unpublished report from UiTM Shah Alam's Infostructure Division, the total number of active academicians in the UiTM system was 8975. To ensure that the findings could be generalized to the target population, the study employed simple random sampling techniques. An online survey was emailed to all non-accounting academicians through the UiTM system. The data collection was carried out over a period of three weeks in September 2021. In total, 358 responses were collected. Of these, 9 questionnaires were partially completed and were discarded. Hence, 349 questionnaires were usable for data analysis.

The survey was designed by using a Google Form to collect the data. The survey was divided into four sections. Section A focused on the demographics of respondents. Section B discussed the significance of tax education in the non-accounting curriculum. Section C, on the other hand, was concerned with academicians' tax knowledge. Finally, Section D focused on preferred tax topics that should be taught if tax subjects are introduced.

3. Results and Discussion

3.1. The Respondents' Profiles

Table 1 shows the demographics of the respondents. Female academicians made up around 74% of the responses, while male academicians made up 26%. More than 55% of respondents were between the ages of 31 and 40, with 29.4% being between the ages of 41 and 50. There were two responses that were over the age of 60. The majority of respondents were aged 31 and above (57%), suggesting that they may have had previous experience as taxpayers. Approximately 78% of respondents were married, while 18.5% were single. In terms of education, over 70% of respondents said they held a master's degree. Around 28% of respondents had a PhD. When asked about their working experience, around 26% said they had 6 to 10 years, while 34.2 percent said they had more than 15 years. Meanwhile, around 40% of those surveyed earned more than RM 8000 per month, with 35.2% earning between RM 6000 and RM 8000 per month. According to a cross-tabulation study, 89 respondents with a monthly salary of more than RM 8001 had worked for more than 15 years. Further research showed that these 89 respondents were Malaysian taxpayers who may have had some knowledge of and experience with tax laws. Regarding the cluster, the largest group of respondents (40.1%) were from Science and Technology, followed by 31.2% from Business and Management, 16.9% from the Academy of Language Studies, 10.3% from Social Sciences and Humanities, and 1.4% from the Academy of Contemporary Studies.

Table 1. The respondents' profiles.

Characteristics		Frequency (*n*)	Percentage (%)
Gender	Male	97	27.8
	Female	252	72.2
Age	≤30 years old	9	2.6
	31–40 years old	199	57.0
	41–50 years old	97	27.8
	51–60 years old	42	12.0
	>61 years old	2	0.6
Marital Status	Single	65	18.6
	Married	273	78.2
	Others	11	3.2
Education	Bachelor's degree	4	1.1
	Master's degree	243	69.6
	Doctor of Philosophy	101	28.9
	Doctor in Business Administration	1	0.3
Work Experience	≤5 years	37	10.6
	Between 6–10 years	90	25.8
	Between 11–15 years	110	31.5
	≥15 years	112	32.1
Monthly Salary	≤RM 4000	16	4.6
	Between RM 4001–RM 6000	69	19.8
	Between RM 6001–RM 8000	123	35.2
	≥RM 8001	141	40.4
Cluster	Business & Management	109	31.2
	Social Sciences & Humanities	36	10.3
	Sciences & Technologies	140	40.1
	Academy of Contemporary Studies	5	1.4
	Academy of Language Studies	59	16.9

3.2. *Opinion on Introducing Tax Subjects into Non-Accounting Curriculum*

The first objective of the current study was to analyze the need for introducing tax subjects in non-accounting curriculums among academicians from non-accounting fields. Three questions were designed to assess the opinions. All items were rated on a five-point Likert scale, with one representing "strongly disagree" and five representing "strongly agree". Table 2 below highlights the mean scores and standard deviation of the three opinions examined in this study. The findings show that most respondents were likely to agree (mean = 3.95) that tax subjects should be introduced to all faculties, and most respondents (mean = 3.93) believed that tax subjects should be offered as an elective subject in their faculty. This implies that the respondents agree that tax subjects should be introduced to all faculties as an elective subject. Still, the respondents did not have a strong opinion (mean = 3.03) about whether tax subjects should be made a required course in their faculty.

Table 2. Opinion on introducing tax subject into non-accounting curriculum.

Statement	Strongly Disagree %	Disagree %	Neutral %	Agree %	Strongly Agree %	Mean	Standard Deviation
Tax subjects should be introduced to all faculties.	4.3	8.0	13.8	36.7	37.2	3.95	1.104
Tax subjects should be introduced as an elective course in my faculty.	3.7	9.5	14.0	36.1	36.7	3.93	1.104
Tax subjects should be introduced as a compulsory course in my faculty.	14.0	22.1	28.9	16.6	18.3	3.03	1.298

The mean of 3.03 was predicted since these respondents are working taxpayers, with some having more than 15 years of experience, and their tax knowledge may have improved because of their experience as taxpayers. Upon checking, it was discovered that 325 of the 349 respondents had been taxpayers for the previous five years. Perhaps this is a reason why some respondents believed it was unnecessary to make tax subjects a compulsory course for their faculty. In line with [3]'s study, tax knowledge may be gained through self-study, formal education, and informal education.

Furthermore, respondents were surveyed on what level they think is best for implementing tax education. Figure 1 demonstrates that most respondents (78%) believed that tax subjects should be introduced at university level. Perhaps this is due to their limited opportunities to learn tax as undergraduate students. This is because many Malaysian higher learning institutions offer tax subjects to accounting and business undergraduate students. As a result, undergraduates from non-accounting fields need to learn informally, either via IRBM's website, via MyTax, or by seeking a tax agent while completing tax returns. Only 20% of respondents indicated a secondary school level. Perhaps it is because there was no formal tax education in Malaysian secondary schools before 2020. As a result, some taxpayers were only able to take tax subjects at university or through informal tax education. Nonetheless, beginning in 2020, our Ministry of Education has included taxation in the mathematics curriculum for Form 5 students. Surprisingly, only two people (1%) thought tax subjects should be taught in primary school. The National Tax Administration (NTA) suggested that offering tax education at a young age might help primary school pupils understand their role and responsibility as citizens to file proper returns and pay their tax obligations [11]. This is something that our Ministry of Education should also learn.

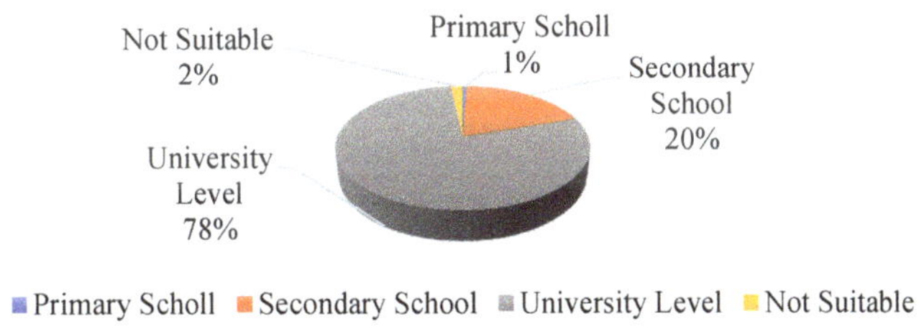

Figure 1. Level for introducing tax subjects.

Next, Table 3 presents opinions on the importance of tax education. On a five-point Likert scale, respondents were asked to assess statements from 1 (strongly disagree) to 5 (strongly agree). With a mean score of above 4.0, the results suggest that all respondents thought that tax education was important. This research shows that tax education has a large and positive impact on the dissemination of tax knowledge. This finding is consistent with that of [8,19]. They discovered that both business and non-business students thought tax was an essential subject to understand at a higher level of schooling. Meanwhile, secondary school students in a study by [20] agreed that taxation should be taught in secondary school, since they saw taxation as general knowledge that everyone should know and understand from the beginning.

Table 3. Opinion on the importance of tax education.

Statement	Strongly Disagree %	Disagree %	Neutral %	Agree %	Strongly Agree %	Mean	Standard Deviation
Learning tax would help me to accurately submit my tax income tax return form.	0.3	1.7	7.2	33.0	57.9	4.46	0.729
Learning tax makes me aware of my responsibility as a taxpayer towards the government.	0.3	2.0	5.2	31.8	60.7	4.51	0.714
Learning tax makes me understand my rights as taxpayer.	0.3	1.1	4.9	31.5	62.2	4.54	0.671
Learning tax makes me aware of any changes or updates about tax relief or rebates.	0.6	0.6	5.4	32.7	60.7	4.52	0.680

3.3. Tax Knowledge on Personal Taxation

The second objective of this study was to measure tax knowledge among non-accounting academicians. Ten questions about rental income, business income, interest income, capital receipts, employment income, and individual personal income were used to test respondents' tax knowledge. Table 4 shows how much the academicians know about tax law. According to the data, 58.5% of the respondents were aware that rental income is taxed. Furthermore, most respondents (87.1%) were aware that business income is taxed. Meanwhile, around 68.5% of respondents were aware that an individual taxpayer is entitled to RM 9000 in personal relief. Approximately 78.5% of respondents were aware that parents may claim child relief of up to RM 8000 per child for unmarried children aged 18 and above who are enrolled in tertiary education. However, more than half of the respondents (54.7%) were unaware that if his wife does not have a source of income, a husband might claim RM 4000 as wife's relief. Furthermore, just 38.4% of those surveyed were aware that interest generated on fixed deposits held with any Malaysian bank, including Maybank, is tax-free. A little more than 30% of those surveyed were aware that cash prizes earned in newspaper contests, TV game shows, and other events are tax-free. Furthermore, the research revealed that respondents have limited tax awareness regarding interest income and capital revenues, with 40.7% being unsure. Only 18.3% of those surveyed were aware of the tax rate. In conclusion, there were some respondents who still had little or no understanding of certain aspects of personal taxation.

Table 4. Tax knowledge on personal taxation.

Statement	Yes Frequency %	No Frequency %	Not Sure Frequency %
Do you know that rental income received from letting of your real property is chargeable for income tax?	204 (58.5)	53 (15.2)	92 (26.4)
Do you know that your business income is chargeable for income tax?	304 (87.1)	15 (4.3)	30 (8.6)
Do you know that your interest income earned from investment in fixed deposits from any local banks in Malaysia, such as Maybank, is exempted from income tax?	134 (38.4)	73 (20.9)	142 (40.7)
Do you know that cash money won from newspaper contests, tv game shows, and other competitions is exempted from income tax?	118 (33.8)	89 (25.5)	142 (40.7)
Do you know that monies received as death gratuity are fully exempted from income tax?	144 (41.3)	79 (22.6)	126 (36.1)
Do you know that an individual taxpayer is entitled to personal relief of RM 9000?	239 (68.5)	43 (12.3)	67 (19.2)
Do you know that parents can claim child relief amounting RM 8000 per child for unmarried children aged 18 years and above and pursuing study in tertiary education?	274 (78.5)	25 (7.2)	50 (14.3)
Do you know that if a wife does not have a source of income, the husband is entitled to claim RM 4000 as wife relief?	191 (54.7)	65 (18.6)	93 (26.6)
Do you know that if a person's taxable income exceeds RM 2,000,000, the tax rate is 30%?	64 (18.3)	122 (35.0)	163 (46.7)
Do you know that if a wife chooses a joint assessment, the husband is entitled to an additional RM 4000 as spouse relief?	83 (23.8)	113 (32.4)	153 (43.8)

This research adopted a knowledge score developed by [10]. Those who answered "Yes" scored a 3 (well-informed), while those who answered "Not Sure" received a 2 (uninformed). Those who answered "No" received a score of 1 (misinformed). The overall score received by the respondents is shown in Table 5. Only 6.3 percent (22/349) of the respondents received full marks, while 0.3 percent (1/349) had a minimum score of 11 points. Most respondents (39/349) received a score of 26.

Table 5. Tax knowledge scores of the respondents.

Score	Frequency	Percentage	Score	Frequency	Percentage
11	1	0.3	21	18	5.2
12	3	0.9	22	37	10.6
13	2	0.6	23	27	7.7
14	9	2.6	24	32	9.2
15	7	2	25	31	8.9
16	5	1.4	26	39	11.2
17	10	2.9	27	21	6
18	21	6	28	23	6.6
19	7	2	29	8	2.3
20	26	7.4	30	22	6.3

The respondents were divided into three groups based on their tax knowledge scores: high, medium, and low tax knowledge. According to Table 6, 50.4% of respondents had a high level of tax knowledge. The results show that the respondents are tax literate. The findings contradict with past studies [5–9] which found that Malaysian taxpayers appear to be tax illiterate. The most likely explanation is that most respondents (89.6%) had more than 5 years of work experience and a salary of RM 4000 or more. Note that single salaried taxpayers with a monthly income of RM 3111 after deducting EPF are required to register for income tax in Malaysia. As a result, the researchers believe that as time passed and they were required to report and pay income tax on a yearly basis, these respondents began to understand informal taxes and gradually improved their tax knowledge. Despite the fact that Malaysia's self-assessment system has been in place for over two decades, 41.8% of respondents had a moderate understanding of taxes and 7.7% of respondents only had a low level of tax knowledge. The findings somewhat indicate that there is a need to introduce tax subjects to all non-accounting curriculums to ensure undergraduates who will become future taxpayers gain some tax knowledge before they become real taxpayers. The findings offer insight into how the Malaysian Ministry of Higher Education and the Malaysian Inland Revenue Board may work together to deliver tax education to non-accounting students at the tertiary level.

Table 6. Level of tax knowledge.

Level of Tax Knowledge	Range of Score	Frequency n	Percentage %
Low	11–16	27	7.7
Medium	17–23	146	41.8
High	24–30	176	50.4

3.4. *Tax Topics Preference*

The third objective was to determine the preferences of non-accounting academicians on taxation topics if tax subjects are integrated into non-accounting curriculums. Ten topics were asked about: real property and gains tax; tax planning for small business; small company tax; tax audit and investigation; budgets; sales and services tax; basic principles and policies; Malaysian taxation system; and tax planning for individual and personal taxation. This study found that real property and gains tax, tax planning for small business, and small company tax were the three most preferred tax topics, with mean scores of 3.31, 3.23, and 3.22, respectively. Research by [7] contradicts the conclusions. According to [7], basic tax concepts, personal taxes, tax planning for individuals, and taxation for small businesses and companies were the three most popular tax topics among students. Nonetheless, based on a survey by [8], personal taxes and individual tax planning were the two most popular tax issues among undergraduates. This revealed that academicians and students have different preferences when it comes to tax topics. Their respondents were students; however, this study surveyed academicians who were already mature and had some assets. Perhaps some of our respondents had prior experience buying and selling real estate, and some of them may have had family members who worked in small businesses.

4. Conclusions

This study aimed to analyze the need for introducing tax subjects in non-accounting curriculums among academicians from non-accounting fields. By surveying 349 academicians from non-accounting disciplines through the UiTM system, the findings suggest that tax education is an essential issue that non-accounting students should be taught. Business students who have studied taxation at universities have a greater degree of understanding of their tax obligations as a result [19]. Tax education has an influence on tax awareness, tax justice, and tax compliance [7,16–18]. They are more likely to comply with tax rules and to be influenced with regards to tax compliance if they have tax knowledge. As a result, non-accounting students should be taught about taxes. An early and formal tax education will help future taxpayers understand the taxation system and their tax responsibilities [15,20]. The authors of [3] found a significant difference in tax knowledge between Indonesian students who received tax education and those who did not. As a result, the Malaysian Ministry of Higher Education and the Inland Revenue Board must collaborate to introduce tax subjects to non-accounting students at the tertiary level. This study has several limitations. To begin, the scope of this study was confined to assessing respondents' perspectives on tax education, the importance of tax education, tax knowledge, and tax topics' preference. Second, the sample was limited to all UiTM non-accounting academicians. Third, tax knowledge level was assessed based on personal taxation. As a result, care should be exercised in the interpretation and generalization of results. Future research could use non-accounting undergraduate students as participants, as well as surveying other specific topics on taxation.

Author Contributions: Conceptualization, N.A.M.Y.; methodology, C.C.L.; software, N.A.M.Y.; validation, R.S.; formal analysis, N.A.M.Y.; investigation, R.S.; resources, C.C.L.; data curation, R.S.; writing—original draft preparation, N.A.M.Y.; writing—review and editing, N.A.M.Y. and R.S.; visualization, R.S.; supervision, N.A.M.Y.; project administration, N.A.M.Y., R.S. and C.C.L.; funding acquisition, N.A.M.Y. All authors have read and agreed to the published version of the manuscript.

Funding: This research was funded by International Matching Grant UiTM-UNAIR, Universiti Teknologi MARA, Malaysia and Universitas Airlangga Indonesia. Ref: 100-TNCPI/INT 16/6/2 (029/2020).

Institutional Review Board Statement: The study was conducted in accordance with the Declaration of Helsinki and approved by the Research Ethics Committee of Universiti Teknologi MARA in accordance with the ICH Good Clinical Practice Guidelines, Malaysian Good Clinical Practice Guidelines (REC/07/2021 (MR/628)—2 August 2021) for studies involving humans.

Informed Consent Statement: Informed consent was obtained from all subjects involved in the study.

Data Availability Statement: Not applicable.

Acknowledgments: The authors would like to thank both the panel of reviewers and the participants for their valuable comments during the presentation at the International Academic Symposium of Social Science 2022 (IASCC 2022), held at Perdana Hotel, Kota Bharu, Malaysia on 3 July 2022.

Conflicts of Interest: The authors declare no conflict of interest.

References

1. OECD. *Revenue Statistics in Asian and Pacific Economies*; OECD Publishing: Paris, France, 2020.
2. Sharifah, S.J. Lots More to Do to Increase Voluntary Compliance—IRB Kuala Lumpur: Theedgemarkets.com. 2019. Available online: https://www.theedgemarkets.com/article/lots-more-do-increase-voluntary-compliance-%E2%80%94-irb (accessed on 30 May 2022).

3. Putro, B.B.R.P.; Tjen, C. Analysis of tax education and tax knowledge: Survey on University Students in Indonesia. *J. Aust. Tax Teach. Assoc.* **2020**, *15*, 232–250.
4. Palil, M.R. Taxpayers Knowledge: A Descriptive Evidence on Demographic Factors in Malaysia. *J. Akunt. Keuang.* **2005**, *7*, 11–21.
5. Barjoyai, B. Tax illiteracy in Malaysia: Problems and solution. *J. Sch. Account. MARA Inst.* **1992**, *13*, 135–143.
6. Loo, E.C.; Ho, J.K. Competency of Malaysian salaried individuals in relation to tax compliance under self assessment. *Ejournal Tax Res.* **2005**, *20*, 45–62.
7. Mat Bahari, A.B.; Lai, M.L. Introducing tax education in non-accounting curriculum in higher education: Survey evidence. *J. Financ. Report. Acounting* **2009**, *7*, 37–51. [CrossRef]
8. Mahat, M.A.; Lai, M.L. Featuring tax education in non-accounting curriculum: Survey evidence. In *Econstor Conference Papers*; ZBW, Leibniz Information Centre for Economics: Kiel, Hamburg, 2011; 56066.
9. Lai, M.L.; Yaacob, Z.; Mahat, M.A.; Choong, K.F. Quest for tax education in non-accounting curriculum: A Malaysian study. *AsianSoc. Sci.* **2013**, *9*, 154–162. [CrossRef]
10. Eriksen, K.; Fallan, L. Tax knowledge and attitudes towards taxation: A report on a quasi experiment. *J. Econ. Psychol.* **1996**, *17*, 387–402. [CrossRef]
11. Sarker, T.K. Improving tax compliance in developing countries via Self-Assessment Systems–what could Bangladesh learn from Japan. *Asia-Pac. Tax Bull.* **2003**, *9*, 1–34.
12. Koster, L. The Incorporating of Basic Tax Education in the Secondary School Curriculum. Master's Thesis, University of Pretoria, Hatfield, South Africa, 2012.
13. Oberholzer, R.; Nel, M. The road to an improved tax curriculum in South African Schools. *Educ. Chang.* **2006**, *10*, 105–119. [CrossRef]
14. Abbas, Y.; Tjen, C.; Wicaksono, T. *Tax Education and Tax Awareness: An Analysis on Indonesian Tax Education Program*; Institute for Economic and Social Research: Jakarta, Indonesia, 2021.
15. Rani Diana, O.; Zubaidah, I.; Nurul Hidayah, A.N. (Eds.) Introducing formal tax education in secondary school: A survey on Malaysian Public's Perception. In Proceedings of the International Conference on Applied Social Sciences, Business and Humanity, ICO-ASCNITY 2019, Padang, Indonesia, 1–3 November 2019.
16. Kasipillai, J.; Noor Afza, A.; Norhani, A. The influence of education on tax avoidance and tax evasion. *eJ. Tax Res.* **2003**, *1*, 134–146.
17. Loo, E.C. Tax knowledge, tax structure and compliance: A report on a quasi-experiment. *N. Z. J. Tax. Law Policy* **2006**, *12*, 117–140.
18. Kurniawan, D. The influence of tax education during higher education on tax knowledge and its effect on personal tax compliance. *J. Indones. Econ. Bus.* **2020**, *35*, 57–72. [CrossRef]
19. Hastuti, R. Tax awareness and tax education: A perception of potential taxpayers. *Int. J. Bus. Econ. Law* **2014**, *5*, 83–91.
20. Mohd Faizal, S.; Mohd Zaini, R.; Somasundaram, N.R. Tax awareness and tax knowledge of students at secondary school level in Malaysia. *Int. Bus. Educ. J.* **2021**, *14*, 65–70.

proceedings

MDPI

Proceeding Paper

Digital Food Supply Chain Traceability Framework †

Pratyusha Reddy [1,*], Sherah Kurnia [1] and Guilherme Luz Tortorella [2]

[1] School of Computing and Information Systems, University of Melbourne, Parkville 3010, Australia
[2] Department of Mechanical Engineering, University of Melbourne, Parkville 3010, Australia
* Correspondence: reddy.p@unimelb.edu.au

† Presented at the International Academic Symposium of Social Science 2022, Kota Bharu, Malaysia, 3 July 2022.

Abstract: The growing importance of digitalising food traceability systems can be attributed to mounting food safety incidents and customer demand for sustainable products. However, most previous studies focused on a specific technology in supply chain management and lacked a holistic approach that could help organisations implement digital food supply chain traceability (DFSCT) systems. This study aims to synthesise the key elements of DFSCT identified in the existing literature and develop a framework to guide future research and practice in DFSCT. The proposed framework captures five dimensions of DFSCT—organisational capacity, enabling technology, the traceability process, expected benefits, and external factors influencing DFSCT system adoption. This study offers important implications for research and practice in DFSCT.

Keywords: digital traceability; food supply chain; food traceability

check for updates

Citation: Reddy, P.; Kurnia, S.; Tortorella, G.L. Digital Food Supply Chain Traceability Framework. *Proceedings* **2022**, *82*, 9. https://doi.org/10.3390/proceedings2022082009

Academic Editor: Mohamad Rahimi Mohamad Rosman

Published: 7 September 2022

1. Introduction

Foodborne illnesses, food fraud, and food scares are rising with the globalisation of supply chains and market diversification [1]. Food traceability has significant potential to protect consumers as it enables food recall, promotes food quality, safety and defence, and eliminates non-consumable food items. Food supply chain traceability (FSCT) is the ability to keep track of a product's movement and retain its recorded information throughout the supply chain [2,3]. Efficiencies achieved through FSCT help organisations gain competitive advantages in the market as they achieve better product quality, safety, regulation and compliance, sustainable business performance, and operational efficiency [4]. Therefore, academics and practitioners recognise the significance of FSCT and are working to improve the current practice.

Technology is one of the critical drivers that open opportunities for food business operators (FBOs) regarding FSCT [5,6]. Tracing and tracking products in real-time are challenging processes due to their dynamic and complex natures. The digitalisation of traceability eases the process, and enhances supplier communication and decision making. Moreover, advanced technologies hold the potential to change the future of the food industry. Barcodes and RFID are the most used technologies in FSCT [7]. Emerging technologies such as the Internet of Things (IoT), big data analytics, and Blockchain offer innovative and robust solutions to improve food supply chain traceability [6,8]. Finally, the industry will benefit from improved consumer confidence and increased efficiency and innovation [9,10].

The apparent positive impacts of digitalising food traceability are insufficient for its adoption because it lacks a commonly agreed-upon lexicon and implementation guidelines. A comprehensive guide to adopting digital food supply chain traceability (DFSCT) is required. The literature presents extensive studies regarding the traceability process [2,3,11–14] and application of specific technologies such as RFID [15], the IoT [9], or Blockchain [16]. However, few studies acknowledge the importance and role of organisational capabilities such as IT infrastructure [17], IT governance [18], strong leadership [8,19,20], and technical and managerial skills [21]. Moreover, data quality management [12] and DFSCT interoperability [22] are seldom considered. Thus, this study aims to address this gap and accelerate the adoption of DFSCT by proposing a holistic framework of DFSCT to guide future research in DFSCT and facilitate its implementation.

The proposed DFSCT framework was developed through an extensive literature review. First, research papers were selected from the domains of supply chain management, food traceability, food traceability technology, and information systems. Previous studies of the adoption of DFSCT were also explored. Second, papers were thematically analysed to identify key concepts and organise them according to DFSCT dimensions and capabilities. Third, key factors affecting the adoption of digital food traceability were identified to provide a holistic view of DFSCT, its potential impacts, and relationships between key concepts related to DFSCT. The proposed DFSCT framework highlights implications regarding the important fields of DFSCT research and practice.

2. Defining Food Traceability

Academics, supply chain professionals, and regulators have tried to define food traceability but have not reached a consensus. This study follows the definition proposed by Bosona and Gebresenbet [11]:

Food traceability is a type of logistics management that captures, stores, and transmits adequate information about a food, feed, food-producing animal, or substance at all stages in the food supply chain so that a product can be checked for safety and quality control, and traced upward, or tracked downward, at any time required. (p. 35).

Figure 1 presents a schematic representation of this concept. The three components of tracing are backward tracing, forward tracing, and product history information. Backward (upward) traceability facilitates finding the source of a problem by finding its origin and characteristics at every point in the supply chain. In contrast, forward (downward) tracing, or tracking, is the ability to follow a product to gather information at any point in the flow [2]. Both types of traceability are essential for effective FSCT, and companies should aim to perfect each of them. Finally, product history information details the movement, time, inputs identification, and operations that a product experienced in a supply chain [23,24]. Additional notable classifications are internal and external traceability. Internal traceability occurs inside a company or production unit, whereas external traceability tracks physical movement between companies [14,22].

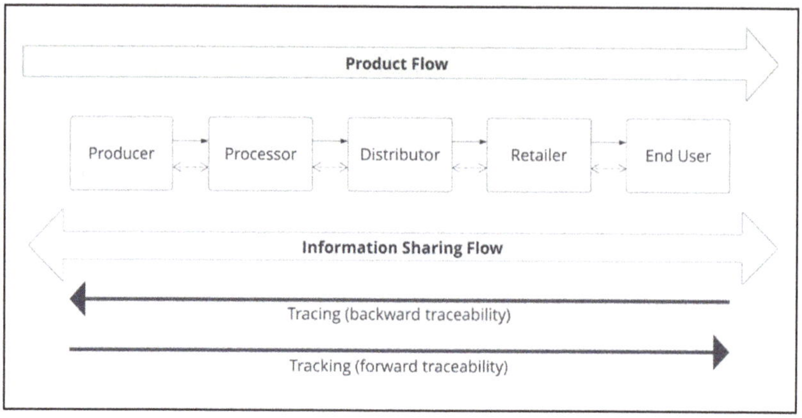

Figure 1. Material and information flow in food supply chain traceability (Adapted with permission from Bosona and Gebresenbet [11], p. 35).

3. Synthesis of Key Concepts

The building blocks of the DFSCT framework, based on our literature analysis and synthesis, are presented below.

3.1. DFSCT Principles

In the literature, traceability principles describe "how" DFSCT is implemented for effective operations. Alternate terminologies found in the literature include components, characteristics, and elements. Various scholars describe DFSCT principles differently across studies, based on their commonalities. In this study, we grouped them into four major principles: (i) identification of traceable resource unit (TRU), (ii) data recording, (iii) data exchange, and (iv) data management. Each principle is briefly described below.

3.1.1. Identification of Traceable Resource Unit (TRU)

TRU is the smallest traceable product or lot [3]; European Council Directive 91/238 defines a lot as "a batch of sales units of foodstuff produced, manufactured or packaged under the same conditions" [25]. In a food crisis, a lot is recalled instead of all affected food instances, allowing targeted operations. Furthermore, TRU simplifies goods semantics, positively impacting the visibility of goods and operational efficiency [22]. Granularity and uniqueness are the two most important factors of TRU identification.

3.1.2. Data Recording

The three types of data recorded are product identification, tracing data, and transformation data, depending on drivers and beneficiaries. The first type is mandatory data for product identification. It includes all information enabling traceability of a specific product [3,25]. The second type is data required to satisfy regulations, standards, or certifications. The third is additional information requested for operational purposes [26]. Captured data are stored in various recording mediums, including simple paper-based systems, computer-based database management systems, Enterprise Resource Planning (ERP), and complex cloud systems [27].

3.1.3. Data Exchange

Data exchange (or integration) enables internal and external traceability by linking, merging, and sharing information [3]. Internal traceability data exchange considers both a product and process data regarding that product moving within the organisation. Data at critical traceability points (CTPs) are mapped using the same TRU identification numbers [13]. A paper-based system is the most widely used for recording data. However, emerging technologies such as RFID, Wireless Sensor Network (WSN), and the Internet of Things (IoT) provide error-free systems and enable explicit linking of necessary data in real-time [28].

3.1.4. Data Management

Information effectiveness, integrity, authenticity, and standardisation are critical to the success of data management [12]. While identification of TRUs, data recording, capturing, and exchange explain the process of traceability, data management focuses on the characteristics of data parsing through the supply chain. It is essential that this information be carefully managed to build and protect trust with supply chain partners and consumers. Therefore, data must be in the correct format, complete, accurate, and credible [29].

Perfecting interoperability, transparency, and accessibility can further improve FSCT. Interoperability allows different technologies to communicate and share information seamlessly [30]. Although this is essential for DFSCT, achieving it could be a challenge, as DFSCT lacks standardisation. Furthermore, transparency between suppliers and customers is vital for adequate informational and knowledge exchange. While visibility poses a risk, the benefits of selective data sharing are worth considering [22]. Finally, confidentiality can be maintained by maintaining different levels of access to data. The control of visibility can build trust [3].

3.2. Technologies

This section provides an overview of the most-used technologies frequently discussed in the literature; their shortcomings and standards are addressed, and followed by an overview of the scope of emerging technologies.

A barcode is a machine-readable pattern of bars and spaces of varying widths. Barcode structure permits accurate, simple, and economical traceability as it is automatic, fast, and precise [15]. GS1 is a common global standard affiliated with Uniform Code Council (UCC) and European Article Numbering (EAN). GS1 Databar and multidimensional barcodes allow barcodes to carry information (including weight, batch numbers, and best-before-dates) in addition to essential identification [10,11]. QR codes are 2D barcodes, which are widely used for automated product tracking and customer reference. However, barcodes require human intervention for positioning and scanning, which introduces errors and inefficiency [3,29].

Radiofrequency Identification is the most widely used food-tracking technology [3,5,31]. RFID tags are electronic labels with a microchip that identifies and tracks tags wirelessly. They help overcome problems associated with traditional solutions (alphanumerical codes and barcode labels) [26,29]. RFID is an effective tool because it supports no-line-of-sight reading. Scanners can read multiple tags simultaneously, have large memories, and allow storage and manipulation of a wide variety of data. Furthermore, automation makes the process of storing and manipulating information in the database error-free and fast [32,33]. The limitations of RFID are high costs and sensitivity towards certain weather conditions and materials [2,3]. The standard for using RFID is EPCglobal Network Standards, launched by GS1, which enables information integration and real-time product visibility in the supply chain. According to the literature, RFID and EPCglobal Network Standards are critical technologies for automating data capture and integrating traceability data [26].

The application of advanced IT in connection to the Internet has become important with regards to information sharing among members of food supply chains [11]. There are two types of traceability information flow models. The first is the 'one step up one step down flow model', where information is filtered at each stage of the supply chain—only certain information is shared with immediate suppliers/customers. The second is the 'aggregated information flow model', where no information filters are applied. Both EU and US regulations have adopted this model for food traceability [2]. Thakur and Donnelly [34] discuss using Electronic Data Interchange (EDI) technology and data format standards, such as Extensible Markup Language (XML), for information exchange in digital traceability systems. EDI enables firms with mature IT capabilities to efficiently exchange standardised and structured data, and XML facilitates the sharing of structured data, mainly via the Internet [3].

Industry 4.0 technologies such as big data analytics, cloud computing, cybersecurity, the Internet of Things (IoT), and Blockchain have the potential to revolutionise food traceability [5]. Research studies regarding these technologies are in nascent stages, but are becoming more frequent in the traceability literature. Alfian et al. [9] used IoT sensors and a machine learning model to improve an RFID-based food traceability system. Lin et al. [35] proposed a framework integrating Blockchain and IoT technologies and demonstrated the capacity of technologies to build a trusted agro-traceability system with all parties in the supply chain, even in the absence of trust between them [36]. However, the high implementation cost, a lack of expertise, and technological immaturity are seen as significant barriers to adopting these modern technologies [5].

3.3. Factors Affecting Adoption of Digital Food Traceability

The literature discusses digital food traceability adoption factors in depth. The following sections elaborate on the benefits and barriers as perceived by FBOs. Drivers are the motivating factors that a company must oblige to successfully establish FSCT systems. Barriers are hurdles that might demotivate an organisation from adopting DFSCT. Table 1 provides an overview of DFSCT key drivers and barriers.

Table 1. Drivers and barriers affecting the adoption of digital food traceability.

Drivers	Barriers
Environment	
• Compliance with regulations, legislation, and certification requirements • Government support and funding • Increase food safety control, reduce food crisis risk, and control disease outbreaks • Mitigate information asymmetry and improve supply chain coordination	• Lack of coordination, trust, confidence, and liability among supply chain partners • Concerns with ethics, privacy, security, reliability, and data protection issues while sharing information
Organisation	
• Decrease the severity, volume, frequency, and cost of a product recall • Improve consumer perception and confidence in brands • Gain competitive advantage, better market access, brand value, and market share • Verify sustainability claims and mitigate potential reputational risk	• Lack of knowledge and awareness • Shortage of skilled staff • Cultural barriers and internal resistance to change • Lack of resources and capital • High initial investment
Technology	
• Availability of digital technologies	• Unreliability of technologies • Lack of interoperability • Lack of uniformity in implementing traceability systems as there are no common standards

4. Proposed Digital Food Supply Chain Traceability Framework

A DFSCT framework (Figure 2) is proposed based on literature analysis and synthesis. This framework captures the critical dimensions of DFSCT, including the required organisational capability, enabling technology, traceability process, and expected benefits. These dimensions are influenced by external factors which either support or hinder the adoption of DFSCT by the food industry. Factors such as access to national IT infrastructure, government provisions, and industry/retailer standards define the operating environment, affecting DFSCT adoption and implementation by FBOs.

DFSCT dimensions have a cause-and-effect relationship in which organisational capability supports the other three dimensions during the development, implementation, and maintenance of DFSCT. Furthermore, technology moderates traceability processes, which results in benefits. Each dimension is further explained regarding specific capabilities in the following subsections.

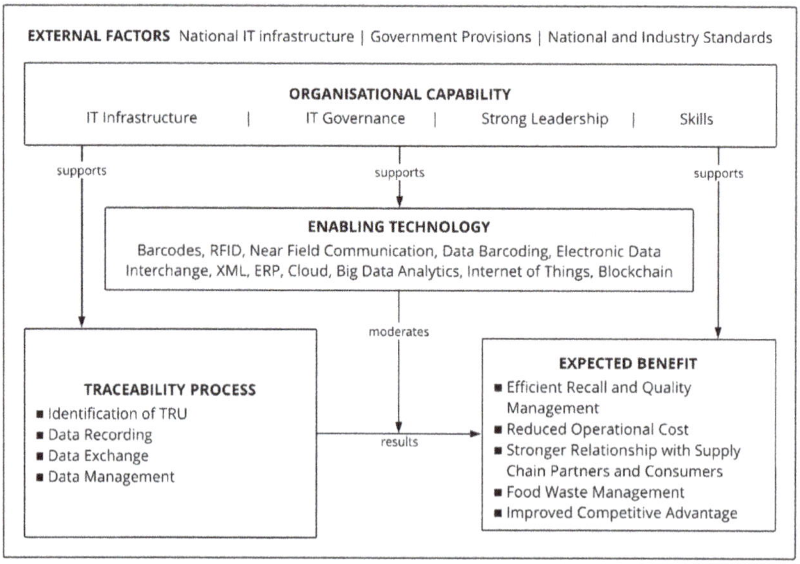

Figure 2. Digital food supply chain traceability framework.

4.1. *Organisational Capability*

The organisational capability dimension consists of four capabilities that aim to establish the right tools, people, processes, and skillsets. Furthermore, this dimension is the foundation for the success of DFSCT as it supports the other three dimensions. The four capabilities drive efficiency and effectiveness; thus, they are vital for creating a robust and agile DFSCT [37], as explained below.

IT Infrastructure: Appropriate applications, databases, systems, and resources are paramount for the development, implementation, and management of DFSCT; together, these form a technical base for evolving technologies and ongoing management [17].

IT Governance: This capability enables organisations to manage strategic business–IT alignment, IT performance, IT resources, IT risk management, and IT value delivery. IT governance ensures appropriate processes are in place to optimise the application of IT to achieve organisational goals by providing direction to leadership, fostering commitment, and establishing accountability [18].

Strong Leadership: The involvement of upper management is critical to the success of DFSCT. Upper management requires a clear understanding of DFSCT's importance and its adoption [8]. Furthermore, upper management is accountable for engaging with internal and external stakeholders to promote productivity and limit change resistance by improving awareness of DFSCT's functions and importance [20]. Moreover, they must establish organisational policies regarding traceability processes to standardise workflow. Finally, performance measurement is the constant evaluation of traceability processes using metrics. It is a quintessential managerial step for warranting the company's ability to stay on track and use its resources effectively [19].

Skills: This capability focuses on the technical and managerial skills of food organisations' employees [20]. It is an essential dimension because successfully enabling technologies requires new skills. Furthermore, organisations may need to hire new employees or train their existing staff to engage in DFSCT [21].

4.2. *Enabling Technology*

The eleven most-cited technologies which enable DFSCT included in this dimension are barcodes, RFID, near field communication, data barcoding, electronic data interchange,

XML, ERP, cloud, big data analytics, the Internet of Things, and Blockchain. Many other advanced technologies are available for industry-specific applications not mentioned here, such as pH indicators for real-time fish spoilage monitoring. These technologies will optimise traceability processes and improve trust between supply chain partners, leading to better recall efficiencies and cost reductions [3,11].

4.3. Traceability Process

The four capabilities essential for the traceability process are identification of TRU, data recording, data exchange, and data management. The first three capabilities are interlinked. Identification is the first step in the DFSCT process, which enables data recording and data exchange. The data recording capability facilitates recording the product information necessary for traceability. Data exchange facilitates continuous traceability information flow, both internally and externally [2].

The fourth capability, data management, includes all processes associated with managing digital data, including data governance, data quality assurance, master data management, data security management, and data analysis. This capability secures data format, completeness, accuracy, and reliability [12,22], and is vital for establishing standardisation, interoperability, accessibility, transparency, integrity, and authenticity.

4.4. Expected Benefit

This dimension represents the cascading effects of DFSCT gained by a food business organisation. The notable benefits are greater efficiency in recalling products and response to quality management, reduced operational costs, stronger relationships with supply chain partners and consumers, food waste management, and improved competitive advantage. Therefore, along with achieving food safety and quality goals, DFSCT plays a crucial role in providing value, improving responsiveness to customer requirements, and increasing sustainable profit margins [38].

5. Discussion

In this study, a comprehensive DFSCT framework is proposed. The framework extends the existing literature by synthesising and drawing out DFSCT's key elements and capabilities from multiple authors' research [2,3,11,12]. Studies regarding organisational capability [8,17–19], enabling technology [9,15,16], the traceability process [2], and benefits [11,38] were brought together to form one robust DFSCT framework with four dimensions. Thus, an aggregate representation of the managerial, process, and technical aspects of DFSCT was compiled by synthesising the traceability frameworks of numerous authors who explored various DFSCT aspects under different lenses. Future research can focus on selected dimensions of the proposed DFSCT framework.

In terms of implications for practice, this study helps identify essential factors that FBOs should focus on to successfully digitalise their food supply chain operations. First, attention should be given to the proper application and customisation of technology, by prioritising what is best suited for business needs over the latest or trending technology. Second, strong leadership is of utmost importance to ensure the company has the right skills and a smooth digital transition with minimal employee resistance. The human factor can be easily overlooked, but it is one of the most challenging barriers to overcome. Finally, FBOs should establish interoperability by standardising data formats and data exchange methods to improve internal and external information sharing between systems. Consolidated efforts from different departments to improve DFSCT could allow an organisation to leverage it for profits and achieve goals sustainably.

6. Conclusions

The importance of digital food supply chain traceability is increasing because it helps food businesses combat and prevent food fraud, improve consumer safety, and achieve sustainability goals. Furthermore, food organisations can actualise sustainable economic, social, and environmental benefits, consequently building a stronger relationship with their supply chain partners and consumers. However, they cannot adopt DFSCT effectively because traceability practices vary across food sectors and there is no standard DFSCT process. Although researchers and practitioners are giving more attention to digital traceability and supply chain management, there is no mutual agreement or holistic discussion regarding DFSCT. Therefore, this research proposes a DFSCT framework to address this gap.

This study contributes to research and practice. In terms of research, a novel comprehensive DFSCT framework is proposed, which contributes to the literature by introducing a holistic overview of the capabilities that food organisations need to build sophisticated digital supply chain traceability. In term of practice, it provides a practical framework to guide DFSCT implementation. It identifies important factors and organisational capabilities that FBOs' senior management need to manage, develop, and support DFSCT implementation. It also helps senior management understand the expected benefits of DFSCT implementation.

As this study is limited to the discussion in the literature, further theoretical and empirical effort is needed to assess and improve the framework. Future research could enhance the findings by conducting in-depth case analyses of multiple food companies. The engagement of diverse food industries and participants with various backgrounds could further evoke new insights and improve the generalisability of the framework.

Future research to develop a cross-disciplinary multi-level digital food supply chain traceability maturity model using the proposed DFSCT framework as a foundation would be useful for both research and practice. A common DFSCT framework and maturity model could standardise the process of digital food traceability, expediting its implementation. Finally, a user-friendly tool for assessing a company's maturity level and planning its next steps could enhance food product traceability along extended supply chains nationally and globally.

Author Contributions: Conceptualization, P.R., S.K. and G.L.T.; methodology, P.R. and S.K.; validation, S.K. and G.L.T.; formal analysis, P.R.; investigation, P.R.; resources, P.R., S.K. and G.L.T.; data curation, P.R.; writing—original draft preparation, P.R.; writing—review and editing, S.K.; visualization, P.R.; supervision, S.K. and G.L.T.; project administration, S.K. All authors have read and agreed to the published version of the manuscript.

Funding: This research received no external funding.

Institutional Review Board Statement: Not applicable.

Informed Consent Statement: Not applicable.

Data Availability Statement: Not applicable.

Conflicts of Interest: The authors declare no conflict of interest.

References

1. Qian, J.; Dai, B.; Wang, B.; Zha, Y.; Song, Q. Traceability in Food Processing: Problems, methods, and performance evaluations—A review. *Crit. Rev. Food Sci. Nutr.* **2020**, *62*, 679–692. [CrossRef] [PubMed]
2. Aung, M.M.; Chang, Y.S. Traceability in a Food Supply Chain: Safety and Quality Perspectives. *Food Control* **2014**, *39*, 172–184. [CrossRef]
3. Islam, S.; Cullen, J.M. Food Traceability: A generic theoretical framework. *Food Control* **2021**, *123*, 107848. [CrossRef]
4. Garcia-Torres, S.; Albareda, L.; Rey-Garcia, M.; Seuring, S. Traceability for Sustainability—Literature review and conceptual framework. *SCM* **2019**, *24*, 85–106. [CrossRef]
5. Kayikci, Y.; Subramanian, N.; Dora, M.; Bhatia, M.S. Food Supply Chain in the Era of Industry 4.0: Blockchain technology implementation opportunities and impediments from the perspective of people, process, performance, and technology. *Prod. Plan. Control* **2020**, *33*, 301–321. [CrossRef]
6. Yu, Z.; Jung, D.; Park, S.; Hu, Y.; Huang, K.; Rasco, B.A.; Wang, S.; Ronholm, J.; Lu, X.; Chen, J. Smart traceability for food safety. *Crit. Rev. Food Sci. Nutr.* **2020**, *62*, 905–916. [CrossRef] [PubMed]
7. Kumperščak, S.; Medved, M.; Terglav, M.; Wrzalik, A.; Obrecht, M. Traceability systems and technologies for better food supply chain management. *Qual. Prod. Improv.-QPI* **2019**, *1*, 567–574. [CrossRef]

8. Wong, L.-W.; Leong, L.-Y.; Hew, J.-J.; Tan, G.W.-H.; Ooi, K.-B. Time to seize the digital evolution: Adoption of blockchain in operations and supply chain management among Malaysian SMEs. *Int. J. Inf. Manag.* **2020**, *52*, 101997. [CrossRef]
9. Alfian, G.; Syafrudin, M.; Farooq, U.; Ma'arif, M.R.; Syaekhoni, M.A.; Fitriyani, N.L.; Lee, J.; Rhee, J. Improving efficiency of RFID-Based traceability system for perishable food by utilizing IoT sensors and machine learning model. *Food Control* **2020**, *110*, 1070161. [CrossRef]
10. Chrysochou, P.; Chryssochoidis, G.; Kehagia, O. Traceability information carriers. The technology backgrounds and consumers' perceptions of the technological solutions. *Appetite* **2009**, *53*, 322–331. [CrossRef]
11. Bosona, T.; Gebresenbet, G. Food traceability as an integral part of logistics management in food and agricultural supply Chain. *Food Control* **2013**, *33*, 32–48. [CrossRef]
12. Duan, Y.; Miao, M.; Wang, R.; Fu, Z.; Xu, M. A framework for the successful implementation of food traceability systems in China. *Inf. Soc.* **2017**, *33*, 226–242. [CrossRef]
13. Karlsen, K.M.; Dreyer, B.; Olsen, P.; Elvevoll, E.O. Literature Review: Does a common theoretical framework to implement food traceability exist? *Food Control* **2013**, *32*, 409–417. [CrossRef]
14. Olsen, P.; Borit, M. How to define traceability. *Trends Food Sci. Technol.* **2013**, *29*, 142–150. [CrossRef]
15. Fan, B.; Qian, J.; Wu, X.; Du, X.; Li, W.; Ji, Z.; Xin, X. Improving continuous traceability of food stuff by using barcode-RFID bidirectional transformation equipment: Two field experiments. *Food Control* **2019**, *98*, 449–456. [CrossRef]
16. Casino, F.; Kanakaris, V.; Dasaklis, T.K.; Moschuris, S.; Stachtiaris, S.; Pagoni, M.; Rachaniotis, N.P. Blockchain-Based food supply chain traceability: A Case Study in the Dairy Sector. *Int. J. Prod. Res.* **2020**, *59*, 5758–5770. [CrossRef]
17. Rahim, M.; Kurnia, S.; Singh, P.; Samson, D. Developing an it maturity model for sustainable supply chain management implementation. In Proceedings of the PACIS 2014 Proceedings, Chengdu, China, 24–28 June 2014.
18. Weill, P.; Ross, J.W. *IT Governance: How Top Performers Manage IT Decision Rights for Superior Results*; Harvard Business School Press: Boston, MA, USA, 2004; ISBN 978-1-59139-253-8.
19. Fagioli, F.F.; Rocchi, L.; Paolotti, L.; Słowiński, R.; Boggia, A. From the farm to the agri-food System: A multiple criteria framework to evaluate extended multi-functional value. *Ecol. Indic.* **2017**, *79*, 91–102. [CrossRef]
20. Queiroz, M.M.; Telles, R. Big data analytics in supply chain and logistics: An empirical approach. *Int. J. Logist. Manag.* **2018**, *29*, 767–783. [CrossRef]
21. Karippacheril, T.G.; Rios, L.D.; Srivastava, L. Global markets, global challenges: Improving food safety and traceability while empowering smallholders Through ICT. In *ICT in Agriculture (Updated Edition): Connecting Smallholders to Knowledge, Networks, and Institutions*; The World Bank: Washington, DC, USA, 2017; pp. 283–308. ISBN 978-1-4648-1002-2.
22. Ringsberg, H. Perspectives on food traceability: A systematic literature review. *Supply Chain. Manag. Int. J.* **2014**, *19*, 558–576. [CrossRef]
23. Manos, B.; Manikas, I. Traceability in the Greek fresh produce sector: Drivers and constraints. *Br. Food J.* **2010**, *112*, 640–652. [CrossRef]
24. Moe, T. Perspectives on traceability in food manufacture. *Trends Food Sci. Technol.* **1998**, *9*, 211–214. [CrossRef]
25. Bendaoud, M.; Lecomte, C.; Yannou, B. A methodological framework to design and assess food traceability systems. *Int. Food Agribus. Manag. Rev.* **2012**, *15*, 103–125.
26. Jakkhupan, W.; Arch-Int, S.; Li, Y. An RFID-Based traceability system: A case study of rice supply chain. *Telecommun. Syst.* **2015**, *58*, 243–258. [CrossRef]
27. Olsen, P.; Borit, M. The components of a food traceability system. *Trends Food Sci. Technol.* **2018**, *77*, 143–149. [CrossRef]
28. Chen, L.; Lee, H.L. Sourcing under supplier responsibility risk: The effects of certification, audit, and contingency payment. *Manag. Sci.* **2017**, *63*, 2795–2812. [CrossRef]
29. Regattieri, A.; Gamberi, M.; Manzini, R. Traceability of food products: General framework and experimental evidence. *J. Food Eng.* **2007**, *81*, 347–356. [CrossRef]
30. Hardt, M.J.; Flett, K.; Howell, C.J. Current barriers to large-scale interoperability of traceability technology in the seafood sector. *J. Food Sci.* **2017**, *82*, A3–A12. [CrossRef]
31. Zhang, A.; Mankad, A.; Ariyawardana, A. Establishing confidence in food safety: Is traceability a solution in consumers' eyes? *J. Consum. Prot. Food Saf.* **2020**, *15*, 99–107. [CrossRef]
32. Scholten, H.; Verdouw, C.N.; Beulens, A.; van der Vorst, J.G.A.J. Defining and analyzing traceability systems in food supply chains. In *Advances in Food Traceability Techniques and Technologies*; Elsevier: Amsterdam, The Netherlands, 2016; pp. 9–33, ISBN 978-0-08-100310-7.
33. Azuara, G.; Luis Tornos, J.; Luis Salazar, J. Improving RFID traceability systems with verifiable quality. *Ind. Manag. Data Syst.* **2012**, *112*, 340–359. [CrossRef]
34. Thakur, M.; Donnelly, K.A.-M. Modeling traceability information in Soybean Value Chains. *J. Food Eng.* **2010**, *99*, 98–105. [CrossRef]
35. Lin, J.; Shen, Z.; Zhang, A.; Chai, Y. Blockchain and IoT based food traceability for smart agriculture. In Proceedings of the 3rd International Conference on Crowd Science and Engineering, Singapore, 28–31 July 2018; Association for Computing Machinery: New York, NY, USA, 2018; pp. 1–6.
36. Azzi, R.; Chamoun, R.K.; Sokhn, M. The power of a blockchain-based supply chain. *Comput. Ind. Eng.* **2019**, *135*, 582–592. [CrossRef]

37. Büyüközkan, G.; Göçer, F. Digital supply chain: Literature review and a proposed framework for future research. *Comput. Ind.* **2018**, *97*, 157–177. [CrossRef]
38. Mattevi, M.; Jones, J.A.; Griffith, C. Food supply chain: Are UK SMEs aware of concept, drivers, benefits and barriers, and frameworks of traceability? *Br. Food J.* **2016**, *118*, 1107–1128. [CrossRef]

proceedings

MDPI

Proceeding Paper

A Ringgit Saved Is a Ringgit Earned: Saving Habits among Malaysian Cadets †

Hafizah Mat Nawi [1,*], Nur Shaliza Sapiai [2], Wan Nur Shawatul Aswal Zulkefle [3], Mohd Nor Yahaya [1] and Ariffin Ismail [1]

1 Faculty of Defence Studies & Management, National Defence University of Malaysia, Kuala Lumpur 53000, Malaysia
2 Faculty of Information Management, Universiti Teknologi Mara Cawangan Kelantan, Machang 18500, Malaysia
3 Faculty of Business and Management, Quest International University, Ipoh 30250, Malaysia
* Correspondence: hafizah.matnawi@upnm.edu.my
† Presented at the International Academic Symposium of Social Science 2022, Kota Bharu, Malaysia, 3 July 2022.

Abstract: Saving money has an important role in both individual lives and economic prosperity. It offers individuals an escape from life's uncertainties and encourages investment, which could indirectly contribute to the economic progress of a nation. This study investigates the factors that influence the saving habits of Malaysian military cadets. The study focuses on the impact of parental socialization, peer influence, financial literacy, and self-control on saving habits. Adopting a positivist paradigm, the research design was drawn from a survey of 400 military cadets studying at the National Defense University of Malaysia. The data was analyzed using multiple regression analysis. The results reveal substantial correlations between the variables except for financial literacy; hence, all of the proposed hypotheses are supported except the hypothesis related to financial literacy. Both the life cycle theory and the theory of planned behavior are supported by the findings. This study provides valuable information about the saving habits of military cadets and soldiers and the importance of savings plans.

Keywords: saving habits; parental socialization; peer influence; financial literacy; self-control; military; cadets

check for updates

Citation: Nawi, H.M.; Sapiai, N.S.; Zulkefle, W.N.S.A.; Yahaya, M.N.; Ismail, A. A Ringgit Saved Is a Ringgit Earned: Saving Habits among Malaysian Cadets. *Proceedings* **2022**, *82*, 20. https://doi.org/10.3390/proceedings2022082020

Academic Editor: Mohamad Rahimi Mohamad Rosman

Published: 9 September 2022

1. Introduction

Saving is defined as expenses subtracted from disposable income. The surplus funds after eliminating personal expenses can be saved by financially savvy individuals. Meanwhile, individuals who rely on credit and loans to meet their requirements will have no funds left for savings. Despite the important role of domestic savings mobilization, the savings rate remains low in most developing countries [1,2].

According to the RinggitPlus Malaysian Financial Literacy Survey 2019 [3], 31 per cent of Malaysians lack financial self-control. A total of 37 per cent of those with incomes below RM2000 do not save for the future. More than half of Malaysians lack sufficient funds to last for three months. The findings of the poll also found that 43 per cent of Malaysians admitted to spending over their means. This indicates that Malaysians live from paycheck to paycheck, preventing them from saving.

We must ask the question, who is a cadet officer? A cadet officer is a young person or youth undergoing training to become an officer in the armed forces. Youth embodies a significant percentage of the population. The National Youth Development Policy of Malaysia [4] defines youth as those between the ages of 15 and 40. According to the Malaysian Department of Statistics [5], the youth population in 2018 is 14.6 million, representing 45.36 per cent of the total population. This is a rise of 160,700, or 1.11 per cent, compared to the year 2017. According to the World Population Review [6], Malaysia's population will gradually grow from year to year. The number implies that this group constitutes nearly half of the population, and this proportion will grow substantially in the future. This demonstrates the impact that exceptional youth could have on the nation.

Unfortunately, financial troubles are on the rise as people, particularly young people, cannot manage their financial situation. Inadequate savings could result in bankruptcy. The young's inability to contribute their full potential to the development of Malaysia has been hampered by the alarming rise in personal bankruptcy. According to the Department of Insolvency Malaysia [7] report, the overall number of bankruptcies to 31 December 2021 was 289,766. Consequently, it was emphasized that over 65 per cent of those declared bankrupt were young adults.

Savings could allow youngsters to prepare for emergencies, short or long-term financial obligations, and retirement. As a result, comprehending the significance of conserving behaviors and having a clear image of the situation should be emphasized. Therefore, this study focuses on the relationship between the primary factors influencing the saving habits of Malaysian cadets. Specifically, the study seeks to determine the impact of parental socialization, peer influence, financial literacy, and self-control, on the saving habits of Malaysian cadets.

2. Review of Literature

2.1. Theories

The study used the life cycle theory and theory of planned behavior as a theoretical lens.

i. *Life-cycle Theory*

The earliest formulation of life-cycle theory by Penrose (1952), as referenced by Mat Nawi [8], focuses on the evolution of an individuals' consumption and saving behaviors. The idea aids in comprehending how individuals spend and save based on their income level, retirement goals, life expectancy, and intergenerational transfer motivations [9]. Throughout the periods of a person's life, the theory predicts that an individual's consumption and savings behaviors would vary considerably based on subjective (i.e., age, marital status, demographic conditions, and socioeconomic) and objective (i.e., income, wealth rates of return) factors. They should accumulate sufficient funds throughout their active career for future use, especially after they retire and become financially inactive.

ii. *Theory of planned behavior*

The theory of planned behavior (TPB), initially proposed by Fishbein and Ajzen [10], emphasizes the behavior of individuals as a result of their intentions to behave; they act in accordance with what they consider to be the most suitable. The objective of the TPB is to facilitate the evaluation of factors that may contribute to financial behaviors associated with the management of an emergency fund and overspending. According to the theory, background factors influence beliefs, which in turn influence attitudes, perceived behavioral control and intent, which eventually influence behavior [11].

Before engaging in particular acts, individuals will have the intent to do so [12]. According to Ajzen and Fishbein [11], attitude towards behavior, subjective norm, and perceived behavioral control are the three aspects that influence the intention of individuals. Attitude towards behavior refers to the extent to which an individual forms a favorable or unfavorable opinion of a given behavior. On the other hand, subjective norms concern a person's perception of social pressure from parents, spouse, classmates, and coworkers, to engage in particular conduct. Alternatively, behavioral control refers to an individual's impression of his ability to accomplish a particular behavior. In other words, it indicates that a person's intention to conduct particular acts can be predicted based on his views of his obstacles in carrying out the behavior.

It is assumed that the individual's impression of his difficulties in completing the behavior reflects both his prior experience and his current impediments. This remark is supported by the findings of Furnham's [13] study, which indicates that highly educated individuals are unlikely to save because they believe that doing so will not result in financial gain.

This study employs two criteria in explaining the influence of financial literacy on cadets' saving habits, i.e., attitude toward behavior and perceived behavioral control. Those with higher financial awareness will see the advantages of saving. This behavioral concept will encourage cadets to develop a favorable attitude towards saving habits. In addition, perceived behavioral control is utilized to explain cadets' self-control. It will be easier for students with greater self-control to save money because they are better able to control their desires and delay pleasure. The subjective norm will be utilized to illustrate how parents and peers influence the saving habit of children. This is because social pressure exerted by cadets' parents and peers has a substantial impact on their propensity to save.

In conclusion, the TPB can explain the independent variables (parental socialization, peer influence, financial literacy, and self-control) in determining cadets' saving habits in Malaysia more visibly.

2.2. *Saving Habits*

The term "saving" encompasses a variety of meanings and connotations. It can be explicated from both economic and psychological perspectives. According to Browning and Lusardi [14], savings are the amount of income remaining after subtracting current spending over a certain period. In contrast, psychologically speaking, saving is defined as the state of not spending money now to use it in the future. Sistiani and Prajawati [15] asserted that savings could provide emergency financial protection. When a person lacks sufficient cash, he or she will be subjected to a large amount of strain and difficulties. Inadequate savings not only causes potential problems in terms of long-term financial security, but would also deter individuals from making vital financial decisions, which can have a severe impact on their lives [16].

On the other hand, saving is essential for sustaining economic growth [17]. Households, businesses, and the government, all play crucial roles in saving due to their interconnected nature. For instance, households with fewer savings and limited emergency money may have financial difficulties that may cause anxiety and stress. From a broader perspective, fewer savings may contribute to insufficient finances for the government to invest in both physical and social infrastructure. In turn, financial institutions will invest money in financial assets to increase the nation's economic growth and productivity.

As the majority of us base our financial decisions on the money we have saved to date, saving habit is regarded as a factor that influences an individual's financial management. In simplest terms, a saving habit is the integration of a future needs perspective, a saving decision, and an action of saving. To have sufficient long-term funds, it is necessary to develop a consistent saving habit.

2.3. *Contributing Factors to Saving Habits*

This section presents the literature on determinants of saving habits. The factors had been identified through past literature. This study focuses on four contributing factors, namely, parental socialization, peer influence, financial literacy, and self-control.

2.3.1. Parental Socialization

Parental advice and influence cultivate higher self-esteem among children [18]. Kim and Jang emphasized that socialization with parents could stimulate the child's views about materialism. In their research, Alekam et al. [19] asserted that parents can instill financial literacy in early childhood by serving as excellent role models. This was proved by Jorgensen's study [20], who found that parental socialization had a significant effect on children's comprehension and financial behavior.

Good examples portrayed by the parents to their children will enable them (i.e., the children) to understand the principles of financial literacy and, consequently, practice prudent spending in their everyday lives [21,22]. Bucciol and Veronesi [23] found that if parents teach their children to save money, their propensity to save will increase by around 16 per cent. Another interesting study related to this issue was conducted by

Otto [24] in England. Empirical evidence showed that parents can foster the development of important saving abilities in children. The results reveal that the majority of the younger generation's saving habits were impacted by parental guidance. Similar conclusions were drawn by [25–27]. They concluded that Generation Y's savings or investment practices are mainly influenced by their parents.

Based on the above evidence, this study proposes the following hypothesis:

H1. ***There is a positive association between parental socialization and saving habits.***

2.3.2. Peer-Influenced

Mangleburg et al. [28] define peer-influenced as the extent to which peers influence an individual's perception and attitude. In addition to parental influence, peer influence also becomes one of the main factors that determine an individual's saving behavior. Franzoi [29] defined both parental socialization and peer influence as social influence. According to Franzoi [29], the behaviors of individuals could be reformed through the use of social power, i.e., social influence.

Individuals with the same preferences are expected to associate with a similar group, hence, establishing a relationship between individual and group attitude. Duflo and Saez [30], who have conducted a study on retirement planning among employees at universities in the U.S., found that peers influence the retirement planning of the samples. Likewise, Jamal et al. [31] asserted that a relatively large number of the young population in Malaysia are poor in managing their finances, which was mainly influenced by their peers.

In addition, Erskine et al. [32], who investigated the saving behaviors of the younger generation in Toronto, found that peers influenced a person's saving behavior. They discovered that those who placed high on the academic-oriented dimension are more likely to practice saving, and they are more patient. Alternatively, categories that score well on the peer-oriented component are typically less patient and less likely to practice saving.

Similarly, Alwi et al. [33] discovered that Generation Y is frequently impacted by peers, while making decisions. Their behaviors could be influenced by direct and indirect interactions with their social networks. A comparable finding was evidenced by Noor Zaihan [34]. She argued that despite the fact that the financial attitude of individuals was fostered by their parents, peer influence still affects their saving behaviors as most individuals spend social time and share thoughts about financial management issues with their peers [22].

Moreover, a significant role of peers in one's retirement planning decisions was studied by Duflo and Saez [30] among university staff in the U.S. The results showed that members of a comparable group share an atmosphere that may influence their perceptions and attitudes. This was because persons with equivalent preferences tend to affiliate with the same group. These two criteria will generate a connection between the group and individual attitudes, which will undoubtedly affect their saving habits. Although most of the previous studies proved a significant relationship between peer-influenced and saving habits, Htet [21] and Suryanti et al. [25] found an opposite finding. They found that the influence of peers on saving behavior is less likely to be agreed upon by respondents (i.e., soldiers and the millennial generation, respectively).

The above discussions lead to the following hypothesis:

H2. ***There is a positive association between peer-influenced and saving habits.***

2.3.3. Financial Literacy

Financial literacy is the capacity of an individual to develop well-informed judgements and to make effective decisions regarding consumption and money management [35]. It refers to the capacity to analyze economic information and make informed decisions regarding debt management, wealth accumulation, financial planning, and retirement planning [36]. Individuals who acquire financial knowledge are better at handling their finances, whether they are in a situation of surplus or deficiency. Personal finance knowledge is necessary for effective personal financial management [37]. Young generations in particular need to make financial decisions that can improve future planning and retirement preparations.

Several empirical studies have been conducted by previous researchers, which found an association between financial literacy and saving habits. For example, a study by Suryanti et al. [25] and Sabri and MacDonald [38] found a favorable effect of financial literacy on millennials' and college students' saving habits, respectively. The results showed that respondents with a deeper understanding of personal finances would engage in prudent saving practices.

Correspondingly, Delafrooz and Paim [39], who studied the association between financial literacy and saving behavior in Malaysia, also found a significant association between saving behavior and financial literacy. It implies that those with insufficient financial literacy will not take the initiative to save and will face financial troubles in the future.

The above discussion suggests the following hypothesis:

H3. ***There is a positive association between financial literacy and saving habits.***

2.3.4. Self-Control

Self-control, according to Ling [40], is the capacity to govern emotions, thoughts, and behavior, in response to temptations and impulses. It is a problem of intrapersonal choice inconsistency, a conflict between 'multiple selves' or cue-triggered errors, among others [41]. Alternatively, Baumeister et al. [42] described self-control as the capacity to identify and control one's feelings and desires. It could be characterized by the initiation of will, self-discipline, and the capacity for delayed satisfaction. Strict self-restraint is one of several prevalent ideas of self-control. Individuals with strict self-restraint are more proficient than their imprudent counterparts at managing their behavioral and emotional urges to achieve long-term objectives [25].

In other words, self-control is the difficulty that arises between an individual's desire and their fulfilment. It is uncommon to discover people that share the same beliefs and perspectives. Some individuals desire to enjoy their current life to the fullest extent possible and tend to spend as much as possible. Others, meanwhile, desire a simple and contented way of life, they manage their time and financial resources appropriately and save money for the future. Cowen [43] believes that individuals cannot achieve their goals in the way they desire unless they manage their financial obligations. For example, education, family, and other incidental expenditures [44].

For example, the previous researcher, Otto [24], discovered in his study that the psychological aspects associated with saving emphasize the importance of self-control and the capacity to delay desire as essential abilities for saving money. Further, Esenvalde [45] has demonstrated through empirical study that self-control is positively associated with saving practices. He asserted that self-control is a powerful and persistent factor used to explain saving practices.

On the other hand, Lim et al. [46], who conducted a study in Malaysia, concluded that an individual's ability to maintain self-control to save money depends on the strength of two competing factors: willpower and desire. Individuals are more likely to save if they have self-control and practice efficient economic cost assessment and budgeting. A similar finding was found by Htet [21], who researched the saving behaviors among army officers.

Based on the evidence, this study proposes the following hypotheses:

H4. ***There is a positive association between self-control and saving habits.***

Based on a literature review, the present study will answer the following research questions:

1. What are the factors that influenced saving habits among Malaysian cadets?
2. Do parental socialization, peer influence, financial literacy, and self-control, influence the saving habits of Malaysian cadets?

The conceptual framework that integrates the aforementioned theories and variables is depicted in Figure 1.

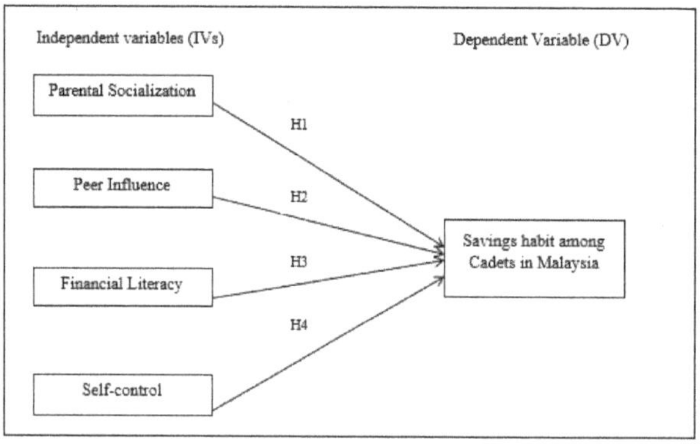

Figure 1. Conceptual model of saving habits.

3. Methodology of Research

This research employed a positivism paradigm. Surveys had been conducted among cadets who are students at the National Defense University of Malaysia. The study used the most convenient way of sampling technique study; i.e., convenience random sampling, for data collection. The study received 400 usable responses, giving a response rate of 92 per cent. The number of samples achieves the minimum sample size for generalization.

A pilot test had been conducted before the main survey to check the reliability and clarity of the questionnaire. With regards to data analysis for the main survey, it began with data screening to identify likely problems from ordinal and ratio variables, by using scatter diagrams. No data had been excluded in this screening process. The initial analysis also showed no issue with multicollinearity since the correlation coefficients are less than 0.7. Cronbach's alpha for this composite measure showed an internal consistency of 0.83. Further, the study performed a multiple regression analysis to test the hypotheses.

Description of the Variables

Tables 1 and 2 present the scale measurement of variables for this study and the construct measurement of each variable, respectively.

Table 1. Scale Measurement of Variables.

Label	Measurement of Variables	Expected Sign	Hypothesis Tested
SavHabit	'Saving Habits' is measured by looking at the Level of agreement (1 = Completely disagree; 0 = Completely agree)	Dependent variable	
ParenSoc	'Parental Socialization' is measured by looking at the Level of agreement (1 = Completely disagree; 0 = Completely agree)	+ve	H1
PEER	'Peer Influence' is measured by looking at the Level of agreement (1 = Completely disagree; 0 = Completely agree)	+ve	H2
FinLit	'Financial Literacy' is measured by looking at the Level of agreement (1 = Completely disagree; 0 = Completely agree)	+ve	H3
SelfCon	'Self-Control' is measured by looking at the Level of agreement (1 = Completely disagree; 0 = Completely agree)	+ve	H4

Table 2. Construct measurement of variables.

Label	Items
SavHabit	I consistently save aside funds for future requirements.
	Saving objectives are not tough for me to attain.
	Before making a purchase, I frequently question whether I truly need the item.
	To save money, I frequently evaluate costs and choose the less expensive one.
	I intend to cut my spending to save money.
	When hazards arise, I have the financial resources to deal with them.
FinLit	I have a habit of keeping accounts.
	I am good at managing my credit consumption.
	I can prepare a weekly/monthly budget myself.
	I know how to invest my money.
	I understand all financial instruments.
	I am well aware of my financial needs.
ParenSoc	My parents are my role models when it comes to managing money.
	I like to discuss with my parents about money management.
	I value my parents' guidance on how to manage my finances.
	I prefer my parents to control my spending and savings.
	My parents are satisfied with my saving.
PEER	I frequently discuss money matters with my friends.
	I constantly compare my savings and expenditures with those of my peers.
	My leisure time is always spent with friends.
	I am constantly engaged in money-spending activities with my buddies.
	I enjoy imitating my friends' spending habits.
SelfCon	I do not save since it is very difficult for me to do so.
	I am susceptible to temptation.
	Every time I receive a lot of money, I immediately spend it (within 1 or 2 days).
	I am more concerned with the immediate future than the distant future.
	"Just do it" is my motto when purchasing something.
	I appreciate spending money on impractical items.

4. Results and Discussion

This study analyzed the reliability of each variable and its internal consistency using the reliability analysis, in both the pilot study and the main study. The items were retained as the Cronbach alpha for all independent variables are above 0.70 (SavHabit = 0.886; ParenSoc = 0.891; PEER = 0.867; FinLit = 0.778; SelfCon = 0.943). The values indicate all variables have achieved the degree of reliability that exceeds what is considered acceptable.

This study also generates a 2-tailed Pearson correlation matrix for each set of the independent variables. The results indicate that the strength of the associations between saving habits (SavHabit) and each independent variable are high as the correlation values fall under the range of 0.70 to 0.90. The significance of the relationship between variables is less than the alpha value of 0.01.

Further, the multiple regression analysis was used to test the hypotheses. The value of R-Squared indicates that 76.5% of the variations in saving habits (SavHabit) are explained by the independent variables (i.e., parents' socialization, peer-influenced, financial literacy, and self-control). In other words, less than 23.5% of the variations are explained by other predictors that are not included in the model. Table 3 summarizes the results of the multiple regression analysis. The table also includes the results and remarks.

Table 3. Regression results.

Independent Variable	Dependent Variable (SavHabit)	Hypothesis	Result	Remark
(Constant) (β)	1.455 (7.182) *			
ParenSoc	0.562 (4.894) ***	H1 Positive	Positive; significant	Supported
PEER	0.650 (7.197) ***	H2 Positive	Positive; significant	Supported
FinLit	0.318 (1.559)	H3 Positive	Not significant	Failed to reject the Null Hypothesis
SelfCon	0.575 (5.436) ***	H4 Positive	Positive; significant	Supported
R-Squared; Adjusted R-square	0.765; 0.656			
F- (p-value)	78.759 (0.000)			

Note: SavHabit—Saving habits; ParenSoc—Parents socialization; PEER—Peer-influenced; FinLit—Financial literacy; SelfCon—Self-control (Absolute value of t-statistics in parentheses, asterisks denote the level of significance * $p < 0.05$; *** $p < 0.001$).

The following is the current regression equation.

SavHabit = 1.455 + 0.650 (PEER) + 0.575 (SelfCon) + 0.562 (ParenSoc) + 0.318 (FinLit)

The equation shows that peer-influenced becomes the most influential factor towards saving habits ($\beta = 0.650$, $p = 0.000$), followed by self-control ($\beta = 0.575$) and parents' socialization ($\beta = 0.562$). The least influential factor is financial literacy ($\beta = 0.318$). Since the significant value of financial literacy is less than 0.05, the study thus failed to reject the null hypothesis 3. The other three hypotheses are significantly associated with saving habits. Hence, the study accepted the three alternate hypotheses and rejected the null hypotheses. It can be concluded that the current regression equation meaningfully explains the relationship between saving habits and the four predictors. The regression model overall predicts SavHabit well as the p-value is below 0.05.

In summary, the findings indicate that the saving habit (SavHabit) of Malaysian cadets is influenced by peer-influenced, parents' socialization, and self-control. Remarkably, their saving habits did not have any relation to their financial literacy. The findings on the significant association of the variables support most of the previous studies, see [21–26,38].

5. Implications and Contributions of the Study

The study's findings aid in the promotion of a comprehensive model of saving habits from the perspective of military trainees, i.e., cadet officers. Essentially, the study is the first of its kind in Malaysia, and it aims to look into cadets' saving practices. Furthermore, it is intended that the findings of this study will contribute to the literature on personal finance and help to raise awareness and comprehension of effective financial planning and management among the general public and, specifically, military personnel. Military personnel should be fully aware that a lack of savings will have an impact on their financial behavior and will prevent them from making sound financial decisions. This, in turn, will harm the personnel's financial situation.

An understanding of the relevance of good financial behavior and financial resilience among Malaysian Armed Forces (MAF) members can aid policymakers in developing appropriate policies for the MAF and its personnel. The outcomes of this study should provide the Ministry of Defense Malaysia with a clearer picture and better understanding of the true saving and financial difficulties experienced by MAF personnel. The existence of contributing factors such as self-control will inform the public about the effectiveness of the government's efforts to assist military personnel in achieving peace of mind and improved concentration, while performing their duties in maintaining law and order and protecting the public. Furthermore, the findings should help the government achieve its goal of financial literacy, as outlined in the national strategy for financial literacy 2019–2023. The findings should also aid in raising financial literacy among Malaysians, as well as fostering responsible behavior and sensible attitudes to ensure people's overall well-being, particularly from a financial standpoint.

6. Conclusions and Recommendations

As previously mentioned, this study investigates how each independent variable influences the saving habits of cadet officers. The data would help banks better comprehend the saving habits of Malaysian cadets. Despite the absence of a correlation between cadets' financial knowledge and their saving habits, financial institutions can provide financial products and services that meet or surpass the requirements of cadet officers. Financial institutions should provide officers with increased financial understanding and assistance with financial literacy, as well as encourage them to save money. Once cadet officers learn how to save, they begin to do so.

Moreover, as a civic responsibility, parents must instill the habit of saving in their children. Retail banks must recognize that parents can play a significant role in encouraging children to save. Furthermore, as a result of peer pressure, one may claim that peers shape the way individuals think and act. Character tends to be shaped by the company one keeps or by the influence of those with whom one spends the most time. According to this current study, cadet officers who have friends with saving habits are more likely to save because they believe saving practices will help them swiftly integrate into the group. Therefore, financial institutions must attract their peers and educate them on saving knowledge. Regarding self-control and peer pressure, the Ministry of Defense could devise a way to save money for cadet officers. The outcomes of the study are essential for policymakers in protecting cadet officers' effective saving practices.

Nevertheless, the limitations do not lessen the significance of the findings; rather, they serve as a foundation for future research to validate and develop the theoretical framework presented in this study. Future research is needed to compare Malaysian cadets' saving practices to those of cadets in other developed and developing nations, and to investigate the differences in order to increase our understanding of finance needs. It would also be beneficial to undertake comparison research of cadets', senior officers', and other ranks' saving habits, as these three groups have distinct traits deriving from their authority and powers. Future studies should also survey a military audience across service branches including the army, air force and navy.

Author Contributions: Conceptualization, H.M.N. and M.N.Y.; methodology, H.M.N. and N.S.S.; software, H.M.N. and N.S.S.; validation, H.M.N. and A.I.; formal analysis, H.M.N.; investigation, H.M.N., M.N.Y. and A.I.; resources, W.N.S.A.Z.; data curation, W.N.S.A.Z.; writing—original draft preparation, H.M.N.; writing—review and editing, H.M.N.; visualization, H.M.N. and W.N.S.A.Z.; supervision, H.M.N.; project administration, H.M.N. and A.I.; funding acquisition, W.N.S.A.Z. and N.S.S. All authors have read and agreed to the published version of the manuscript.

Funding: This research received no external funding.

Institutional Review Board Statement: Not applicable.

Informed Consent Statement: Not applicable.

Data Availability Statement: Not applicable.

Acknowledgments: Paper presented by Hafizah Mat Nawi at the International Academic Symposium of Social Science 2022, which was organized by UiTM Kelantan, Malaysia.

Conflicts of Interest: The authors declare no conflict of interest.

References

1. Sakyi, D.; Onyinah, P.O.; Baidoo, S.T.; Ayesu, E.K. Empirical determinants of saving habits among commercial drivers in Ghana. *J. Afr. Bus.* **2021**, *22*, 106–125. [CrossRef]
2. Amakor, I.C.; Eneh, O.M.R. Fund Mobilization and Sustainable Economic Growth; the Nigerian's Experience. *Int. J. Trend Sci. Res. Dev. (IJTSRD)* **2021**, *5*, 1026.
3. RinggitPlus Malaysia Financial Literacy Survey—RMFLS 2019. 2019. Available online: https://media.ringgitplus.com/pdf/RP+2019+Financial+Survey+Report_Full+Report.pdf (accessed on 2 February 2022).
4. National Youth Development Policy of Malaysia. 2022. Available online: https://www.youthpolicy.org/factsheets/country/malaysia/#:~{}:text=The%20national%20youth%20policy\T1\textquoterights%20objective,a%20focus%20on%20skill%20development (accessed on 2 February 2022).
5. Malaysian Department of Statistics. 2018. Available online: https://www.dosm.gov.my/v1/index.php?r=column/coneandmenu_id=eGUyTm9RcEVZSllmYW45dmpnZHh4dz09 (accessed on 2 February 2022).
6. World Population Review. 2022. Available online: https://worldpopulationreview.com/countries/malaysia-population (accessed on 2 February 2022).
7. Department of Insolvency Malaysia. 2022. Available online: https://docs.google.com/viewerng/viewer?url=https://www.mdi.gov.my/images/documents/Statistics/Bankrupcy/BRIEFNOTEDIS.pdf (accessed on 2 February 2022).
8. Mat Nawi, H. Determinants of Capital Structure in Small and Medium Sized Enterprises in Malaysia. Doctoral Dissertation, Brunel University, London, UK, 2015.
9. Chaffin, C.R. *The Financial Planning Competency Handbook*; John Wiley& Sons, Inc.: Hoboken, NJ, USA, 2013.
10. Fishbein, M.; Ajzen, I. *Belief, Attitude, Intention and Behavior: An Introduction to Theory and Research*; Addison-Wesley: Reading, MA, USA, 1975.
11. Ajzen, I.; Fishbein, M. The influence of attitudes on behavior. *Handb. Attitudes* **2005**, *173*, 31.
12. Ajzen, I. The theory of planned behavior. *Organ. Behav. Hum. Decis. Process.* **1991**, *50*, 179–211. [CrossRef]
13. Furnham, A. Why do people save? Attitudes to, and habits of saving money in Britain. *J. Appl. Soc. Psychol.* **1985**, *15*, 354–373.
14. Browning, M.; Lusardi, A. Household saving: Micro theories and micro facts. *J. Econ. Lit.* **1996**, *34*, 1797–1855.
15. Sistiani, F.; Prajawati, M.I. Saving behavior of female workers in new normal. *Int. J. Soc. Sci. Hum. Res.* **2021**, *4*, 2390–2398. [CrossRef]
16. Babiarz, P.; Robb, C.A. Financial literacy and emergency saving. *J. Fam. Econ. Issues* **2014**, *35*, 40–50. [CrossRef]
17. Prawitz, A.; Garman, E.T.; Sorhaindo, B.; O'Neill, B.; Kim, J.; Drentea, P. In charge financial distress/financial wellbeing scale: Development, administration, and score interpretation. *J. Financ. Couns. Plan.* **2006**, *17*, 34–50.
18. Kim, T.E.; Jang, C.Y. The relationship between children's flourishing and being overweight. *J. Exerc. Rehabil.* **2018**, *14*, 598. [CrossRef]
19. Alekam, J.M.E. The effect of family, peer, behavior, saving and spending behavior on financial literacy among young generations. *Int. J. Organ. Leadersh.* **2018**, *7*, 309–323. [CrossRef]
20. Jorgensen, B.L. Financial Literacy of College Students: Parental and Peer Influences. Doctoral Dissertation, Virginia Tech, Blacksburg, VA, USA, 2007.
21. Htet, K.M. Influencing Factors on Saving Behaviour among Army Officers in Yangon Centonment. Doctoral Dissertation, MERAL, University of Yangon, Yangon, Myammar, 2020.
22. Dangol, J.; Maharjan, S. Parental and peer influence on the saving behavior of the youth. *Int. Res. J. Manag. Sci.* **2018**, *3*, 42–63. [CrossRef]
23. Bucciol, A.; Veronesi, M. Teaching children to save: What is the best strategy for lifetime savings? *J. Econ. Psychol.* **2014**, *45*, 1–17. [CrossRef]
24. Otto, A.M.C. The Economic Psychology of Adolescent Saving. Doctoral Dissertation, University of Exeter, Exeter, UK, 2009.
25. Suryanti, R.; Setyawanand, W.; Nopiana, U.D. Faktor-faktor yang mempengaruhi saving behavior generasi milenial. *J. Akunt. Bisnis Pelita Bangsa* **2021**, *6*, 47–58. [CrossRef]
26. Hasni, H. Factors that Influence Saving Habits among Gen Y: Case Study on Students of Politeknik Sultan Azlan Shah. Doctoral Dissertation, Universiti Utara Malaysia, Kedah, Malaysia, 2014.
27. Ganesan, A.S. Consumption, Spending and Investment Behaviour of Malaysia Generation Y. Doctoral Dissertation, Universiti Tunku Abdul Rahman (UTAR), Kampar, Malaysia, 2012.
28. Mangleburg, T.F.; Doney, P.M.; Bristol, T. Shopping with friends and teens' susceptibility to peer influence. *J. Retail.* **2004**, *80*, 101–116. [CrossRef]
29. Franzoi, S.L. *Social Psychology*, 4th ed.; McGraw-Hill: Boston, MA, USA, 2006.
30. Duflo, E.; Saez, E. Participation and investment decisions in a retirement plan: The influence of colleagues' choices. *J. Public Econ.* **2002**, *85*, 121–148. [CrossRef]
31. Jamal, A.A.A.; Ramlan, W.K.; Karim, M.A.; Osman, Z. The effects of social influence and financial literacy on savings behavior: A study on students of higher learning institutions in Kota Kinabalu, Sabah. *Int. J. Bus. Soc. Sci.* **2015**, *6*, 110–119.

32. Erskine, M.; Kier, C.; Leung, A.; Sproule, R. Peer crowds, work experience, and financial saving behaviour of young Canadians. *J. Econ. Psychol.* **2006**, *27*, 262–284. [CrossRef]
33. Alwi, S.; Amir Hashim, I.Z.; Ali, M.S. Factors affecting savings habits within millennials in Malaysia: Case study on students of Taylor's university. In Proceedings of the Fourth Asia-Pacific Conference on Global Business, Economics, Finance and Social Sciences, Kuala Lumpur, Malaysia, 7–9 August 2015; pp. 1–10.
34. Noor Zaihan, D. Determinants of Saving Behavior among Generation Y Students in Universiti Utara Malaysia. Doctoral Dissertation, Universiti Utara Malaysia, Kedah, Malaysia, 2016.
35. Albeerdy, M.I.; Gharleghi, B. Determinants of the financial literacy among college students in Malaysia. *Int. J. Bus. Adm.* **2015**, *6*, 15–24. [CrossRef]
36. Lusardi, A.; Mitchell, O.S. The economic importance of financial literacy: Theory and evidence. *J. Econ. Lit.* **2014**, *52*, 5–44. [CrossRef]
37. Garman, T.E.; Forgue, R.E. *Personal Finance*, 8th ed.; Houghton Mifflin Company: Boston, MA, USA, 2006.
38. Sabri, M.F.; MacDonald, M. Savings behavior and financial problems among college students: The role of financial literacy in Malaysia. *Cross-Cult. Commun.* **2010**, *6*, 103–110.
39. Delafrooz, N.; Paim, L.H. Determinants of financial wellness among Malaysia workers. *Afr. J. Bus. Manag.* **2011**, *5*, 10092–10100.
40. Ling, H. Determinants of Saving Behaviour among Universities Students in Guangdong Province. *Front. Soc. Sci. Technol.* **2021**, *3*, 51–70.
41. Gathergood, J. Self-control, financial literacy and consumer over-indebtedness. *J. Econ. Psychol.* **2012**, *33*, 590–602. [CrossRef]
42. Baumeister, R.F. Yielding to temptation: Self-control failure, impulsive purchasing, and consumer behavior. *J. Consum. Res.* **2002**, *28*, 670–676. [CrossRef]
43. Cowen, E.L. In pursuit of wellness. *Am. Psychol.* **1991**, *46*, 404. [CrossRef]
44. Awais, M. Self-control & Saving Habits. *Asian J. Eng. Sci. Technol.* **2014**, *4*, 71–77.
45. Esenvalde, I. *Psychological Predictors of Savings Behavior: Contrasting the Impact of Optimism and Burnout on Self-Control, Achievement Motivation and Savings Behavior*; Alliant International University: Los Angeles, CA, USA, 2011.
46. Lim, C.S.; Sia, B.K.; Gan, G.J. The analysis of psychological factors affecting savers in Malaysia. *Middle Easter Financ. Econ.* **2011**, *12*, 77–85.

proceedings

Proceeding Paper

Exploratory Factor Analysis of Financial Literacy in the Malaysian Context †

Julaina Baistaman *, Farahiyah Akmal Mat Nawi, Wan Masnieza Wan Mustapha and Mazlina Mamat

Faculty of Business and Management, Universiti Teknologi MARA Kelantan Branch, Bukit Ilmu, Machang 18500, Kelantan, Malaysia
* Correspondence: julaina@uitm.edu.my
† Presented at the International Academic Symposium of Social Science 2022, Kota Bharu, Malaysia, 3 July 2022.

Abstract: Financial literacy has become one of the most important factors in personal financial management. Low financial literacy among Malaysians is an issue that leads to various problems such as bankruptcy, poor financial planning, inability to save for emergencies, and many more. This study aims to validate the measurement model for a Financial Literacy construct. Generally, the Exploratory Factor Analysis (EFA) is required to expose the appropriate items for use in research instruments. Thus, this study takes steps to develop and validate the instrument of financial literacy in the Malaysian context. This study adopted a cross-sectional research design, with data collected from 100 school teachers in the state of Kelantan, Malaysia. Items with a 10-interval scale were used. Using the Principal Component extraction method with Varimax Rotation, the researchers performed the EFA procedure on construct elements using SPSS software. Bartlett's Test of Sphericity and the Kaiser–Meyer–Olkin (KMO) sampling adequacy were also performed. Cronbach's alpha was used to test the reliability of the retained items. The items of financial literacy initially consist of 15 items. Based on the factor analysis, this study finalized the instrument of 11 financial literacy items, yielding three dimensions, i.e., numeracy (four items), time money value (four items), and inflation (three items). This study explained in detail the procedures for carrying out EFA analysis for the Financial Literacy construct. Due to the limited knowledge regarding the dimensions of the Financial Literacy construct, this study adds an important survey instrument to the literature that will help to measure financial literacy dimensions.

Keywords: financial literacy; personal financial management; exploratory factor analysis; Malaysian context

Citation: Baistaman, J.; Nawi, F.A.M.; Mustapha, W.M.W.; Mamat, M. Exploratory Factor Analysis of Financial Literacy in the Malaysian Context. *Proceedings* **2022**, *82*, 24. https://doi.org/10.3390/proceedings2022082024

Academic Editor: Mohamad Rahimi Mohamad Rosman

Published: 13 September 2022

1. Introduction

OECD [1] examined cross-country financial literacy levels across 30 countries using financial knowledge, financial attitude, and financial behavior as a combined measure of financial literacy. It was found that overall financial literacy levels are low. According to cross-country evidence provided by [2], the financial literacy rates among adults are at least 65% in countries such as Australia, Canada, Denmark, Finland, Germany, Israel, The Netherlands, Norway, Sweden, and the United Kingdom, whereas they are only 25% or less in South Asia. In contrast, some countries in South Asia have the lowest financial literacy levels, where only a quarter of adults or fewer are financially literate.

Globally, financial literacy has become one of the major financial management concerns in many countries. The authors in [2] illustrated the financial literacy breakdown to demonstrate the literacy scores among countries, as shown in Figure 1. The finding shows that Malaysia has produced an average score of 36% from all countries, indicating that financial illiteracy is prevalent in many countries, including Malaysia.

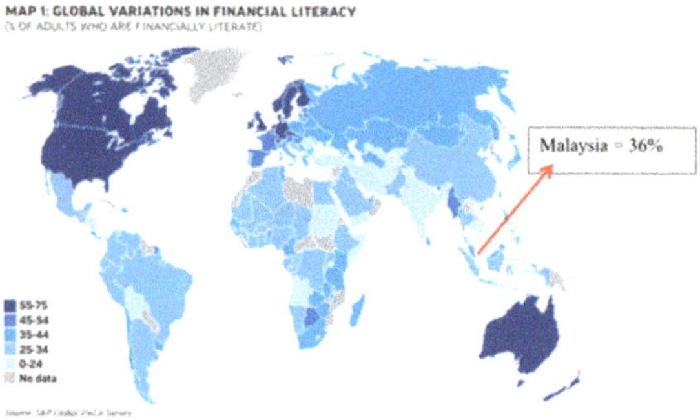

Figure 1. Global Variations in Financial Literacy by Klapper et al. [2].

One of the main concerns among Malaysians is the low financial literacy issue caused by financial illiteracy. In connection with this issue, a survey by Bank Negara Malaysia (BNM) exposed that three out of four Malaysians find it difficult to raise even RM1000 for an emergency. Furthermore, 9 out of every 10 Malaysian households have no emergency funds, besides having significant debts [3].

According to [4], young Malaysians possess poor financial knowledge, which causes them to be heavily in debt [5]. Furthermore, less than a quarter of people have any kind of investment to help them financially. The survey also showed that a majority of Malaysians tend to spend money for instant gratification instead of planning for the long term. This can be seen when only 40% of Malaysians considered themselves as financially ready for retirement, despite the steadily increasing life expectancy of Malaysians. Most people in Malaysia have insufficient information on the importance of having strong finances, including retirement savings, to avoid long-term financial commitment. Therefore, it is important to investigate the level of financial literacy among Malaysians.

The aim of this paper is to explore different dimensions and factors of financial literacy in Malaysian working adults. This study would generate items to measure the construct in this study using an appropriate quantitative approach.

2. Literature Review

Financial literacy typically refers to a capacity to use a knowledge of finance and to be able to use it to make wise decisions for oneself [6]. More specifically, it refers to the abilities and information that enable people to make wise financial decisions [7]. According to [8], the capacity to comprehend and utilize financial matters is known as financial literacy. A person's ability to manage their finances and to prevent future large debt obligations will be aided by their understanding of financial literacy.

Financial literacy is formed with two types of measures, objective and subjective [9]. Objective financial literacy can be measured through questions related to various financial issues such as interest rates, inflation, and the time value of money [10]. The "Big Three" [11] are three questions about compound interest, inflation, and risk diversification developed by [12] to create a measure of financial literacy on an objective measure. The authors in [13] measured financial literacy by asking basic questions related to knowledge about numeracy (interest), compound interest, inflation, and risk diversification.

Consumers can handle their financial affairs autonomously if they have a basic understanding of financial concepts and the capacity to apply their numeracy abilities in a financial environment. They will react appropriately to news and events that could affect their financial situation [14]. Financial inclusion, financial development, and ultimately financial stability are all thought to benefit from financial literacy [15]. In addition, financially smart investors are also more likely to divide their money over numerous projects in order to diversify risk [16]. According to [17], better financial decision-making is supported by increased financial capability and literacy. Thus, this facilitates better management and planning of life events, including retirement, housing purchases, and education. These planning elements must become a top priority and one of the key factors influencing financial stability.

Exploratory Factor Analysis

Typically, in social science studies, there are two main classes of factor analysis: Exploratory Factor Analysis (EFA) and Confirmatory Factor Analysis (CFA). Prior to doing the subsequent CFA, the EFA is conducted first [18]. In general, EFA is heuristic. EFA has been one of the statistical techniques that is used the most frequently, particularly in social science research. Research suggests that the EFA technique produces more precise results when multiple measurable variables that are either endogenous or exogenous constructs in the analysis are used to represent each common factor [19,20]. In addition, EFA outlines the fundamental relationship among the studied variables and cannot be measured directly, but is represented as a group of items [21]. According to [22], EFA is used when the number of factors included in a set of variables is unclear.

As the name implies, EFA is exploratory, and researchers have no assumptions regarding the quantity or kind of variables. That is, it enables the researchers to investigate the key dimensions in order to develop a theory or model from a sizable collection of latent constructs, frequently represented by a set of items. Principal component analysis (PCA), which is utilized for data reduction in EFA, does not distinguish between common and unique variance [23]. As advised by [24], in the EFA technique, the value is suppressed at the threshold of 0.60 or higher once the EFA procedure is implemented. It is suggested that the high factor loading is a crucial signal. Moreover, EFA proposes the factor loading into the same component, in addition to minimizing the number of variables used in this investigation. Indicators included in the same component demonstrate that this outer loading aims to represent the measurement model. This component will be used in structural equation modeling after the researchers complete the EFA technique (SEM).

3. Methodology

To achieve the research objectives, this study used a cross-sectional research design to create a valid and reliable measure for the Financial Literacy construct, particularly in the working adult context. The participants in this study were school educators in Kelantan, Malaysia, and 100 randomly selected school educators were used as the sample. Data were collected through a structured questionnaire and analyzed using Exploratory Factor Analysis (EFA) to establish the rotated component analysis used for significant items in the model.

To determine the suitable Financial Literacy measures among working adults in Malaysia, the researchers developed a structured questionnaire that comprised 15 items, which were measured using a 10-point interval scale, from "1 = Strongly Disagree" to "10 = Strongly Agree". The measurement of financial literacy was adopted and adapted from [25–27].

4. Results

Calculating the coefficient alpha allows one to assess how reliable a scale is. Therefore, the reliability of the study's items was assessed using the conventional Cronbach's alpha approach. Based on suggestions by [28], a more favorable coefficient alpha is above 0.70, as stated by [29]. He states that Cronbach's alpha is a consistent coefficient, indicating that the relationships among the item sets are proportionally correlated. In addition, in his opinion, a reliability value below 0.70 is considered a weak model. Moreover, the 15 items in the questionnaire have a Cronbach's alpha value of 0.819, demonstrating the suitability and reliability of the Financial Literacy construct items in measuring the response.

4.1. Kaiser–Meyer–Olkin (KMO) and Bartlett's Test of Sphericity

The researchers conducted EFA using principal component analysis with the varimax rotation method on 100 datasets to evaluate and purify the scale items, and to separate which items should be classified in the same components. According to [30], a Kaiser–Meyer–Olkin (KMO) value of greater than 0.50 is used to purify the measurement items. However, this study used factor loadings above 0.60, so that only the remaining high factor loading can be processed for the subsequent step. In this case, the KMO value of 0.868 is excellent, as it exceeds the recommended value of 0.6. Additionally, the value of Bartlett's Test of Sphericity must be less than 0.05 (p-value < 0.05) for factor analysis to be acceptable. The Bartlett's Test significance value for the present study is 0.000, which meets the required significance value of less than 0.05 [31,32].

Both Bartlett's Test of Sphericity is significant (p-value < 0.05) and the sampling adequacy by Kaiser–Meyer–Olkin (KMO) is excellent, since it exceeded the required value of 0.6. The authors in [33] suggested that the data are adequate to proceed further with the data reduction procedure in EFA [34].

The total variance explained is also an extraction process of the items, to reduce them into a manageable number before further analysis. In this process, components with eigenvalues exceeding 1.0 are extracted into different components [35]. The outcome showed that the EFA process produced three components based on a computed Eigenvalue of greater than 1.0. The eigenvalues ranged between 2.866 and 2.105. Meanwhile, the variance explained for component 1 is 26.051%, for component 2 is 24.205%, and for component 3 is 19.134. This construct's measurement yields a total explained variance of 69.390%. Given that it surpasses the required minimum of 60%, the total variance explained is acceptable [36]. This indicates that the items are grouped into three dimensions with a total explained variance of 69.390%, which would be considered for further analysis.

4.2. Scree Plot

The Exploratory Factor Analysis (EFA) was employed for data analysis, to determine the number of components of the Financial Literacy construct. As illustrate in Figure 2, the screen plot has identified three factors, while three factors have also been identified based on the initial eigenvalues and the total variance explained.

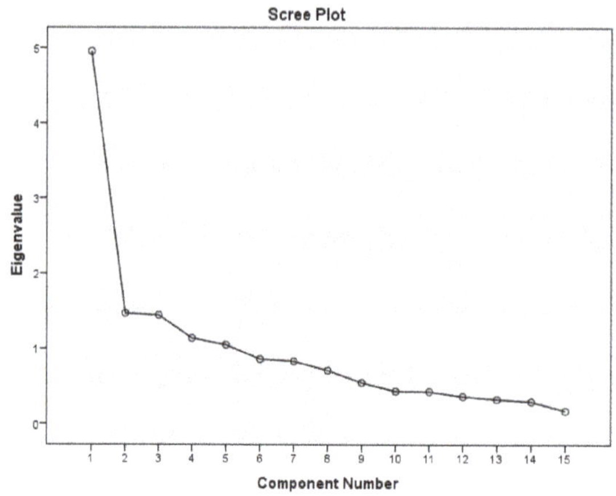

Figure 2. Scree plot for Financial Literacy construct.

4.3. Component Matrix

According to [31], the rotated component matrix was examined, and only items with a factor loading of above 0.6 were retained for further analysis.

Table 1 presents the three dimensions or components that emerged, and their respective items resulting from the EFA procedure. The factor loading for every item should be greater than 0.6 in order to be retained, since it indicates the usefulness of items in measuring the particular construct [34,35]. As a result, Table 2 shows all 11 retained items, which will be considered for further analysis under the three dimensions of the Financial Literacy construct.

Table 1. Rotated Component Matrix of Financial Literacy Construct.

Item	Item Statement	Comp 1	Comp 2	Comp 3	
1.	RM100 today will be worth less than RM100 in the future.			0.743	
2.	High-risk investment offers higher returns.		0.717		
3.	Time horizon is an important consideration in investment decisions.		0.779		
4.	Inflation reduces my buying power.			0.826	
5.	Inflation is an important consideration in my retirement planning.			0.794	
6.	Diversifying investment is important to reduce the risk of losing money.				Drop
7.	I understand the term, "Don't put all eggs in one basket".				Drop
8.	The interest rate influences the saving value in the future to come.		0.797		
9.	Safe investment offers lower returns.		0.805		
10.	Given the saving interest is 4% per annum, saving on RM1000 will accumulate to RM1040 at the end of the first year.	0.768			
11.	If monthly income is RM1000, it is advisable to save RM100-RM300 per month.	0.792			
12.	Given current monthly income is RM1000 and for comfortable living, it is advisable to have more than RM600 per month after retirement.				Drop
13.	Given the monthly income is RM1000, the total installment payments of all loans must not exceed RM400.	0.818			
14.	Given a loan of RM1000 within a year, (5% interest per annum), the amount payable is RM1050.	0.841			
15.	Given the monthly income is RM1000, the amount that can be spent on entertainment is RM50.				Drop

Extraction Method: Principal Component Analysis. Rotation Method: Varimax with Kaiser Normalization.

Table 2. Items retained for Financial Literacy Construct.

Rotated Component Matrix [a]						
No.	Sub-Construct	Item Label	Item Statement	Component		
				1	2	3
	Financial Literacy					
1	Numeracy	FL11		0.768		
2		FL12		0.792		
3		FL13		0.818		
4		FL14		0.841		
5	Time Money Value	FL21			0.717	
6		FL22			0.779	
7		FL23			0.797	
8		FL24			0.805	
9	Inflation	FL31				0.743
10		FL32				0.826
11		FL33				0.794

Extraction Method: Principal Component Analysis. Rotation Method: Varimax with Kaiser Normalization.
[a]. Rotation converged in 5 iterations.

Next, the researchers constructed the latent variables for each variable, based on the EFA report. In this study, financial literacy is a second-order construct, measured by the following three components, namely numeracy, time money value, and inflation. Finally, the researchers computed Cronbach's alpha using the internal reliability statistics test as presented in Table 3. Cronbach's alpha (α) was used to measure the internal consistency reliability of the selected items in measuring the construct. The value of Cronbach's alpha should be greater than 0.7 for the items to achieve Internal Reliability [31].

Table 3. Internal Reliability for Financial Literacy Construct.

Component	Rename Components	N of Items	Cronbach's Alpha (α)
1	Numeracy	4	0.873
2	Time Money Value	4	0.834
3	Inflation	3	0.754

5. Conclusions

The present study contributes to the measurement of the Financial Literacy construct, particularly among school teachers in Kelantan, Malaysia. The EFA results of this study produced a structure that extracts three dimensions of the Financial Literacy construct, which are numeracy, time money value, and inflation. These dimensions can be measured using the 11 items developed in this study. This is because all of the reliability measures for the three dimensions showed a high Cronbach's alpha value, hence meeting Bartlett's Test achievements (significant), acceptable KMO values (>0.6), and factor loadings exceeding the minimum threshold of 0.6. This reflects that the items are applicable in this study [36].

This study adds an important survey instrument to the literature that will help to measure the financial literacy dimensions. Without a proper definition and dimension available for financial literacy, there is a need for the present study to explore and examine the reliability of the Financial Literacy construct for working adults in the Malaysian context. Thus, using the final validated version of this dimension will be useful for future research. The finding of this study can be applied to the structural model in future research, and could benefit applied researchers who are interested in financial literacy research.

Author Contributions: Conceptualization, J.B.; methodology, J.B. and F.A.M.N.; investigation, J.B. and W.M.W.M.; validation, J.B., F.A.M.N., W.M.W.M. and M.M.; formal analysis, J.B.; writing—original draft preparation, J.B., F.A.M.N., W.M.W.M. and M.M.; writing—review and editing, J.B., F.A.M.N., W.M.W.M. and M.M.; supervision, J.B.; project administration, J.B. All authors have read and agreed to the published version of the manuscript.

Funding: This research received no external funding.

Institutional Review Board Statement: Not applicable.

Informed Consent Statement: Not applicable.

Data Availability Statement: Not applicable.

Acknowledgments: This research was partially supported by Universiti Teknologi MARA Kelantan Branch, Malaysia.

Conflicts of Interest: The authors declare no conflict of interest.

References

1. OECD. OECD/INFE International Survey of Adult Financial Literacy Competencies. 2016. Available online: https://www-oecd-org.ezaccess.library.uitm.edu.my/finance/oecd-infe-survey-adult-financial-literacy-competencies.htm (accessed on 5 March 2020).
2. Klapper, L.; Lusardi, A.; Van Oudheusden, P. *Financial Literacy Around the World*; World Bank: Washington, DC, USA, 2015.
3. Ngui, N. Malaysians Are Borrowing Too Much, Not Saving Enough, The Star Online (30 August 2016). Available online: https://www.thestar.com.my/business/businessnews/2016/08/30/khazanah-malaysians-borrowing-too-much/ (accessed on 12 January 2018).

4. Asian Institute of Finance, Finance Matters: Understanding Gen Y- Bridging the Gap of Malaysian Millennials. 2015. Available online: https://www.aif.org.my/clients/aif_d01/assets/multimediaMS/publication/Finance_Matters_Understaning_Gen_Y_Bridging_the_Knowledge_Gap_of_Malaysias_Millennials.pdf (accessed on 7 February 2019).
5. Murugiah, L. The level of understanding and strategies to enhance financial literacy among Malaysians. *Int. J. Econ. Financ. Issues* **2016**, *6*, 130–139.
6. Hogarth, J.M. Financial Literacy and Family and Consumer Sciences. *J. Fam. Consum. Sci.* **2002**, *94*, 15–28.
7. Norman, A.S. Importance of financial education in making informed decision on spending. *J. Econ. Int. Financ.* **2010**, *2*, 199–207.
8. Remund, D.L. Financial literacy explicated: The case for a clearer definition in an increasingly complex economy. *J. Consum. Aff.* **2010**, *44*, 276–295. [CrossRef]
9. Allgood, S.; Walstad, W.B. The effects of perceived and actual financial literacy on financial behaviors. *Econ. Inq.* **2016**, *54*, 675–697. [CrossRef]
10. Khan, M.N.; Rothwell, D.W.; Cherney, K.; Sussman, T. Understanding the financial knowledge gap: A new dimension of inequality in later life. *J. Gerontol. Soc. Work* **2017**, *60*, 487–503. [CrossRef] [PubMed]
11. Agnew, J.R.; Bateman, H.; Thorp, S. Financial literacy and retirement planning in Australian. *Numeracy* **2013**, *6*, 7. [CrossRef]
12. Lusardi, A.; Mitchell, O.S. Financial literacy and retirement planning in the United States. *J. Pension Econ. Financ.* **2011**, *10*, 509–525. [CrossRef]
13. Hasler, A.; Lusardi, A. *The Gender Gap in Financial Literacy: A Global Perspective*; Global Financial Literacy Excellence Center, The George Washington University School of Business: Washington, DC, USA, 2017.
14. Morgan, P.J.; Trinh, L.Q. Determinants and impacts of financial literacy in Cambodia and Viet Nam. *J. Risk Financ. Manag.* **2019**, *12*, 19. [CrossRef]
15. Ramakrishnan, D. Financial Literacy-The Demand Side of Financial Inclusion. 2011. Available online: https://ssrn.com/abstract=1958417 (accessed on 22 April 2019).
16. Abreu, M.; Mendes, V. Financial literacy and portfolio diversification. *Quant. Financ.* **2010**, *10*, 515–528. [CrossRef]
17. Mahdzan, N.S.; Tabiani, S. The Impact of Financial Literacy on Individual Saving: An Exploratory Study in the Malaysian Context. *Transform. Bus. Econ.* **2013**, *12*, 4155.
18. Afthanorhan, W.M.A.B.W.; Ahmad, S.; Mamat, I. Pooled confirmatory factor analysis (PCFA) using structural equation modeling on volunteerism program: A step-by-step approach. *Int. J. Asian Soc. Sci.* **2014**, *4*, 642–653.
19. MacCallum, R.C.; Widaman, K.F.; Zhang, S.; Hong, S. Sample size in factor analysis. *Psychol. Methods* **1999**, *4*, 84. [CrossRef]
20. Velicer, W.F.; Fava, J.L. Effects of variable and subject sampling on factor pattern recovery. *Psychol. Methods* **1998**, *3*, 231. [CrossRef]
21. Hair, J.F.; Sarstedt, M.; Hopkins, L.; Kuppelwieser, V.G. Partial least squares structural equation modeling (PLS-SEM): An emerging tool in business research. *Eur. Bus. Rev.* **2014**, *26*, 106–121. [CrossRef]
22. Nayak, T.; Sahoo, C.K. Quality of work life and organizational performance: The mediating role of employee commitment. *J. Health Manag.* **2015**, *17*, 263–273. [CrossRef]
23. Bentler, P.M.; Kano, Y. On the equivalence of factors and components. *Multivar. Behav. Res.* **1990**, *25*, 67–74. [CrossRef]
24. Hair, J.F.; Ringle, C.M.; Sarstedt, M. PLS-SEM: Indeed, a silver bullet. *J. Mark. Theory Pract.* **2011**, *19*, 139–152. [CrossRef]
25. Potrich, A.C.G.; Vieira, K.M.; Mendes-Da-Silva, W. Development of a financial literacy model for university students. *Manag. Res. Rev.* **2015**, *39*, 356–376. [CrossRef]
26. Van Rooij, M.; Lusardi, A.; Alessie, R. Financial literacy and stock market participation. *J. Financ. Econ.* **2011**, *101*, 449–472. [CrossRef]
27. AKPK Financial Behaviour Survey 2018 (AFBeS'18). Financial Behaviour and State of Financial Well-being of Malaysian Working Adults. 2018. Available online: https://www.akpk.org.my/sites/default/files/AKPK_Financial%20Behaviour%20and%20State%20of%20Finanical%20Well-being%20of%20Malaysian%20Working%20Adult.pdf (accessed on 29 January 2021).
28. Nunnally, J.C. An overview of psychological measurement. In *Clinical Diagnosis of Mental Disorders*; Wolman, B.B., Ed.; Springer: New York, NY, USA, 1978; pp. 97–146.
29. Sekaran, U.; Bougie, R. *Research Methods for Business: A Skill Building Approach*; Wiley: Hoboken, NJ, USA, 2010.
30. Hair, J.F.; Wolfinbarger, M.F.; Ortinau, D.J.; Bush, R.P. *Essentials of Marketing Research*; McGraw-Hill/Higher Education: New York, NY, USA, 2008.
31. Awang, Z. *Structural Equation Modeling Using AMOS Graphic*; Penerbit Universiti Teknologi MARA: Kuala Lumpur, Malaysia, 2012.
32. Hoque, A.S.M.M.; Awang, Z.; Baharu, S.M.A.T.; Siddiqui, B.A. Upshot of Generation 'Z' Entrepreneurs' E-lifestyle on Bangladeshi SME Performance in the Digital Era. *Int. J. Entrep. Small Medium Enterp. (IJESME)* **2018**, *5*, 97–118.
33. Pallant, J. *SPSS Survival Manual. A Step-by-Step Guide to Data Analysis Using SPSS*, 4th ed.; Allen & Unwin: Crows Nest, Australia, 2011; Available online: www.allenandunwin.com/spss (accessed on 28 March 2021).
34. Yahaya, T.; Idris, K.; Suandi, T.; Ismail, I. Adapting instruments and modifying statements: The confirmation method for the inventory and model for information sharing behavior using social media. *Manag. Sci. Lett.* **2018**, *8*, 271–282. [CrossRef]
35. Awang, Z. *SEM Made Simple: A Gentle Approach to Learning Structural Equation Modelling*; MPWS Rich Resources: Bandar Baru Bangi, Malaysia, 2015.
36. Awang, Z.; Afthanorhan, A.; Mamat, M.; Aimran, N. Modeling Structural Model for Higher Order Construct (HOC) using Marketing Model. *World Appl. Sci. J.* **2017**, *35*, 1434–1444.

proceedings

MDPI

Proceeding Paper

Alliance Management Capabilities and Enterprise Resilience—The Mediating Role of Information Technology Capabilities: The Case of Indonesia's State-Owned Enterprises †

Purwatiningsih Lisdiono [1,*], Jamaliah Said [2], Haslinda Yusoff [3], Ancella A. Hermawan [1] and Dewi Izzwi binti Abdul Manan [4]

1 Faculty of Economics and Business, Universitas Indonesia, Depok City 16424, Indonesia
2 Accounting Research Institute, Universiti Teknologi MARA, Shah Alam 40450, Malaysia
3 Faculty of Accountancy, Universiti Teknologi MARA, Selangor Branch, Puncak Alam Campus, Puncak Alam 42300, Malaysia
4 Faculty of Business, Accountancy and Social Science, Kolej Universiti Poly-Tech MARA, Kuala Lumpur 56100, Malaysia
* Correspondence: purwatiningsih.mba@ui.ac.id
† Presented at the International Academic Symposium of Social Science 2022, Kota Bharu, Malaysia, 3 July 2022.

Abstract: Based on dynamic capability theory, this study examines Alliance Management Capabilities (AMCs) in fostering the enterprise resilience (ER) of Indonesia's State-Owned Enterprises (SOEs) and their subsidiaries. Information Technology Capabilities (ITCs) are used as mediating variables. The study was administered through an online survey using questionnaires to the board of director members and senior managers. There were 322 valid responses received. The findings of the PLS-SEM analysis show that AMCs influence the creation of ER and ITCs. ITCs mediate the relationship between AMCs and ER. The findings could provide the management and policymakers the ability to develop a strategy for building and improving ER. This study broadens the scope of prior research in ER and dynamic capability theory in SOEs in emerging economies. The findings offer novelty to the limited literature on enterprise resilience in public organizations from dynamic capability theory operating in emerging markets.

Keywords: alliance management capabilities; dynamic capability theory; enterprise resilience; sustainability; Indonesia; information technology capabilities; state-owned enterprises

Citation: Lisdiono, P.; Said, J.; Yusoff, H.; Hermawan, A.A.; Dewi-Izzwi, A.M. Alliance Management Capabilities and Enterprise Resilience—The Mediating Role of Information Technology Capabilities: The Case of Indonesia's State-Owned Enterprises. *Proceedings* **2022**, *82*, 28. https://doi.org/10.3390/proceedings2022082028

Academic Editor: Mohamad Rahimi Mohamad Rosman

Published: 13 September 2022

1. Introduction

The rapid changes have increased the importance of the issue of growing uncertainty that emphasizes the critical role of organizational resilience in surviving, adjusting, and thriving in an unpredictable world [1,2]. Resilience comprises pre-event preparation for hardships, a post-event response that enables effective and prompt turnaround, and the capabilities for novel regeneration through creativity [3]. Enterprises must cultivate resilience to succeed in dynamic situations [4,5]. A growing number of companies are focusing on developing dynamic capabilities to stay ahead of the competition and be resilient [6].

AMCs are the ability to develop, expand, or rearrange the enterprise's asset value by creating collaboration with an ally [7] and the capacity to handle several collaborations. The research on AMCs is quite new [8]. AMCs help elucidate alliance success, which may be essential for surviving in a continually shifting ecosystem [9,10].

Companies must be flexible, agile, and resilient in a dynamic business context [11] and should rely on solid information structures and capabilities in this digital era. The technology captures data and converts them into information, including alliance information, as bases for decision-making [12]. Organizations with excellent information collection, analysis, and usage capabilities could foresee developments and take proactive modifications to reduce the negative impacts of uncertainties [13]. Thus, enterprise resilience depends on the capacity to gather and process resources to deal with the environment and utilize information, knowledge, experiences, and other resources [14].

This study integrates dynamic capabilities and ER to understand better how dynamic capabilities (in this case, AMCs and ITCs) may better assist Indonesia's SOEs in building and improving enterprise resilience. The research questions in this study are: (1) Do AMCs influence enterprise resilience? (2) Do ITCs influence enterprise resilience? (3) Do AMCs influence ITCs? And (4) do ITCs mediate the relation between AMCs and enterprise resilience? The article consists of the following structure: first, a discussion of the theoretical foundation which leads to the development of hypotheses; second, it explains the methodology; the third part presents empirical study and findings; then, it will be summarized with the discussion of the key results of the study.

2. Literature Review and Hypothesis Development

2.1. Enterprise Resilience and Dynamic Capabilities

Firms may encounter difficulties achieving their objectives in a business setting that is dynamic and uncertain [15]. Companies must be flexible, agile, and resilient [11]. According to [16], resilience is formed through dynamic development that necessitates continuous activity. Resilience is defined as the processes of reasoning, anticipating, adapting, improvising, and recovering after coping with unexpected adverse incidences [1]. Developing or creating resilience occurs before, during, and following an adverse event with minimal impact on organizational effectiveness [2]. Thus, resilient firms are not only reactive but also proactive in preventing disruptions before they occur. Refs. [17,18] propose a dynamic capabilities perspective on organizational resilience processes that happen prior to, throughout, and after an adverse event [19]. Thus, resilience is dynamic in nature, demonstrating businesses' responses to a dynamic landscape [1].

This dynamic approach has already been mentioned by [20], who state that resilience is about surviving and taking proactive measures to identify and prepare mitigation strategies that will ensure stability and victory in the face of adverse shocks. Dynamic capabilities [21] are the outcomes of two separate components of capability, namely dynamic capability and operational capability [22], that strengthen a firm's resilience [23]. Dynamic capabilities include sensing, capturing, and transforming the company's resources according to the need. In this regard, technology, especially information technology, plays a crucial role [24,25].

SOEs have played a crucial part in the world economy, despite the wave of privatization over the past two decades [26]. Due to implicit government support and their frequently dominant market positions, SOEs may be less concerned with operational efficiency, including resilience and sustainability, than private companies [27]. SOEs are companies that are either wholly or partially managed by the state [28]. They exist in both developing and industrialized nations [29]. Due to the essential character of their role, the poor performance of SOEs is an issue of concern for stakeholders because their failure could pose a threat to national economic development [30]. Therefore, improving the performance of SOEs remains crucial for many governments [29]. Despite emphasizing resilience in public policy and management, little research has been conducted on how public sector companies attain resilience [31].

Figure 1 demonstrates the research framework presented for this study.

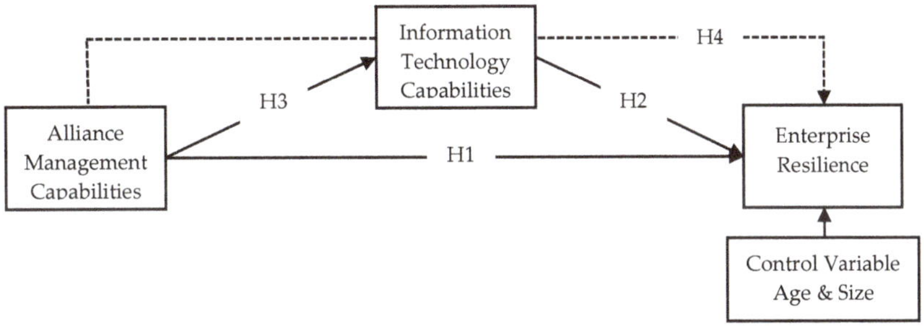

Figure 1. Research framework.

2.2. Relationship between Alliance Management Capabilities and Enterprise Resilience

In a dynamic business climate, it is essential to cultivate dynamic core capabilities [25]. AMCs could be chosen as the capabilities that should be implemented to support business results [8]. Ref. [32] found that strategic alliances are a significant source of assets and know-how and could create competitiveness. Ref. [33] cite alliance development as a typical illustration of dynamic capability. Alliance capabilities will assist businesses, including SOEs, in developing, sharing, and accessing knowledge from partners and investigating network use [34]. The vast majority of research on AMCs is conceptual studies [8]. Alliance is one of the strategy options that could be chosen to shine a spotlight on opportunities and make the most of favorable conditions to be resilient [35]. Businesses must have sufficient resources to create resilience capabilities, such as solid alliances [36].

Hypothesis 1 (H1). *Alliance Management Capabilities significantly influence enterprise resilience.*

2.3. Relationship between Information Technology Capabilities and Enterprise Resilience

High levels of uncertainty in the economic landscape necessitate adaptability on the part of enterprises, and information technology capabilities are viewed as a means of reacting more quickly to a changing environment [37]. A study by [38] defines Information Technology Capability (ITC) as an organizational capability and empirically investigates the relationship between IT capability and business performance. She categorized Organization-specific IT assets as IT infrastructure, human IT resources and intangibles enabled by IT [38]. Ref. [39] defined ITCs as a company's capacity to expand competitive agility through the delivery of better products and services within short production cycle times and to cultivate a high-skilled and dynamic team. Ref. [40] states that ITCs are vital for the agility and robustness of enterprises. Numerous firms view ITCs as significant and distinctive assets that give them competitiveness [41]. Ref. [12] ITCs may assist businesses in identifying and capitalizing on opportunities in a volatile environment.

Companies that successfully take advantage of the situation and capitalize on it may receive support from technological capabilities that allow them to deploy, reconfigure, and protect their assets during and after the crisis and provide firms with immediate and long-term competitive advantages [42]. ITCs improve the company's responsiveness to interruptions and unforeseen shifts [43]. ITC is highly correlated with agility, contributing to the development or enhancement of organizational resilience [44]. Thus, enterprise resilience depends on the capacity to gather and process resources in order to deal with environmental instability [14]. Better ITCs enhance a company's sensing and information processing capabilities, enabling it to react quickly to unanticipated events and successfully compete in a dynamic situation. They also improve the company's responsiveness to interruptions and unforeseen shifts [40,45–47].

Hypothesis 2 (H2). *Information Technology Capabilities significantly influence enterprise resilience.*

2.4. Relationship between Alliance Management Capability and Information Technology Capabilities

Enterprises are not similarly effective at foreseeing and handling the complex nature of a changing business environment, given the presence of substantial inter-firm disparities in their capacity to generate value through collaboration [10]. Ref. [6] found that AMCs significantly influence ITCs. Ref. [48] discovered that companies with superior ITCs could benefit more from an alliance and boosts alliance performance. Strategic alliances are fragile and necessitate communication and coordination. ITCs play their role in such situations [49] and in integrating alliance activities and understanding better partner's initiatives [48].

Hypothesis 3 (H3). *Alliance Management Capabilities significantly influence Information Technology Capabilities.*

2.5. Information Technology Capabilities Mediate the Relationship between Alliance Management Capabilities and Enterprise Resilience

The dynamic and unsure business climate has compelled the organization to determine its means of survival. This dynamic business climate necessitates adaptation and prospering in an unknown environment, where organizational agility and resiliency are essential and provide the foundation for survival [50]. Ref. [51] revealed that collaborative knowledge facilitates the development of proactive e-business responses. In addition, they discovered the mediation role of ITCs in the form of e-business proactiveness in the relationship between collaborative knowledge development and firm agility, which is a component of enterprise resilience. A study by [14] shows that organizations could improve their resilience by using digital technology in their operations and alliances. The deployment of digitalization enhances the productivity of processing information, allows enterprises to manage resources, including alliances, better, and respond quickly to disasters [52]. Xie et al. [14] identify that digital technology proficiency moderates the business network-resilience relationship in SOEs more than in private enterprises.

Hypothesis 4 (H4). *Information Technology Capabilities mediate the relationship between Alliance Management Capabilities and enterprise resilience.*

3. Methodology

This quantitative research approach employs a cross-sectional time frame and survey methodology. The unit analysis of this study is Indonesia's SOEs and their subsidiaries because of their vital role and contribution to the national economy and social welfare [53]. We targeted the Board of Directors (BOD) and senior management of SOEs listed on the website of the Indonesia Ministry of State-Owned Enterprises (https://bumn.go.id; accessed on 30 April 2020) and the website of each SOEs as respondents. The survey instrument was pre-tested to assure the accuracy and applicability of the measuring instruments. We conducted the common method bias test using the procedures mentioned by [54]. Our cover letter also mentioned that we maintained their answers and comments as private and anonymous. We also conducted a pilot test with thirty-six (36) board members and senior managers from the SOEs and subsidiaries. SPSS was utilized to assess the reliability and validity of the pilot test's results. The outcome is satisfactory. Firm size (average revenue over the past three years) and firm age (number of years since inception) served as control variables in this study, since these two variables impact enterprise performance or survivability [55].

A survey used a questionnaire to collect data. There were 114 SOEs and 530 SOE subsidiaries in total in the period of survey (June 2020–August 2020). Because to the participants' tight schedules and the nature of the field research that was conducted during Indonesia's lockdown regulation because of the COVID-19 pandemic, this study used an online questionnaire using Google Forms.

4. Results and Discussion

The study examined 322 instances of valid data from online surveys. Then, the data were analyzed using SPSS 23 and the partial least squares structural equation modeling (PLS-SEM) method with reflective models. Seventy-seven percent of the enterprises are more than ten years old. Over fifty-two percent of enterprises have had more than 1000 billion IDR average revenue in the last three years. Meanwhile, the informant's characteristics are dominated by men (85%), 86% over 40 years old, and 66.5% on the company board.

4.1. Assessment of Measurement Model

We performed four preliminary tests before conducting the PLS-SEM assessment analysis using SmartPLS 3 [56]. The preliminary results are non-normal distributed data, with no evidence of common method bias. There are also no collinearity and non-response bias issues. Table 1 shows the measurement model assessment. Internal reliability Cronbach's Alpha (CA) and Composite Reliability (CR), and validity (convergent and discriminant) met the criteria [57,58]. SRMR value as the goodness of measure for PLS-SEM is less than 0.08 and is considered a good fit [59]. The Variance Inflation Factors (VIF) of the variables in the datasets are below 3, so collinearity is not an issue among the predictor constructs [60].

Table 1. Measurement model assessment.

	Internal Consistency		Convergent Validity	Discriminant Validity	Model Fit	Collinearity
	CA	CR	AVE	HTMT	SRMR	VIF
AMCs	0.936	0.947	0.666			
ITCs	0.936	0.946	0.663	<0.90	0.064	<3.3
ER	0.940	0.948	0.568			

4.2. Assessment of Structural Model and Hypotheses Test

PLS-SEM is a nonparametric method; thus, bootstrapping is employed to determine statistical significance [57,61]. Shown in Table 2, the R^2 (explained variance of the endogenous variables), f^2 (the effect size of predictor relationship), and Q^2 (the predictive relevance of the model) [57,61].

Table 2. Structural model assessment.

Variable and Relationship	R^2	f^2	Q^2
ER	0.698 (substantial)		0.388 (medium)
ITC	0.481 (moderate)		0.314 (medium)
AMCs → ITCs		0.686 (substantial)	
AMCs → ER		0.927 (substantial)	
ITCs → ER		0.115 (small)	

Table 3 exhibits the hypotheses test with all hypotheses supported.

Table 3. Hypotheses test.

Hypotheses	B	*SD*	*t*-Value *	*p*-Value *	Decision
H1: AMCs → ER	0.635	0.044	14.467	0.000	Supported
H2: ITCs → ER	0.262	0.050	5.278	0.000	Supported
H3: AMCs → ITCs	0.694	0.041	17.017	0.000	Supported
H4: AMCs → ITCs → ER	0.182	0.034	5.420	0.034	Supported

* *Supported:* *t*-value > 1.96 (two-tailed), $p < 0.05$.

5. Discussion

This research investigates the relationship between AMCs and ITCs with enterprise resilience. This study also examines the mediating role of ITCs on the relationship between AMCs and enterprise resilience. The findings indicate that AMC significantly affects ER, as suggested by previous research [35,44,47]. The findings suggest that having alliances and the competence to handle those alliances could allow businesses to reach their ultimate objective securely and successfully. The Ministry of SOEs could utilize this outcome to draft appropriate regulations, processes, and guidelines for SOEs to form reciprocal and long-lasting alliances. The results of the analysis support the hypothesis that ITCs have a significant effect on ER (as reported in earlier research). This finding indicated that SOEs and their subsidiaries must devote, build, and strengthen ITCs in order to proactively detect, capture opportunities, and transform enterprises' resources to create more agile enterprises. This research demonstrates the evidence to support the hypothesis that AMC has a significant effect on ITCs [6,48]. The results indicate that ITCs mediate the relationship between AMCs and enterprise resilience. This result suggests that businesses can create and enhance their resilience by forming alliances, managing them effectively, and increasing their ITCs.

6. Conclusions

This research extends the literature on AMCs, ITCs, and organizational resilience from the perspective of dynamic capability theory, particularly for state-owned firms in emerging economies which is still limited. The size and age of SOEs do not influence their capability to build their resilience. The results may be used to study SOE resilience in other developing countries and to compare private and public enterprise resilience that helps decision-making processes. These study results may convince the Ministry of SOEs to build and enhance the dynamic capabilities of all SOEs, including subsidiaries, to increase resilience. The Ministry of SOEs may create resiliency standards and measures.

This research has flaws. First, perceptual self-rating assessments can be worrisome. Second, a cross-sectional temporal horizon was used, and data were collected during COVID-19, and thus could have been affected by the conditions and environments, which could have been different if the data had been collected earlier. Third, this study only focuses on two dynamic capabilities (ITCs and AMCs). Fourth, it relies on a single informant, a board member, or a senior manager. Personal bias can occur. Nevertheless, these limits could open avenues for new research.

Author Contributions: Conceptualization, P.L. and J.S.; methodology, P.L. and J.S.; software, P.L.; validation, P.L., J.S.; formal analysis, P.L.; investigation, P.L.; resources, J.S. and D.I.b.A.M.; data curation, P.L.; writing—original draft preparation, P.L.; writing—review and editing, P.L., J.S., H.Y., A.A.H.; visualization, P.L.; supervision, J.S., H.Y., A.A.H.; project administration, P.L.; funding acquisition, J.S. and D.I.b.A.M. All authors have read and agreed to the published version of the manuscript.

Funding: This research received no external funding.

Institutional Review Board Statement: Not applicable.

Informed Consent Statement: Informed consent was stated at the cover letter of the questionnaire and being read by the participants. Anonymity is assured.

Data Availability Statement: The data available are not publicly available due to the agreement of confidentiality with the respondents and The Ministry of State-Owned Enterprises Republic of Indonesia. Further information on the data can be addressed to the authors.

Acknowledgments: Thank you to the Ministry of Higher Education Malaysia for HICoE Research Funding, Accounting Research Institute Universiti Teknologi MARA Malaysia, Research Management Center, Kolej University Poly-Tech Mara, Malaysia, and Faculty Economics and Business, Universitas Indonesia, Indonesia, for all support and resources.

Conflicts of Interest: The authors declare no conflict of interest.

References

1. Williams, T.A.; Gruber, D.A.; Sutcliffe, K.M.; Shepherd, D.A.; Zhao, E.Y. Organizational response to adversity: Fusing crisis management and resilience research streams. *Acad. Manag. Ann.* **2017**, *11*, 733–769. [CrossRef]
2. Linnenluecke, M.K. Resilience in business and management research: A review of influential publications and a research agenda. *Int. J. Manag. Rev.* **2017**, *19*, 4–30. [CrossRef]
3. Kantur, D.; İşeri-Say, A. Organizational resilience: A conceptual integrative framework. *J. Manag. Organ.* **2012**, *18*, 762–773. [CrossRef]
4. Heredia, J.; Rubiños, C.; Vega, W.; Heredia, W.; Flores, A. New strategies to explain organizational resilience on the firms: A cross-countries configurations approach. *Sustainability* **2022**, *14*, 1612. [CrossRef]
5. Wieczorek-Kosmala, M. A study of the tourism industry's cash-driven resilience capabilities for responding to the COVID-19 shock. *Tour. Manag.* **2022**, *88*, 104396. [CrossRef]
6. Dubey, R.; Bryde, D.J.; Graham, G.; Foropon, C.; Kumari, S.; Gupta, O. The role of alliance managementt, big data analytics and information visibility on new-product development capability. *Ann. Oper. Res.* **2021**, 1–25. [CrossRef]
7. Helfat, C.E.; Finkelstein, S.; Mitchell, W.; Peteraf, M.; Singh, H.; Teece, D.; Winter, S.G. *Dynamic Capabilities: Understanding Strategic Change in Organizations*; Blackwell: London, UK, 2007.
8. Kohtamäki, M.; Rabetino, R.; Möller, K. Alliance capabilities: A systematic review and future research directions. *Ind. Mark. Manag.* **2018**, *68*, 188–201. [CrossRef]
9. Schilke, O.; Goerzen, A. Alliance management capability: An investigation of the construct and its measurement. *J. Manag.* **2010**, *36*, 1192–1219. [CrossRef]
10. Sluyts, K.; Matthyssens, P.; Martens, R.; Streukens, S. Building capabilities to manage strategic alliances. *Ind. Mark. Manag.* **2011**, *40*, 875–886. [CrossRef]
11. Ciampi, F.; Marzi, G.; Rialti, R. Artificial Intelligence, Big Data, Strategic Flexibility, Agility, and Organizational Resilience: A Conceptual Framework Based on Existing Literature. In Proceedings of the International Conferences on WWW/Internet 2018 and Applied Computing, Budapest, Hungary, 21–23 October 2018; Volume 15, pp. 51–58.
12. Cetindamar, D.; Phaal, R.; Probert, D. Understanding technology management as a dynamic capability: A framework for technology management activities. *Technovation* **2009**, *29*, 237–246. [CrossRef]
13. Srinivasan, R.; Swink, M. Leveraging supply chain integration through planning comprehensiveness: An organizational information processing theory perspective. *Decis. Sci.* **2015**, *46*, 823–861. [CrossRef]
14. Xie, X.; Wu, Y.; Palacios-Marqués, D.; Ribeiro-Navarrete, S. Business networks and organizational resilience capacity in the digital age during COVID-19: A perspective utilizing organizational information processing theory. *Technol. Forecast. Soc. Change* **2022**, *177*, 121548. [CrossRef]
15. Almeida, R.; Teixeira, J.M.; Mira da Silva, M.; Faroleiro, P. A conceptual model for enterprise risk management. *J. Enterp. Inf. Manag.* **2019**, *32*, 843–868. [CrossRef]
16. Hamel, G.; Välikangas, L. The quest for resilience. *Harv. Bus. Rev.* **2003**, *81*, 52–63, 131. Available online: http://www.ncbi.nlm.nih.gov/pubmed/12964393 (accessed on 2 January 2022). [PubMed]
17. Van der Vegt, G.S.; Essens, P.; Wahlström, M.; George, G. Managing risk and resilience. *Acad. Manag. J.* **2015**, *58*, 971–980. [CrossRef]
18. Helfat, C.E.; Martin, J.A. Dynamic managerial capabilities: Review and assessment of managerial impact on strategic change. *J. Manag.* **2015**, *41*, 1281–1312. [CrossRef]
19. Conz, E.; Magnani, G. A dynamic perspective on the resilience of firms: A systematic literature review and a framework for future research. *Eur. Manag. J.* **2019**, *38*, 400–412. [CrossRef]
20. Somers, S. Measuring resilience potential: An adaptive strategy for organizational crisis planning. *J. Contingencies Cris. Manag.* **2009**, *17*, 12–23. [CrossRef]
21. Teece, D.J.; Pisano, G.; Shuen, A. Dynamic capabilities and strategic management. *Strateg. Manag. J.* **1997**, *18*, 509–533. [CrossRef]
22. Cepeda, G.; Vera, D. Dynamic capabilities and operational capabilities: A knowledge management perspective. *J. Bus. Res.* **2007**, *60*, 426–437. [CrossRef]
23. Zhao, E.Y.; Fisher, G.; Lounsbury, M.; Miller, D. Optimal distinctiveness: Broadening the interface between institutional theory and strategic management. *Strateg. Manag. J.* **2016**, *38*, 93–113. [CrossRef]
24. Augier, M.; Teece, D.J. Dynamic capabilities and the role of managers in business strategy and economic performance. *Organ. Sci.* **2009**, *20*, 410–421. [CrossRef]
25. Teece, D.J. Explicating dynamic capabilities: The nature and microfoundations of (sustainable) enterprise performance. *Strateg. Manag. J.* **2007**, *28*, 1319–1350. [CrossRef]
26. Asian Development Bank. State-Owned Enterprises and Economic Development in Asia. In *Reforms, Opportunities, and Challenges for State-Owned Enterprises*; Ginting, E., Naqvi, K., Eds.; Asian Development Bank: Metro Manila, Philippines, 2020; pp. 1–43.
27. Okhmatovskiy, I. Performance implications of ties to the government and SOEs: A political embeddedness perspective. *J. Manag. Stud.* **2010**, *47*, 1020–1047. [CrossRef]
28. PwC's Public Sector Research Centre. State-Owned Enterprises Catalysts for Public Value Creation? 2015. Available online: https://www.pwc.com/gr/en/publications/assets/state-owned-enterprises-catalysts-for-public-value-creation.pdf (accessed on 11 January 2022).

29. Heo, K. *Effects of Corporate Governance on the Performance of State-Owned Enterprises*. Washington, D.C., WPS8555. 2018. Available online: http://documents.worldbank.org/curated/en/523421534424982014/Effects-of-corporate-governance-on-the-performance-of-state-owned-enterprise (accessed on 14 December 2020).
30. Okhmatovskiy, I.; Grosman, A.; Sun, P. Hybrid governance of state-owned enterprises. In *Oxford Handbook of State Capitalism and the Firm*; Wright, M., Wood, G., Cuervo-Cazurra, A., Sun, P., Okhmatovskiy, I., Grosman, A., Eds.; Oxford University Press: Oxford, UK, 2021.
31. Van der Stede, W.A.; Chow, C.W.; Lin, T.W. Strategy, choice of performance measures, and performance. *Behav. Res. Account.* **2006**, *18*, 185–205. [CrossRef]
32. Ireland, R.D.; Hitt, M.A.; Vaidyanath, D. Alliance management as a source of competitive advantage. *J. Manag.* **2002**, *28*, 413–446. [CrossRef]
33. Eisenhardt, K.M.; Martin, J.A. Dynamic capabilities: What are they? *Strateg. Manag. J.* **2000**, *21*, 1105–1121. [CrossRef]
34. Abdullah, N.H.N.; Said, J. Enhancing the governance of government linked companies via strategic management accounting practices and value creation. *Procedia Econ. Financ.* **2015**, *28*, 222–229. [CrossRef]
35. Inigo, E.A.; Ritala, P.; Albareda, L. Networking for sustainability: Alliance capabilities and sustainability-oriented innovation. *Ind. Mark. Manag.* **2017**, *89*, 550–565. [CrossRef]
36. Välikangas, L.; Romme, A.G.L. Building resilience capabilities at Big Brown Box, Inc. *Strateg. Leadersh.* **2012**, *40*, 43–44. [CrossRef]
37. Cai, Z.; Liu, H.; Huang, Q.; Liang, L. Developing organizational agility in product innovation: The roles of IT capability, KM capability, and innovative climate. *R D Manag.* **2017**, *49*, 421–438. [CrossRef]
38. Bharadwaj, A.S. A resource-based perspective on information technology capability and firm performance: An empirical investigation. *MIS Q.* **2000**, *24*, 169–196. [CrossRef]
39. Clark, C.E.; Cavanaugh, N.C.; Brown, C.V.; Sambamurthy, V. Readiness change- capabilities in organization: Insights from the bell atlantic experience. *Manag. Inf. Syst. Q.* **2010**, *21*. [CrossRef]
40. Mandal, S. Exploring the influence of IT capabilities on agility and resilience in tourism: Moderating role of technology orientation. *J. Hosp. Tour. Technol.* **2019**, *10*, 401–414. [CrossRef]
41. Mikalef, P.; Pateli, A. Information technology-enabled dynamic capabilities and their indirect effect on competitive performance: Findings from PLS-SEM and fsQCA. *J. Bus. Res.* **2017**, *70*, 1–16. [CrossRef]
42. Gu, M.; Yang, L.; Huo, B. The impact of information technology usage on supply chain resilience and performance: An ambidexterous view. *Int. J. Prod. Econ.* **2021**, *232*, 107956. [CrossRef] [PubMed]
43. Wamba, S.F.; Gunasekaran, A.; Akter, S.; fan Ren, S.J.; Dubey, R.; Childe, S.J. Big data analytics and firm performance: Effects of dynamic capabilities. *J. Bus. Res.* **2017**, *70*, 356–365. [CrossRef]
44. Sabahi, S.; Parast, M.M. Firm innovation and supply chain resilience: A dynamic capability perspective. *Int. J. Logist. Res. Appl.* **2019**, *23*, 254–269. [CrossRef]
45. Chae, H.C.; Koh, C.E.; Park, K.O. Information technology capability and firm performance: Role of industry. *Inf. Manag.* **2018**, *55*, 525–546. [CrossRef]
46. Dubey, R.; Gunasekaran, A.; Childe, S.J.; Blome, C.; Papadopoulos, T. Big data and predictive analytics and manufacturing performance: Integrating institutional theory, resource-based view and big data culture. *Br. J. Manag.* **2019**, *30*, 341–361. [CrossRef]
47. Wamba, S.F.; Dubey, R.; Gunasekaran, A.; Akter, S. The performance effects of big data analytics and supply chain ambidexterity: The moderating effect of environmental dynamism. *Int. J. Prod. Econ.* **2020**, *222*, 107498. [CrossRef]
48. Lioukas, C.S.; Reuer, J.J.; Zollo, M. Effects of information technology capabilities on strategic alliances: Implications for the resource-based View. *J. Manag. Stud.* **2016**, *53*, 161–183. [CrossRef]
49. Schreiner, M.; Kale, P.; Corsten, D. What really is alliance management capability and how does it impact alliance outcomes and success? *Strateg. Manag. J.* **2009**, *30*, 1395–1419. [CrossRef]
50. Altschuller, S.; Gelb, D.S.; Henry, T.F. IT as a resource for competitive agility: An analysis of firm performance during industry turbulence. *J. Int. Technol. Inf. Manag.* **2010**, *19*. Available online: https://scholarworks.lib.csusb.edu/jitim/vol19/iss1/1 (accessed on 22 March 2022).
51. Al-Omoush, K.S.; Simón-Moya, V.; Sendra-García, J. The impact of social capital and collaborative knowledge creation on e-business proactiveness and organizational agility in responding to the COVID-19 crisis. *J. Innov. Knowl.* **2020**, *5*, 279–288. [CrossRef]
52. Joseph, J.; Gaba, V. Organizational structure, information processing, and decision making: A Retrospective and roadmap for research. *Acad. Manag. Ann.* **2020**. [CrossRef]
53. Khatri, Y.; Ikhsan, M. Enhancing the development contribution of indonesia's state-owned enterprises. In *Reforms, Opportunities, and Challenges For State-Owned Enterprises*; Ginting, E., Naqvi, K., Eds.; Asian Development Bank: Mandaluyong, Philippines, 2020; pp. 88–132.
54. Podsakoff, P.M.; MacKenzie, S.B.; Podsakoff, N.B. Sources of method bias in social science research and recommendations on how to control it. *Annu. Rev. Psychol.* **2012**, *63*, 539–569. [CrossRef]
55. Yasuda, T. Firm growth, size, age and behavior in Japanese manufacturing. *Small Bus. Econ.* **2005**, *24*, 1–15. [CrossRef]
56. Ringle, C.M.; Wende, S.; Becker, J.M. SmartPLS 3. 2015. Available online: https://www.smartpls.com (accessed on 20 March 2022).

57. Hair Jr, J.; Hult, G.T.; Ringle, C.; Sarstedt, M. *A Primer on Partial Least Squares Structural Equation Modeling (PLS-SEM)*, 2nd ed.; SAGE Publications, Inc.: Thousand Oaks, CA, USA, 2017; ISBN 9781483377445.
58. Henseler, J.; Ringle, C.M.; Sarstedt, M. A New criterion for assessing discriminant validity in variance-based structural equation modeling. *J. Acad. Mark. Sci.* **2015**, *43*, 115–135. [CrossRef]
59. Henseler, J.; Dijkstra, T.K.; Sarstedt, M.; Ringle, C.M.; Diamantopoulos, A.; Straub, D.W.; Ketchen David, J., Jr.; Hair, J.F.; Hult, G.T.M.; Calantone, R.J. Common beliefs and reality about PLS: Comments on rönkkö and evermann (2013). *Organ. Res. Methods* **2014**, *17*, 182–209. [CrossRef]
60. Diamantopoulos, A.; Siguaw, J.A. Formative versus reflective indicators in organizational measure development: A comparison and empirical illustration. *Br. J. Manag.* **2006**, *17*, 263–282. [CrossRef]
61. Chin, W.W. Commentary issues and opinion on structural equation modeling. *MIS Q.* **1998**, *22*, vii–xvi.

proceedings

MDPI

Proceeding Paper

Socio-Economic Differences in Response to the COVID-19 Pandemic: A Case in Malaysia †

Nor Salwati Othman [1], Zuraidah Mohamed Isa [2,*], Norhidayah Ali [2], Dahlia Ibrahim [2], Azyyati Anuar [2] and Suhaida Abu Bakar [3]

1 Department of Accounting and Finance, College of Business Management and Accounting, Universiti Tenaga Nasional, Muadzam Shah 26700, Pahang, Malaysia
2 Faculty of Business and Management, Universiti Teknologi MARA Kedah Branch, Merbok 08400, Kedah, Malaysia
3 Academy of Contemporary Islamic Studies, Universiti Teknologi MARA Kedah Branch, Merbok 08400, Kedah, Malaysia
* Correspondence: zuraidah588@uitm.edu.my
† Presented at the International Academic Symposium of Social Science 2022, Kota Bharu, Malaysia, 3 July 2022.

Abstract: The COVID-19 pandemic has drastically transformed the human lifestyle and world socioeconomic conditions. It has attracted many scholars to assess the consequences of this pandemic. From a different angle, this study aims to compare the socioeconomic conditions before and during the COVID-19 pandemic; and to assess which socioeconomic indicator was most impacted by the pandemic. This empirical study employed a valid and reliable questionnaire that was randomly distributed to 516 respondents aged 18 to 65 years old in Peninsular Malaysia. Paired sample t-test and frequency analysis were utilised for the data analysis. Results show a significant difference in terms of socioeconomic conditions, namely income, saving, job security, health conditions, security/personal safety, emotional condition, spirituality, and work productivity was observed before and during the COVID-19 pandemic. Furthermore, spirituality (99.4%), food security (96.9%), health (88%), and personal safety (83.5%) were the most affected indicators during the COVID-19 pandemic. The outcomes of this study could provide policymakers with a clear picture of the effective strategy for achieving Sustainable Development Goals (SDG) and aiding Malaysia in its economic and social recovery after the pandemic.

Keywords: COVID-19 pandemic; socioeconomic impact; Malaysia

Citation: Othman, N.S.; Mohamed Isa, Z.; Ali, N.; Ibrahim, D.; Anuar, A.; Abu Bakar, S. Socio-Economic Differences in Response to the COVID-19 Pandemic: A Case in Malaysia. *Proceedings* **2022**, *82*, 32. https://doi.org/10.3390/proceedings2022082032

Academic Editor: Mohamad Rahimi Mohamad Rosman

Published: 13 September 2022

1. Introduction

The COVID-19 pandemic is known as one of the biggest shocks with a remarkable impact on economic, social, health, political, and technological conditions worldwide. The disease has spread to every continent and case numbers continue to rise [1]. The World Health Organization (WHO) reported 212 million infections and about 4.4 million deaths worldwide from the pandemic since its discovery until August 2021. As for Malaysia, 1.5 million infections and 14 thousand deaths have been reported as of 24th August 2021 [1,2]. This never-ending story has created anxiety, tension, and disrupted everyone's life including children and aged individuals. The human lifestyle has changed 360 degrees as a result of this pandemic. Many countries have responded with one or more successive lockdowns (including school closures, workplace, closures, travel bans, and stay-at-home requirements) particularly to avoid overloading the health system which in turn slows down the productivity of affected countries.

Malaysia is no exception. In mid-March 2020, at the beginning of the pandemic in Malaysia, national and federal governments implemented lockdown measures, which resulted in a full halt to public life. These measures included stay-at-home orders, banning outdoor activities, including interstate travel, and shutting down all businesses except a few designated essential services and the natural resource sectors. The lockdown is turning into an economic knockout. The economy is plummeting with growing negative impacts on jobs, incomes, and livelihoods, disrupting supply chains and upending businesses, and exacerbating inequalities, poverty, and hardships, especially among the most vulnerable [3].

Several explanatory or descriptive studies have been conducted to elucidate the socioeconomic impact of previous pandemics. A previous study [4] reported contradicting results regarding the impact of previous pandemics on different socioeconomic groups. For example, some authors recorded higher illiteracy rates to be associated with an increased risk of mortality during the 1918 pandemic in the USA, whereas other researchers found no differences in the socioeconomic status in New Zealand during the same pandemic. Similarly, the impact of the 2009 influenza pandemic was reportedly higher in lower socioeconomic groups in England but not in France. For COVID-19, several studies have analysed the effects of the pandemic on socioeconomic status in different populations and have yielded mixed results (the detail are reported in the next section). However, there is limited information regarding the impact of the COVID-19 pandemic in developing countries such as Malaysia. This article attempts to fill this research gap by answering the following questions: (1) Is there any difference in terms of socioeconomic conditions before and during the COVID-19 pandemic? (2) Which socioeconomic condition was impacted most by the pandemic? To answer these questions, the current study aims to compare the socioeconomic conditions before and during the COVID-19 pandemic; and to assess the socioeconomic components that were highly impacted by the pandemic. The results will contribute to a rapidly growing body of literature on the effects of the pandemic and assist in elucidating the implications of the policy responses to mitigate them at a critical time. Meanwhile, the main contribution of this study lies precisely in the identification of key socioeconomic and demographic factors and the possible regional mechanisms of action through which COVID-19 spreads at an ecological level.

This article will be structured as follows: Section 2 presents the past literature on event analysis; Section 3 draws the methodology step; Sections 4 and 5 comprise the findings and discussion respectively; and finally, Section 6 presents the conclusion and policy implications.

2. Literature Review

With the rising number of COVID-19 cases, many researchers worldwide have attempted to explore the cause and effect of this pandemic from diverse perspectives. An extensive review [5] was provided on the impact of the COVID-19 outbreak, stating that the pandemic has had a remarkable impact on the environment, economics, mental wellbeing, human health, and socioeconomic conditions. In addition, COVID-19 affected global trade [6]. Nevertheless, the economic consequences of the outbreak are underestimated due to over-reliance on historical parallels of the SARS pandemic or the financial crisis of 2008/2009.

In the case of Bangladesh, there was an immediate impact of the COVID-19 lockdown order on women and families in rural areas, which included a reduction in paid work, increased food insecurity, and heightened levels of depression and anxiety symptoms [7]. People's perceptions of the socioeconomic crisis and human stress in resource-limited settings in Bangladesh during the COVID-19 outbreak were assessed [8]. They revealed that food and nutritional deficiency were present among the vulnerable poorest sections due to the loss of livelihood. Additionally, high mental stress was reported among front-liners such as doctors, healthcare staff, police forces, volunteer organisations, and bankers. In Italy, the impact of the pandemic on hematologic patients (HP) was measured [9]. The results showed that HP experienced extremely severe depression, anxiety, and stress. In fact, 3.7% of HP did not work, and the main reasons were layoffs and lack of jobs. Meanwhile, the association between deprivation and COVID-19 incidence (case-hospitalisation and case-fatality) during pre-lockdown, lockdown and post-lockdown in Italy was investigated [4]. No significant differences were documented in case-hospitalisation and case-fatality according to deprivation in any of the observation periods.

In Brazil, the impact of the COVID-19 pandemic on the Brazilian electricity distribution market applying a socioeconomic regulatory model was evaluated [10]. The researchers found that both consumer and power distribution companies were significantly affected during the pandemic. Distributors at the concession area that had no access to a COVID-19 account (mitigation policy) were impacted the most. However, a report [11] has identified the most significant risk factors for the spread of COVID-19 in 29 countries (16 in Europe; 8 in Asia; 2 in Australia; and 1 each in South America and North America). Results showed that government interventions and the number of days to impose lockdowns, overweight population, and the presence of air pollution were significantly associated with the spreading rate of the novel virus in these countries. In North Africa, COVID-19 had a strong influence on social contact and economic activities through enforced policies on social isolation in the travel sector, financial market, and health system [12]. Specifically, the global demand for air travel, including travel in and out of Africa, dropped significantly.

Meanwhile, the socioeconomic determinants of COVID-19 for 42 Asian countries were analysed [13]. Net migration and higher economic activities were identified as predictors of the occurrence of COVID-19 cases in Asian countries. The knowledge, behaviour, health, and socioeconomic circumstances in response to the COVID-19 outbreak were assessed [14]. People in south Asian countries displayed good knowledge of COVID-19 symptoms and transmission but access to hygiene and personal protection resources was low. Additionally, the prevalence of unemployment rose and household income declined during the lockdown. The determinants of employee engagement during the COVID-19 pandemic in Vietnam were analysed [15]. A significant and positive effect of perceived organisational support and perceived family support on employee engagement was highlighted. In the same vein, customer purchasing intention on healthcare products during the pandemic in Indonesia was examined and revealed that social influence was a major indicator of consumer intention [16].

Several studies have been conducted in Malaysia to determine the impact of COVID-19. An overview of the measurement taken by the Malaysian government in response to COVID-19 and the effectiveness of the Movement Control Order (MCO) were provided [17]. The MCO and its compliance in mid-April effectively reduced the new number of active COVID-19 cases. A previous study [18] has provided a conceptual discussion and analysis of the impact of COVID-19 on financial crime and regulatory compliance. The number of financial crimes reportedly reduced during the pandemic; however, cybercrime cases increased during the same period. The regulatory compliance was unsatisfactory before and during the COVID-19 outbreak. Financial development scaled down during the COVID-19 outbreak in the Malaysian tourism industry [19]. Likewise, another study [20] investigated the psychosocial impact of COVID-19 on Malaysian families and found that most respondents with no permanent employment status faced a high psychological impact. The association between depression, anxiety, stress, and perceived quality of life among the Malaysia B40 urban community during the COVID-19 lockdown was measured [21]. Significant negative associations were reported between depression, anxiety, stress, and perceived quality of life, with the strongest correlation being observed between depression and psychological domains.

In summary, the effects of the COVID-19 pandemic on socioeconomic conditions have yielded mixed results. The effects of the pandemic appear to vary from one country to another. Nevertheless, no study has investigated the socioeconomic differences in response to the COVID-19 pandemic in Malaysia to date. In other words, the socioeconomic perspective of the impact of the pandemic needs to be elucidated. This study tends to fill the gap in the literature; hence, it was hypothesized that socioeconomic conditions were significantly different before and during the COVID-19 pandemic in Malaysia.

3. Methodology

Quantitative data were used to measure the variables in this study. The Malaysian population between the ages of 18 and 65 years was targeted for the survey. The minimum sample size for this study was determined as a rule of thumb [22]. A 95% confidence level was employed for the sample size calculation using Raosoft software, which was estimated as 385. To avoid bias and incomplete responses, a calculated working sample of 516 was drawn for the survey. Convenience sampling was used in this study due to the enforcement of the MCO in Malaysia to reduce the spread of COVID-19. Convenience sampling is a non-probability sampling technique that is used to select respondents based on availability [23]. Data were collected using Google forms, which has a flexible and practical web interface for designing and developing a web-based survey or questionnaire [24]. Respondents were sent a URL link via a WhatsApp group and an email in order to access and complete the questionnaire.

This study used a combination of multiple-choice and scaled questionnaires to answer the research questions and achieve the research objectives. COVID-19 was also examined in relation to socioeconomic factors in Malaysia. The researchers also collected demographic information, including age, gender, marital status, ethnicity, academic qualification, type of residence, and household income. Information was gathered on respondents' income, savings, job security, food, health, personal safety, emotion, spirituality, and productivity, among other socioeconomic factors. The questionnaire for the current study comprised five sections; demographic items were presented in section A, while section B contained yes/no closed questions relating to pre- and post-COVID-19 socioeconomic conditions. Meanwhile, section C included closed and multiple-choice questions on socioeconomic factors before and during the COVID-19 pandemic with "yes" or "no" responses. Sections D and E were presented using a 5-point Likert scale. Examples of studies that used 5-item Likert scales are [25–30]. Data analysis was performed using paired sample T-tests to detect if significant differences existed in the respondents' socioeconomic status before and during the COVID-19 outbreak. Frequency analysis was utilised to determine the socioeconomic component that was significantly impacted by the pandemic.

4. Findings

4.1. Descriptive Analysis

This section describes the unit of analysis in this case the Peninsular Malaysian population aged 18 and 65 years old. The survey included 34.1% and 65.9% of male and female respondents, respectively. The majority of respondents were between 40 and 49 years old (65.9%) and 52.7% of them were married. Meanwhile, Malay represented the largest ethnic group percentage and academic qualifications at 96.5% and 41.9%, respectively. A higher proportion of the respondents (40.1%) had an income less than RM 4360, whereas 31.2% earned a monthly income between RM 4361 and RM 9619. Descriptive analysis of respondents' socioeconomic factors revealed that emotion (with the highest mean scores; M = 3.77, SD = 1.081) and personal safety (M = 3.76, SD = 1.095) were the most dominant socioeconomic factors to a significant degree. Following that, socioeconomic moderate factors included productivity (M = 3.72, SD = 1.119), health (M = 3.68, SD = 1.161), savings (M = 3.64, SD = 1.163), and spirituality (M = 3.61, SD = 1.186). Income (M = 3.54, SD = 1.200), food (M = 3.54, SD = 1.207), and job (M = 3.52, SD = 1.236) recorded the lowest mean scores. Overall, all dimensions in this study achieved acceptable or satisfactory levels of implementation.

4.2. Frequency Analysis

For robustness of the earlier findings, frequency analysis was conducted. The frequency analysis indicated that there was a significant difference in terms of socioeconomic conditions, such as income, savings, job stability, health, security/personal safety, emotion, spirituality, and working productivity, before and after the COVID-19 pandemic. This means that the COVID-19 pandemic has the potential to affect the above-mentioned socioeconomic indicators, as well as cause other types of socioeconomic failures. Specifically, COVID-19 negatively impacted 26.6% of the 516 respondents' income through job loss (25%), pay loss (36.8%), overtime loss (24.3%), pay cut (20.8%), and allowance cut (24.3%). Further, 42.1% of respondents expressed that their savings were affected during the COVID-19 pandemic due to employment cutback (14.5%), loss of pay (20.4%), loss of overtime (13.1%), pay cut (11.8%), allowance cut (13.6%), support for food (64.7%) and an increase in current bills (53.8). In summary, the portion of job losses before and during COVID-19 increased from 34.2% to 52.2%, which highlights the impact of the pandemic on respondents' income and savings.

The majority of respondents' food security was also affected by the COVID-19 pandemic (96.9%) resulting from a lack of funding (50%), health concerns (34.2%), and food supply difficulties (34.2%). In terms of health conditions, 88% of respondents were impacted by the COVID-19 pandemic, comprising lack of sleep (55.1%), family problems (22.4%), loss of appetite (14.3%), shortage of money to buy medicines (10.2%), shortage of medical supply (10.2%), and others (34%). Approximately 83.5% of respondents felt the MCO (66.7%), social media attack (28.7%), fraud (16.1%), theft (9.2%), burglaries (4.6%), and others (21%) affected their personal safety. In addition, 39.7% had their emotion disturbed (ranging from slightly to frequently) due to the MCO (57.4%), domestic violence (0.6%), lack of social activities (51.1%), concern of infection (64.6%), loss of income (23.4%), and others (35.8%). Almost all (99.4%) of the respondents experienced an increase in their spiritual activities, where 97.8% prayed or worshipped God, 76.9% read a religious book, 70.4% did zikr, 19.7% performed song/nasyeed, 52.9% did meditation, 60% made sadaqah, and 2.4% performed activities. As for productivity, 57.4% of respondents felt there were changes in productivity due to the MCO (76.7%), lack of social activities (57.3%), the concern of infection (45.8%), loss of income (18.5%), and others (5.2%).

In conclusion, the socioeconomic factor that was most impacted was spirituality (99.4%), followed by food security (96.9%), health (88%), and personal safety (83.5%), which is consistent with the findings of [7,20,21]. (See Table 1, for details).

Table 1. Frequency analysis: Socioeconomic factors.

No.	Socioeconomic Factors	Is This Factor Affected by COVID-19?		If Yes, What Are the Causes of the DV?	%
		YES	NO		
1	Income	26.6% (137)	73.4% (379)	Loss of job	25% (36)
				Loss of pay	36.8% (54)
				Loss of overtime	24.3% (36)
				Pay cut	20.8% (30)
				Allowance cut	24.3% (35)
2	Savings	42.1% (217)	57.9% (299)	Loss of job	14.5% (32)
				Loss of pay	20.4% (46)
				Loss of overtime	13.1% (30)
				Pay cut	11.8% (26)
				Allowance cut	13.6% (30)
				Support for food	64.7% (144)
				Increase in current bill	53.8% (120)
3	Job	Yes, affected. The percentage not working has increased from 34.2% (177) to 52.2% (269).			

Table 1. *Cont.*

No.	Socioeconomic Factors	Is This Factor Affected by COVID-19?		If Yes, What Are the Causes of the DV?	%
		YES	NO		
4	Food	96.9% (500)	3.1% (16)	Insufficient money	50% (19)
				Concern of affected health	34.2% (13)
				Difficulty in getting supply	21.1% (8)
				Others	31% (12)
5	Health	88% (454)	12% (62)	Lack of sleep	55.1% (27)
				Family problem	22.4% (11)
				Loss of appetite	14.3% (7)
				Shortage of money to buy medicines	10.2% (5)
				Shortage of medical supply	10.2% (5)
				Others	34% (17)
6	Personal Safety	83.5% (431)	16.5% (85)	MCO	66.7% (58)
				Social media attack	28.75% (25)
				Fraud	16.1% (14)
				Theft	9.2% (8)
				Burglaries	4.6% (4)
				Others	21% (19)
7	Emotion	Normal/Stable 60.3 Slightly disturbed 36.4 Frequently disturbed 3		MCO	57.4
				Domestic violence	0.6
				Lack of social activities	51.1
				Concern of infection	64.6
				Loss of income	23.4
				Others	35.8
8	Spirituality	99.40%	0.60%	Pray or worship God	97.8
				Read religious books	76.9
				Zikr	70.4
				Spiritual song/Nasyeed	19.7
				Meditation/Self-reflection	52.9
				Sadaqah	60
				Others	2.4
9	Productivity	57.4	42.6	MCO	76.7
				Domestic violence	0
				Lack of social activities	57.3
				Concern of infection	45.8
				Loss of income	18.5
				Others	5.2

4.3. *Paired T-Test Analysis*

Based on these findings, the t-test demonstrates robust validity relative to previous findings. A total of nine factors, namely income, savings, job security, food, health, personal safety, emotion, spirituality, and productivity, were tested to determine if they were

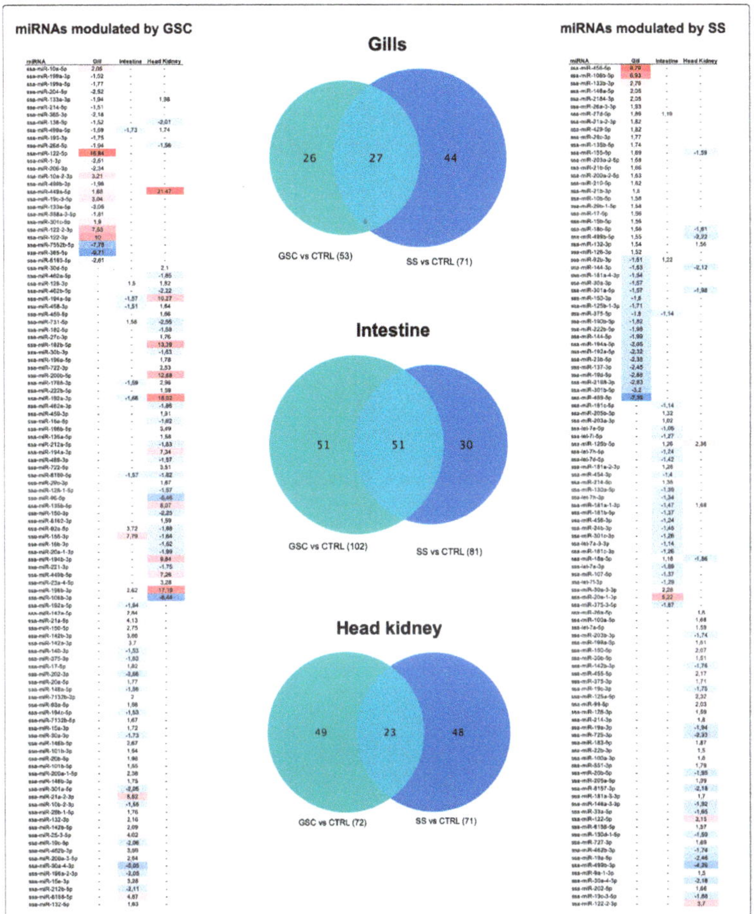

Figure 7. miRNA differential expression analysis of Atlantic salmon tissues under both conditions, GSC and SS. Tables show the fold-change values of miRNA for each tissue; red: upregulated, blue: downregulated.

3.7. GO Enrichment of miRNA Target Genes

The differentially expressed mRNAs in each tissue of fish groups exposed to GSC and SS were evaluated as target genes of differently modulated miRNAs. Notably, GO enrichment analysis of putative mRNAs suggests that gradual salinity changes trigger a greater number of regulatory responses than the salinity shock in the evaluated tissues. This is reflected in BP number, which seems to be modulated by differential transcribed miRNAs in the GSC fish group (Table 1). Response to stimulus, cell communication, response to stress, and immune system process were found among the Biological Process (BPs) putatively modulated by miRNAs in the three tissues (Table 1). The fish group exposed to SS exhibits a reduced number of BPs putatively modulated by differentially expressed miRNAs (Table 1). Furthermore, among the tissues, different target processes were identified. For instance, among the most representative processes in gills were pattern-recognition receptor-signaling pathways and response to ATP. On the other hand, regulation of mononuclear cell proliferation and antigens processing a presentation by MHC class II were observed in the intestine. While, the BPs found in head kidney were glycosylation and transposition (Table 1).

Table 1. GO enrichment analysis of differently modulated putative miRNA target genes in Atlantic salmon exposed to gradual salinity changes (GSC) and salinity shock (SS).

GSC Atlantic Salmon		**SS Atlantic Salmon**	
Gill Tissue Biological Process	**N° GO Term**	**Gill Tissue Biological Process**	**N° GO Term**
Cellular process	269	Pattern recognitio receptor signaling pathway	126
Metabolic prosess	259	Regulation of Pattern recognitio receptor signaling pathway	121
Response to stimilus	243	Truptophanil-tRNA aminoacylation	59
Macromolecule metabolic process	241	Protein ADP-ribosylation	53
Biological regulation	239	Regulation od protein ADP-ribosylation	51
Regulation of biological process	234	Spindle assembly	43
Cellular biosynthetic process	233	Mitotic spindle assembly	32
Regulation of cellular process	232	Response to ATP	30
Organic cyclic compound metabolic process	229	Regulation of meiotic cell cycle	28
Celullar response to stimulus	226	Microtube nucleation	23
Gene expression	221	Carbon utilization	22
Regulation of metabolic process	218	Regulation of mitotic spindle assembly	20
Localization	215		
Cell communication	214		
Transport	212		
Organic ciclyc compound biosynthetic process	210		
Regulation of cellular metabolic process	210		
Cellular component organization	209		
Cellular component organization or biogenesis	209		
Portein metabolic process	209		
Cellular macomolecule biosynthetic process	208		
Response to stress	203		
Immune system process	202		
Small molecules metabolic process	202		
Regulation of gene expression	202		
Intestine Tissue Biological Process	**N° GO Term**	**Intestine Tissue Biological Process**	**N° GO Term**
Cellular process	269	Regulation of mononuclear cell proliferation	8
Metabolic prosess	259	Regulation of lymphocyte proliferation	8
Organic substance metabolic process	256	Antigen processing and presentation of peptide	8
Cellular metabolic process	255	Hydrogen peroxide catabolic process	7
Prymary metabolic process	253	Antigen processing and presentation of peptide antigen	7
Response to stimilus	243	MCH class II	6
Macromolecule metabolic process	241	Fertilization	5
Biological regulation	239	Cell-cell recognition	4
Organic substance biosynthetic process	237	Sperm-egg recognition	3
Biosynthetic process	237	Lung epithelium development	3
Regulation of biological process	234	Hemolysis in other organims	3
Cellular macomolecule metabolic process	233	Microtubule polymerization	3
Regulation of cellular process	233	Regulation of platelet activation	3
Cellular response to stimulus	232	Forebrain neuron differentation	3
Gene expression	226	Forebrain generation of neurons	3
Nucleobase-containing compund metabolic process	221		
Regulation of metabolic process	221		
Cellular nitrogen compound bisynthetic process	218		
Localization	215		
Cell communication	214		
Transport	212		
Response to stress	203		
Immune system process	202		
Regulation of gene expression	202		
Head Kidney Tissue Biological Process	**N° GO Term**	**Head Kidney Tissue Biological Process**	**N° GO Term**
Cellular process	169	DNA integration	82
Metabolic prosess	165	Transposition	55
Cellular metabolic process	163	Transposition, DNA-mediated	55
Organic substance metabolic process	163	Macromolecule glycosylation	42
Primary metabolic process	158	Protein glycosilation	42
Nitrogen compund metabolic process	154	Pryrimidine nucleotide-sugar transmembrane transport	19
Cellular biosynthetic process	153	Nucelotide-sugar transmembrane transport	19
Biosyntetic process	153	ARF protein signal transduction	10
Organic substance biosynthetic process	152	Regulation of ARF protein signal transduction	10
Macromolecule metabolic process	151		
Biological regulation	149		
Organonitrogen compound metabolic process	149		
Response to stimilus	147		
Cellular macomolecule metabolic process	147		
Regulation of biological process	146		
Localization	144		
Cellular nitrogen compound metabolic process	143		
Regulation of cellular process	143		
Organic cyclic compound metabolic process	143		
Transport	142		
Establishment of localization	142		
Immune system process	124		
Immune response	119		
Defense response	108		
Response to biotic stimilus	103		

In particular, the evaluation of expression changes of miRNAs and their putative targets by tissue suggested a putative regulatory role of ssa-miR-205-5p in the regulation serine/threonine kinase or the regulation of the HSP70 gene by ssa-moR-19c-3-5p in gills exposed to GSC (Table 2, Figure S6). On the other hand, in the case of SS fish gills, a putative regulatory role of ssa-miR-19d-5p and ssa-miR-222b-5p with ATPase inhibitor can be mentioned and transposase, respectively (Table 2). Moreover, a putative regulation of *myosin* and *annexin A2-like* genes in the intestine tissue of GSC fish was observed by SSA-miR-92a-5p and ssa-miR-15a-3p, respectively. In addition, in SS fish group was observed a negative correlation expression between *free fatty acid receptor* and *low-density lipoprotein receptor* genes, with miRNAs 30a-3p and 125b-5p, respectively (Table 2). In the case of head kidney tissue of GSC fish, ssa-miR-128-1-5p showed a negative correlation expression with the *SPATA5*gene, and *laminin* gene exhibited a binding site to ssa-miR-194a-3p in fish exposed to GSC (Table 2).

Finally, expression-correlation analysis was conducted among the differently expressed miRNAs and smoltification-related genes and TEs to evidence the investigated role observed during smoltification. Expression-correlation analysis exhibited a negative correlation value between ssa-miRNA-204-5p with the *TE Tcb1* and the *sodium/potassium-transporting ATPase subunit alpha and beta.* Additionally, TEs showed negative correlation values with miRNAs ssa-miRNA-19c-3-5p, ssa-miRNA-138-5p, and ssa-miRNA-10a-2-3p (Figure 8). Interestingly, positive correlation values were found in some isoforms of *sodium/potassium-transporting ATPase subunit alpha and beta* and the miRNAs belonging to the families ssa-miRNA-206, ssa-miRNA-214, and ssa-miRNA-219. This result could be associated with specific regulatory roles between miRNAs and transcript isoforms.

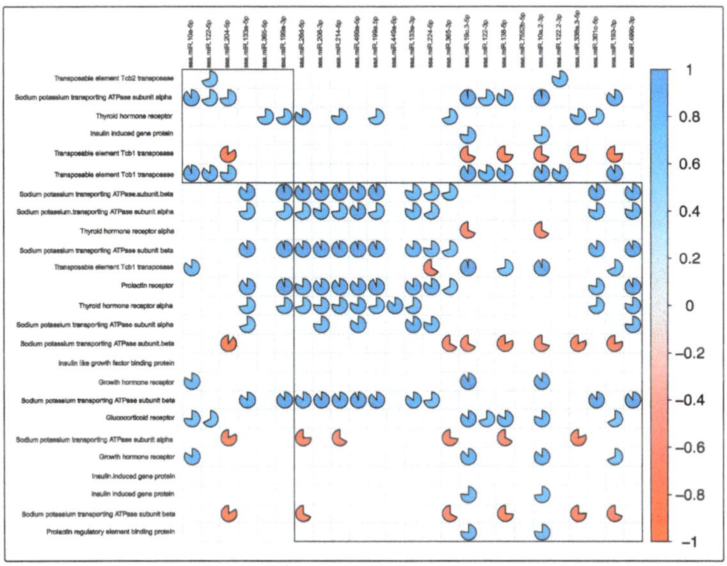

Figure 8. Correlation analysis of expression pairs among smoltification-related genes, TEs, and differential expressed miRNAs. TPM values. Corrplot analyses were conducted on differentially expressed smoltification-related genes, TEs, and differential expressed miRNAs (fold change > |4| and *p*-value < 0.01) in the combination of all the data, including exposure to the three tissue and GSC and SS conditions. Only Pearson's correlation values that were significant (*p*-value > 0.01) are shown in the plot. Red pies correspond to significant negative correlations, and blue pies correspond to significant positive correlations. The completeness of the pies corresponds to the correlation level, where pies closer to circular shape correspond to values more proximal to |1| in Pearson's calculation.

Table 2. Expression modulation of miRNAs and their putative target genes in Atlantic salmon exposed to GSC and SS.

Tissue	miRNA	Fold Change GSC	Fold Change SS	*De nodo* Assembly Contig	Fold Change GSC	Fold Change SS	Delta G	Description
Gills	ssa-miR-204-5p	−2, 52	0	contig_33389	2, 14	0	−12.00	PREDICTED: serine/threonine-protein kinase WNK2-like isoform X2 [Salmo salar]
	ssa-miR-19c-3-5p	3, 04	0	contig_41523	−2, 29	0	−12.00	PREDICTED: heat shock 70 kDa protein 12B-like [Salmo salar]
	ssa-miR-214-5p	−1, 51	0	contig_50634	2, 59	0	−12.00	PREDICTED: creatine kinase S-type, mitochondrial isoform X1 [Salmo salar]
	ssa-miR-214-5p	−1, 51	0	contig_29197	2, 47	0	−12.20	PREDICTED: MAPK/MAK/MRK overlapping kinase-like isoform X1 [Salmo salar]
	ssa-miR-199a-3p	−1, 52	0	contig_12464	2, 54	0	−12.00	PREDICTED: interleukin-20 receptor subunit beta-like [Salmo salar]
	ssa-miR-19c-3-5p	3, 04	0	contig_3735	−2, 11	0	−12.00	PREDICTED: myosin light chain kinase, smooth muscle-like isoform X3 [Salmo salar]
	ssa-miR-456-5p	0	8, 79	contig_63289	0	−3, 32	−12.00	PREDICTED: fibroblast growth factor 10-like [Salmo salar]
	ssa-miR-150-3p	0	−1, 6	contig_15517	0	2, 02	−12.10	ATPase inhibitor, mitochondrial precursor [Salmo salar]
	ssa-miR-19d-5p	0	−2, 68	contig_13064	0	2, 07	−12.00	ATP-binding cassette sub-family F member 2 [Salmo salar]
	ssa-miR-222b-5p	0	−1, 98	contig_78633	0	3, 86	−12.00	transposase [Salmo salar]
	ssa-miR-204-5p	−2, 52	0	contig_66355	2, 96	0	−12.00	transposase [Salmo salar]
	ssa-miR-18b-5p	0	1, 56	contig_19956	0	−17, 31	−12.00	PREDICTED: haptoglobin-like [Salmo salar]
	ssa-miR-456-5p	0	8, 79	contig_9077	0	−3, 2	−12.00	SPATA5 [Salmo salar]
	ssa-miR-456-5p	0	8, 79	contig_52147	0	−2, 72	−12.10	PREDICTED: MAPK/MAK/MRK overlapping kinase-like isoform X3 [Salmo salar]
	ssa-miR-456-5p	0	8, 79	contig_32835	0	−2, 64	−12.00	interleukin-17A/F3 [Salmo salar]
Intestine	ssa-miR-25-3-5p	4, 02	0	contig_19100	−2, 43	0	−30.60	PREDICTED: non-syndromic hearing impairment protein 5-like isoform X2 [Salmo salar]
	ssa-miR-30a-3-3p	0	2, 28	contig_58533	−3, 12	0	−30.50	PREDICTED: kynureninase-like [Salmo salar]
	ssa-miR-92b-3p	0	1, 22	contig_2303	2, 07	0	−30.40	PREDICTED: coronin-1B-like [Salmo salar]
	ssa-miR-214-5p	0	1, 35	contig_3322	2, 75	0	−30.10	PREDICTED: fibroblast growth factor receptor substrate 2-like [Salmo salar]
	ssa-miR-19c-5p	−2, 06	0	contig_51580	3, 29	0	−30.10	PREDICTED: protein-tyrosine kinase 6-like [Salmo salar]
	ssa-miR-92a-5p	3, 72	0	contig_4409	−4, 12	0	−30.00	PREDICTED: dihydropyrimidinase-related protein 4-like [Salmo salar]
	ssa-miR-125b-5p	0	1, 26	contig_4543	0	−1, 97	−29.80	PREDICTED: low-density lipoprotein receptor-related protein 2-like [Salmo salar]
	ssa-miR-30a-3-3p	0	2, 28	contig_3114	0	−1, 94	−29.00	PREDICTED: free fatty acid receptor 3-like [Salmo salar]
	ssa-miR-30a-3p	−1, 73	0	contig_44524	2, 17	0	−29.00	PREDICTED: annexin A2-like [Salmo salar]
	ssa-miR-92a-5p	3, 72	0	contig_51439	−3, 15	0	−28.90	PREDICTED: myosin heavy chain, fast skeletal muscle-like [Salmo salar]
	ssa-miR-15a-3p	1, 72	0	contig_9149	−2, 1	0	−28.40	PREDICTED: annexin A2-like [Salmo salar]
	ssa-miR-140-3p	−1, 53	0	contig_25304	4, 52	0	−28.40	PREDICTED: leucine-rich repeat-containing protein 58-like [Salmo salar]
	ssa-miR-456-3p	3, 35	0	contig_24874	−1, 73	0	−17.60	PREDICTED: sialic acid-binding Ig-like lectin 5 [Salmo salar]
	ssa-let-7i-5p	0	5, 51	contig_66659	0	−1, 37	−16.80	PREDICTED: dedicator of cytokinesis protein 3-like, partial [Salmo salar]
	ssa-let-7d-5p	4	0	contig_28976	−5, 05	0	−16.00	PREDICTED: fibroblast growth factor receptor substrate 2-like [Salmo salar]
Head kidney	ssa-miR-8157-3p	0	−2, 18	contig_122554	0	6, 77	−30.20	PREDICTED: E3 ubiquitin-protein ligase TRIP12-like isoform X1 [Salmo salar]
	ssa-miR-128-1-5p	−1, 57	0	contig_72217	6, 3	0	−29.60	SPATA5 [Salmo salar]
	ssa-miR-214-3p	0	1, 8	contig_21752	0	−1, 47	−29.50	PREDICTED: BTB/POZ domain-containing adapter for CUL3-mediated RhoA degradation protein 2 isoform X1 [Salmo salar]
	ssa-miR-8157-3p	0	−2, 18	contig_46746	0	1, 67	−29.50	PREDICTED: phospholipid-transporting ATPase 11C-like isoform X1 [Salmo salar]
	ssa-miR-205a-5p	0	1, 99	contig_45675	0	−1, 03	−29.40	PREDICTED: von Willebrand factor A domain-containing protein 7-like [Salmo salar]
	ssa-miR-214-3p	0	1, 8	contig_95191	0	−1, 84	−28.90	PREDICTED: fibrinogen beta chain-like [Salmo salar]
	ssa-miR-194a-3p	7, 34	0	contig_28663	−3, 31	0	−28.60	PREDICTED: laminin subunit alpha-4-like isoform X1 [Salmo salar]
	ssa-miR-462a-3p	−1, 86	0	contig_22955	2, 08	0	−34.50	PREDICTED: 3-keto-steroid reductase-like isoform X2 [Salmo salar]
	ssa-miR-122-5p	0	3, 15	contig_68369	0	−1, 89	−29.80	PREDICTED: transcription factor E2F7-like [Salmo salar]
	ssa-miR-125a-5p	0	2, 32	contig_31191	0	−1, 03	−27.70	PREDICTED: tectonic-2 [Salmo salar]
	ssa-miR-449a-5p	21, 47	0	contig_44769	−2, 24	0	−20.40	PREDICTED: guanine nucleotide-binding protein G(I)/G(S)/G(T) subunit beta-3-like isoform X1 [Salmo salar]
	ssa-miR-192b-5p	13, 39	0	contig_65873	−2, 18	0	−20.00	PREDICTED: E3 ubiquitin-protein ligase RNF144A-like [Salmo salar]
	ssa-miR-212a-5p	−1, 83	0	contig_49183	2, 56	0		PREDICTED: collagen EMF1-alpha-like [Salmo salar]

4. Discussion

This study proposes a novel approach to evaluate transcriptome data, where time-series of gene-expression profiling from different tissues and experimental conditions are evaluated. Here, the whole-genome expression-profiling of mRNAs and miRNAs of Atlantic salmon during the smoltification process were explored. Furthermore, the methodology to determine the chromosome gene expression (CGE index) was described. CGE index indicates the coverage differences among two conditions along the Atlantic salmon genome, reflecting differences in the number of mapped reads in specific genomic regions. The study included three different Atlantic salmon tissues with osmoregulation function —gills, intestine, and head kidney. Fish were exposed to gradual salinity changes and salinity shock.

Interestingly, synteny analysis shows high homology among chromosomes with a high CGE index. Furthermore, these chromosomes showed a large number of mobile genetic elements as transposable elements (TE) from Tc1 family and transposase genes. The transposable elements are highly represented in the Atlantic salmon genome. For instance, only the transposable elements of the Tc1-mariner family class represent the 12.89% of the genome [1]. Moreover, this group of transposable elements was associated with signal transduction, regulation of transcription, and defense response [39]. Additionally, it has been reported that transposon expression in rainbow trout is triggered by an external stimulus, such as stress, toxicity, or bacterial antigens [40]. In addition, the growth hormone gene highly associated with the smoltification process, a transposon insertion in a specific isoform of growth hormone gene (gh2) promotor of Atlantic and chinook salmon, has been described. Furthermore, the authors identified a second Tc1 transposon only in the gh2 gene of Atlantic salmon, associated with speciation events [41]. Notably, a previous study performed for our group reported a high presence of TE located near lncRNAs with a putative role in Atlantic salmon seawater adaptation [9]. We suggest a strong regulatory event associated with TEs during Atlantic salmon smoltification. Moreover, this study reports upregulation of TEs in head kidney tissue of Atlantic salmon exposed to GSC and SS. In contrast, gill tissue showed upregulation of TEs in fish exposed to SS, suggesting a putative role of TEs in salmon seawater adaptation. However, additional functional studies are needed to demonstrate the role of TEs during the smoltification process.

Differential expression patterns during parr-smolts transition have been reported by cDNA microarray analysis, highlighting up-modulation of genes associated with growth, metabolism, oxygen transport, and osmoregulation [7]. Additionally, upregulation of genes related to transcription, oxygen transport, electron transport, and protein biosynthesis has been reported [8]. Furthermore, an RNA-Seq study of Atlantic salmon gills showed expression variation during gradual salinity changes, showing higher modulation of processes associated with stress response, cell division and proliferation, tissue development, and collagen catabolic process [9]. Additionally, significant enrichment of genes related to immune response, response to stress, and growth have been described in Atlantic salmon smolt head kidney tissue following seawater transfer [18]. In addition, the authors reported an immune response downregulation after a week in seawater [42]. Moreover, a downregulation of the immune system has been suggested by Shwe, Østbye, Krasnov, Ramberg, and Andreassen [18], associated with the pathogen susceptibility reported for Atlantic salmon during the first period in seawater [42].

Interestingly, a remarkable difference in the number of transcripts differently expressed (DE) among fish exposed to GSC and SS was observed in the present study. It is possible to suggest that salinity shock triggers a higher transcription response than GSC, where the number of DE transcripts was low. GO enrichment analysis of evaluated tissues exhibit BPs associated with protein metabolic process, localization and cell motility, biosynthesis, metabolism, immune response, response to oxidative stress, and response to growth factor. Interestingly, eye morphogenesis was also identified among GO terms identified in gills tissue. In vertebrates, the eyes have a relevant osmoregulatory role [43]. In Coilia nasus, expression variation of eye transcriptomes was observed between hyperosmotic and hypoosmotic conditions. Among the differentially regulated genes were annotated genes associated with immune response, metabolism, and transport [43].

Previously, our group characterized long, non-coding RNAs of Atlantic salmon gills during the smoltification process and reported on the relevance of non-coding RNAs in regulation of this biological process [9]. Interestingly, the study observed that highly regulated lncRNAs were located near genes associated with the process as growth, cell death, catalytic activity, and apoptotic process [9]. The role of miRNAs during smoltification and the early seawater phase was described by Shwe, Østbye, Krasnov, Ramberg, and Andreassen [18] in head kidney of Atlantic salmon. The authors reported 71 DE miRNAs during their evaluation time. Among the relevant miRNAs identified by the authors, the miRNA from family mir-192 has been associated with hypoxia [44]. In the present study, the mir-192 family was overexpressed in response to GSC in head kidney and down modulated in gills of the SS fish group. Another miRNA family associated with hypoxia [44], and downregulated in the intestine in our study was mir-181. Notably, in the present study, SSA-miR-499a-5p was up-regulated in gills and head kidney tissues of Atlantic salmon exposed to GSC. Moreover, from the target analysis, it is possible to suggest that miRNA has a role in fish growth, with gene collagen as its target. In contrast, miRNA was downregulated in head kidney of the SS fish group. Similar results were reported by Shwe, Østbye, Krasnov, Ramberg, and Andreassen [18], who observed a decrease in ssa-miR-499a-5p expression in Atlantic salmon head kidney during transfer from FW to SW.

Finally, GO enrichment analysis of putative target genes of differently modulated miRNAs from the GSC group showed a large number of transcripts associated with cellular process, response to stimulus, cell communication, response to stress, and immune system process in gills, intestine, and head kidney tissues. A similar process was reported to be associated with target genes of miRNAs differently expressed in Atlantic salmon head kidney during seawater adaptation [18]. In contrast, in this study, a small number of target genes was annotated in response to salinity shock. For instance, biological process as a response to A4TP, antigens processing a presentation by MHC class, and glycosylation were identified in this study. These results suggest bounded Atlantic salmon miRNA modulation in response to salinity shock, compared with gradual salinity change, where Atlantic salmon miRNAs display regulation of a large number of biological processes. Functional analysis will be conducted to demonstrate the regulatory role of the highlight miRNAs identified in this study.

5. Conclusions

The CGE index proposed in this study explains differences in expression profiling among fish exposed to gradual salinity change and salinity shock during seawater transfer. Furthermore, a great abundance of transposable elements was identified among the chromosomes, with significant expression differences between groups. This is the first study showing the harmony among mRNA and miRNA transcription profiles at the genome level of Atlantic salmon during the smoltification process. The proposed transcriptome analysis revealed significant differences among tissues, time, and experimental conditions. In addition, the relevance of miRNA function in the regulation different biological process is suggested, such as growth, stress response, catabolism, and immune response.

Supplementary Materials: The following are available online at https://www.mdpi.com/article/10.3390/biology11010001/s1. Figure S1. Experimental design diagram showing the GSC and SS conditions for Atlantic salmon smolts (n = 30 fish/tank). Sample points (S) are indicated with red arrow. T1-T4 indicate the sample points. For GSC Atlantic salmon group T1, T2, and T3 correspond a week after salmons were at 10, 20, and 32 PSU, respectively. For the SS group, the T1 and T2 are 2 and 3 weeks before salinity shock, and T3 is a week after salinity shock (32 PSU). Figure S2. Histochemistry analyses showing the localization of chloride cells in the gill filament epithelium. (**A**) FW gills sample; (**B**) GSC 10 PSU; (**C**) GSC 20 PSU; (**D**) GSC 32 PSU; (**E**) GSC 32 PSU. (**A**) Na+/K+-ATPase positive cells located in the middle region of the superficial interlamellar space and near the lamellar vascular axis. B-C Na+/K+-ATPase positive cell deeper in the epithelium.

Na+/K+-ATPase positive cells are indicated with black arrows. Magnification 60×. Figure S3. RT-qPCR analysis of ATPase-α and ATPase-β subunits in Atlantic salmon gills exposed to GSC and SS. Significant differences between experimental conditions are indicate with asterisk ($p < 0.05$). Figure S4. Heatmap representation of Atlantic salmon transcriptome for gills, intestine and head kidney tissues exposed GSC and SS conditions. Red and blue colors represent the gene expression levels from high to low transcription values. Figure S5. Cluster gene expression analysis of transposable elements (TEs) during SW transfer in Atlantic salmon gills. (**A**) Expression values of TE transcripts in Atlantic salmon gradually transferred from FW to SW (GSC) and from FW to salinity shock (SS), respectively. The four gene clusters were detected by K-means algorithm using TPM values. (**B**) RT-qPCR validation for transposable elements Tcb1 and Tcb2 for GSC and SS, respectively. For GSC Atlantic salmon group T1, T2, and T3 correspond a week after salmons were at 10, 20, and 32 PSU, respectively. For the SS group, the T1 and T2 are 2 and 3 weeks before salinity shock, and T3 is a week after salinity shock (32 PSU). Elongation factor was used as endogenous control for normalize data. Figure S6. Cluster gene expression analysis of miRNAs during SW transfer in Atlantic salmon gills. (**A**) Expression values of TE transcripts in Atlantic salmon gradually transferred from FW to SW (GSC) and from FW to salinity shock (SS), respectively. The four gene clusters were detected by K-means algorithm using TPM values. (**B**) RT-qPCR validation for ssa-miR-143, ssa-miR-21b, ssa-miR-10d expressed in response to GSC, and ssa-miR-181, ssa-miR-10b and ssa-miR-30d expressed in gills samples exposed to SS. For GSC Atlantic salmon group T1, T2, and T3 correspond a week after salmons were at 10, 20, and 32 PSU, respectively. For the SS group, the T1 and T2 are 2 and 3 weeks before salinity shock, and T3 is a week after salinity shock (32 PSU). Ssa-mir-455-5p was used as endogenous control for normalize data. Figure S7. RT-qPCR validation of candidate miRNAs and their putative target gene. Fold-changes (Log2) of gene expression were calculated using the FW condition as control group. Elongation factor and Ssa-mir-455-5p were used as endogenous control for mRNAs and miRNAs, respectively. Table S1. chr clusters. Table S2. DEG transcripts annotation. Table S3. miRNA prediction. Table S4. Primer list.

Author Contributions: V.V.-M. and C.G.-E. designed the experiment; B.P.B. performed RNA extraction and RT-qPCR analysis. H.E.-S. software analysis. V.V.-M. coordinated the experimental trials. V.V.-M., C.G.-E., B.P.B., G.N.-A., D.V.-M. and J.A.V. participated in writing of the manuscript. All authors have read and agreed to the published version of the manuscript.

Funding: ANID-Chile funded this study through the Postdoctoral grant FONDECYT (3190320), grants FONDAP (15110027) and FONDECYT (1210852).

Institutional Review Board Statement: Not applicable.

Informed Consent Statement: Not applicable.

Data Availability Statement: Not applicable.

Conflicts of Interest: The authors declare no conflict of interest.

References

1. Lien, S.; Koop, B.F.; Sandve, S.R.; Miller, J.R.; Kent, M.P.; Nome, T.; Hvidsten, T.R.; Leong, J.S.; Minkley, D.R.; Zimin, A.; et al. The Atlantic salmon genome provides insights into rediploidization. *Nature* **2016**, *533*, 200–205. [CrossRef]
2. McCormick, S.D.; Björnsson, B.T. Physiological and hormonal differences among Atlantic salmon parr and smolts reared in the wild, and hatchery smolts. *Aquaculture* **1994**, *121*, 235–244. [CrossRef]
3. McCormick, S.D.; Hansen, L.P.; Quinn, T.P.; Saunders, R.L. Movement, migration, and smolting of Atlantic salmon (Salmo salar). *Can. J. Fish. Aquat. Sci.* **1998**, *55*, 77–92. [CrossRef]
4. McCormick, S.D.; O'Dea, M.F.; Moeckel, A.M.; Björnsson, B.T. Endocrine and physiological changes in Atlantic salmon smolts following hatchery release. *Aquaculture* **2003**, *222*, 45–57. [CrossRef]
5. Ebbesson, L.O.E.; Ekström, P.; Ebbesson, S.O.E.; Stefansson, S.O.; Holmqvist, B. Neural circuits and their structural and chemical reorganization in the light–brain–pituitary axis during parr–smolt transformation in salmon. *Aquaculture* **2003**, *222*, 59–70. [CrossRef]
6. Stefansson, S.O.; Björnsson, B.T.; Ebbesson, L.O.E.; McCormick, S.D. Smoltification. In *Fish Larval Physiology*; Finn, R.N., Kapoor, B.G., Eds.; CRC Press: Boca Raton, FL, USA, 2008; pp. 639–681.
7. Marshall, W.S. Na^+, Cl^-, Ca^{2+} and Zn^{2+} transport by fish gills: Retrospective review and prospective synthesis. *J. Exp. Zool.* **2002**, *293*, 264–283. [CrossRef]
8. Seear, P.J.; Carmichael, S.N.; Talbot, R.; Taggart, J.B.; Bron, J.E.; Sweeney, G.E. Differential gene expression during smoltification of Atlantic salmon (*Salmo salar* L.): A first large-scale microarray study. *Mar. Biotechnol.* **2010**, *12*, 126–140. [CrossRef] [PubMed]

9. Robertson, L.S.; McCormick, S.D. Transcriptional profiling of the parr–smolt transformation in Atlantic salmon. Comp. *Biochem. Physiol. Part D Genom. Proteom.* **2012**, *7*, 351–360.
10. Valenzuela-Muñoz, V.; Váldes, J.A.; Gallardo-Escárate, C. Transcriptome Profiling of Long Non-coding RNAs During the Atlantic Salmon Smoltification Process. *Mar. Biotechnol.* **2021**, *23*, 308–320. [CrossRef] [PubMed]
11. Bartel, D.P. MicroRNAs: Genomics, Biogenesis, Mechanism, and Function. *Cell* **2004**, *116*, 281–297. [CrossRef]
12. Bartel, D.P. MicroRNAs: Target recognition and regulatory functions. *Cell* **2009**, *136*, 215–233. [CrossRef] [PubMed]
13. Bi, Y.; Liu, G.; Yang, R. MicroRNAs: Novel regulators during the immune response. *J. Cell. Physiol.* **2009**, *218*, 467–472. [CrossRef]
14. Judice, C.C.; Bourgard, C.; Kayano, A.C.; Albrecht, L.; Costa, F.T. MicroRNAs in the host-apicomplexan parasites interactions: A review of immunopathological aspects. *Front. Cell. Infect. Microbiol.* **2016**, *6*. [CrossRef]
15. Lindsay, M.A. MicroRNAs and the immune response. *Trends Immunol.* **2008**, *29*, 343–351. [CrossRef] [PubMed]
16. Sonkoly, E.; Ståhle, M.; Pivarcsi, A. MicroRNAs and immunity: Novel players in the regulation of normal immune function and inflammation. *Semin. Cancer Biol.* **2008**, *18*, 131–140. [CrossRef] [PubMed]
17. Kozomara, A.; Griffiths-Jones, S. MiRBase: Annotating high confidence microRNAs using deep sequencing data. *Nucleic Acids Res.* **2014**, *42*, D68–D73. [CrossRef] [PubMed]
18. Andreassen, R.; Worren, M.M.; Høyheim, B. Discovery and characterization of miRNA genes in Atlantic salmon (*Salmo salar*) by use of a deep sequencing approach. *BMC Genom.* **2013**, *14*, 1–11. [CrossRef]
19. Shwe, A.; Østbye, T.-K.K.; Krasnov, A.; Ramberg, S.; Andreassen, R. Characterization of differentially expressed MiRNAs and their predicted target transcripts during smoltification and adaptation to seawater in head kidney of Atlantic salmon. *Genes* **2020**, *11*, 1059. [CrossRef] [PubMed]
20. Deng, X.; Zhironkina, O.A.; Cherepanynets, V.D.; Strelkova, O.S.; Kireev, I.I.; Belmont, A.S. Cytology of DNA Replication Reveals Dynamic Plasticity of Large-Scale Chromatin Fibers. *Curr. Biol.* **2016**, *26*, 2527–2534. [CrossRef] [PubMed]
21. Wilson, R.H.C.; Coverley, D. Relationship between DNA replication and the nuclear matrix. *Genes Cells* **2013**, *18*, 17–31. [CrossRef] [PubMed]
22. Nagano, T.; Lubling, Y.; Stevens, T.J.; Schoenfelder, S.; Yaffe, E.; Dean, W.; Laue, E.D.; Tanay, A.; Fraser, P. Single-cell Hi-C reveals cell-to-cell variability in chromosome structure. *Nature* **2013**, *502*, 59–64. [CrossRef] [PubMed]
23. Schwarz-Finsterle, J.; Scherthan, H.; Huna, A.; Gonzalez, P.; Mueller, P.; Schmitt, E.; Erenpreisa, J.; Hausmann, M. Volume increase and spatial shifts of chromosome territories in nuclei of radiation-induced polyploidizing tumour cells. *Mutat. Res.Genet. Toxicol. Environ. Mutagenesis* **2013**, *756*, 56–65. [CrossRef] [PubMed]
24. Mehta, I.S.; Kulshreshtha, M.; Chakraborty, S.; Kolthur-Seetharam, U.; Rao, B.J. Chromosome Territories Reposition During DNA Damage-Repair Response. *Biophys. J.* **2014**, *106*, 79A. [CrossRef]
25. Zheng, H.; Xie, W. The role of 3D genome organization in development and cell differentiation. *Nat. Rev. Mol. Cell Biol.* **2019**, *20*, 535–550. [CrossRef] [PubMed]
26. Balajee, A.S.; Sanders, J.T.; Golloshi, R.; Shuryak, I.; McCord, R.P.; Dainiak, N. Investigation of Spatial Organization of Chromosome Territories in Chromosome Exchange Aberrations After Ionizing Radiation Exposure. *Health Phys.* **2018**, *115*, 77–89. [CrossRef] [PubMed]
27. Fritz, A.J.; Barutcu, A.R.; Martin-Buley, L.; van Wijnen, A.J.; Zaidi, S.K.; Imbalzano, A.N.; Lian, J.B.; Stein, J.L.; Stein, G.S. Chromosomes at Work: Organization of Chromosome Territories in the Interphase Nucleus. *J. Cell Biochem.* **2016**, *117*, 9–19. [CrossRef] [PubMed]
28. Fritz, A.J.; Sehgal, N.; Pliss, A.; Xu, J.H.; Berezney, R. Chromosome territories and the global regulation of the genome. *Gene Chromosome Canc.* **2019**, *58*, 407–426. [CrossRef] [PubMed]
29. Kinney, N.A.; Sharakhov, I.V.; Onufriev, A.V. Chromosome-nuclear envelope attachments affect interphase chromosome territories and entanglement. *Epigenetics Chromatin* **2018**, *11*, 3. [CrossRef] [PubMed]
30. Sehgal, N.; Fritz, A.J.; Vecerova, J.; Ding, H.; Chen, Z.H.; Stojkovic, B.; Bhattacharya, S.; Xu, J.H.; Berezney, R. Large-scale probabilistic 3D organization of human chromosome territories. *Hum. Mol. Genet.* **2016**, *25*, 419–436. [CrossRef] [PubMed]
31. Sole, M.; Blanco, J.; Gil, D.; Valero, O.; Fonseka, G.; Frodsham, R.; Vidal, F.; Sarrate, Z. Chromosome territories in mice spermatogenesis: A new three-dimensional methodology. *Mol. Cytogenet.* **2017**, *10*, 236.
32. Krzywinski, M.I.; Schein, J.E.; Birol, I.; Connors, J.; Gascoyne, R.; Horsman, D.; Jones, S.J.; Marra, M.A. Circos: An information aesthetic for comparative genomics. *Genome Res.* **2009**, *19*, 1639–1645. [CrossRef] [PubMed]
33. Yu, G.; Wang, L.-G.; Han, Y.; He, Q.-Y. clusterProfiler: An R package for comparing biological themes among gene clusters. *Omics A J. Integr. Biol.* **2012**, *16*, 284–287. [CrossRef] [PubMed]
34. Griffiths-Jones, S.; Saini, H.K.; Van Dongen, S.; Enright, A.J. MiRBase: Tools for microRNA genomics. *Nucleic Acids Res.* **2007**, *36*, D154–D158. [CrossRef]
35. Valenzuela-Muñoz, V.; Novoa, B.; Figueras, A.; Gallardo-Escárate, C. Modulation of Atlantic salmon miRNome response to sea louse infestation. *Dev. Comp. Immunol.* **2017**, *76*, 380–391. [CrossRef]
36. Miranda, K.C.; Huynh, T.; Tay, Y.; Ang, Y.S.; Tam, W.L.; Thomson, A.M.; Lim, B.; Rigoutsos, I. A pattern-based method for the identification of MicroRNA binding sites and their corresponding heteroduplexes. *Cell* **2006**, *126*, 1203–1217. [CrossRef]
37. R Core Team. *R: A Language And Environment For Statistical Computing*; R Foundation for Statistical Computing: Vienna, Austria, 2021.
38. Wei, T.; Simko, V.; Levy, M.; Xie, Y.; Jin, Y.; Zemla, J. Package 'corrplot'. *Statistician* **2017**, *56*, 316–324.

39. Johansen, I.; Andreassen, R. Validation of miRNA genes suitable as reference genes in qPCR analyses of miRNA gene expression in Atlantic salmon (*Salmo salar*). *BMC Res. Notes* **2014**, *8*, 945. [CrossRef] [PubMed]
40. Krasnov, A.; Koskinen, H.; Afanasyev, S.; Mölsä, H. Transcribed Tc1-like transposons in salmonid fish. *BMC Genom.* **2005**, *6*, 107. [CrossRef]
41. Von Schalburg, K.R.; Yazawa, R.; De Boer, J.; Lubieniecki, K.P.; Goh, B.; Straub, C.A.; Beetz-Sargent, M.R.; Robb, A.; Davidson, W.S.; Devlin, R.H. Isolation, characterization and comparison of Atlantic and Chinook salmon growth hormone 1 and 2. *BMC Genom.* **2008**, *9*, 522. [CrossRef] [PubMed]
42. Johansson, L.-H.; Timmerhaus, G.; Afanasyev, S.; Jørgensen, S.M.; Krasnov, A. Smoltification and seawater transfer of Atlantic salmon (*Salmo salar* L.) is associated with systemic repression of the immune transcriptome. *Fish. Shellfish Immunol.* **2016**, *58*, 33–41. [CrossRef]
43. Gao, J.; Xu, G.; Xu, P. Full-length transcriptomic analysis reveals osmoregulatory mechanisms in *Coilia nasus* eyes reared under hypotonic and hyperosmotic stress. *Sci. Total Environ.* **2021**, *799*, 149333. [CrossRef] [PubMed]
44. Jacobs, L.A.; Bewicke-Copley, F.; Poolman, M.G.; Pink, R.C.; Mulcahy, L.A.; Baker, I.; Beaman, E.-M.; Brooks, T.; Caley, D.P.; Cowling, W.; et al. Meta-Analysis Using a Novel Database, miRStress, Reveals miRNAs That Are Frequently Associated with the Radiation and Hypoxia Stress-Responses. *PLoS ONE* **2013**, *8*, e80844. [CrossRef] [PubMed]

biology

MDPI

Article

Expression Analysis in Atlantic Salmon Liver Reveals miRNAs Associated with Smoltification and Seawater Adaptation

Alice Shwe [1], Aleksei Krasnov [2], Tina Visnovska [3], Sigmund Ramberg [1], Tone-Kari K. Østbye [2] and Rune Andreassen [1,*]

[1] Department of Life Science and Health, Faculty of Health Sciences, OsloMet-Oslo Metropolitan University, 0167 Oslo, Norway; aliceshw@oslomet.no (A.S.); sigmundr@oslomet.no (S.R.)

[2] Nofima (Norwegian Institute of Food, Fisheries and Aquaculture Research), 1430 Ås, Norway; aleksei.krasnov@nofima.no (A.K.); tone-kari.ostbye@nofima.no (T.-K.K.Ø.)

[3] Bioinformatics Core Facility, Oslo University Hospital, 0372 Oslo, Norway; martvis@ifi.uio.no

* Correspondence: rune.andreassen@oslomet.no

Simple Summary: Smoltification is a developmental process that preadapts Atlantic salmon for a life in seawater. Suboptimal smoltification and poor timing of transfer to seawater is associated with increased mortality. MicroRNAs (miRNAs) are small non-coding genes. They regulate gene expression post-transcriptionally as part of the miRNA induce silencing complex (miRISC) where they guide miRISC to particular mRNAs (target genes). The aim of this study was to identify Atlantic salmon miRNAs expressed in liver that are associated with smoltification and adaptation to seawater as well as to predict their target genes. In total, 62 guide miRNAs were identified, and by their expression patterns they were clustered into three groups. Target gene predictions followed by gene enrichment analysis of the predicted targets indicated that the guide miRNAs were involved in post-transcriptional regulation of important smoltification associated biological processes. Some of these were energy metabolism, protein metabolism and transport, circadian rhythm, stress and immune response. Together, the results indicate that certain miRNAs are involved in the regulation of many of the important changes occurring in the liver during this developmental transition.

Abstract: Optimal smoltification is crucial for normal development, growth, and health of farmed Atlantic salmon in seawater. Here, we characterize miRNA expression in liver to reveal whether miRNAs regulate gene expression during this developmental transition. Expression changes of miRNAs and mRNAs was studied by small-RNA sequencing and microarray analysis, respectively. This revealed 62 differentially expressed guide miRNAs (gDE-miRNAs) that could be divided into three groups with characteristic dynamic expression patterns. Three of miRNA families are known as highly expressed in liver. A rare arm shift was observed during smoltification in the Atlantic salmon-specific novel-ssa-miR-16. The gDE-miRNAs were predicted to target 2804 of the genes revealing expression changes in the microarray analysis. Enrichment analysis revealed that targets were significantly enriched in smoltification-associated biological process groups. These included lipid and cholesterol synthesis, carbohydrate metabolism, protein metabolism and protein transport, immune system genes, circadian rhythm and stress response. The results indicate that gDE-miRNAs may regulate many of the changes associated with this developmental transition in liver. The results pave the way for validation of the predicted target genes and further study of gDE-miRNA and their targets by functional assays.

Keywords: smoltification; seawater adaptation; microRNAs; small-RNA sequencing; liver; Atlantic salmon; microarray transcriptome

Citation: Shwe, A.; Krasnov, A.; Visnovska, T.; Ramberg, S.; Østbye, T.-K.K.; Andreassen, R. Expression Analysis in Atlantic Salmon Liver Reveals miRNAs Associated with Smoltification and Seawater Adaptation. *Biology* **2022**, *11*, 688. https://doi.org/10.3390/biology11050688

Academic Editor: Patricia Pereiro

Received: 25 March 2022
Accepted: 24 April 2022
Published: 30 April 2022

1. Introduction

Atlantic salmon is one of the most successful aquaculture species. The total worldwide production of farmed Atlantic salmon was around 2638 kilotons in 2020 and 1821 kilotons of this was produced in the North Atlantic area [1]. The majority of farmed salmon in the North Atlantic area is produced by Norway and UK (Scotland) (77% and 11%, respectively) [1]. The rapid growth in the salmon aquaculture industry is due to continuous improvement in production practices [2]. However, salmon aquaculture is still facing several challenges including high mortality after sea transfer associated with deficient smoltification and higher susceptibility to infection diseases [3–5]. Studies of salmonids smoltification have also revealed changes in expression of genes involved in immune responses [6–9].

Smoltification (also known as Parr–Smolt transformation) is a complex preparatory developmental process that transforms parr to smolt for a successful life in the marine environment [10–12]. Smoltification involves changes in physiology (e.g., increased metabolic rate, increased seawater tolerance and alterations in lipid metabolism) [13], morphology (e.g., smolts acquire silver skin pigmentation and more streamlined body shape) and behavior (e.g., downstream movement and the loss of territorial behavior) [12,14,15]. Smoltification is a highly energy-demanding process and it is associated with decrease in liver glycogen, whole-body lipid content and muscle content in smolt [16,17]. An increase in mitochondrial enzymes activity [18], enzymes of glycolysis, fatty acid and lactate metabolism (i.e., phosphofructokinase (PFK), β-hydroxyacyl-coenzyme A dehydrogenase (HOAD) and lactate dehydrogenase (LDH)) in the liver of Atlantic salmon have also been reported. Elevated levels of thyroid hormones could be responsible for this increase [18,19]. It has also been reported that β-oxidation capacity in liver increased significantly prior to seawater transfer which gives liver an important role in energy production during this period [16]. Citrate synthase activity in liver, gill and kidney was also enhanced during smoltification [18,20] and did not change after seawater transfer [21].

In nature, smolting is stimulated by the increasing day length (photoperiod) in spring and seasonal temperature fluctuations [10]. The aquaculture industry takes advantage of this photoperiod-dependence in the production of seawater-tolerant juvenile salmon [22]. Smolting in salmon aquaculture is artificially achieved by exposing parr reaching a desirable size to a short photoperiod for several weeks and then returning them to continuous light [23]. Optimal smoltification and correct timing of seawater transfer (SWT) are crucial for normal development, growth and health of farmed salmon [24,25]. Various tests for assessing smolt readiness and quality are employed, including salinity tolerance test, seawater challenge hypo-osmoregulatory test (SCHT), mRNA expression and enzymatic activity of gill Na^+/K^+-ATPase (NKA) and measurement of hormones involved in smolting [24,26–28]. Salinity tolerance test and SCHT are inexpensive and relative quick to perform. The SCHT is used by many commercial hatcheries in Norway and Canada [24]. Studies showed that despite satisfactory reactions to salinity tolerance test and SCHT, some hatchery-reared smolts perform poorly at sea in terms of survival and growth rate compared to their naturally produced smolts [29]. Additionally, gill NKA activity in freshwater at the peak of smolting does not always predict long-term growth in seawater [30].

In recent years, incomplete smoltification is reported as one of the most important causes of mortality and contributor to disease development in the period following sea transfer in Norwegian salmon farms [5,25]. According to the annual summary of fish health in Norway [25] from the Norwegian Veterinary Institute, 52.1 million of 290 million smolts transferred to sea died prior to harvest in 2020. The high mortality rate of farmed salmon in the period following sea transfer represents a major fish welfare problem as well as a major economic loss for the aquaculture industry. Better measures for optimal smoltification and transfer to sea at the correct time and good follow up during the first period in seawater are required to reduce mortalities and improve welfare of farmed salmon [25]. Research on smoltification in the past 30 years was particularly directed to gain a better understanding of optimal timing for transfer of smoltified juveniles into ocean net pens [3,26]. In recent years, more studies of gene expression associated with smoltification have been carried out to better understand this developmental transition as well as to search for new biomarkers [7,22,31].

MicroRNAs (miRNAs) are single-stranded, non-coding RNAs (typically ~22 nucleotides in length) that regulate a large variety of biological processes at the post-transcriptional level [32,33]. Mature miRNAs are processed by a cascade of several nuclear and cytoplasmic enzymatic processing steps [34]. Most miRNAs are transcribed into long primary miRNAs (pri-miRNAs) that are processed further into hairpin-structured precursor miRNAs (pre-miRNAs) and finally short miRNA duplexes [35]. At the final step, one of the strands of the miRNA duplex (termed as "mature miRNA" or "guide miRNA") is loaded into and retained in Argonaute (AGO) proteins, forming the miRNA-induced silencing complex (miRISC). The other strand of the miRNA duplex (termed as "passenger miRNA") is released from AGO proteins and degraded. The guide miRNAs direct the miRISCs to their target messenger RNAs (mRNAs), usually by binding partially to the 3′-untranslated region (3′-UTR) of the target transcripts. This leads to translational repression or degradation of target mRNAs [36]. The guide miRNA may originate from the 5′ end of the pre-miRNA (referred to as "5p") or the 3′end of the pre-miRNA (referred to as "3p") [34,37]. Atlantic salmon miRNAs are among the best characterized in teleost [38]. Due to the salmonid specific genome duplication [39,40], the number of miRNA gene families is larger in salmonids than in any other teleost [38,41]. Recently, the transcriptomes from liver, head kidney and gills from different stages of the smoltification process were characterized by full-length error corrected mRNA sequencing. This provided a catalog of genes and splice variants that are expressed in liver during this developmental process [42]. Furthermore, the 3′UTRs from the full-length sequenced mRNAs were analyzed regarding their potential as miRNA targets. This resulted in a comprehensive resource of predicted mRNA targets for any of the mature miRNAs in Atlantic salmon [43].

The miRNA expression analysis often shows that one of the mature miRNAs appears as the highly expressed one compared to the other mature miRNA processed from same precursor. This is a consistent pattern shown in several studies [38,41]. The explanation to this difference in abundance of the two mature miRNAs from one precursor is that the more abundant one is the biologically important guide miRNA incorporated into miRISC while the other one with the much lower abundancy, is the passenger miRNAs that are degraded [37]. Here, in cases where there is a large difference in abundance between miRNAs from same precursor, we refer to the abundant mature miRNAs (10 times higher read counts than its corresponding mature from same precursor) as the biologically important guide miRNAs. Previous mammalian studies have characterized some miRNAs that were highly expressed in liver [44,45], suggesting that they are important regulators in liver development [46], liver homeostasis, ion metabolism and lipid, and cholesterol biosynthesis [47–49]. Studies in teleost fish have also reported that miRNAs are involved in immune response [50–52] and response to environmental stimuli [53]. A study of hepatic miRNAs in rainbow trout and farmed carp indicated that miRNAs are involved in hepatic energy metabolism [54,55] while a study in Atlantic salmon revealed several miRNAs associated with lipid metabolism in liver [56]. Furthermore, miR-122-5p, miR-8163-3p, miR-148-5p and miR-101-3p have been reported as highly expressed in the liver of marine teleost, indicating that they serve a common important function in this organ [38,57].

Our recent study conducted in the head kidney of Atlantic salmon indicated that miRNAs are involved in post-transcriptional regulation of gene expression during smoltification and adaptation to seawater in this organ [58]. Gene ontology analysis of the predicted target genes showed that they were enriched in head kidney specific biological processes associated with smoltification and seawater adaptation [58]. So far, similar studies have not been carried out in liver, an organ that plays an important role in storage (lipids, carbohydrates, vitamin A and iron) and detoxification in teleost [59]. In addition, the liver has a central position in amino acid and carbohydrate metabolism and in the synthesis and export of many proteins [60].

The aim of the present study was to characterize miRNA expression changes associated with smoltification and adaptation to seawater in the first month following seawater transfer. Investigating the post-transcriptional interaction between miRNAs and their predicted target mRNAs during this critical period of Atlantic salmon life may provide a better understanding of how the fine-tuning of gene expression in liver may help in facilitating this developmental transition.

2. Materials and Methods

2.1. Experimental Fish Trial and Samplings

The experiment was carried out at the Nofima's Research Station for Sustainable Aquaculture (Sunndalsøre, Norway) in accordance with the Guidelines of the EU-legislation (2010/63/EU), as well as with the Norwegian legislation on animal experimentation. The experimental fish were not exposed to any pain or distress. They were solely killed for the use of their tissues in this project and, thus, approval from the Norwegian Food Safety Authority was not required.

The experimental fish were from SalmonBreed commercial strain SalmonBreed-Model SB-Optimal. The most important characteristic of this strain is good growth combined with a balanced weighting for health properties through family selection. A total of 70 fish were selected for this experiment. The fish were kept in one tank supplied with running water throughout the experimental period and they were fed commercial dry feed (Skretting, Norway). The fish used in this study were the same experimental fish as described in Shwe et al. [58]. From the start of feeding, the fish were kept in freshwater with an average water temperature of 13 °C and 24 h continuous light. The average water temperature was then dropped to 8 °C 2 weeks before the start of smoltification process. The initial smoltification process started by decreasing daylight from 24 h to 12 h and increasing water temperature from 8 °C to 13 °C for 5 days, followed by 12 °C for 41 days. Subsequently, the daylight was increased to 24 h and the water temperature lowered to 8 °C for the final stage of smoltification (Table 1). The seawater challenge test was performed once a week in the last 3 weeks before SWT using a salinity of 35‰. Seawater challenge test, blood plasma ions (Cl^-, Na^+ and Mg^{2+}) level and the change to silvery skin color indicated that the experimental fish were smoltified 81 days after onset of the experiment. The smoltified smolts with average weight 72.4 ± 8.7 g were then transferred to seawater. The average weight at 1-week post-SWT and one-month post-SWT was 63.2 ± 8.5 g and 98.4 ± 14.9, respectively. No mortality of smolts was observed during smoltification or after SWT.

Table 1. Photoperiod and water temperature during the experimental trial.

Experimental Days	Hours of Light per Day (h)	Water Temperature (°C)	Water Type
Day 0	24	8	Fresh water
Day 1–5	12	13	Fresh water
Day 6–47	12	12	Fresh water
Day 48–60	24	12	Fresh water
Day 61–81	24	8	Fresh water
Day 82–111	24	8	seawater

The liver samples were collected at six time points (Table 2), T1: parr, 1 day prior to light treatment, T2: halfway through light treatment at the change from 24 h to 12 h light and temperature to 12 °C (47 days post-onset of light treatment (POL)), T3: three quarters into the light treatment period (67 days POL), T4: smolt, 1 day prior to SWT (81 days POL), T5: 1 week after SWT (88 days POL), and T6: 1 month after SWT (111 days POL). The experimental conditions such as day light, water temperature, average weight of experimental fish and water type at each sampling points are provided in Table 2. Ten fish were euthanized with an overdose of anesthetic metacain (MS-222; 0.1 g/L) and killed by a blow to the head prior to weighing and sampling at each time point. The collected liver samples were frozen immediately in liquid hydrogen and stored at −80 °C. Seven fish from each time points were used for RNA extraction and small RNA sequencing, while RNA from five of these fish was also used for microarray analysis.

Table 2. Time-points and conditions where liver samples were collected.

Group	Sample Collection Time Points	Light [1]	Temp. [2]	Weight [3]	Water Type	Sampling [4]
T1	Parr, 1 day prior to light treatment	24	8	29.4 ± 5.6	Fresh water	Day 0
T2	Halfway into light treatment	12	12	52.6 ± 5.9	Fresh water	Day 47
T3	Three quarters into light treatment	24	8	63.9 ± 10.1	Fresh water	Day 67
T4	Smolt, 1 day prior to SWT	24	8	72.4 ± 8.7	Fresh water	Day 81
T5	One week after SWT	24	8	63.2 ± 8.5	Seawater	Day 88
T6	One month after SWT	24	8	98.4 ± 14.9	Seawater	Day 111

[1] Hours with day light. [2] Water temperature in degree Celsius (°C). [3] Average weight in gram of the experimental fish collected at each time points. [4] Sampling day within the experimental period.

2.2. Total RNA Extraction for Sequencing and Microarray Analysis

Total RNA was isolated using the mirVanaTM miRNA Isolation Kit (Ambion, Life Technologies, Carlsbad, CA, USA) according to the manufacturer's protocol. The RNA quality and quantity were determined using NanoDropTM1000 Spectrophotometer (Nanodrop ND-1000, Thermo Fisher Scientific, Wilmington, DE, USA). The integrity of total RNA (RIN value) was measured using the Agilent 2100 Bioanalyzer in combination with an Agilent 6000 Nano Chip (Agilent Technologies, Santa Clara, CA, USA). Extracted total RNA was stored at −80 °C.

2.3. Small-RNA Library Preparation and Sequencing

Library construction and sequencing of 42 liver samples were carried out at the Norwegian High-Throughput Sequencing Centre (NSC; Oslo, Norway). Seven liver samples from each of the time points T1-T6 were selected for sequencing. The NEBnext® multiplex small RNA Library Prep Set (New England Biolabs, Inc., Ipswich, MA, USA) was used to construct libraries for 42 liver samples in accordance with manufacturers protocol. One µg total RNA from each sample were used as input for preparation of the libraries followed by 5′ and 3′ adapter ligation, reverse transcription, PCR amplification and size selection of 140–150 bp fragments using 6% polyacrylamide gel. Sequencing was performed on a NextSeq 500 from Illumina (Illumina, Inc, San Diego, CA, USA), producing 75 bp single end reads. All sequenced samples have been submitted to the NCBI Sequence Read Archieve Centre (SRA) (https://www.ncbi.nlm.nih.gov/sra accessed 20 April 2022) with accession bioproject number PRJNA665200 and will automatically be released by NCBI at the publication of this study.

2.4. Processing of Small-RNA Reads and DESeq2 Expression Analysis

Data processing and quality control of small-RNA reads were performed according to the procedure described in Shwe et al. [58]. The quality of raw reads was checked using FASTQC software (v.0.11.8). The raw reads that passed quality control were trimmed using Cutadapt (v.2.3) Python package (v.3.7.3) [61]. Trimmed reads were processed further by size filtering to discard all reads that were shorter than 18 nucleotides (nts) or longer than 25 nts. An additional FASTQC analysis was performed to make sure that there were no adapter sequences or poor-quality reads in our final data set of clean reads.

Two samples from smoltified fish (T4) and saltwater-adapted fish (T6) were used in mirDeep2 analysis (v.0.0.7) [62] for miRNA discovery as described in Woldemariam et al. [38], but no Atlantic salmon miRNAs other than those already described in the Atlantic salmon miRNAome [38] were discovered. Subsequently, clean reads were aligned to the reference index of all known *Salmo salar* mature miRNAs [38] using STAR aligner software with default parameters except modified with parameter—alignIntronMax 1 (v.2.5.2b) [63]. The output files of STAR alignment (BAM format) were processed further in R-studio using the feature Counts function from Rsubread package (v.1.34.2) to produce matrices [64]. These count tables were used as input in the DESeq2 R package (v.1.24.0) for miRNA differential expression analysis by comparing each of the time points T2, T3, T4, T5 and T6 with T1. DESeq2 performs an internal normalization by estimating the size factor for each sample. The size factor is estimated by first calculating geometric mean for each gene across all samples. The counts for a gene in each sample is then divided by this geometric mean. The median of these ratios in a sample corresponds to the size factor for that sample [65].

All miRNAs with log2 fold-change ≤ -1.0 or ≥ 1.0, Benjamini-Hochberg adjusted p-value ≤ 0.05 and with average normalized read counts > 30 in at least at one comparison were defined as differentially expressed miRNAs (DE-miRNAs). Subsequently, read counts of DE-miRNAs originating from the same precursor were compared. If one showed >10 times more reads than the other mature from the same precursor, this one was assumed to be the biologically active guide miRNA and subsequently used in the in silico target analysis and gene enrichment analysis. Additional unsupervised hierarchical clustering with complete linkage and spearman correlation was performed with DE-miRNA log2-fold changes as input using *hclust* function from the stats package (v.3.6.1) in R. Heatmap2 from R-package gplots (v.3.0.1.1) was used to plot heatmaps of DE-miRNAs grouped by the hierarchical clustering analysis.

2.5. Microarray Analysis

The expression profiling of differentially expressed mRNAs (DE-mRNAs) in the liver of Atlantic salmon was performed at NOFIMA (Ås, Norway) using 44 k DNA oligonucleotide microarray containing 60-mer probes to protein coding genes (Salgeno-2, GPL28080). The oligonucleotide microarray for Atlantic salmon were designed at Nofima and annotated with bioinformatics package STARS [66]. Microarrays were manufactured by Agilent Technologies (Inc., Cedar Creek, TX, USA), and the reagents and equipment were from the same source. One-color hybridization was used, and each sample was analyzed with separate array. Microarray was performed on 28 of the 42 liver samples selected for small-RNA sequencing, representing six time points (T1–T6), with five fish per time point except T5 and T6 where four fish were analyzed.

Total RNA (220 ng) from each sample was used as input for cDNA synthesis, amplification and Cy3 labeling of cRNA using a LowInput QuickAmp Labeling Kit according to the manufacturer's protocol. The labeled/amplified cRNA were purified using Qiagen's RNeasy Mini Kit (QIAGEN group, Hilden, Germany). The quantity and quality of the purified cRNA was assessed by NanoDropTM1000 Spectrophometer (Nanodrop ND-1000, Thermo Fisher Scientific, Wilmington, DE, USA). Cy3-labeled cRNA (1650 ng) was used as input to prepare the hybridization mix for each sample using Gene Expression Hybridization Kit. Slides were hybridized in oven (17 h, 65 C, rotation speed 0.01 g). The hybridized slides were washed and scanned with SureScan Microarray Scanner (Agilent Technologies, Santa Clara, CA, USA). Nofima's bioinformatic package STARS [66] was used for subsequent data processing of mRNA array data. Differential expression analyses were carried out by comparing each of the time points (T2–T6) with T1. The transcripts/mRNAs with log2 fold-changes ≤ -0.80 or ≥ -0.80 and $p < 0.05$ (t-test) were defined as significantly changed and termed differentially expressed mRNAs (DE-mRNAs).

Enrichment analysis of DE-mRNAs was performed as described in Krasnov et al. [67] by comparing the numbers of DE-mRNAs per functional category using GO and STARS annotation data sets. Additional pathway analysis was carried out using KEGG annotation data set. Functional categories or pathways with ratio ≥ 2 and Yates' corrected chi-square ($p \leq 0.05$) were defined as overrepresented in DE-mRNA enrichment analysis.

2.6. In Silico Target Gene Predictions and Enrichment Analysis of Predicted miRNA Targets

Prediction of guide DE-miRNA targets was carried out against the DE-mRNAs. Firstly, to provide 3′UTR sequences for in silico prediction, probe sequences of the 5708 DE-mRNAs on the microarray were utilized to identify matching transcripts in the Atlantic salmon full-length (FL) mRNA transcriptome [42]. Analysis was performed with the BLASTN tool in the BLAST+ package (v.2.9.0+) [68]. Match criteria of 90% sequence identity and 91% query coverage were used to classify an alignment of a probe sequence as identifying its matching transcript in the FL-transcriptome. The set of FL-transcripts that were classified as matches to at least one DE-mRNA was used as input in the MicroSalmon GitHub repository (http://github.com/AndreassenLab/MicroSalmon/ accessed 20 April 2022) [43]. This produced a list of transcripts that were identified both as being DE-mRNAs and predicted as targets of the gDE-miRNAs revealed in this study (FL-targets).

The gene symbols of FL-targets were retrieved from both MicroSalmon [43] and the Universal Protein Resource (UniProt) (https://beta.uniprot.org/ accessed 20 April 2022) using the FL-target transcripts annotation information as input. Gene ontology enrichment analysis was performed using PANTHER Overrepresentation Test (version 16.0) (http://pantherdb.org/ accessed 20 April 2022) [69]. *Homo sapiens* was chosen as reference gene list for the enrichment analysis as this is the most complete functional annotated database while teleost is rather incompletely annotated. The annotation data sets GO biological process complete and Reactome pathways were used to identify enriched biological processes (BP) and gene pathways associated with DE-miRNA targets, respectively. Fold enrichment (FE) ≥ 2 and Fisher's Exact test with False Discovery Rate (FDR) less than 0.05 as calculated by the Benjamini–Hochberg procedure were used as thresholds in the enrichment analysis. Subclasses that were related to the same functional category or pathway category were grouped together in the PANTHER analysis and sorted by most specific subclasses. The grouped outputs also show the related biological processes ranked from general to specific biological processes (see Supplementary Files in Section 3.6).

3. Results

3.1. RNA Library Preparation and Small RNA Sequencing

Total RNA from 42 liver samples collected before, during smoltification and post-SWT was successfully extracted, and all samples were subsequently small-RNA sequenced. The number of raw reads obtained from small-RNA sequencing of liver samples (n = 42) ranged from 5.7 to 12.3 million. The quality-filtered (Phred score >32), adapter-trimmed and size-filtered reads for each sample ranged from 3.3 million to 8.4 million. The clean reads uniquely mapped as mature *Salmo salar* miRNAs in each sample ranged from 49.9 to 79.6%. An overview of all samples including their RNA concentration, RIN value, read numbers, reads uniquely mapped as mature miRNAs and the SRA accession numbers are given in Table S1.

3.2. miRNAs with Differential Expression Changes in Liver during Smoltification and Post SWT

To identify DE-miRNAs, we compared the miRNA expression before smoltification (parr) (T1) against the ongoing smoltification period (T2, T3), smolt (T4) and samples from post-SWT (T5 and T6). An overview of conditions at different sampling points is given in Table 1. A total of 88 miRNAs (Benjamini–Hochberg adjusted p-value ≤ 0.05, log2 fold-change ≤ -1.0 or ≥ 1.0 and average normalized read counts >30) belonging to 47 miRNA families were differentially expressed relative to T1 on at least one of the five time points. The relative expression changes of the 88 DE-miRNAs at each timepoint and the mature miRNA sequences of each of the DE-miRNAs is given in Table S2. A heatmap illustrating changes in expression pattern of all 88 DE-miRNAs at all time points is given in Figure S1.

There were 62 among these 88 that were identified as guide DE-miRNAs (gDE-miRNAs), meaning that they were the highly abundant mature miRNA of the two miRNAs originated from the same precursor or both mature miRNAs if present in similar amounts (see definition in introduction and methods). Read counts of the gDE-miRNAs are given in Table S3. The annotated gDE-miRNAs in this study agreed with the previously characterized differences in abundance between miRNAs from same precursor [38,41]. Hierarchical clustering analysis of these 62 gDE-miRNAs revealed 3 major clusters. The results are illustrated in the heatmap in Figure 1.

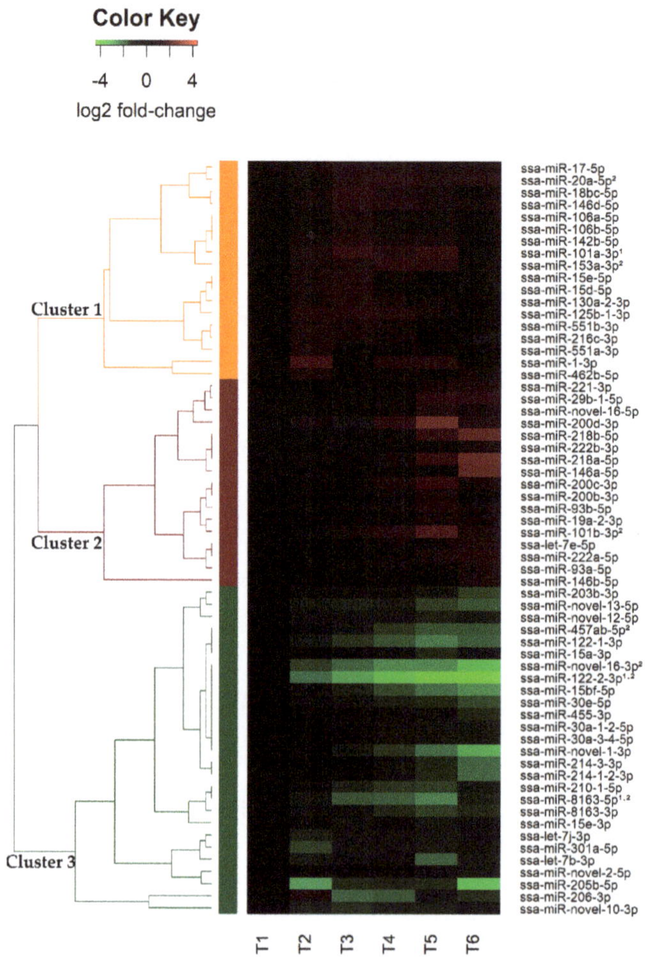

Figure 1. Heatmap and hierarchical clustering of the 62 guide DE-miRNAs. Each row represents a miRNA, and each column represents the expression changes at each time points relative to T1 (pre-smolt, one day before smoltification). T2–T4 and T5–T6 are relative expression changes during smoltification period and post-SWT period, respectively. The dendrogram and the row side colors on the left show the three major clusters of DE-miRNAs (Cluster 1-orange, Cluster 2-red, Cluster 3-green). The direction of expression changes in terms of log2 fold-change is illustrated by the color key above the heatmap. The annotation (1) indicates liver-specific miRNAs and (2) indicates some miRNAs with large changes from T1 to T4 (smoltified fish).

Cluster 1 consisted of 18 gDE-miRNAs belonging to 15 miRNA families (Table 3). This cluster was characterized by larger increases in miRNA expression occurring during smoltification including some gDE-miRNAs with modest changes 1 week post-SWT (Figure 1). The second cluster consisted of 17 gDE-miRNAs from 11 miRNA families including ssa-miR-novel-16 which has so far only been discovered in Atlantic salmon (Table 3). The expression pattern common to cluster 2 gDE-miRNAs was a small upregulation of their expression during smoltification (T2–T4) that continued in a larger increase and peaked post-SWT (T5–T6). The gDE-miRNAs included in cluster 3 were those characterized by a decrease in their expression relative to T1. The expression of some miRNAs in this cluster decreased gradually from T1 to T6 while others decreased at later time points (Figure 1).

Common to all these, except for ssa-miR-206-3p, were that the downregulation peaked after SWT. There were 27 miRNAs in this cluster belonging to 19 miRNA families. Notably, six of these miRNAs have only been discovered in Atlantic salmon (ssa-miR-novel-1, ssa-miR-novel-2-5p, ssa-miR-novel-10-3p, ssa-miR-novel-12-5p, ssa-miR-novel-13-5p and ssa-miR-novel-16-3p) (Table 3).

Table 3. Overview of differentially expressed guide miRNAs in cluster 1, 2 and 3.

Cluster 1.	Cluster 2	Cluster 3
ssa-miR-1-3p	ssa-let-7e-5p	ssa-let-7b-3p
ssa-miR-15d-5p	ssa-miR-19a-2-3p	ssa-let-7j-3p
ssa-miR-15e-5p	ssa-miR-29b-1-5p	ssa-miR-15a-3p
ssa-miR-17-5p	ssa-miR-93a-5p	ssa-miR-15bf-5p
ssa-miR-18bc-5p	ssa-miR-93b-5p	ssa-miR-15e-3p
ssa-miR-20a-5p [2]	ssa-miR-101b-3p [2]	ssa-miR-30a-1-2-5p
ssa-miR-101a-3p [1]	ssa-miR-146a-5p	ssa-miR-30a-3-4-5p
ssa-miR-106a-5p	ssa-miR-146b-5p	ssa-miR-30e-5p
ssa-miR-106b-5p	ssa-miR-200b-3p	ssa-miR-122-1-3p
ssa-miR-125b-1-3p	ssa-miR-200c-3p	ssa-miR-122-2-3p [1,2]
ssa-miR-130a-2-3p	ssa-miR-200d-3p	ssa-miR-203b-3p
ssa-miR-142b-5p	ssa-miR-218a-5p	ssa-miR-205b-5p
ssa-miR-146d-5p	ssa-miR-218b-5p	ssa-miR-206-3p
ssa-miR-153a-3p [2]	ssa-miR-221-3p	ssa-miR-210-1-5p
ssa-miR-216c-3p	ssa-miR-222a-5p	ssa-miR-214-1-2-3p
ssa-miR-462b-5p	ssa-miR-222b-3p	ssa-miR-214-3-3p
ssa-miR-551a-3p	ssa-miR-novel-16-5p	ssa-miR-301a-5p
ssa-miR-551b-3p		ssa-miR-455-3p
		ssa-miR-457ab-5p [2]
		ssa-miR-8163-3p
		ssa-miR-8163-5p [1,2]
		ssa-miR-novel-1-3p
		ssa-miR-novel-2-5p
		ssa-miR-novel-10-3p
		ssa-miR-novel-12-5p
		ssa-miR-novel-13-5p
		ssa-miR-novel-16-3p [2]

The annotation ([1]) indicates liver-specific miRNAs and ([2]) indicates miRNAs with large changes from T1 to T4.

3.3. *Liver Specific DE-miRNAs, ARM-SHIFT and Potential Biomarker miRNAs*

Three of the DE-miRNAs (ssa-miR-101a-3p, ssa-miR-122-2-3p and ssa-miR-8163-5p, Table 3) are among those highly expressed in the liver of teleost fish and are assumed to have liver-specific functions [38,57]. The expression of the liver-specific ssa-miR-101a-3p (cluster 1, Figure 1) showed an increase during smoltification and 1 week post-SWT followed by a slight decrease in expression 1-month post-SWT. The liver-specific ssa-miR-122-2-3p (cluster 3, Figure 1), on the other hand, showed a decrease in expression during smoltification followed by an even larger decrease in expression after SWT. The liver-enriched miRNA, ssa-miR-8163-5p, also belonged to cluster 3. This miRNA was characterized by a decrease in expression but differed from ssa-miR-122-2-3p by peaking 1 week post-SWT (T5, Figure 1).

Ssa-miR-novel-16-5p increased in expression during smoltification and post-SWT while ssa-miR-novel-16-3p decreased in its expression across all time points. Interestingly, the mature 5p and 3p showed a significantly inverse correlation relationship from T1 to T6 with Spearman's rho coefficient of -0.83 and $p = 0.04$. Taken together, this revealed that there was a change of arm dominance of the mature miRNAs processed from the miR-novel-16 precursor. The change was from 3p being the major expressed mature miRNA at pre-smolt stage (433 normalized read counts at T1, 69 normalized read counts at T6) to 5p being the major expressed miRNA post-SWT (85 normalized read counts at T1,

197 normalized read counts at T6). Such changes are referred to as arm shifts. In this case, the arm shift was occurring gradually over the experimental timepoints measured.

The expression changes of some gDE-miRNAs (miR-20a-5p, miR-153a-3p, miR-101b-3p, miR-457ab-5p, miR-novel-16-3p, miR-122-2-3p and miR-8163-5p) were rather large from T1 to T4 (smoltified fish, 1 day prior to SWT) (Figure 1), indicating potential biomarkers for identifying smoltified fish. On the other hand, eight miRNAs (miR-140-5p, miR-107-3p, miR-103-3p, miR-130a-3p, miR-148a-3p, miR-199a-3p, miR-22a-3p and miR-21a-5p) showed stable expression across all time points. These eight miRNAs would be suitable reference miRNAs in RT-qPCR analysis of miRNAs changing expression over this developmental transition. Three of these miRNAs miR-183, miR-140 and miR-107 have previously been validated as stable miRNAs suitable as references in RT-qPCR analysis [57].

3.4. Identification of DE-mRNAs and Enrichment Analysis of DE-mRNAs

The same liver samples and time-points used for characterization of miRNA expression were used in the microarray analysis. This analysis revealed 5708 mRNAs that were differentially expressed relative to T1 on at least one of the time points compared. The expression changes of all DE-mRNAs over the experimental period and p-values are provided in Table S4.

To identify whether these DE-mRNAs were associated with particular biological processes or gene pathways, we performed enrichment analysis. STARS and GO annotation data sets were used for finding enriched functional categories and KEGG was used for gene pathways enrichment analysis. Enriched biological categories and pathways with a ratio ≥ 2 and $p \leq 0.05$ are given in Table S5. Several biological processes are associated with smoltification in liver, including glycogen metabolic process [16], fatty acid metabolism, lactate regulation [19], stress response [67], mitochondrial matrix and energy metabolism [18] were overrepresented. Immune-related functional groups have previously been reported in head kidney [7,58]. Additionally, in the liver samples investigated here, there were several immune-related functional groups that were enriched (immune antigen presentation and immune complement).

3.5. In Silico Prediction Revealed 2804 gDE-miRNA Target Genes

A common approach to understanding the regulatory role of the miRNAs showing differential expression is to predict their target genes. We, therefore, carried out in silico target prediction with the 62 gDE-miRNAs (Figure 1 and Table 3) against all the 5708 DE-mRNAs identified by microarray analysis (Section 3.4).

The probe sequences from the microarray were used to identify their matching FL-transcripts (described in Section 2.6). Applying the 3′UTRs from these transcripts as input showed that there were DE-mRNAs from 4188 unique loci that were predicted as targets of the gDE-miRNAs. The number of genes were further reduced, as many of the genes from the 4188 loci were duplicates (isoforms or paralogs). The final target gene set consisted of 2804 different genes. A complete overview of predicted mRNAs targets along with their targeting miRNAs, gene names and symbols, transcript accession numbers and GO terms are given in Table S6.

3.6. Enriched Biological Processes and Pathways Associated with the Predicted Target Genes

Enrichment analysis was carried out to determine enriched biological processes and gene pathways associated with the 2804 predicted target genes (Table S7). The specific subclasses of enriched biological processes are shown in Figure 2 while Figure 3 shows the enriched gene pathways. The complete outputs from gene ontology (GO) enrichment analysis with term biological process (BP) and gene pathways enrichment analysis are provided in Tables S8 and S9, respectively.

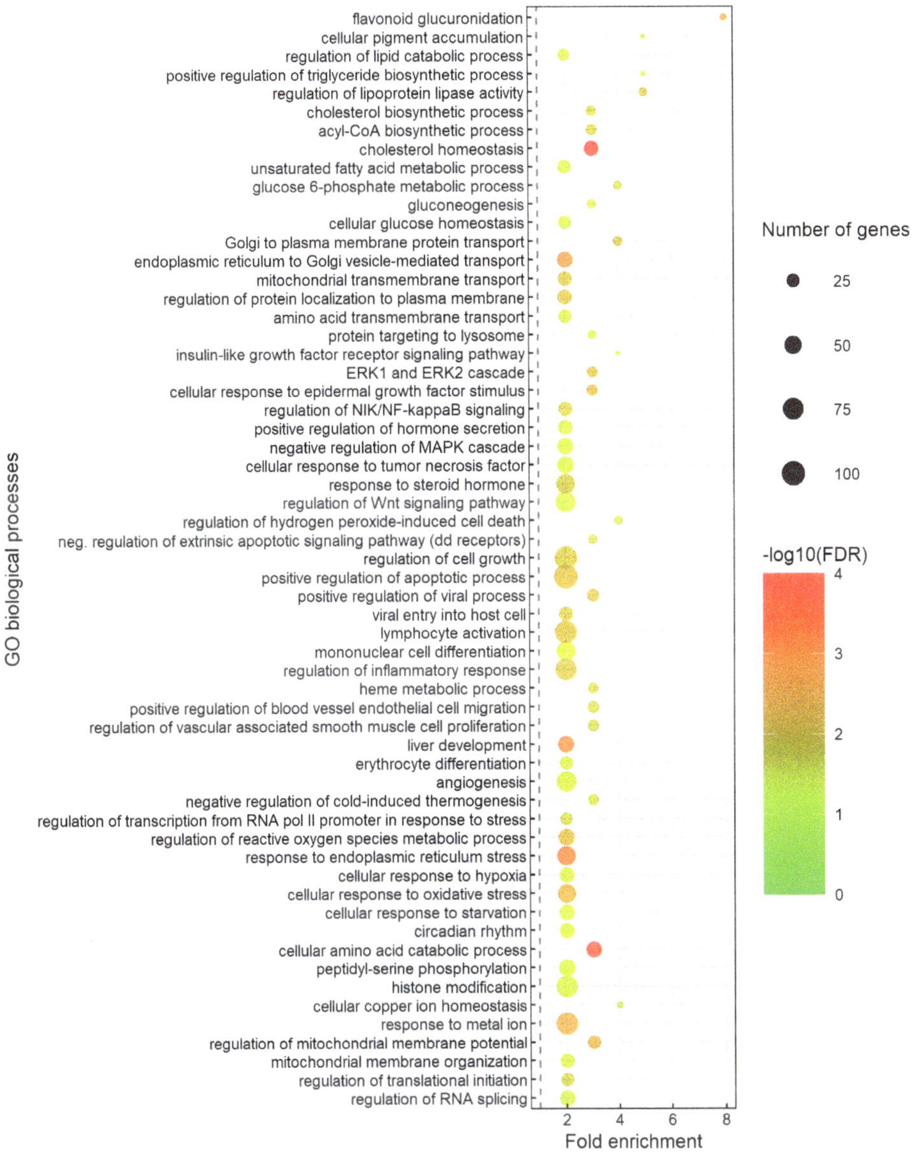

Figure 2. Significantly overrepresented biological processes in the predicted target genes of gDE-miRNAs dataset. The specific and representative subclasses of biological processes are shown here while the complete results from analysis of enriched biological processes are given in Table S8. The dot size indicates the number of DE-miRNA target genes associated with the process and the dot color indicates the significance of the enrichment (−log10 (FDR-corrected P-values)). The vertical grey dashed line represents a fold enrichment of 1.

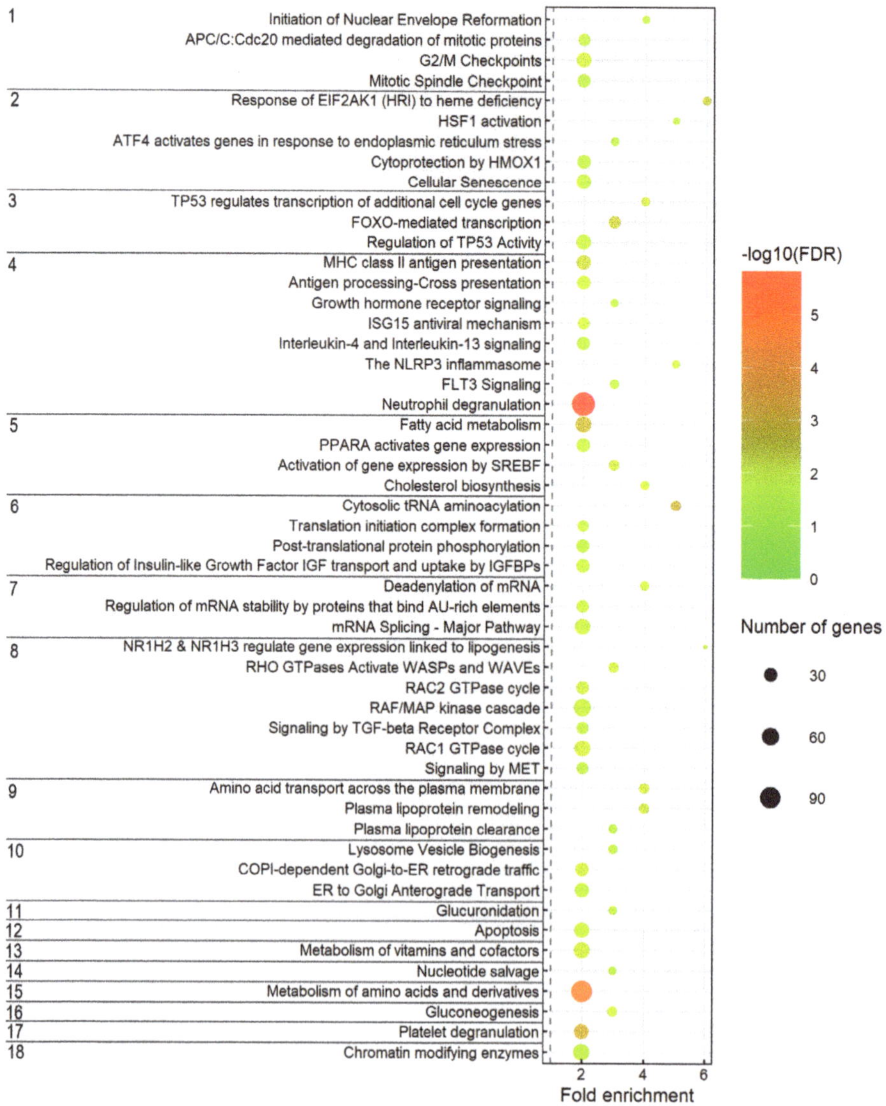

Figure 3. Enriched gene pathways associated with predicted target genes. The specific and representative subclasses of pathways are shown on the y-axis while the complete results from the analysis of enriched Reactome pathways are given in Table S9. The enriched pathways were grouped into categories and each category were assigned a number where each number represents the following categories, 1. Cell cycle, 2. Cellular responses to stress, 3. Gene expression (Transcription), 4. Immune system, 5. Metabolism of lipids, 6. Metabolism of proteins, 7. Metabolism of RNA, 8. Signal Transduction, 9. Transport of small molecules, 10. Vesicle-mediated transport, 11. Biological oxidations, 12. Programmed cell death, 13. Metabolism of vitamins and cofactors, 14. Metabolism of nucleotides, 15. Metabolism of amino acids and derivatives, 16. Metabolism of carbohydrates, 17. Hemostasis, 18. Chromatin organization). The dot size indicates the number of DE-miRNA target genes associated with the pathway and the dot color indicates the significance of the enrichment (−log10 (FDR-corrected p-values). The vertical grey dashed line represents a fold enrichment of 1.

The enriched biological processes included carbohydrate biosynthesis, amino acid metabolism, lipid biosynthesis, steroid metabolic process (cholesterol biosynthesis), apoptotic process and protein transport (Figure 2). Biological processes related to response to environmental stimuli that are associated with the developmental changes during smoltification were also enriched. This included response to stress, external stimuli (circadian rhythm) and organ developmental processes (liver development and angiogenesis).

The gene pathways analysis (Figure 3), also revealed enriched pathways associated with those revealed in the enrichment analysis of biological processes. Interestingly, several pathways related to immune-system-like MHC class II antigen presentation and antigen-processing–cross presentation, immune system by Cytokine signaling-like growth hormone receptor signaling, ISG15 antiviral mechanism, Interleukin-4 and Interleukin-13 signaling including the NLRP3 inflammasome, FLT3 signaling and Neutrophil degranulation were also revealed in the gene pathways enrichment analysis (Figure 3). The other significantly enriched pathways were mainly related to cell cycle, cellular responses to stress, transcription, signal transduction, small molecules and vesicle-mediated transport, biological oxidation, cell death, hemostasis and metabolism (lipids, proteins, RNA, nucleotides, amino acids, derivates, vitamins and cofactors).

4. Discussion

4.1. Differential Expression of 62 Guide miRNAs Indicates That They Are Involved in Post-Transcriptional Gene Regulation during Smoltification and Seawater Adaptation

The differential expression analysis of miRNAs investigated here allowed us to identify 88 miRNAs that changed their expression during smoltification. Sixty-two of them were annotated as the biologically important guide miRNAs (Table 3) by examining their level of average normalized read counts compared to their corresponding mature miRNAs originated from the same precursor (Table S3). In a few cases, DE-miRNAs from the same precursor (e.g., ssa-miR-8163-3p and ssa-miR-8163-5p, Table S3) revealed similar mature miRNA read counts. In cases like this, both mature miRNAs were regarded as biologically important gDE-miRNAs. The hierarchical analysis of the 62 gDE-miRNAs also revealed that DE-miRNAs in the same family showed similar expression pattern. Consequently, same family DE-miRNAs usually clustered together as either upregulated or downregulated. This is also consistent with expected results for mature miRNAs from same family having the same regulatory function [70].

Among the gDE-miRNAs, there were three miRNA families (miRNA-101, miRNA-122 and miRNA-8163) that have previously been reported as liver-specific in teleost [38,57]. Two of these, miR-101a-3p and ssa-miR-122-2-3p, were predicted to target genes involved in fatty acid metabolism (e.g., FAM3A-like protein, non-specific lipid-transfer protein and diacylglycerol O-acyltransferase 2-like) and xenobiotic glucuronidation (e.g., UDP-glucuronosyltransferase-like isoform and multidrug resistance-associated protein 4-like), respectively. The third one, miR-8163-5p, was predicted to target genes that are associated with liver regeneration (e.g., hepatocyte growth factor-like protein and hepatocyte nuclear factor 3-beta-like) [71]. Another predicted target gene of miR-8163-5p was insulin-like growth factor-binding protein 1 (IGFBP-1), a regulator that is mainly produced in liver [72]. IGFBP-1 is known to bind to insulin-like growth factor (IGF-I) [73], an important regulator for smoltification [26]. In addition, fatty acid synthase was also a predicted target of miR-8163-5p. This is a housekeeping protein that is involved in production of fat for storage when nutrients are present in excess in liver [74]. Altogether, these observations were in agreement with these three liver-specific miRNAs being involved in regulation of biological processes associated with liver-specific functions affected by smoltification and adaptation to seawater.

When comparing results from differential expression analysis of miRNAs in head kidney (HK) [58] and liver (this study) sampled from the same experimental fish group, there were 12 miRNAs (ssa-let-7b-3p, ssa-miR-1-3p, ssa-miR-29b-1-5p, ssa-miR-125b-1-3p, ssa-miR-146a-5p, ssa-miR-153a-3p, ssa-miR-200b-3p, ssa-miR-205b-5p, ssa-miR-218a-5p, ssa-miR-301a-5p, ssa-miR-novel-2-5p and ssa-miR-novel-12-5p) that were gDE-miRNAs in both head kidney and liver. Four of these miRNAs (ssa-miR-29b-1-5p, ssa-miR-146a-5p, ssa-miR-153a-3p and ssa-miR-218a-5p) showed an increasing expression during smoltification and SWT while six miRNAs (ssa-miR-1-3p, ssa-let-7b-3p, ssa-miR-125b-1-3p, ssa-miR-205b-5p, ssa-miR-novel-2-5p and ssa-miR-novel-12-5p) showed a decreasing expression during smoltification and SWT. The fact that they revealed same expression profiles in different organs indicates that they are involved in regulation of basic cellular homeostasis, most likely not associated with organ specific functions in kidney or liver. The two remaining miRNAs (ssa-miR-200b-3p and ssa-miR-301a-5p) revealed differences in expression profiles between liver and HK, showing an increased expression level in liver and a decreased expression level in HK.

One DE-miRNA, miR-novel-16 revealed an arm-shift gradually changing from a dominant 3p at pre-smolt stage to a dominant 5p at post-SWT stage. miRNA arm shifting is a change in the preferential enrichment of 5′ or 3′ mature miRNA from a precursor. This has previously been reported as highly dynamic, tissue-specific and could also be time-dependent [75,76]. This finding indicates that both miR-novel-16-3p and miR-novel-16-5p were biologically functional, but the preferences of 3p or 5p as guide miRNAs depended on the developmental stage of the fish.

Seven miRNAs showed rather large changes from T1 (parr) to T4 (smolt) (miR-20a-5p, miR-153a-3p, miR-101b-3p, miR-457ab-5p, miR-novel-16-3p, miR-122-2-3p and miR-8163-5p). Three of these miRNAs (miR-20a-5p, miR-8163-5p and miR-novel-16-3p) predicted to target genes (Na^+,K^+-ATPase(NKA) α- and β-subunit isoforms) that are often considered as indicator of smolt development in gill [24]. These gDE-miRNAs with large expression changes from T1 (parr) to T4 (smolt) point themselves out as potential novel biomarkers for assessing smoltification status in Atlantic salmon. An RT-qPCR assay applying these as biomarkers would require reference miRNAs that are stably expressed. Eight miRNAs (miR-140-5p, miR-107-3p, miR-103-3p, miR-130a-3p, miR-148a-3p, miR-199a-3p, miR-22a-3p and miR-21a-5p) could fit as such reference genes since their expression were stable across all the six time points. In particular, miR-183, miR-140 and miR-107 are good candidates, as they previously have been validated as stable and worked well (efficiency and specificity) in RT-qPCR assays [57]. Although appearing as promising biomarkers, they need to be further validated experimentally. In summary, three different expression profiles of gDE-miRNAs identified here show that some miRNAs expressed in liver are associated with smoltification and seawater adaptation.

4.2. Enriched Biological Processes and Pathways Associated with Predicted DE-miRNA Targets

The enriched biological categories observed in the microarray analysis were, in general, in agreement with findings observed in previous studies of differentially expressed genes [7,58] and physiological and biochemical changes [16,18,19] associated with smoltification. The DE-mRNAs discovered in the microarray analysis were used as input in the in silico target prediction analysis. The predictions were carried out in this manner to make the predictions relevant to genes important in this developmental transition. Aiming to further minimize false positives, only predicted target genes supported by RNAhybrid and at least two of the prediction tools (PITA, miRanda, TargetSpy) were included as putative target genes [43]. Applying this strategy, the in silico target predictions revealed 2804 unique genes as potential targets.

Enrichment analysis is an approach to better understand the general biological processes that are affected when a gene set shows differential expression in the condition studied. Applying this approach using the predicted miRNA targets could indicate if particular biological processes and gene pathways were likely controlled by miRNA guided post-transcriptional regulation. Therefore, gene enrichment analysis was performed with the predicted targets. This analysis revealed many enriched biological processes and gene pathways that were previously reported as smoltification related. Functional groups of lipid homeostasis, lipid synthesis and cholesterol biosynthesis (Figures 2 and 3) were among those observed as enriched. In addition, regulation of lipoprotein lipase activity and plasma lipoprotein remodeling was also enriched. Enriched lipid metabolism related biological processes are in agreement with previous studies reporting that lipid synthesis and liver glycogen (a major storage form of energy) decrease during smoltification [13,16]. Changes in lipid metabolism may affect tissue fatty acid compositions as previous study suggested that fatty acid compositions changed as a pre-adaptive response to seawater entry [77,78]. Functional groups related to energy metabolism including amino acid catabolic process, fatty acid metabolic process and acyl-CoA biosynthetic process (Figures 2 and 3) were also enriched. Acyl-CoA enters the citric acid cycle in the mitochondrion by combining with oxaloacetate to form citrate catalyzed by citrate synthase [16,79]. Subsequent oxidation of citrate provides the major source of cellular ATP production which is essential in the smoltification as it is a highly energy-demanding process [80]. The enrichment of the functional groups related to energy metabolism observed shows that miRNAs may fine tune the genes important in energy metabolism which is essential to success in this developmental transition [16,18,20].

Other enriched biological processes and pathways that were specifically related to smoltification were metabolism of proteins and amino acids (e.g., cytosolic tRNA aminoacylation, cellular amino acid catabolic process, metabolism of amino acids and derivatives). The contents of free amino acids have previously been reported to reorganize during smoltification. This was suggested to be linked to changes in cellular osmoregulatory gradients [17]. Glucuronidation which is an important detoxification process occurring in fish liver [59] was also enriched, indicating that miRNAs may involve in regulation of detoxification in liver during smoltification.

Another interesting enriched biological process was circadian rhythm, an endogenous time-keeping system that controls and coordinates metabolism, physiology and behavior to adapt to variations within the day and the seasonal environmental cycles [81]. This indicates that miRNAs may also regulate genes involved in circadian rhythm which is likely to be affected by changes in photoperiod during smoltification. Circadian rhythm is important for smoltification as they translate changes in daylength into physiological response [26].

Various signaling pathways related to signal transduction, processes that regulate overall growth and cellular proliferation, differentiation and survival [82] were also enriched among miRNA targets. The enriched signaling pathways observed here, e.g., signaling by TGF-beta receptor complex, NR1H2 and NR1H3 regulate gene expression linked to lipogenesis and signaling by MET, may play a crucial role in affecting cellular metabolism during smoltification and in the adaptation to seawater.

The results showed that gene pathways associated with immune system were enriched among the gDE-miRNA target genes. This indicates that miRNAs may be involved in the regulation of immune system changes during smoltification and adaptation to seawater. It has also been reported that decreased immune defenses to pathogens was correlated with the high energetic cost due to smoltification related changes [7,83]. Both of these biological processes are associated with the gDE-miRNAs by their targets being enriched in these groups. Biological processes related to stress response (e.g., cellular response to oxidative stress, response to endoplasmic reticulum stress and regulation of transcription from RNA pol II promoter in response to stress) as well as liver development and angiogenesis were enriched. Together, this shows that miRNAs may influence genes important in stress response and liver development associated with smoltification and adaptation to seawater.

Interestingly, regulation of Insulin-like Growth Factor (IGF) transport and uptake by Insulin-like Growth Factor Binding Proteins (IGFBPs) was also enriched. Previous studies reported that growth hormone (GH) increased during smoltification [84] and GH stimulates IGF-1 and IGF-2, leading to modulation of intermediary metabolism and osmoregulatory mechanism in fish [85,86]. Furthermore, the interaction of cortisol with GH, IGF-1 and corticosteroid receptors promotes salinity tolerance, changes in growth and metabolism [26]. The enrichment of these groups indicates that miRNAs may be involved in the IGF-associated changes occurring during smoltification.

5. Conclusions

In conclusion, enrichment analysis of the predicted target genes revealed that they were significantly enriched in important biological processes groups and pathways associated with smoltification and adaptation to seawater. Whether any of the gDE-miRNAs are themselves among molecular initiators of the smoltification process or they regulate the downstream cellular processes triggered by this developmental transition, is uncertain. Nevertheless, this approach allowed us to identify miRNAs associated with smoltification and predict high confident target genes. The findings in this study pave the way for validation of the predicted target genes and further studies of such miRNA–mRNA interactions by experimental assays.

Supplementary Materials: The following are available online at https://www.mdpi.com/article/10.3390/biology11050688/s1, Figure S1: Heatmap of 88 DE-miRNAs, Table S1: Overview of the 42 experimental liver samples, Table S2: DE-miRNAs with their family, mature sequences and relative expression values, Table S3: Complete overview of DE-miRNAs and their corresponding 3p or 5p mature sequences, Table S4: Overview of differentially expressed mRNAs, Table S5: Overview of enriched DE-mRNAs, Table S6: List of all predicted targets of guide DE-miRNAs, Table S7: List of unique gene symbols from DE-miRNAs predicted target genes, Table S8: List of enriched GO biological processes with fold enrichment over 2, Table S9: List of enriched pathways with fold enrichment over 2.

Author Contributions: Conceptualization, R.A. and T.-K.K.Ø.; methodology, A.S., R.A. and T.-K.K.Ø.; software, A.S., A.K., T.V. and T.-K.K.Ø.; and S.R.; validation, A.S. and A.K.; formal analysis, A.S., A.K., T.V. and S.R.; investigation, A.S.; resources, R.A., T.-K.K.Ø. and A.K.; data curation, A.S.; writing—original draft preparation, A.S.; writing—review and editing, A.S., R.A., T.-K.K.Ø. and A.K.; visualization, A.S.; supervision, R.A. and T.-K.K.Ø.; project administration, R.A.; funding acquisition, R.A. All authors have read and agreed to the published version of the manuscript.

Funding: This research was funded by THE RESEARCH COUNCIL OF NORWAY, grant number 280839/E40.

Institutional Review Board Statement: The fish material which this manuscript is based on originated form the Nofima Research station for sustainable aquaculture in Sunndalsora Norway. These research facilities are licensed under Norwegian law to do experimental work on fish. The regulation in question is FOR-2015-06-18-761 «Forskrift om bruk av dyr i forsok», and is a national adaptation of (and in compliance with) the "Directive 2010/63/EU of the European Parliament an of the Council on the protection of animals used for scientific purposes". The fish were euthanized before any sampling of tissue. No experimental procedures were applied in vivo either to the fish group or to the sampled individuals. Thus, the activities are covered by the exemption given in the abovementioned regulation, §6 section 1, which states that a specific license is not required if the experiment only include euthanasia followed by harvest of tissues or other materials from the animals.

Data Availability Statement: All sequenced samples have been submitted to the NCBI Sequence Read Archive Centre (SRA) (https://www.ncbi.nlm.nih.gov/sra accessed 20 April 2022) with accession bioproject number PRJNA665200.

Acknowledgments: The authors would like to thank staff at the Sunndalsøra Research Station (Sunndalsøra, Møre og Romsdal, Norway) for their assistance with fish husbandry and sampling. We are also grateful to Marianne Helén Selander Hansen (NOFIMA, Ås) for her assistance with microarray analysis.

Conflicts of Interest: The funders had no role in the design of the study; in the collection, analyses, or interpretation of data; in the writing of the manuscript, or in the decision to publish the results. All authors declare no conflict of interest.

References

1. Julien, A.; Hlynur, B.; Ida, A.-B.; Geir, H.B.; Cindy, B.; Mathieu, B.; Karin, C.; Gerald, C.; Anne, C.; Guillaume, D. *Working Group on North Atlantic Salmon (WGNAS)*; ICES Publication: Copenhagen V, Denmark, 2021; p. 417.
2. Asche, F. Farming the Sea. *Mar. Resour. Econ.* **2008**, *23*, 527–547. [CrossRef]
3. Morera, F.J.; Castro-Guarda, M.; Nualart, D.; Espinosa, G.; Muñoz, J.L.; Vargas-Chacoff, L. The biological basis of smoltification in Atlantic salmon. *Austral J. Vet. Sci.* **2021**, *53*, 73–82. [CrossRef]
4. SERNAPESCA. *Informe Sanitario de Salmonicultura en Centros Marinos 1er Semestre año 2019*; Ministerio de Economía, Fomento y Turismo: Santiago, Chile, 2019.
5. Jensen, B.B.; Lillehaug, A.; Oliveria, V.; Jansen, M.D.; Nilsen, A.; Gismervik, K.; Gåsnes, S.K.; Nielsen, K.V.; Mejdell, C.M.; Sindre, H.; et al. *The Health Situation in Norwegian Aquaculture 2019*; Veterinærinstituttet/Norwegian Veterinary Institute 2020: Oslo, Norway, 2019.
6. Boulet, M.; Normandeau, É.; Bougas, B.; Audet, C.; Bernatchez, L. Comparative transcriptomics of anadromous and resident brook charr Salvelinus fontinalis before their first salt water transition. *Curr. Zool.* **2012**, *58*, 158–170. [CrossRef]
7. Johansson, L.-H.; Timmerhaus, G.; Afanasyev, S.; Jørgensen, S.M.; Krasnov, A. Smoltification and seawater transfer of Atlantic salmon (Salmo salar L.) is associated with systemic repression of the immune transcriptome. *Fish Shellfish. Immunol.* **2016**, *58*, 33–41. [CrossRef] [PubMed]
8. Seear, P.J.; Carmichael, S.N.; Talbot, R.; Taggart, J.B.; Bron, J.E.; Sweeney, G.E. Differential gene expression during smoltification of Atlantic salmon (Salmo salar L.): A first large-scale microarray study. *Mar. Biotechnol.* **2010**, *12*, 126–140. [CrossRef] [PubMed]
9. Maryoung, L.A.; Lavado, R.; Bammler, T.; Gallagher, E.; Stapleton, P.; Beyer, R.; Farin, F.; Hardiman, G.; Schlenk, D. Differential Gene Expression in Liver, Gill, and Olfactory Rosettes of Coho Salmon (Oncorhynchus kisutch) After Acclimation to Salinity. *Mar. Biotechnol.* **2015**, *17*, 703–717. [CrossRef] [PubMed]
10. Hoar, W.S. 4 The Physiology of Smolting Salmonids. In *Fish Physiology*; Hoar, W.S., Randall, D.J., Eds.; Academic Press: Cambridge, MA, USA, 1988; Volume 11, pp. 275–343.
11. Björnsson, B.T.; Bradley, T.M. Epilogue: Past successes, present misconceptions and future milestones in salmon smoltification research. *Aquaculture* **2007**, *273*, 384–391. [CrossRef]
12. McCormick, S.D.; Sheehan, T.F.; Björnsson, B.T.; Lipsky, C.; Kocik, J.F.; Regish, A.M.; O'Dea, M.F. Physiological and endocrine changes in Atlantic salmon smolts during hatchery rearing, downstream migration, and ocean entry. *Can. J. Fish. Aquat. Sci.* **2013**, *70*, 105–118. [CrossRef]
13. Sheridan, M.A. Alterations in lipid metabolism accompanying smoltification and seawater adaptation of salmonid fish. *Aquaculture* **1989**, *82*, 191–203. [CrossRef]
14. McCormick, S.D.; Hansen, L.P.; Quinn, T.P.; Saunders, R.L. Movement, migration, and smolting of Atlantic salmon (Salmo salar). *Can. J. Fish. Aquat. Sci.* **1998**, *55*, 77–92. [CrossRef]
15. Björnsson, B.T.; Stefansson, S.O.; McCormick, S.D. Environmental endocrinology of salmon smoltification. *Gen. Comp. Endocrinol.* **2011**, *170*, 290–298. [CrossRef] [PubMed]
16. Stubhaug, I.; Lie, Ø.; Torstensen, B.E. β-Oxidation capacity in liver increases during parr-smolt transformation of Atlantic salmon fed vegetable oil and fish oil. *J. Fish Biol.* **2006**, *69*, 504–517. [CrossRef]
17. Finn, R.N.; Kapoor, B.G. *Fish Larval Physiology*; Taylor & Francis Group: Enfield, IL, USA, 2008.
18. McCormick, S.D.; Saunders, R.L.; MacIntyre, A.D. Mitochondrial enzyme and Na+, K+-ATPase activity, and ion regulation during parr-smolt transformation of Atlantic salmon (Salmon salar). *Fish Physiol. Biochem.* **1989**, *6*, 231–241. [CrossRef] [PubMed]
19. Leonard, J.B.K.; McCormick, S.D. Metabolic enzyme activity during smolting in stream- and hatchery-reared Atlantic salmon (Salmo salar). *Can. J. Fish. Aquat. Sci.* **2001**, *58*, 1585. [CrossRef]
20. Mizuno, S.; Urabe, H.; Aoyama, T.; Omori, H.; Iijima, A.; Kasugai, K.; Torao, M.; Misaka, N.; Koide, N.; Ueda, H. Changes in activity and transcript level of liver and gill metabolic enzymes during smoltification in wild and hatchery-reared masu salmon (Oncorhynchus masou). *Aquaculture* **2012**, *362–363*, 109–120. [CrossRef]
21. McCormick, S.D.; Moyes, C.D.; Ballantyne, J.S. Influence of salinity on the energetics of gill and kidney of Atlantic salmon (Salmo salar). *Fish Physiol. Biochem.* **1989**, *6*, 243–254. [CrossRef]
22. Iversen, M.; Mulugeta, T.; Gellein Blikeng, B.; West, A.C.; Jørgensen, E.H.; Rød Sandven, S.; Hazlerigg, D. RNA profiling identifies novel, photoperiod-history dependent markers associated with enhanced saltwater performance in juvenile Atlantic salmon. *PLoS ONE* **2020**, *15*, e0227496. [CrossRef]
23. McCormick, S.D.; Shrimpton, J.M.; Moriyama, S.; Björnsson, B.T. Differential hormonal responses of Atlantic salmon parr and smolt to increased daylength: A possible developmental basis for smolting. *Aquaculture* **2007**, *273*, 337–344. [CrossRef]
24. Craig Clarke, W.; Saunders, R.L.; McCormick, S.D. Chapter 8—Smolt Production. In *Developments in Aquaculture and Fisheries Science*; Pennell, W., Barton, B.A., Eds.; Elsevier: Edinburgh, UK, 1996; Volume 29, pp. 517–567.
25. De Oliveira, V.H.S.; Nilsen, A.; Jensen, B.B.; Sommerset, I.; Jansen, M.D.; Brun, E.; Grismervik, K.; Gåsnes, S.K.; Nielsen, K.V.; Mejdell, C.M.; et al. *The Health Situation in Norwegian Aquaculture 2020*; Norwegian Veterinary Institute: Ås, Norway, 2020.
26. McCormick, S.D. 5—Smolt Physiology and Endocrinology. In *Fish Physiology*; McCormick, S.D., Farrell, A.P., Brauner, C.J., Eds.; Academic Press: Cambridge, MA, USA, 2012; Volume 32, pp. 199–251.
27. McCormick, S.D.; Regish, A.M.; Christensen, A.K. Distinct freshwater and seawater isoforms of Na+/K+-ATPase in gill chloride cells of Atlantic salmon. *J. Exp. Biol.* **2009**, *212*, 3994–4001. [CrossRef]

28. Nilsen, T.O.; Ebbesson, L.O.E.; Madsen, S.S.; McCormick, S.D.; Andersson, E.; Björnsson, B.T.; Prunet, P.; Stefansson, S.O. Differential expression of gill Na+,K+-ATPaseα- and β-subunits, Na+,K+,2Cl-cotransporter and CFTR anion channel in juvenile anadromous and landlocked Atlantic salmon Salmo salar. *J. Exp. Biol.* **2007**, *210*, 2885–2896. [CrossRef]
29. Jensen, A.J.; Berg, M.; Bremset, G.; Finstad, B.; Hvidsten, N.A.; Jensås, J.G.; Johnsen, B.O.; Lund, E. Passing a seawater challenge test is not indicative of hatchery-reared Atlantic salmon Salmo salar smolts performing as well at sea as their naturally produced conspecifics. *J. Fish Biol.* **2016**, *88*, 2219–2235. [CrossRef] [PubMed]
30. Zydlewski, G.B.; Zydlewski, J. Gill Na+,K+-ATPase of Atlantic salmon smolts in freshwater is not a predictor of long-term growth in seawater. *Aquaculture* **2012**, *362–363*, 121–126. [CrossRef]
31. Houde, A.L.S.; Günther, O.P.; Strohm, J.; Ming, T.J.; Li, S.; Kaukinen, K.H.; Patterson, D.A.; Farrell, A.P.; Hinch, S.G.; Miller, K.M. Discovery and validation of candidate smoltification gene expression biomarkers across multiple species and ecotypes of Pacific salmonids. *Conserv. Physiol.* **2019**, *7*, coz051. [CrossRef]
32. Bartel, D.P. MicroRNAs: Genomics, Biogenesis, Mechanism, and Function. *Cell* **2004**, *116*, 281–297. [CrossRef]
33. Chekulaeva, M.; Filipowicz, W. Mechanisms of miRNA-mediated post-transcriptional regulation in animal cells. *Curr. Opin. Cell Biol.* **2009**, *21*, 452–460. [CrossRef] [PubMed]
34. Bartel, D.P. Metazoan MicroRNAs. *Cell* **2018**, *173*, 20–51. [CrossRef]
35. O'Brien, J.; Hayder, H.; Zayed, Y.; Peng, C. Overview of MicroRNA Biogenesis, Mechanisms of Actions, and Circulation. *Front. Endocrinol.* **2018**, *9*, 402. [CrossRef]
36. Bartel, D.P. MicroRNAs: Target Recognition and Regulatory Functions. *Cell* **2009**, *136*, 215–233. [CrossRef]
37. Kobayashi, H.; Tomari, Y. RISC assembly: Coordination between small RNAs and Argonaute proteins. *Biochimica et Biophysica Acta (BBA)-Gene Regul. Mech.* **2016**, *1859*, 71–81. [CrossRef]
38. Woldemariam, N.T.; Agafonov, O.; Høyheim, B.; Houston, R.D.; Taggart, J.B.; Andreassen, R. Expanding the miRNA Repertoire in Atlantic Salmon; Discovery of IsomiRs and miRNAs Highly Expressed in Different Tissues and Developmental Stages. *Cells* **2019**, *8*, 42. [CrossRef]
39. Allendorf, F.W.; Thorgaard, G.H.; Turner, B. *Evolutionary Genetics of Fishes*; Turner, B., Ed.; Springer: Boston, MA, USA, 1984. [CrossRef]
40. Lien, S.; Koop, B.F.; Sandve, S.R.; Miller, J.R.; Kent, M.P.; Nome, T.; Hvidsten, T.R.; Leong, J.S.; Minkley, D.R.; Zimin, A.; et al. The Atlantic salmon genome provides insights into rediploidization. *Nature* **2016**, *533*, 200–205. [CrossRef] [PubMed]
41. Andreassen, R.; Worren, M.M.; Høyheim, B. Discovery and characterization of miRNA genes in Atlantic salmon (Salmo salar) by use of a deep sequencing approach. *BMC Genom.* **2013**, *14*, 482. [CrossRef] [PubMed]
42. Ramberg, S.; Høyheim, B.; Østbye, T.-K.K.; Andreassen, R. A de novo Full-Length mRNA Transcriptome Generated From Hybrid-Corrected PacBio Long-Reads Improves the Transcript Annotation and Identifies Thousands of Novel Splice Variants in Atlantic Salmon. *Front. Genet.* **2021**, *12*, 656334. [CrossRef] [PubMed]
43. Ramberg, S.; Andreassen, R. MicroSalmon: A Comprehensive, Searchable Resource of Predicted MicroRNA Targets and 3'UTR Cis-Regulatory Elements in the Full-Length Sequenced Atlantic Salmon Transcriptome. *Non-Coding RNA* **2021**, *7*, 61. [CrossRef] [PubMed]
44. Chang, J.; Nicolas, E.; Marks, D.; Sander, C.; Lerro, A.; Buendia, M.A.; Xu, C.; Mason, W.S.; Moloshok, T.; Bort, R.; et al. miR-122, a mammalian liver-specific microRNA, is processed from hcr mRNA and may downregulate the high affinity cationic amino acid transporter CAT-1. *RNA Biol.* **2004**, *1*, 106–113. [CrossRef] [PubMed]
45. Lagos-Quintana, M.; Rauhut, R.; Yalcin, A.; Meyer, J.; Lendeckel, W.; Tuschl, T. Identification of tissue-specific microRNAs from mouse. *Curr. Biol.* **2002**, *12*, 735–739. [CrossRef]
46. Shah, N.; Nelson, J.E.; Kowdley, K.V. MicroRNAs in Liver Disease: Bench to Bedside. *J. Clin. Exp. Hepatol.* **2013**, *3*, 231–242. [CrossRef]
47. Roderburg, C.; Trautwein, C. Cell-specific functions of miRNA in the liver. *J. Hepatol.* **2017**, *66*, 655–656. [CrossRef]
48. Girard, M.; Jacquemin, E.; Munnich, A.; Lyonnet, S.; Henrion-Caude, A. miR-122, a paradigm for the role of microRNAs in the liver. *J. Hepatol.* **2008**, *48*, 648–656. [CrossRef]
49. Castoldi, M.; Vujic Spasic, M.; Altamura, S.; Elmén, J.; Lindow, M.; Kiss, J.; Stolte, J.; Sparla, R.; D'Alessandro, L.A.; Klingmüller, U.; et al. The liver-specific microRNA miR-122 controls systemic iron homeostasis in mice. *J. Clin. Investig.* **2011**, *121*, 1386–1396. [CrossRef]
50. Andreassen, R.; Høyheim, B. miRNAs associated with immune response in teleost fish. *Dev. Comp. Immunol.* **2017**, *75*, 77–85. [CrossRef] [PubMed]
51. Andreassen, R.; Woldemariam, N.T.; Egeland, I.Ø.; Agafonov, O.; Sindre, H.; Høyheim, B. Identification of differentially expressed Atlantic salmon miRNAs responding to salmonid alphavirus (SAV) infection. *BMC Genom.* **2017**, *18*, 349. [CrossRef] [PubMed]
52. Smith, N.C.; Christian, S.L.; Woldemariam, N.T.; Clow, K.A.; Rise, M.L.; Andreassen, R. Characterization of miRNAs in Cultured Atlantic Salmon Head Kidney Monocyte-Like and Macrophage-Like Cells. *Int. J. Mol. Sci.* **2020**, *21*, 3989. [CrossRef] [PubMed]
53. Bizuayehu, T.T.; Johansen, S.D.; Puvanendran, V.; Toften, H.; Babiak, I. Temperature during early development has long-term effects on microRNA expression in Atlantic cod. *BMC Genom.* **2015**, *16*, 305. [CrossRef] [PubMed]
54. Kostyniuk, D.J.; Mennigen, J.A. Meta-analysis of differentially-regulated hepatic microRNAs identifies candidate post-transcriptional regulation networks of intermediary metabolism in rainbow trout. *Comp. Biochem. Physiol. Part D Genom. Proteom.* **2020**, *36*, 100750. [CrossRef] [PubMed]

55. Rasal, K.D.; Iquebal, M.A.; Pandey, A.; Behera, P.; Jaiswal, S.; Vasam, M.; Dixit, S.; Raza, M.; Sahoo, L.; Nandi, S.; et al. Revealing liver specific microRNAs linked with carbohydrate metabolism of farmed carp, Labeo rohita (Hamilton, 1822). *Genomics* **2020**, *112*, 32–44. [CrossRef] [PubMed]
56. Østbye, T.K.; Woldemariam, N.T.; Lundberg, C.E.; Berge, G.M.; Ruyter, B.; Andreassen, R. Modulation of hepatic miRNA expression in Atlantic salmon (Salmo salar) by family background and dietary fatty acid composition. *J. Fish Biol.* **2021**, *98*, 1172–1185. [CrossRef]
57. Andreassen, R.; Rangnes, F.; Sivertsen, M.; Chiang, M.; Tran, M.; Worren, M.M. Discovery of miRNAs and Their Corresponding miRNA Genes in Atlantic Cod (Gadus morhua): Use of Stable miRNAs as Reference Genes Reveals Subgroups of miRNAs That Are Highly Expressed in Particular Organs. *PLoS ONE* **2016**, *11*, e0153324. [CrossRef]
58. Shwe, A.; Østbye, T.-K.K.; Krasnov, A.; Ramberg, S.; Andreassen, R. Characterization of Differentially Expressed miRNAs and Their Predicted Target Transcripts during Smoltification and Adaptation to Seawater in Head Kidney of Atlantic Salmon. *Genes* **2020**, *11*, 1059. [CrossRef]
59. Dutta, H.; Puslé, J.; Anadon, G.G.i. The Structure and Function of Fish Liver. In *Fish Morphology*; Dutta, H., Ed.; Routledge: London, UK, 2017; pp. 77–93. [CrossRef]
60. Carter, C.G.; Houlihan, D.F. Protein synthesis. In *Fish Physiology*; Academic Press: Cambridge, MA, USA, 2001; Volume 20, pp. 31–75.
61. Martin, M.J.E.j. Cutadapt removes adapter sequences from high-throughput sequencing reads. *EMBnet.journal* **2011**, *17*, 10–12. [CrossRef]
62. Friedländer, M.R.; Mackowiak, S.D.; Li, N.; Chen, W.; Rajewsky, N. miRDeep2 accurately identifies known and hundreds of novel microRNA genes in seven animal clades. *Nucleic Acids Res.* **2011**, *40*, 37–52. [CrossRef] [PubMed]
63. Dobin, A.; Davis, C.A.; Schlesinger, F.; Drenkow, J.; Zaleski, C.; Jha, S.; Batut, P.; Chaisson, M.; Gingeras, T.R. STAR: Ultrafast universal RNA-seq aligner. *Bioinformatics* **2012**, *29*, 15–21. [CrossRef] [PubMed]
64. Liao, Y.; Smyth, G.K.; Shi, W. featureCounts: An efficient general purpose program for assigning sequence reads to genomic features. *Bioinformatics* **2013**, *30*, 923–930. [CrossRef] [PubMed]
65. Love, M.I.; Huber, W.; Anders, S. Moderated estimation of fold change and dispersion for RNA-seq data with DESeq2. *Genome Biol.* **2014**, *15*, 550. [CrossRef]
66. Krasnov, A.; Timmerhaus, G.; Afanasyev, S.; Jørgensen, S.M. Development and assessment of oligonucleotide microarrays for Atlantic salmon (Salmo salar L.). *Comp. Biochem. Physiol. Part D Genom. Proteom.* **2011**, *6*, 31–38. [CrossRef]
67. Krasnov, A.; Johansen, L.-H.; Karlsen, C.; Sveen, L.; Ytteborg, E.; Timmerhaus, G.; Lazado, C.C.; Afanasyev, S. Transcriptome Responses of Atlantic Salmon (Salmo salar L.) to Viral and Bacterial Pathogens, Inflammation, and Stress. *Front. Immunol.* **2021**, *12*, 705601. [CrossRef]
68. Camacho, C.; Coulouris, G.; Avagyan, V.; Ma, N.; Papadopoulos, J.; Bealer, K.; Madden, T.L. BLAST+: Architecture and applications. *BMC Bioinform.* **2009**, *10*, 421. [CrossRef]
69. Mi, H.; Muruganujan, A.; Thomas, P.D. PANTHER in 2013: Modeling the evolution of gene function, and other gene attributes, in the context of phylogenetic trees. *Nucleic Acids Res.* **2013**, *41*, D377–D386. [CrossRef]
70. Kamanu, T.K.K.; Radovanovic, A.; Archer, J.A.C.; Bajic, V.B. Exploration of miRNA families for hypotheses generation. *Sci. Rep.* **2013**, *3*, 2940. [CrossRef]
71. Nakamura, T.; Sakai, K.; Nakamura, T.; Matsumoto, K. Hepatocyte growth factor twenty years on: Much more than a growth factor. *J. Gastroenterol. Hepatol.* **2011**, *26*, 188–202. [CrossRef]
72. Garcia de la Serrana, D.; Macqueen, D.J. Insulin-Like Growth Factor-Binding Proteins of Teleost Fishes. *Front. Endocrinol.* **2018**, *9*, 80. [CrossRef] [PubMed]
73. Grimberg, A.; Cohen, P. Role of insulin-like growth factors and their binding proteins in growth control and carcinogenesis. *J. Cell Physiol.* **2000**, *183*, 1–9. [CrossRef]
74. Jensen-Urstad, A.P.; Semenkovich, C.F. Fatty acid synthase and liver triglyceride metabolism: Housekeeper or messenger? *Biochim. Biophys. Acta* **2012**, *1821*, 747–753. [CrossRef] [PubMed]
75. Kuo, W.-T.; Su, M.-W.; Lee, Y.L.; Chen, C.-H.; Wu, C.-W.; Fang, W.-L.; Huang, K.-H.; Lin, W.-C. Bioinformatic Interrogation of 5p-arm and 3p-arm Specific miRNA Expression Using TCGA Datasets. *J. Clin. Med.* **2015**, *4*, 1798–1814. [CrossRef]
76. Kern, F.; Amand, J.; Senatorov, I.; Isakova, A.; Backes, C.; Meese, E.; Keller, A.; Fehlmann, T. miRSwitch: Detecting microRNA arm shift and switch events. *Nucleic Acids Res.* **2020**, *48*, W268–W274. [CrossRef]
77. Bell, J.G.; Tocher, D.R.; Farndale, B.M.; Cox, D.I.; McKinney, R.W.; Sargent, J.R. The effect of dietary lipid on polyunsaturated fatty acid metabolism in Atlantic salmon (Salmo salar) undergoing parr-smolt transformation. *Lipids* **1997**, *32*, 515–525. [CrossRef]
78. Ota, T.; Takagi, T.; Terao, T. Changes in Fatty Acid Composition of Masu Salmon, Oncorhynchus masou, Reared in Sea Water. *Hokkaidō Daigaku Suisan Gakubu Kenkyū Ihō* **1978**, *29*, 155–163.
79. Murakami, M.; Kouyama, T. Crystal Structures of Two Isozymes of Citrate Synthase from Sulfolobus tokodaii Strain 7. *Biochem. Res. Int.* **2016**, *2016*, 7560919. [CrossRef]
80. Iacobazzi, V.; Infantino, V. Citrate-new functions for an old metabolite. *Biol. Chem.* **2014**, *395*, 387–399. [CrossRef]
81. Arafa, K.; Emara, M. Insights About Circadian Clock and Molecular Pathogenesis in Gliomas. *Front. Oncol.* **2020**, *10*, 199. [CrossRef]

82. Kramer, I.M. Chapter 2—An Introduction to Signal Transduction. In *Signal Transduction*, 3rd ed.; Kramer, I.M., Ed.; Academic Press: Boston, MA, USA, 2016; pp. 53–183. [CrossRef]
83. Pontigo, J.P.; Agüero, M.J.; Sánchez, P.; Oyarzún, R.; Vargas-Lagos, C.; Mancilla, J.; Kossmann, H.; Morera, F.J.; Yáñez, A.J.; Vargas-Chacoff, L. Identification and expressional analysis of NLRC5 inflammasome gene in smolting Atlantic salmon (Salmo salar). *Fish Shellfish. Immunol.* **2016**, *58*, 259–265. [CrossRef] [PubMed]
84. McCormick, S.D. Endocrine Control of Osmoregulation in Teleost Fish. *Integr. Comp. Biol.* **2001**, *41*, 781–794. [CrossRef]
85. McCormick, S.; Sakamoto, T.; Hasegawa, S.; Hirano, T. Osmoregulatory actions of insulin-like growth factor-I in rainbow trout (Oncorhynchus mykiss). *J. Endocrinol.* **1991**, *130*, 87–92. [CrossRef] [PubMed]
86. McCormick, S.D. Effects of Growth Hormone and Insulin-like Growth Factor I on Salinity Tolerance and Gill Na+, K+-ATPase in Atlantic Salmon (Salmo salar): Interaction with Cortisol. *Gen. Comp. Endocrinol.* **1996**, *101*, 3–11. [CrossRef]

biology

MDPI

Article

Whole-Transcriptome Analysis Identifies Gender Dimorphic Expressions of Mrnas and Non-Coding Rnas in Chinese Soft-Shell Turtle (*Pelodiscus sinensis*)

Junxian Zhu [1,2,†], Luo Lei [1,3,†], Chen Chen [1], Yakun Wang [1], Xiaoli Liu [1], Lulu Geng [1,3], Ruiyang Li [1], Haigang Chen [1], Xiaoyou Hong [1], Lingyun Yu [1], Chengqing Wei [1], Wei Li [1,*] and Xinping Zhu [1,2,3,*]

1 Key Laboratory of Tropical & Subtropical Fishery Resource Application & Cultivation of Ministry of Agriculture and Rural Affairs, Pearl River Fisheries Research Institute, Chinese Academy of Fishery Sciences, Guangzhou 510380, China; zhujunxian_1994@163.com (J.Z.); 2019213005@stu.njau.edu.cn (L.L.); chenchen@prfri.ac.cn (C.C.); wangyk@prfri.ac.cn (Y.W.); liuxl@prfri.ac.cn (X.L.); gengll97@163.com (L.G.); a3021893023@163.com (R.L.); zjchenhaigang@prfri.ac.cn (H.C.); hxy@prfri.ac.cn (X.H.); yuly@prfri.ac.cn (L.Y.); zjweichengqing@prfri.ac.cn (C.W.)
2 College of Fisheries and Life Science, Shanghai Ocean University, Shanghai 201306, China
3 Wuxi Fisheries College, Nanjing Agricultural University, Wuxi 214081, China
* Correspondence: liwei@prfri.ac.cn (W.L.); zhuxinping@prfri.ac.cn (X.Z.)
† These authors contributed equally to this work.

Simple Summary: *Pelodiscus sinensis* has significant gender dimorphism in its growth patterns. Nevertheless, its underlying molecular mechanisms have not been elucidated well. Here, a whole-transcriptome analysis of the female and male gonads was performed. Altogether, 7833 DE mRNAs, 619 DE lncRNAs, 231 DE circRNAs, and 520 DE miRNAs were identified and enriched in the pathways related to sex differentiation. Remarkably, the competing endogenous RNA interaction network was constructed, including the key genes associated with sex differentiation. Collectively, this study provides comprehensive data, contributing to the exploration of master genes during sex differentiation and highlighting the potential functions of non-coding RNAs in *P. sinensis* as well as in other turtles.

Abstract: In aquaculture, the Chinese soft-shelled turtle (*Pelodiscus sinensis*) is an economically important species with remarkable gender dimorphism in its growth patterns. However, the underlying molecular mechanisms of this phenomenon have not been elucidated well. Here, we conducted a whole-transcriptome analysis of the female and male gonads of *P. sinensis*. Overall, 7833 DE mRNAs, 619 DE lncRNAs, 231 DE circRNAs, and 520 DE miRNAs were identified. Some "star genes" associated with sex differentiation containing *dmrt1*, *sox9*, and *foxl2* were identified. Additionally, some potential genes linked to sex differentiation, such as *bmp2*, *ran*, and *sox3*, were also isolated in *P. sinensis*. Functional analysis showed that the DE miRNAs and DE ncRNAs were enriched in the pathways related to sex differentiation, including ovarian steroidogenesis, the hippo signaling pathway, and the calcium signaling pathway. Remarkably, a lncRNA/circRNA–miRNA–mRNA interaction network was constructed, containing the key genes associated with sex differentiation, including *fgf9*, *foxl3*, and *dmrta2*. Collectively, we constructed a gender dimorphism profile of the female and male gonads of *P. sinensis*, profoundly contributing to the exploration of the major genes and potential ncRNAs involved in the sex differentiation of *P. sinensis*. More importantly, we highlighted the potential functions of ncRNAs for gene regulation during sex differentiation in *P. sinensis* as well as in other turtles.

Keywords: *Pelodiscus sinensis*; whole-transcriptome sequencing; sex differentiation; non-coding RNAs; ceRNA

Citation: Zhu, J.; Lei, L.; Chen, C.; Wang, Y.; Liu, X.; Geng, L.; Li, R.; Chen, H.; Hong, X.; Yu, L.; et al. Whole-Transcriptome Analysis Identifies Gender Dimorphic Expressions of Mrnas and Non-Coding Rnas in Chinese Soft-Shell Turtle (*Pelodiscus sinensis*). *Biology* **2022**, *11*, 834. https://doi.org/10.3390/biology11060834

Academic Editor: Patricia Pereiro

Received: 26 April 2022
Accepted: 26 May 2022
Published: 29 May 2022

1. Introduction

Gender dimorphism of morphology and physiology is widely present in the animal kingdom, which contributes to making the world complicated and wonderful [1]. In aquaculture, the outstanding gender dimorphism of growth rate and body size has drawn special attention from scientists [2–4], as these factors can be manipulated to achieve higher economic value and broader consumer demand. Consequently, the breeding of a monosexual population has been extensively used in aquaculture species [5–7]. The Chinese soft-shell turtle (*Pelodiscus sinensis*) is a vital economic species in aquaculture, with an annual production of over 330,000 tons [8] and outstanding sexual dimorphism in body growth, calipash, and fat content between males and females, with males having higher commercial value than females [9]. Therefore, there is also demand for the single-sex breeding of *P. sinensis*.

Studies on the regulation mechanism of sex determination and sex differentiation in *P. sinensis* are the basis of unisexual breeding. However, achieving consensus on the sex determination mechanism of *P. sinensis* is complicated. Early studies have shown that the sex of *P. sinensis* embryos was affected by the incubation temperature, so researchers inferred that the *P. sinensis* had temperature-dependent sex determination (TSD) [10–12]. With the discovery of the ZZ/ZW micro-chromosome, genetic sex determination (GSD) was verified as the sex-determination system of *P. sinensis* [13,14]. Additionally, via molecular cytogenetics and incubation experiments, TSD was ruled out in *P. sinensis* [15]. Currently, although researchers have conducted relevant studies and achieved some success [16,17], they are not sufficient to completely reveal the mechanisms of sex determination and sex differentiation in *P. sinensis*. Accordingly, massive studies on molecular mechanisms, particularly the genes and the regulatory pathways, could promote the mechanistic comprehension of sex differentiation and/or sex determination in *P. sinensis*.

The regulation of sex determination and sex differentiation is complex and diverse, and the master genes play a key role. Sex-determining genes (SDGs) drive the sex differentiation of bipotential gonads into the ovaries or the testes [18]. The sex-determining region Y (*Sry*) gene [19] and the W-linked DM-domain (*Dm-w*) gene [20] were the first SDGs identified in the XX/XY system and ZZ/ZW system, respectively. Subsequently, other SDGs and candidates, including *Zglp1* [21] and *Sox9* [22] in mammals, *Dmy* [23] and *Foxl3* [24] in fishes, *Dmrt1* [25] and *Dsx* [26] in birds, and *Dmrt1* [16,27] and *Amh* [28] in reptiles, were cognized in GSD species. Therefore, more information about these candidates or their homologs should be provided to help us understand their potential roles in the sex differentiation of *P. sinensis*.

Compared to DNA with coding genes, mammalian genomes have large amounts of non-coding DNAs, known as "junk DNA" or "genomic dark matter", which can be transcribed into non-coding RNAs (ncRNAs) containing long ncRNAs (lncRNAs), microRNAs (miRNAs), and circular RNAs (circRNAs), and although most of them do not encode proteins, they are involved in multiple biological processes [29,30]. More recently, studies have reported that miR-124 repressed the expression of *Sox9* in mice (*Mus musculus*) ovaries, indicating a major role during ovarian differentiation [31]. In addition, in chicken (*Gallus gallus*) male hypermethylation (MHM), a Z sex chromosome-linked locus adjacent to *dmrt1* was methylated and transcriptionally silent in male cells (ZZ), while it was hypomethylated in female cells (ZW) and then transcribed a long non-coding RNA (MHM). Then, injecting the expression plasmids of MHM into adult chicken testes could inhibit *Dmrt1* expression [32]. Moreover, a circular *Sry* transcript was produced in the testes of adult mice [33,34] and functioned as an miR-138 sponge, resulting in a more than tenfold enrichment of AGO2 (Argonaute 2) [35]. The roles of mRNAs and non-coding RNAs in sex differentiation are not independent of each other, and the theory of competing endogenous RNAs (ceRNA) unifies them as a system. Briefly, the lncRNAs and circRNAs function as a sponge to competitively adsorb miRNAs, thereby promoting mRNA expression [36,37]. Despite the fact that the studies about non-coding RNA are a popular topic in the field of sex differentiation, to our knowledge, studies on *P. sinensis* and turtles, especially the interaction between ncRNAs and mRNAs, are far from adequate.

In the present study, we conducted a transcriptome analysis on the ovaries and testes of *P. sinensis* using Illumina RNA Sequencing (RNA-Seq) technology, established differentially expressed (DE) profiles, and analyzed the pathways of DE mRNAs, lncRNAs, miRNAs, and circRNAs. Furthermore, the correlation networks of lncRNA/circRNA–miRNA–mRNA were constructed to disclose the interactions among them. The relative expressions of candidate mRNAs, lncRNAs, miRNAs, and circRNAs were validated using quantitative reverse-transcription PCR (qRT-PCR). Collectively, the comprehensive data contribute to the exploration of master genes during the sex-differentiation process and highlight the potential functions of non-coding RNAs in *P. sinensis* as well as in other turtles.

2. Materials and Methods

2.1. *P. sinensis* Collection and Total RNA Isolation

The Animal Care and Ethics Committee of the Pearl River Fisheries Research Institute, Chinese Academy of Fishery Sciences approved this research, and all experimental protocols and methods were performed in accordance with the relevant guidelines and regulations. A total of 18 healthy 6-month-old *P. sinensis* subjects, 9 males and 9 females, were collected and sacrificed humanely. Testes and ovaries were obtained and immediately frozen in liquid nitrogen for storage at −80 °C for RNA isolation. Total RNA was isolated using TRIzol reagent (Ambion, Carlsbad, CA, USA). The consistency and purity were tested by the NanoQ™ (Thermo Scientific, Madison, WI, USA), and the quality was evaluated using an Agilent 2100 bioanalyzer (Agilent Technologies, Santa Clara, CA, USA). Three biological replicates were obtained from the 9 males and 9 females, which were pooled with 3 turtles, respectively.

2.2. Library Construction and Sequencing

High-quality RNAs were used for library construction. The mRNAs were first enriched using oligo (dT) magnetic beads. The rRNAs were removed using a Ribo-Zero™ rRNA Removal kit (Epicenter, Madison, WI, USA). The rRNA-depleted RNAs were fragmented using fragmentation buffer and then the first-strand cDNA was synthesized using random hexamer primers. Subsequently, the second-strand cDNA was progressed, adding buffer, dNTPs, RNase H, and DNA polymerase I. The double-stranded cDNA was converted into blunt ends and 3′ ends were adenylated and then purified by AMPure XP (Beckman Coulter, Brea, CA, USA) beads to select insert fragments. After PCR and ligation of the sequencing adapters, the final library was estimated with the Agilent Bioanalyzer 2100 (Agilent Technologies, Santa Clara, CA, USA) and sequenced at Vazyme Biotech Co., Ltd. (Nanjing, China) using an Illumina HiSeq 4000™ System (Illumina, San Diego, CA, USA).

2.3. Quality Control and Transcriptome Assembly

The reads, including adapters, >10% of ploy-N, and those of low quality (>50% of the bases had quality scores of ≤5), were removed to obtain clean data. Meanwhile, the Q20, Q30, and GC content of the clean data were calculated and used for further bioinformatics analysis. The clean reads from each library were mapped to the reference genome of *P. sinensis* (PRJNA221645, NCBI, accessed on 10 January 2022) using Hisat2 v2.1.0 [38]. The mapped reads were assembled with StringTie v1.3.5 [39] and then annotated by the gffcompare program.

2.4. Differential Expression Analysis and Functional Annotation

For each library, the fragments per kilobase per million reads (FPKM) and the transcripts per kilobase of exon model per million mapped reads (TPM) scores were calculated using StringTie v1.3.5 [39]. The differentially expressed levels of mRNAs and ncRNAs were evaluated using edgeR (accessed on 13 January 2022) [40]. The critical values of DE mRNAs and DE ncRNAs were given as a false discovery rate (FDR) of <0.05

and an absolute value of the log2 (fold change) of >2. The Gene Ontology project (GO, http://www.geneontology.org, accessed on 13 January 2022) and Kyoto Encyclopedia of Genes and Genomes (KEGG, https://www.kegg.jp, accessed on 13 January 2022) were used for the functional annotation of DE mRNAs and DE ncRNAs. A p value of <0.05 was considered significantly enriched in GO terms and KEGG pathway analyses.

2.5. Validation of Candidate mRNAs and ncRNAs by qRT-PCR

Eight of the DE mRNAs, lncRNA, and circRNAs, and four of the DE miRNAs were randomly selected to validate the RNA-seq data using qRT-PCR. All primers used in this study are provided in Table S1. The reaction program of the qRT-PCR was set as follows: 95 °C pre-denaturation for 10 min; 95 °C for 15 s, 60 °C for 20 s, and 72 °C for 20 s for 40 total cycles. The expression levels of the DE mRNAs, lncRNAs, and circRNAs were normalized with the ef1α gene [41], while the U6 gene [42] was applied for the normalization of the DE miRNAs. The relative expressions of the DE mRNAs and DE ncRNAs were calculated using the $2^{-\Delta\Delta Ct}$ method [43]. All data are shown as the means $\pm$ standard error of the mean (SEM) and were analyzed using ANOVA (accessed on 6 March 2022). $p < 0.05$ was considered to be significantly different.

2.6. Construction of the ceRNA Interaction Network

To better elucidate the potential role of DE ncRNAs in the sex differentiation of *P. sinensis*, a lncRNA/circRNA–miRNA–mRNA interaction network was constructed. miRanda software (Memorial Sloan Kettering Cancer Center, New York, NY, USA) was used to forecast the interaction relationships of the mRNAs, lncRNAs, and circRNAs with miRNAs, and Cytoscape 3.2 (National Institute of General Medical Sciences, Bethesda, MD, USA) was applied for visualization.

3. Results

3.1. Overview of Transcriptome Sequencing

After removing low-quality reads, we obtained the totally clean reads of the mRNAs, lncRNAs, and circRNAs, ranging from 42,487,292 to 46,871,481, and the miRNAs ranging from 9,765,869 to 14,021,878 via the whole-transcriptome analysis of the early female and male gonads in *P. sinensis*. The GC contents of the mRNAs, lncRNAs, and circRNAs were between 45.70 and 47.94%, and those of the miRNAs were between 45.99 and 50.54%. The Q30 bases of the mRNAs, lncRNAs, and circRNAs were greater than 93.98%, and those of the miRNAs were greater than 97.76% (Table S2). The results of principal component analysis (PCA) showed good similarity among the three biological replicates (Figure S1). The length distributions of the mRNAs and ncRNAs are also provided in Table S3.

3.2. Identification of DE mRNAs and ncRNAs

A volcano plot and heat map were used to exhibit the differentially expressed results (fold change of $\geq$2 and FDR of <0.05) of the mRNAs and ncRNAs. In total, 7833 DE mRNAs were identified, of which 3192 were up-regulated and 4641 were down-regulated (Figure 1A, Table S4); 619 DE lncRNAs were identified, of which 385 were up-regulated and 234 were down-regulated (Figure 1C, Table S4); 231 DE circRNAs were identified, of which 39 were up-regulated and 192 were down-regulated (Figure 1E, Table S4); and 520 DE miRNAs were identified, of which 219 were up-regulated and 301 were down-regulated (Figure 1G, Table S4). Likewise, the hierarchical clustering of the DE mRNAs and DE ncRNAs was divided into two distinct clusters, revealing remarkable differences between the ovaries and testes of *P. sinensis* (Figure 1B,D,F,H).

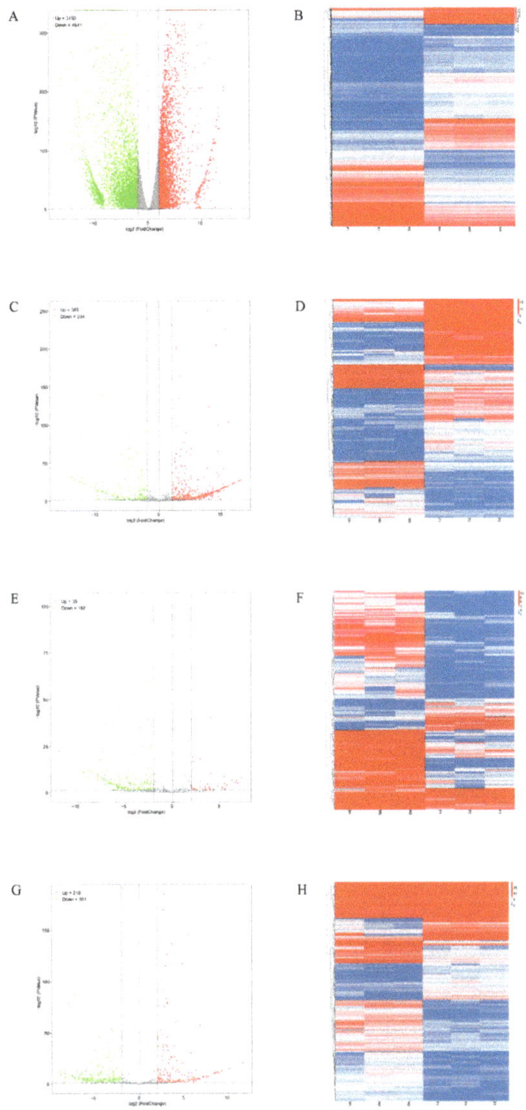

Figure 1. The volcano plots and heat maps of DE mRNAs and ncRNAs. (**A**,**C**,**E**,**G**) indicate the volcano plots of DE mRNAs, DE lncRNAs, DE circRNAs, and DE miRNAs, respectively. (**B**,**D**,**F**,**H**) show the heat maps of DE mRNAs, DE lncRNAs, DE circRNAs, and DE miRNAs, respectively.

3.3. GO and KEGG Analysis of DE mRNAs and ncRNAs

GO and KEGG analyses were applied to annotate the biological functions and enriched pathways of the DE mRNAs and ncRNAs. GO terms were mainly classified into three categories, including biological process, cellular component, and molecular function. For the DE mRNAs and ncRNAs, 7833 DE mRNAs, 619 DE lncRNAs, 231 DE circRNAs, and 520 DE miRNAs were used for GO annotation, respectively (Table S5), and all of them were mainly enriched in the cellular process of the biological process, the cellular anatomical entity of the cellular component, and binding of the molecular function (Figure 2A,C,E,G). In addition, a total of 25 KEGG signaling pathways of DE mRNAs, 37 KEGG signaling pathways of DE lncRNAs, 16 KEGG signaling pathways of DE circRNAs, and 30 KEGG signaling pathways of DE miRNAs were significantly enriched, respectively (Table S6). The top three significantly enriched KEGG signaling pathways were related to the neuroactive ligand-receptor interaction, the olfactory transduction, and the calcium signaling pathway in DE mRNAs (Figure 2B); the Ras signaling pathway, the rap1 signaling pathway, and the regulation of actin cytoskeleton in DE lncRNAs (Figure 2D); the phosphatidylinositol signaling system, the glycerolipid metabolism, and the hippo signaling pathway in DE circRNAs (Figure 2F); and the neuroactive ligand-receptor interaction, the Ras signaling pathway, and the regulation of actin cytoskeleton in DE miRNAs (Figure 2H). The KEGG signaling pathways associated with sex differentiation were also annotated, such as the ovarian steroidogenesis, the hippo signaling pathway, and the calcium signaling pathway.

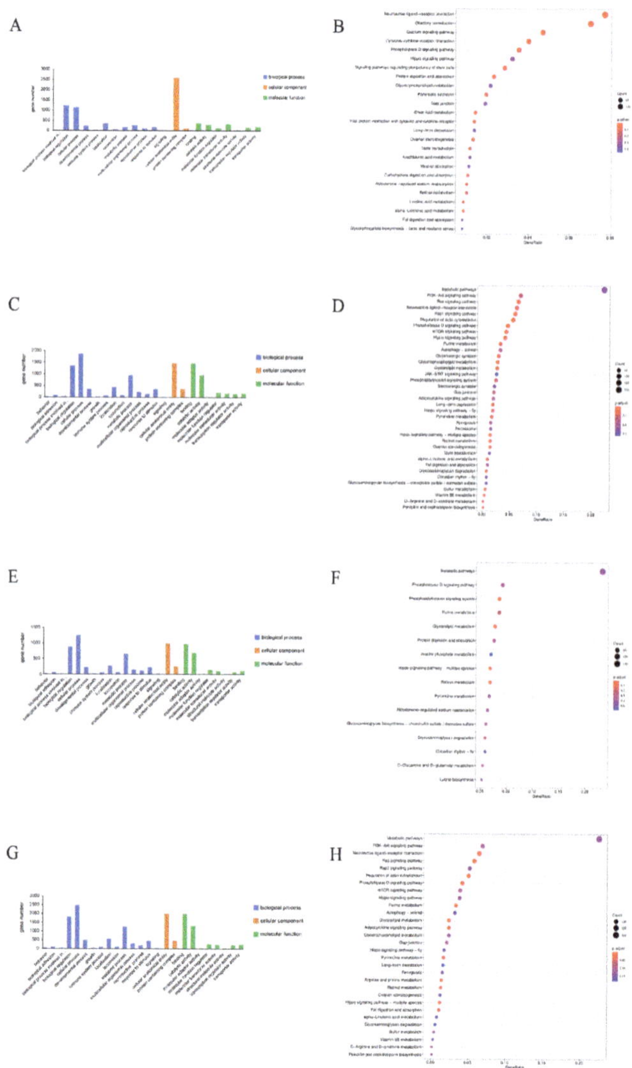

Figure 2. Analyses of functions and pathways of DE mRNAs and ncRNAs. (**A**,**C**,**E**,**G**) signify the GO terms of DE mRNAs, DE lncRNAs, DE circRNAs, and DE miRNAs, respectively. (**B**,**D**,**F**,**H**) illustrate the KEGG signaling pathways of DE mRNAs, DE lncRNAs, DE circRNAs, and DE miRNAs, respectively.

3.4. Validation of DE mRNAs and ncRNAs

Eight DE mRNAs (*foxl2*, *bmp2*, *ran*, *sox3*, *dmrt1*, *dkkl1*, *spo11*, and *sox9*), lncRNA (lncRNA_21915, lncRNA_34729, lncRNA_27020, lncRNA_52064, lncRNA_32111, lncRNA_1-6229, lncRNA_15672, and lncRNA_55405), circRNAs (circRNA_1083, circRNA_1243, circRNA_1021, circRNA_546, circRNA_464, circRNA_389, circRNA_823, and circRNA_668), and four DE miRNAs (Chr1_880, Chr20_20074, Chr9_14462, and Chr22_21171) were chosen for qRT-PCR analysis to evaluate the reliability of the RNA-seq data presented with the TPM value [44]. The results show that the data of the qRT-PCR (Figure 3B,D,F,H) were consistent with the RNA-seq data (Figure 3A,C,E,G), affirming the accuracy of the RNA-seq data and indicating the existence of sexually dimorphic expression profiles between the female and male gonads of *P. sinensis*.

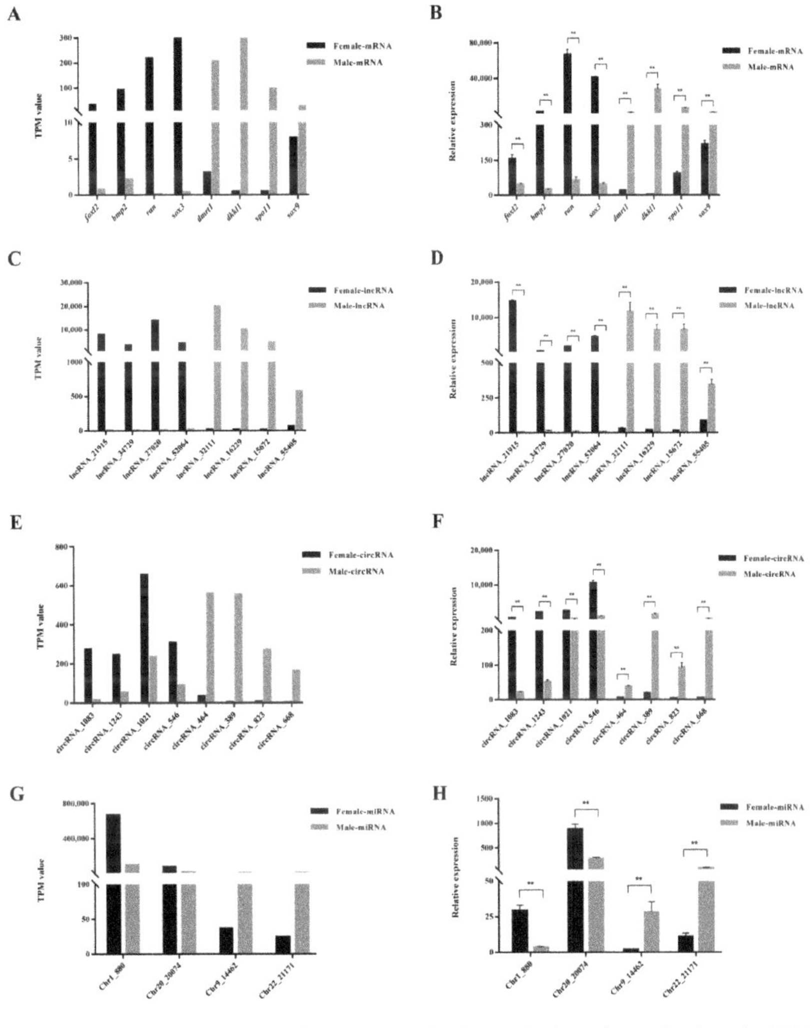

Figure 3. Validation of RNA-seq data by qRT-PCR. (**A**,**C**,**E**,**F**) show the tendencies of mRNAs, lncRNAs, circRNAs, and miRNAs, respectively, in RNA-seq data using the TPM value. (**B**,**D**,**F**,**H**) represent the relative expressions of candidate mRNAs, lncRNAs, circRNAs, and miRNAs, respectively, by qRT-PCR. The TPM value is represented as the means, while the data of the relative expressions shown are the means ± SEM. The asterisk indicates significant differences. ** $p < 0.01$.

3.5. Construction of the lncRNA/circRNA–miRNA–mRNA Network

The interaction network was predicted using miRanda software with the default parameters [45]. Fold changes greater than 2 and less than -2 indicate up- and down-regulation, and q value < 0.05. Altogether, 131 DE lncRNAs, 40 DE miRNAs, and 1690 DE mRNAs were used to establish the lncRNA–miRNA–mRNA interaction network (Figure 4A, Table S7), and 17 DE circRNAs, 16 DE miRNAs, and 749 DE mRNAs were applied to build the circRNA–miRNA–mRNA interaction network (Figure 4B, Table S8). The circles, triangles, and squares represent the miRNAs, lncRNAs or circRNAs, and mRNAs, respectively. Red indicates up-regulation and green indicates down-regulation. The gray lines illustrated their interactions with each other. Remarkably, we identified some pivotal genes, such as *fgf9*, *foxl3*, and *dmrta2*, in the two interaction networks that participate in sex differentiation. *Fgf9* could be regulated with lncRNA_3086, lncRNA_34437, and lncRNA_42520 by the miRNAs of Chr19_19503, Chr19_19503, and Chr1_3952, respectively. Additionally, *Fgf9* could also be regulated with circRNA_302 and circRNA_1030 through the miRNAs of Chr21_20215. *Foxl3* could be regulated with six lncRNAs (lncRNA_21044, lncRNA_35038, lncRNA_37410, lncRNA_44739 lncRNA_47846, and lncRNA_49172), and all by an miRNA of Chr24_22338. *Dmrta2* was regulated with lncRNA_62580 via the miRNAs of Chr6_12669, Chr1_3133, and Chr1_3134 (Figure 4, Tables S7 and S8).

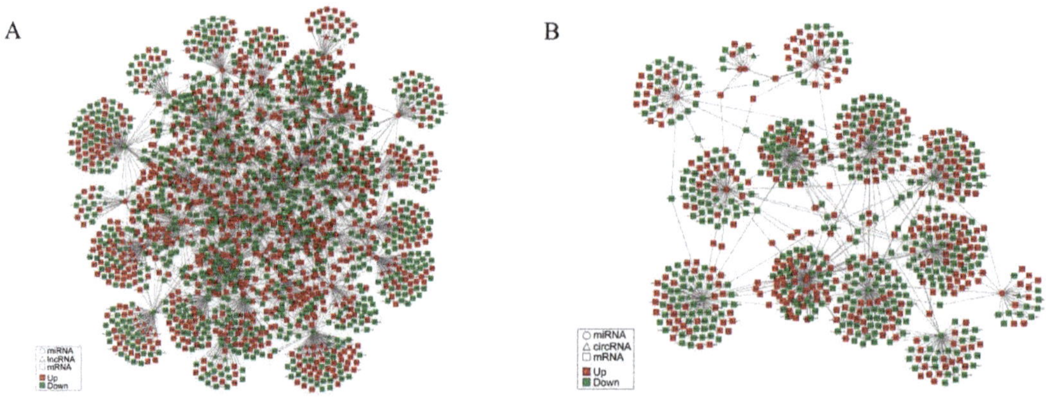

Figure 4. (**A**) The interaction network of lncRNA/circRNA–miRNA–mRNA, and (**B**) the interaction network of circRNA–miRNA–mRNA.

4. Discussion

In this study, we performed a whole-transcriptome analysis of the early female and male gonads in *P. sinensis*. The results of the quality of each library and the verification of qRT-PCT demonstrated that we established a high-quality and reliable transcriptome database.

Subsequently, we identified 7833 DE mRNAs. Undoubtedly, the "star genes" associated with sex differentiation were identified, including *dmrt1*, *foxl2*, and *sox9*. *Dmrt1* was the first gene identified to be required for the male sex differentiation of *P. sinensis* and exhibited a sex-dimorphic expression pattern throughout the embryonic period. An RNA interference experiment showed that the knockdown of *dmrt1* in male embryos of *P. sinensis* led to sex reversal [16]. Moreover, *Dmrt1* was also the first master gene identified for sex determination in *Trachemys scripta elegans* (a TSD species), and *dmrt1* was directly promoted for transcription by KDM6B to eliminate the trimethylation of H3K27 near its promoter, thereby opening the pathway of male sex differentiation [27]. In the XY system, *sox9* was a downstream target [46] of *sry*, which was located on the Y chromosome, initiating the development of a bipotential primordial gonad in the testes [47]. *sox9* is essential for male sex differentiation in the red-eared slider turtle [48]. *Foxl2* is a member of the Wnt pathway and was the downstream target of *wnt4* [49]. A loss of *foxl2* suppressed *sox9* expression, resulting in a partial reversal of the ovaries [50].

Notably, some novel genes in *P. sinensis* related to sex differentiation were also identified, including *bmp2*, *sox3*, *dkkl1*, *spo11*, and *ran*. *Zglp1* was a downstream effector of *bmp2*, activating the oogenic program and determining the oogenic fate [21]. *Bmp2* and retinoic acid synergistically affirmed the female pathway [51] and responded to Wnt4, a classical feminizing signal [52]. The deletion of *sox9* induced the expression of *bmp2* in XY gonads [53,54]. In addition, *bmp2* presented gender dimorphic expression in *White leghorns* gonads, and was upregulated and hypermethylated in the ovary [55,56]. *Sox3* was the evolutionary ancestor of *sry* and could functionally substitute *sry*, opening the male pathway [57]. In *Oryzias dancena*, the functional deficiency of *sox3* resulted in sex reversal, and *sox3* started male sex differentiation via upregulating the expression of *gsdf* [58]. In *Epinephelus coioides*, a hermaphroditic fish, the continuous expression of *sox3* developed the primordial germ cells toward oogonia; however, when *sox3* ceased expression, the primordial germ cells developed toward spermatogonia [59]. In *Rana rugosa*, *sox3* activated the transcription of *cyp19* by binding to the promoter region, driving the development of gonads into the ovaries [60]. *Dkkl1* was only detected in the testes of mice and humans [61,62], and was mainly distributed in mice spermatocytes and the acrosomes of sperms [63], and in human spermatocytes and round spermatids [61] involved in meiosis and spermatogenesis. Moreover, *dkkl1* could decrease the transcriptional activity of *sf1*, a key sex differentiation gene [64], thereby inhibiting the expression of *cyp17* and *cyp11* and reducing testosterone synthesis in Leydig cells [65]. *Spo11*, a meiosis-specific protein, initiated meiotic recombination in multiple species [66] and the knockout of *spo11* caused meiotic arrest, the apoptosis of spermatocytes during early prophase, and follicle deficiency [67]. *Spo11* was also identified as a male-biased gene in *Danio rerio* [68] and *Anguilla japonica* [69]. Males of *spo11* knockout were completely sterile, but females were fertile [68]. *Ran* could function as an androgen receptor coactivator and weak coactivation of *ran* might result in partial androgen insensitivity [70,71]. The nuclear entry pathway of the sex-determining factor *sry* was demonstrated to require the canonic ran-dependent pathway [72,73]. Likewise, the significant differences of these novel genes in the sex-dimorphic expression patterns of female and male gonads may suggest that their potential function in the sex differentiation of *P. sinensis* has not yet been annotated.

Recently, studies have shown that ncRNAs are extensively involved in the process of sex differentiation [31–35]. Thus, in all of the RNA-seq data, we identified 619 DE lncRNAs, 231 DE circRNAs, and 520 DE miRNAs. The DE mRNAs and DE ncRNAs were enriched in the sex differentiation-associated pathways. Estradiol (E2) is an essential steroid hormone in the ovaries, which has been confirmed to participate in the sex differentiation of turtles. E2 treatment caused the ovarian differentiation of female embryos and sex reversal of male embryos in *P. sinensis* [74], resulting in a rapid up-regulation of *cyp19a1* and completing the sex reversal of male embryos in *Mauremys reevesii* [75]. *Yap* and *wwtr1* were two downstream effectors of the hippo signaling pathway that were inactivated, leading to the up-regulation of female sex-differentiation genes, such as *foxl2* and *wnt4*, and the down-regulation of male sex-differentiation genes, such as *dmrt1* and *sox9* [76]. The calcium signaling pathway was broadly connected to diverse biological processes [77]. Recently, a study found that high temperatures promoted an influx in Ca^{2+} in the somatic cells of gonads, thereby activating pSTAT3 and inhibiting Kdm6b and then Dmrt1 expression, and then ultimately turning on the female pathway [78].

Meanwhile, based on RNA-seq data, we constructed a lncRNA/circRNA–miRNA–mRNA interaction network. Particularly, sex differentiation-related genes *fgf9*, *foxl3*, and *dmrta2* were predicted to be regulated by three pairs of lncRNA–miRNA and two pairs of circRNA–miRNA, and six pairs of lncRNA–miRNA and three pairs of lncRNA–miRNA, respectively. Previous studies have suggested that *fgf9* is a downstream target of *sry*, sufficiently inducing the male sex fate in XX individuals [79]. The lncRNA *LOC105611671* was an upstream regulator of *fgf9* and regulated *fgf9* expression by targeting oar-miR-26a, thereby enhancing testicular steroidogenesis in Hu sheep [80]. A member of the fibroblast growth factor (FGF) family could interact with miRNA-541 to suppress androgen receptor signals in prostate cancer [81]. In *Oryzias latipes*, *foxl3* expressed in germ cells took part in sex determination, and the knockdown of *foxl3* produced functional sperms in females [24]. A study showed that miR-9 targeted the 3′ untranslated region of *foxl3*, promoting oocyte degeneration and meiosis in *Monopterus albus*. miR-430 directly interacted with a member of the forkhead box (FOX) family and repressed its expression [82]. In *Scophthalmus maximus*, *dmrta2* was also identified as a candidate gene related to sex differentiation [83]. A member of the doublesex and mab-3-related transcription factor (DMRT) family was transcriptionally promoted by a transcription factor of the lncRNA DMRT2-AS (referred to as *dmrt2* antisense), participating in the sex differentiation of *Cynoglossus semilaevis* [84]. Analogously, these ncRNAs, which were predicted to regulate the sex differentiation-related genes, may play a vital role in the sex differentiation process of *P. sinensis* and should be further studied.

5. Conclusions

In summary, we have provided complete transcriptome data that reveal a gender-dimorphic expression profile in the female and male gonads of *P. sinensis*, which contributes to exploring the master genes and the potential ncRNAs closely related to sex differentiation, as well as understanding the relationship between genes and ncRNAs in the regulatory mechanism of sex differentiation in *P. sinensis*.

Supplementary Materials: The following supporting information can be downloaded at: https://www.mdpi.com/article/10.3390/biology11060834/s1, Figure S1: Principal component analysis of each sample; Figure S2: Pearson's correlation coefficient of each sample; Table S1: All primers used in qRT-PCR analysis; Table S2: Summary of RNA-Seq data; Table S3: Length distribution of mRNAs and ncRNAs; Table S4: All differentially expressed mRNAs and ncRNAs; Table S5: GO analysis; Table S6: KEGG analysis; Table S7: The lncRNA-miRNA-mRNA network; Table S8: The circRNA-miRNA-mRNA network; Table S9: The correlation coefficient of the results between qRT-PCR and RNA-seq.

Author Contributions: X.Z. and W.L. designed the experiments. C.C., Y.W., X.L., L.G., R.L., H.C., X.H., L.Y. and C.W. collected samples. J.Z. and L.L. performed the experiments and analyzed the data; J.Z. wrote the main manuscript. X.Z. and W.L. revised the manuscript. All authors have read and agreed to the published version of the manuscript.

Funding: This work was funded by the National Key R&D Program of China (2018YFD0900201), the National Natural Science Foundation of China (32102792), The Project Supported by Guangdong Natural Science Foundation (2018A030313520), the Central Public-Interest Scientific Institution Basal Research Fund, CAFS (2020TD35), the National Freshwater Genetic Resource Center (NFGR-2022), the Guangdong Agricultural Research System (2019KJ150), and the China-ASEAN Maritime Cooperation Fund (CAMC-2018F).

Institutional Review Board Statement: The Animal Care and Ethics Committee of the Pearl River Fisheries Research Institute, Chinese Academy of Fishery Sciences (Guangzhou, China) approved this research (Approval Code: LAEC-PRFRI-20210148; Approval Date: January 10, 2021), and all experimental protocols and methods were performed in accordance with the relevant guidelines and regulations.

Informed Consent Statement: Not applicable.

Data Availability Statement: The raw sequence reads used for analysis in this study can be obtained from the Sequence Read Archive (SRA) under BioProject PRJNA838782. All data are available upon request from the corresponding author.

Conflicts of Interest: All authors declare no conflict of interest.

References

1. Mei, J.; Gui, J.F. Genetic basis and biotechnological manipulation of sexual dimorphism and sex determination in fish. *Sci. China Life Sci.* **2015**, *58*, 124–136. [CrossRef] [PubMed]
2. Hidir, A.; Aaqillah-Amr, M.A.; Azra, M.N.; Shahreza, M.S.; Abualreesh, M.H.; Peng, T.H.; Ma, H.Y.; Waiho, K.; Fazhan, H.; Ikhwanuddin, M. Sexual dimorphism of mud crab, genus *Scylla* between sexes based on morphological and physiological characteristics. *Aquac. Res.* **2021**, *52*, 5943–5961. [CrossRef]
3. Liu, X.; Zhu, Y.; Wang, Y.; Li, W.; Hong, X.; Zhu, X.; Xu, H. Comparative transcriptome analysis reveals the sexual dimorphic expression profiles of mRNAs and non-coding RNAs in the Asian yellow pond turtle (*Meauremys mutica*). *Gene* **2020**, *750*, 144756. [CrossRef] [PubMed]
4. Ou, M.; Chen, K.; Gao, D.; Wu, Y.; Chen, Z.; Luo, Q.; Liu, H.; Zhao, J. Comparative transcriptome analysis on four types of gonadal tissues of blotched snakehead (*Channa maculata*). *Comp. Biochem. Physiol. Part D Genom. Proteom.* **2020**, *35*, 100708. [CrossRef]
5. Liu, H.; Guan, B.; Xu, J.; Hou, C.; Tian, H.; Chen, H. Genetic manipulation of sex ratio for the large-scale breeding of YY super-male and XY all-male yellow catfish (*Pelteobagrus fulvidraco* (Richardson)). *Mar. Biotechnol.* **2013**, *15*, 321–328. [CrossRef]
6. Angienda, P.O.; Aketch, B.O.; Waindi, E.N. Development of all-male fingerlings by heat treatment and the genetic mechanism of heat induced sex determination in nile tilapia (*Oreochromis niloticus* L.). *Adv. Mater.* **2013**, *21*, 3727–3729.
7. Zhao, J.; Ou, M.; Wang, Y.; Liu, H.; Chen, K. Breeding of YY super-male of blotched snakehead (*Channa maculata*) and production of all-male hybrid (*Channa argus* ♀× *C. maculata* ♂). *Aquaculture* **2021**, *538*, 736450. [CrossRef]
8. Bureau of Fisheries, Ministry of Agriculture and Rural Affairs of PR China. *China Fisheries Statistics Yearbook*; China Agriculture Press: Beijing, China, 2021; p. 35.
9. Zhou, X.; Zhu, Y. Sex-specific growth characteristics of Chinese soft-shelled turtle, *Pelodiscus sinensis*. *Chin. Aquac.* **2011**, *32*, 11–13.
10. Nie, L.; Guo, C.; Wang, M.A.; Wang, Q. Sex determination mechanism of *Trionyx Sinensis*. *Chin. J. Appl. Environ. Biol.* **2001**, *7*, 258–261.
11. Zheng, J.; Zhu, M. Isolation and sequence analysis of the *Sox*-1, -2, -3 homologs in *Trionyx sinensis* and *Alligator sinensis* having temperature-dependent sex determination. *Biochem. Genet.* **2006**, *44*, 101–112. [CrossRef]
12. Zhu, D.; Sun, X. Sex determination in *Trionyx sinensis*. *Chin. J. Zool.* **2000**, *6*, 37–38.
13. Kawagoshi, T.; Uno, Y.; Matsubara, K.; Matsuda, Y.; Nishida, C. The ZW micro-sex chromosomes of the Chinese soft-shelled turtle (*Pelodiscus sinensis*, Trionychidae, Testudines) have the same origin as chicken chromosome 15. *Cytogenet. Genome Res.* **2009**, *125*, 125–131. [CrossRef] [PubMed]
14. Badenhorst, D.; Stanyon, R.; Engstrom, T.; Valenzuela, N. A ZZ/ZW microchromosome system in the spiny softshell turtle, *Apalone spinifera*, reveals an intriguing sex chromosome conservation in Trionychidae. *Chromosome Res.* **2013**, *21*, 137–147. [CrossRef]
15. Mu, Y.; Zhao, B.; Tang, W.Q.; Sun, B.J.; Zeng, Z.G.; Valenzuela, N.; Du, W.G. Temperature-dependent sex determination ruled out in the Chinese soft-shelled turtle (*Pelodiscus sinensis*) via molecular cytogenetics and incubation experiments across populations. *Sex. Dev.* **2015**, *9*, 111–117. [CrossRef]
16. Sun, W.; Cai, H.; Zhang, G.; Zhang, H.; Bao, H.; Wang, L.; Ye, J.; Qian, G.; Ge, C. *Dmrt1* is required for primary male sexual differentiation in Chinese soft-shelled turtle *Pelodiscus sinensis*. *Sci. Rep.* **2017**, *7*, 4433. [CrossRef] [PubMed]
17. Zhang, Y.; Xiao, L.; Sun, W.; Li, P.; Zhou, Y.; Qian, G.; Ge, C. Knockdown of *R-spondin1* leads to partial sex reversal in genetic female Chinese soft-shelled turtle *Pelodiscus sinensis*. *Gen. Comp. Endocrinol.* **2021**, *309*, 113788. [CrossRef] [PubMed]
18. Capel, B. Vertebrate sex determination: Evolutionary plasticity of a fundamental switch. *Nat. Rev. Genet.* **2017**, *18*, 675–689. [CrossRef]
19. Hiramatsu, R.; Matoba, S.; Kanai-Azuma, M.; Tsunekawa, N.; Katoh-Fukui, Y.; Kurohmaru, M.; Morohashi, K.; Wilhelm, D.; Koopman, P.; Kanai, Y. A critical time window of *Sry* action in gonadal sex determination in mice. *Development* **2009**, *136*, 129–138. [CrossRef]
20. Yoshimoto, S.; Okada, E.; Umemoto, H.; Tamura, K.; Uno, Y.; Nishida-Umehara, C.; Matsuda, Y.; Takamatsu, N.; Shiba, T.; Ito, M. A W-linked DM-domain gene, DM-W, participates in primary ovary development in *Xenopus laevis*. *Proc. Natl. Acad. Sci. USA* **2008**, *105*, 2469–2474. [CrossRef]
21. Nagaoka, S.I.; Nakaki, F.; Miyauchi, H.; Nosaka, Y.; Ohta, H.; Yabuta, Y.; Kurimoto, K.; Hayashi, K.; Nakamura, T.; Yamamoto, T.; et al. *ZGLP1* is a determinant for the oogenic fate in mice. *Science* **2020**, *367*, 6482. [CrossRef]
22. Vidal, V.P.; Chaboissier, M.C.; de Rooij, D.G.; Schedl, A. *Sox9* induces testis development in XX transgenic mice. *Nat. Genet.* **2001**, *28*, 216–217. [CrossRef] [PubMed]
23. Matsuda, M.; Nagahama, Y.; Shinomiya, A.; Sato, T.; Matsuda, C.; Kobayashi, T.; Morrey, C.E.; Shibata, N.; Asakawa, S.; Shimizu, N.; et al. *DMY* is a Y-specific DM-domain gene required for male development in the medaka fish. *Nature* **2002**, *417*, 559–563. [CrossRef] [PubMed]
24. Nishimura, T.; Sato, T.; Yamamoto, Y.; Watakabe, I.; Ohkawa, Y.; Suyama, M.; Kobayashi, S.; Tanaka, M. *Foxl3* is a germ cell-intrinsic factor involved in sperm-egg fate decision in medaka. *Science* **2015**, *349*, 328–331. [CrossRef] [PubMed]
25. Smith, C.A.; Roeszler, K.N.; Ohnesorg, T.; Cummins, D.M.; Farlie, P.G.; Doran, T.J.; Sinclair, A.H. The avian Z-linked gene *DMRT1* is required for male sex determination in the chicken. *Nature* **2009**, *461*, 267–271. [CrossRef]

26. Ioannidis, J.; Taylor, G.; Zhao, D.; Liu, L.; Idoko-Akoh, A.; Gong, D.; Lovell-Badge, R.; Guioli, S.; McGrew, M.J.; Clinton, M. Primary sex determination in birds depends on DMRT1 dosage, but gonadal sex does not determine adult secondary sex characteristics. *Proc. Natl. Acad. Sci. USA* **2021**, *118*, e2020909118. [CrossRef]
27. Ge, C.; Ye, J.; Weber, C.; Sun, W.; Zhang, H.; Zhou, Y.; Cai, C.; Qian, G.; Capel, B. The histone demethylase KDM6B regulates temperature-dependent sex determination in a turtle species. *Science* **2018**, *360*, 645–648. [CrossRef]
28. Zhou, Y.; Sun, W.; Cai, H.; Bao, H.; Zhang, Y.; Qian, G.; Ge, C. The role of anti-mullerian hormone in testis differentiation reveals the significance of the TGF-beta pathway in reptilian sex determination. *Genetics* **2019**, *213*, 1317–1327. [CrossRef]
29. Rastetter, R.H.; Smith, C.A.; Wilhelm, D. The role of non-coding RNAs in male sex determination and differentiation. *Reproduction* **2015**, *150*, R93–R107. [CrossRef]
30. Burgos, M.; Hurtado, A.; Jiménez, R.; Barrionuevo, F.J. Non-Coding RNAs: lncRNAs, miRNAs, and piRNAs in sexual development. *Sex. Dev.* **2021**, *15*, 335–350. [CrossRef]
31. Real, F.M.; Sekido, R.; Lupiáñez, D.G.; Lovell-Badge, R.; Jiménez, R.; Burgos, M. A microRNA (mmu-miR-124) prevents *Sox9* expression in developing mouse ovarian cells. *Biol. Reprod.* **2013**, *89*, 78. [CrossRef]
32. Roeszler, K.N.; Itman, C.; Sinclair, A.H.; Smith, C.A. The long non-coding RNA, MHM, plays a role in chicken embryonic development, including gonadogenesis. *Dev. Biol.* **2012**, *366*, 317–326. [CrossRef] [PubMed]
33. Hacker, A.; Capel, B.; Goodfellow, P.; Lovell-Badge, R. Expression of *Sry*, the mouse sex determining gene. *Development* **1995**, *121*, 1603–1614. [CrossRef] [PubMed]
34. Jeske, Y.W.; Bowles, J.; Greenfield, A.; Koopman, P. Expression of a linear *Sry* transcript in the mouse genital ridge. *Nat. Genet.* **1995**, *10*, 480–482. [CrossRef] [PubMed]
35. Hansen, T.B.; Jensen, T.I.; Clausen, B.H.; Bramsen, J.B.; Finsen, B.; Damgaard, C.K.; Kjems, J. Natural RNA circles function as efficient microRNA sponges. *Nature* **2013**, *495*, 384–388. [CrossRef]
36. Siniscalchi, C.; Di Palo, A.; Russo, A.; Potenza, N. The lncRNAs at X chromosome inactivation center: Not just a matter of sex dosage compensation. *Int. J. Mol. Sci.* **2022**, *23*, 611. [CrossRef] [PubMed]
37. Kartha, R.V.; Subramanian, S. Competing endogenous RNAs (ceRNAs): New entrants to the intricacies of gene regulation. *Front. Genet.* **2014**, *5*, 8. [CrossRef]
38. Kim, D.; Langmead, B.; Salzberg, S.L. HISAT: A fast spliced aligner with low memory requirements. *Nat. Methods* **2015**, *12*, 357–360. [CrossRef]
39. Pertea, M.; Pertea, G.M.; Antonescu, C.M.; Chang, T.C.; Mendell, J.T.; Salzberg, S.L. StringTie enables improved reconstruction of a transcriptome from RNA-seq reads. *Nat. Biotechnol.* **2015**, *33*, 290–295. [CrossRef]
40. Robinson, M.D.; McCarthy, D.J.; Smyth, G.K. edgeR: A Bioconductor package for differential expression analysis of digital gene expression data. *Bioinformatics* **2010**, *26*, 139–140. [CrossRef]
41. Li, W.; Zhang, P.; Wu, X.; Zhu, X.; Xu, H. A novel dynamic expression of vasa in male germ cells during spermatogenesis in the Chinese soft-shell turtle (*Pelidiscus sinensis*). *J. Exp. Zool. B Mol. Dev. Evol.* **2017**, *328*, 230–239. [CrossRef]
42. Farhadi, A.; Lv, L.; Song, J.; Zhang, Y.; Ye, S.; Zhang, N.; Zheng, H.; Li, S.; Zhang, Y.; Ikhwanuddin, M.; et al. Whole-transcriptome RNA sequencing revealed the roles of chitin-related genes in the eyestalk abnormality of a novel mud crab hybrid (*Scylla serrata* ♀× *S. paramamosain* ♂). *Int. J. Biol. Macromol.* **2022**, *208*, 611–626. [CrossRef] [PubMed]
43. Livak, K.J.; Schmittgen, T.D. Analysis of relative gene expression data using real-time quantitative PCR and the 2(-Delta Delta C(T)) Method. *Methods* **2001**, *25*, 402–408. [CrossRef] [PubMed]
44. Zhao, S.; Ye, Z.; Stanton, R. Misuse of RPKM or TPM normalization when comparing across samples and sequencing protocols. *RNA* **2020**, *26*, 903–909. [CrossRef] [PubMed]
45. Enright, A.J.; John, B.; Gaul, U.; Tuschl, T.; Sander, C.; Marks, D.S. MicroRNA targets in *Drosophila*. *Genome Biol.* **2003**, *5*, R1. [CrossRef]
46. Bishop, C.E.; Whitworth, D.J.; Qin, Y.; Agoulnik, A.I.; Agoulnik, I.U.; Harrison, W.R.; Behringer, R.R.; Overbeek, P.A. A transgenic insertion upstream of *sox9* is associated with dominant XX sex reversal in the mouse. *Nat. Genet.* **2000**, *26*, 490–494. [CrossRef]
47. Koopman, P.; Gubbay, J.; Vivian, N.; Goodfellow, P.; Lovell-Badge, R. Male development of chromosomally female mice transgenic for *Sry*. *Nature* **1991**, *351*, 117–121. [CrossRef]
48. Hui, H.B.; Xiao, L.; Sun, W.; Zhou, Y.J.; Zhang, H.Y.; Ge, C.T. *Sox9* is indispensable for testis differentiation in the red-eared slider turtle, a reptile with temperature-dependent sex determination. *Zool. Res.* **2021**, *42*, 721–725. [CrossRef]
49. Warr, N.; Siggers, P.; May, J.; Chalon, N.; Pope, M.; Wells, S.; Chaboissier, M.C.; Greenfield, A. *Gadd45g* is required for timely *Sry* expression independently of RSPO1 activity. *Reproduction* **2022**, *163*, 333–340. [CrossRef]
50. Uhlenhaut, N.H.; Jakob, S.; Anlag, K.; Eisenberger, T.; Sekido, R.; Kress, J.; Treier, A.C.; Klugmann, C.; Klasen, C.; Holter, N.I.; et al. Somatic sex reprogramming of adult ovaries to testes by FOXL2 ablation. *Cell* **2009**, *139*, 1130–1142. [CrossRef]
51. Miyauchi, H.; Ohta, H.; Nagaoka, S.; Nakaki, F.; Sasaki, K.; Hayashi, K.; Yabuta, Y.; Nakamura, T.; Yamamoto, T.; Saitou, M. Bone morphogenetic protein and retinoic acid synergistically specify female germ-cell fate in mice. *EMBO J.* **2017**, *36*, 3100–3119. [CrossRef]
52. Kashimada, K.; Pelosi, E.; Chen, H.; Schlessinger, D.; Wilhelm, D.; Koopman, P. FOXL2 and BMP2 act cooperatively to regulate follistatin gene expression during ovarian development. *Endocrinology* **2011**, *152*, 272–280. [CrossRef] [PubMed]
53. Bernard, P.; Ryan, J.; Sim, H.; Czech, D.P.; Sinclair, A.H.; Koopman, P.; Harley, V.R. Wnt signaling in ovarian development inhibits *Sf1* activation of *Sox9* via the Tesco enhancer. *Endocrinology* **2012**, *153*, 901–912. [CrossRef] [PubMed]

54. Chaboissier, M.C.; Kobayashi, A.; Vidal, V.I.; Lützkendorf, S.; Van de Kant, H.J.; Wegner, M.; De Rooij, D.G.; Behringer, R.R.; Schedl, A. Functional analysis of *Sox8* and *Sox9* during sex determination in the mouse. *Development* **2004**, *131*, 1891–1901. [CrossRef] [PubMed]
55. Carré, G.A.; Couty, I.; Hennequet-Antier, C.; Govoroun, M.S. Gene expression profiling reveals new potential players of gonad differentiation in the chicken embryo. *PLoS ONE* **2011**, *6*, e23959. [CrossRef]
56. Li, J.; Zhang, X.; Wang, X.; Sun, C.; Zheng, J.; Li, J.; Yi, G.; Yang, N. The m6A methylation regulates gonadal sex differentiation in chicken embryo. *J. Anim. Sci. Biotechnol.* **2022**, *13*, 52. [CrossRef]
57. Bergstrom, D.E.; Young, M.; Albrecht, K.H.; Eicher, E.M. Related function of mouse SOX3, SOX9, and SRY HMG domains assayed by male sex determination. *Genesis* **2000**, *28*, 111–124. [CrossRef]
58. Takehana, Y.; Matsuda, M.; Myosho, T.; Suster, M.L.; Kawakami, K.; Shin, I.T.; Kohara, Y.; Kuroki, Y.; Toyoda, A.; Fujiyama, A.; et al. Co-option of *Sox3* as the male-determining factor on the Y chromosome in the fish *Oryzias dancena*. *Nat. Commun.* **2014**, *5*, 4157. [CrossRef]
59. Zhou, L.; Gui, J.F. Molecular mechanisms underlying sex change in hermaphroditic groupers. *Fish. Physiol. Biochem.* **2010**, *36*, 181–193. [CrossRef]
60. Oshima, Y.; Naruse, K.; Nakamura, Y.; Nakamura, M. *Sox3*: A transcription factor for *Cyp19* expression in the frog *Rana rugosa*. *Gene* **2009**, *445*, 38–48. [CrossRef]
61. Yan, Q.; Wu, X.; Chen, C.; Diao, R.; Lai, Y.; Huang, J.; Chen, J.; Yu, Z.; Gui, Y.; Tang, A.; et al. Developmental expression and function of *DKKL1/Dkkl1* in humans and mice. *Reprod. Biol. Endocrinol.* **2012**, *10*, 51. [CrossRef]
62. Kaneko, K.J.; DePamphilis, M.L. *Soggy*, a spermatocyte-specific gene, lies 3.8 kb upstream of and antipodal to *TEAD-2*, a transcription factor expressed at the beginning of mouse development. *Nucleic Acids Res.* **2000**, *28*, 3982–3990. [CrossRef] [PubMed]
63. Kohn, M.J.; Kaneko, K.J.; DePamphilis, M.L. DkkL1 (Soggy), a Dickkopf family member, localizes to the acrosome during mammalian spermatogenesis. *Mol. Reprod. Dev.* **2005**, *71*, 516–522. [CrossRef] [PubMed]
64. Sekido, R.; Lovell-Badge, R. Sex determination involves synergistic action of SRY and SF1 on a specific *Sox9* enhancer. *Nature* **2008**, *453*, 930–934. [CrossRef] [PubMed]
65. Dakhova, O.; O'Day, D.; Kinet, N.; Yucer, N.; Wiese, M.; Shetty, G.; Ducy, P. Dickkopf-like1 regulates postpubertal spermatocyte apoptosis and testosterone production. *Endocrinology* **2009**, *150*, 404–412. [CrossRef]
66. Smirnova, N.A.; Romanienko, P.J.; Khil, P.P.; Camerini-Otero, R.D. Gene expression profiles of *Spo11*$^{-/-}$ mouse testes with spermatocytes arrested in meiotic prophase I. *Reproduction* **2006**, *132*, 67–77. [CrossRef]
67. Baudat, F.; Manova, K.; Yuen, J.P.; Jasin, M.; Keeney, S. Chromosome synapsis defects and sexually dimorphic meiotic progression in mice lacking Spo11. *Mol. Cell.* **2000**, *6*, 989–998. [CrossRef]
68. Yunsheng, Z.; Zhiping, L.; Yuan, N.; Guixiu, O.; Changzhi, C.; Shangye, C.; Liangguo, L.; Pinhong, Y. Sexually dimorphic reproductive defects in zebrafish with *spo11* mutation. *Aquac. Res.* **2020**, *51*, 4916–4924.
69. Ozaki, Y.; Miura, C.; Miura, T. Molecular cloning and gene expression of *Spo11* during spermatogenesis in the Japanese eel, *Anguilla japonica*. *Comp. Biochem. Physiol. B Biochem. Mol. Biol.* **2006**, *143*, 309–314. [CrossRef]
70. Hsiao, P.W.; Lin, D.L.; Nakao, R.; Chang, C. The linkage of Kennedy's neuron disease to ARA24, the first identified androgen receptor polyglutamine region-associated coactivator. *J. Biol. Chem.* **1999**, *274*, 20229–20234. [CrossRef]
71. Harada, N.; Ohmori, Y.; Yamaji, R.; Higashimura, Y.; Okamoto, K.; Isohashi, F.; Nakano, Y.; Inui, H. ARA24/Ran enhances the androgen-dependent NH2- and COOH-terminal interaction of the androgen receptor. *Biochem. Biophys. Res. Commun.* **2008**, *373*, 373–377. [CrossRef]
72. Kaur, G.; Jans, D.A. Dual nuclear import mechanisms of sex determining factor SRY: Intracellular Ca^{2+} as a switch. *FASEB J.* **2011**, *25*, 665–675. [CrossRef] [PubMed]
73. Hanover, J.A.; Love, D.C.; Prinz, W.A. Calmodulin-driven nuclear entry: Trigger for sex determination and terminal differentiation. *J. Biol. Chem.* **2009**, *284*, 12593–12597. [CrossRef] [PubMed]
74. Toyota, K.; Masuda, S.; Sugita, S.; Miyaoku, K.; Yamagishi, G.; Akashi, H.; Miyagawa, S. Estrogen receptor 1 (ESR1) agonist induces ovarian differentiation and aberrant müllerian duct development in the Chinese soft-shelled turtle, *Pelodiscus sinensis*. *Zool. Stud.* **2020**, *59*, e54. [PubMed]
75. Wu, P.F.; Wang, X.F.; Gao, F.; Du, W.G. Role of *Cyp19a1* in the female pathway of a freshwater turtle species (*Mauremys reevesii*) with temperature-dependent sex determination. *Zool. Res.* **2022**, *43*, 4. [CrossRef]
76. Levasseur, A.; Paquet, M.; Boerboom, D.; Boyer, A. Yes-associated protein and WW-containing transcription regulator 1 regulate the expression of sex-determining genes in Sertoli cells, but their inactivation does not cause sex reversal. *Biol. Reprod.* **2017**, *97*, 162–175. [CrossRef]
77. Shah, K.; Seeley, S.; Schulz, C.; Fisher, J.; Gururaja Rao, S. Calcium channels in the heart: Disease states and drugs. *Cells* **2022**, *11*, 943. [CrossRef]
78. Weber, C.; Zhou, Y.; Lee, J.G.; Looger, L.L.; Qian, G.; Ge, C.; Capel, B. Temperature-dependent sex determination is mediated by pSTAT3 repression of *Kdm6b*. *Science* **2020**, *368*, 303–306. [CrossRef]
79. Li, Y.H.; Chen, T.M.; Huang, B.M.; Yang, S.H.; Wu, C.C.; Lin, Y.M.; Chuang, J.I.; Tsai, S.J.; Sun, H.S. *FGF9* is a downstream target of SRY and sufficient to determine male sex fate in ex vivo XX gonad culture. *Biol. Reprod.* **2020**, *103*, 1300–1313. [CrossRef]

80. Gao, X.; Zhu, M.; An, S.; Liang, Y.; Yang, H.; Pang, J.; Liu, Z.; Zhang, G.; Wang, F. Long non-coding RNA *LOC105611671* modulates fibroblast growth factor 9 (*FGF9*) expression by targeting oar-miR-26a to promote testosterone biosynthesis in Hu sheep. *Reprod. Fertil. Dev.* **2020**, *32*, 373–382. [CrossRef]
81. Hu, S.; Li, L.; Yeh, S.; Cui, Y.; Li, X.; Chang, H.C.; Jin, J.; Chang, C. Infiltrating T cells promote prostate cancer metastasis via modulation of FGF11→miRNA-541→androgen receptor (AR)→MMP9 signaling. *Mol. Oncol.* **2015**, *9*, 44–57. [CrossRef]
82. Zhang, L.; Yang, Q.; Xu, W.; Wu, Z.; Li, D. Integrated analysis of miR-430 on steroidogenesis-related gene expression of larval rice field eel *Monopterus albus*. *Int. J. Mol. Sci.* **2021**, *22*, 6994. [CrossRef] [PubMed]
83. Viñas, A.; Taboada, X.; Vale, L.; Robledo, D.; Hermida, M.; Vera, M.; Martínez, P. Mapping of DNA sex-specific markers and genes related to sex differentiation in turbot (*Scophthalmus maximus*). *Mar. Biotechnol.* **2012**, *14*, 655–663. [CrossRef] [PubMed]
84. Feng, B.; Li, S.; Wang, Q.; Tang, L.; Huang, F.; Zhang, Z.; Mahboobe, S.; Shao, C. lncRNA DMRT2-AS acts as a transcriptional regulator of *dmrt2* involving in sex differentiation in the Chinese tongue sole (*Cynoglossus semilaevis*). *Comp. Biochem. Physiol. B Biochem. Mol. Biol.* **2021**, *253*, 110542. [CrossRef] [PubMed]

biology

MDPI

Article

The Important Role of Sex-Related *Sox* Family Genes in the Sex Reversal of the Chinese Soft-Shelled Turtle (*Pelodiscus sinensis*)

Yubin Wang [1,2], Xiangzhong Luo [2], Chunjuan Qu [3], Tao Xu [4], Guiwei Zou [2,*] and Hongwei Liang [2,*]

1 College of Fisheries and Life Science, Shanghai Ocean University, Shanghai 201306, China; wybingo992@126.com
2 Key Laboratory of Aquatic Genomics, Ministry of Agriculture and Rural Affairs, Yangtze River Fisheries Research Institute, Chinese Academy of Fisheries Science, Wuhan 430223, China; lxz@yfi.ac.cn
3 Bengbu Aquatic Technology Promotion Center, Bengbu 233000, China; cjqubbsc@outlook.com
4 College of Biology & Pharmacy, China Three Gorges University, Yichang 443002, China; xutaoxinong@126.com
* Correspondence: zougw@yfi.ac.cn (G.Z.); lianghw@yfi.ac.cn (H.L.)

Simple Summary: *Pelodiscus sinensis* is an important aquatic economic species in China with sexual dimorphism. All-male breeding is becoming a research hotspot. Here, comparative transcriptome analyses of female, male, and pseudo-female gonads were performed. We found that the differences between males and pseudo-females were mainly related to steroid hormone synthesis at the transcriptome level. When it comes to the *sox* family genes, *sox3* may have a role in the process of sex reversal from male to pseudo-female, when *sox8* and *sox9* were inhibited by exogenous estrogen.

Abstract: The Chinese soft-shelled turtle *Pelodiscus sinensis* shows obvious sexual dimorphism. The economic and nutrition value of male individuals are significantly higher than those of female individuals. Pseudo-females which are base to all-male breeding have been obtained by estrogen induction, while the gene function and molecular mechanism of sex reversal remain unclear in *P. sinensis*. Here, comparative transcriptome analyses of female, male, and pseudo-female gonads were performed, and 14,430 genes differentially expressed were identified in the pairwise comparison of three groups. GO and KEGG analyses were performed on the differentially expressed genes (DEGs), which mainly concentrated on steroid hormone synthesis. Furthermore, the results of gonadal transcriptome analysis revealed that 10 sex-related *sox* genes were differentially expressed in males vs. female, male vs. pseudo-female, and female vs. pseudo-female. Through the differential expression analysis of these 10 *sox* genes in mature gonads, six *sox* genes related to sex reversal were further screened. The molecular mechanism of the six *sox* genes in the embryo were analyzed during sex reversal after E_2 treatment. In mature gonads, some *sox* family genes, such as *sox9 sox12*, and *sox30* were highly expressed in the testis, while *sox1*, *sox3*, *sox6*, *sox11*, and *sox17* were lowly expressed. In the male embryos, exogenous estrogen can activate the expression of *sox3* and inhibit the expression of *sox8*, *sox9*, and *sox11*. In summary, *sox3* may have a role in the process of sex reversal from male to pseudo-female, when *sox8* and *sox9* are inhibited. *Sox* family genes affect both female and male pathways in the process of sex reversal, which provides a new insight for the all-male breeding of the Chinese soft-shelled turtle.

Keywords: transcriptome; *sox* family genes; *Pelodiscus sinensis*; estradiol; pseudo-female; sex-related

Citation: Wang, Y.; Luo, X.; Qu, C.; Xu, T.; Zou, G.; Liang, H. The Important Role of Sex-Related *Sox* Family Genes in the Sex Reversal of the Chinese Soft-Shelled Turtle (*Pelodiscus sinensis*). *Biology* **2022**, *11*, 83. https://doi.org/10.3390/biology11010083

Academic Editor: Patricia Pereiro

Received: 30 November 2021
Accepted: 2 January 2022
Published: 6 January 2022

1. Introduction

Pelodiscus sinensis, known as the Chinese soft-shelled turtle, is widely distributed in many freshwater areas, such as rivers and lakes in China, Korea, Russia, Thailand, Vietnam, and Japan [1]. This turtle shows obvious sexual dimorphism: males have a larger size, faster growth rate, and wider and thicker calipash than females. Furthermore, the male juvenile is more popular for aquaculture practices because it is priced higher than the female juvenile. Therefore, all-male breeding of *P. sinensis* by using sex control approaches has become important [2].

In aquaculture, unisexual offspring cannot be obtained by controlling the incubation temperature during the embryo development of *P. sinensis*. In addition, some studies have identified ZZ/ZW micro-sex chromosomes in *P. sinensis*, which is significantly different from the typical temperature-dependent sex determination (TSD) in *Trachemys scripta* [3–5]. Sex-specific markers have been developed to accurately identify the genetic sex of a turtle by using RAD-Seq technology [1]. These studies suggest that genetic sex determination can be used for *P. sinensis*. Pseudo-female turtles (ΔZZ) with a female phenotype and male genotype can be obtained by using estradiol (E_2) to induce male embryos (ZZ) in the sex determination stage to differentiate into physiological females [6]. The pseudo-female turtle is used as the female parent (ΔZZ) and the male turtle is used as the male parent (ZZ) when they reach sexual maturity. All their offspring should be males (ZZ). Therefore, it is important to study the sex determination mechanism of *P. sinensis* for the all-male aquaculture of this species.

Unlike the typical TSD mechanism of turtles such as *T. scripta*, the sex determination mechanism of *P. sinensis* is a more complex process that involves genes and hormones [7,8]. Estrogen is a gonadal steroid hormone that plays a key role in female sex determination in vertebrates [9]. E_2 can induce the expression of *cyp19a1* in an embryo, promote ovary development, and even induce the sex reversal of *P. sinensis* [10]. Some sex-specific genes commonly found in other species, such as *dmrt1* [11,12], *cyp19a1* [13], *foxl2* [14], and *rspo1* [6,15], have been reported in the preliminary studies of *P. sinensis*. These genes were not only directly involved in the sex determination of *P. sinensis* but also affected by exogenous E_2, which was significantly changed during the sex determination period [6,10,13,14,16]. These studies have not formed a systematic molecular mechanism, and further analysis will provide a new understanding for the sex differentiation and reversal of *P. sinensis*.

A series of SRY-related high-mobility group (HMG)-box (SOX) transcription factors with an HMG box DNA-binding domain are called *sox* family genes, which play important roles in embryonic development, neurogenesis, and other aspects [17]. *Sry* was the first sox transcription factor to be identified, and it is involved in male sex determination in mammals. When *sry* is absent, male XY mice develop into females [18]. *Sox9* was specifically expressed in the early stage of gonadal differentiation in male *P. sinensis* embryos, and it is an important gene involved in male sex determination in vertebrates [19]. The loss of *sox9* resulted in the reversal from male to female in mice [20]. In vertebrates, *sox3* inhibited the expression of *sox9* in the ovaries and promoted ovary development and even directly activated the transcription of *cyp19* [21–23]. In addition, *sry* can inhibit the expression of *sox3* and promote the expression of *sox9*, ensuring that mice can be differentiated into males [24]. Other *sox* genes are also involved in sex differentiation and gonadal development [25–31]. To date, the interaction between sex-related *sox* genes and estrogen in the sex differentiation and reversal of *P. sinensis* has not yet been elucidated.

With the rapid development of omics research, high-throughput and high-sensitivity second-generation transcriptome sequencing technologies have been widely used to breed aquaculture animals, such as *Cynoglossus semilaevis* [32] and *Oreochromis niloticus* [33]. Currently, transcriptome studies on *P. sinensis* mainly focus on growth and immunity [34,35], and there is a lack of transcriptome studies on gonad differentiation and sex reversal. In this study, firstly, a comparative transcriptome analysis was performed using the gonadal tissues of E_2-induced pseudo-female and female and male *P. sinensis*. The expression profiles of differentially expressed genes (DEGs) in the gonads of *P. sinensis* were established. Candidate genes and signaling pathways related to gonad differentiation and development were analyzed. Then, the differential expressions of significantly different *sox* family genes in pseudo-female, female, and male gonads were analyzed. The *sox* genes which may be involved in gonad development and function maintenance during the sex reversal were further screened and their expression patterns were analyzed after E_2 treatment. These results provided transcriptome resources for analyzing the molecular mechanism of gonad differentiation and sex reversal of *P. sinensis*.

2. Materials and Methods

2.1. Ethical Approval

The procedures in this study were performed according to the Animal Experimental Ethical Inspection of Laboratory Animal Centre of the Yangtze River Fisheries Research Institute, Chinese Academy of Fishery Sciences (Wuhan, China; ID Number: 20200118).

2.2. Sample Collection

Two-year-old *P. sinensis*, 3 males (mean weight 1075 $\pm$ 126 g, recorded as M-1, M-2, and M-3), 3 females (mean weight 816 $\pm$ 72 g, recorded as F-1, F-2, and F-3), and 3 pseudo-females (mean weight 929 $\pm$ 77g, female phenotype and male genotype, recorded as PF-1, PF-2, and PF-3), were collected from Anhui Xijia Agricultural Development Co. Ltd. (Bengbu, Anhui Province, China). The pseudo-female turtles were obtained by treating the eggs with 30 mg/mL E_2 at the stage 12 of embryo development which was the critical period of sex differentiation [36]. The biological sex and genetic sex of juvenile turtles were identified by phenotypic and sex-specific markers, respectively after they were cultured in greenhouse for 8 months [37]. The treated embryos at gonadal differentiation period (stage 12, 13, 14, 15, 16, and 17 of embryonic development) were collected [36], and the sex of the embryos was identified using sex-specific markers. [1]. All turtles were anesthetized with 0.05% MS-222 (Sigma, St. Louis, MO, USA), and the gonad tissues were collected and stored in liquid nitrogen.

2.3. RNA Extraction, Library Preparation and Transcriptome Sequencing

The total RNA was extracted from the gonads by using TRIzol reagent (Invitrogen, Carlsbad, CA, USA), according to the manufacturer's instructions. The RNA quality was monitored using 1.5% agarose gels. The RNA purity was checked with the NanoPhotometer® spectrophotometer (Implen, Westlake Village, CA, USA). The RNA integrity was tested with the RNA Nano 6000 Assay Kit of the Agilent Bioanalyzer 2100 system (Agilent Technologies, Santa Clara, CA, USA). Then, RNA concentration was measured using the Qubit® RNA Assay Kit in Qubit® 3.0 Fluorometer (Life Technologies, Carlsbad, CA, USA). The NEBNext® UltraTM RNA Library Prep Kit for Illumina® (NEB, Ipswich, MA, USA) was used to generate the sequencing libraries, and the Agilent Bioanalyzer 2100 System was used to assess library quality. The libraries were sequenced on an Illumina Hiseq X Ten platform, and 150 bp paired-end reads were obtained. The raw reads were filtered to remove the low-quality reads and reads with the adapter and N content more than 10% and obtain clean reads. Then, FastQC v1.2 was used to evaluate the quality of the sequencing data.

2.4. Identification of the Differentially Expressed Genes (DEGs)

The clean reads were aligned to the *P. sinensis* reference genome (https://www.ncbi.nlm.nih.gov/genome/?term=Pelodiscus+sinensis, PRJNA221645, Pelsin_1.0) by using the software Tophat2 v2.1.1 [38] and mapped to the coding sequences with bowtie2 v2.2.2 [39]. The gene and transcript expression levels were calculated using fragments per kilobase of transcripts per million bases [40] values in RSEM with default settings [41]. By using fragments per kilobase per million bases (FPKM) transformation, the paired-end reads from the same fragment were used as a fragment to obtain gene and transcription levels. Principal component analysis (PCA) was used to detect the similarity detection of three biological repeats. The DEGs were identified using R package DEseq2 [42], with false discovery rate (FDR) < 0.05 and $\log_2$FC (fold change (condition 2/condition 1) > 1 or $\log_2$FC < -1. The upregulated DEGs showed FDR < 0.05 and $\log_2$FC > 1, and the downregulated DEGs, FDR < 0.05 and $\log_2$FC < -1.

2.5. GO and KEGG Pathway Enrichment Analysis of DEGs

GOseq v1.22 was used for the GO enrichment analysis, which is based on the algorithm of hypergeometric distribution. The GO term of FDR < 0.05 was considered as a significantly enriched term. The KEGG enrichment analysis was used as a hypergeometric test to identify significantly enriched pathways relative to the annotated genes. KOBAS v3.0 was used for the KEGG pathway enrichment analysis. A pathway with FDR < 0.05 was defined as significantly enriched with DEGs.

2.6. Validation of the Transcriptome with RT-qPCR

To verify the accuracy of the transcriptomic data, 13 DEGs related to gonadal differentiation and development were randomly selected for RT-qPCR. All the selected DEGs showed significantly different expressions in different samples. *Gapdh* was used as the endogenous reference gene, and RT-qPCR primers for the selected DEGs were designed using Primer Premier 5 (Table S1). The HiScript® III 1st Strand cDNA Synthesis Kit (+gDNA wiper) (Vazyme, Wuhan, China) was used to synthesize the template cDNA. The ChamQ™ Universal SYBR® qPCR Master Mix (Vazyme, Wuhan, China) was used to establish the reaction system (total volume, 20 μL): 10 μL of 2 × Master Mix, 0.4 μL of each primer (total, 10 μM), 1 μL of template cDNA, and 8.2 μL of RNase-free ddH_2O. The reaction was performed using the QuantStudio® 5 Real-Time PCR Instrument (Applied Biosystems, Thermo Fisher Scientific, Waltham, MA, USA), and the qPCR program was as follows: 95 °C for 3 min, followed by 40 cycles of 95 °C for 15 s and 60 °C for 34 s. The relative gene expression levels were calculated using the $2^{-\Delta\Delta CT}$ method [43], and $\log_2$ (fold change) was used for comparison with the RNA-seq data. The Duncan method of SPSS 22 was used for the significance analysis.

2.7. Expression Patterns of Sox Family Genes during Sex Reversal

To our knowledge, *sox* family genes play important roles in sex differentiation. Ten sex-related *sox* family genes were screened from the transcriptomic data on the basis of $p < 0.05$ and $\log_2FC > 1$ or $\log_2FC < -1$ to analyze their molecular functions in sex reversal. Furthermore, the expression patterns of the identified genes were analyzed in the male, female, and pseudo-female gonads. Next, six *sox* genes with significantly different expressions between pseudo-female and female or male were screened, and qPCR was used to detect the expression levels of the selected genes during E_2-induced embryonic sex reversal.

3. Results

3.1. Quality Assessment of the Sequencing Data

The study was conducted according to the experimental process (Figure 1A). Transcriptome sequencing was performed using the gonads of the female (F), male (M), and pseudo-female (PF) of *P. sinensis* (Table 1). The number of clean reads in all samples ranged from 42,130,694 to 50,909,630. The GC content of each sample was between 48% and 51%. Q30 bases were more than 92%, indicating high sequencing quality. The clean reads were aligned to the reference genome of *P. sinensis*, and the results showed that 67.34–72.69% of the clean reads were successfully mapped (Table 2). The similarity between the three biological replicates was tested by principal component analysis, and the results showed good similarity between the samples (Figure 1B). These results showed that the sequencing data can be further analyzed.

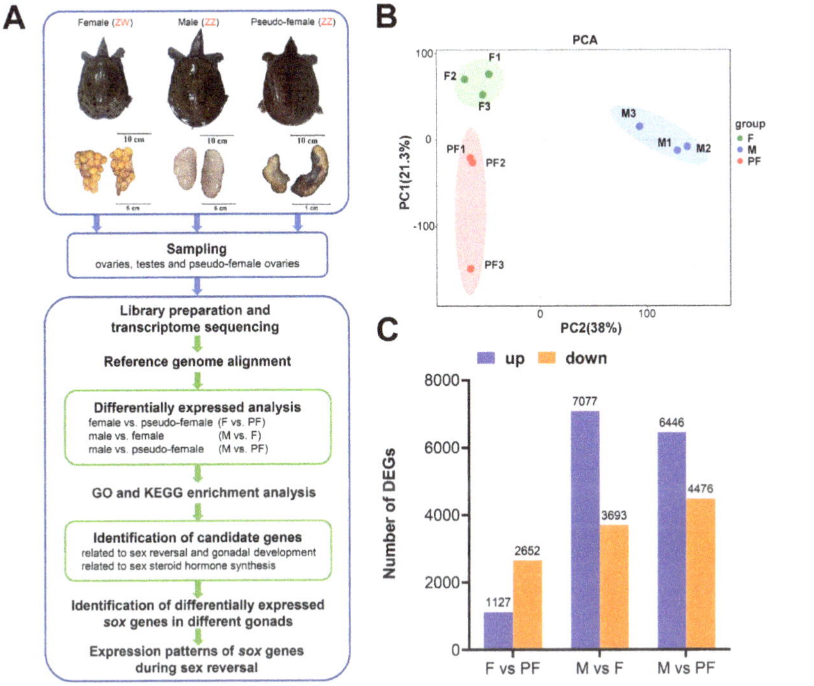

Figure 1. Experimental design and data analysis. (**A**) Diagram of the experiment. (**B**) Principal component analysis between three sets of data. (**C**) The number of DEGs in the three comparisons of *P. sinensis*. Up-regulated DEGs (blue), and down-regulated DEGs (orange) are presented by a histogram. Filter threshold is FDR < 0.05, $\log_2$FoldChange > 1 or $\log_2$FC < −1.

Table 1. Summary of the sequencing data quality.

Sample	Clean Reads	Clean Bases	Clean Reads Pair	Q20 (%)	Q30 (%)	GC (%)
F-1	43,494,664	6,524,199,600	21,747,332	97	93	48
F-2	48,689,394	7,303,409,100	24,344,697	97	92	51
F-3	50,909,630	7,636,444,500	25,454,815	97	92	48
M-1	48,936,618	7,340,492,700	24,468,309	97	92	48
M-2	44,953,598	6,743,039,700	22,476,799	97	93	49
M-3	42,130,694	6,319,604,100	21,065,347	97	92	48
PF-1	45,059,436	6,758,915,400	22,529,718	97	92	48
PF-2	49,918,734	7,487,810,100	24,959,367	97	93	48
PF-3	49,677,990	7,451,698,500	24,838,995	97	93	49

Table 2. Summary of the clean reads mapped to the reference genome.

Sample	Clean Reads Pair	Mapped Reads	Uniquely Mapped Reads	Multiple Mapped Reads
F-1	21,747,332 (100.00%)	15,704,473 (72.21%)	13,724,054 (63.11%)	1,980,419 (9.11%)
F-2	24,344,697 (100.00%)	16,394,180 (67.34%)	14,155,999 (58.15%)	2,238,181 (9.19%)
F-3	25,454,815 (100.00%)	18,295,164 (71.87%)	15,903,317 (62.48%)	2,391,847 (9.40%)
M-1	24,468,309 (100.00%)	17,785,438 (72.69%)	15,500,943 (63.35%)	2,284,495 (9.34%)
M-2	22,476,799 (100.00%)	16,292,285 (72.48%)	14,220,854 (63.27%)	2,071,431 (9.22%)
M-3	21,065,347 (100.00%)	15,184,447 (72.08%)	13,244,886 (62.88%)	1,939,561 (9.21%)
PF-1	22,529,718 (100.00%)	16,110,869 (71.51%)	14,009,770 (62.18%)	2,101,099 (9.33%)
PF-2	24,959,367 (100.00%)	17,777,029 (71.22%)	15,329,064 (61.42%)	2,447,965 (9.81%)
PF-3	24,838,995 (100.00%)	18,011,332 (72.51%)	15,716,312 (63.27%)	2,295,020 (9.24%)

3.2. Analysis of DEGs

The pairwise comparisons of F and PF, M and F, and M and PF were used to identify the DEGs. Genes with $|\log_2 FC| \geq 1$ and FDR < 0.05 were determined to be DEGs. In the present study, a total of 14,430 DEGs were obtained from the three comparisons after filtration. In F vs. PF, 1127 upregulated DEGs and 2652 downregulated DEGs were identified in the female (Figure 1C). According to the results of M vs. F, 7077 DEGs were upregulated in the male and 3693 DEGs were upregulated in the female. When compared with PF, M showed 6446 upregulated and 4476 downregulated DEGs. Of the 14,430 DEGs, 3017 and 975 sex different genes were specifically expressed in the males and females. In addition, 147 genes expressed in only the pseudo-female but not in both female and male were screened (Table S2).

3.3. GO and KEGG Enrichment Analysis of DEGs

To investigate the potential functions of genes in *P. sinensis*, the DEGs were annotated in the GO database. In F vs. PF, 55, 29, and 41 GO terms were significantly enriched in biological process (BP), cellular component (CC), and molecular function (MF), respectively. In M vs. F, 63, 46, and 74 GO terms were significantly enriched in BP, CC, and MF, respectively. In M vs. PF, 86, 43, and 57 GO terms were significantly enriched in BP, CC, and MF, respectively (Table S3). The significantly enriched GO terms related to sexual reversal can be found in F vs. PF and M vs. PF, of which the DEGs enriched in BP were mainly associated with metabolism and cell cycle, such as metabolic process (GO: 0008152), primary metabolic process (GO: 0044238), and DNA replication (GO: 0006260) (Figure 2). On the other hand, reproduction (GO: 0000003), reproductive process (GO: 0022414), and other reproductive activities were significantly enriched between males and females (Figure S1). The three groups were all significantly enriched in terms related to chromosome replication, catalytic activity, and molecular binding.

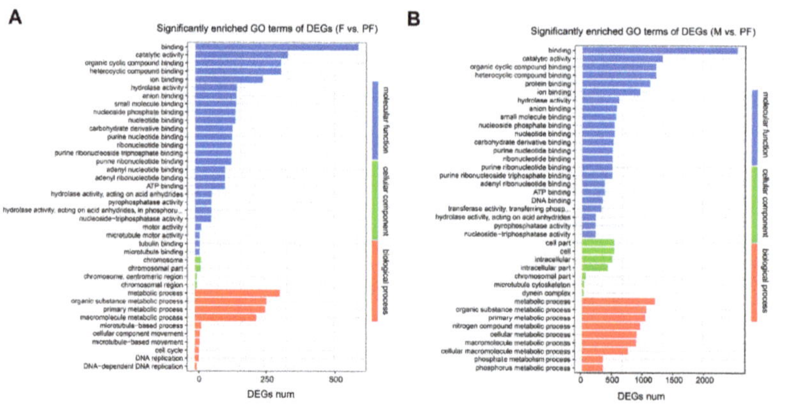

Figure 2. Significantly enriched GO terms of DEGs comparison among the groups. (**A**) F vs. PF. (**B**) M vs. PF. Statistical significance GO terms were determined based on FDR < 0.05. The x-axis indicates the number of genes, and the y-axis indicates the second-level GO term.

KEGG enrichment analysis was performed to reveal the functional characteristics of the DEGs. In this study, a total of 340 signaling pathways were found, and phenylpropanoid biosynthesis (ko00940) was only significantly enriched between males and females. Indole alkaloid biosynthesis (ko00901) and betalain biosynthesis (ko00965) were only enriched between males and pseudo-females (Table S4). In F vs. PF, 1652 DEGs were mainly involved in cell cycle (ko04110), cell cycle—yeast (ko04111), and meiotic—yeast (ko04113) (Figure 3A). In M vs. PF, 3469 DEGs were observed in 335 signaling pathways, and gap junction (ko04540), phosphatidylinositol signaling system (ko04070), and purine metabolism (ko00230) were the most prominent (Figure 3B). Furthermore, most DEGs were enriched in cell cycle (ko04110), cell cyclic–yeast (ko04111), and oocyte meiosis (ko04114) in M vs. F (Figure S2). Among these pathways, those involved in physiological activities, such as cell cycle (ko04110) and purine metabolism (ko00230), were significantly enriched. However, great differences existed in the pathways involved in the synthesis and metabolism of steroid hormones between males and pseudo-females. Metabolism of xenobiotics by cytochrome P450 (ko00980), drug metabolism–cytochrome P450 (ko00982), and other steroid metabolic pathways were significantly enriched in the pseudo-females. Reproductive-related pathways such as oocyte meiosis (ko04114), meiosis–yeast (ko04113), and progesterone-mediated oocyte maturation (ko04914) were significantly enriched in the males and females.

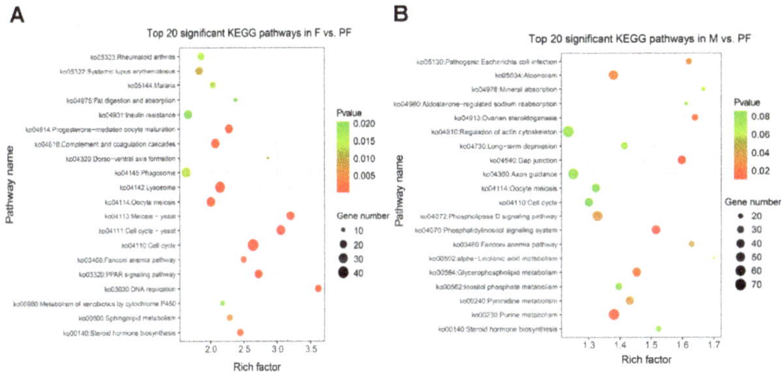

Figure 3. Top 20 KEGG enrichment significant pathways. (**A**) Female vs. Pseudo-female. (**B**) Male vs. Pseudo-female. Rich factor is the ratio of DEGs and back genes in the pathway, the closer *p* value is to zero, the more significant is the enrichment.

3.4. Screening of Candidate DEGs Related to Sex Reversal and Gonadal Development

In this study, sex-related GO terms and KEGG signaling pathways were screened out, e.g., meiotic cell cycle (GO: 0051321), sexual reproduction (GO: 0019953), steroid hormone biosynthesis (ko00140), ovarian steroidogenesis (ko04913), and progesterone-mediated oocyte maturation (ko04914) (Table S5). They were mainly related to reproductive activities, such as steroid hormone synthesis, gonadal development, oocyte maturation, gametogenesis, and binding. Twenty-eight candidate DEGs involved in gonadal development and sex reversal were mainly screened from steroid synthesis and gonadal development pathways and genes significantly expressed between pseudo-females and common sex types (Table 3). Some genes, such as corticosteroid 11-β-dehydrogenase isozyme 2 (*hsd11b2*) and 17-β-Hydroxysteroid dehydrogenase type 7 (*hsd17b7*), were differentially expressed in steroid hormone biosynthesis. Some genes showed sex-specific expression patterns. *Foxl2*, *fgf8*, *fgf9*, *bmp15*, and *gdf9* were highly expressed in the pseudo-females, and *dmrt1*, *klhl10*, *theg*, and *fam71d* were specifically expressed in the males. Moreover, Genes (*wnt1*, *wnt2*, *rspo1*, and *rspo2*) involved in wnt signaling pathway (ko04310) were highly expressed in the ovaries. Some *sox* family genes (*sox1*, *sox2*, *sox3*, *sox11*, *sox12*, and *sox17*) were highly expressed in the pseudo-female ovary, but the expression of *sox30* was higher in the testis. Among them, *sox17* was enriched in wnt signaling pathway of female pathway. A total of 17 *sox* family genes were obtained from the transcriptome data, most of which were differentially expressed in pseudo-female ovaries (Table S6). Sex-related *sox* genes will be further screened and analyzed for their roles in the sex reversal.

Table 3. Candidate differentially expressed genes (DEGs) putatively related to steroid synthesis and gonadal development.

Gene	Description	Log₂FoldChange		
		F vs. PF	M vs. F	M vs. PF
dmrt1	Double sex and mad-3 related transcription factor 1	−2.97	9.92	6.96
sox8	Sry-like HMG box 8	1.64	0.04	−0.88
sox30	Sry-like HMG box 30	−1.65	7.39	5.74
klhl10	Kelch-like protein 10	0.12	13.98	14.11
fam71d	Family with sequence similarity 71, member D	−1.35	12.74	11.38
theg	Testicular haploid expressed gene protein	2.03	8.56	10.58
hsd11b2	Corticosteroid 11-β-dehydrogenase isozyme 2	−0.98	3.97	2.99
hsd17b7	17-β-Hydroxysteroid dehydrogenase type 7	0.55	2.10	2.65
LOC106731888	17-β-Hydroxysteroid dehydrogenase type 8 like (*hsd17b8*)	0.62	3.09	3.71
LOC102455057	Cytochrome P450 cholesterol side-chain cleavage (*cyp11a*)	−2.81	3.99	1.16
LOC102453952	3-β-hydroxysteroid dehydrogenase/Delta 5 (*hsd3b*)	1.24	4.34	5.57
LOC102459111	Cytochrome P450 1A1 (*cyp19a1*)	−1.55	−6.93	−8.48
LOC112546066	Bone morphogenetic protein 15-like (*bmp15*)	−1.51	−9.91	−11.42
sox1	Sry-like HMG box 1	−0.34	0.82	−1.53
sox2	Sry-like HMG box 2	−0.24	0.84	−1.60
sox3	Sry-like HMG box 3	−0.78	−4.09	−4.87
sox11	Sry-like HMG box 11	−0.71	−1.33	−2.04
sox12	Sry-like HMG box 12	0.21	0.97	−2.49
sox17	Sry-like HMG box 17	−0.13	−2.00	−2.13
foxl2	Forkhead box L2	0.31	−7.43	−7.12
fgf8	Fibroblast growth factor 8	−5.80	−0.73	−6.53
fgf9	Fibroblast growth factor 9	−1.79	−0.73	−2.52
gdf9	Growth/differentiation factor 9	−2.61	−10.57	−13.17
wnt1	Wingless-type MMTV integration site family, member 1	−2.00	−3.20	−5.18
wnt2	Wingless-type MMTV integration site family, member 2	−1.38	−3.59	−4.97
rspo2	R-spondin-2	−0.46	−1.26	−1.72
rspo3	R-spondin-3	−0.78	−1.47	−2.25
hsd17b1	17-β-Hydroxysteroid dehydrogenase type 1	−2.71	−5.42	−8.13

3.5. DEGs Were Verified with RT-qPCR

Ten sex-related DEGs were selected randomly from the candidate sex-related genes for RT-qPCR verification. Five genes (*hsd3b*, *hsd11b2*, *hsd17b7*, *hsd17b8*, *cyp19a1*) were involved in sex steroid hormone synthesis, three were *sox* family genes (*sox3*, *sox17*, *sox30*), and two female-specific genes (*bmp15* and *gdf9*). The validation results were generally consistent with the transcriptomic data, which confirmed the reliability of the transcriptomic data (Figure 4).

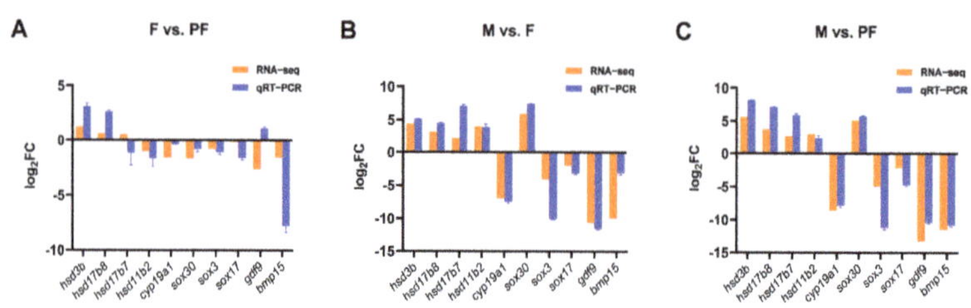

Figure 4. Validation of the RNA-seq data by RT-qPCR. (**A**) F vs. PF. (**B**) M vs. F. (**C**) M vs. PF. The x-axis presents the gene name, and the y-axis presents a fold change in gene expression.

3.6. Identification of Sex-Related Sox Genes in Different Gonads

Of the sex-related *sox* family genes, *sox1*, *sox2*, *sox3*, *sox6*, *sox8*, *sox9*, *sox11*, *sox12*, *sox17*, and *sox30* were screened on the basis of the transcriptomic data to analyze their molecular functions during E_2-induced sex reversal of *P. sinensis*. The sex-related *sox* family genes were screened from the DEGs of F vs. PF and M vs. PF, on the basis of $p < 0.05$ and $\log_2FC > 1$ or $\log_2FC < -1$. The expression patterns of these genes were analyzed in the normal ovary, testis, and pseudo-female ovary (Figure 5). In this study, *sox1*, *sox3*, *sox6*, *sox9*, *sox11*, *sox12*, *sox17*, and *sox30* showed sex specificity. The expression levels of *sox9*, *sox12* and *sox30* were higher in the testis than in the ovary, whereas *sox1*, *sox3*, *sox6*, *sox11*, and *sox17* showed the opposite trend. No significant differences were observed in the expression levels of *sox2*, and *sox8* in the females and males. Furthermore, the expression levels of *sox3*, *sox8*, *sox11*, and *sox17* were significantly higher in the pseudo-female gonads than in the males and females, whereas the expression level of *sox9* was significantly lower. The results suggest that these genes may play an important role in the development and maturation of pseudo-female gonads, even during sexual reversal and differentiation. Therefore, *sox3*; *sox8*, *sox9*, *sox11*, *sox17*, and *sox30* were selected for further analyses during the sex reversal of exogenous estrogen treatment. These genes showed significant differences between pseudo-female ovary and testis.

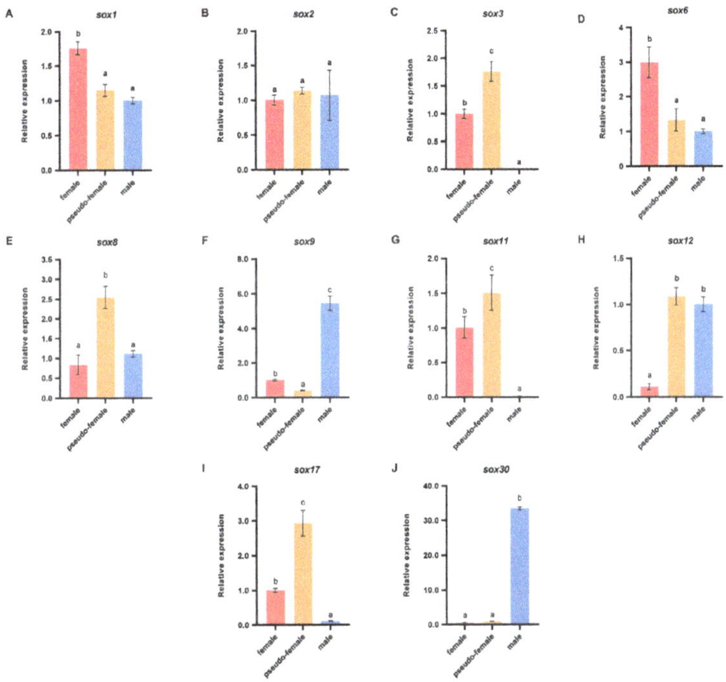

Figure 5. Differential expressions of sex-related *sox* genes in different gonads of *P. sinensis*. (**A**) *sox1*. (**B**) *sox2*. (**C**) *sox3*. (**D**) *sox6*. (**E**) *sox8*. (**F**) *sox9*. (**G**) *sox11*. (**H**) *sox12*. (**I**) *sox17*. (**J**) *sox30*. Each value is presented as the mean ± SD of three repetitions. One-way ANOVA with Tukey post-hoc tests were used to analyze the means. Different letters indicate significant differences.

3.7. Expression Patterns of Sox Genes in the Embryonic Sex Reversal after E_2 Treatment

The expression patterns of *sox3*, *sox8*, *sox9*, *sox11*, *sox17*, and *sox30* in the embryo were recorded during sex reversal after E_2 treatment. In the female embryos treated with E_2, the expression pattern of *sox3* was significantly upregulated and peaked at stage 13 and then decreased, but it was still higher than that in the untreated embryo ($p < 0.05$, Figure 6). *Sox8* and *sox17* were definitely inhibited from stage 13, and their levels then remained low. Although the expression level of *sox30* was significantly different at stage 13 and 17, it was not affected by E_2 on the whole. No arresting changes were observed in the expression patterns of *sox9*, and *sox11*. In the male embryos, the expression level of *sox3* was higher than that in the control, and it reached peaked at stage 15 and then decreased (Figure 7). However, *sox8*, *sox9*, and *sox11* were dramatically inhibited during sex differentiation. During the differentiation of the primordial gonads into ovaries, the expression level of *sox3* was obviously increased by exogenous estrogen. At this point, *sox8*, *sox9*, and *sox11* were inhibited in the male embryos. It was suggested that *sox3* may play an important role in the sex reversal from male to pseudo-female. No effect of exogenous estrogen on the expression level of *sox30* was found in male embryos.

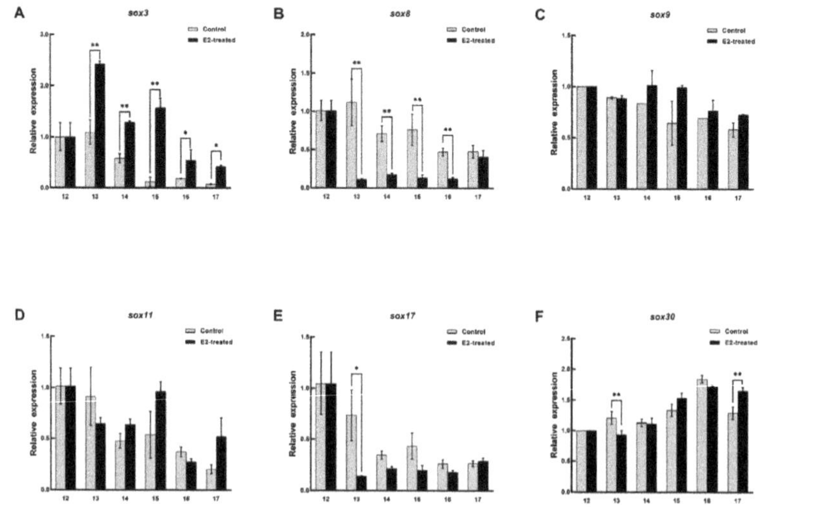

Figure 6. Expression changes of six *sox* genes during the sex differentiation of female embryo after estradiol treatment. (**A**) *sox3*. (**B**) *sox8*. (**C**) *sox9*. (**D**) *sox11*. (**E**) *sox17*. (**F**) *sox30*. The x axis represents the embryonic development stage, and the y axis represents the relative expression level. Each value is presented as the mean ± SD of three repetitions. One-way ANOVA with Tukey post-hoc tests were used to analyze the means. * $p < 0.05$ and ** $p < 0.01$.

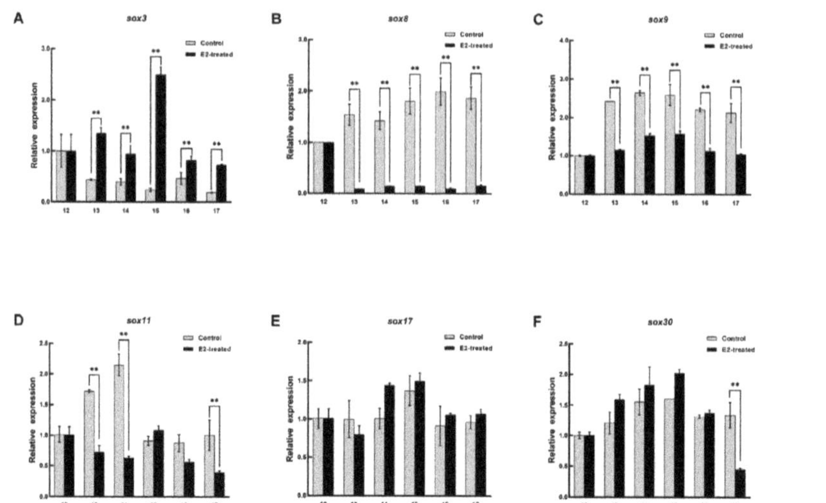

Figure 7. Expression changes of six *sox* genes during the sex differentiation of male embryo after estradiol treatment. (**A**) *sox3*. (**B**) *sox8*. (**C**) *sox9*. (**D**) *sox11*. (**E**) *sox17*. (**F**) *sox30*. The x axis represents the embryonic development stage, and the y axis represents the relative expression level. Each value is presented as the mean ± SD of three repetitions. One-way ANOVA with Tukey post-hoc tests were used to analyze the means. ** $p < 0.01$.

4. Discussion

The research field of sex determination and gonadal development mechanism of *P. sinensis* is widely concerned because of the economic characteristics associated with significant sexual dimorphism. In order to obtain all-male offspring, the pseudo-females (ΔZZ) after sex reversal will be reproduced as the female parent. Pseudo-females resemble females in gonadal morphology. Our transcriptome results showed that pseudo-females were closer to females at mRNA level. During the sex differentiation of vertebrates, exogenous sex steroids can influence the phenotypic sex greatly [44] and *sox* family genes play a crucial role in the process. In this study, a comparative transcriptome analysis was performed using the gonadal tissues of *P. sinensis* males, females, and pseudo-females. The objective of this study was to identify DEGs in the gonads of the different sex types of *P. sinensis*. Differentially expressed genes between male and pseudo-female gonads may be the key genes during the sex reversal. Further, expression patterns of *sox* family genes were analyzed during sex reversal after E_2 treatment to explore the role of *sox* family genes in sex reversal.

Sex steroid hormones, especially androgen and E_2, play an important regulatory role in reproductive activities such as sex determination, gametogenesis, and storage in turtles and other vertebrates [45–47]. In the gonadal transcriptome of *P. sinensis*, *cyp19a1*, *cyp11a*, *hsd3b*, *hsd11b2*, and *hsd17b7* were found to be significantly enriched in steroid hormone biosynthesis and ovarian steroid hormone genesis pathways. Previous studies have shown that *star*, *cyp11a1*, and *hsd3b* are closely related to gonadal development and gametogenesis in fish [45,48,49]. *Hsd11b2* is involved in the synthesis of androgen 11-kT, and it plays an important role in the male sex differentiation of vertebrates such as *Epinephelus coioides* [50] and *Cynoglossus semilaevis* [51]. *Hsd17b7* could convert estrone to E_2 and played an important role in mouse embryonic development [52]. The expression levels of *cyp11a*, *hsd3b*, and *hsd11b2* were observably higher in male *P. sinensis* than in the female, which was consistent with the results of *Oryzias latipes* [53]. However, the expression level of *hsd17b7* was inconsistent with that reported in previous studies. The expression level of *hsd17b7* was higher in the males than in the females and pseudo-females. This may be because, when *cyp19a1* is inhibited, the male turtle upregulates the expression of *hsd17b7* to maintain life activities. On the other hand, steroid biosynthesis pathways, such as steroid hormone biosynthesis (ko00140), ovarian steroidogenesis (ko04913), and progesterone-mediated oocyte maturation (ko04914), were enriched in pseudo-female, suggesting that the pseudo-females could maintain ovarian development and maturation through these pathways as females do.

In addition, several female-specific genes have been identified. *Fgf9* is a downstream target of the male sex-determining gene *sox9*, and it participates in male sex determination by positive feedback regulation of *sox9*. However, *fgf9* inhibits the activation of the wnt signaling pathway and expression of *foxl2* [54]. Mice that lacked *fgf9* showed sex reversal from male to female [55]. Our results show some differences: the expression levels of *fgf8* and *fgf9* were significantly higher in the pseudo-females than in the males and females. Some studies have shown that *fgf8* and *fgf9* promote follicular maturation during gonadal development [56]. Therefore, *fgf8* and *fgf9* may be key genes during estrogen-induced sex reversal of *P. sinensis*, but this needs further experimental verification.

The SOX transcription factors play a vital role in the gonadal development of many animals [29]. Not all *sox* genes in *P. sinensis* have been found to be involved in sex determination, especially in the males [57]. In this study, *sox9* and *sox17* were enriched in cAMP signaling pathway (ko04024) and wnt signaling pathway (ko04310) related to gender determination, respectively. As a result of RT-qPCR, *sox1*, *sox3*, *sox6*, *sox11*, and *sox17* exhibited female-specific expression in the gonads, whereas *sox9*, *sox12*, and *sox30* exhibited male specificity. Among the male-specific genes, *sox12* and *sox30* exhibited different mRNA levels in pseudo-female ovaries. *Sox30* was almost not expressed in both females and pseudo-females and was hardly affected by exogenous estradiol in the sex reversal of *P. sinensis*. It has been reported that the silencing of *sox30* in the common carp (*Cyprinus carpio*) decreased the expression level of *sox9* and significantly decreased serum testosterone [31]. Contrary to previous studies, the expression level of *sox12* in pseudo-female was similar to that in male, but higher than that in female [58]. Researches in mice have also shown that *sox12* can regulate gonad morphogenesis and germ cell differentiation [59]. All in all, *sox12* and *sox30* are male-specific genes involved in the maturation and maintenance of the testis in *P. sinensis*, and not involved in sex differentiation and sex reversal. It has been reported that *sox9* is an important male sex-determining gene in *P. sinensis* [16]. Our study confirmed that estrogen inhibited the expression of *sox9* in embryos.

Among the female-specific genes, *sox1* and *sox6* were highly expressed in the ovary with no difference between pseudo-female ovary and testis. Its expression pattern was consistent with that of *Acipenser sinensis* [60]. The mRNA expression level of *sox17* was higher in the pseudo-females than in the females. However, *sox17* was not affected by exogenous estrogen in the sex reversal, which was different from the increased *sox17* expression level reported in *Dicentrarchus labrax* [30] during gonadal differentiation. These results suggest that the molecular functions of *sox1, sox6,* and *sox17* may be related to ovarian development and maintenance rather than sex reversal.

The expression levels of *sox3, sox8,* and *sox11* were higher in the pseudo-females than in the males and females. The expression level of *sox3* increased in the embryos after E_2 treatment, whereas *sox8, sox9,* and *sox11* decreased during the sex differentiation period in the males. Previous studies have revealed that E_2 can cause sex reversal in *P. sinensis* [13]. In this process, *sox3* may promote the sex reversal of male to pseudo-female, and *sox8, sox9* and *sox11* were inhibited by E_2. Interestingly, *sox11* is female-specific but inhibited by estrogen during gonadal differentiation, suggesting that *sox11* is related to ovarian development and does not participate in the sex differentiation of *P. sinensis*. Both *sox8* and *sox9* were inhibited by exogenous estrogen during sex reversal, but the expression level of *sox8* was higher in the pseudo-female ovaries than in the males. The expression level of *sox9* showed the opposite trend. Previous studies have showed that the cooperative functions of *sox9* and *sox8* play an important role in the maintenance of testicular function in mice [61]. Our results may indicate that *sox8* promoted the development of pseudo-female ovaries, but this needs to be studied further.

In vertebrates, such as *Xenopus laevis* [21], *Rana rugosa* [23], and *Mus musculus* [24], *sox3* inhibited the expression of *sox9* in the ovaries, promoted the development of ovaries, and even directly activated the transcription of *cyp19*. Deletion and overexpression of *sox3* can lead to sex reversal in *Oryzias dancena* [22]. Our studies showed that *sox3* was increased by exogenous estrogen during sex differentiation in both female and male embryos. Therefore, it was speculated that *sox3* may have a key role in the regulation of female sex differentiation in *P. sinensis* through the estrogen pathway.

5. Conclusions

In conclusion, in this study, gonadal transcriptomic differences between E_2-induced pseudo-female, male, and female *P. sinensis* were investigated, and *sox* family genes were analyzed after E_2 treatment. The results showed that the pseudo-females were more similar to the females with respect to mRNA expression levels. The important genes during sex reversal were identified, especially *sox3, sox8,* and *sox9,* and they may play a vital role in the sex reversal of male to pseudo-female. *Sox3* may promote male-to-female sex reversal, and *sox8* and *sox9* were inhibited by E_2 during the sex reversal (Figure 8). This study provides a reference for further investigations of the molecular mechanism of sex regulation and all-male breeding of *P. sinensis*.

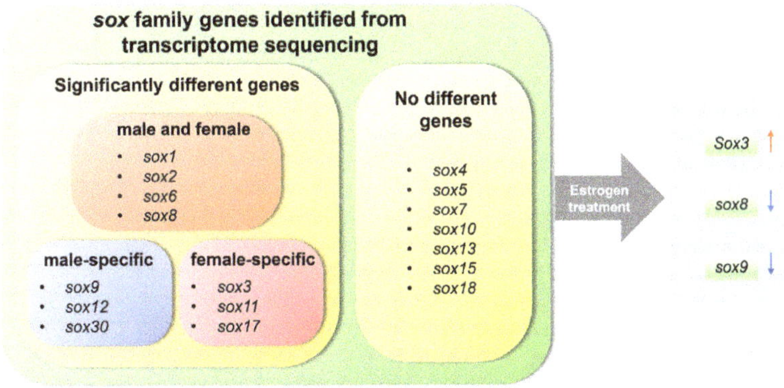

Figure 8. Schematic diagram illustrating the role of *sox* genes in gonad and exogenous estrogen-induced gonadal development and sex reversal in Chinese soft-shelled turtle.

Supplementary Materials: The following supporting information can be downloaded at: https://www.mdpi.com/article/10.3390/biology11010083/s1, Table S1: Primer sequences used in this experiment; Table S2: Differentially expressed genes in the transcriptome; Table S3: GO enrichment analysis of DEGs; Table S4: KEGG enrichment analysis of DEGs; Table S5: Significantly enriched GO terms and KEGG pathways related to sex; Table S6: *Sox* family genes identified form the transcriptome. Figure S1: Significantly enriched GO terms in M vs. F. Figure S2: Top 20 KEGG enrichment significant pathways in M vs. F.

Author Contributions: Conceptualization, Y.W., G.Z. and H.L.; methodology, Y.W. and H.L.; validation, Y.W., X.L. and T.X.; formal analysis, Y.W.; investigation, Y.W., X.L., C.Q. and T.X.; resources, C.Q., G.Z. and H.L.; data curation, Y.W.; writing—original draft preparation, Y.W.; writing—review and editing, X.L., T.X., G.Z. and H.L.; visualization, Y.W.; supervision, G.Z. and H.L.; project administration, H.L.; funding acquisition, G.Z. and H.L. All authors have read and agreed to the published version of the manuscript.

Funding: This research was funded by National Freshwater Aquatic Germplasm Resource Center (grant NO. FGRC18537) and Central Public-interest Scientific Institution Basal Research Fund, CAFS (grant NO. 2020TD33).

Institutional Review Board Statement: The study was conducted according to the appropriate Animal Experimental Ethical Inspection of Laboratory Animal Centre of the Yangtze River Fisheries Research Institute, Chinese Academy of Fishery Sciences (Wuhan, China) (ID Number: 20200118).

Informed Consent Statement: Not applicable.

Data Availability Statement: The raw sequence data reported in this study were deposited in the Genome Sequence Archive (GSA) in National Genomics Data Center (https://ngdc.cncb.ac.cn/gsa/), China National Center for Bioinformation/Beijing Institute of Genomics, Chinese Academy of Sciences under BioProject PRJCA007752, with accession number CRA005737. The data presented in this study are also available on request from the corresponding author.

Conflicts of Interest: The authors declare no conflict of interest.

References

1. Liang, H.W.; Wang, L.H.; Sha, H.; Zou, G.W. Development and Validation of Sex-Specific Markers in *Pelodiscus Sinensis* Using Restriction Site-Associated DNA Sequencing. *Genes* **2019**, *10*, 302. [CrossRef]
2. Liang, H.W.; Cao, L.H.; Li, X.; Tong, M.M.; Jiang, Y.L.; Li, Z.; Luo, X.Z.; Zou, G.W. Morphological differences analysis of three strains of *Pelodiscus sinensis*. *Freshw. Fish.* **2017**, *47*, 91–96.
3. Mu, Y.; Zhao, B.; Tang, W.Q.; Sun, B.J.; Zeng, Z.G.; Valenzuela, N.; Du, W.G. Temperature-dependent sex determination ruled out in the Chinese soft-shelled turtle (*Pelodiscus sinensis*) via molecular cytogenetics and incubation experiments across populations. *Sex. Dev.* **2015**, *9*, 111–117. [CrossRef]
4. Kim, S.; Kettlewell, J.R.; Anderson, R.C.; Bardwell, V.J.; Zarkower, D. Sexually dimorphic expression of multiple doublesex-related genes in the embryonic mouse gonad. *Gene Expr. Patterns GEP* **2003**, *3*, 77–82. [CrossRef]

5. Kawagoshi, T.; Uno, Y.; Matsubara, K.; Matsuda, Y.; Nishida, C. The ZW micro-sex chromosomes of the Chinese soft-shelled turtle (*Pelodiscus sinensis*, Trionychidae, Testudines) have the same origin as chicken chromosome 15. *Cytogenet. Genome Res.* **2009**, *125*, 125–131. [CrossRef]
6. Liang, H.W.; Meng, Y.; Cao, L.H.; Li, X.; Zou, G.W. Effect of exogenous hormones on R-spondin 1 (*RSPO1*) gene expression and embryo development in *Pelodiscus sinensis*. *Reprod. Fertil. Dev.* **2019**, *31*, 1425–1433. [CrossRef]
7. Zhao, L.; Wang, X.; Wan, Q.H.; Fang, S.G. Identification of sex using *SBNO1* gene in the Chinese softshell turtle, *Pelodiscus sinensis* (Trionychidae). *J. Genet.* **2019**, *98*, 36. [CrossRef]
8. Wang, L. The Effect of *DMRT1* Gene on Male Gonadal Development in *Pelodiscus sinensis*. Master's Thesis, Shanghai Ocean University, Shanghai, China, 2014.
9. Toyota, K.; Masuda, S.; Sugita, S.; Miyaoku, K.; Yamagishi, G.; Akashi, H.; Miyagawa, S. Estrogen Receptor 1 (*ESR1*) Agonist Induces Ovarian Differentiation and Aberrant Mullerian Duct Development in the Chinese Soft-shelled Turtle, *Pelodiscus sinensi*. *Zool. Stud.* **2020**, *59*, e54. [CrossRef]
10. Bao, H.S.; Cai, H.; Han, W.; Zhang, H.Y.; Sun, W.; Ge, C.T.; Qian, G.Y. Functional characterization of *Cyp19a1* in female sexual differentiation in *Pelodiscus sinensis*. *Sci. Sin. Vitae* **2017**, *47*, 640–649.
11. Sun, W.; Cai, H.; Zhang, G.; Zhang, H.; Bao, H.; Wang, L.; Ye, J.; Qian, G.; Ge, C. *Dmrt1* is required for primary male sexual differentiation in Chinese soft-shelled turtle *Pelodiscus sinensis*. *Sci. Rep.* **2017**, *7*, 4433. [CrossRef]
12. Sun, W.; Shi, S.R.; Cai, H.; Mi, M.D.; Qian, G.Y.; Ge, C.T. Function Analysis of *Dmrt1* in Male Sexual Differentiation in *Pelodiscus sinensis*. *Sci. Sin. Vitae* **2015**, *45*, 881–889.
13. Liang, H.W.; Meng, Y.; Cao, L.H.; Li, X.; Zou, G.W. Expression and characterization of the *cyp19a* gene and its responses to estradiol/letrozole exposure in Chinese soft-shelled turtle (*Pelodiscus sinensis*). *Mol. Reprod. Dev.* **2019**, *86*, 480–490. [CrossRef]
14. Gao, L.L.; Diao, X.M.; Li, Y.; Zhai, X.L.; Zhou, C.L. Molecular cloning and expression of *FOXL2* gene induced by exogenous hormone in the *Pelodiscus sinensis*. *Acta Hydrobiol. Sin.* **2019**, *43*, 45–51.
15. Zhang, Y.; Xiao, L.; Sun, W.; Li, P.; Zhou, Y.; Qian, G.; Ge, C. Knockdown of R-spondin1 leads to partial sex reversal in genetic female Chinese soft-shelled turtle *Pelodiscus sinensis*. *Gen. Comp. Endocrinol.* **2021**, *309*, 113788. [CrossRef]
16. Wang, L.; Hu, T.; Mi, M.D.; Yang, K.Z.; Qian, G.Y.; Ge, C.T. The effects of estrogen on gonadal differentiation and expressions of *DMRT1* and *SOX9* in *Pelodiscus sinensis*. *Acta Hydrobiol. Sin.* **2014**, *38*, 467–473.
17. Denny, P.; Swift, S.; Brand, N.; Dabhade, N.; Barton, P.; Ashworth, A. A conserved family of genes related to the testis determining gene, *SRY*. *Nucleic Acids Res.* **1992**, *20*, 2887. [CrossRef]
18. Gubbay, J.; Collignon, J.; Koopman, P.; Capel, B.; Economou, A.; Münsterberg, A.; Vivian, N.; Goodfellow, P.; Lovell-Badge, R. A gene mapping to the sex-determining region of the mouse Y chromosome is a member of a novel family of embryonically expressed genes. *Nature* **1990**, *346*, 245–250. [CrossRef] [PubMed]
19. Wang, L.; Sun, W.; Chu, J.Q.; Liu, Y.; Shi, S.R.; Ge, C.T.; Qian, G.Y. The expression pattern of *SOX9* gene during embryonic development and its expression changes in sex reversal in *Pelodiscus sinensis*. *J. Fish. China* **2014**, *38*, 1286–1293.
20. Bishop, C.E.; Whitworth, D.J.; Qin, Y.; Agoulnik, A.I.; Agoulnik, I.U.; Harrison, W.R.; Behringer, R.R.; Overbeek, P.A. A transgenic insertion upstream of *sox9* is associated with dominant XX sex reversal in the mouse. *Nat. Genet.* **2000**, *26*, 490–494. [CrossRef]
21. Koyano, S.; Ito, M.; Takamatsu, N.; Takiguchi, S.; Shiba, T. The *Xenopus Sox3* gene expressed in oocytes of early stages. *Gene* **1997**, *188*, 101–107. [CrossRef]
22. Takehana, Y.; Matsuda, M.; Myosho, T.; Suster, M.L.; Kawakami, K.; Shin, I.T.; Kohara, Y.; Kuroki, Y.; Toyoda, A.; Fujiyama, A.; et al. Co-option of *Sox3* as the male-determining factor on the Y chromosome in the fish *Oryzias dancena*. *Nat. Commun.* **2014**, *5*, 4157. [CrossRef] [PubMed]
23. Oshima, Y.; Naruse, K.; Nakamura, Y.; Nakamura, M. *Sox3*: A transcription factor for *Cyp19* expression in the frog *Rana rugosa*. *Gene* **2009**, *445*, 38–48. [CrossRef] [PubMed]
24. Bergstrom, D.E.; Young, M.; Albrecht, K.H.; Eicher, E.M. Related function of mouse *SOX3*, *SOX9*, and *SRY* HMG domains assayed by male sex determination. *Genesis* **2000**, *28*, 111–124. [CrossRef]
25. Schartl, M.; Schories, S.; Wakamatsu, Y.; Nagao, Y.; Hashimoto, H.; Bertin, C.; Mourot, B.; Schmidt, C.; Wilhelm, D.; Centanin, L.; et al. *Sox5* is involved in germ-cell regulation and sex determination in medaka following co-option of nested transposable elements. *BMC Biol.* **2018**, *16*, 16. [CrossRef] [PubMed]
26. Butler, A.M.; Owens, D.A.; Wang, L.; King, M.L. A novel role for *sox7* in *Xenopus* early primordial germ cell development: Mining the PGC transcriptome. *Development* **2018**, *145*, dev155978. [CrossRef]
27. O'Bryan, M.K.; Takada, S.; Kennedy, C.L.; Scott, G.; Harada, S.; Ray, M.K.; Dai, Q.; Wilhelm, D.; de Kretser, D.M.; Eddy, E.M.; et al. *Sox8* is a critical regulator of adult Sertoli cell function and male fertility. *Dev. Biol.* **2008**, *316*, 359–370. [CrossRef]
28. Polanco, J.C.; Wilhelm, D.; Davidson, T.L.; Knight, D.; Koopman, P. Sox10 gain-of-function causes XX sex reversal in mice: Implications for human 22q-linked disorders of sex development. *Hum. Mol. Genet.* **2010**, *19*, 506–516. [CrossRef] [PubMed]
29. Zhang, J.; Yu, P.; Liu, T.T.; Qiao, D.; Hu, Q.T.; Su, S.P. Identification and functional analysis of SOX transcription factors in the genome of the Chinese soft-shell turtle (*Pelodiscus sinensis*). *Comp. Biochem. Physiol. Part B Biochem. Mol. Biol.* **2020**, *242*, 110407. [CrossRef]
30. Navarro-Martín, L.; Galay-Burgos, M.; Sweeney, G.; Piferrer, F. Different *sox17* transcripts during sex differentiation in sea bass, *Dicentrarchus labrax*. *Mol. Cell. Endocrinol.* **2009**, *299*, 240–251. [CrossRef]

31. Anitha, A.; Senthilkumaran, B. Role of *sox30* in regulating testicular steroidogenesis of common carp. *J. Steroid Biochem. Mol. Biol.* **2020**, *204*, 105769. [CrossRef]
32. Lin, G.; Gao, D.; Lu, J.; Sun, X. Transcriptome Profiling Reveals the Sexual Dimorphism of Gene Expression Patterns during Gonad Differentiation in the Half-Smooth Tongue Sole (*Cynoglossus semilaevis*). *Mar. Biotechnol.* **2021**, *23*, 18–30. [CrossRef] [PubMed]
33. Teng, J.; Zhao, Y.; Chen, H.J.; Wang, H.; Ji, X.S. Transcriptome Profiling and Analysis of Genes Associated with High Temperature-Induced Masculinization in Sex-Undifferentiated Nile Tilapia Gonad. *Mar. Biotechnol.* **2020**, *22*, 367–379. [CrossRef] [PubMed]
34. Tong, Z.F.; Hu, Y.Z.; Tan, J.; Wang, P.; Chen, Z.N.; Tian, Y.; Wang, X.Q. Comparative Analysis Transcriptome of Spleen of Chinese Soft-shelled Turtle *Pelodiscus sinensis*. *Genom. Appl. Biol.* **2020**, *39*, 5449–5456.
35. Zhang, H.Q.; Xu, X.J.; He, Z.Y.; Zheng, T.L.; Shao, J.Z. De novo transcriptome analysis reveals insights into different mechanisms of growth and immunity in a Chinese soft-shelled turtle hybrid and the parental varieties. *Gene* **2017**, *605*, 54–62. [CrossRef]
36. Wang, L.H.; Zhang, Y.P.; Li, C.; Zou, G.W.; Liang, H.W. Chinese Softshelled Turtle *Pelodiscus sinensis*: Embryonic Development and Embryo Staging. *Chin. Agric. Sci. Bull.* **2020**, *36*, 152–158.
37. Zhou, T.; Sha, H.; Chen, M.; Chen, G.B.; Zou, G.W.; Liang, H.W. MicroRNAs May Play an Important Role in Sexual Reversal Process of Chinese Soft-Shelled Turtle, *Pelodiscus sinensis*. *Genes* **2021**, *12*, 1696. [CrossRef]
38. Trapnell, C.; Pachter, L.; Salzberg, S.L. TopHat: Discovering splice junctions with RNA-Seq. *Bioinformatics* **2009**, *25*, 1105–1111. [CrossRef] [PubMed]
39. Langmead, B. Aligning short sequencing reads with Bowtie. *Curr. Protoc. Bioinform.* **2010**, *32*, 11.7.1–11.7.14. [CrossRef] [PubMed]
40. Li, B.; Dewey, C.N. RSEM: Accurate transcript quantification from RNA-Seq data with or without a reference genome. *BMC Bioinform.* **2011**, *12*, 323. [CrossRef]
41. Trapnell, C.; Williams, B.A.; Pertea, G.; Mortazavi, A.; Kwan, G.; van Baren, M.J.; Salzberg, S.L.; Wold, B.J.; Pachter, L. Transcript assembly and quantification by RNA-Seq reveals unannotated transcripts and isoform switching during cell differentiation. *Nat. Biotechnol.* **2010**, *28*, 511–515. [CrossRef]
42. Robinson, M.D.; McCarthy, D.J.; Smyth, G.K. edgeR: A Bioconductor package for differential expression analysis of digital gene expression data. *Bioinformatics* **2010**, *26*, 139–140. [CrossRef]
43. Schmittgen, T.D.; Livak, K.J. Analyzing real-time PCR data by the comparative C(T) method. *Nat. Protoc.* **2008**, *3*, 1101–1108. [CrossRef]
44. Pan, Z.; Zhu, C.; Chang, G.; Wu, N.; Ding, H.; Wang, H. Differential expression analysis and identification of sex-related genes by gonad transcriptome sequencing in estradiol-treated and non-treated Ussuri catfish Pseudobagrus ussuriensis. *Fish Physiol. Biochem.* **2021**, *47*, 565–581. [CrossRef]
45. Akhavan, S.R.; Salati, A.P.; Jalali, A.H.; Falahatkar, B. Expression profile of *star* and *cyp19* and plasma sex steroid during gonad development from previtellogenesis to early atresia in captive Sterlet sturgeon, *Acipenser ruthenus*. *J. Appl. Ichthyol.* **2018**, *35*, 249–256. [CrossRef]
46. Katsu, Y.; Ichikawa, R.; Ikeuchi, T.; Kohno, S.; Guillette, L.J.J.; Iguchi, T. Molecular cloning and characterization of estrogen, androgen, and progesterone nuclear receptors from a freshwater turtle (*Pseudemys nelsoni*). *Endocrinology* **2008**, *149*, 161–173. [CrossRef]
47. Schroeder, A.; Rhen, T. Role for androgens in determination of ovarian fate in the common snapping turtle, *Chelydra serpentina*. *Gen. Comp. Endocrinol.* **2019**, *281*, 7–16. [CrossRef]
48. Hsu, H.J.; Lin, J.C.; Chung, B.C. Zebrafish *cyp11a1* and *hsd3b* genes: Structure, expression and steroidogenic development during embryogenesis. *Mol. Cell. Endocrinol.* **2009**, *312*, 31–34. [CrossRef] [PubMed]
49. Nakamoto, M.; Fukasawa, M.; Tanaka, S.; Shimamori, K.; Suzuki, A.; Matsuda, M.; Kobayashi, T.; Nagahama, Y.; Shibata, N. Expression of 3β-hydroxysteroid dehydrogenase (*hsd3b*), *star* and *ad4bp/sf-1* during gonadal development in medaka (*Oryzias latipes*). *Gen. Comp. Endocrinol.* **2012**, *176*, 222–230. [CrossRef]
50. Peng, C.; Wang, Q.; Shi, H.; Chen, J.; Li, S.; Zhao, H.; Lin, H.; Yang, J.; Zhang, Y. Natural sex change in mature protogynous orange-spotted grouper (*Epinephelus coioides*): Gonadal restructuring, sex hormone shifts and gene profiles. *J. Fish Biol.* **2020**, *97*, 785–793. [CrossRef]
51. Hao, X.C.; Feng, B.; Shao, C.W.; Wang, Q. Molecular Characterization and Expression Patterns of *hsd11b1l* and *hsd11b2* and Their Response to High Temperature Stress in Chinese Tongue Sole *Cynoglossus semilaevis*. *Prog. Fish. Sci.* **2021**, *42*, 45–54.
52. Shehu, A.; Mao, J.; Gibori, G.B.; Halperin, G.B.; Le, J.; Devi, Y.S.; Merrill, B.; Kiyokawa, H.; Gibori, G. Prolactin receptor-associated protein/17beta-hydroxysteroid dehydrogenase type 7 gene (*Hsd17b7*) plays a crucial role in embryonic development and fetal survival. *Mol. Endocrinol.* **2008**, *22*, 2268–2277. [CrossRef] [PubMed]
53. Chen, H.; Li, Z.; Wang, Y.; Huang, H.; Yang, X.; Li, S.; Yang, W.; Li, G. Comparison of Gonadal Transcriptomes Uncovers Reproduction-Related Genes with Sexually Dimorphic Expression Patterns in *Diodon hystrix*. *Animals* **2021**, *11*, 1042. [CrossRef] [PubMed]
54. Kim, Y.; Kobayashi, A.; Sekido, R.; DiNapoli, L.; Brennan, J.; Chaboissier, M.C.; Poulat, F.; Behringer, R.R.; Lovell-Badge, R.; Capel, B. *Fgf9* and *Wnt4* act as antagonistic signals to regulate mammalian sex determination. *PLoS Biol.* **2006**, *4*, e187. [CrossRef] [PubMed]
55. Cen, C.H.; Chen, M.; Jiang, L.; Hou, X.H.; Gao, F. The regulation of gonadal somatic cell differentiation during sex determination in mice. *Acta Physiol. Sin.* **2020**, *72*, 20–30.

56. Price, C.A. Mechanisms of fibroblast growth factor signaling in the ovarian follicle. *J. Endocrinol.* **2016**, *228*, R31–R43. [CrossRef]
57. Li, S.F.; Li, Y.P. Roles of SOX Transcription Factors on Fate Decision of Germ Cells. *Chin. J. Cell Biol.* **2019**, *41*, 961–966.
58. Anitha, A.; Senthilkumaran, B. Role of *sox* family genes in teleostean reproduction-an overview. *Reprod. Breed.* **2021**, *1*, 22–31. [CrossRef]
59. Zhao, L.; Arsenault, M.; Ng, E.T.; Longmuss, E.; Chau, T.C.; Hartwig, S.; Koopman, P. *SOX4* regulates gonad morphogenesis and promotes male germ cell differentiation in mice. *Dev. Biol.* **2017**, *423*, 46–56. [CrossRef] [PubMed]
60. Yang, J.; Hu, Y.C.; Han, J.L.; Xiao, K.; Liu, X.Q.; Tan, C.; Zeng, Q.K.; Du, H.J. Genome-wide analysis of the Chinese sturgeon sox gene family: Identification, characterisation and expression profiles of different tissues. *J. Fish Biol.* **2020**, *96*, 175–184. [CrossRef] [PubMed]
61. Barrionuevo, F.; Georg, I.; Scherthan, H.; Lécureuil, C.; Guillou, F.; Wegner, M.; Scherer, G. Testis cord differentiation after the sex determination stage is independent of *Sox9* but fails in the combined absence of *Sox9* and *Sox8*. *Dev. Biol.* **2009**, *327*, 301–312. [CrossRef]

biology

MDPI

Article

Comparative Transcriptome Analysis Revealed Genes Involved in Sexual and Polyploid Growth Dimorphisms in Loach (*Misgurnus anguillicaudatus*)

Li-Fei Luo [1,2], Zi-Sheng Xu [1,2], Eman Abdelwareth Baioumy Elsayed Elgazzar [1,2], Hang Du [1,2], Dan-Yang Li [1,2], Xiao-Yun Zhou [1,2,3,*] and Ze-Xia Gao [1,2,3,*]

1 Key Lab of Freshwater Animal Breeding, Ministry of Agriculture, Key Lab of Agricultural Animal Genetics, Breeding and Reproduction of Ministry of Education, Engineering Research Center of Green Development for Conventional Aquatic Biological Industry in the Yangtze River Economic Belt, Ministry of Education, College of Fisheries, Huazhong Agricultural University, No. 1 Shizishan Street, Hongshan District, Wuhan 430070, China; luolifei@mail.hzau.edu.cn (L.-F.L.); xu-zisheng@webmail.hzau.edu.cn (Z.-S.X.); xiaoruDong@webmail.hzau.edu.cn (E.A.B.E.E.); hbdh1996@163.com (H.D.); lidanyang20210727@163.com (D.-Y.L.)
2 Hubei Hongshan Laboratory, Wuhan 430070, China
3 Engineering Technology Research Center for Fish Breeding and Culture in Hubei Province, Wuhan 430070, China
* Correspondence: zhouxy@mail.hzau.edu.cn (X.-Y.Z.); gaozx@mail.hzau.edu.cn (Z.-X.G.); Tel.: +86-2787282113 (Z.-X.G.); Fax: +86-2787282114 (Z.-X.G.)

Citation: Luo, L.-F.; Xu, Z.-S.; Elgazzar, E.A.B.E.; Du, H.; Li, D.-Y.; Zhou, X.-Y.; Gao, Z.-X. Comparative Transcriptome Analysis Revealed Genes Involved in Sexual and Polyploid Growth Dimorphisms in Loach (*Misgurnus anguillicaudatus*). *Biology* **2021**, *10*, 935. https://doi.org/10.3390/biology10090935

Academic Editor: Patricia Pereiro

Received: 26 July 2021
Accepted: 13 September 2021
Published: 18 September 2021

Simple Summary: *Misgurnus anguillicaudatus* not only exhibits sexual size dimorphism, but also shows polyploid size dimorphism. Here, we performed comparative transcriptome integration analysis of multiple tissues of diploid and tetraploid *M. anguillicaudatus* of both sexes. We found that differences in energy metabolism and steroid hormone synthesis levels may be the main causes of sexual and polyploidy growth dimorphisms of *M. anguillicaudatus*. Fast-growing *M. anguillicaudatus* (tetraploids, females) have higher levels of energy metabolism and lower steroid hormone synthesis and fatty acid degradation abilities than slow-growing *M. anguillicaudatus* (diploids, males).

Abstract: Sexual and polyploidy size dimorphisms are widespread phenomena in fish, but the molecular mechanisms remain unclear. Loach (*Misgurnus anguillicaudatus*) displays both sexual and polyploid growth dimorphism phenomena, and are therefore ideal models to study these two phenomena. In this study, RNA-seq was used for the first time to explore the differentially expressed genes (DEGs) between both sexes of diploid and tetraploid loaches in four tissues (brain, gonad, liver, and muscle). Results showed that 21,003, 17, and 1 DEGs were identified in gonad, liver, and muscle tissues, respectively, between females and males in both diploids and tetraploids. Regarding the ploidy levels, 4956, 1496, 2187, and 1726 DEGs were identified in the brain, gonad, liver, and muscle tissues, respectively, between tetraploids and diploids of the same sex. When both sexual and polyploid size dimorphisms were considered simultaneously in the four tissues, only 424 DEGs were found in the gonads, indicating that these gonadal DEGs may play an important regulatory role in regulating sexual and polyploid size dimorphisms. Regardless of the sex or ploidy comparison, the significant DEGs involved in glycolysis/gluconeogenesis and oxidative phosphorylation pathways were upregulated in faster-growing individuals, while steroid hormone biosynthesis-related genes and fatty acid degradation and elongation-related genes were downregulated. This suggests that fast-growing loaches (tetraploids, females) have higher energy metabolism levels and lower steroid hormone synthesis and fatty acid degradation abilities than slow-growing loaches (diploids, males). Our findings provide an archive for future systematic research on fish sexual and polyploid dimorphisms.

Keywords: *Misgurnus anguillicaudatus*; sexual size dimorphism; polyploid size dimorphism; growth; comparative transcriptome; gene expression

1. Introduction

Sexual size dimorphism, which is the relative difference in body size and growth rate between males and females of the same species, has been observed in many cultivable fish species, and this phenomenon varies widely across species. Some species, such as carp (*Cyprinus carpio*), rainbow trout (*Oncorhynchus mykiss*), Japanese flounder (*Paralichthys olivaceus*), and half-smooth tongue sole (*Cynoglossus semilaevis*), show sexual size dimorphism, with females being larger and growing faster than males. In contrast, some other species, such as Nile tilapia (*Oreochromis niloticus*), yellow catfish (*Pelteobagrus fulvidraco*), and channel catfish (*Ictalurus punctatus*), show that males have a faster growth rate and larger body size than females [1]. It is often assumed that the growth difference is mediated by differences in the expression of genes present in both sexes [2,3]. For example, *growth hormone* (*gh*) expression level of females was higher than that of males in *C. semilaevis* and spotted scat (*Scatophagus argus linnaeus*), in which females grow faster than males [4,5]. The sex phenotype is established by several genes that act together to form a complicated regulatory network [6]. Comparative transcriptome analysis can provide a basis for exploring the genetic and molecular mechanisms of sexual dimorphism. With the rapid development of next-generation sequencing (NGS) technologies, RNA-seq technology has become an attractive, low-cost, and highly accurate method that has revealed an increasing number of novel transcripts and sequence variations [7,8], and has also been used to analyze important trait-related gene pathways in organisms, even those without genomic information. Recently, RNA-Seq has been used to identify sex-biased genes in many fish species, such as platy fish (*Xiphophorus maculatus*) [9], *O. mykiss* [10], *O. niloticus* [11], rockfish (*Sebastiscus marmoratus*) [12], catfish (*Ictalurus punctatus*) [13], and turbot (*Scophthalmus maximus*) [14]. Such studies have provided some overview of sex-biased gene expression in fish and offer more useful information about sexual dimorphism.

Some fish not only exhibit sexual size dimorphism, but also show polyploid size dimorphism. Polyploids tend to be larger in size and grow faster than diploids in multiple fish species. For example, polyploids in the genus *Barbus* and the Japanese spined loach (*Cobitis biwae*) are larger than those in diploids [15]. Polyploidy in fish is also associated with improved longevity and better ecological adaptability compared to diploids [16]. Currently, most studies on different ploidy individuals in animals have focused on biological characteristics [17], genes related to sex differentiation [18], and comparison of genomic methylation [19]. There are only a few studies on polyploid growth dimorphism in fish, mainly focusing on the differential expression of growth axis-related genes and their regulatory hormone genes in fishes with different ploidy. By comparing the expression of GH/IGF axis genes of *Carassius* carp with different ploidy, it has been revealed that elevated expression of GH/IGF axis genes in triploids plays a crucial role in the faster growth rate of triploids [20]. Tao et al. carried out studies on cDNA cloning and expression of the *cyp19a1a* gene in *Carassius* carp with different ploidy [21]. Polyploids are generally known to have higher levels of genetic diversity than diploid species, owing to the larger total number of chromosome copies [22]. Although there have been some reports on the transcriptome of different ploidy fish [23,24], the molecular regulatory mechanisms of the growth difference between different ploidy fish are still unclear.

Individual growth traits in fish are an important economic characteristic, and is closely linked with fish farming production in aquatic products. In addition to being regulated by the well-known "hypothalamus-pituitary-liver" axis, the growth of vertebrates is also regulated by several energy metabolic processes, such as glycolysis/gluconeogenesis and oxidative phosphorylation. Glycolysis/gluconeogenesis plays an important role in biological synthesis and catabolism [25]. The main physiological functions of glycolysis and gluconeogenesis are the breakdown and synthesis of sugars, respectively. They are two major metabolic pathways that provide precursors and rapid energy sources for cell growth [26]. In Drosophila and zebrafish, glycolysis is located downstream of the GH/IGF axis and plays an important regulatory role in muscle development and animal growth [27]. Oxidative phosphorylation, like glycolysis, is the main source of cellular energy [28]. Oxidative phosphorylation not only supports the growth of tumor cells, but also plays an important role in the growth of normal differentiated somite cells [29,30]. In animals, glucose homeostasis is mutually controlled by catabolic glycolysis/oxidative phosphorylation and anabolic gluconeogenesis pathways [31], thus providing energy for growth.

To date, studies on the regulation of sexual and polyploid growth dimorphisms-related genes in fish are limited. Loach (*Misgurnus anguillicaudatus*), a small freshwater fish species widely distributed in eastern Asia, including China, Korea and Japan, possesses both sexual and polyploid growth dimorphisms [32]. In loach, females grow faster than males and tetraploids grow faster than diploids; therefore, this species is a suitable model to clarify whether fish sexual and polyploid growth dimorphisms are regulated through the same signaling pathways and genes. In this study, comparative transcriptome profiling of brain, gonad, liver, and muscle tissues of 18-month-old diploid and tetraploid males and females was performed to identify the DEGs related to the growth difference between male and female individuals of the same ploidy and between different ploidies of the same sex. Through comparative transcriptome integration analysis of multiple tissues of diploids and tetraploids of both sexes, the key genes and pathways that regulate sexual and polyploid growth dimorphisms of loach are revealed. Our study provides basic information for breeding loach strains with fast growth and uniform size.

2. Materials and Methods

2.1. Ethic Statement

All experimental protocols were approved by the Animal Experimental Ethical Inspection of Laboratory Animal Center, Huazhong Agricultural University, Wuhan, China (HZAUDO-2016-005, 2016-10-26). All surgery was performed under MS-222 (Sigma, Saint Louis, MO, USA; 100 mg/L) anesthesia, and all efforts were made to minimize suffering. All experiments were performed in accordance with relevant guidelines and regulations.

2.2. Sampling and RNA Extraction

Loaches for sampling were collected from the Fisheries Experimental Station of Huazhong Agricultural University. A total of 100 loaches were separately selected from previously established 18-month-old diploid and tetraploid loach populations. The ploidy of fish was determined by flow cytometry analysis as previously described [33], and *Paramisgurnus dabryanus*, which is a known diploid, was used as a control (Figure S1). Forty-eight healthy loaches (24 diploids and 24 tetraploids, containing both females and males) were randomly selected and acclimatized in two tanks with adequate aeration at 24–26 °C for one week before processing. Four tissues (brain, gonad, liver and muscle) of female and male of diploid and tetraploid individuals were immediately sampled separately. Total RNA from each sample was isolated using RNAiso Plus (TaKaRa, Dalian, China) according to the manufacturer's instructions. RNA quality and concentration were determined using an Agilent Bioanalyzer 2100 (Agilent Technologies, Santa Clara, CA, USA) and Qubit® RNA Assay Kit in Qubit® 2.0 Fluorometer (Life Technologies, Carlsbad, CA, USA), respectively. High-quality RNA was subjected to further libraries construction.

2.3. Library Construction and Sequencing

Equal amounts of RNA from four individual samples were mixed as one sequencing sample. A total of 1.5 μg RNA per sample was used as input material for the RNA sample preparations following the removal of possible genomic DNA. Sequencing libraries were generated using NEBNext® Ultra™ RNA Library Prep Kit for Illumina® (NEB, Ipswich, MA, USA) following the manufacturer's recommendations and index codes were added to attribute sequences to each sample. Subsequently, the library preparations were sequenced on an Illumina HiSeq 2500 platform and 150 bp paired-end reads were generated. Library construction and sequencing were performed by Novogene (Beijing, China). In total, 48 sequencing libraries were constructed and transcriptome data were obtained in this study. All sequencing data were uploaded to the Sequence Read Archive (SRA) of the National Center for Biotechnology Information (NCBI accession number PRJNA678824).

2.4. Preliminary Analysis of the Transcriptome Data

Raw data (raw reads) in fastq format were first processed using in-house Perl scripts. At this step, clean data (clean reads) were obtained after filtering out reads with adapters and poly-N, as well as low-quality reads (<Q20) from the raw data. Simultaneously, Q20, Q30, GC-content and sequence duplication level of the clean data were calculated.

All subsequent analyses were based on high-quality clean data. The clean reads from the high-quality clean data were then de novo assembled using Trinity software with min_kmer_cov set to 2 as default and all other parameters set to default [34]. The expression level of each transcript was calculated by the expected number of fragments per kilobase of transcript sequence per million base pairs sequenced (FPKM) method [35].

2.5. Screening and Functional Analysis of Differentially Expressed Genes (DEGs)

Differential expression analysis of the samples was performed using DEGseq2 [36]. Genes with an adjusted *p*-value (padj) < 0.05 and $|\log_2(\text{fold-change})| > 1$ were set as the threshold for significantly differential expression.

Gene ontology (GO) enrichment analysis of the differentially expressed genes (DEGs) was implemented in the GOseq R packages based on Wallenius non-central hyper-geometric distribution, which can adjust for gene length bias in DEGs. In GO enrichment analysis, only categories with a corrected *p*-value < 0.05 were considered as enriched in the network. The Kyoto Encyclopedia of Genes and Genomes (KEGG) pathway analysis was used to determine the location of the DEGs in the different pathways. In this study, KOBAS software was used to test the statistical enrichment of DEGs in the KEGG pathways [37]. Pathways with a corrected *p*-value < 0.05 were considered as the enriched items.

2.6. Validation of RNA-Seq Data by Quantitative Real-Time PCR

To confirm the reliability of the data obtained with RNA-Seq, four genes (*pgk, gpi, ldh,* and *eno*) involved in the glycolytic pathway, three estrogen receptor genes (*erα, erβ1,* and *erβ2*), and several reported growth-related genes (*gh, igf1, igf2, igf1r,* and *igf2r*) were selected for validation by quantitative real-time PCR (qPCR) using the primers listed in Table S1. qPCR was performed on the Applied Biosystems QuantStudio 6 Flex Real-time PCR System (Applied Biosystems, Carlsbad, CA, USA) with SYBR® Premix DimerEraser™ (TaKaRa, Shiga, Japan) according to the manufacturer's instructions. Five biological replicates were performed, with *β-actin* serving as the reference for internal standardization. The PCR cycling conditions were as follows: 95 °C for 30 s, followed by 40 cycles of 95 °C for 5 s, 60 °C for 30 s, and 72 °C for 30 s. The qPCR results were analyzed using the $2^{-\Delta\Delta Ct}$ method [38]. All data were expressed as mean ± standard deviation (S.D.) values of five replicates. Statistical significance was determined using one-way analysis of variance (ANOVA) followed by multiple comparison testing with the least significant distance (LSD) *t*-test using SPSS 17.0 software. Statistical significance was set at *p*-value < 0.05. The entire research scheme diagram is shown in Figure 1.

3. Results

3.1. De novo Transcriptome Assembly and Quality Evaluation

Before de novo transcriptome assembly, paired-end raw reads were processed to remove adapter fragments and low-quality bases, generating clean reads from raw reads. As shown in Table S2, the raw reads, clean reads, clean bases, Q20, Q30, and the GC content were recorded for 48 libraries. For all libraries, Q20 $\geq$ 90%, and Q30 $\geq$ 89%, showing the high-quality of the sequencing data and, thus, could be used for subsequent analyses. After using the Trinity software, de novo sequence assembly resulted in 351,696 unigenes with 956 bp N50 and an average length of 791 bp ranging from 301 bp to 42,939 bp (Tables S3 and S4). The number of unigenes and their length distribution indicated that

68,606 unigenes were >1000 bp in length (Table S3). The functions of unigenes were annotated using GO and KEGG databases. According to GO annotations, the unigenes were involved in 25 biological processes, 16 cell components, and 9 molecular functions (Figure 2A). These unigenes participated in 32 KEGG pathways (Figure 2B).

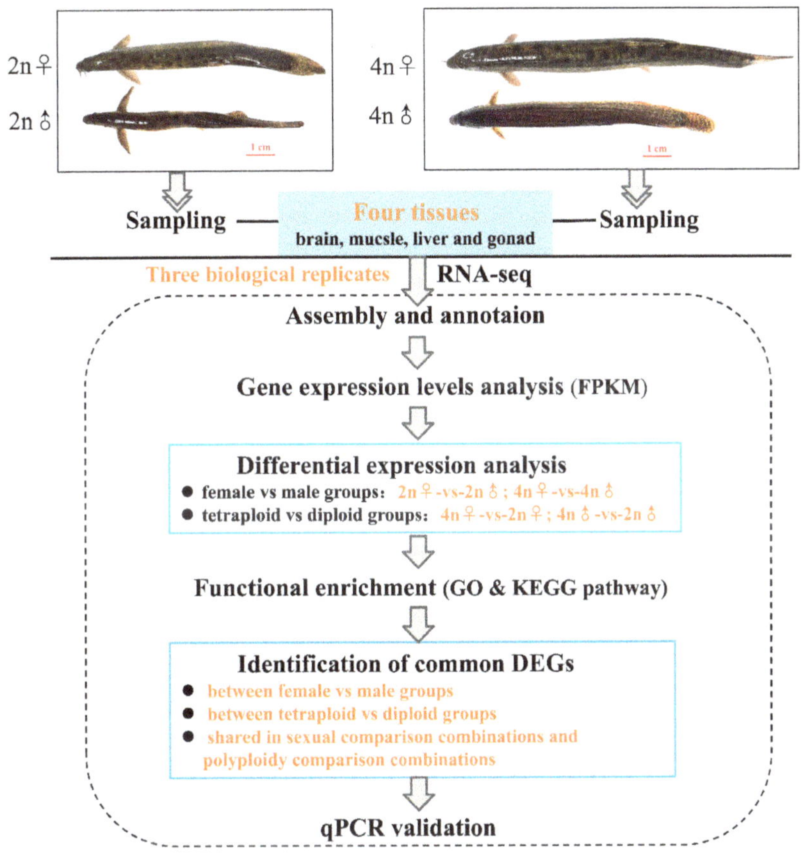

Figure 1. Flow chart of transcriptome sequencing analysis.

3.2. Analysis of DEGs between Female and Male Loaches

Genes were considered DEGs if they showed a padj value < 0.05 and $|\log_2(\text{fold-change})| > 1$. Among the comparisons of diploid female and male loaches (brain, gonad, liver, and muscle samples from diploid females were compared with those from diploid males, comparisons were named D-B-F vs. D-B-M, D-G-F vs. D-G-M, D-L-F vs. D-L-M, and D-M-F vs. D-M-M), 126 (68 upregulated and 58 downregulated), 36,765 (12,622 upregulated and 24,143 downregulated), 257 (131 upregulated and 126 downregulated), and 150 (65 upregulated and 85 downregulated) DEGs were identified in brain, gonad, liver, and muscle tissues, respectively (Figure 3A). Based on the results of GO annotation, the DEGs in the brain were enriched in three molecular function items, including oxygen binding, heme binding, and tetrapyrrole binding (Table S5). In gonads, 305, 101, and 125 DEGs were enriched in biological processes, cell components, and molecular functions, respectively (Table S6), among which most of the DEGs were involved in protein binding, ion binding, oxidoreductase activity, protein-disulfide reductase activity, and cellular protein modification processes. In the liver, DEGs were significantly enriched in two biological processes and two cell components, which were mainly involved in lipid transport, lipid localization and substrate characteristic transport activity. The DEGs in the muscle were mainly enriched on four molecular functional related items, which were associated with oxidoreductase activity, phosphorylase activity, and glycogen phosphorylase activity (Table S5). Furthermore, according to the KEGG pathway analysis results, the DEGs identified in the brain were mainly involved in steroid hormone biosynthesis and a series of metabolic processes including porphyrin and chlorophyll, amino acids and arachidonic acid metabolism. Meanwhile, the DEGs in gonad, liver, and muscle were mainly associated with cell cycle and PI3K-Akt signaling pathway, phagosome and antigen processing and presentation, glycolysis/gluconeogenesis and HIF-1 signaling pathway, respectively (Table S7).

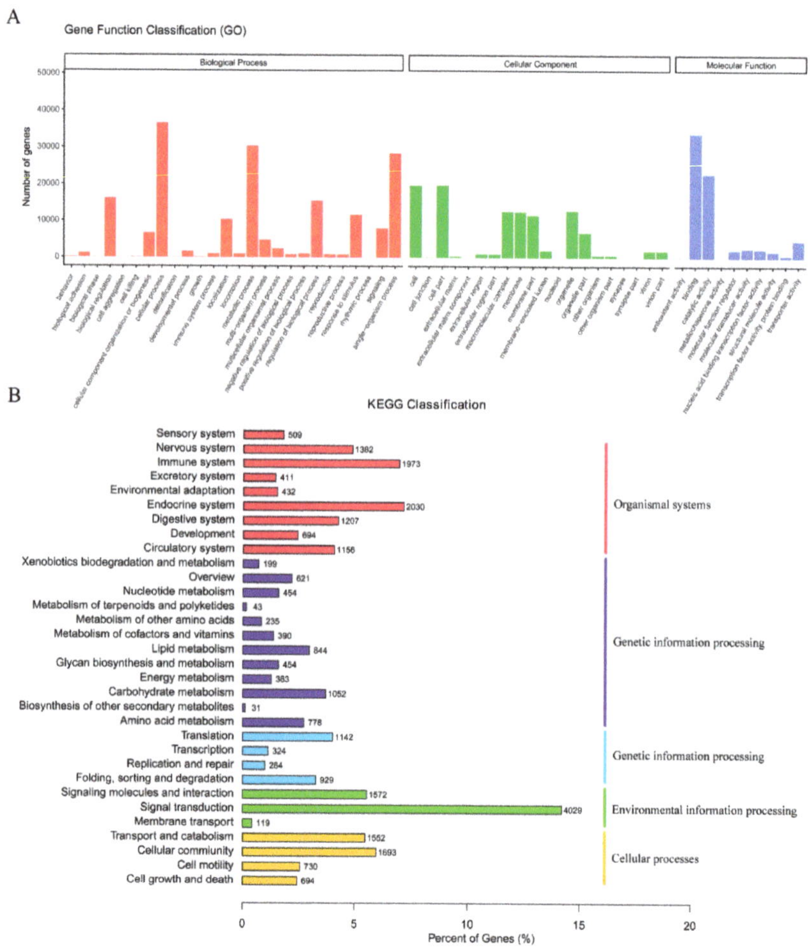

Figure 2. (**A**) GO categorization of the unigenes in the transcriptome of *M. anguillicaudatus*. (**B**) KEGG assignment of unigenes in the transcriptome of *M. anguillicaudatus*.

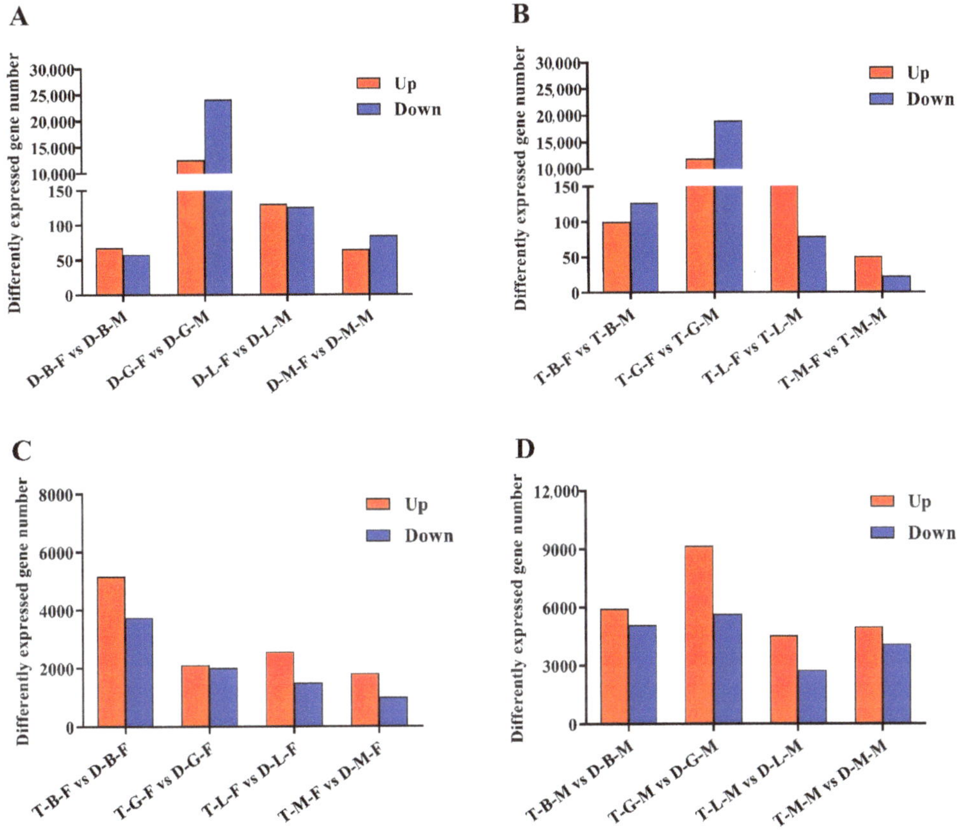

Figure 3. The number of DEGs in four tissues identified in the comparisons of: (**A**) diploid females and diploid males; (**B**) tetraploid females and tetraploid males; (**C**) tetraploid females and diploid females; and (**D**) tetraploid males and diploid males. The letters B, G, L, and M in the middle of each group represent brain, gonad, liver, and muscle, respectively.

Among the comparisons of tetraploid female and male loaches (brain, gonad, liver and muscle samples from tetraploid females were compared with those from tetraploid males, and comparisons were named T-B-F vs. T-B-M, T-G-F vs. T-G-M, T-L-F vs. T-L-M, and T-M-F vs. T-M-M), a total of 227 (100 upregulated and 127 downregulated), 30,970 (11,896 upregulated and 19,074 downregulated), 253 (174 upregulated and 79 downregulated), and 74 (51 upregulated and 23 downregulated) DEGs were identified in brain, gonad, liver, and muscle tissues, respectively (Figure 3B). Further GO annotation results showed that there were no GO items in the brain. In gonads, 216, 82, and 116 DEGs were enriched in biological processes, cell components, and molecular functions, respectively (Table S8), among which most of the DEGs were involved in protein binding, ion binding, and oxidoreductase activity (Table S5). In the liver, similar to the GO items of DEGs identified in the D-L-F vs. D-L-M group, the DEGs identified in the T-L-F vs. T-L-M group were also mainly involved in lipid transport, lipid location and matrix characteristic transport activity. In the muscle, unlike the enriched items of DEGs identified in the D-M-F vs. D-M-M group, the DEGs identified in the T-M-F vs. T-M-M group were enriched in the biological processes related to cell cycle and the cell component related to skeletal muscle. The KEGG pathways mainly enriched in the brain, gonad, liver, and muscle were involved in some metabolic processes (protein digestion and absorption, taurine and hypotaurine metabolism), protein processing in endoplasmic reticulum, biosynthesis, and tight junction, respectively (Table S7).

3.3. Analysis of DEGs between Tetraploid and Diploid Loaches

To characterize the differences between tetraploid and diploid loaches, the transcriptional levels of the brain, gonad, liver, and muscle tissues of tetraploids and diploids of the same sex were compared. Among the comparisons of tetraploid and diploid female loaches (brain, gonad, liver, and muscle samples from tetraploid females were compared with those from diploid females, and comparisons were named T-B-F vs. D-B-F, T-G-F vs. D-G-F, T-L-F vs. D-L-F, and T-M-F vs. D-M-F, respectively), a total of 8913 (5156 upregulated and 3757 downregulated), 4136 (2114 upregulated and 2022 downregulated), 4069 (2563 upregulated and 1506 downregulated) and 2832 (1820 upregulated and 1012 downregulated) DEGs were identified in brain, gonad, liver, and muscle tissues, respectively (Figure 3C). Among the comparisons of tetraploid and diploid male loaches (T-B-M vs. D-B-M, T-G-M vs. D-G-M, T-L-M vs. D-L-M, and T-M-M vs. D-M-M represent the comparison of brain, gonad, liver, and muscle samples between tetraploid and diploid males, respectively), a total of 11,014 (5931 upregulated and 5083 downregulated), 14,868 (9173 upregulated and 5675 downregulated), 7307 (4559 upregulated and 2748 downregulated), and 9090 (5000 upregulated and 4090 downregulated) DEGs were identified in the brain, gonad, liver, and muscle tissues, respectively (Figure 3D).

GO annotation of the DEGs among the comparisons of tetraploids and diploids of both sexes indicated that the DEGs identified in the brain were associated with DNA integration, DNA metabolic process and oxidoreductase activity. In addition, the DEGs identified in the gonad mainly performed the molecular function of nickel cation binding; those in the liver participated in DNA integration, binding and metabolic processes. Meanwhile, the DEGs in the muscle were enriched in terms of cell components, among which microtubule and actin cytoskeleton were the most represented (Table S5). KEGG pathway analysis further indicated that DEGs between the tetraploids and diploids of both sexes were mainly involved in pathways related to a series of metabolic processes in the brain, cell cycle and oxidative phosphorylation in the gonad, and related to some bacterial infections and their phagocytosis in the liver (Table S7).

3.4. Identification of Common DEGs Shared in Different Comparison Groups in the Same Tissue

In order to obtain the common DEGs shared in different comparison combinations within the same tissue, the DEGs identified in four types of comparative combinations (diploid females vs. diploid males (D-B/G/L/M-F vs. D-B/G/L/M-M), tetraploid females vs. tetraploid males (T-B/G/L/M-F vs. T-B/G/L/M-M), tetraploid females vs. diploid females (T-B/G/L/M-F vs. D-B/G/L/M-F), and tetraploid males vs. diploid males (T-B/G/L/M-M vs. D-B/G/L/M-M) were comprehensively analyzed. When considering the common DEGs between D-B/G/L/M-F vs. D-B/G/L/M-M and T-B/G/L/M-F vs. T-B/G/L/M-M comparison combinations, Venn diagram showed that there were 21,003, 17, and 1 DEGs in the gonad, liver, and muscle, respectively (Figure 4), which were only related to sexes but not ploidy. Surprisingly, there were no common DEGs shared in the comparison combinations of brain (D-B-F vs. D-B-M and T-B-F vs. T-B-M). In addition, the overlapping DEGs in the comparison combinations of T-B/G/L/M-F vs. D-B/G/L/M-F and T-B/G/L/M-M vs. D-B/G/L/M-M were also identified. As shown in Figure 4, there were 4956, 1496, 2187, and 1726 common DEGs shared in the brain, gonad, liver, and muscle, respectively, which were only related to ploidy but not sex. According to KEGG enrichment result, whether it was only related to ploidy or only related to sexes, the DEGs were mainly enriched in the oxidative phosphorylation pathway, metabolic pathways, and cardiac muscle contraction process, among which most of the genes, such as cytochrome c oxidase subunit 2 (*cox2*), *cox3*, *nd4*, *nd5*, and ubiquinol-cytochrome c reductase cytochrome b (*cytb*), were upregulated genes, whereas *cox1* was the most enriched downregulated gene.

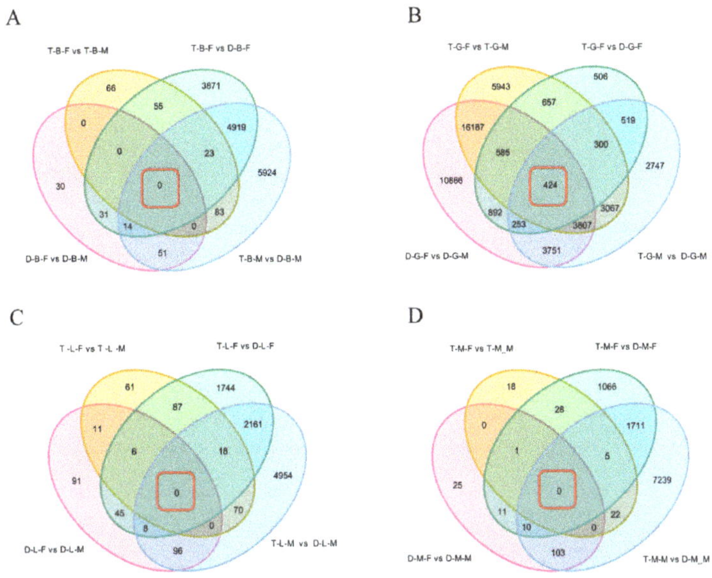

Figure 4. Venn diagram of shared and unique DEGs in brain (**A**), gonad (**B**), liver (**C**), and muscle (**D**) of four comparison groups. The red box indicates the number of DEGs overlapping among the four comparison groups.

Moreover, the overlapping DEGs in the four comparison combinations (D-B/G/L/M-F vs. D-B/G/L/M-M, T-B/G/L/M-F vs. T-B/G/L/M-M, T-B/G/L/M-F vs. D-B/G/L/M-F, and T-B/G/L/M-M vs. D-B/G/L/M-M) were genes related to sexual and polyploidy size dimorphisms of loaches. The Venn diagram showed that there were no common DEGs in the four comparison combinations of brain, liver, and muscle tissues, but there were 424 common DEGs in gonad tissue. KEGG pathway enrichment analysis showed that the common DEGs of different comparison combinations of gonads were mainly involved in transcriptional misregulation in cancer, EGFR tyrosine kinase inhibitor resistance, ErbB signaling pathway, endocrine resistance, and phospholipase D signaling pathway (Table S9), which were associated with a series of cellular processes, including growth, proliferation, and differentiation.

3.5. Analysis of KEGG Pathways Related to Sexual and Polyploid Growth Dimorphisms of Loaches

In order to understand the pathways involved in the regulation of sexual growth dimorphism, we focused on the growth-related pathways enriched by DEGs in two types of comparative combinations (D-B/G/L/M-F vs. D-B/G/L/M-M and T-B/G/L/M-F vs. T-B/G/L/M-M). Among the comparisons of diploid females and males (D-B/G/L/M-F vs. D-B/G/L/M-M), the growth-related pathways enriched by DEGs in the brain were steroid hormone biosynthesis and a series of metabolic processes including porphyrin and chlorophyll, amino acids and arachidonic acid metabolism (Table 1). Among them, metabolic pathway genes (E2.3.1.37, 5-aminolevulinate synthase and *Alox5*, arachidonate 5-lipoxygenase) and protein digestion and absorption-related genes (*cola1* and *cola2*, collagen type I alpha 1/2 chain) were upregulated in females, whereas the steroid hormone biosynthesis-related genes *hsd17b3* (hydroxysteroid 17-beta dehydrogenase 3) and *srd5a1* (3-oxo-5-alpha-steroid 4-dehydrogenase 1) were downregulated in females (Figure 5A). In the gonads of diploid females and males, the DEGs related to growth were mainly enriched in the cell cycle and DNA replication pathways. Most of the genes involved in the cell cycle (such as *cdc45*, *ccne*, *apc*, *orc1*, and *p53*) and DNA replication (*pcna*, *ssb*, *mcm2*, and *RNase-HI*) were expressed at higher levels in females than in males. The DEGs between diploid females and males in the liver were not significantly enriched in any growth-related pathways, but were mainly enriched in pathogenic infection and immune-related pathways, whereas those in the muscle were mainly enriched in glycolysis/gluconeogenesis and glucagon signaling pathways. These pathways-related genes (*gpi, pgk, pgam, eno,* and *gapdh*) were all upregulated in females (Figures 5B and 6A). In the comparison of tetraploid females and males (T-B/G/L/M-F vs. T-B/G/L/M-M), the DEGs involved in protein digestion and absorption and drug metabolism-cytochrome P450 pathways enriched in the brain were upregulated in females, while those involved in cardiac muscle contraction, insulin secretion, MAPK signaling pathway, and calcium signaling pathway were downregulated. Interestingly, in gonads, the mRNA expression levels of ribosome biogenesis in eukaryotes and protein processing in endoplasmic reticulum process-related genes (*utp22, utp6, emg1, ssr1, eif2α,* and *ire1*) were also upregulated in females. In the liver, the expression levels of the enriched ribosome-related genes (*lp1, l11e,* and *l23e*) and glycolysis/gluconeogenesis-related genes (*pfk* and *pgk*) were upregulated in females, while the biosynthesis of unsaturated fatty acids pathway gene *scd* and steroid hormone biosynthesis-related genes *cyp7a1* and *hsd11b2* were downregulated. In the muscle, the expression of the DEGs involved in tight junction (*myl6, myh9,* and *myl12*) and fatty acid degradation (*cpt1*) processes were higher in males than in females. However, the expression of genes related to glycolysis and linoleic acid metabolism was reversed, whereby it was higher in females than in males.

Table 1. Growth-related KEGG pathways in each comparison group.

Group	KEGG Terms	Upregulated Genes	Downregulated Genes	Corrected *P*-Value
D-B-F vs. D-B-M	Porphyrin and chlorophyll metabolism	E2.3.1.37(5-aminolevulinate synthase)		4.36×10^{-2}
	Glycine, serine and threonine metabolism	E2.3.1.37(5-aminolevulinate synthase)		4.36×10^{-2}
	Arachidonic acid metabolism	*alox5*		4.36×10^{-2}
	Protein digestion and absorption	*cola1, cola2*		8.82×10^{-2}
	Ovarian steroidogenesis	*alox5*		4.36×10^{-2}
	Steroid hormone biosynthesis		*hsd17b3, srd5a1*	4.36×10^{-2}
D-G-F vs. D-G-M	Cell cycle	*cdc45, ccne, apc1, orc1, p53, mcm2*		4.63×10^{-2}
	Spliceosome	*cdc5, cype, prp17, prp18, eif4a1*		7.83×10^{-2}
	DNA replication	*pcna, ssb, RNaseHI, mcm2, mcm4*		8.47×10^{-2}
D-L-F vs. D-L-M	Taurine and hypotaurine metabolism	*csad*		3.39×10^{-1}
	Arginine and proline metabolism	*prodh2*	*aoc1*	3.39×10^{-1}
D-M-F vs. D-M-M	Glycolysis/gluconeogenesis	*gpi, pgk, pgam, ldh*		2.58×10^{-7}
	Glucagon signaling pathway	*pgam, ldh*		9.64×10^{-3}
	Starch and sucrose metabolism	*gpi*		2.30×10^{-2}
T-B-F vs. T-B-M	Glycolysis/gluconeogenesis	*pgk*		4.88×10^{-1}
	Protein digestion and absorption	*col1a, cola2*		3.31×10^{-1}
	Drug metabolism-cytochrome P450	*fmo*		3.54×10^{-1}
	Taurine and hypotaurine metabolism	*ggt1-5*		4.88×10^{-1}
	Cardiac muscle contraction		*cacna1c, cacna1d*	4.94×10^{-1}
	GnRH signaling pathway		*cacna1c, cacna1f*	4.88×10^{-1}
	Insulin secretion		*cacna1c*	4.88×10^{-1}
	MAPK signaling pathway		*cacna1a*	5.77×10^{-1}
	Calcium signaling pathway		*cacna1c*	5.77×10^{-1}
T-G-F vs. T-G-M	Ribosome biogenesis in eukaryotes	*utp22, utp6, imp3, ck2a, emg1*		2.77×10^{-1}
	Protein processing in endoplasmic reticulum	*atf6, ire1, eif2ak1, eif2α, ssr1*		3.17×10^{-1}
T-L-F vs. T-L-M	Ribosome	*lp1, lp2, l11e, l23e*		4.92×10^{-11}
	Protein processing in endoplasmic reticulum	*ssr1*	*hsp70*	8.01×10^{-2}
	Glycolysis/gluconeogenesis	*pfka, pgk*		2.93×10^{-1}
	Biosynthesis of unsaturated fatty acids		*△9-desaturase*	2.26×10^{-2}
	Steroid hormone biosynthesis		*cyp7a1, hsd11b2*	7.18×10^{-2}
T-M-F vs. T-M-M	Tight junction		*myl6, myl12, myh9*	3.85×10^{-2}
	DNA replication	*rfc3/5*		1.98×10^{-1}
	Linoleic acid metabolism	*cyp2j*		1.98×10^{-1}
	Fatty acid degradation		*cpt1*	2.16×10^{-1}

Table 1. *Cont.*

Group	KEGG Terms	Upregulated Genes	Downregulated Genes	Corrected *P*-Value
T-B-F vs. D-B-F	Oxidative phosphorylation	*cox3, cox5b, cytb, nd3, nd4*	*cox1, atpef0a*	1.04×10^{-4}
	Cell cycle	*pcna, cyca, cycb, apc/c*		4.07×10^{-1}
	PI3K-Akt signaling pathway	*creb, akt, pp2a, ras*	*itga, pkc, bim*	7.58×10^{-1}
	PPAR signaling pathway	*lpl, acs, scp*		8.70×10^{-1}
	Protein digestion and absorption	*slc15a1, mme*		1.20×10^{-1}
	Regulation of actin cytoskeleton	*actb-g1, myl2, myl5*	*egfr, fgfr2, fgfr3*	9.99×10^{-1}
	Drug metabolism-cytochrome P450	*mao, ugt, gst*		6.32×10^{-1}
	GnRH signaling pathway	*slc5a6, rft2*	*cubn, abcc1*	9.99×10^{-1}
	Estrogen signaling pathway	*gper, mmp, creb1*		7.58×10^{-1}
	Fatty acid degradation		*mecr, fada, acaa2*	9.99×10^{-1}
	Fatty acid elongation		*elovl1, mecr, acaa2, ter*	8.70×10^{-1}
	Cardiac muscle contraction	*myh6/7, myl2, myl3, atp1a, atp1b*		3.82×10^{-1}
T-G-F vs. D-G-F	Oxidative phosphorylation	*cox3, nd4, nd4l, cytb*	*cox1, sdhb*	6.35×10^{-1}
	Arachidonic acid metabolism	*prxl2b, cbr1, cbr2, ptgds, cyp4f, alox12*		6.35×10^{-1}
	p53 signaling pathway	*igf, tsp, mdm2, cyclinb, casp8, perp*		6.35×10^{-1}
	Ascorbate and aldarate metabolism			6.35×10^{-1}
T-L-F vs. D-L-F	Oxidative phosphorylation	*cox3, nd4, nd4l, cytb*	*cox1*	1.77×10^{-6}
	Mineral absorption	*zip4, atpase*		1.04×10^{-5}
	Vitamin digestion and absorption	*apob-48, lrat, rft2*		1.64×10^{-4}
T-M-F vs. D-M-F	Tight junction	*actin*	*myl6, myl1, myh9*	5.02×10^{-5}
	Cardiac muscle contraction	*actin, tpm1*		4.64×10^{-13}
	Oxidative phosphorylation	*cox3, nd4, nd4l, cytb, qcr2, qcr10*	*cox1*	1.00×10^{-9}
	Protein digestion and absorption	*slc8a, mme*	*slc3a2*	6.82×10^{-5}
T-B-M vs. D-B-M	Oxidative phosphorylation	*cox3, cox5, cytb, nd3, nd4*	*cox1, atpef0a*	2.30×10^{-2}
	Glutathione metabolism	*gpx, anpep, rrm1*	*gpx4, ggct*	1.45×10^{-2}
	Steroid hormone biosynthesis		*cyp11b1, cyp11b2, srd5a1*	1.88×10^{-1}
T-G-M vs. D-G-M	Cell cycle	*cdc45, ccne, apc1, orc1, p53, mcm2*		3.50×10^{-6}
	DNA replication	*pcna, mcm2, mcm4*		4.25×10^{-7}
	Regulation of actin cytoskeleton	*iqgap, fak, rho, pak1, f-actin, drf3*		1.04×10^{-1}
T-L-M vs. D-L-M	Mineral absorption	*mt, fpn1*	*hmox, znt1*	1.51×10^{-3}
	Cardiac muscle contraction	*oplah, gpx, ggt*	*ggct, g6pd*	5.45×10^{-1}
	Steroid hormone biosynthesis		*cyp2r1, cyp51, dwf, fdft1*	1.18×10^{-6}
	Fat digestion and absorption	*abca1, apoa, dgat*		2.44×10^{-3}
	Oxidative phosphorylation	*cox3, cox5, cytb, nd3, nd4*		1.40×10^{-2}
	Glycolysis/gluconeogenesis	*fbp, tpi, eno, adh1, g6pc*		1.93×10^{-1}
T-M-M vs. D-M-M	Glycolysis/gluconeogenesis	*gpi, pgk, pgam, ldh, tpi*		6.20×10^{-7}
	Tight junction	*actin, lgl1*	*claudin, pp2a, myl2*	1.00×10^{-9}
	Cardiac muscle contraction	*actin, tpm, atp*	*tnt, tnc*	7.34×10^{-14}
	Oxidative phosphorylation	*cox3, nd4, nd4l, cytb, qcr2, qcr10*	*cox1*	7.34×10^{-14}

Note: brain, gonad, liver, and muscle samples from diploid females were compared with those from diploid males, abbreviated as D-B-F vs. D-B-M, D-G-F vs. D-G-M, D-L-F vs. D-L-M, and D-M-F vs. D-M-M, respectively, and T-B-F vs. T-B-M, T-G-F vs. T-G-M, T-L-F vs. T-L-M, and T-M-F vs. T-M-M represent the comparisons of those samples between females and males in tetraploids. T-B-F vs. D-B-F, T-G-F vs. D-G-F, T-L-F vs. D-L-F, and T-M-F vs. D-M-F represent the comparison of brain, gonad, liver, and muscle samples of tetraploid and diploid females, respectively, and T-B-M vs. D-B-M, T-G-M vs. D-G-M, T-L-M vs. D-L-M, and T-M-M vs. D-M-M represent comparisons between tetraploid and diploid males.

In addition, in order to reveal the pathways regulating polyploid growth dimorphism, we selected the growth-related pathways enriched by the DEGs in T-B/G/L/M-F vs. D-B/G/L/M-F and T-B/G/L/M-M vs. D-B/G/L/M-M for analysis. According to the results of KEGG pathway enrichment analysis, most of the DEGs identified in the four tissues in the comparison of tetraploid and diploid females (T-B/G/L/M-F vs. D-B/G/L/M-F) were significantly enriched in the oxidative phosphorylation pathway (Table 1). Most of the genes (NADH-ubiquinone oxidoreductase chain 3 (*nd3*), *nd4*, *nd4l*, *cytb*, cytochrome c oxidase subunit 3 (*cox3*), *cox4*, and *cox5*) exhibited higher expression in tetraploids than in diploids (Figures 5C and 6B), whereas the expression of the DEGs related to steroid hormone biosynthesis and fatty acid degradation pathways was generally lower in tetraploids than in diploids. Similarly, in the comparison of tetraploid and diploid males (T-B/G/L/M-M vs. D-B/G/L/M-M) in multiple tissues, DEGs that were relatively highly expressed in tetraploid males were also significantly enriched in oxidative phosphorylation and glycolysis pathways, while those with lower expression in tetraploid males were mainly enriched in steroid hormone synthesis (Table 1). A heat-map was constructed by filtering some key genes related to glycolysis and oxidative phosphorylation pathways, and the results showed that genes related to glycolysis had high expression levels in both diploid and tetraploid muscle tissues, while most of the genes related to oxidative phosphorylation had high expression levels in multiple tissues (Figure 6).

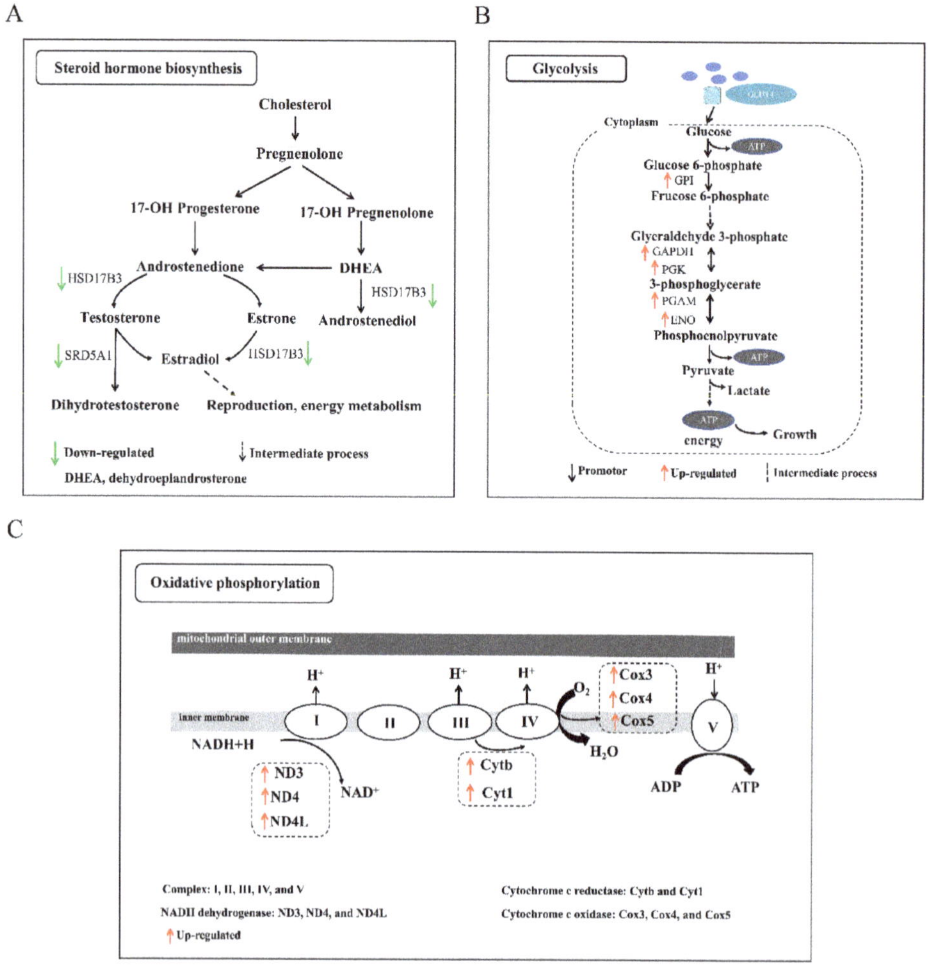

Figure 5. The schematic diagram of key pathways. (**A**) Steroid hormone biosynthesis. Note: HSD17B3, hydroxysteroid 17-beta dehydrogenase 3; SRD5A1, 3-oxo-5-alpha-steroid 4-dehydrogenase 1; DHEA, dehydroepiandrosterone. (**B**) Glycolysis. Note: GPI, glucose-6-phosphate isomerase; GAPDH, glyceraldehyde 3-phosphate dehydrogenase; PGK, phosphoglycerate kinase; PGAM, 2,3-bisphosph-oglycerate-dependent phosphoglycerate mutase; ENO, enolase. (**C**) Oxidative phosphorylation pathway. Note: I, II, III, IV, and V represent different complexes. ND3/4/4L, NADH-ubiquinone oxidoreductase chain 3/4/4L; Cytb/1, ubiquinol-cytochrome c reductase cytochrome b/1; Cox3/4/5, cytochrome c oxidase subunit 3/4/5. The green arrow represents the downregulated, and the dotted arrow represents the intermediate process. The black and red arrow represents the promotor and upregulated DEGs, respectively.

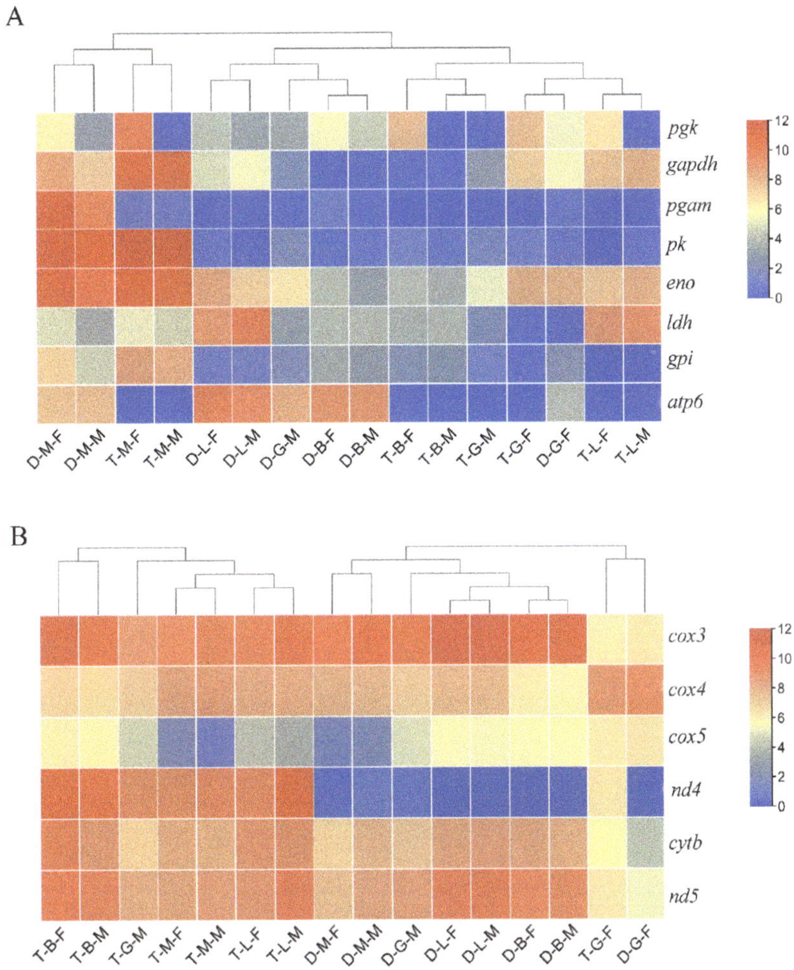

Figure 6. Heat-maps of the key genes related to growth regulation in *M. anguillicaudatus*. (**A**) The heat-map of the key genes related to glycolysis in each group. (**B**) The heat-map of the key genes related to oxidative phosphorylation pathway in each group.

3.6. Verification of RNA-seq Data by qPCR

To confirm the reliability of the RNA-seq data, four genes (*pgk*, *gpi*, *ldh*, and *eno*) involved in the glycolytic pathway, three estrogen receptor genes (*erα*, *erβ1*, and *erβ2*) and several growth-related genes (*gh*, *igf1*, *igf2*, *igf1r*, and *igf2r*) were selected for qPCR analysis. The expression levels of these genes are shown in Figure 7. Overall, the differential expression patterns detected by the two methods showed good consistency, although the exact Log2 (fold-change) in the genes at several data points varied between the RNA-seq and qPCR analyses.

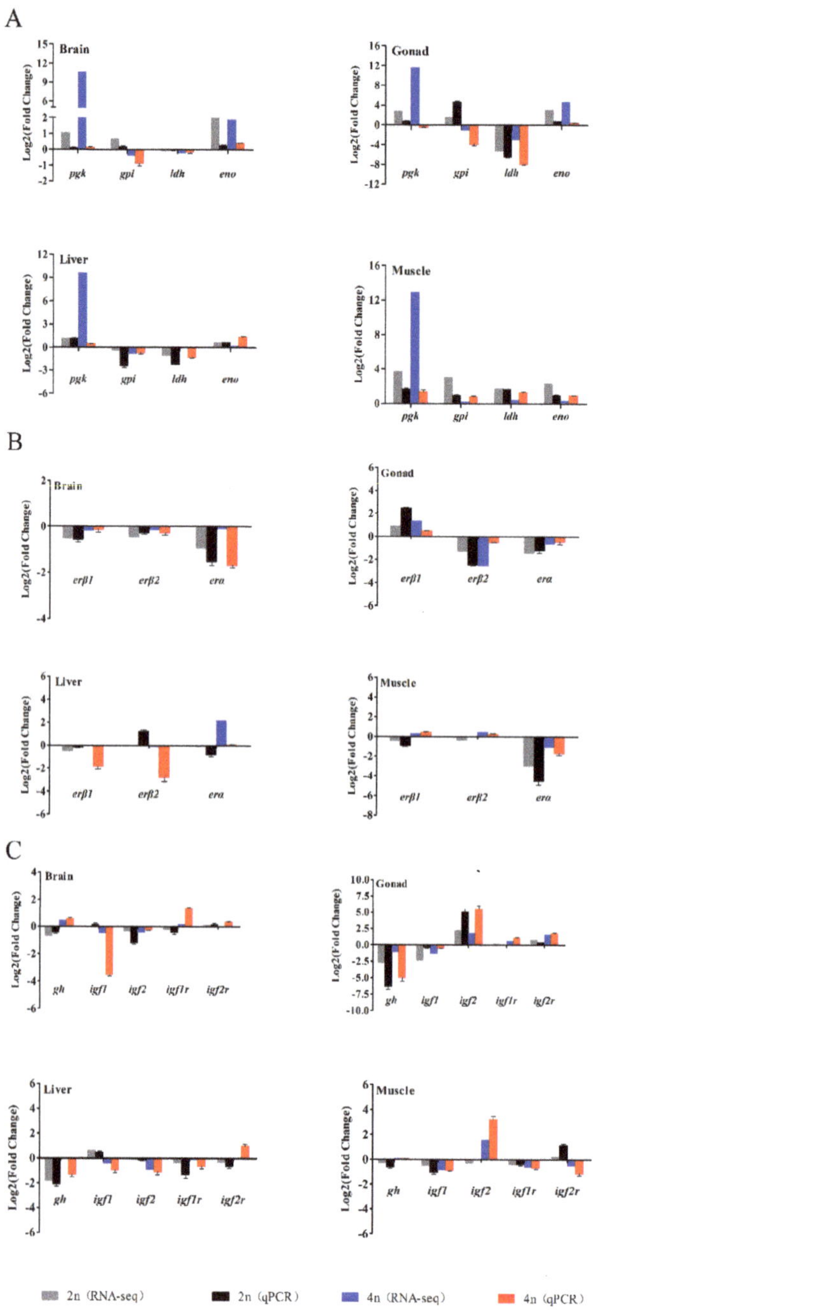

Figure 7. Comparison of the expression profiles of genes involved in the glycolytic pathway. (**A**) Three estrogen receptor genes (**B**) and several reported growth-related genes (**C**) determined by RNA-seq and qPCR in four tissues of female loaches. Each bar represents the expression fold change in a gene compared to that in the male loaches.

4. Discussion

Growth is one of the most important economic traits for aquaculture fish, and is of great significance to the development of the aquaculture industry [39]. Understanding the molecular mechanisms of sexual and polyploid growth dimorphisms will provide a theoretical basis for the cultivation of fast-growing populations with stable and uniform growth performance in aquaculture. So far, research on the growth differences between female and male fish has mainly focused on the aspects of feeding and digestion [5], growth, reproductive energy allocation [40], inheritance, genotypes and phenotypes [41], steroid hormone levels [42], and growth axis genes [43,44]. Moreover, past research on polyploid growth dimorphism has focused on the expression of a single gene [18]. Unlike most previous studies, we performed comparative transcriptome integration analysis of multiple tissues of diploids and tetraploids in both sexes. This is the first report of a combined analysis of sexual and polyploid growth dimorphisms in fish, and it will help us understand the regulatory mechanisms involved and provide basic information for breeding the loach strains with fast growth and uniform size.

In the comparison of male and female loach transcriptomes, our results demonstrated that the main pathways related to growth differences between diploid females and males were glycolysis, hormone synthesis, and cell cycle. Specifically, the expression levels of unigenes related to metabolism (*gpi*, *pgk*, *pgam*, etc.) and cell cycle (*cdc45*, *cyce*, *apc*, etc.) were higher in females than in males, while the expression level of hormone synthesis-related genes (such as *hsd17b3*) in females was lower than that in males. Similar to the male and female diploid comparison group, the male and female tetraploid comparison group was also significantly enriched in glycolysis/gluconeogenesis, and the glycolysis/gluconeogenesis level was also higher in females than in males. Glycolysis/gluconeogenesis is the central pathway of most organisms, providing energy in the form of ATP and reducing power, pyruvate to fuel for the tricarboxylic acid cycle, and precursors for secondary metabolism, amino acid and fatty acid biosynthesis [45]. PGK is a key enzyme for adenosine triphosphate (ATP) generation in glycolysis. Taken together, our results showed that female individuals had higher metabolic levels than male individuals, but lower levels of steroid hormone synthesis than males. On theoretical grounds, there is a link between growth rate and metabolic rate, with fast growth requiring fast metabolism to provide the growing organism with the necessary materials and energy [46]. Glycolysis has also been found up-regulated in the muscle of domestic rainbow trout compared to their wild counterparts [47], and also in faster-growing fish of the same species; this up-regulation has been associated with an increased muscle energy demand in fast-growing fish [48,49]. Studies in humans have also shown that men grow faster than women and have higher metabolic levels [50]. However, in cows where males grow faster than females, the total glucose metabolism is two-fold higher in males than in females, but the activity of the pentose phosphate pathway is four times higher in females than in males [51]. The molecular mechanisms underlying these effects are unclear. The earlier view is that females may have increased metabolic levels to reduce the effects of oxygen free radicals, allowing them to grow faster and have higher viability [52]. In our study, we speculate that the higher metabolic levels of female individuals provide more energy for growth, and the reduction in hormone synthesis levels also saves energy. Therefore, female individuals grow faster than male individuals and are larger in size.

In addition, protein digestion and absorption-related genes were also significantly upregulated in females, among which *cola1* and *cola2* were the most enriched. COLA1 and COLA2 are members of the type I collagen family. Collagen is not only an important protein in animal connective tissue, but also the main component of the extracellular matrix (*ECM*), which can provide support for cell growth [53]. Other collagens have also been found to be closely related to the development of osteogenic cell [54,55]. Our results indicated that the expression level of type I collagen is higher in females with faster growth than in males with slower growth, which is consistent with the reported growth-promoting effect of collagens. However, among the DEGs, the most downregulated genes in females were *cacna1c* and *cacna1d*, which were involved in multiple pathways, including insulin secretion, calcium signaling pathway and MAPK signaling pathway. It has been reported that mice with *cacna1c* and *cacna1d* deficiencies often lack of exercise and social interaction skills [56]. Female individuals may reduce exercise and social skills by regulating the decreased expression of *cacna1c* and *cacna1d* in the body, thus reducing energy expenditure and maintaining a larger body size.

In the comparative analysis of brain, liver, muscle, and gonad tissues of male and female individuals, both DEGs in male and female brains of diploids and tetraploids were enriched in metabolic processes, such as amino acid metabolism, glycolysis/gluconeogenesis, and oxidative phosphorylation. The DEGs in the muscle were also significantly enriched in metabolic processes, whereas those in the liver were mainly related to *Staphylococcus aureus* infection. Growth is regulated by the growth axis "hypothalamic-pituitary-target organ" [57]. DEGs in both male and female brains and muscles were enriched in the same processes, suggesting that the hypothalamus mainly acts on the target organ of muscle to regulate the growth differences between male and female loaches. It is not surprising that the DEGs in the liver are enriched in immune-related processes, as the liver is an important immune organ [58]. Although gonads can regulate growth, they mainly play a role in reproduction.

The transcriptome integration analysis results of tetraploids and diploids showed that regardless of sex, oxidative phosphorylation was the most significantly enriched in multiple tissues. Genes related to this pathway (such as *cox3*, *nd4*, *nd4l*, *cytb*, *qcr2*, and *qcr10*) were upregulated in tetraploids. In Wistar rats and monkeys, where males have a growth advantage, the levels of oxidative phosphorylation were higher in females than in males [59,60]. Our results showed that oxidative phosphorylation levels were higher in fast-growing individuals, which is contrary to the results in mammals. The level of oxidative phosphorylation reflects the mitochondrial activity [61]. Oxidative phosphorylation provides most of the ATP for animals and plants to maintain homeostasis in life and energy metabolism [62]. Moreover, oxidative phosphorylation is an important form of energy metabolism. This suggests that the difference in growth of tetraploids and diploids is closely related to energy metabolism. Our results indicate that the mechanisms regulating energy metabolism and growth in fish may be different from those in mammals. Studies have shown that estrogen is involved in the regulation of mitochondrial activity in animals, including regulation of mitochondrial biogenesis, oxygen consumption, and energy production [63,64]. Estrogen receptor is the key to the physiological role of estrogen. There are two subtypes of estrogen receptor genes (*erα* and *erβ*) in mammals, while three subtypes of estrogen receptor genes (*erα*, *erβ1*, and *erβ2*) in teleost fish [65,66]. Therefore, we speculate that the different estrogen receptor gene types between mammals and fish may result in different ways in which they regulated energy metabolism.

In general, regardless of sexual or polyploidy growth dimorphisms, processes related to growth differences were significantly enriched in pathways related to energy metabolism, including glycolysis/gluconeogenesis, oxidative phosphorylation, hormone synthesis, and fat metabolism. Among them, glycolysis/gluconeogenesis and oxidative phosphorylation pathways were upregulated in fast-growing loach individuals (females, tetraploids), while hormone synthesis and fat metabolism pathways were upregulated in slow-growing loach individuals (males, diploids). Studies have demonstrated that glucose oxidative metabolism reduces the formation of free radicals, while fat metabolism promotes the formation of free radicals [67]. Free radicals play an important role in the entire process of life activities. Thus, it is suggested that the reduction of free radicals weakens the activity of the entire body and provides more energy for growth, while the increase in free radicals accelerates the body's activities and increases energy consumption, leaving less energy for growth. Further research is needed to be undertaken to explore the molecular mechanisms by which females and males regulate energy metabolism balance and growth in fish.

5. Conclusions

In conclusion, by using comparative transcriptome analysis, we successfully identified the DEGs related to sexual and polyploid growth dimorphisms in *M. anguillicaudatus*. Based on transcriptome integration analysis, we found that regardless of the comparison of sexes or ploidy, the DEGs were mainly involved in glycolysis/gluconeogenesis, oxidative phosphorylation, steroid hormone biosynthesis, fatty acid degradation, and elongation. Our results suggest that the differences in energy metabolism levels, steroid hormone synthesis, and fatty acid degradation abilities might be important reasons for the sexual and polyploidy dimorphisms in loaches. Specifically, fast-growing loaches (tetraploids, females) have higher levels of energy metabolism and lower steroid hormone synthesis and fatty acid degradation abilities than slow-growing loaches (diploids, males). Our study not only indicates the direction for us to further investigate the molecular mechanisms of sexual and polyploidy growth dimorphisms of loaches, but also provides essential gene information for future functional studies.

Supplementary Materials: The following are available online at https://www.mdpi.com/article/10.3390/biology10090935/s1, Table S1: Primers sequences used in qPCR experiments, Table S2: Information and quality of RNA-seq. Table S3: Assembly statistics of the transcriptome data, Table S4: Assembly length distribution, Table S5: Top six enriched GO terms for the DEGs, Table S6: Significantly enriched GO annotations for DEGs between diploid female and male gonads, Table S7: Top six enriched KEGG terms for the DEGs, Table S8: Significantly enriched GO annotations for DEGs between tetraploid female and male gonads, Table S9: KEGG enrichment results of common DEGs in gonads, Figure S1: DNA content of diploid and tetraploid cells of loaches measured by flow cytometry. (A), *Paramisgurnus dabryanus*, a known diploid, was used as a control. (B), diploid *M. anguillicaudatus*, which have the same DNA content as the diploid control; (C), tetraploid *M. anguillicaudatus*, which have doubled DNA content compared with diploids.

Author Contributions: Z.-X.G. and X.-Y.Z. designed and supervised the study. L.-F.L., E.A.B.E.E., Z.-S.X., H.D., and D.-Y.L. performed the sample collection, RNA isolation, and qPCR validation experiments. L.-F.L. performed the data curation and prepared and wrote the original draft. Z.-X.G. and X.-Y.Z. revised the manuscript. All authors have read and agreed to the published version of the manuscript.

Funding: This work was financially supported by the National Key Research and Development Program (grant no. 2018YFD0900205), National Natural Science Foundation of China (grant no. 31872559), and the China Agriculture Research System of MOF and MARA (grant no. CARS-46-08).

Institutional Review Board Statement: Not applicable.

Informed Consent Statement: Not applicable.

Data Availability Statement: All sequencing data have been uploaded to the Sequence Read Archive (SRA) of the National Center for Biotechnology Information (NCBI accession number PRJNA678824).

Conflicts of Interest: The authors declare no conflict of interest.

References

1. Mei, J.; Gui, J. Genetic basis and biotechnological manipulation of sexual dimorphism and sex determination in fish. *Sci. China Life Sci.* **2015**, *58*, 124–136. [CrossRef]
2. Connallon, T.; Knowles, L. Intergenomic conflict revealed by patterns of sex-biased gene expression. *Trends Genet.* **2005**, *21*, 495–499. [CrossRef] [PubMed]
3. Rinn, J.; Snyder, M. Sexual dimorphism in mammalian gene expression. *Trends Genet.* **2005**, *21*, 298–305. [CrossRef]
4. Ji, X. *Artificial Gynogenesis, Genetic Analysis and Differential Expression of Growth-Related Genes in Half-Smooth Tongue Sole*; Shandong Agricultural University: Shandong, China, 2009. (In Chinese)
5. Wu, B. *Difference of Growth Performance, Digestive Enzyme Activities and Growth Hormone (GH) Expressions between Male and Female Scatophagus argus Linnaeus*; Guangdong Ocean University: Guangdong, China, 2013.
6. Cutting, A.; Chue, J.; Smith, C. Just how conserved is vertebrate sex determination? *Dev. Dyn.* **2013**, *242*, 380–387. [CrossRef]
7. Wang, B.; Guo, G.; Wang, C.; Lin, Y.; Wang, X.; Zhao, M.; Guo, Y.; He, M.; Zhang, Y.; Pan, L. Survey of the transcriptome of *Aspergillus oryzae* via massively parallel mRNA sequencing. *Nucleic Acids Res.* **2010**, *38*, 5075–5087. [CrossRef]

8. Li, Y.; Wang, G.; Tian, J.; Liu, H.; Yang, H.; Yi, Y.; Wang, J.; Shi, X.; Jiang, F.; Yao, B.; et al. Transcriptome analysis of the silkworm (*Bombyx mori*) by high-throughput RNA sequencing. *PLoS ONE* **2012**, *7*, e43713. [CrossRef] [PubMed]
9. Zhang, Z.; Wang, Y.; Wang, S.; Liu, J.; Warren, W.; Mitreva, M.; Walter, R. Transcriptome analysis of female and male *Xiphophorus maculatus* Jp 163 A. *PLoS ONE* **2011**, *6*, e18379. [CrossRef] [PubMed]
10. Salem, M.; Rexroad, C.; Wang, J.; Thorgaard, G.; Yao, J. Characterization of the rainbow trout transcriptome using Sanger and 454-pyrosequencing approaches. *BMC Genom.* **2010**, *11*, 564. [CrossRef] [PubMed]
11. Tao, W.; Yuan, J.; Zhou, L.; Sun, L.; Sun, Y.; Yang, S.; Li, M.; Zeng, S.; Huang, B.; Wang, D. Characterization of gonadal transcriptomes from nile tilapia (*Oreochromis niloticus*) reveals differentially expressed genes. *PLoS ONE* **2013**, *8*, e63604. [CrossRef]
12. Sun, L.; Wang, C.; Huang, L.; Wu, M.; Zuo, Z. Transcriptome analysis of male and female *Sebastiscus marmoratus*. *PLoS ONE* **2012**, *7*, e50676. [CrossRef] [PubMed]
13. Sun, F.; Liu, S.; Gao, X.; Jiang, Y.; Perera, D.; Wang, X.; Li, C.; Sun, L.; Zhang, J.; Kaltenboeck, L.; et al. Male-biased genes in catfish as revealed by RNA-seq analysis of the testis transcriptome. *PLoS ONE* **2013**, *8*, e68452. [CrossRef]
14. Ribas, L.; Pardo, B.; Fernandez, C.; Antonio Alvarez-Dios, J.; Gomez-Tato, A.; Isabel Quiroga, M.; Planas, J.; Sitja-Bobadilla, A.; Martinez, P.; Piferrer, F. A combined strategy involving Sanger and 454 pyrosequencing increases genomic resources to aid in the management of reproduction, disease control and genetic selection in the turbot (*Scophthalmus maximus*). *BMC Genom.* **2013**, *14*, 180. [CrossRef]
15. Krysanov, E.; Golubtsov, A. Karyotypes of four fish species from the Nile and Omo-Turkana basins in Ethiopia. *J. Ichthyol.* **2014**, *54*, 889–892. [CrossRef]
16. Comber, S.; Smith, C. Polyploidy in fishes: Patterns and processes. *Biol. J. Linn. Soc.* **2004**, *82*, 431–442. [CrossRef]
17. Xiao, J.; Zou, T.; Chen, Y.; Chen, L.; Liu, S.; Tao, M.; Zhang, C.; Zhao, R.; Zhou, Y.; Long, Y.; et al. Coexistence of diploid, triploid and tetraploid crucian carp (*Carassius auratus)* in natural waters. *BMC Genet.* **2011**, *12*, 20. [CrossRef]
18. Tao, M.; Liu, S.; Long, Y.; Zeng, C.; Liu, J.; Liu, L.; Zhang, C.; Duan, W.; Liu, Y. The cloning of Dmc1 cDNAs and a comparative study of its expression in different ploidy cyprinid fishes. *Sci. China Ser. C Life Sci.* **2008**, *51*, 38–46. [CrossRef] [PubMed]
19. Zhou, H.; Ma, T.Y.; Zhang, R.; Xu, Q.; Shen, F.; Qin, Y.; Xu, W.; Wang, Y.; Li, Y. Analysis of different ploidy and parent-offspring genomic DNA methylation in the loach *Misgurnus anguillicaudatus*. *Int. J. Mol. Sci.* **2016**, *17*, 1299. [CrossRef]
20. Zhong, H.; Yi, Z.; Liu, S.; Tao, M.; Long, Y.; Liu, Z.; Zhang, C.; Duan, W.; Hu, J.; Song, C.; et al. Elevated expressions of GH/IGF axis genes in triploid crucian carp. *Gen Comp. Endocr.* **2012**, *178*, 291–300. [CrossRef] [PubMed]
21. Tao, M.; Liu, S.; Zhan, Z.; Chen, J.; Liu, W.; Liu, Y. Molecular cloning and comparative expression patterns of cyp19a1a of gene in different ploidy cyprinid fishes. *J. Fish. China* **2014**, *38*, 1201–1210. (In Chinese)
22. Meirmans, P.; Liu, S.; Van Tienderen, P. The Analysis of Polyploid Genetic Data. *J. Hered.* **2018**, *109*, 283–296. [CrossRef] [PubMed]
23. Odei, D.; Hagen, O.; Peruzzi, S.; Falk-Petersen, I.; Fernandes, J. Transcriptome sequencing and histology reveal dosage compensation in the liver of triploid pre-smolt Atlantic salmon. *Sci. Rep.* **2020**, *10*, 16836. [CrossRef] [PubMed]
24. Zhou, R.; Wu, Y.; Tao, M.; Zhang, C.; Liu, S. MicroRNA profiles reveal female allotetraploid hybrid fertility. *BMC Genet.* **2015**, *16*, 119. [CrossRef]
25. Michaeloudes, C.; Kuo, C.; Haji, G.; Finch, D.; Halayko, A.; Kirkham, P.; Chung, K.; Adcock, I. Metabolic re-patterning in COPD airway smooth muscle cells. *Eur. Respir. J.* **2017**, *50*, 1700202. [CrossRef] [PubMed]
26. Roosterman, D.; Meyerhof, W.; Cottrell, G. Proton Transport Chains in Glucose Metabolism: Mind the Proton. *Front. Neurosci.* **2018**, *12*, 404. [CrossRef]
27. Tixier, V.; Bataillé, L.; Etard, C.; Jagla, T.; Weger, M.; Daponte, J.; Strähle, U.; Dickmeis, T.; Jagla, K. Glycolysis supports embryonic muscle growth by promoting myoblast fusion. *Proc. Natl. Acad. Sci. USA* **2013**, *110*, 18982–18987. [CrossRef] [PubMed]
28. Yuan, Q.; Miao, J.; Yang, Q.; Fang, L.; Fang, Y.; Ding, H.; Zhou, Y.; Jiang, L.; Dai, C.; Zen, K.; et al. Role of pyruvate kinase M2-mediated metabolic reprogramming during podocyte differentiation. *Cell Death Dis.* **2020**, *11*, 355. [CrossRef]
29. Jiang, D.; LaGory, E.; Kenzelmann Brož, D.; Bieging, K.; Brady, C.; Link, N.; Abrams, J.; Giaccia, A.; Attardi, L. Analysis of p53 transactivation domain mutants reveals Acad11 as a metabolic target important for p53 pro-survival function. *Cell Rep.* **2015**, *10*, 1096–1109. [CrossRef] [PubMed]
30. Lin, T.; Wu, S. Reprogramming with Small Molecules instead of Exogenous Transcription Factors. *Stem Cells Int.* **2015**, 794632. [CrossRef]
31. Dong, C.; Yuan, T.; Wu, Y.; Wang, Y.; Fan, T.; Miriyala, S.; Lin, Y.; Yao, J.; Shi, J.; Kang, T.; et al. Loss of FBP1 by Snail-mediated repression provides metabolic advantages in basal-like breast cancer. *Cancer Cell* **2013**, *23*, 316–331. [CrossRef]
32. Feng, B.; Yi, S.; Li, R.; Zhou, X. Comparison of age and growth performance of diploid and tetraploid loach *Misgurnus anguillicaudatus* in the Yangtze River basin, China. *Environ. Biol. Fish* **2017**, *100*, 815–828. [CrossRef]
33. Zhong, J.; Yi, S.; Ma, L.; Wang, W. Evolution and phylogeography analysis of diploid and polyploid *Misgurnus anguillicaudatus* populations across China. *Proc. Biol. Sci.* **2019**, *286*, 20190076. [CrossRef] [PubMed]
34. Grabherr, M.; Haas, B.; Yassour, M.; Levin, J.; Thompson, D.; Amit, I.; Adiconis, X.; Fan, L.; Raychowdhury, R.; Zeng, Q.; et al. Full-length transcriptome assembly from RNA-Seq data without a reference genome. *Nat. Biotechnol.* **2011**, *29*, 644–652. [CrossRef]
35. Trapnell, C.; Williams, B.; Pertea, G.; Mortazavi, A.; Kwan, G.; van Baren, M.; Salzberg, S.; Wold, B.; Pachter, L. Transcript assembly and quantification by RNA-seq reveals unannotated transcripts and isoform switching during cell differentiation. *Nat. Biotechnol.* **2010**, *28*, 511–515. [CrossRef]

36. Love, M.; Huber, W.; Anders, S. Moderated estimation of fold change and dispersion for RNA-seq data with DESeq2. *Genome Biol.* **2014**, *15*, 550. [CrossRef] [PubMed]
37. Kanehisa, M.; Araki, M.; Goto, S.; Hattori, M.; Hirakawa, M.; Itoh, M.; Katayama, T.; Kawashima, S.; Okuda, S.; Tokimatsu, T.; et al. KEGG for linking genomes to life and the environment. *Nucleic Acids Res.* **2008**, *36*, D480–D484. [CrossRef] [PubMed]
38. Livak, K.; Schmittgen, T. Analysis of Relative Gene Expression Data Using Real-Time Quantitative PCR and the $2^{-\Delta\Delta CT}$ Method. *Methods* **2001**, *25*, 402–408. [CrossRef]
39. Blay, C.; Haffray, P.; Bugeon, J.; D'Ambrosio, J.; Dechamp, N.; Collewet, G.; Enez, F.; Petit, V.; Cousin, X.; Corraze, G.; et al. Genetic parameters and genome-wide association studies of quality traits characterised using imaging technologies in rainbow trout, *Oncorhynchus mykiss*. *Front. Genet.* **2021**, *12*, 639223. [CrossRef]
40. Wang, X.; Li, C.; Xie, Z.; Fan, W.; Zhang, J. Studies on the growth difference of the male and female *Siniperca chuatsi*. *Freshw. Fish.* **2006**, *36*, 34–37.
41. Toguyeni, A.; Fauconneau, B.; Fostier, A.; Abucay, J.; Mair, G.; Baroiller, J. Influence of sexual phenotype and genotype, and sex ratio on growth performances in tilapia, *Oreochromis niloticus*. *Aquaculture* **2002**, *207*, 249–261. [CrossRef]
42. Zhang, J.; Ma, W.; He, Y.; Wu, J.; Dawar, F.; Ren, F.; Zhao, X.; Mei, J. Sex biased expression of ghrelin and GHSR associated with sexual size dimorphism in yellow catfish. *Gene* **2016**, *578*, 169–176. [CrossRef]
43. Chatchaiphan, S.; Srisapoome, P.; Kim, J.; Devlin, R.; Na-Nakorn, U. De novo transcriptome characterization and growth-related gene expression profiling of diploid and triploid bighead catfish (*Clarias macrocephalus Gunther*, 1864). *Mar. Biotechnol.* **2017**, *19*, 36–48. [CrossRef]
44. Ma, W.; Wu, J.; Zhang, J.; He, Y.; Gui, J.; Mei, J. Sex differences in the expression of GH/IGF axis genes underlie sexual size dimorphism in the yellow catfish (*Pelteobagrus fulvidraco*). *Sci. China-Life Sci.* **2016**, *59*, 431–433. [CrossRef] [PubMed]
45. Plaxton, W. The organization and regulation of plant glycolysis. *Annu. Rev. Plant Physiol.* **1996**, *47*, 185–214. [CrossRef] [PubMed]
46. Montes, L.; Le Roy, N.; Perret, M.; de Buffrenil, V.; Castanet, J.; Cubo, J. Relationships between bone growth rate, body mass and resting metabolic rate in growing amniotes: A phylogenetic approach. *Biol. J. Linn. Soc.* **2007**, *92*, 63–76. [CrossRef]
47. Tymchuk, W.; Sakhrani, D.; Devlin, R. Domestication causes large-scale effects on gene expression in rainbow trout: Analysis of muscle, liver and brain transcriptomes. *Gen. Comp. Endocrinol.* **2009**, *164*, 175–183. [CrossRef] [PubMed]
48. Danzmann, R.; Kocmarek, A.; Norman, J.; Rexroad, C.; Palti, Y. Transcriptome profiling in fast versus slow-growing rainbow trout across seasonal gradients. *BMC Genom.* **2016**, *17*, 60. [CrossRef]
49. Robledo, D.; Rubiolo, J.; Cabaleiro, S.; Martínez, P.; Bouza, C. Differential gene expression and SNP association between fast- and slow-growing turbot (*Scophthalmus maximus*). *Sci. Rep.* **2017**, *7*, 12105. [CrossRef]
50. Ray, P.; Conaghan, J.; Winston, R.; Handyside, A. Increased number of cells and metabolic activity in male human preimplantation embryos following in vitro fertilization. *J. Reprod. Fertil.* **1995**, *104*, 165–171. [CrossRef]
51. Tiffin, G.; Rieger, D.; Betteridge, K.; Yadav, B.; King, W. Glucose and glutamine metabolism in pre-attachment cattle embryos in relation to sex and stage of development. *J. Reprod. Fertil.* **1991**, *93*, 125–132. [CrossRef]
52. Rieger, D. Relationships between energy metabolism and development of early mammalian embryos. *Theriogenology* **1992**, *37*, 75–93. [CrossRef]
53. Sorushanova, A.; Delgado, L.; Wu, Z.; Shologu, N.; Kshirsagar, A.; Raghunath, R.; Mullen, A.; Bayon, Y.; Pandit, A.; Raghunath, M.; et al. The Collagen Suprafamily: From Biosynthesis to Advanced Biomaterial Development. *Adv. Mater.* **2019**, *31*, e1801651. [CrossRef]
54. Izu, Y.; Ezura, Y.; Koch, M.; Birk, D.; Noda, M. Collagens VI and XII form complexes mediating osteoblast interactions during osteogenesis. *Cell Tissue Res.* **2016**, *364*, 623–635. [PubMed]
55. Wen, Y.; Yang, H.; Wu, J.; Wang, A.; Chen, X.; Hu, S.; Zhang, Y.; Bai, D.; Jin, Z. COL4A2 in the tissue-specific extracellular matrix plays important role on osteogenic differentiation of periodontal ligament stem cells. *Theranostics* **2019**, *9*, 4265–4286. [CrossRef]
56. Redecker, T.; Kisko, T.; Schwarting, R.; Wohr, M. Effects of Cacna1c haploinsufficiency on social interaction behavior and 50-kHz ultrasonic vocalizations in adult female rats. *Behav. Brain Res.* **2019**, *367*, 35–52. [CrossRef] [PubMed]
57. Chen, J.; Cao, M.; Zhang, A.; Shi, M.; Tao, B.; Li, Y.; Wang, Y.; Zhu, Z.; Trudeau, V.L.; Hu, W. Growth Hormone Overexpression Disrupts Reproductive Status Through Actions on Leptin. *Front. Endocrinol.* **2018**, *9*, 131. [CrossRef]
58. Wang, B.; Wangkahart, E.; Secombes, C.J.; Wang, T. Insights into the Evolution of the Suppressors of Cytokine Signaling (SOCS) Gene Family in Vertebrates. *Mol. Biol. Evol.* **2019**, *36*, 393–411. [CrossRef]
59. Colom, B.; Oliver, J.; Roca, P.; Garcia-Palmer, F. Caloric restriction and gender modulate cardiac muscle mitochondrial H2O2 production and oxidative damage. *Cardiovasc. Res.* **2007**, *74*, 456–465. [CrossRef]
60. Yan, L.; Ge, H.; Li, H.; Lieber, S.; Natividad, F.; Resuello, R.; Kim, S.; Akeju, S.; Sun, A.; Loo, K.; et al. Gender-specific proteomic alterations in glycolytic and mitochondrial pathways in aging monkey hearts. *J. Mol. Cell. Cardiol.* **2004**, *37*, 921–929. [CrossRef]
61. Brand, S.; Ebner, K.; Mikoteit, T.; Lejri, I.; Gerber, M.; Beck, J.; Holsboer-Trachsler, E.; Eckert, A. Influence of Regular Physical Activity on Mitochondrial Activity and Symptoms of Burnout-An Interventional Pilot Study. *J. Clin. Med.* **2020**, *9*, 667. [CrossRef] [PubMed]
62. Wilson, D. Oxidative phosphorylation: Unique regulatory mechanism and role in metabolic homeostasis. *J. Appl. Physiol.* **2017**, *122*, 611–619. [CrossRef]
63. Klinge, C. Estrogenic control of mitochondrial function and biogenesis. *J. Cell. Biochem.* **2008**, *105*, 1342–1351. [CrossRef] [PubMed]

64. Palmer, B.; Clegg, D. The sexual dimorphism of obesity. *Mol. Cell. Endocrinol.* **2015**, *402*, 113–119. [CrossRef] [PubMed]
65. Meng, X.; Bartholomew, C.; Craft, J. Differential expression of vitellogenin and oestrogen receptor genes in the liver of zebrafish, Danio Rerio. *Anal. Bioanal. Chem.* **2010**, *396*, 625–630. [CrossRef] [PubMed]
66. Zhang, Y.; Wang, H.; Qin, F.; Liu, S.; Wu, T.; Li, M.; Xu, P.; Zhang, X.; Wang, X.; Hu, G.; et al. Molecular characterization of estrogen receptor genes in loach Paramisgurnus dabryanus and their expression upon 17 alpha-ethinylestradiol exposure in juveniles. *Gen. Comp. Endocr.* **2012**, *178*, 194–205. [CrossRef]
67. Speijer, D.; Manjeri, G.; Szklarczyk, R. How to deal with oxygen radicals stemming from mitochondrial fatty acid oxidation. *Philos. Trans. R. Soc. Lond. B Biol. Sci.* **2014**, *369*, 20130446. [CrossRef]

biology

MDPI

Article

Environment-Dependent Heterosis and Transgressive Gene Expression in Reciprocal Hybrids between the Channel Catfish *Ictalurus punctatus* and the Blue Catfish *Ictalurus furcatus*

Haolong Wang [1,2], Timothy J. Bruce [2,3], Baofeng Su [2,3], Shangjia Li [2,3], Rex A. Dunham [2,3] and Xu Wang [1,2,4,*]

[1] Department of Pathobiology, College of Veterinary Medicine, Auburn University, Auburn, AL 36849, USA; hzw0088@auburn.edu

[2] Alabama Agricultural Experiment Station, Auburn, AL 36849, USA; tjb0089@auburn.edu (T.J.B.); BZS0014@auburn.edu (B.S.); szl0164@auburn.edu (S.L.); dunhara@auburn.edu (R.A.D.)

[3] School of Fisheries, Aquaculture and Aquatic Sciences, Auburn University, Auburn, AL 36849, USA

[4] HudsonAlpha Institute for Biotechnology, Huntsville, AL 35806, USA

* Correspondence: xzw0070@auburn.edu; Tel.: +1-344-844-7511

Simple Summary: The hybrid catfish, generated by crossing female channel catfish and male blue catfish, has occupied the majority of the market share due to superior performance in growth rate, yield, and disease resistance in pond culture. However, we found that channel catfish have the best growth performance in tank units of smaller size, indicating that the heterosis is environment-dependent. To investigate the mechanisms of this intriguing phenomenon, hematological assays and transcriptome analysis were performed in the parental species and hybrid crosses. Lower levels of innate immunity activity, stress, as well as lowered blood glucose/lactate were found in channel catfish, which are associated with superiority in growth. Functional enrichment analysis revealed that genes involved in fatty acid metabolism/transport pathways are significantly upregulated in channel catfish. The results provide insights into the molecular mechanisms of heterosis and will inform the development of new strategies for genetic enhancement through hybrid breeding.

Abstract: The hybrid between female channel catfish (*Ictalurus punctatus*) and male blue catfish (*Ictalurus furcatus*) is superior in feed conversion, disease resistance, carcass yield, and harvestability compared to both parental species. However, heterosis and heterobeltiosis only occur in pond culture, and channel catfish grow much faster than the other genetic types in small culture units. This environment-dependent heterosis is intriguing, but the underlying genetic mechanisms are not well understood. In this study, phenotypic characterization and transcriptomic analyses were performed in the channel catfish, blue catfish, and their reciprocal F_1s reared in tanks. The results showed that the channel catfish is superior in growth-related morphometrics, presumably due to significantly lower innate immune function, as investigated by reduced lysozyme activity and alternative complement activity. RNA-seq analysis revealed that genes involved in fatty acid metabolism/transport are significantly upregulated in channel catfish compared to blue catfish and hybrids, which also contributes to the growth phenotype. Interestingly, hybrids have a 40–80% elevation in blood glucose than the parental species, which can be explained by a phenomenon called transgressive expression (overexpression/underexpression in F_1s than the parental species). A total of 1140 transgressive genes were identified in F_1 hybrids, indicating that 8.5% of the transcriptome displayed transgressive expression. Transgressive genes upregulated in F_1s are enriched for glycan degradation function, directly related to the increase in blood glucose level. This study is the first to explore molecular mechanisms of environment-dependent heterosis/heterobeltiosis in a vertebrate species and sheds light on the regulation and evolution of heterosis vs. hybrid incompatibility.

Keywords: heterosis; heterobeltiosis; environment; RNA-Seq; transcriptomics; transgressive genes; aquaculture

Citation: Wang, H.; Bruce, T.J.; Su, B.; Li, S.; Dunham, R.A.; Wang, X. Environment-Dependent Heterosis and Transgressive Gene Expression in Reciprocal Hybrids between the Channel Catfish *Ictalurus punctatus* and the Blue Catfish *Ictalurus furcatus*. *Biology* **2022**, *11*, 117. https://doi.org/10.3390/biology11010117

Academic Editor: Patricia Pereiro

Received: 30 November 2021
Accepted: 6 January 2022
Published: 12 January 2022

1. Introduction

Interspecific hybrids are formed by crossing two distinct species. The hybrid offspring are sometimes superior in yield, size, growth rate, strength, fertility, or longevity traits than their parents. This phenomenon was first reported by Charles Darwin [1], and it is described by the following three terms depending on which trait it is referring to and how it is calculated: (1) hybrid vigor, when referring to fitness or reproductive traits leading to increased output of offspring [2]; (2) heterosis, the superiority of hybrids in production traits over the parental mean, which is measured by the average of the reciprocal F1s minus the parental mean divided by the parental mean; (3) heterobeltiosis, a special case of heterosis when the hybrids' traits exceed the best performing parent [3], which is also known as Dunham's practical heterosis [4,5]. In agricultural practices, many plant and animal breeds exhibit heterosis through breeding practices [6]. One of the most successful applications of selective breeding is the crossbred maize (*Zea mays*), with a higher grain yield [7]. In mammals, the benefits of heterosis are significant. For instance, post-weaning body weight gain in cattle increased by 1.49 kg in the hybrid of Continental (*Bos taurus*) × Zebu (*Bos indicus*) and by 14.68 kg in the British (*Bos taurus*) × Zebu (*Bos indicus*) hybrids [8]. For fertility traits, hybrid vigor was around 10–25% in crossbred dairy cattle [9]. Sheep survival was improved from 8.8% to 14.6% by crossbreeding among 14 different breeds [10]. In poultry, bodyweight at different development stages and egg production in F1 crossbred chicken were increased by 3.76–22.33% and 8.25%, respectively [11].

From an evolutionary perspective, the formation of new species occurs as populations diverge. Before speciation, there is a short period (heterosis phase) in which hybrid fitness is higher than that of the two parental species [12]. Thus, hybrid vigor may facilitate speciation by increasing adaptation to hostile environments, which has important implications in evolution and speciation. Variations in adaptive traits are heavily influenced by changes in gene expression [13]. Cumulative genetic variation and stabilizing selection of gene expression lead to coevolution, and thus molecular functions are preserved [14]. Crosses between species can disturb this coevolution and result in hybrid incompatibility (also called genetic incompatibility) [15]. In the hybrids, the expression level of most genes is close to the mid-parent value (additive effect) or near the level of one parent (dominance or partial dominance). Hybrid incompatibility is the manifestation of gene expression misregulation, resulting in expression levels higher (positive overdominance) or lower (negative overdominance) than both parental species [16]. Genetic loci with such expression patterns are called transgressive genes [17], which directly contribute to the superior phenotypes in heterosis or misregulation of gene expression in the hybrid breakdown or outbreeding depression [18].

The effect of gene expression regulation is often asymmetric in the two reciprocal F_1 hybrid crosses. For example, gene regulatory evolution studies on frogs (crosses from *Xenopus laevis* and *X. muelleri*) [19], fish (reciprocal crosses among centrarchid species) [20], and bird (zebra finch subspecies) [21] showed asymmetric patterns of heterobeltiosis in one but incompatibility in the other reciprocal hybrids. The reduction in the fitness of hybrids is generally caused by Dobzhansky–Muller incompatibilities in the evolution field [22]. For example, the interaction between Lethal hybrid rescue (*Lhr*) and Hybrid male rescue (*Hmr*) in F_1 hybrid males leads to lethality in the interspecific crosses between *Drosophila melanogaster* and *D. simulans* [23]. Interestingly, the genetic diversity between two parental species has a considerable effect on the performance of hybrid offspring. As the genetic distance increases, stronger heterosis effects may be observed in some cases [24]. However, when the two parental species are too distantly related, detrimental effects surpass heterosis, resulting in genetic incompatibility [25]. A further study shows the advantage of heterobeltiosis is partially countered by the disadvantage of genetic incompatibility [26].

Although the phenomenon of heterosis has been reported for more than a century, the underlying genetic mechanisms are still poorly understood. Three theories about heterosis have been proposed, which were widely discussed and debated: (1) Dominance hypothesis, the dominant allele suppresses the expression of the recessive allele according to Mendel's theory [27]. The deleterious effects from recessive alleles carried by parental gametes were suppressed, and only the effects of beneficial dominant alleles were maintained; (2) Overdominance hypothesis, the expression of heterozygotes outperformed the highest parental expression values [28]; (3) Epistatic hypothesis, gene or allele interactions are existing and result in two conditions of epistasis: positive epistasis which causes the phenotype to be better than predicted, and negative epistasis in which performance is lower than expected. Positive epistatic interactions between non-allelic genes were thought to contribute to heterosis [29,30]. According to studies on gene transcription, expression differences between hybrid progeny and the parents, especially non-additive expression, contribute to heterosis [31]. However, existing theories of genetics basic do not give a unified explanation for both heterosis and hybrid incompatibility, which appear to be contradictory [32]. To understand the underlying genetic mechanisms, additional studies in different hybrid systems are needed.

Interspecific hybridization can mix genetic materials from two different species, which was proven to be an effective way to increase phenotypic variability and achieve genetic improvement in F_1s [33]. In the breeding of aquatic species, interspecific hybridization, and crossbreeding within species have been applied as an effective means of genetic enhancement in a few cases. For example, crossbreeding has successfully improved the genetic stocks of common carp [34]. The sturgeon aquaculture industry was greatly promoted by the interspecific cross of Beluga (*Huso huso*) and sterlet (*Acipenser ruthenus*) because the hybridization reduced the sexual maturity time from 20 years to 6–9 years [35,36]. The growth rate of interspecific hybrids between the black drum (*Pogonias cromis*) and red drum (*Sciaenops ocellatus*) was significantly higher than both parents [37]. Hybrid progenies of the red abalone (female *Haliotis rufescens*) and the pink abalone (male *H. corrugate*) have significantly elevated growth and survival during conditions of thermal stress (22 °C) [38]. Hybrid striped bass was found to be superior in survival rate than the two parent species, white bass (*Morone chrysops*) and striped bass (*Morone saxatilis*). The hybrid bass has an intermediate growth rate (better than white bass) and thermal tolerance (better than stripe bass), and it is preferred for aquaculture [39]. Thus, aquaculture can greatly benefit from heterobeltiosis or overall greater performance in controlled hybridizations, allowing producers to use these enhanced attributes to increase production metrics and yields.

As the largest finfish aquaculture in the US [40], catfish accounts for 70% of total US freshwater production, and it is one of the most successful examples of the application of heterosis in aquaculture. Researchers evaluated 42 different interspecific ictalurid catfish hybrids by crossing two distinct species. Only the combination of channel catfish (*Ictalurus punctatus*) female × blue catfish (*Ictalurus furcatus*) male (C×B hybrid) has superior feed conversion efficiency [41,42], higher carcass yield [43], better tolerance to low oxygen [44], improved disease resistance [45], and enhanced harvestability [46] to market size. Collectively, these characteristics exhibiting heterobeltiosis enable a commercial production rate of 13,000 kg ha^{-1}, which doubles the yield of traditional channel catfish farming [42,47,48]. Nowadays, the C×B hybrid catfish constitutes more than 50% of the total catfish harvest in the US [49].

In this study, the characteristics in C×B hybrid were found to be not heterotic in tank culture. Instead, the channel catfish parent was the superior genetic type in the aquarium environment. This finding suggested that the C×B heterobeltiosis was only observed in pond culture, which was an instance of environment-dependent heterobeltiosis/incompatibility. This phenomenon was previously reported in *Drosophila*, in which hybrid heterosis was higher in optimal density than in a crowded environment, and much higher at a lower than optimal temperature [50]. To understand the molecular basis of environment-dependent heterosis, growth-related morphometric traits were measured, including total length, standard length, body depth, body weight, head length, head depth, head width, and caudal depth, as well as physiological and immune parameters in channel catfish, blue catfish, and their reciprocal crosses under the tank culture. Liver transcriptome analyses were conducted to identify differentially expressed and transgressive genes, which provide insights into the molecular basis of environment-dependent heterosis.

2. Materials and Methods

2.1. Fish Maintenance and RNA-Seq Sample Collection

All experimental animal protocols, including animal care and tissue sample collections, were approved by the Auburn University Institutional Animal Care and Use Committee (AU-IACUC). Blue catfish (PB), channel catfish (PC), B×C hybrid catfish (F_1BC), and C×B hybrid catfish (F_1CB) were reared at the Auburn University Fish Genetics Research Unit (Auburn, AL, USA; Figure 1A). The indoor unit has a recirculatory aquaculture system (RAS) equipped with mechanical and biological filters to clean and recycle rearing water to the fish culture tanks. Dissolved oxygen was maintained above 5 mg L^{-1}, pH between 7.0 and 7.3, and water temperatures between 25 and 27 °C. One hundred fish from each group were maintained in separate 60 L rectangular tanks. At 12 months of age, two randomly selected fish from each genetic type were euthanized with buffered tricaine methanesulfonate (MS-222, Syndel Inc., Ferndale, WA, USA) for RNA-seq experiments. Liver tissues were immediately dissected, flash-frozen in liquid nitrogen, and stored in a −80 °C freezer.

2.2. Morphometric Measurements

Six fish (fingerlings in July 2020) were randomly selected from each genetic type, and morphometric traits were measured at 10 months of age, including total length, standard length, body depth, body weight, head length, head depth, head width, and caudal depth (Figure 1B). Statistical significance among PB, PC, F_1BC, and F_1CB was assessed using the non-parametric Mann–Whitney U test. The significant p-value was shown by using asterisk rating system: *, $p < 0.05$; **, $p < 0.01$; ***, $p < 0.001$.

2.3. Biochemical and Immunological Assays

At 10 months of age, six fish for each genetic type (PB, PC, F_1BC, and F_1CB) were randomly selected. The total length of the fish ranged from 11.9 to 18.4 cm. Catfish were anesthetized with buffered MS-222 (100 mg L^{-1}), and blood samples were collected from the caudal vasculature with BD U-100 syringes and transferred to lithium heparin-containing blood collection tubes (Becton Dickinson and Company, Franklin Lakes, NJ, USA). To obtain plasma, blood samples were immediately centrifuged at 1000× g for 10 min. Plasma glucose concentration was determined using Liquid Glucose (Oxidase) Reagent Set (Pointe Scientific Inc, Canton, MI, USA) using a 450 nm wavelength, with an input of 10 μL plasma per replicate. Lactate level was quantified by the Lactate (Liquid) Reagent Set (Pointe Scientific Inc, Canton, MI, USA) using a 595 nm wavelength and a sample input volume of 10 μL plasma per replicate. Lysozyme activity was determined based on the lysis of lysozyme-sensitive Gram-positive bacterium *Micrococcus lysodeikticus* (Sigma, St. Louis, MO, USA) by lysozyme present in the plasma according to Sankaran and Gurnani [51]. A total of 10 μL plasma was used as input, and 250 μL of bacterial cell suspension was added. The initial and final (after 30 min incubation at 37 °C) absorbances of the samples were measured at 450 nm. The rate of reduction in absorbance of samples was converted to lysozyme concentration (μg mL^{-1}) using a standard curve. Alternative complement hemolytic activity (ACH50) was detected following a microplate protocol previously described by Welker [52]. The input of plasma was 25 μL per replicate, and the absorbance of the samples was measured at 405 nm.

significantly different before and during the COVID-19 pandemic. The results depicted that respondents' income (M = 0.347, SD = 0.640, t = 12.314, p = 0.00) 0, savings (M = 0.372, SD = 0.627, t = 13.477, p = 0.000), and job security (M = 0.180, SD = 0.414, t = 9.891, p = 0.000) significantly differed between the two periods. In other words, the pandemic had a significant impact on respondents' income, savings, and job security. Further analysis revealed that there was a statistically significant difference in respondents' food security (M = 0.019, SD = 0.196, t = −2.245, $p < 0.0$) before and during COVID-19 with an effect size of 0.0969 (small effect). Meanwhile, a large effect size (2.2700 and 1.3607) was recorded for the significant differences in health (M = 0.849, SD = 0.374, t = −51.494, p = 0.05) and personal safety (M = 0.713, SD = 0.524, t = −30.901, p = 0.05) before and during the pandemic. Subsequent dimensions demonstrated significant differences in emotion (M = −0.614, SD = 0.547, t = −25.499, $p < 0.0$), spirituality (M = 0.029, SD = 0.179, t = 3.682, $p < 0.0$), and productivity (M = −0.368, SD = 0.179, t = −16.395, $p < 0.0$) between the respondent' situations before and during COVID-19. Specifically, the levels of emotion, spirituality, and productivity were significantly lower before COVID-19 than during the pandemic. The detailed results are presented in Table 2.

Table 2. Paired Samples Test.

Before and After COVID-19		Paired Differences					t	df	Sig. (2-Tailed)
		Mean (M)	Std. Deviation (SD)	Std. Error Mean	95% Confidence Interval of the Difference				
					Lower	Upper			
Pair 1	Income	−0.347	0.640	0.028	−0.402	−0.292	−12.314	515	0.000
Pair 2	Saving	−0.372	0.627	0.028	−0.426	−0.318	−13.477	515	0.000
Pair 3	Job Security	−0.180	0.414	0.018	−0.216	−0.144	−9.891	515	0.000
Pair 4	Food	−0.019	0.196	0.009	−0.036	−0.002	−2.245	515	0.025
Pair 5	Health	−0.849	0.374	0.016	−0.881	−0.816	−51.494	515	0.000
Pair 6	Personal Safety	−0.713	0.524	0.023	−0.759	−0.668	−30.901	515	0.000
Pair 7	Emotion	−0.614	0.547	0.024	−0.662	−0.567	−25.499	515	0.000
Pair 8	Spirituality	0.029	0.179	0.008	0.014	0.045	3.682	515	0.000
Pair 9	Productivity	−0.368	0.510	0.022	−0.412	−0.324	−16.395	515	0.000

5. Discussion

The results from paired t-tests on income, savings, and job security revealed that there were significant differences between respondents' conditions before and during COVID-19. This crisis has disrupted not only many lives but also socioeconomic factors such as income, savings, and job stability. Some actions are needed to counter the negative impact of COVID-19 to elevate the socioeconomic status of those affected by the crisis, especially the poor. The research findings on income, savings, and job security are consistent with a previous study [31] in which individuals and households faced high economic risk as their financial sources and savings were severely affected during the crisis. According to them, some did experience a cut to their basic income while others were forced to depend on their savings to pay for necessary items, especially among the B40 households. Some Malaysians were reported to have experienced a loss of income due to COVID-19 [32]. Additionally, movement restrictions have affected daily incomes and some of them have lost their jobs. These findings are pertinent for the relevant agencies to introduce appropriate plans or policies to ensure the well-being of those impacted by the pandemic or any future crisis. COVID-19 has impacted individuals and businesses, especially small and medium enterprises following massive lay-offs and loss of income [33]. The researcher concluded that government intervention is vital to ensure recovery and growth so that Malaysia can be sustainable and resilient for any upcoming crisis. In brief, COVID-19 has impacted income, savings, and job security, especially during the event of the crisis. Therefore, relevant government agencies should undertake proactive actions and policies to mitigate the impacts of COVID-19 or any upcoming crisis.

Although there was a small difference in respondents' food security before and during COVID-19, the current data emphasises the need to investigate the issue to implement adequate measures in the future to alleviate hunger, reduce malnutrition, and other serious impacts of food insecurity on the poor. This finding is also in line with another study [8] which reported the prevalence of food and nutritional deficiencies during the COVID-19 pandemic and measures to address the issue. The government and other relevant authorities should support affected individuals in diversifying their sources of income in order to strengthen food security. The government's national food security programmes should be well-planned and focused to reduce the burden of food insecurity among affected groups. Integrated farming entrepreneurship and rural empowerment programmes should be established and intensified to eliminate food insecurity of the public. Admittedly, COVID-19 remains a public health emergency worldwide. The current study revealed that there is a remarkable difference in public health before and during the pandemic. This finding is consistent with research [12] in which COVID-19 was reported to have a significant impact on healthcare systems. This study highlights the need for more comprehensive health education, focusing on information consistency from the government to the general public. Health educational efforts are urgently needed to reach the general population. New education systems need to be deployed to increase general population awareness of COVID-19 and its preventative practices to reach its elimination targets. Knowledge and awareness of the disease are crucial characteristics for the implementation of protective measures to reduce the risk of illness exposure. This study also revealed that personal safety demands urgent intention. The current study depicted a large difference in the personal safety of the public before and during COVID-19. Most respondents in the present study asserted that the pandemic affected their personal safety. In other words, their safety has declined since the MCO was implemented. There are also issues of social media attacks, fraud, and theft, which make them feel traumatised. Another previous study [20] reported the psychosocial impact of COVID-19 on Malaysian families. The term "safety" refers to a situation in which there is no threat, thus, meaning it is described as the ability of an individual to go about their daily lives without fear of psychological, emotional, or physical harm from others. It is necessary pertaining to one's personal experience. This sense of safety is important to individuals who are substantially not at risk of being a victim due to the pandemic. Therefore, it is pertinent for the government to consider how to improve individual safety during the pandemic. A personal safety plan could assist the victims in individual protection and aid in the preparation for future violence and abuse. The sample safety plans should be adapted to fit the specific needs during the pandemic, for example, a self-guide for domestic violence, safety planning for children and youth, programmes for domestic abuse, an educational toolkit for domestic abuse victims and survivors, and many more.

The current findings have elucidated the impact of COVID-19 on socioeconomic factors, comprising emotion, spirituality, and productivity. Regarding emotion, the majority of respondents agreed that one of the reasons their emotions were disturbed was their fear of contracting COVID-19. Infection with COVID-19 might disrupt individual health, limit daily routine and activities, and the most feared outcome is death. In parallel, it has been reported by Forbes that lockdown and social distancing measures to prevent the spread of COVID-19 have heightened fears of increasing levels of domestic violence, which includes physical, emotional, and sexual abuse [34]. Meanwhile, a study [7] found that over half of the women experiencing emotional or moderate physical violence reported an increase in these events since the lockdown. With no job, no income, and limited mobility, individuals' emotions are affected, as highlighted by the respondents of this study. This can be seen where, on average, they consider this pandemic to be unexpected. Therefore, it is a great challenge for the respondents to continue surviving, especially during the pandemic. Although COVID-19 had a significant impact on the majority of the respondents, it is interesting to note that they strive to be closer to their God in terms of spirituality. These people may believe that the pandemic is a test from their God to see how strong and patient they are in the face of adversity. In a previous study [35], spirituality or faith was one of the tools for dealing with stress and the negative effects of life problems and illnesses. Indeed, a prior study demonstrated that families rely on their spirituality for emotional, mental, and physical well-being [36]. In fact, spiritual practices have long been recognised as an effective coping mechanism for dealing with life-altering and traumatic events [36]. In Indonesia, the rise in COVID-19-related anxiety cases necessitated increased advocacy for holistic mental health services, with spirituality being recommended among people suffering from anxiety [37]. Moreover, in the present study, most respondents stated that practicing prayer, dhikr, and reading religious books could help their emotions to be calmer and more stable. These respondents have been able to reflect on why COVID-19 is occurring as a result of the emphasis on spirituality. Consequently, productivity among respondents was also affected during the COVID-19 outbreak. Following the MCO announcement by the government, respondents were unable to perform their usual activities. This has resulted in declining productivity specifically if working from home (WFH). In the same vein, workers who worked from home had lower productivity than those who did not [38]. Second, the same study also revealed that poor WFH setups and communication issues are the primary causes of productivity losses. According to Stanford economist, Nicholas Bloom, the global WFH attempts to maintain output and efficiency during the COVID-19 pandemic may cause a worldwide productivity slump and threaten economic growth for many years [39]. WFH is undeniably not the same as working in an office or outside. The respondents were in a similar situation, where they felt unproductive when sitting at home alone given that most of them were self-employed while others run their own businesses to make ends meet.

6. Conclusions and Policy Implications

This research aimed at comparing the socioeconomic conditions before and during the COVID-19 pandemic and determining the most affected socioeconomic indicator. This empirical study used a valid and reliable questionnaire that was purposely distributed to 516 respondents aged 18 to 65 years in Peninsular Malaysia. The findings revealed a significant difference in socioeconomic conditions, such as income, savings, job stability, health, security/personal safety, emotion, spirituality, and working productivity. Other conditions included spirituality, food security, health, and personal safety. The results impys that the COVID-19 pandemic did affect the socioeconomic conditions of most of the respondents due to loss of income and challenges in paying for their expenses. Nevertheless, the findings are inconclusive since no causal relationship or effect could be determined on respondents' socioeconomic conditions. In the future, advanced techniques and methodologies, such as the structural equation model (SEM) and a combination of primary and secondary data analysis could be employed. Such analyses will assist in yielding advanced recommendations, primarily related to the post-COVID-19 challenges and prospects of the Malaysian economy. Therefore, to overcome the impact of COVID-19 on socioeconomic conditions among Malaysians, the government and NGOs need to come up with post-pandemic policies and initiatives, which include providing constant support to enterprises, jobs, and incomes through appropriate fiscal and monetary policies. Additionally, Malaysian firms can increase OSH safeguards, reschedule or reorganise working hours, and change their working styles to protect their employees in the workplace. For example, keep a specific number of employees at their workstations while allowing others to telework or work from home. By improving the employees' working environment, it is expected that productivity will increase and contribute to a better economic rebound.

Author Contributions: N.S.O. and Z.M.I. contributed to the conceptualization; N.S.O. wrote the introduction and literature review; Z.M.I., D.I., and S.A.B. contributed to the methodology; Z.M.I., N.A., A.A. and N.S.O. worked on analysis, findings, and discussion; N.S.O. and D.I. dealt with the conclusion and policy implications; N.S.O., Z.M.I., N.A., D.I., A.A. and S.A.B. contributed to reviewing, editing, and project administration; Z.M.I., N.A., D.I., A.A. and S.A.B. were involved in funding acquisition. All authors have read and agreed to the published version of the manuscript.

Funding: This research was funded by UiTM Kedah Branch under *Geran Dalaman Negeri (GDN) Tahun 2020*, grant number 600-UiTMKDH (PJI.5/4/1)(4/2020).

Institutional Review Board Statement: This research was carried out with the consent of the UiTM Research Ethics Committee (Ref. number: REC/08/2021 (MR/692)).

Informed Consent Statement: Not applicable.

Data Availability Statement: The data presented in this study are available upon request to the corresponding author.

Acknowledgments: The authors would like to thank UiTM Kedah Branch for awarding the *Geran Dalaman Negeri (GDN) Tahun 2020* with the research titled "Impact of COVID-19 Spread on Malaysia's Socioeconomic," with the project identification number 600-UiTMKDH (PJI.5/4/1)(4/2020).

Conflicts of Interest: The authors declare no conflict of interest.

References

1. Pettersson, H.; Manley, B.; Hernandez, S. Tracking COVID-19's Global Spread. CNN Health. Available online: https://www.edition.cnn.com/interactive/2020/health/coronavirus-maps-and-cases/ (accessed on 15 June 2022).
2. Ministry of Health Malaysia. COVIDNOW in Malaysia. Available online: https://covidnow.moh.gov.my/ (accessed on 15 June 2022).
3. Lim, L.L. The Socioeconomic Impacts of COVID-19 in Malaysia: Policy Review and Guidance for Protecting the Most Vulnerable and Supporting Enterprises. 2020. International Labour Organisation. Available online: http://ilo.org/wcmsp5/groups/public/---asia/---ro-bangkok/documents/publication/wcms_751600.pdf (accessed on 15 June 2022).
4. Mateo-Urdiales, A.; Fabiani, M.; Rosano, A.; Vescio, M.F.; Del Manso, M.; Bella, A.; Riccardo, F.; Pezzotti, P.; Regidor, E.; Andrianou, X. Socioeconomic patterns and COVID-19 outcomes before, during and after the lockdown in Italy. *Health Place* **2020**, *71*, 102642. [CrossRef] [PubMed]
5. Chung, C.; FSingau, J.B.; Pazim, K.H.; Mansur, K. The repercussions of covid-19 pandemic on the wellbeing of olderpeople in Malaysia: A literature review. *Int. J. Stud. Child. Women Elderly Disabled* **2020**, *11*, 17–22. Available online: https://www.ijcwed.com/wp-content/uploads/2020/09/IJCWED11_017.pdf (accessed on 15 June 2022).
6. Fernandes, N. Economic Effects of Coronavirus Outbreak (COVID-19) on Theworld Economy. 2020. Available online: https://mediaroom.iese.edu/wp-content/uploads/2020/03/Fernandes-Nuno_20200322-Global-Recession-is-inevitable.pdf (accessed on 15 June 2022).
7. Hamadani, J.D.; Hasan, M.I.; Baldi, A.J.; Hossain, S.J.; Shiraji, S.; Bhuiyan, M.S.; Mehrin, S.F.; Fisher, J.; Tofail, F.; Tipu, S.M.; et al. Immediate impact of stay-at-home orders to control COVID-19 transmission on socioeconomic conditions, food insecurity, mental health, and intimate partner violence in Bangladeshi women and their families: An interrupted time series. *Lancet Global Health* **2020**, *8*, E1380–E1389. [CrossRef]
8. Shammi, M.; Bodrud-Doza, M.; Towfiqul Islam, A.R.M.; Rahman, M.M. COVID-19 pandemic, socioeconomic crisis and human stress in resource-limited settings: A case from Bangladesh. *Heliyon* **2020**, *6*, e04063. [CrossRef] [PubMed]
9. De Muro, M.; Amadori, S.; Ardu, N.R.; Cerchiara, E.; De Fabritiis, P.; Niscola, P.; Dante, S.; Schittone, V.; Tesei, C.; Trawinska, M.M.; et al. Impact on Mental Health, Disease Management and Socioeconomic Modifications in Hematological Patients during COVID-19 Pandemia in Italy. *Blood* **2020**, *136*, 35–37. [CrossRef]
10. Costa, V.B.F.; Bonatto, B.D.; Pereira, L.C.; Silva, P.F. Analysis of the impact of COVID-19 pandemic on the Brazilian distribution electricity market based on a socioeconomic regulatory model. *Int. J. Elec. Power Energ. Syst.* **2021**, *132*, 107172. [CrossRef]
11. Mashrur, F.R.; Roy, A.D.; Chhoan, A.P.; Sarker, S.; Saha, A.; Hasan, S.N.; Saha, S. Impact of demographic, environmental, socioeconomic, and government intervention on the spreading of COVID-19. *Clin. Epidemiol. Glob. Health* **2021**, *12*, 100811. [CrossRef]
12. Ozili, P. COVID-19 in Africa: Socio-economic impact, policy response and opportunities. *Int. J. Sociol. Soc. Policy* **2020**, *42*, 177–200. [CrossRef]
13. Varkey, R.S.; Joy, J.; Sarmah, G.; Panda, P.K. Socioeconomic determinants of COVID-19 in Asian countries: An empirical analysis. *Public Aff.* **2020**, *21*, e2532. [CrossRef]
14. Kusuma, D.; Pradeepa, R.; Khawaja, K.I.; Hasan, M.; Siddiqui, S.; Mahmood, S.; Shah, S.M.; De Silva, C.K.; de Silva, L.; Gamage, M.; et al. Low uptake of COVID-19 prevention behaviours and high socioeconomic impact of lockdown measures in South Asia: Evidence from a large-scale multi-country surveillance programme. *SSM-Popul. Health* **2021**, *13*, 100751. [CrossRef]
15. Nguyen, H.N.; TRAN, M.D. The effect of perceived organizational support on employee engagement during the COVID-19 Pandemic: An empirical study in Vietnam. *J. Asian Financ. Econ. Bus.* **2021**, *8*, 415–426. [CrossRef]
16. Hidayat, S.; Wibowo, W.; Gunawan, Y.E.; Dewi, G.C.; Wijayaningtyas, M. FactorsInfluencing Purchase Intention of Healthcare Products During the COVID-19 Pandemic:An Empirical Study in Indonesia. *J. Asian Finance Econ. Bus.* **2021**, *8*, 337–345. [CrossRef]
17. Tang, K.H.D. Movement control as an effective measure against Covid-19 spread in Malaysia: An overview. *Z Gesundh Wiss.* **2022**, *30*, 583–586. [CrossRef] [PubMed]
18. Jamil, A.H.; Mohd Sanusi, Z.; Yaacob, N.M.; Mat Isa, Y.; Tarjo, T. The Covid-19 impact on financial crime and regulatory compliance in Malaysia. *J. Financ. Crime* **2022**, *29*, 491–505. [CrossRef]

19. Shakeel, S.; Ahmed Hassali, M.A.; Abbas Naqvi, A. Health and Economic Impact of COVID-19: Mapping the Consequences of a Pandemic in Malaysia. *Malays. J. Med. Sci.* **2020**, *27*, 159–164. [CrossRef]
20. Zainudeen, Z.T.; Abd Hamid, I.J.; Azizuddin, M.N.A.; Abu Bakar, F.F.; Sany, S.; Zolkepli, I.A.; Mangantig, E. Psychosocial impact of COVID-19 pandemic on Malaysian families: A cross-sectional study. *BMJ Open* **2021**, *11*, e050523. [CrossRef]
21. Sing Joo, G.; Owen Devan, D.M.; Shao Qi, C.; Patil, S.S. Association between depression, anxiety, stress and perceived quality of life in a Malaysian B40 urban community during the COVID-19 lockdown: A cross-sectional study. *F1000Research* **2021**, *10*, 693. [CrossRef]
22. Krejcie, R.V.; Morgan, D.W. Determining sample size for research activities. *Educ. Psychol. Meas.* **1970**, *30*, 607–610. [CrossRef]
23. Taherdoost, H. Sampling methods in research methodology; How to choose a sampling technique for research. *Int. J. Acad. Res. Manag.* **2016**, *5*, 18–27. [CrossRef]
24. Michaelidou, N.; Dibb, S. Using email questionnaires for research: Good practice in tackling non-response. *J. Target. Meas. Anal. Mark.* **2006**, *14*, 289–296. [CrossRef]
25. Lee, S.A. Coronavirus Anxiety Scale: A brief mental health screener for COVID-19 related anxiety. *Death Stud.* **2020**, *44*, 393–401. [CrossRef] [PubMed]
26. Liu, N.; Zhang, F.; Wei, C.; Jia, Y.; Shang, Z.; Sun, L.; Wu, L.; Sun, Z.; Zhou, Y.; Wang, Y.; et al. Prevalence and predictors of PTSS during COVID-19 outbreak in China hardest-hit areas: Gender differences matter. *Psychiatry Res.* **2020**, *287*, 112921. [CrossRef] [PubMed]
27. Oosterhoff, B.; Palmer, C.A. Psychological correlates of news monitoring, social distancing, disinfecting, and hoarding behaviors among US Adolescents during the COVID-19 Pandemic. *PsyArXiv* **2020**. *Reprint*. Available online: https://psyarxiv.com/rpcy4/ (accessed on 13 September 2022). [CrossRef]
28. Kelly, B.; Squiers, L.; Bann, C.; Stine, A.; Hansen, H.; Lynch, M. Perceptions and plans for prevention of Ebola: Results from a national survey. *BMC Public Health* **2015**, *15*, 1136. [CrossRef] [PubMed]
29. Richards, P.; Amara, J.; Ferme, M.C.; Kamara, P.; Mokuwa, E.; Sheriff, A.I.; Suluku, R.; Voors, M. Social pathways for Ebola virus disease in rural Sierra Leone, and some implications for containment. *PLoS Negl. Trop. Dis.* **2015**, *9*, e0003567. [CrossRef]
30. Yang, Z.J. Altruism during Ebola: Risk perception, issue salience, cultural cognition, and information processing. *Risk Anal.* **2016**, *36*, 1079–1089. [CrossRef]
31. Ismail, M.K.; Kumaran, V.V.; Munawwarah, S.N.; Muhamad, M.Z.; Sarifuddin, S. COVID-19 Outbreak: An analysis of Malaysian household income class during Movement Control Orders (MCO). *Asia Proc. Soc. Sci.* **2021**, *7*, 26–29. [CrossRef]
32. Saari, M.Y.; Ibrahim, K.M.; Habibullah, M.S. Assessing the income distributional effect of lockdowns in Malaysia. *J. Econ. Impact* **2022**, *4*, 132–138. [CrossRef]
33. Daud, S. The COVID-19 pandemic crisis in Malaysia and the social protection program. *J. Dev. Soc.* **2021**, *37*, 480–501. [CrossRef]
34. Gaskell, A. How Productive Have Remote Workers Been during COVID? 2021. Forbes. Available online: https://www.forbes.com/sites/adigaskell/2021/05/31/how-productive-have-remote-workers-been-during-covid/ (accessed on 15 June 2022).
35. Ribeiro, M.R.C.; Damiano, R.F.; Marujo, R.; Nasri, F.; Lucchetti, G. The role of spirituality in the COVID-19 pandemic: A spiritual hotline project. *J. Public Health.* **2020**, *42*, 855–856. [CrossRef]
36. Roman, N.V.; Mthembu, T.G.; Hoosen, M. Spiritual care—'A deeper immunity'—A response to Covid-19 pandemic. *Afr. J. Prim. Health Care Fam. Med.* **2020**, *12*, e1–e3. [CrossRef] [PubMed]
37. Rias, Y.A.; Rosyad, Y.S.; Chipojola, R.; Wiratama, B.S.; Safitri, C.I.; Weng, S.F.; Yang, C.Y.; Tsai, H.T. Effects of spirituality, knowledge, attitudes, and practices toward anxiety regarding COVID-19 among the general population in INDONESIA: A cross-sectional study. *J. Clin. Med.* **2020**, *9*, 3798. [CrossRef] [PubMed]
38. Kitagawa, R.; Kuroda, S.; Okudaira, H.; Owan, H. Working from home and productivity under the COVID-19 pandemic: Using survey data of four manufacturing firms. *PLoS ONE* **2021**, *16*, e0261761. [CrossRef] [PubMed]
39. Adam, G. The Productivity Pitfalls of Working from Home in the Age of COVID-19. 2022. Stanford News. Available online: https://news.stanford.edu/2020/03/30/productivity-pitfalls-working-home-age-covid-19/ (accessed on 15 June 2022).

proceedings

MDPI

Proceeding Paper

The E-Servqual Effect on Mobile Stickiness Intention of E-Commerce Marketplace †

Azlin Zanariah Bahtar *, Geetha Muthusamy, Zarinah Abu Yazid and Shahreena Daud

Faculty of Business and Management, Universiti Teknologi MARA, Cawangan Melaka Kampus Bandaraya Melaka, Melaka 75300, Malaysia

* Correspondence: azlin_bahtar@uitm.edu.my

† Presented at the International Academic Symposium of Social Science 2022, Kota Bharu, Malaysia, 3 July 2022.

Abstract: Digital devices and the internet have created a new lifestyle among Malaysian consumers, especially after COVID-19. Online transactions have become common. Shopping on e-commerce marketplaces has become inseparable from the life of Malaysian consumers. The most popular e-commerce marketplace applications used here are Shopee and Lazada. The purpose of this study is to determine the effect of e-service quality on stickiness intention in Malaysia. This study obtained 251 responses and uses the Statistical Package for Social Sciences (SPSS) to test all four hypotheses. The results showed that three relationships (reliability, website design, and customer service and support) were found to have significant effects on mobile stickiness intention while the second hypothesis (privacy and security) was found to have an insignificant negative effect on mobile stickiness intention. The findings of this study provide insight into consumer behaviour, e-commerce, marketing, and technology. This study contributes to the growing literature on E-Servqual and stickiness intention.

Keywords: E-Servqual; e-commerce; online marketing

Citation: Bahtar, A.Z.; Muthusamy, G.; Yazid, Z.A.; Daud, S. The E-Servqual Effect on Mobile Stickiness Intention of E-Commerce Marketplace. *Proceedings* **2022**, *82*, 34. https://doi.org/10.3390/proceedings2022082034

Academic Editor: Mohamad Rahimi Mohamad Rosman

Published: 13 September 2022

1. Introduction

In the past two years, online digital platforms have become the busiest they have ever been in terms of both sellers and buyers. The usage of the internet and online selling have greatly accelerated and created massive shifts in consumer buying behaviour [1–3]. Sellers have embraced and utilised these platforms in an effort to remain relevant and to re-penetrate the market as well as increase effectiveness and efficiency [3–5]. Meanwhile, buyers have chosen to conduct transactions via these platforms with the intention to minimise movements and physical contacts [1,3,6]. The pattern of online buying and selling is expected to not return to how things were experienced before COVID-19 [7,8]. Although many operations have resumed normally, due to the high increase in COVID-19 cases reported in early 2022 [9], the majority of Malaysians still prefer to purchase their non-complex items via digital platforms.

Generally, there are many e-commerce platforms where buyers can conduct transactions such as sellers' own websites, mobile applications, social media platforms, or even e-commerce marketplace [5]. In Malaysia, e-commerce marketplaces (e.g., Shopee, Lazada, and Happy Fresh) are booming and receiving many visitors, especially during their special promotion days. Moreover, the e-commerce marketplaces are now available on mobile applications, thus, giving consumers more benefits and privileges such as sufficient information, convenience, and a great number of choices [2,10,11]. With the choices and options available on these platforms, consumers could purchase many items under one roof with minimum disruptions [5]. This has created a positive impact where an increase in sales between 10% and 15% on the e-commerce platforms in Malaysia has been recorded [2]. Other than these benefits, the quality of online services has also become a major factor contributing to consumers' decisions to purchase from or revisit the platforms. Good and consistent E-Service Quality (E-Servqual) provided by the e-commerce marketplace is able to increase the trust and satisfaction level among the customers [11,12]. This will enhance the possibility of platform stickiness intention [4,12].

Stickiness is important specifically for e-commerce platforms as it has become one of the success determinants. According to previous studies, no solid factors have been identified on how to make visitors stick with a platform [4,12]. Nevertheless, findings from [4] showed that good practice in E-Servqual (especially during this COVID-19 period) managed to increase the stickiness intention among visitors and users. E-Servqual is considered fundamental and must be taken seriously. Close monitoring and improvement from time to time are required to ensure the services provided will always be high quality and satisfying. This will encourage the return of visitors and repurchase activities [11,12]. As mentioned earlier, a high number of buyers are purchasing from this digital marketplace; we do know that stickiness intention will affect sustainability, and E-Servqual is said to influence the stickiness intention. Thus, the purpose of this study is to examine the effect of E-Servqual on the e-commerce marketplace.

2. Literature Review

2.1. Stickiness Intention

As stated earlier, stickiness intention is important in determining the future of a business, and this is not a new area of study as it has been widely discussed for years. It is defined as the repetitive visits to and use of a certain e-commerce platform. It is also about the willingness to spend an extensive period of time browsing and scrolling the platform [4]. Reference [13] viewed stickiness as the ability of a platform to attract and develop visitors' behaviour to revisit and retain them. Stickiness will only take place if the users and visitors are comfortable, satisfied, and trust the platforms they visited [12].

Wu, Liu, & Cui (2021) summarised the factors discussed in previous studies that influence stickiness intention. The factors are website content/information, situation, infrastructure, usefulness, system quality, product options, trust, satisfaction, belongingness, and positive attitudes [12]. Studies conducted by [4,13] showed that E-Servqual (or service quality) do affect and influence stickiness intention (or user stickiness).

2.2. E-Service Quality (E-Servqual)

E-Servqual dimensions are not a new area to discuss and study. However, due to the drastic changes and accelerating digital business resulting from the COVID-19 pandemic, quality of service should be at the centre of the discussion again. This is because more sellers have redeveloped and enhanced their online marketing strategies as digital marketing has changed tremendously [14,15]. As one of the tactics to be competitive on the digital platform, the services provided by digital sellers must be of the very best quality [16]. It is believed that E-Servqual is potent in influencing online users' behaviour including sticking to and remaining with digital platforms [4,17].

Over the decades, the dimensions of E-Servqual have changed and been reconceptualised to fit with the situation and surroundings that are still based on the first service quality measurements [16,18]. According to [19], providing stable service quality to customers is a complex activity. One model used for one platform or industry may not be suitable for another [18] as the aim of providing E-Serqual is to satisfy platform users [20,21]. In this study, researchers adapted the E-Servqual dimensions from a study conducted by [21] comprising reliability, privacy and security, website design, and customer service and support. Many past studies have conducted E-Servqual research using purchase intention, customer satisfaction, and customer loyalty. However, very few have undertaken research with stickiness intention as the dependent variable. Hence, the hypotheses of this study are as follows:

Hypothesis 1 (H1). *Reliability has a positive significant relationship with stickiness intention.*

Hypothesis 2 (H2). *Privacy and security have a positive significant relationship with stickiness intention.*

Hypothesis 3 (H3). *Website design has a positive significant relationship with stickiness intention.*

Hypothesis 4 (H4). *Customer service and support have a positive significant relationship with stickiness intention.*

3. Methodology

In this study, a quantitative technique was employed for data collection, with the researcher distributing questionnaires specially prepared with research-related questions to the target respondents. Initially, a series of questions pertaining to the study variables were created to elicit responses from the target respondents. The objective is to determine which factors of online retail service quality impact consumer stickiness intention. The questionnaire set is divided into three sections, which are Section A, Section B and Section C. Section A, the first part of the questionnaire, consists of questions regarding the demographic background of the respondents.

There were three questions under Section B, where researchers asked about respondents' online purchasing behaviour. Meanwhile, the last section comprised five subsections focusing on four independent variables of the study, which are reliability, privacy and security, website design, customer service and support that influence the dependent variable—stickiness intention. The results were utilised to analyse the most important factors influencing consumer stickiness intention. In this research, the independent variables and dependent variables were measured using a five-point Likert scale ranging from 1 to 5 as follows: 1 represents Strongly Disagree, 2 represents Disagree, 3 represents Neutral, 4 represents Agree, and 5 represents Strongly Agree.

The researcher utilised a non-probability sampling technique for this study. The G*Power 3 calculator was used in this investigation to estimate the sample size from the general population. The choice of convenience sampling and the selection of respondents were based on the researcher's ease of accessibility. Data were gathered from online shoppers after the survey link was sent through messages to Shopee and Lazada account users. The total number of replies obtained was 251 sets.

The data were analysed using both descriptive and inferential methods. Methods of analysis such as reliability analysis, Pearson's correlation coefficient, and multiple regression analysis were employed in this research. The Statistical Package for Social Sciences (SPSS) Version 25.0 software was used to evaluate and analyse the data that were collected from the questionnaire. The statistical analysis was used to determine the relationships between E-Servqual dimensions and customer stickiness intention.

4. Findings

4.1. Characteristics of Respondents

In this study, a total of 251 responses were obtained. The respondents are individuals who had shopped online using E-commerce marketplaces, specifically Shopee and Lazada. Of the total number of responses, 80.5% are female and only 49 respondents are male. As for the age group, the majority of the respondents are 18–25 years old (60.6%), followed by 17.5% who are above 45 years old. A total of 13.5% are those between 35 and 45 years old. Lastly, only 8.4% are people aged between 26 and 35 years old. From the obtained data, most of the respondents hold a bachelor's degree (61.4%). Out of 251 responses, 41.8% are employed in various sectors and industries, while the remaining individuals are students or recently graduated. As the majority of the respondents are students and unemployed, hence 66.6% of the total respondents earn less than RM2500.00 monthly. Due to that, only 24.8% out of the 251 respondents have conducted online purchases more than 15 times in the past 2 years. A total of 47% of the respondents prefer to pay their purchases using online banking transfers rather than other options available.

4.2. Data Analysis

A reliability test was conducted to determine the items' Cronbach alpha figures and range. There are five variables altogether (reliability, privacy and security, website design, and customer support and service) used and tested in this study. The Cronbach alpha ranges are between 0.808 and 0.912. This indicates the items and variables are all reliable and can be further analysed.

Data then were tested under the Pearson correlation analysis. This test was conducted to identify the direction and strength of the relationship between the independent variables and the dependent variable. The final analysis conducted was the multiple regression analysis. The purpose of this analysis is to prove the hypotheses and to determine which E-Servqual determinants impacted the stickiness intention the most, by calculating the coefficient of determination (R^2). The results are shown in Table 1.

Table 1. Result.

Variables	Cronbach Alpha	Path Coefficient	T-Value	Sig.
Stickiness Intention	0.906	-	-	-
Reliability	0.808	0.631	4.928	0.000
Privacy and Security	0.882	−0.492	−0.072	0.943
Website Design	0.885	0.679	6.008	0.000
Customer Service and Support	0.912	0.559	2.116	0.000
	R Square			0.538
	Adjusted R Square			0.530
	N			251

Dependent Variable: Stickiness Intention.

5. Conclusions

The results in Table 1 showed three hypotheses (reliability, website design, and customer support and service) were found to have positive significant relationships with stickiness intention. However, the second hypothesis, which was on privacy and security in relation to stickiness intention was found to be insignificant and had a negative direction.

In digital practices, especially involving transactions between customers and merchants, privacy and security should be the very most important element to pay attention to. Platforms that are incompetent in protecting customer data definitely will receive fewer visitors and users resulting from customers' reluctance to trust the platforms. Therefore, merchants and platforms these days invest a great deal in digital security to ensure every customer will be protected as they deserve. E-Commerce marketplaces such as Shopee provided a clear privacy policy (last modified on 29 October 2021) [22]. In the privacy policy, Shopee mentioned:

> *"We implement a variety of security measures and strive to ensure the security of your personal data on our systems. User personal data is contained behind secured networks and is only accessible by a limited number of employees who have special access rights to such systems. However, there can inevitably be no guarantee of absolute security"*

Although Shopee does not guarantee absolute security, we can still conclude from the finding that users and customers are comfortable and trust the platform. With such confidence, the customers shall continue to visit and repurchase from the platform without ignoring other service quality factors, especially how the platform is designed.

From the table above, website design is found to be the most influential factor in customer stickiness intention. With that, all major e-commerce marketplaces are suggested to continuously monitor their layouts by keeping them neat and clean. A neat and well-structured layout will definitely deliver a positive emotion to the visitors and soon will encourage them to revisit the page.

Referring to the results, E-Servqual variables used in this research are viewed to be important in leading customers to stick to a platform. Continuing to maintain a good quality of services can definitely bring visitors to return to Shopee. However, it is highly recommended for future researchers to include emotions as mediators or geography as a moderator. Findings with such variables will make the results more interesting and enhance the body of knowledge as a whole.

Author Contributions: Conceptualization, A.Z.B.; methodology, G.M.; software, A.Z.B. and G.M.; validation, A.Z.B. and G.M.; formal analysis, A.Z.B.; investigation, A.Z.B., G.M., Z.A.Y. and S.D.; resources, A.Z.B. and G.M.; data curation, A.Z.B. and G.M.; writing—original draft preparation, A.Z.B.; writing—review and editing, Z.A.Y. and S.D.; visualization, A.Z.B.; supervision, A.Z.B.; project administration, Z.A.Y. and S.D.; funding acquisition, A.Z.B., G.M., Z.A.Y. and S.D. All authors have read and agreed to the published version of the manuscript.

Funding: This research received no external funding.

Institutional Review Board Statement: Not applicable.

Informed Consent Statement: Not applicable.

Data Availability Statement: Not applicable.

Acknowledgments: The authors would like to express their gratitude to all respondents who have participated in this study and to everyone who has contributed to the idea of getting this paper completed. Moreover, we would like to thank the Faculty of Business and Management UiTM Cawangan Melaka for supporting us in this study and the Faculty of Business and Management Cawangan Kelantan for the opportunity to present this paper at the International Academic Symposium of Social Science 2022, online, 3 July 2022.

Conflicts of Interest: The authors declare no conflict of interest.

References

1. Halik, A.; Arif, D.; Ekasari, R.; Yucha, N.; Pramudita, R.A. Economic Impacts of COVID-2019 and its Implications on e-Commerce Buying Interest: A Case Study in Shopee Online-shop. *Int. J. Adv. Sci. Technol.* **2020**, *29*, 2530–2538.
2. Isa, K.; Shah, J.M.; Palpanadan, S.T.; Isa, F. Malaysians' Popular Online Shopping Websites during Movement Control Order (MCO). *Int. J. Adv. Trends Comput. Sci. Eng.* **2020**, *9*, 2154–2158.
3. Koch, J.; Frommeyer, B.; Schewe, G. Online Shopping Motives during the COVID-19 Pandemic—Lessons from the Crisis. *Sustainability* **2020**, *12*, 10247. [CrossRef]
4. Kusumawati, A.; Augustinah, F.; Alhabsyi, T.; Suharyono, S. The E-servqual effect on the stickiness intention of marketplace during COVID-19 pandemic: An empirical study in Indonesia. *J. Asian Financ. Econ. Bus.* **2021**, *8*, 573–581.
5. Ratmono, R.; Djuniardi, D.; Musdalipa, M.; Nurhaida, T.; Saati, E.A.; Rahayu, S.; Andiyan, A. In the Middle of the COVID-19 Pandemic, Using the Marketplace as a Promotional Medieum to Increase Sales. *Acad. Strateg. Manag. J.* **2022**, *21*, 1–5.
6. Ali, B.J. Impact of COVID-19 on consumer buying behavior toward online shopping in Iraq. *Econ. Stud. J.* **2020**, *18*, 267–280.
7. Butu, A.; Brumă, I.S.; Tanasă, L.; Rodino, S.; Vasiliu, C.D.; Doboș, S.; Butu, M. The impact of COVID-19 crisis upon the consumer buying behavior of fresh vegetables directly from local producers. Case study: The quarantined area of Suceava County, Romania. *Int. J. Environ. Res. Public Health* **2020**, *17*, 5485. [CrossRef] [PubMed]
8. Sharma, A. Changing consumer behaviours towards online shopping-an impact of COVID 19. *Acad. Mark. Stud. J.* **2020**, *24*, 1–10.
9. Salim, S. *Malaysia's COVID-19 Cases Hit Record High of 162,793 in Week of Feb 11–17*; The Edge Markets: Petaling Jaya, Malaysia, 18 February 2022.
10. Martinez, B.M.; McAndrews, L.E. The influence of mobile application design features on users' stickiness intentions as mediated by emotional response. *Int. J. Retail Distrib. Manag.* **2021**, *49*, 1497–1511. [CrossRef]
11. Saraswati, A.; Indriani, F. Effect E-service quality on customer satisfaction and impact on Repurchase in Lazada Indonesia's online selling in the city of Semarang. *Int. J. Econ. Bus. Account. Res.* **2021**, *5*, 1203–1215.
12. Wu, J.; Liu, L.; Cui, T. What drives consumer website stickiness intention? The Role of website service quality and website involvement. *Int. J. Serv. Technol. Manag.* **2021**, *27*, 189–208. [CrossRef]
13. Lijuan, H.; Lin, J.; Jiahe, S. Antecedents of user stickiness and loyalty and their effects on users' group-buying repurchase intention. In Proceedings of the 21st Americas Conference on Information Systems, Fajardo, Puerto Rico, 13–15 August 2015; pp. 1–17.
14. Khan, A.; Zubair, S.S.; Khurram, S.; Khan, M.A. Service quality dimensions and customer satisfaction in online shopping: A customer's perspective. *J. Appl. Econ. Bus. Stud.* **2020**, *4*, 53–76. [CrossRef]
15. Tata, S.V.; Prashar, S.; Parsad, C. Examining the Influence of Satisfaction and Regret on Online Shoppers' Post-Purchase Behaviour. *Benchmarking* **2021**, *28*, 1987–2007. [CrossRef]

16. Saadah, A.; Ying, T.Y. The impact of online retail service quality on customer satisfaction. *J. Technol. Manag. Technopreneurship* **2020**, *8*, 32–51.
17. Al Ardi, A.N.; Yulisetiarini, D. The effect of lazada website quality to satisfaction and consumer loyalty. *Int. J. Res. Sci. Manag.* **2018**, *5*, 11–15.
18. Kowalska, N.; Ostrega, A. Using SERVQUAL method to assess tourist service quality by the example of the Silesian Museum established on the post-mining area. *Land* **2020**, *9*, 333. [CrossRef]
19. Ko, C.-H.; Chou, C.-M. Apply the SERVQUAL instrument to measure service quality for the adaptation of ICT technologies: A case study of nursing homes in Taiwan. *Healthcare* **2020**, *8*, 108. [CrossRef] [PubMed]
20. Ogiemwonyi, O.; Harun, A.; Rahman, A.; Alam, M.N.; Hamawandy, N.M. The relationship between service quality dimensions and customer satisfaction towards hypermarket in Malaysia. *Int. J. Psychosoc. Rehabil.* **2020**, *24*, 2062–2071. [CrossRef]
21. Ul Haq, I.; Awan, T.M. Impact of e-banking service quality on e-loyalty in pandemic times through interplay of e-satisfaction. *Vilakshan—XIMB J. Manag.* **2020**, *17*, 39–55. [CrossRef]
22. Shopee Privacy Policy. Available online: https://help.shopee.com.my/portal/article/77216 (accessed on 21 April 2022).

proceedings

Proceeding Paper

Intelligence-Led Policing Acceptance and Policing Effectiveness: The Roles of Organizational Change, Innovative Behavior and Knowledge Sharing †

Mohammed Ibrahim Mohammed Ibrahim Alblooshi [1,2] and Erne Suzila Kassim [3,*]

1 Department of Postgraduate & Professional Studies, Faculty of Business and Management, Universiti Teknologi MARA, Shah Alam 40150, Selangor, Malaysia
2 Abu Dhabi Police General Headquarters, Muroor Road, Abu Dhabi 253, United Arab Emirates
3 Faculty of Business and Management, Universiti Teknologi MARA Cawangan Selangor, Puncak Alam 42300, Selangor, Malaysia
* Correspondence: ernekassim@uitm.edu.my
† Presented at the International Academic Symposium of Social Science 2022, Kota Bharu, Malaysia, 3 July 2022.

Abstract: The study was conducted to examine the linkage between ILP acceptance and policing effectiveness, and to examine the relationships between organizational change, innovative behavior and knowledge sharing practices. Using the quantitative approach, data was collected through a survey from 400 police officers in Abu Dhabi Police (ADP). The data was analyzed for the measurement and structural model, using Smart PLS software. The results of the hypothesis testing suggest the relationship between ILP acceptance and policing effectiveness is significant, and readiness of change is significantly related to ILP acceptance. Furthermore, innovative behavior, knowledge sharing, climate of change and process of change are related to readiness of change. The study concludes ILP enforcement is fundamental for public safety, but the implementation model should consider the factors of organizational change.

Keywords: intelligence-led policing (ILP); Abu Dhabi Police (ADP); policing and organizational change; intelligence-led policing and policing effectiveness

Citation: Mohammed Ibrahim Alblooshi, M.I.; Kassim, E.S. Intelligence-Led Policing Acceptance and Policing Effectiveness: The Roles of Organizational Change, Innovative Behavior and Knowledge Sharing. *Proceedings* **2022**, *82*, 57. https://doi.org/10.3390/proceedings2022082057

Academic Editor: Mohamad Rahimi Mohamad Rosman

Published: 16 September 2022

1. Introduction

Intelligence-led policing (ILP) is a policing philosophy that emerged aggressively as a consequence to the 9/11 attack [1], although it has been used since 1990s. ILP, according to [2], refers to a model which pertains to managerial law enforcement with a purpose to bring the crime intelligence approach to the front position, particularly in decision making. Thus, it is imperative to further comprehend the structure as an operational process within the sphere of the enforcement of the law units, which include both the similarities and differences. On a same note, intelligence-led policing encompasses the analysis of information for the development and implementation of strategic actions for the purpose of efficiency towards diminishing the occurrence of crimes [3]. ILP offers innovative approach in the law enforcement functioned by the police force heading towards a better and more precise and systematic policing. However, utilizing the ILP as a means of governance faces a great challenge because they will be forced to ensure that their respective countries do not bear the losses, in terms of resources that are necessary to ensure appropriate use of identified ILP systems. The Organization for Security and Co-operation in Europe (2017) suggests that the ILP model does not only offer a modernized approach in identifying a countermeasures plan against the threats of terrorism and organized crime, rather ILP could also be applicable to routine proactive police management.

Over the past decade, significant changes have transpired in the Abu Dhabi Police (ADP) modernization program [4]. As the program continues, the United Arab Emirates

(UAE) government has provided sponsorship to ADP, which eventually has fostered several projects, including the ILP. Giving the citizens fast responses in safety is the priority of the ADP [5]. Moreover, safeguarding public safety is fundamental; integrated, simplified and effective system [6]. Thus, provision of ILP comes with the mission to create positive impacts on the citizens, including public trust.

However, ILP is an innovation that shall take some time to go with the traditional factors and restrictions of policing operations. It is a transformation that requires changes to the traditional practices. Hence, understanding the police institutions change strategy is fundamental. One of the prominent models that explains changes in organization is Korler's. It has become a subject of investigation within the policing institution by [6]. However, the use of Kotler's leading change model, although significant, does not provide a generalizability to the policing institutions because of the method selection. In addition, the change climate, process and readiness have not been addressed and emphasized in past ILP change management studies [7] Likewise, ILP is a system that requires acceptance among the police officers. Therefore, from the lens of behavioral studies, it raises the question of whether innovative behavior and knowledge-sharing practices determine the police officers' readiness to accept ILP. One of the foremost concerns in the formation and implementation of ILP is the knowledge sharing. Nevertheless, challenges are faced in this area, given the fact that both the information as well as the importance of knowledge is considered significant in the operation of policing. However, it is observed that behaviours with regards to maladaptive information, including values and cultures of policing, persists to influence the gathering and sharing of information and knowledge to every unit of the police department. Research related to ILP has not given sufficient discussion on how knowledge sharing and police officer's innovative behavior are playing the role in ILP adoption intention and practices. Thus, it opens an avenue for further examinations. As a result, the study was conducted with the following objectives. First, is to examine the relationship between ILP acceptance and policing effectiveness in ADP. Second, is to determine the factors that shape the ILP acceptance and police officers' readiness from the lens of organizational change theory. The introduction should briefly place the study in a broad context and define the purpose of the work and its significance.

2. Review of Literature

2.1. *Intelligence-Led Policing (ILP)*

Intelligence-led policing, according to [2], refers to a model which pertains to managerial law enforcement of which the purpose is the positioning of a crime intelligence approach at the front position, particularly in decision making. Thus, it is imperative to further comprehend the structure as well as the operational processes in the sphere of the enforcement of the law units, which include both the similarities and differences within the organizational units. The purpose is to better understand the influence of the nuances in shaping up the approaches of ILP. [8] also highlighted ILP is a concept in which intelligence and data are used to objectively detect and efficiently handle crime threats. ILP uses intelligence and data to prioritize issues (which may be more subtle), such as a human trafficking ring operating in the neighborhood that the general public is unaware of. To put it another way, although public input is greatly sought and valued in both COP and ILP, objective analysis utilized to establish the most significant criminal threats is the prime priority and mechanism used to proactively police under the ILP model [1]. In order to better comprehend ILP, awareness of who in particular is going to use the term is considered crucial, verification about the context of the application of ILP and where it shall be applied should be sorted out, why is it necessary to apply, and what are the anticipated outcomes for its use. In the absence of the foregoing fundamental frame of reference, ILP from the general perspective is going to be an insignificant concept and of no logical application.

2.2. Organizational Change, Innovation Behavior, Knowledge Sharing and ILP

ILP acceptance entails obtaining and analyzing crime data while taking into account the elements that contribute to crime with the objective of providing actionable information to assist law enforcement in formulating tactical and/or strategic responses to threats in the face of new or changing threats [1]. Policing service effectiveness (PSE) refers to the state of police public services as an institution, with a focus on its efficacy in reducing crime [9]. PSE ensures that civilians are enlisted to assist ADP in critical areas. Crime analysts, for example, assist in identifying the persons who commit the most crimes, as well as crime patterns, so that ADP may better target recurrent and repeat offenders.

Knowledge sharing is the voluntary transmission or dissemination of information from one individual to another or a group within an organization [10]. As one of the major variables in this study, the process of knowledge sharing shall be investigated to assert its relationship to ILP. Brodeur and Dupont's definition of knowledge-based policing or knowledge as it applies to police, according to [11], may be divided into old and new knowledge. Old knowledge may be defined as information that has historically been acknowledged within the area of law enforcement and is also linked to criminal activity. It is a knowledge that is assembled by way of traditional law enforcement approaches in resemblance to handling of informant, investigation, as well as investigative interrogation. Conversely, Ratcliffe goes on to say that, as a result of rapid technological progress, a new kind of knowledge has emerged, with a deeper awareness of crime and police efficiency in combatting and reducing crime, as well as a new era of accountability. Thus, this so-called new knowledge encompasses a wider perspective and interpretation pertaining to intelligence involving approaches pertaining to crime mapping, trend, along with demographic analysis and strategic intelligence with the use of open-source information.

When comparing old and new intelligence, the new intelligence is conducted by people who spend the most of their time at the police station and may or may not be civilians working in a sworn setting. In the realm of policing, the slow paradigm shift with regards to the information that is supposedly significant and of value has taken most law enforcers into uncharted territory; furthermore, it is perceived to have potential for the alteration of dynamics in relation to the value within policing. [12] claimed that knowledge sharing plays a significant role in the development of positive attitudes with respect to diversity within a team. Additionally, knowledge sharing is also causative to a variety of non-traditional work-related effects, such as those that affect team atmosphere and employees' readiness to change. Hence the following hypothesis is proposed:

H1. *There is a significant relationship between police officers' knowledge-sharing behavior and readiness of change.*

Innovative behavior refers to the use of a new concept, product, procedures in accomplishing tasks at work or in the company [13]. Workplace innovation is defined as a social process that includes cooperation with workers and corporate peers, as well as the support and refinement of new ideas in order for them to be accepted. As a result, employee satisfaction was associated with organizational support for innovation, which was linked to innovative behavior. As a result, workplace support and opportunity increase the effectiveness of achievement and encourage creative behavior [14]. Employees develop creative solutions in the organization [15]. Hence, employee behaviors are important to the organizational innovation goal. Furthermore, they assert that good ideas are built on creativity. As a result, it is important to raise awareness of the importance of people and their unique characteristics in the success of creative initiatives. Hence, the following hypothesis is offered:

H2. *There is a significant relationship between police officers' innovative behavior and readiness to change.*

Organizational change (OC) is the process wherein a company or organization changes its working strategies or aim in order to maintain a competitive advantage [16]. In this study, OC plays a major component, especially in the innovative approaches that shall ensure policing effectiveness is delivered for the interest of the people. Readiness to change deals with the commitment of every member of the organization along with a shared belief to implement and perform changes [17]. As one of the key factors in this research, being ready to change shall be examined to establishment law enforcement accountabilities and steadfastness towards the enforcement of the law in the perspective of organizational change. Climate change is defined as any job task being changed due to an indirect or direct variables that affect their work [18]. In this study, climate change refers to companies that want to prove that they satisfy society standards through social and environmental reporting, which helps them manage stakeholder perceptions. Process change is a set of procedures wherein top leaders are positioned in order to guarantee an effectual operation, and further refrain from losses that may result from fraud, malfunction of technological aspect, and also error [19]. Thus, based on the discussion, the following hypotheses are offered:

H3. *There is a significant relationship between climate of change and readiness of change.*

H4. *There is a significant relationship between process of change and readiness of change.*

H5. *There is a significant relationship between readiness of change and ILP acceptance.*

H6. *There is a significant relationship between ILP acceptance and policing effectiveness.*

3. Research Method

3.1. The Survey Instrument

The survey instrument consisted of items from seven measures, which were adopted from past studies. Items to measure knowledge sharing behavior were adopted from [20], innovation behavior [21], ILP acceptance [22] and policing service effectiveness [23]. Organizational change was measured as climate of change, process of change and readiness of change and adopted from [24] Two dimensions were used to measure climate of change, which are trust and cohesion. There are four dimensions to measure process of change: participation, supervisor support, quality of communication and top management attitude. On the other hand, there are three dimensions to measure readiness of change: emotion, cognitive and intentional. The items were measured as a five-point-Likert scale and the choices of the responses were listed in accordance with the appropriateness of the variables.

3.2. Participation and Data Collection Procedure

The instrument was verified and validated by two academics who were the expert in the field of organizational policy and information system, and also by the Abu Dhabi Police Strategic Advisor. The instrument was translated to the Arabic language using the back translation method. Then, a pilot study was conducted with 40 police officers that took part in the procedure. The goals of the pilot test were to assess the item clarity, the common instructions used and the survey length. The results of the internal consistency scores for all variables were above 0.700 which suggests a good scale reliability. The population of the study was the police officers in Abu Dhabi. The sample was selected based on the convenient sampling technique. Although the technique is least preferred in any social science study, the convenient sampling technique was considered as the best strategy, given the police officers' hectic schedule and the impossibility to simply interfere in their daily duties. The sample size was set based on the G Power calculator, which suggested for 129 samples. However, given the higher the sample size, the lower the sampling error, data was collected from 400 police officers. A permission to collect the data was obtained from the Abu Dhabi Police authority. The collected responses were screened for missing values, and the data was also assessed for the common method bias. The respondents were officers from the Abu Dhabi Police Headquarters (31%), Allain (28%). Al Dhafa (22%) and other external areas (19%). The designations included Search and Investigation Officers (40%),

Interrogation Officers (32%), Patrol Officers (9%) and others including Operation Officers, Cyber Crime and Traffic.

4. Results and Discussion

4.1. Measurement Reliability and Validity

The data was analysed using Smart PLS 3.3.2. The results of the reflective measurement are shown in Table 1. They indicate the composite reliability scores were higher than the cut-off value of 0.70, whereas the average variance extracted (AVE) were above 0.50 [25].

Table 1. Measurement Reliability and Validity.

Construct	Cronbach's Alpha	Composite Reliability	Average Variance Extracted (AVE)
Knowledge Sharing	0.947	0.957	0.789
Innovative Behavior	0.931	0.948	0.783
Climate of Change	0.902	0.862	0.758
Process of Change	0.955	0.893	0.677
Readiness of Change	0.952	0.913	0.777
ILP Acceptance	0.859	0.899	0.640
Policing Effectiveness	0.908	0.932	0.732

In addition, the heterotrait-monotrait criterion (HTMT) procedure was also conducted as a strategy to test for the discriminant validity. The results are in Table 2, and it indicated the HTMT values fulfill the guideline, which suggests the values should be less than 0.90 [26].

Table 2. Heterotrait-Monotrait ratio of correlations, HTMT.

Construct	1	2	3	4	5	6	7
Climate of Change							
Effectiveness	0.459						
Innovative Behavior	0.645	0.443					
Knowledge Sharing	0.544	0.362	0.602				
ILP Acceptance	0.423	0.537	0.327	0.326			
Process of Change	0.660	0.399	0.409	0.449	0.447		
Readiness	0.608	0.443	0.501	0.530	0.397	0.623	

4.2. Structural Model and Discussion

The structural model was evaluated by following the procedure of (1) the lateral collinearity (VIF), (2) the path coefficients, and t-value and p-value, (3) in-sample predictive power (R^2), and (4) the effect size (f^2) and (5) the predictive accuracy (Q^2). The results are shown in Table 3. The hypotheses were tested by using a bootstrap re-sampling technique with an iteration of 5000 sub-sample.

The scores of the variance inflation factor (VIF) show all values were between 1.00 and 2.135, which suggests multicollinearity is not an issue [27]. The results also suggest all hypotheses were supported and the relationships were all significant. For the predictive power of coefficient determination (R^2), the results show 47.4% of the variance in readiness of change is explained by innovative behavior, knowledge sharing, climate of change and process of change. Subsequently, 13.1% of the variance in ILP acceptance is explained by readiness of change, whereas 22.8% of the variance on policing effectiveness is explained by ILP acceptance. Based on [28], the predictions were substantial, moderate, and substantial accordingly. Next, the effect size (f^2) values based on [28] suggest that process of change ($f^2 = 0.156$), readiness of change ($f^2 = 0.151$) and ILP acceptance ($f^2 = 0.295$) demonstrated a medium effect size in generating the respective R^2. Likewise, the other effects sizes were small. The predictive accuracy was assessed based on the blindfolding approach [29]. The results show the endogenous variables (readiness of change = 0.295, ILP acceptance = 0.081, and policing effectiveness = 0.163) exhibited predictive accuracy of the model (with $Q^2 > 0$).

Table 3. Results of Hypothesis Testing.

Hypothesis	Beta	Standard Error (SE)	*t*-Value	*p*-Value	VIF	f^2
Innovative Behavior → Readiness	0.114	0.055	2.064	0.020	1.837	0.013
Knowledge Sharing → Readiness	0.199	0.054	3.699	0.000	1.635	0.046
Climate of Change → Readiness	0.171	0.062	2.765	0.003	2.135	0.026
Process of Change → Readiness	0.368	0.057	6.492	0.000	1.647	0.156
Readiness → ILP Acceptance	0.362	0.050	7.265	0.000	1.000	0.151
Acceptance → Policing Effectiveness	0.477	0.046	10.380	0.000	1.000	0.295

R^2: Readiness = 0.474, Acceptance = 0.131, Policing Effectiveness = 0.228, Q^2: Readiness = 0.295, Acceptance = 0.081, Policing Effectiveness = 0.163.

The findings reveal that readiness of change is very important to the decision to accept the ILP practices. In addition, what predicts the readiness includes how the process of change is governed, the degree of the climate of change, and equally important is the innovative behavior of the police officers and also the willingness to share knowledge among the police officers. Employee innovative behavior (e.g., inventing, adopting, and implementing new ideas for goods and work techniques) is a significant asset that helps an organization to prosper in a dynamic business environment.

In addition, the findings suggest there is a significant relationship between knowledge sharing and readiness to change. What this suggests is, to encourage knowledge-sharing readiness, individuals must be aware of the need to share knowledge and the benefits of doing so. Furthermore, if their personal or their organization's social norms require them to share their knowledge, they are more ready for any changes [30]. Workplace knowledge sharing is influenced by employees' interactions with others, which shows that employees' willingness to share knowledge are influenced by their relationships with others. The relationships between organizational change, measured as climate of change and process of change, were found to be significant to readiness to change. The employees' readiness to change is influenced by the environment and process of organizational transformation [31]. Likewise, the significant relationship between climate of change and readiness exemplifies the proposition that changes in work procedure, policies and reward systems influence behavior.

5. Conclusions

The main objective of the study is to bridge the gap between ILP acceptance and policing effectiveness in the policing institution of the Abu Dhabi Police. ILP is a transformation towards a more secured, efficient and robust policing operation. The transformation requires changes not only to the institution, but studies have shown changes were affecting the motivation, readiness and acceptance from the police employees. The study also concluded that ILP is predominantly significant in addressing the need to effectively control crimes and wrongdoings. ILP is noted as encompassing strategic policing such as knowledge-driven policing techniques, data sharing among linked and relevant departments, along with hazard appraisal, which are all necessary to ensure the safety and security of the citizenry. For future research, it is suggested for exploring how the citizens react to the changes and accept the service quality of the policing system. Finally, as technological changes are robust, a future study should also evaluate the policing services across time and to consider the changes within the technological and innovation landscape.

Author Contributions: M.I.M.I.A.: Conceptualization, methodology, formal analysis, investigation, data curation, visualization, writing—original draft preparation; E.S.K.: validation, formal analysis, writing—review and editing, supervision. All authors have read and agreed to the published version of the manuscript.

Funding: This research received no external funding.

Informed Consent Statement: Informed consent was obtained from all subjects involved in the study.

Data Availability Statement: Not applicable.

Acknowledgments: The authors would like to thank Mark Hartley and Keith Deakin, Strategic Advisor of Abu Dhabi Police, Sheeren Noranee, and Universiti Teknologi MARA.

Conflicts of Interest: The authors declare no conflict of interest.

References

1. Carter, J.G.; Foz, B. Community policing and intelligence-led policing: An examination of convergent or discriminant validity. *Polic. Int. J.* **2018**, *42*, 43–58. [CrossRef]
2. Burcher, M.; Whelan, C. Intelligence-led policing in practice: Reflections from intelligence analysts. *Police Q.* **2019**, *22*, 139–160. [CrossRef]
3. Summers, L.; Kim Rossmo, D. Offender interviews: Implications for intelligence-led policing. *Polic. Int. J.* **2019**, *42*, 31–42. [CrossRef]
4. Al Karaeen, B. effective leadership during strategic change: An investigative study of Abu Dhabi Police. Ph.D. Thesis, University of East Anglia, Norwich, UK, 2016.
5. Mohammed Ibrahim Alblooshi, M.I.; Kassim, E.S. Exploring citizens' perception of Abu Dhabi policing e-service quality. *Int. J. Acad. Res. Bus. Soc. Sci.* **2021**, *11*, 519. [CrossRef]
6. Lambert, D.E. Addressing challenges to homeland security information sharing in American policing: Using Kotter's leading change model. *Crim. Justice Policy Rev.* **2019**, *30*, 1250–1278. [CrossRef]
7. Gemke, P.; Den Hengst, M.; Van Rosmalen, F.; De Boer, A. Towards a maturity model for intelligence-led policing A case study research on the investigation of drugs crime and on football and safety in the Dutch police. *Police Pract. Res.* **2021**, *22*, 190–207. [CrossRef]
8. Bullock, K. Community, intelligence-led policing and crime control. *Polic. Soc.* **2013**, *23*, 125–144. [CrossRef]
9. Ullah, F.; Ibrar, M. Effectiveness of police public services mechanism in crime reduction in Khyber Pakhtunkhwa, Pakistan. *Pak. J. Criminol.* **2019**, *11*, 131–140.
10. Adeyelure, T.S.; Kalema, B.M.; Motlanthe, B.L. An empirical study of knowledge sharing: A case of South African healthcare system. *Knowl. Manag. E-Learn. Int. J.* **2019**, *11*, 114–128.
11. Ratcliffe, J. Intelligence-led policing and the problems of turning rhetoric into practice. *Polic. Soc.* **2002**, *12*, 53–66. [CrossRef]
12. Ahmad, F.; Karim. M. Impacts of knowledge sharing: A review and directions for future research. *J. Workplace Learn.* **2019**, *31*, 207–230. [CrossRef]
13. West, M.A.; Farr, J.L. Innovation at work: Psychological perspectives on social behavior. *Commun. Monogr.* **1989**, *58*, 273–288.
14. Riaz, S.; Xu, Y.; Hussain, S. Understanding employee innovative behavior and thriving at work: A Chinese perspective. *Adm. Sci.* **2018**, *8*, 46. [CrossRef]
15. Purc, E.; Laguna, M. Personal values and innovative behavior of employees. *Front. Psychol.* **2019**, *10*, 865. [CrossRef]
16. Grimsley, S.; Allison, J. What Is Organizational Change? Theory & Example. 2015. Available online: https://study.com/academy/lesson/what-is-organizational-change-theory-example-quiz.html (accessed on 28 May 2022).
17. Shea, C.M.; Jacobs, S.R.; Esserman, D.A.; Bruce, K.; Weiner, B.J. Organizational readiness for implementing change: A psychometric assessment of a new measure. *Implement. Sci.* **2014**, *9*, 7. [CrossRef]
18. Ahmad, N.N.N.; Hossain, D.M. Exploring the meaning of climate change discourses: An impression management exercise? *Account. Res. J.* **2019**, *32*, 113–128. [CrossRef]
19. Aguirre, D.; Alpern, M. 10 Principles of Leading Change Management. Strategy+ Business. 2014. Available online: https://www.strategy-business.com/article/00255?gko=6c601 (accessed on 28 May 2022.).
20. Van Den Hooff, B.; Ridder. J.A. Knowledge sharing in context: The influence of organizational commitment, communication climate and CMC use on knowledge sharing. *J. Knowl. Manag.* **2004**, *8*, 117–130. [CrossRef]
21. Janssen, O.; Van Yperen, N.W. Employees' goal orientations, the quality of leader-member exchange, and the outcomes of job performance and job satisfaction. *Acad. Manag. J.* **2004**, *47*, 368–384.
22. Carter, J.G. Institutional pressures and isomorphism: The impact on intelligence-led policing adoption. *Police Q.* **2016**, *19*, 435–460. [CrossRef]
23. Tilley, N. *Problem-Oriented Policing, Intelligence-Led Policing and the National Intelligence Model*; Jill Dando Institute of Crime Science, University College London: London, UK, 2003.

24. Bouckenooghe, D.; Devos, G.; Van den Broeck, H. Organizational change questionnaire–climate of change, processes, and readiness: Development of a new instrument. *J. Psychol.* **2009**, *143*, 559–599. [CrossRef]
25. Hair, J.F.; Risher, M.S.; Ringle, C.M. When to use and how to report the results of PLS-SEM. *Eur. Bus. Rev.* **2019**, *31*, 2–24. [CrossRef]
26. Henseler, J.; Ringle, C.M.; Sarstedt, M. A new criterion for assessing discriminant validity in variance-based structural equation modeling. *J. Acad. Mark. Sci.* **2015**, *43*, 115–135. [CrossRef]
27. Becker, J.-M.; Ringle, C.M.; Sarstedt, M.; Völckner, F. How collinearity affects mixture regression results. *Mark. Lett.* **2015**, *26*, 643–659. [CrossRef]
28. Cohen, J. *Statistical Power Analysis for the Behavioral Sciences*; Routledge: London, UK, 2013.
29. Geisser, S. The predictive sample reuse method with applications. *J. Am. Stat. Assoc.* **1975**, *70*, 320–328. [CrossRef]
30. Small, C.T.; Sage, A.P. Knowledge management and knowledge sharing: A review. *Inf. Knowl. Syst. Manag.* **2005**, *5*, 153–169.
31. Holt, D.T.; Achilles, A.S.; Harris, S.G.; Field, H.S. Toward a comprehensive definition of readiness for change: A review of research and instrumentation. *Res. Organ. Chang. Dev.* **2007**, *16*, 289–336.

proceedings

Proceeding Paper

The Impact of Human Capital Index on Economic Growth in Malaysia †

Noorziah Mohd Salleh [1,*], Imbarine Bujang [1], Christina Andin [2] and Mohammad Nur Azhar Mazlan [3]

1 Faculty of Business Management, Universiti Teknologi MARA Cawangan Sabah, Kota Kinabalu 88770, Sabah, Malaysia
2 Faculty of Psychology and Education, Universiti Malaysia Sabah, Kota Kinabalu 88400, Sabah, Malaysia
3 Faculty of Cognitive Sciences and Human Development, Universiti Malaysia Sarawak, Kota Samarahan 94300, Sarawak, Malaysia
* Correspondence: noorziah@uitm.edu.my
† Presented at the International Academic Symposium of Social Science 2022, Kota Bharu, Malaysia, 3 July 2022.

Abstract: The purpose of this study was to investigate the impact of the human capital index (HCI) on economic growth (EG). Despite a consistent pattern between human capital capabilities and economic growth, there was no significant correlation between HCI and EG, while employment was positively correlated. HCI improves both labor force education and the health care commitment of a nation, two important factors affecting employment. Accordingly, the country's human capital development plan should improve the health and education of its citizens, including monitoring their health closely and providing quality education to students. In this study, we explore the relationship between human capital index accumulation (health and education) and economic growth in Malaysia by using a Dickey–Fuller test with an Augmented protocol from 1979 and Philperrons tests from 1988, which have been conducted.

Keywords: human capital index in Malaysia; economic growth; human capital; HCI in Malaysia

Citation: Salleh, N.M.; Bujang, I.; Andin, C.; Mazlan, M.N.A. The Impact of Human Capital Index on Economic Growth in Malaysia. *Proceedings* **2022**, *82*, 71. https://doi.org/10.3390/proceedings2022082071

Academic Editor: Mohamad Rahimi Mohamad Rosman

Published: 21 September 2022

1. Introduction

In Malaysia, the rapid advancement of the country requires high investments in human capital that will strengthen industrial growth, serve as a backbone for the country's development plans, and most importantly, accelerate its readiness for the Fourth Industrial Revolution (4IR). Therefore, human capital intellectual development is crucial for identifying a country's key investments as a means of better mobilizing economic growth, especially in a rapidly evolving business environment. Investing in people is crucial to maintain a competitive advantage. Although managing human capital has been widely discussed in the literature, there appears to have been no suitable method for forecasting human resources potential until the World Bank introduced the human capital index (HCI) [1]. Human capital index focuses on education quality and health outcomes for individuals (and society as a whole) [1,2]. According to [1], improving health and education among the nation's citizens would benefit the nation by the age of 18. In light of the comprehensive nature of the concept, this research is heavily based on the seminal work by the [2] and [1].

A long-term investment in people will yield positive results by highlighting the issues and potential of human capital, and these efforts will increase economic growth and eradicate poverty. However, since COVID-19 started in 2020, Malaysia's economy has deteriorated and faces extensive health and education challenges. Generally, the experience of the COVID-19 outbreaks throughout 2020 indicates that health and education have a significant impact on economic growth. World Bank data shows that the Malaysian economy was healthy between 1982 and 2019, as the gross domestic product (GDP) of the country averaged between 4% and more than 5% per year but dropped drastically to –5.6% during the pandemic in 2020 (GDP per capita growth (annual percentage %)-Malaysia, 2021).

Nevertheless, managing human resources presents a major challenge for the country. According to [3], Malaysia will not face a lack of jobs, but a shortage of skills required for future jobs. Therefore, the country should increase its labor force output and calibrate the development of its people skills in accordance with the labour market's demands [4]. According to previous studies, Malaysian workers, particularly graduates, lack critical thinking skills, communication skills, language proficiency, and positive personality traits [5]; there is consequently a mismatch between employers' requirements and product offerings of higher education institutions. These situations may have contributed to the high rate of youth unemployment and, consequently, the moderate performance of the national economy. In spite of positive economic growth in 2017, the unemployment rate for graduates remained high (10.5%)-around three times the national unemployment rate in 2016. One article [6] cited Malaysia as having the third-highest youth unemployment rate in the Asia-Pacific region after Indonesia and South Korea. However, the country still maintains a lower unemployment rate compared to the other two countries due to its relatively stronger economy, lower population density, and lower corruption level [7].

Malaysia is ranked 62nd in the World Bank's 2018 human capital index, behind Singapore, which is ranked first; Malaysia scored 0.62 against Singapore's 0.88. The lowest HCI was found in Chad (0.27%) (Human Capital Index (HCI) (scale 0-1)-Chad | Data, 2021) with South Korea (0.84%), Japan (0.84%) and Hong Kong (0.82%) way ahead [8]. Even before the pandemic, in the fourth quarter of 2017, skilled employed persons decreased by 0.8 percentage points to 27.2% and the number of semi-skilled persons also decreased by 0.5 percentage points to 59.2 percent while the employed persons with secondary education has increased by 1.0 percentage point to 56.2% [9]. The employed persons with tertiary education decreased by 0.8 percentage points to 28.0% [9].

It is important to note that in Malaysia, the majority of employed persons do not have tertiary education. This may be caused by the fact that more low-skilled jobs are being created, since Malaysia does not pursue knowledge-intensive and innovation-driven activities. Today's generation should pay more attention to their education in order to increase the number of people with tertiary education in the workforce. As a result, it is crucial that intellectual development begins immediately by focusing on the health and education of the new generation in order to improve the capabilities of its workforce. However, Malaysia appears to place more emphasis on physical investment and less on people's well-being. Currently, there are several major infrastructure projects underway, such as the Pan Borneo Highway and the East Coast Rail Link (ECRL), which cost USD7.05 and USD10.5 billion respectively and are expected to promote economic growth [10]. The outcome of these rapid infrastructure advances certainly promotes development, but human capital remains the main driving force behind large-scale infrastructure projects.

Malaysia's budget for health and education is relatively small compared to the sums allocated for infrastructure. Although the health budget allocation for the federal government was RM31 billion, an increase of 6.7% or 10% of the total RM297 billion, it covered the entire federal government. According to the budget, development expenditure was RM56 billion in 2020. The budget plan was heavily criticized as it again assigned more funding for construction and infrastructure projects than for its people. RM69 billion was allocated for the purposes of infrastructure and only RM50.4 billion for education.

The potential for Malaysia's development is excellent. Global investors make up the largest portion of the economy's support, led by China (16.7%), Singapore (12.9%), the United States (8.3%), Japan (7.1%) Thailand (5.6%); its main markets, which contribute 51.3% of its exports, are China (13.9%), Singapore (13.9%), the United States (9.1%), Hong Kong (7.5%) and Japan (6.9%) [11]. The country earned a gross domestic product (GDP) of US364.7 billion from its economy with a population of more than 32.7 million people [12]. Compared with its neighbours such as Singapore, Thailand, and the Philippines, the country is doing well [13]. The country's economic growth, high labor participation rate (68.4%), and low unemployment rate (3.3%) [14] have helped Malaysia rank 17th globally for manufacturing competitiveness (see Deloitte Touche Tohmatsu's Global Manufacturing Competitiveness Index 2016-Deloitte.com). In terms of technology and innovation, Malaysia ranked 37th out of 127 countries, and 8th in Asia in the Global Innovation Index 2017 conducted by Cornell University, INSEAD and WIPO [15]. In addition, the Readiness for the Future of Production Report (2018) identified Malaysia as a well-positioned country to benefit from Industry 4.0 [16]. As such, Malaysia has a huge potential for development in the near future, but it must focus on improving its greatest asset, human capital, rather than just its physical assets.

In this study, the aim is to investigate the relationship between education and health, which are the elements in the HCI, with economic growth in Malaysia. Robust studies have examined how human resources affect firm performance and its ability to establish competitive advantages [17–19]. However, a study on the correlation between health and economic performance remains largely contested, despite the fact that health plays an important role in EG. The reason for this is that the data is too complex to quantify [20–23]; several recent studies have attempted to explain the link of health and education with EG [24,25], however, they were less comprehensive as they were done within a specific country or region. (This is claimed as less effective for many reasons: for instance, a study used nonparametric and a semiparametric analysis on a sample of 100 countries over the period from 1970 to 2014).

Currently, HCI is being actively studied, so there is only a limited amount of evidence discussing the index. Therefore, this study aims to fill the gap in the literature by exploring the effects of the human capital index through education and health on economic growth in Malaysia.

2. Literature Review

The following literature discusses the HCI concept, the education and health elements of the HCI and the relationship between human capital (HC) and economic growth (EG).

Defining HCI: In 1990, Amartya Sen and Mahbub ul Haq, together with Yale University professor Gustav Ranis and Lord Meghnad Desai, established the HCI. Continuing this work, health and education are used to predict what an 18-year-old born today might achieve, given the risks of a country where education and health are poor. HCI measures productivity in relation to a benchmark of a complete education and full health. It ranges from 0 to 1. A child born today can expect to be only ×*100 percent as productive as a future worker as they would be if they had a complete education and full health [1]. Using HCI, a standardised index can be used to measure HC capabilities effectiveness in a country, specifically a country's productivity. It is also can be used as an indicator of a country's HC growth as claimed; a person with a good education and good health benefits society and is the main factor behind economic growth [26,27]. Thus, education, health, and productivity are the main direct mechanisms through which human capital influences economic growth. According to the HCI model, investing in human capital can have significant economic benefits in the long run, especially when it comes to economic growth, particularly in the employment sector [1]. The relationship between employment and economic growth is clear, which was evident during the recent pandemic in 2020/21 [28]; Malaysia's unemployment increased by 4.5% from 3.3% (2019) when the EG declined in 2020 [12].

The health aspect of the HCI: several studies have found a reciprocal relationship between economic growth and healthcare spending. A country with high healthcare expenditures sees high economic growth; but poor countries with low GDP ratios cannot spend as much on healthcare, eventually causing them to become poorer [29,30]. Studies show varying relationships between healthcare spending and economic growth. A study by [31] examined the stationarity and cointegration of variables across 13 MENA countries with respect to economic growth (GDP) and healthcare spending over the period of 1995 to 2005. It was found that healthcare expenditure and economic growth are closely related, with a negative relationship between healthcare spending and gross domestic product, and that healthcare spending is a necessity in the countries studied [29,32]. Measuring the health spending is challenging; hence, the correlation between health and EG is hard to demonstrate as it requires long-term observation and possibly requires both qualitative and quantitative methods [29,32].

The education aspect of the HCI: there have been many empirical studies that have shown education has a strong effect on economic growth and productivity in many countries, including [33] who found primary and secondary education have a causal impact on economic growth and productivity in India, [34] who found primary and secondary education have a causal influence on productivity, and [35] who found higher and secondary education had positive effects on economic growth. According to [36], their study looked at 14 countries in the Middle East and North Africa region and found that average human capital influences growth but does not affect productivity. In a study by [37], 94 developed and developing country educational attainment levels were examined. The author found that human capital positively impacted productivity growth. In the study by [38], an increase of 1% in school enrollment rates led to an increase between 1 and 3% in GDP per capita. The study by [39] indicated that current educational expenditures lead to economic growth in the future. According to [40], education in Malaysia has a positive significant effect on growth in the long run, with secondary education having a significant positive effect compared with primary and tertiary education.

The HCI and Economic Growth: the majority of research on HC and EG shows a positive correlation [41,42] and a variety of HC variables were analyzed for their impact on EG, including human resource functions, company size and culture, corporate strategy, customer satisfaction, and company structure; the results mostly showed indirect effects on EG. In order to gain more direct and conclusive evidence, it was claimed that the HC and EG correlation required quantitative and qualitative analyses [43,44]. Recently, many studies have shown some positive results in terms of accounting models, and the studies have stressed the importance of measuring changes in the quality of the labour force, measured by qualifications, health and higher skills, when attempting to account for economic growth over the long run.

3. Material and Methods

According to the discussion, it is difficult to prove the relationship between education, health and EG. In this study, time series analysis was used to test for stationarity in order to provide insight into the relationship in Malaysia. A Dickey–Fuller test with an Augmented protocol from 1979 and Philperrons tests from 1988 have been conducted. The only level at which the data were stationary was at the level associated with economic growth. At the second and first differences, the human capital index (HCI) and employment rate (UE) were sufficient. The following criteria were used to test the empirical models:

$$\text{Model 1: } LnEG_t = \alpha + \beta_1 D_2 HCI_t + \varepsilon_t$$

$$\text{Model 2: } D_1 LnEM_t = \alpha + \beta_1 D_2 HCI_t + \varepsilon_t$$

where the variables are measured based on Table 1.

Table 1. Measurement of variable.

Variables	Definition	Sources
Index of Human Capital (HCI)	Index of Human Capital per Person for Malaysia, Index, Annual, Not Seasonally Adjusted from FREDD economic database	[45]
Economic growth (EG)	Economic growth based on % change of Gross Domestic Product from World Bank data	[13]
Employment rate (UE)	Employment rate based on total labour from World Bank data	[46]

As part of the procedure to validate the results, a diagnostic test called an autocorrelation test used the Breusch Godfrey and Serial Correlation LM test [47,48]. Time series analysis was used to perform the Eviews 9 statistical software, and the findings are presented in the following section.

4. Result

A total of 37 observations are included in this study that extracts data from 1982 until 2019. All the necessary estimation procedures had been completed in order to report a meaningful discussion. In Table 1 below, a descriptive statistic is presented to illustrate the characteristics of the data used in this study. According to the table, the data for HCI were normally distributed. Economic growth and unemployment have been transformed using the log transformation to standardize their values. All preliminary descriptives that meet the criteria qualify for the next step of the analysis. And the descriptive statistics result is shown in Table 2.

Table 2. Results for descriptive statistics.

Variables	Mean	Jarque Bera Test
EG	5.7377	33.23007 (0.0001) ***
EM	62.91	28.11942 (0.0001) ***
HCI	2.5547	2.6474 (0.2752)

*** denotes a significant level at 99% confidence level.

The final results were presented in Table 3 below. There was insufficient evidence to support the relationship between the human capital index (HCI) and economic growth (EG). This indicates that the data failed to reject the null hypothesis between the two variables. Interestingly, there is a significant positive relationship between the human capital index and the employment rate at a 5% level. The results suggested that as the human capital index increases by 1%, the employment rate increase by 0.0089%. However, the effect of the human capital index on economic growth failed to capture a significant relationship. Diagnostic testing shows that both estimated models did not suffer from the serial correlation, as the *p*-value was significantly greater than 0.05 on the Breusch–Godrey test.

Table 3. Results for simple linear regression analysis.

Variables	Model 1 (Economic Growth)	Model 2 (Employment Rate)
D2HCI		
β	8.6818	0.0089
t-value	0.934963	2.4319
p-value	(0.3562)	(0.02) **
Constant		
β	2.2538	4.1182
t-value	7.048221	4.192
p-value	(0.0001) ***	(0.001) ***
R-Squared	0.0243	0.1445
Breusch Godfrey serial correlation LM test		
p-values chi-square	(0.5212)	(0.360)
No of observations	37	37

*** and ** denotes a significant level at 99% and 95% confidence level respectively.

5. Discussion

Human capital index (HCI) and economic growth (EG) have no significant relationship in this study; however, there is a significant relationship between HCI and employment. The results indicate adverse consequences, despite the importance of HCI and EG being discussed here. The role of the community itself, the well-being of friends and family, and other factors can adversely affect healthy and educated individuals. HCI was also found to be based on years spent in school, and returns to education, based on the index of human capital per person extracted from the Fredd database in this study. Fredd evaluated the HCI solely on the basis of the number of years of education, without considering educational quality or healthcare quality. It would be useful to include these data in the future. A number of studies have found that healthy human capital is positively correlated with economic growth; however, as discussed, establishing a cause-and-effect relationship between the two is challenging. For instance, when a company invests in the health and education of its employees, this will likely result in greater employee productivity, and this in turn will improve the company's productivity. It is, therefore, possible that this effort will have an effect on EG if it is successful in increasing the output of the company's industry. Investment in training helps to create a skills and knowledge base and thus helps with the absorption of new technologies which leads to higher production and thus economic growth. With highly qualified workers contributing to high productivity and strong local economic growth, it may increase the GDP of a country. Furthermore, more research is required, but the findings of this study indicate that trained and educated populations can contribute positively to economic growth when they act according to sound principles.

The findings are in line with the concept of employment being closely linked to a country's performance and could thus be used to eradicate poverty. A country's performance is heavily dependent on employment and would alleviate poverty, thus enhancing life quality. It supports the literature review's assertion that if an individual has acquired skills before entering the labor market, they are capable of contributing to the overall productivity of the country. As a result, education and health are regarded as being the two key aspects of HCI that should be improved in order to increase HC productivity. These findings are consistent with the results of previous studies indicating that HCI has a positive and significant relationship with employment. Several studies have demonstrated that human capital can significantly contribute to economic growth when health and education are improved [49–51].

6. Conclusions

Several important points are highlighted in this study; by exploring the connection between human capital accumulation (health and education) and economic growth, this study contributes to the ongoing debate. Human capital has a limited impact on economic growth, necessitating further evidence that HCI has a significant effect on EG and employment. As discussed, the two important areas of human capital development and employment are vital to the economic development of the country. Therefore, the study concludes that urgent interventions are needed to improve research and development, as well as investments in the well-being of the people and the effective use of human capital to promote growth while reducing externalities or spillovers. Malaysia's human capital development (HCD) plans should be taken more seriously, and it should improve the health and education of its citizens to become a developed country—a vision of 2020 that is considered a failure. According to previous studies discussed in this paper, physical assets are necessary to sustain economic progress, but the economy itself must also have a more skilled labor force. No matter how important physical capital is, the development of a developed nation that has been dreamed of for so long will not be achieved without skilled human capital.

Author Contributions: Conceptualization, N.M.S.; methodology, I.B. and N.M.S.; software, M.N.A.M.; validation, N.M.S., C.A.; formal analysis, I.B., N.M.S.; investigation, N.M.S., I.B.; resources, M.N.A.M., C.A.; data curation, I.B., N.M.S.; Original draft preparation, N.M.S.; writing—N.M.S., C.A. and M.N.A.M.; Review and editing, N.M.S., C.A., M.N.A.M.; visualization, I.B.; supervision, N.M.S. and M.N.A.M.; project administration, N.M.S.; funding acquisition. All authors. All authors have read and agreed to the published version of the manuscript.

Funding: This work was supported by Sustainable Development Goal (SDG)-Borneo [600-RMC/SDG-BORNEO 5/3 (007/2020)].

Institutional Review Board Statement: Not applicable.

Informed Consent Statement: Not applicable.

Data Availability Statement: Not applicable.

Conflicts of Interest: The authors declare no conflict of interest.

References

1. Kraay, A. Methodology for a World Bank Human Capital Index. *Policy Res. Work. Pap.* **2018**. Available online: https://openknowledge.worldbank.org/handle/10986/30466 (accessed on 14 September 2021).
2. Investing in People to Build Human Capital. Available online: https://www.worldbank.org/en/news/immersive-story/2018/08/03/investing-in-people-to-build-human-capital (accessed on 12 May 2021).
3. How Universities Should Prepare for Industry 4.0. Available online: https://www.monash.edu.my/news-and-events/pages/latest/articles/2018/how-universities-should-prepare-for-industry-4.0 (accessed on 4 April 2019).
4. Intense job competition spurring demand for graduate degrees in Malaysia. *ICEF Monit.* **2017**. Available online: https://monitor.icef.com/2017/10/intense-job-competition-spurring-demand-graduate-degrees (accessed on 21 February 2018).
5. Mohd Salleh, N.; Ab Rahman, B. Malaysia. In *The Future of Work in Asia and Beyond: A Technological Revolution or Evolution?* 1st ed.; Nankervis, A.R., Connell, J., Burgess, J., Eds.; Routledge: Oxfordshire, UK, 2020.
6. Youth Unemployment Rate Remains High as Skills Mismatch Stay Prevalent. Available online: https://www.midf.com.my/sites/corporate/files/2022-01/Econs-Msia-2017-Youth-UE-MIDF-030518.pdf (accessed on 14 January 2022).
7. Gupta, A. Changing forms of corruption in India. *Mod. Asian Stud.* **2017**, *51*, 1862–1890. [CrossRef]
8. Human Capital Index (HCI). Available online: https://data.worldbank.org/indicator/HD.HCI.OVRL?locations=TD (accessed on 11 September 2021).
9. Principal Statistics of Labour Force, Malaysia, Fourth Quarter (Q4) 2017. Available online: https://www.dosm.gov.my/v1/ (accessed on 9 September 2021).
10. Mahalingam, E. Is Sabah Highway Next To Get the Chop? Available online: https://www.thestar.com.my/business/business-news/2018/05/30/sabah-highway-next-to-get-the-chop/ (accessed on 11 April 2019).
11. Malaysia External Trade Development Corporation (MATRADE). Available online: https://twitter.com/matrade/media (accessed on 7 February 2019).
12. Unemployment 1982–2020 and Jan 2020–May 2021. Available online: https://www.dosm.gov.my/v1/ (accessed on 9 September 2021).
13. GDP per Capita (Current US$). Available online: https://data.worldbank.org/indicator/NY.GDP.PCAP.CD?locations=MY (accessed on 3 September 2021).
14. Labour Force Survey Report, Malaysia, 2019. Available online: https://www.dosm.gov.my/v1/ (accessed on 9 September 2021).
15. Malaysia Ranks 37th in Global Innovation Index. Available online: https://www.thestar.com.my/business/business-news/2017/06/15/malaysia-ranks-37th-in-global-innovation-index (accessed on 19 August 2019).

16. Industry 4WRD: National Policy on Industry 4.0. Available online: https://www.miti.gov.my/miti/resources/STA%20Folder/PDF%20file/Industry4WRD_-_National_Policy_on_Industry_4.0_.pdf (accessed on 19 August 2019).
17. Sehgal, K.; Malati, N. Employer branding: A potent organizational tool for enhancing competitive advantage. *IUP J. Brand Manag.* **2013**, *10*, 51–65.
18. Browning, V.; Edgar, F.; Gray, B.; Garrett, T. Realising competitive advantage through HRM in New Zealand service industries. *Serv. Ind. J.* **2009**, *29*, 741–760. [CrossRef]
19. Wolfe, R.; Wright, P.M.; Smart, D.L. Radical HRM innovation and competitive advantage: The moneyball story. *Hum. Resour. Manag.* **2006**, *45*, 111–145. [CrossRef]
20. Lopreite, M.; Zhu, Z. The effects of ageing population on health expenditure and economic growth in China: A Bayesian-VAR approach. *Soc. Sci. Med.* **1982**, *265*, 113513. [CrossRef]
21. Mariana, I. Consequences of the Investment in Education as Regards Human Capital. *Procedia-Econ. Financ.* **2015**, *23*, 362–370. [CrossRef]
22. Ogundari, K.; Awokuse, T. Human capital contribution to economic growth in Sub-Saharan Africa: Does health status matter more than education? *Econ. Anal. Policy* **2018**, *58*, 131–140. [CrossRef]
23. Wilson, R.A.; Briscoe, G. The impact of human capital on economic growth: A review. In *Impact of Education and Training, Third Report on Vocational Training Research in Europe: Background Report*; Descy, P., Tessaring, M., Eds.; Cedefop Reference Series, 54; Luxembourg Office for Official Publications of the European Communities: Luxembourg, 2004.
24. Widarni, E.L.; Wilantari, R.N. The relationship between human capital development and economic growth: Evidence from Malaysia. *J. Asian Financ. Econ. Bus.* **2021**, *8*, 641–650.
25. Abdullah, A.; Doucouliagos, H.; Manning, E. Does education reduce income inequality? A meta-regression analysis. *J. Econ. Surv. Wiley Blackwell* **2015**, *29*, 301–316. [CrossRef]
26. Yahong, W.; Khan, S. A cross-sectional analysis of employment returns to education and health status in China: Moderating role of gender. *Front. Psychol.* **2021**, *12*, 129. [CrossRef]
27. Hatak, I.; Zhou, H. Health as human capital in entrepreneurship: Individual, extension, and substitution effects on entrepreneurial success. *Entrep. Theory Pract.* **2021**, *45*, 18–42. [CrossRef]
28. Sipalan, J. Update 2-Malaysia's Economy Posts Biggest Annual Decline since 1998 Crisis. Available online: https://www.reuters.com/article/malaysia-economy-gdp-idUSL4N2KG1HZ (accessed on 8 April 2021).
29. Atilgan, E.; Kilic, D.; Ertugrul, H. The dynamic relationship between health expenditure and economic growth: Is the health-led growth hypothesis valid for Turkey? *Eur. J. Health Econ.* **2016**, *18*, 567–574. [CrossRef] [PubMed]
30. Lee, H.; Oh, D.Y.; Meng, M. Stationarity and cointegration of health care expenditure and GDP: Evidence from tests with smooth structural shifts. *Empir. Econ.* **2019**, *57*, 631–652. [CrossRef]
31. Omri, A.; Nguyen, D.K.; Rault, C. Causal interactions between CO_2 emissions, FDI, and economic growth: Evidence from dynamic simultaneous-equation models. *Econ. Model* **2014**, *42*, 382–389. [CrossRef]
32. Abdouli, M.; Omri, A. Exploring the Nexus among FDI Inflows, environmental quality, human capital, and economic growth in the Mediterranean region. *J. Knowl. Econ.* **2021**, *12*, 788–810. [CrossRef]
33. Self, S.; Grabowski, R. Does education at all levels cause growth? India, a case study. *Econ. Educ. Rev.* **2004**, *23*, 47–55. [CrossRef]
34. Bratti, M.; Bucci, A.; Moretti, E. *Demographic Trends, Human Capital and Economic Growth in Developing Countries: Theory and Evidence*; University of Ancona, Department of Economics: Ancona, Italy, 2004.
35. Villa, S. *Determinants of Growth in Italy. A Time Series Analysis*; Working Paper; University of Foggia: Foggia, Italy, 2005.
36. Haouas, I.; Yagoubi, M. *Openness and Human Capital as Sources of Productivity Growth: An Empirical Investigation from the MENA Countries*; IZA Discussion Paper No. 1461: Bonn, Germany, 2005.
37. Park, J. Dispersion of Human Capital and Economic Growth. *EconsPapers JEL-Codes* **2004**, *12*, J2. [CrossRef]
38. Sianesi, B.; van Reenen, J. *The Returns to Education: A Review of the Macro-Economic Literature*; CEE Discussion Papers; Centre for the Economics of Education, LSE: London, England, 2000.
39. Sylwester, K. Income inequality, education expenditures, and growth. *J. Dev. Econ.* **2000**, *63*, 379–398. [CrossRef]
40. Shaihani, M.A.L.; Haris, A.; Ismail, N.W.; Said, R. Long run and short run effects on education levels: Case in Malaysia. *Int. J. Econ. Res.* **2011**, *2*, 77–87.
41. Sara, I.; Saputra, K.A.K.; Utama, I. The effects of strategic planning, human resource and asset management on economic productivity: A case study in Indonesia. *J. Asian Financ. Econ. Bus.* **2021**, *8*, 381–389.
42. Chang, W.-J.A.; Huang, T.C. The impact of human resource capabilities on internal customer satisfaction and organisational effectiveness. *Total Qual. Manag. Bus. Excell.* **2010**, *21*, 633–648. [CrossRef]
43. McCourt, W.; Ramgutty-Wong, A. Limits to strategic HRM: The case of the Mauritian civil service. *Int. J. Hum. Resour. Manag.* **2003**, *14*, 600–618. [CrossRef]
44. Barro, R.J. Human Capital and Growth. *Am. Econ. Rev.* **2001**, *91*, 12–17. [CrossRef]
45. Index of Human Capital per Person for Malaysia. Available online: https://fred.stlouisfed.org/series/HCIYISMYA066NRUG (accessed on 5 September 2021).
46. Human Capital Index 2018. Available online: http://pubdocs.worldbank.org/en/625641565356285634/IBRD-Financial-Statements-June-2019.pdf (accessed on 26 September 2021).
47. Breusch, T. Testing for autocorrelation in dynamic linear models. *Aust. Econ. Pap.* **1978**, *17*, 334–355. [CrossRef]

48. Godfrey, L. Testing against general autoregressive and moving average error models when the regressors include lagged dependent variables. *Econometrica* **1978**, *46*, 1293. [CrossRef]
49. Bashir, F.; Farooq, S.; Nawaz, S.; Bagum, M.; Sandila, M.A.; Arshad, M.R. Education, health and employment in Pakistan: A Co-integration analysis. *Res. Humanit. Soc. Sci.* **2012**, *2*, 53–64.
50. Pandey, M.K. On ageing, health and poverty in rural India. *ASARC Work. Paper* **2009**, 1–17. Available online: https://mpra.ub.uni-muenchen.de/15932/ (accessed on 26 September 2021).
51. Laplagne, P.; Glover, M.; Shomos, A. Effects of Health and Education on Labour Force Participation. Available online: https://ideas.repec.org/p/ris/prodsw/0704.html (accessed on 10 September 2021).

proceedings

Proceeding Paper

Repositioning of Planning and Budgeting Functions with Respect to the Chief Financial Officer †

Irwan Suliantoro [1,*], Bambang Soedaryono [2] and Muhammad Zilal Hamzah [2]

1 Department of Public Financial Management, Polytechnic of State Finance STAN, South Tangerang 15222, Banten, Indonesia
2 Faculty of Economics and Business, Trisakti University, West Jakarta 11440, Jakarta, Indonesia
* Correspondence: irwans@pknstan.ac.id
† Presented at the International Academic Symposium of Social Science 2022, Kota Bharu, Malaysia, 3 July 2022.

Abstract: In state budget preparations in Indonesia, the Ministry of Finance has authority over state budgeting functions, and the National Development Planning Agency has authority over planning functions. Prior to 2003, the planning function was dominant. After 2003, the budgeting function was more dominant. After 2017, planning and budgeting functions were synchronized in almost all annual-budget preparation processes. Based on a focus-group discussion, it was shown that by completing the synchronization process through the Memorandum of Understanding, the relationship between planning and budgeting functions remains separated in two different entities. However, the results of the NVivo analysis show that despite the adjustment of interests, institutional pride and institutional competition still exist. To solve this problem, it is necessary to synchronize legal products not only at the level of government regulations but also at the level of laws.

Keywords: planning; budgeting; synchronization

Citation: Suliantoro, I.; Soedaryono, B.; Hamzah, M.Z. Repositioning of Planning and Budgeting Functions with Respect to the Chief Financial Officer. *Proceedings* **2022**, *82*, 78. https://doi.org/10.3390/proceedings2022082078

Academic Editor: Mohamad Rahimi Mohamad Rosman

Published: 23 September 2022

1. Introduction

1.1. Preface

State finances are all rights and obligations of the state that can be valued in currency and everything that can be used as state property in connection with the implementation of these rights and obligations. A state's finance includes monetary authority and fiscal authority. Fiscal authority includes, among others, the management of state revenues and the management of state expenditures. The management of state finances is directed to achieve the goals of the state. To carry out government functions to achieve state goals, a state budget is prepared. A state budget is the state's annual financial plan that is approved by the parliament.

The management of state finances is authorized by the Minister of Finance as the fiscal manager. As the fiscal manager, the Minister of Finance prepares the draft of the state budget. In the context of preparing the draft of the state budget, the line ministries compile budget documents that comprise annual financial plan documents. The process of preparing the budget documents is coordinated by the Directorate General of Budget at the Ministry of Finance. Thus, the Directorate General of Budget has authority over the function of state budgeting. Budgeting can be interpreted as a process for preparing government financial plans that are prepared based on applicable rules to achieve state goals.

The management of state finances is directed at achieving the goals of the state, among others, in the form of the general welfare. To achieve the goals of the state, national development is needed. National development is a series of continuous (development) activities to realize national goals. A planning process is needed to achieve this form of sustainability with respect to short-term, medium-term, and long-term periods. The development planning process involves line ministries, local governments, and community participation. The resulting planning documents are the annual planning document, the five-year planning document, and the twenty-year planning document. The development of planning processes at the line ministries is coordinated by the National Development Planning Agency (Bappenas). Thus, Bappenas has authority over the function of development planning, which is realized by project planning to support the achievement of national development.

If line ministries propose a new project during development planning, the mechanism is carried out through the Proposed New Initiative. A New Initiative is a proposed additional performance plan in the form of a program, activity, or output (or project). The Proposed New Initiative proposed to Bappenas and to the Directorate General of Budget. Concerning the proposed new project, Bappenas conducts a review of the results (projects) from the aspect of performance targets and conformity to policy directions and national development priorities. The Directorate General of Budget examines the efficiency and effectiveness of state spending (Regulation of the Head of Bappenas Number 1/2021 concerning Procedures for Compiling, Reviewing, and Amending Work Plans of the Line Ministries).

In Indonesia, the relationship between the planning function and the budgeting function fluctuates. Before 2003, the planning function was very dominant compared to the budgeting function. After the enactment of Law Number 17 of 2003 concerning State Finances, the budgeting function became relatively more dominant than the planning function. After the enactment of Government Regulation 17/2017 concerning the Synchronization of National Development Planning and Budgeting Processes, to maintain consistency between the planning documents and the budgeting documents, Bappenas was further involved in budget preparation processes. The budget planning process in Indonesia is a unique model because two units with different functions (planning and budgeting functions) synchronize to work together in almost all annual budget preparation processes [1].

1.2. Research Gap

The research objectives achieved are as follows: (i). We analyze and assess whether the current planning process (in Bappenas) conflicts with the authority of the Minister of Finance as the Chief Financial Officer in the preparation of the state budget; (ii). we provide alternative policies so that the relationship between the planning function (at Bappenas) and the budgeting function (at the Ministry of Finance) becomes more constructive in the preparation of the state's budget.

The Synchronization of the Planning and Budgeting Process is a process of harmonizing and strengthening the planning and budgeting process for controlling the achievement of development goals (Government Regulation 17/2017). There were several weaknesses before the enactment of Government Regulation 17/2017 [1]: (i). a weak integration between planning documents and budgeting documents was observed; (ii). the role of Bappenas is limited to the planning aspects for priority programs, while the real budget available to the Ministry of Finance can be diverted from the original plan; (iii). changes in the budgeting document side are not accommodated in the planning document. The implication is that there is a potential for funding changes for priority projects that are not monitored on the planning side. However, there are several weaknesses with the enactment of Government Regulation 17/2017: (i). There is a potential struggle for influence between Bappenas and the Ministry of Finance in the budgeting process; (ii). the function of the Chief Financial Officer is not fully under the authority of the Ministry of Finance, as stipulated in Law Number 17 of 2003; (iii). there is a duplication of planning functions in the budgeting function, which causes inefficiency in organizational functions.

It can be said that Indonesia has experienced three patterns of change (fluctuation) in the relationship between the planning function and the budgeting function. The pattern of changes in the relationship is a form of repositioning in order to respond to the dynamics of existing policies. However, there is no guarantee that the current pattern of relationships will not change in the future. Alternative solutions that are more permanent for the long term are required. Alternative solutions that can be considered include repositioning, which not only comprises a synchronization between the two planning and budgeting entities but also comprises a repositioning in planning and budgeting functions in the perspective of a single entity. With respect to these conditions, for the sake of efficiency, World Bank Indonesia provides options or discourses to combine the two functions so that planning, budget allocation, budget distribution, review, analysis, monitoring, and evaluation processes are in one entity [1].

From the explanation above, there is a practical gap in which the planning function and the budgeting function are two different entities, and Bappenas (to some extent) is included in the area of the budgeting function. In addition to the practical gap, there is an academic gap in which, in some OECD (Organisation for Economic Co-operation and Development) countries, the planning and budgeting functions are in one entity.

1.3. Literature Review

After the 1997/1998 economic crisis, Indonesia reformed its budgeting system, which was marked by the issuance of Law Number 17 of 2003 concerning State Finance. Several points of concern from the law are as follows [2] (p. 7): (i). There is a consensus from the parliament with respect to what a major achievement is after the crisis; (ii). the law regulates in great detail and specifically regarding budgetary control in the context of increasing fiscal responsibility and the precautionary principle of state finances; and (iii). Separation of laws related to budgeting and planning. The Explanation Sheet for Law Number 17 of 2003 somewhat ignores the planning function.

Apart from the legal framework, budget transformation is achieved by integrating the regular budget with the development budget (investment expenditure) and including the non-budget in the statutory budget. Initially, the regular budget was the responsibility of the Ministry of Finance, and the development budget was the responsibility of Bappenas. Double budgets should be avoided by integrating regular and development budgets. A feature of Indonesian budgeting is the national planning function. Regarding the national planning function, which is within territory of Bappenas, there are some issues related to the budgeting process. First, achieving economic growth since independence has been inseparable from the success of development plans. Second, this plan has become an important tool for organizing donor development assistance to Indonesia. Third, the five-year plan runs in parallel with the presidential term and can reflect the political agenda of the presidential term. It can be said that the planning function of the Budget Bureau is outside the Indonesian Budget Bureau, that is, in Bappenas. In OECD countries, this planning function is integrated into a single budget office rather than individually as in Indonesia. In Indonesia, planning and budgeting are inefficient due to their separate structures, but this separation is addressed as part of performance-based budgeting reform [2] (p. 12).

The following is a budget preparation cycle in Indonesia.

1. Preparation of Budget Availability: Budget availability preparations are carried out by the Fiscal Policy Agency and the Directorate General of Budget and usually begin in February to guide the budget preparation process. After the macroeconomic framework is established, the Directorate General of Budget will fund the availability of existing budgets for regular spending (non-discretionary spending) and new programs/activities/projects (discretionary spending).
2. Prioritization of new programs: After the Ministry of Finance prepares the Availability of Budget for development expenditure, Bappenas then distributes the budget into priority projects. This priority project refers to the five-year planning document, which is further elaborated in the annual planning document and becomes the basis for line ministries to prepare budget documents. Furthermore, the Ministry of Finance and Bappenas issued a Joint Letter on the Indicative Ceiling containing the estimated funding for each program at the line ministries.

3. Preliminary Talks with the parliament: The government submits to Parliament a document on the macroeconomic framework and the basics of fiscal policy. This document is basically a pre-budget report that includes a macroeconomic framework, fiscal policy and priorities, deficit targets, and revenue forecasts. The government also submits a plan document to parliament for discussion.
4. Finalization of budget proposals: After the government reaches an agreement with Parliament on budget policies and priorities in mid-June, the Ministry of Finance issues a Finance Ministers' Decree on Budget Caps for line ministries' programs. The line ministry then creates a budget document with a different structure and format than the planning document. As part of the performance-based budgeting reform, synchronization efforts are underway to reconcile the two. Line ministries must submit a budget by July. Bappenas reviews budget documents to ensure consistency with planning documents, and DG Budget reviews budget documents to ensure they comply with budget caps, unit prices, and spending classifications.

In the United States, planning and budgeting processes are carried out by the Office of Management and Budget (OMB). The OMB is the central budgeting office that is part of the Executive Office of the President. The OMB director is equivalent to a minister and is a member of the cabinet. The OMB oversees the implementation of coordination and management of all executive agencies. OMB serves as the central clearing house for all communications between the executive and Congress. All legislation and other submissions to Congress must be reviewed by the OMB [3] (p. 12). In South Korea, the planning and budgeting process is carried out by the Ministry of Economy and Finance. Initially, the Ministry of Economy and Finance were two separate entities, namely the Ministry of Finance and the Ministry of Economic Planning Board. The Ministry of Finance was responsible for the management of financial resources, the development of financial, monetary, and currency exchange policy policies. Meanwhile, the Ministry of Economic Planning Board was responsible for budget planning and economic development [4] (p. 309). In Singapore, the planning and budgeting process is carried out by the Development Planning Committee consisting of the Minister of Finance, Minister for Trade and Industry, and ministers who have sectoral projects. The committee is responsible for all capital expenditures or development expenditures [5] (p. 59). In the United Kingdom, the planning and budgeting functions are performed by the HM Treasury. The HM Treasury comprises the Ministry of Economy and Finance, which compiles public spending, sets the direction of economic policy, and determines strong and sustainable economic growth [4].

2. Research Methodology

In this study, the data collection method was carried out through a Focus Group Discussion (FGD) and literature study. The FGD was conducted by involving experts from the Directorate General of Budget, Bappenas, representatives from the line ministries, and consultants from the World Bank Indonesia: (i). Eko Roestanto, Head of Sub-Directorate of Budget for Public Works, Agrarian, and Spatial Planning (Ministry of Finance); (ii). Anantyo Wahyu Nugroho, Coordinator of Development Planning System (Bappenas); (iii). M. Taufik Kurniawan, Head of Planning and Performance Accountability (Financial Transaction Reporting and Analysis Center); (iv). Agus Haryanto, Sub-Coordinator of Work Plan Alignment, Planning Bureau (Ministry of Marine Affairs and Fisheries); (v). Sudes Nazarudin, Representative of the Association of Budget Analysts (Ministry of Finance); (vi). Hari Purnomo, Senior Public Finance Management Specialist (World Bank Indonesia).

The literature study was conducted by examining articles and books. The FGD was held online on 18 January 2022. The data analysis method used is the NVivo application. NVivo is an application that is used to assist in the processing and analysis of qualitative data. Furthermore, to maintain validity, the triangulation method was carried out. In this study, triangulation was carried out based on the data sources. In particular, the analysis was not only carried out on the results of interview transcripts, the Focus Group

Discussions, and the observations but was supplemented with data sourced from the various documents, such as annual reports, regulations, and journal articles.

Furthermore, the framework of this research is described in the diagram, as shown in Figure 1 below. Government Regulation Number 17 of 2017 concerning the Synchronization of the National Development Planning and Budgeting Process implies the existence of a phenomenon in the form of a duplication of authority between the planning function and the budgeting function, which ultimately leads to inefficiency in organizational functions. To minimize these implications, it is necessary to reposition functions in the planning and budgeting domains.

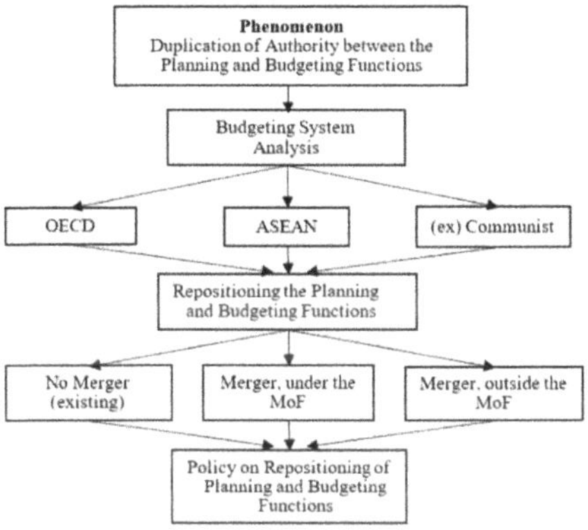

Figure 1. Framework of Thinking.

The design of this research study is as follows. The results of the FGD are recorded in the form of a transcript for each expert (informant) involved. Furthermore, based on the prepared transcript, a systematic coding process was carried out. In this case, coding is intended for drawing the themes contained in the perspective of the sources in the form of coding nodes and cases. In the coding and analysis process, the NVivo application is used. The analysis carried out comprises hierarchical analysis, comparison diagram, matrix code, and concept map.

3. Research Result and Discussion

3.1. Focus Group Discussion Results

In the Focus Group Discussion (FGD), the experts (informants) were given two questions:

1. Does the current planning process not conflict with the authority of the Minister of Finance as the Chief Financial Officer in the preparation of the state budget?
2. What policies need to be taken to make the relationship between planning and budgeting more constructive in the preparation of the state budget? Do the planning and budgeting functions remain separate (model 1)? Do we merge Bappenas with the Ministry of Finance (model 2) or do we merge Bappenas with the Directorate General of Budget (model 3)?

In the first question, based on the results of the FGD, it can be concluded that the current planning process conflicts with the authority of the Minister of Finance as the Chief Financial Officer. The conflict of authority is a form of collaboration and synergy. The current planning process (synchronization of planning with budgeting) is aimed at aligning planning documents with budgeting documents. To synchronize planning documents with budgeting documents, the planning function is included in (part of) the budgeting process. Thus, the conflict of authority is intended to further build synergies. Regarding the second point question, based on the results of the FGD, it can be concluded that the relationship between the planning and budgeting processes leads to model 1 (current conditions, without merging) by perfecting the synchronization process through a Memorandum of Understanding.

Regarding the conflict of interest, viewed from the theory of public policy, the policy of synchronizing planning and budgeting (which causes a conflict of authority) reflects public policy with a combination model of the institutional theory and the group theory. The policy of synchronizing planning and budgeting comprises decision making based on internal bureaucratic actors attached to the organizational structure of the government. On the other hand, the policy of synchronizing planning and budgeting is the result of an alignment of interests between two groups, namely Bappenas and the Ministry of Finance. In group theory, there can be changes in the balance of group interests. In this case, the group's interest is in the form of the inclusion of the planning function in (part of) the budgeting area. The balance in question is in the form of increasing the authority of the entity that carries out the planning function.

Here are some statements from FDG respondents.

> "Regarding the results of the evaluation of the synchronization of planning and budgeting, there are several challenges, namely synchronizing the five-year planning with fiscal capacity, a potential deviation, and an inefficient process. Therefore, the solution is to improve regulations, improve data synchronization ... ". [6]

> "This synergy is quite difficult to challenge, it is easier to say, for example, who leads the (meeting) is a sensitive issue. (In the past) in a meeting when the Bappenas was invited, the Ministry of Finance did not necessarily want to attend in full. On the other hand, if the Ministry of Finance invites it as if the lead is (from) the Minister of Finance ... ". [7]

> " ... what we are doing now (making the Memorandum of Understanding) is something important, so that the planning and budgeting process is really integrated in a real way ... ". [8]

3.2. NVivo Analysis Results

The results of the Focus Group Discussion were processed by using the NVivo application. The final output of the NVivo application is Concept Mapping, which describes interrelationships of various issues/discussion topics. Some of the issues/discussion topics (on the Concept Mapping) reflect the solutions desired by the experts involved. Some of the other issues/topics reflect the obstacles faced in realizing the intended solution. Obstacles that arise in Concept Mapping are then used as a reference to provide recommendations so that the relationship between the planning function and the budgeting function becomes more constructive.

The Concept Mapping shown in Figure 2 below can be interpreted as follows. To improve the quality of planning and budgeting results, the relationship between planning and budgeting functions needs to be more synchronized. In the synchronization process, based on the results of the NVivo application analysis, several interrelated topics/problems require attention, namely Data Integration, Asymmetric Information, Technology Integration, Information Transparency, Institutional Coordinators, Institutional Pride, Institutional Asynchronous, Planning and Budgeting Asynchronous, Institutional Competition, and Supreme Regulation. From the relationship between these problems, the following analysis was obtained:

1. To build a constructive relationship between the planning function and the budgeting function, Synchronization is necessary. We realize Synchronization is not easy. Some obstacles arise in the form of Asymmetric Information, Institutional Pride, Institutional Asynchronous, Planning, and Budgeting Asynchronous, and Institutional Competition.

2. Institutional Competition and Institutional Pride are causes of Institutional Asynchronous and also has an impact on Planning and Budgeting Asynchronous.
3. To overcome this problem (Institutional Pride and Institutional Competition), it is necessary to return it to the Supreme Regulation or through the coordinator of the two institutions, for example, the President or Vice President or possibly another independent institution.

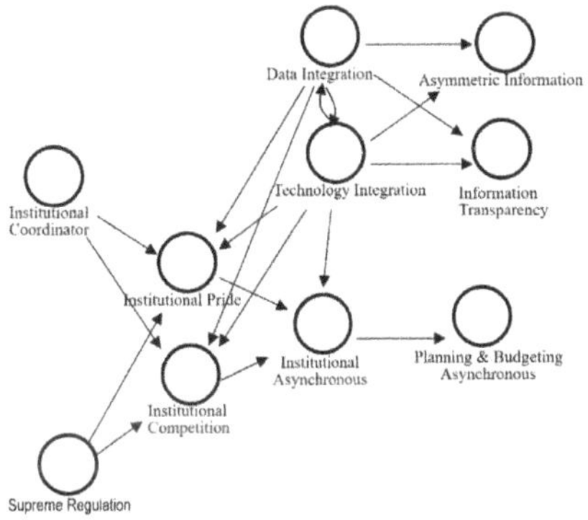

Figure 2. Concept Mapping.

4. Conclusions and Recommendations

4.1. Conclusions

From the results of the discussion in the previous chapter, the following conclusions can be drawn:

1. The current planning process (to some extent) conflicts with the authority of the Minister of Finance as the Chief Financial Officer in the preparation of the state's budget.
2. The relationship between the planning function and the budgeting function leads to model 1 (current conditions, without merger) by perfecting the synchronization process through the Memorandum of Understanding.

4.2. Recommendation

4.2.1. Short-Term Recommendations

- To increase synergy and minimize the impact of Institutional Pride, it is necessary to create an employee exchange program so that Bappenas employees serve for some time at the Directorate General of Budget and vice versa. In addition, a task force can be formed between the Ministry of Finance and Bappenas to discuss or handle strategic issues (for example) in the planning or evaluation of national priority activities.
- To improve Data Integration, Technology Integration, Information Transparency, and overcome Asymmetric Information, it is necessary to perform data integration that is not only limited to reference data but also requires is expanded to output achievements and budget realization data. Thus far, the output achievement data and the budget realization data in the Bappenas eMonev application have been filled in manually by spending units at the line ministries, even though the data are already available at the Ministry of Finance.
- To improve Synchronization/Coordination, there is a need for an additional Memorandum of Understanding outside the Memorandum of Understanding related to the information system's integration.
- There is a need for further research on the relationship between the planning function and the budgeting function with respondents from members of parliament as partners of the government in the preparation of the state budget.

4.2.2. Medium-Term Recommendations

- To support Data Integration, the features in the Bappenas eMonev application need to be integrated with the monitoring and evaluation application at the Directorate General of Budget so that the spending units at the line ministries do not need to use two different applications with the same or similar functions [9]. In addition, it is necessary to distinguish between the information needs of the budgeting unit and the spending unit so that evaluation results are more useful in the budgeting process [10] (p. 29). Furthermore, it is necessary to have a mechanism that allows executive leadership to track the progress of achieving strategic targets (outcomes at the line ministries level) to pursue their strategic goals [11] (p. 18).
- To improve the efficiency of business processes, it is necessary to consider the simplification of business processes so that meetings or discussions regarding planning and budgeting are not carried out repeatedly (bilateral meetings, first-stage trilateral meetings, second-stage trilateral meetings, first-stage budget document reviews, and second-stage budget document reviews).
- To improve synchronization, we should consider strengthening the Ministry of Finance 's role in the process of medium-term planning and budgeting. Thus far, the Ministry of Finance's role has been more focused on the annual budget. In this case, it is necessary to conduct research that examines the integration of medium-term planning (at the Ministry of Finance) and five-year planning (at Bappenas).
- The annual planning document has revised targets and funding needs, but the five-year planning document has not been revised (Presidential Regulation Number 18 of 2020 concerning the 2020–2024 National Medium-Term Development Plan); thus, it needs to be considered so that the revised target and indication of funding needs can be implemented in the five-year planning document. Regarding the 2020–2024 medium-term planning document, Bappenas can make program adjustments and overhaul all programs and targets set based on various assumptions and new developments after COVID-19 and the economic crisis that accompanied it. In this case, all programs that have been set out in the 2020–2024 medium-term planning document are reviewed, the strategy is reformulated, and the implementation period is rescheduled [12] (p. 250).
- To some extent, it is necessary to consider synchronization between the five-year planning document (at Bappenas) and the medium-term expenditure framework documents (at the Ministry of Finance). The expenditure figures in the medium-term expenditure framework document can be used as a reference for the preparation or adjustment of the planning document. In addition, in the preparation of the five-year planning document, it is necessary to review real conditions by involving the independent institution [13] (p. 12). Furthermore, the budget documents containing medium-term information and performance achievements need to be presented to the parliament [14] (p. 58).

4.2.3. Long-Term Recommendations

Since the implementation of Law Number 17 of 2003 regarding the State Finances does not pay attention to planning aspects, the synchronization of planning and budgeting is not only carried out at the government regulation level but also at the legislation level; in this case, it involves Law Number 17 of 2003 and Law Number 25 of 2004 concerning the National Development Planning System. Further research is needed on the synchronization of Law Number 25 of 2004 and Law Number 17 of 2003:

- In Law Number 25 of 2004, it is necessary to consider the chapter on the preparation of the Macroeconomic Framework, Principles of Fiscal Policy, and Budget Availability. In addition, it is necessary to consider synchronization between the development targets in the twenty-year planning document and the development targets in the five-year planning document and the annual planning documents.
- In Law Number 17 of 2003, it is necessary to consider the subject of monitoring and evaluation of the implementation of programs and activities, the subject of consistency

between planning and budgeting, and the subject of medium-term budgeting. The medium-term budgeting concept includes the Medium-Term Fiscal Framework, the Medium-Term Budget Framework, and the Medium-Term Expenditure Framework.

Author Contributions: Conceptualization, I.S.; methodology, I.S. and M.Z.H.; software, I.S.; validation, I.S., B.S. and M.Z.H.; formal analysis, I.S.; investigation, I.S.; resources, M.Z.H.; data curation, I.S.; writing—original draft preparation, I.S.; writing—review and editing, B.S. and M.Z.H.; visualization, I.S.; supervision, B.S. and M.Z.H.; project administration, I.S.; funding acquisition, I.S. All authors have read and agreed to the published version of the manuscript.

Funding: This research received no external funding.

Institutional Review Board Statement: Not applicable.

Informed Consent Statement: Not applicable.

Data Availability Statement: Not applicable.

Acknowledgments: In this study, the data-collection method was carried out through the Focus Group Discussion. Therefore, the authors would like to thank the Focus Group Discussion's participants, namely, Eko Roestanto and Kiswanto (Directorate General of Budget), Anantyo Wahyu Nugroho (Bappenas), M. Taufik Kurniawan (line ministry), Agus Haryanto (line ministry), Sudes Nazarudin (the budget analyst), and Hari Purnomo (the World Bank Indonesia Governance Global Practice Team). In addition, we thank Eleonora Sofilda (Trisakti University), the World Bank, the International Monetary Fund, and the OECD for sharing journals/papers/reports/e-books on related websites.

Conflicts of Interest: The authors declare no conflict of interest.

References

1. World Bank Indonesia Governance Global Practice Team. *Synchronization for the Process of National Development Planning and Budgeting.pptx*; World Bank Indonesia Governance Global Practice Team: Jakarta, Indonesia, 2017.
2. Blöndal, J.R.; Hawkesworth, I.; Choi, H. Budgeting in Indonesia. *OECD J. Budg.* **2009**, *2*, 1–30. [CrossRef]
3. Blöndal, J.R.; Kraan, D.; Ruffner, M. Budgeting in the United States. *OECD J. Budg.* **2003**, *3*, 7–48. [CrossRef]
4. Lienert, I.; Jung, M. *The Legal Framework for Budget Systems—An International Comparison*; OECD Publishing: Paris, France, 2004; p. 309.
5. Blöndal, J.R. Budgeting in Singapore. *OECD J. Budg.* **2006**, *6*, 45–84. [CrossRef]
6. Roestanto, E. *Integrasi Sistem Perencanaan dan Penganggaran*; FGD Reposition of Planning and Budgeting Functions in the Financial Chief Officer: Jakarta, Indonesia, 2022.
7. Purnomo, H. *Indonesia—Synchronization of the Process of National Development Annual Planning and Budgeting*; FGD Reposition of Planning and Budgeting Functions in the Financial Chief Officer: Jakarta, Indonesia, 2022.
8. Nazarudin, S. *Perencanaan dan Penganggaran*; FGD Reposition of Planning and Budgeting Functions in the Financial Chief Officer: Jakarta, Indonesia, 2022.
9. Suliantoro, I. Duplikasi Sistem Monitoring dan Evaluasi Belanja Kementerian/Lembaga. *J. Publ. Fina. Mana.* **2020**, *4*, 16–30. [CrossRef]
10. Robinson, M. *Connecting Evaluation and Budgeting*; World Bank: Washington, DC, USA, 2014; p. 29.
11. Beazley, I. *Incentivising Performance in Public Investment Policies Delivered at National and Subnational Levels: Managing across Temporal and Institutional Horizons*; OECD-European Commission: Paris, France, 2017; p. 18.
12. Muhyiddin, M. COVID-19, New Normal dan Perencanaan Pembangunan di Indonesia. *Indones. J. Dev. Plan.* **2020**, *4*, 240–252.
13. Setiawan, L.; Prasojo, W.; Parulian, A. Refleksi Penerapan Kerangka Fiskal Jangka Menengah di Indonesia. *J. Indones. Stat. Budg. Fina.* **2020**, *2*, 163–174.
14. Allen, R.; Eckardt, S.; Jacobs, D.; Kristensen, J.; Lienert, I.; Schiavo-Campo, S. *Indonesia: Budget Reform Strategy Priorities*; International Monetary Fund & World Bank: Washington, DC, USA, 2007; p. 58.

proceedings

MDPI

Proceeding Paper

Rhetorical Organisation of Risk Management Reports by Malaysian Banks †

Khairul Firhan Yusob [1,2], Afida Mohamad Ali [2,*], Mei Yuit Chan [2] and Geok Imm Lee [2]

1 Akademi Pengajian Bahasa, Universiti Teknologi MARA Pahang, Bandar Jengka 26400, Pahang, Malaysia
2 Faculty of Modern Languages and Communication, Universiti Putra Malaysia, Serdang 43400, Selangor, Malaysia
* Correspondence: afida@upm.edu.my
† Presented at the International Academic Symposium of Social Science 2022, Kota Bharu, Malaysia, 3 July 2022.

Abstract: Risk management reports (henceforth, RMRs) are paramount to an organisation's business processes and are used to inform stakeholders on the current risks experienced by a company. Some reports by companies, such as banks, are receiving more attention due to their importance to the public and current economy. While a plethora of studies have been conducted on RMRs in other disciplines, such as business, accountancy and finance, the same scenario is not apparent in language studies. Therefore, the present study intends to linguistically analyse the RMRs published by Malaysian commercial banks by examining their rhetorical organisation. In doing so, 40 RMRs published between 2016 and 2020 were analysed for their rhetorical moves and steps. The findings recognised five rhetorical moves in the RMRs, of which one was optional, one was conventional, while the rest were obligatory. The present study could assist readers of RMRs to better understand the structure of the reports.

Keywords: commercial banks; genre analysis; rhetorical organisation; risk management reports

Citation: Yusob, K.F.; Ali, A.M.; Chan, M.Y.; Lee, G.I. Rhetorical Organisation of Risk Management Reports by Malaysian Banks. *Proceedings* **2022**, *82*, 80. https://doi.org/10.3390/proceedings2022082080

Academic Editor: Mohamad Rahimi Mohamad Rosman

Published: 23 September 2022

1. Introduction

A risk management report (henceforth, RMR) is a section in corporate annual reports that is used as risk disclosure by companies. It was first included by choice, where companies voluntarily included the section only if there was a need to do so. However, beginning in 2005, the section has been made mandatory by mature capital markets. According to the authors of [1], the United States became the first country to enforce this. The practise was followed by China in 2007. The RMR is regarded as an important section because it informs stakeholders on the current circumstances of a company's risk [2], and it has been the topic of interest of researchers in the fields of finance and accountancy for years [3].

In Malaysia, the way that the RMR has been referred to has gone through several changes. It was first known as Risk Management. However, a few years later, some banks changed it to Managing Risks or the Risk Management Approach. Since 2014, Bursa Malaysia has set guidelines for all publicly listed companies, including banks, to label the section as the Statement on Risk Management and Internal Control. Apart from that, the RMRs by the listed companies in Malaysia are also guided by the Statement on Risk Management and Internal Control: Guidelines for Directors of Listed Issuers [4]. Some of the guidelines that all publicly listed companies must adhere to include:

1. The main features of the company's risk management and internal control system;
2. The ongoing process for identifying, evaluating and managing the significant risks faced by the company in its achievement of objectives and strategies;
3. The process has been in place for the year under review and updated for inclusion in the annual report;

4. The process (or where applicable, through its committees) has been applied in reviewing the risk management and internal control system, confirming that necessary actions have been or are being taken to remedy any significant failings or weaknesses identified from that review;
5. That a review on the adequacy and effectiveness of the risk management and internal control system has been undertaken;
6. Commentary on the adequacy and effectiveness of the risk management and internal control system;
7. The process has been applied to deal with the material internal control aspects of any significant problems disclosed in the annual report and financial statements;
8. Where material joint ventures and associates have not been dealt with as part of the group for the purposes of applying these guidelines, this should be disclosed.

Even though a lot of studies have been conducted so far on RMRs, studies from linguistic perspectives are minimal. It would be interesting to find out how a genre-based methodology can be used to analyse the organisational structure and communicative functions of RMRs. Therefore, the present study is guided by two research questions:

1. How are the RMRs of Malaysian commercial banks rhetorically organised in terms of the move–step structure?
2. To what extent are the moves obligatory, conventional or optional in nature?

2. Risk Management Reports (RMRs)

RMRs are reports about the risks encountered by companies. Because different companies have to deal with different types of risks, they tend to communicate about the risks in different ways. The authors of [5] argued that risk reporting is normally communicated differently using either the internal, external or intermediate level between the two. Internal risk reporting takes place between management and employees, and the information on the risk identification, measurement, performance development and monitoring circulates within the company only to ensure that the company's objective is achieved [5]. The intermediate level refers to a situation in which the communication involves the company's board of directors and is between an internal information channel and the external public disclosure panel, where the board of directors plays a role as a management control measure, and is able to successfully manage risk-related issues in the company [5]. Lastly, external reporting takes place when the communication of the risk information involves the public. It becomes one of the main requirements because the companies are using external financing and should comply with the regulatory agencies, creditors and investors [5]. The focus of the present study is on the external level of risk reporting, which can be found in the annual reports.

In the banking sector, risk reporting is regarded as a "fundamental tenet of a sound banking system" [6]. It reflects the practise of the risk management of banks. Moreover, banks are exposed to different types of risks that may affect their performances. Therefore, it is important for the banks to inform their stakeholders regarding their current risk management status. Furthermore, the information is also important to existing and potential investors prior to making financial decisions. This has become the main factor for banks to practise honesty in disclosing risks. Failure to do so will result in the loss of trust and confidence in the CARs of the companies [5,7].

Hitherto, RMRs have become the interest of researchers in the fields of accountancy and business. Therefore, it is not surprising to see that most of the studies conducted on RMRs are from these disciplines. In recent years, however, some scholars have conducted language studies on RMRs, even though they are not based on linguistic theory. The studies were more concerned with the impact of language on RMRs. The authors of [8], for instance, looked at cybersecurity risk reporting by 112 companies in the United States. The reports were published between 2011 and 2018. The findings show that the length of the reports increased as a result of regulatory enforcement. Using multiple regression analysis, they discovered that the latter reports were more difficult to read. The author of [9] revealed that less specific or boilerplate language in risk reporting is not helpful in making investment decisions. Boilerplate language often relates to using repeated information in disclosures. Some scholars refer to this as disclosure inertia [10]. Both studies highlighted the function of the language used in RMRs, rather than studying the language used itself. Only recently, the authors of [11] investigated the rhetorical moves in RMRs by an Islamic bank in Malaysia. The study, however, lacks diversity, and it is still at the preliminary stage because it only involved five RMRs produced by the same bank. Thus, the results are different from the moves in the present study. Nevertheless, the study has opened opportunities for more studies on similar genres.

3. Rhetorical Organisation

Rhetorical structure is a term that has often been used to refer to the structure of a genre text. Some scholars have been referring to it as structure [12,13], organisation [14], schematic structure [15], cognitive move-structure [16] and rhetorical organisation [17]. Studying the rhetorical structure of a text is common in genre analysis. It describes the structure that constitutes a text, which is often addressed as moves. The author of [18] defines moves as "discoursal or rhetorical units that perform coherent communicative functions in a written or spoken discourse" (p. 228). By identifying moves, readers will be aware of the rhetorical goals of the writers or genre producers. Furthermore, moves are often identified with steps, which are the building blocks of moves. Steps often work in combination to achieve the overall purpose of the moves. While some moves may comprise a few steps, there are also moves that do not require any steps to achieve the communicative functions.

Therefore, to understand the rhetorical structure of a text entails the identification of the moves and steps through a close analysis of the text. The authors of [19] argue that genre knowledge is "best conceptualized as a form of situated cognition embedded in disciplinary activities.... " (p. 3). This shows the importance of knowing how a text is structured. They further illustrated how possessing a genre and knowing its rhetorical structure is beneficial for academic and professional writers.

During its initial years, move analysis was often conducted on the academic genre. The author of [10] explored the moves of the introduction sections of research articles. His analysis discovered four moves that constituted research articles: establishing the field, summarising previous research, preparing for present research and introducing present research. This later became a steppingstone for other scholars to explore the rhetorical structures of more academic texts, such as these [20–22], and research articles [23–25]. There are also a number of studies that have been conducted on the spoken academic genre, such as oral presentations [26,27] and academic lectures [28,29].

With the advancement of knowledge, move analysis has now been widely used as a tool to analyse the rhetorical structures of professional genres. However, the number of studies is comparatively lower than the studies on academic genres. One of the professional genres that has caught the attention of genre analysts is annual reports. Many studies have investigated the moves and steps of different sections of annual reports [30,31]. Some of them include chairmen's statements [17], managerial forewords [32,33] and management discussion and analysis [34]. However, studies on RMRs from a genre standpoint are still limited and will therefore be discussed in the present study.

4. Method

The present study adopted a qualitative research design. It used the Move Analysis by the author of [12,13] as the theoretical framework. The following subsections further describe the methodology used in analysing the data.

4.1. Corpus

A corpus that comprised 40 RMRs by eight commercial banks in Malaysia was built for the present study. The corpus was named the Corpus of Risk Disclosure by Malaysian

Banks (CORDMAB), and it consisted of 155,162 word tokens. All RMRs were published in the banks' annual reports from 2016 to 2020. They were obtained from the Bursa Malaysia website. Only the last five years of RMRs were selected for each bank because they are more relevant to the current practise of the banks' risk reporting. For confidentiality purposes, the names of the banks were changed to alphabet letters from A to H, followed by the last two digits representing the years the RMRs were published. For example, A16 refers to RMRs by Bank A published in 2016.

4.2. Data Analysis

The analysis of the corpus was conducted using Atlas.ti 22, a qualitative analysis software. Prior to the analysis, all RMRs were extracted from annual reports and uploaded onto the programme. Then, codes were created based on the moves and steps identified. The RMRs were later tagged with the codes to ease the process of the frequency count. An example of the tagging process is shown in Figure 1.

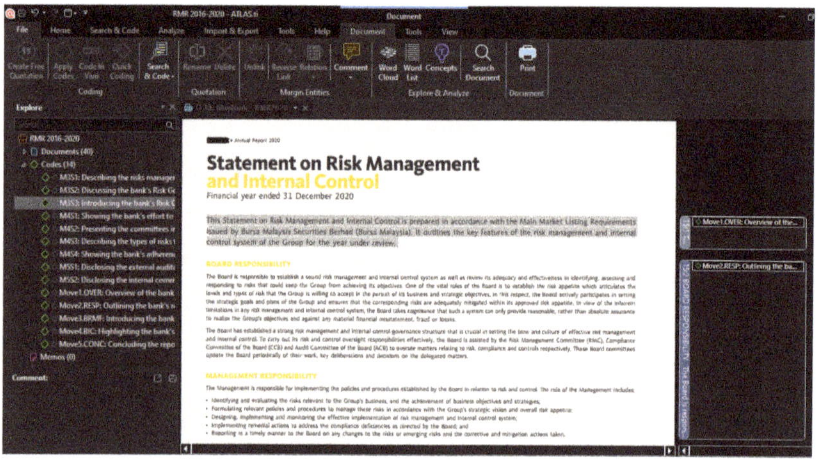

Figure 1. Move annotation using Atlas.ti 22.

To the best of the researchers' knowledge, there have been few studies on RMRs from a genre standpoint. Therefore, this provides more opportunities for research from this perspective. What is the generic structure of RMRs? Is it possible to come up with a reliable move scheme that can be a reference for those whose work contributes to RMRs? These are some instances of questions that genre-based research may be able to answer.

In genre-based studies, it is important to ensure the reliability of the moves and steps found by researchers. Thus, to answer Research Question 1, two intercoders were asked to check the demarcation of the moves and steps. The first intercoder holds a master's degree in ESL and has prior experience in coding moves for genre studies. The second intercoder is currently working on his master's research in the field of genre analysis. The researchers first identified the move scheme, which was later used to train the intercoders. The Statement on Risk Management and Internal Control: Guidelines for Directors of Listed Issuers [4] was also used as a reference in determining the moves and steps. Later, a series of trainings and discussions were held with the intercoders. Prior to finalising the move scheme, the agreement percentage of the demarcation of the moves and steps was sought between the researchers and intercoders. The percentage recorded was more than 80%, and the scheme was later finalised.

Using the finalised move scheme, the researchers coded the rest of the samples. Once the coding process was complete, the annotated moves were counted for their frequency. This was performed to answer Research Question 2. Moves that occurred 100% are categorised as obligatory [16], conventional moves are those with 60% to 99% occurrence and below, and moves that occurred 60% are optional [35].

5. Results and Discussion

The analysis on all 40 RMRs revealed that they are formed of five moves, which could be further described into steps. The RMR section for all banks is named the Statement on Risk Management and Internal Control, and it is divided into two parts. The first part focuses on general information on the management of the risks, while the second part highlights the internal efforts made by the banks in governing the risks. During the analysis, it was found that each bank had similar information from year to year. However, this is not surprising in annual reports because some companies were inclined to practise "disclosure inertia". The term refers to providing little or insignificant changes in disclosures [10]. This practise often takes place in RMRs [36]. Table 1 shows the moves found in the RMRs. The findings in this subsection answer Research Questions 1 and 2.

Table 1. Percentages of occurrences of moves and steps of RMRs by local commercial banks.

Moves	Percentage	Obligatory/Conventional/Optional
Move1.OVER: Overview of the bank's RMR.	47.5	Optional
Move2.RESP: Outlining the bank's responsibilities.	100	Obligatory
Move3.BRMF: Introducing the bank's risk management framework.	87.5	Conventional
M3S1: Describing the risk management framework.	87.5	Conventional
M3S2: Discussing the bank's risk management process.	87.5	Conventional
Move4.BIC: Highlighting the bank's internal control in mitigating risks.	100	Obligatory
M4S1: Showing the bank's effort to stop financial crime.	62.5	Conventional
M4S2: Explicating the types of risks the bank encounters.	55	Optional
M4S3: Displaying the bank's adherence to human resource policies.	100	Obligatory
Move5.CONC: Concluding the report.		
M5S1: Disclosing the internal auditors involved in reviewing the risk	100	Obligatory
management system.	100	Obligatory
M5S2: Disclosing the external committees involved in reviewing	97.5	Conventional
the RMR.		

5.1. Move 1: Overview of the Bank's RMR

The first move begins with a statement that the RMR was written as a requirement set by Bursa Malaysia. However, it was found that Move 1 is optional, as it could only be found in 47.5% of the corpus. Presumably, this low occurrence is because the overview of RMRs was not made compulsory by Bursa Malaysia. An example of Move 1 can be seen in the following excerpt taken from Sample A17: "The Statement on Risk Management and Internal Control is made pursuant to Bursa Malaysia Securities Berhad Listing Requirements which require the Board of Directors ("the Board") to disclose in its Company Annual Report a statement on key features of the risk management and internal control system of the Group during the year under review."

5.2. Move 2: Outlining the Bank's Responsibilities

Move 2 informs the readers of the responsible parties in charge of the risk management of the bank. In the corpus, they are referred to as the Board, which should be understood as the Board of Directors who are responsible for the bank's situation. This is evident in the following example: "The Board is cognisant of its overall responsibility and oversight of the Group's system of internal controls and is constantly keeping abreast with developments in areas of risk and governance." (Sample D18).

5.3. Move 3: Introducing the Bank's Risk Management Framework

Having good risk management requires a good framework, which serves as a strong foundation to support the risk governance. In doing so, banks need to introduce the risk management framework, which is evident in Move 3. Unlike the first two moves, which can stand on their own without any move constituents or steps, Move 3 comprises two steps. As shown in Table 1, the move and its steps are conventional, with 87.5% occurrence, as they are not found in the RMRs of Bank C. The following excerpt serves as an example (Sample H18): "The Group has in place a risk management framework approved by the Board for identifying, measuring, monitoring and reporting of significant risks faced by the Group in the achievement of the Group's business objectives and strategies".

The first step of Move 3, M3S1, describes the main function of the risk management framework. It tells readers how the framework helps the banks in governing the risk. This can be seen in the following example (Sample H18): "The Group's risk management framework ensures that there is an effective on-going process to identify, evaluate and manage risk across the Group". This step is also conventional, with 87.5% occurrence, as it was missing from Bank C.

Following the step is M3S2, which discusses the bank's risk management process. This step often begins with a diagram to help illustrate the process. This step is conventional and appears in 87.5% of the RMRs only. An example of the step is displayed in Figure 2 (Sample H18).

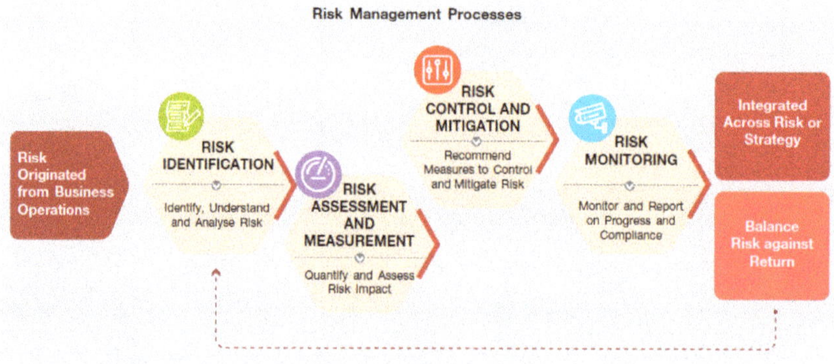

Figure 2. A diagram used to illustrate M3S2.

5.4. *Move 4: Highlighting the Bank's Internal Control in Mitigating Risks*

The second part of RMRs, which emphasises the internal efforts made by the banks, is articulated in Move 4. Because this is an essential part of RMRs, its occurrence was 100%. An example of the move can be seen in the following excerpt (Sample H17): "The Group's system of internal control is designed to manage and reduce risks that will hinder the Group from achieving its goals and objectives". This move is realised by three steps, which will be further described below.

The first step, M4S1, shows the bank's efforts to stop financial crime. The number of financial crime cases, such as financial fraud, has been escalating in recent years [36]. Therefore, it becomes necessary for companies, and especially banks, to prevent it from taking place in their institutions. In doing so, some measures are introduced under this step, such as "Anti-bribery and corruption", and "Anti-money laundering/ counter financing of terrorism". This step is conventional because it only occurred at 62.5% in the corpus. The following excerpt serves as an example of this step: "ANTI-BRIBERY AND CORRUPTION—One of the core values of the Group is integrity, and the Group will not tolerate any acts which are in breach of this value. The Group firmly believes in acting professionally, fairly and with integrity in all business dealings and relationships." (Sample D19).

Following the first step is M4S2, which presents and describes the risks that the banks encounter. However, this step only had 55% occurrence. Presumably, this step is neither compulsory nor requested in the Statement on Risk Management and Internal Control: Guidelines for Directors of Listed Issuers. Some banks also disclosed the risks in other sections of the annual reports rather than the RMR. Some of the risks that the banks encountered include Credit Risk, Operational Risk, Market Risk, Liquidity Risk, IT and Cyber Risk and Environmental Risk. An example of this step is shown in Sample B17: "EMERGING RISKS—The Group has identified several emerging risks, as listed below":

Human resource is an important element because it contributes to the growth of any company, including banks. They must show their adherence to the human resource policies. Therefore, this has become the main purpose of M4S3. In this step, the policies presented are mainly related to staff development. This step is conventional because it occurred in 75% of the corpus. The following excerpt serves as an example of this step: "The Group People Policies serves as a baseline with clarity on the philosophy and principles for People Management and Development in the Group." (Sample F19).

5.5. Move 5: Concluding the Report

Similar to other sections of an annual report, an RMR is also ended with a formal closing [37,38]. This occurred in all the RMRs, and thus the occurrence was 100%, making it an obligatory move. This move is composed of two steps that concern the review of the report. Some banks included the "Conclusion" heading to signal the move, while the rest immediately disclosed the reviewing process.

The first step, M5S1, discloses the internal committees involved in reviewing the risk management system of the banks. It is an obligatory step because it had 100% occurrence. This step assures readers that the risk management and internal control practises have been thoroughly reviewed by the internal auditors appointed by the banks. This can be seen in the following example: "The Board has received assurance from the Group Managing Director/Chief Executive Officer, Chief Financial Officer, Chief Risk Officer, Chief Internal Auditor and Chief Compliance Officer that the Group's risk management and internal control system is operating adequately and effectively, in all material aspects, based on the risk management and internal control system of the Group" (Sample E20).

The last step is M5S2, which discloses the external reviewers of the RMR being published. However, unlike the previous step, its occurrence was only 97.5% because it was not found in C16. This made the step conventional. M5S2 is used to inform readers that the reviewing process of the RMR involved external auditors and followed the guidelines that had been set by Bursa Malaysia. The following excerpt serves as an example of this step: "As required by Bursa Securities' MMLR Paragraph 15.23, the external auditors have reviewed this Statement on Risk Management and Internal Control. Their limited assurance review was in accordance with Audit and Assurance Practice Guide 3: Guidance for Auditors on Engagements to Report on the Statement of Risk Management and Internal Control included in the Annual Report, issued by the Malaysian Institute of Accountants." (Sample E20).

6. Conclusions

The present study investigates the moves and steps of RMRs by Malaysian commercial banks. A total of 40 RMRs by eight Malaysian commercial banks listed in Bursa Malaysia served as the data of the study. The RMRs were found to consist of two parts, which included general information on the management of the risks, and the internal efforts made by the banks in governing the risks. The findings revealed five rhetorical moves used in the reports. One optional move, one conventional move and three obligatory moves were identified in the corpus. It should also be noted that the practise of disclosure inertia [10] was found in the RMRs of all the banks. Through this practise, the information included in the reports was retained every year, with very minimal changes.

Genre-based studies on RMRs are still new, and therefore, this leaves a lot more opportunities for future research. The studies could include a comparative move analysis of the RMRs between countries or different types of companies. Some countries are still new to disclosing risks. Thus, the studies may provide insights into how similar or different the rhetorical moves are that can be found in the RMRs in these countries and countries that have been pioneering RMRs, such as the United States. The findings may help researchers and practitioners to understand the difference from language perspectives.

Author Contributions: Conceptualisation, K.F.Y. and A.M.A.; data curation, K.F.Y. and A.M.A.; formal analysis, K.F.Y., A.M.A., M.Y.C. and G.I.L.; investigation, K.F.Y.; methodology, K.F.Y., A.M.A.; project administration, K.F.Y., A.M.A., M.Y.C. and G.I.L.; software, K.F.Y.; supervision, A.M.A., M.Y.C. and G.I.L.; validation, K.F.Y. and A.M.A.; visualisation, K.F.Y.; writing—original draft, K.F.Y. and A.M.A.; writing—review and editing, A.M.A., M.Y.C. and G.I.L. All authors have read and agreed to the published version of the manuscript.

Funding: This research received no external funding.

Institutional Review Board Statement: Not applicable.

Informed Consent Statement: Not applicable.

Data Availability Statement: Data available in a publicly accessible repository that do not issue DOIs publicly available datasets were analysed in this study. These data can be found here: https://www.bursamalaysia.com/ (accessed on: 1 November 2022).

Conflicts of Interest: The authors declare no conflict of interest.

References

1. Gao, X.; Wang, X.; Tian, F. Do significant risk warnings in CARs increase corporate bond credit spreads? Evidence from China. *China J. Account. Res.* **2019**, *12*, 191–208. [CrossRef]
2. Ng, S.H.; Lee, C.S. Does IPO prospectus in Malaysia disclose relevant risk? *Asia-Pacific J. Bus. Adm.* **2019**, *11*, 301–323. [CrossRef]
3. Szczepankiewicz, E.I. Identification of Going-Concern Risks in CSR and Integrated Reports of Polish Companies from the Construction and Property Development Sector. *Risks* **2021**, *9*, 85. [CrossRef]
4. Bursa Malaysia, Statement on Risk Management & Internal Control: Guidelines for Directors of Listed Issuers. Available online: https://bursa-malaysia.s3.amazonaws.com/reports/listing_requirement_guidelines_guidance_for_directors_plc.pdf (accessed on 15 April 2022).
5. Lajili, K.; Zéghal, D. A content analysis of risk management disclosures in Canadian CARs. *Can. J. Adm. Sci.* **2005**, *22*, 125–142. [CrossRef]
6. Basel Committee on Banking Supervision, Standards: Revised Pillar 3 Disclosure Requirements. Available online: https://www.bis.org/bcbs/publ/d455.pdf (accessed on 16 April 2022).
7. Linsley, P.; Shrives, P. Disclosure of risk information in the banking sectors. *J. Fin. Reg. Compli.* **2005**, *13*, 205–214. [CrossRef]
8. Gao, L.; Calderon, T.G.; Tang, F. Public companies' cybersecurity risk disclosures. *Int. J. Account. Inf. Syst.* **2020**, *38*, 100468. [CrossRef]
9. Arikan, O. The effect of boilerplate language on nonprofessional investors' judgments. *Account. Bus. Res.* **2022**, *52*, 417–442. [CrossRef]
10. Abraham, S.; Shrives, P.J. Improving the relevance of risk factor disclosure in corporate annual reports. *British Account. Rev.* **2013**, *46*, 91–107. [CrossRef]
11. Aziz, R.A.; Abas, N.; Yusob, K.F.; Yusof, N.M. Genre analysis of risk management reports by an islamic financial institution in Malaysia. *GEMA Online J. Lang. Stud.* **2022**, *22*, 63–87. [CrossRef]
12. Swales, J. *Aspects of Article Introductions*; The University of Aston: Birmingham, UK, 1981.
13. Swales, J. *Genre Analysis: English in Academic and Research Settings*, 1st ed.; Cambridge University Press: Cambridge, UK, 1990.
14. Bazerman, C. *Shaping Written Knowledge: The Genre and Activity of the Experimental Article in Science*; University of Wisconsin Press: Madison, WI, USA, 1988.
15. Bloor, T.; Pindi, M. Schematic structure in economics structure. In *The Language of Economics: The Analysis of Economics Discourse*; Dusley-Evans, T., Henderson, W., Eds.; Modern English Publications in Association with British Council: Oxford, UK, 1990.
16. Bhatia, V.K. *Analysing Genre: Language Use in Professional Settings*; Longman Group Ltd.: New York, NY, USA, 1993.
17. Skulstad, A.S. Rhetorical organisation of chairmen's statements. *Int. J. Appl. Ling.* **1996**, *6*, 43–62. [CrossRef]
18. Swales, J. *Research Genres: Explorations and Applications*; Cambridge University Press: Cambridge, UK, 2004.
19. Berkenkotter, C.; Huckin, T.N. Rethinking genre from a sociocognitive perspective. *Written Comm.* **1993**, *10*, 475–509. [CrossRef]
20. Basturkmen, H. A genre-based investigation of discussion sections of research articles in dentistry and disciplinary variation. *J. Engl. Acad. Purposes* **2012**, *11*, 134–144. [CrossRef]
21. Ebadi, S.; Salman, A.R.; Nguyen, T.T.L.; Weisi, H. Rhetorical Structure Variations in Abstracts and Introductions of Applied Linguistics Master's Theses by Iraqi and International Students. *Iran. J. Lang. Teach. Res.* **2019**, *7*, 101–117.
22. Pratiwi, S.N.; Kurniawan, E. Rhetorical move and genre knowledge development of English and Indonesian abstracts: A comparative analysis. *Stud. Eng. Lang. Edu.* **2021**, *8*, 885–900. [CrossRef]

23. Can, S.; Karabacak, E.; Qin, J. Structure of moves in research article abstracts in applied linguistics. *Publications* **2016**, *4*, 23. [CrossRef]
24. Duan, J.; Wei, J. A Genre Analysis of English and Chinese Legal Research Article Abstracts: A Corpus-based Approach. *J. Lang. Teach. Res.* **2021**, *12*, 810–821. [CrossRef]
25. Kanoksilapatham, B. Rhetorical structure of biochemistry research articles. *Engl. Specif. Purp.* **2005**, *24*, 269–292. [CrossRef]
26. Singh, K.K.M.; Mohamad Ali, A.; Chan, M.Y.; Tan, H. A Genre-Based Investigation of the Introduction Sections of Academic Oral Presentations. *Asian J. Uni. Edu.* **2019**, *15*, 94–124. [CrossRef]
27. Hu, G.; Liu, Y. Three-minute thesis presentations as an academic genre: A cross-disciplinary study of genre moves. *J. Eng. Acad. Purp.* **2018**, *35*, 16–30. [CrossRef]
28. Thompson, S. Frameworks and contexts: A genre-based approach to analysing lecture introductions. *Engl. Spec. Purp.* **1994**, *13*, 171–186. [CrossRef]
29. Deng, L.; Wannaruk, A. A Contrastive Study of Rhetorical Move Structure of English Medium Instruction Lectures Given by Native English and Chinese Lecturers. *LEARN J.* **2021**, *14*, 451–477.
30. Hussain, S.S.; Mohamad Ali, A.; Mohd Kasim, Z.; Jalaluddin, I. A review on the rhetorical structure and linguistic features of corporate annual reports. *Int. J. Acad. Res. Busi. Soc. Sci.* **2020**, *10*, 236–256.
31. Hussain, S.S. A Comparative Critical Genre Analysis of Statements on Corporate Governance of Malaysian Corporations between Pre and During Currency Depreciation. Ph.D. Thesis, Universiti Putra Malaysia, Selangor, Malaysia, December 2021.
32. De Groot, E.B.; Korzilius, H.; Nickerson, C.; Gerritsen, M. A corpus analysis of text themes and photographic themes in managerial forewords of Dutch-English and British annual general reports. *IEEE Trans. Prof. Comm.* **2006**, *49*, 217–235. [CrossRef]
33. Mobasher, A.; Mohamad Ali, A. Exploring the Macrostructure of Management Forewords of Corporate Annual Reports. *Int. J. Appl. Ling. Engl. Lit* **2015**, *4*, 14–25.
34. Qian, Y. A critical genre analysis of MD&A discourse in corporate annual reports. *Discourse Comm.* **2020**, *14*, 424–437.
35. Salmani Nodoushan, M.A.; Montazeran, H. The Book Review Genre: A Structural Move Analysis. *Int. J. Lang. Studies* **2012**, *6*, 1–30.
36. Abdullah, M.; Shukor, Z.A.; Mohamed, Z.M.; Ahmad, A. Risk management disclosure: A study on the effect of voluntary risk management disclosure toward firm value. *J. Appl. Account. Res.* **2015**, *16*, 400–432. [CrossRef]
37. Bhatia, V.K. *Worlds of Written Discourse: A Genre-Based View*; Continuum International Publishing Group: New York, NY, USA, 2004.
38. Nickerson, N.; de Groot, E.B. Dear shareholder, dear stakeholder, dear stockholder: The business letter genre in the annual general report. In *Genre Variation in Business Letters*; Gillaerts, P., Gotti, M., Eds.; Peter Lang: Bern, Switzerland, 2005; pp. 324–346.

proceedings

Proceeding Paper

The Mediator Effect of Corporate Social Responsibility Disclosure on the Relationship between Corporate Governance and Bank Performance †

Ibrahim Alnohoud *, Maisarah Binti Mohammed Saat and Suresh A/L Ramakrishnan

Department of Accounting and Finance, Azman Hashim International Business School, Universiti Teknologi Malaysia, Johor Bahru 81310, Malaysia
* Correspondence: akibrahim@graduate.utm.my
† Presented at the International Academic Symposium of Social Science 2022, Kota Bharu, Malaysia, 3 July 2022.

Abstract: The performance of banks is one of the critical indicators for investors and a testament to the success of the economic cycle. Bank performance is the mediator between society and the state through loans to institutions and individuals. Thus, when corporate governance is activated and corporate social responsibility disclosure is disclosed in a banking environment, these are proven factors that lead to better performance. This study used the Baron and Kenny approach to test the mediator effect. The result of this study confirmed that corporate social responsibility disclosure acts as a mediator between corporate governance and bank performance.

Keywords: CSRD; bank performance; corporate governance

Citation: Alnohoud, I.; Saat, M.B.M.; Ramakrishnan, S.A/L. The Mediator Effect of Corporate Social Responsibility Disclosure on the Relationship between Corporate Governance and Bank Performance. *Proceedings* **2022**, *82*, 83. https://doi.org/10.3390/proceedings2022082083

Academic Editor: Mohamad Rahimi Mohamad Rosman

Published: 26 September 2022

1. Introduction

In general, corporate social responsibility disclosure (CSRD) has become a research subject for all organizations worldwide. It has gained increasing interest from scholars and practitioners around the world in recent years. Many strongly argue that corporations should not be judged solely based on their commercial success and economic efficiency, but that they should also be socially responsible. CSRD encompasses the notion that a profit-seeking company should expand the scope of its objectives and priorities beyond increasing the assets of its investors. It implies that corporations should balance their financial and non-financial objectives with their host society's best interests, and in particular the natural environment, workers, and consumers [1].

In the last few decades, companies have been trying to achieve a competitive advantage and overcome competitors by investing in social responsibility. In recent years, this has been enhanced due to changes in investor behavior and attitudes towards society. Thus, corporate social responsibility disclosure (CSRD) has frequently been used as a parameter in research to measure CSR practices. However, CSRD in Jordan has received only a small amount of attention from the Amman Stock Exchange (ASE) companies in their annual reports. The most revealed items were those relating to human resources and involvement in the social community. In contrast, Jordanian companies need for more attention with relation to environmental disclosure [2].

Several factors may influence bank performance, including CSRD, which is the company's responsibility to investors and other stakeholders such as staff, consumers, government, society, and the environment. CSRD can be used as a measure of the willingness of the company to fulfil stakeholder expectations effectively. The considerable emphasis placed on the societal role of business these days is due to the widespread conviction that proportions of corporate achievement must not focus only on profit. The firm should also align with the needs of shareholders and society as a whole; the social part of the duty of any corporation in the 21st century is more critical than ever. In recent times, CSRD has become a characteristic concern of the sector, impacting corporate profits and credibility [3].

There has been extensive research into the relationship between CSRD and bank performance, including samples to date from developed countries and non-financial firms. Findings from studies linked to the CSRD and bank performance relationship can be categorized as positive, negative, or insignificant. The positive relationship between corporate social responsibility disclosure and bank performance has been supported by many researchers [4–7].

2. Literature Review

In this review, we examine research on corporate social responsibility and its impact on bank performance. We also research all decisions about corporate social responsibility and financial efficiency, since we recognize that bank performance is the most important element in strengthening and supporting corporate social responsibility, ensuring profitability for every firm and bank. Corporate social responsibility is often referred to as corporate conscience, corporate citizenship, or responsible business, and is incorporated into business models as a form of corporate self-regulation. Today, everywhere we look, businesses the world must adopt more socially responsible practices. Numerous scholars and authors have carried out studies on corporate social responsibility and bank performance worldwide. The fundamental conclusion of the literature is that corporate social responsibility in every area of life plays a role in financial efficiency. The underlying literature is centered on specific national financial panel research on the financial results of banks. In this article, we analyze various research papers from Jordan, Malaysia, Yemen, United Kingdom, Bangladesh, India, Nigeria, and Pakistan [8] to explore the disclosure of corporate sustainability (CS) practices and analyze the connection between sustainability performance and financial performance in the Asian context, considering companies from India and Japan. The current study was focused on secondary data obtained from the annual reports and CS reports of 28 and 35 Indian and Japanese non-financial firms listed between 2009 and 2014. To measure the sustainability disclosure score based on the context of Global Reporting Initiatives [9], content analysis (binary coding system) was used. To calculate financial efficiency, market-to-book ratios were used. The study shows that in Japanese companies, the average degree of transparency is higher than in Indian companies. The study found that the effect of CS performance on financial performance is positive and significant for both nations, using the regression model.

The authors of [10] reported the relationship between the operations of CSR and the financial results of publicly traded companies in Malaysia, along with data from the 2009–2013 annual reports. With the aid of these results, the researchers clarified that Malaysian top-100 businesses are supportive and active in CSR operations and further found that CSR operations play a critical role in enhancing financial support.

The authors of [11] defined the impact of corporate social responsibility on the performance of Yemeni non-financial organizations. Three components of CSR—economic, legal, and ethical—were shown to have a positive and meaningful relationship with each other using different statistical methods and results. In addition, after review, it was found that there was no statistically significant gap between SOEs and private companies concerning the extent of periodic implementation of corporate social responsibility activities.

The authors of [12] identified the effect of corporate social responsibility adoption on corporate financial performance from Jordanian banks in Jordan. Researchers used the model calculated by multiple regression in this article. According to quantitative analysis, a significant relationship exists between bank levels, bank size, the level of risk, the level of all advertisement expenses, and CSR. In the banking sector of Jordan, CSR is also described as the most important.

The authors of [13] examined the effect of corporate social responsibility on the financial results of Islamic and traditional banks in different Asian countries. The secondary data was obtained from the annual reports of the respective banks. This analysis was quantitative. Correlation and regression methods for data processing purposes were used. The study results indicate that the overall positive and meaningful relationship between corporate social responsibility and financial performance is focused on regression and rigorous research.

On the other hand, some studies have determined the relationship between corporate social responsibility disclosure and financial efficiency to be insignificant. The authors of [14] reported the effect of CSR on the profitability of the business and country in Nigeria had been established. They obtained results through ordinary least square analysis that indicated the negative relationship between the financial performance measure of the business and expenditure on social responsibility and concluded that profitability is high in Nigeria, but organizations do not spend much. The authors of [15] analyzed the effect of corporate governance and corporate social responsibility on the financial performance of the banking sector as reported on the Indonesian stock exchange. Secondary data is used in this descriptive analysis in the form of annual corporate financial reports. Census techniques with an observation period of 5 years, from 2012 to 2016, were chosen for the survey. Multiple regression models were used in the data analysis methodology. The study findings indicate that corporate social responsibility, as calculated by the Return on Investment, has no major impact on financial efficiency (ROA).

Hypothesis H1: *the mediating effect of corporate social responsibility disclosure on the relationship between corporate governance and bank performance.*

3. Materials and Methods

The authors of [16] reported that three different data collection strategies were suggested: (1) quantitative; (2) qualitative; and (3) mixed approaches.

1. Quantitative research is empirical research in which the data is presented in numerical form; this method is appropriate for studying programs, behaviors, and social interactions. It is separated into two categories: primary and secondary data [17].
2. Qualitative research is empirical research in which the data is not presented in numerical form; as a result, qualitative research is well suited to social phenomena and introduces inventive new research methodologies [17].
3. Mixed methods research is a method of gathering and interpreting data that combines qualitative and quantitative methodologies in one study [17].

Data can be analyzed in terms of numbers by using a quantitative analysis method. The result can be evaluated and interpreted using numerical methods. In this process, data is defined, as well as the relationship between the variables and the outcome. Various data analysis tools were used to determine the financial results of banks after the data had been cleaned and pre-processed. These models were used to capture the impact of corporate governance on bank performance. The role of corporate social responsibility disclosure was investigated as a mediator variable in the relationship between corporate governance and bank performance.

3.1. Sampling, Research Population, and Period of Study

The present study analyzes corporate social responsibility in Jordanian bank data collected from the entire population of conventional banks. The study excludes Islamic banks because Islamic banks have different regulations and rules. As a country in the Middle East, Jordan has a diverse culture and different laws from other developed countries [18]. A total of 13 conventional banks in Jordan were assessed through the period of 2011 to 2019. This period is considered important for banking in Jordan for several reasons. Firstly, during this period, problems arose in the Arab countries surrounding Jordan, such as Syria and Iraq, and simultaneously lead to the increased consumption of resources; the Jordanian economy experienced risk exposure, thus leading to few employment opportunities; and a greater increase in the number of poor people than necessary, which caused restraint from the banking sector to stand with the government in minimizing these problems by supporting small and medium enterprises. Initially, this study mainly relies on secondary data, which have been extracted from the annual report.

3.2. Research Process

Preliminary consideration, procedure, and data analysis and interpretation were the three phases of this study's research process. The first step was a preliminary evaluation, which involved deciding the study target. Literature analysis and research design are two critical aspects of any research project. The second stage was concerned with the procedure e the data collection and data were collected from the annual reports of Jordanian banks from 2011 to 2019. Moreover, the researchers used the content analysis of items disclosed in the annual report according to the global report initiatives [9]. In the last stage of data analysis and interpretation, the Baron and Kenny approach was used [19], and a discussion of the result follows according to the objective of this study.

3.3. Baron and Kenny Approach

3.3.1. Mediator Variable

This study utilized the standard procedure of [19] to analyze the mediating effects of corporate social responsibility disclosure on the association between corporate governance and bank performance. It is explained with the following procedure:

In the first stage, Y as the dependent variable is regressed on X as the independent variable.

$$Y = \alpha 0 + \beta 1X + \varepsilon \quad (1)$$

Equation (1) examines the relationship between corporate governance and bank performance, where Y is bank performance and X is corporate governance.

In the second stage, M as the mediator variable regresses on X, which acts as an independent variable.

$$M = \alpha 0 + \beta 2X + \varepsilon \quad (2)$$

Equation (2) examines the relationship between corporate governance on corporate social responsibility disclosure, where M is corporate social responsibility disclosure and X is corporate governance.

At the final stage, Y as the dependent variable is regressed on X, which is the independent variable, and M as the mediator simultaneously.

$$Y = \alpha 0 + \beta 3M + \beta 4X + \varepsilon \quad (3)$$

Equation (3) examines the relationship between corporate governance with corporate social responsibility disclosure on bank performance, where Y is Bank performance, M is corporate social responsibility disclosure, and X is corporate governance.

Equation (3) represents the role of corporate social responsibility disclosure as a mediator in relationship between corporate governance and performance. This equation stands for bank performance and regresses on corporate social responsibility disclosure (M) and corporate governance (X).

3.3.2. Sobel Test

Another common approach to mediation analysis results from an inference coefficient approach, or Sobel Test [20]. However, the authors of [19] proposed the Sobel Test to test the value of the indirect path (a * b) and find out whether the discrepancy between the total effect and the direct effect is statistically significant. This Sobel Test aims to check whether a mediator converts the effect of an independent variable into a dependent variable.

Average error and coefficient among independent variables and mediators and average error and coefficient between the mediator and dependent variable were used for testing the mediating effect by the Sobel Test. After applying the Sobel Test, results would indicate whether corporate social responsibility disclosure is a mediator or not.

Figure 1 shows the effect of corporate social responsibility disclosure as a mediator variable on the relationship between corporate governance and bank performance. X acts as an independent variable, Y shows the dependent, and M stands as a mediator variable. C and SE stand for coefficient and standard error, respectively.

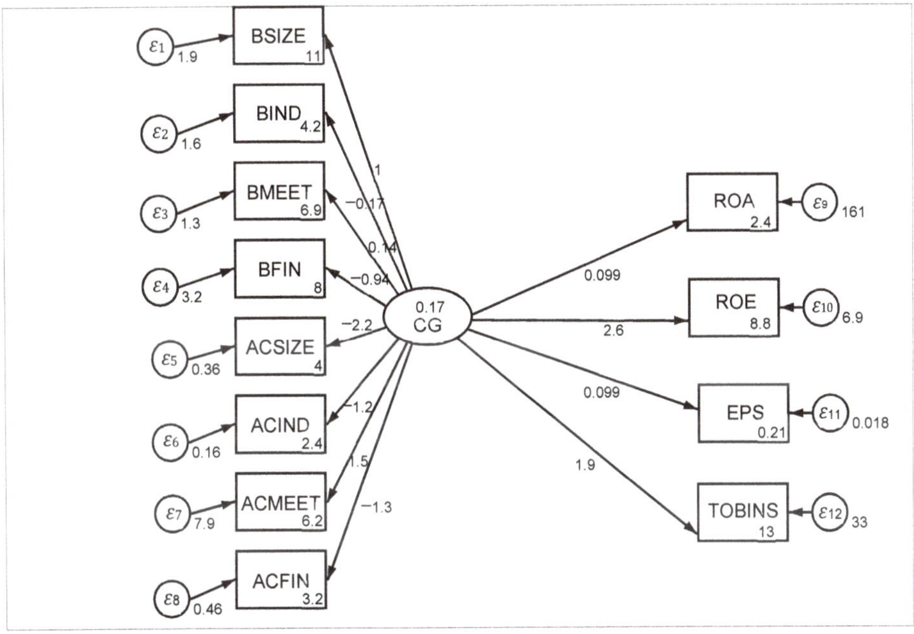

Figure 1. Result of corporate governance and bank performance.

3.3.3. Mediator Variable

This section aims to describe the role of corporate social responsibility disclosure as a mediator variable between corporate governance and bank performance in conventional banks separately. The approach of Baron and Kenny [19] and the Sobel test [20] have been used for this reason.

In the first stage, *Y* as the dependent variable is regressed on *X* as the independent variable.

$$y = \alpha 0 + \beta 1X + \varepsilon \quad (4)$$

In the second stage, *M* as the mediator variable regresses on *X*, which acts as an independent variable.

$$M = \alpha 0 + \beta 2X + \varepsilon \quad (5)$$

At the final stage, *Y* as the dependent variable is regressed on *X*, which is the independent variable, and *M* as mediator simultaneously.

$$Y = \alpha 0 + \beta 3M + \beta 4X + \varepsilon \quad (6)$$

4. Results

The first condition of the Causal Steps Method confirmed that there is a significant relationship between corporate governance and bank performance, implying that companies with strong corporate governance are more likely to have better firm performance. Figure 2 and Table 1 show the results of the first condition of the Causal Steps Method, which confirmed that there is a significant relationship between corporate governance and bank performance (implying that companies with strong corporate governance are more likely to have better firm performance). This finding is in line with prior research, which has found that effective and good corporate governance leads to improved firm performance [21].

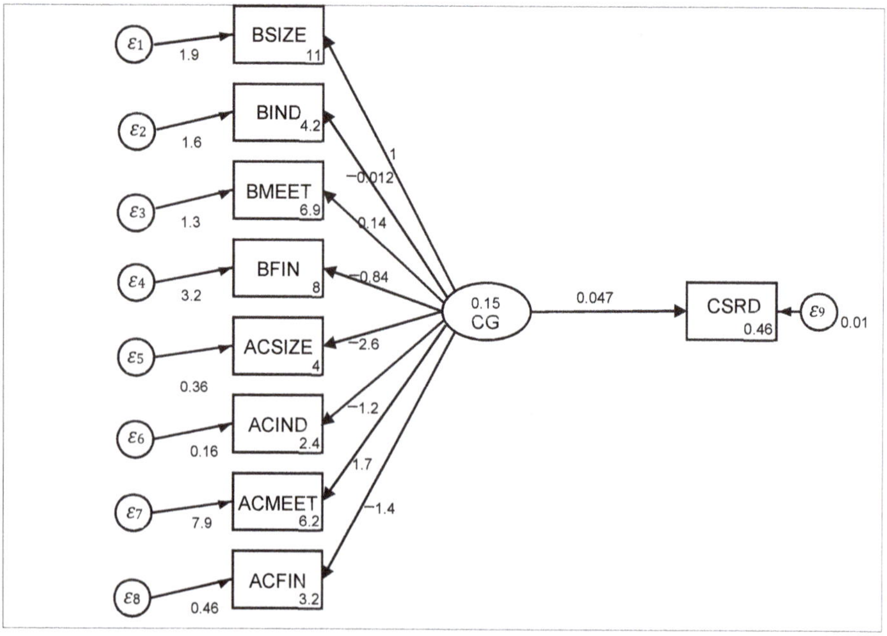

Figure 2. Result of corporate governance and corporate social responsibility disclosure.

Table 1. Result of corporate governance and firm performance.

STRUCTURAL	COEFFICIENT	P > Z
ROA < G	0.0987987	0.975
ROE < CG	2.632945	0.021
EPS < CG	0.099025	0.045
TOBIN'S < CG	1.932218	0.279

The second condition of the Causal Steps Method resulted in a significant association between corporate governance and corporate social responsibility disclosure as shown in Figure 2 and Table 2. This shows that firms with excellent corporate governance are more likely to demand high levels of corporate social responsibility. This finding is in line with legitimacy theory and earlier research [22].

Table 2. Result of corporate governance and corporate social responsibility disclosure.

Structural	Coefficient	P > Z
CSRD < -CG	0.047006	0.0000

The result of the Causal Steps Method's third condition, as shown in Figure 3 and Table 3, confirmed that there is a significant relationship between corporate social responsibility disclosure and bank performance, implying that companies with high corporate social responsibility disclosure are more likely to have better bank performance. This finding suggests that a high level of corporate social responsibility can help financial firms perform better, and is in line with earlier research [23].

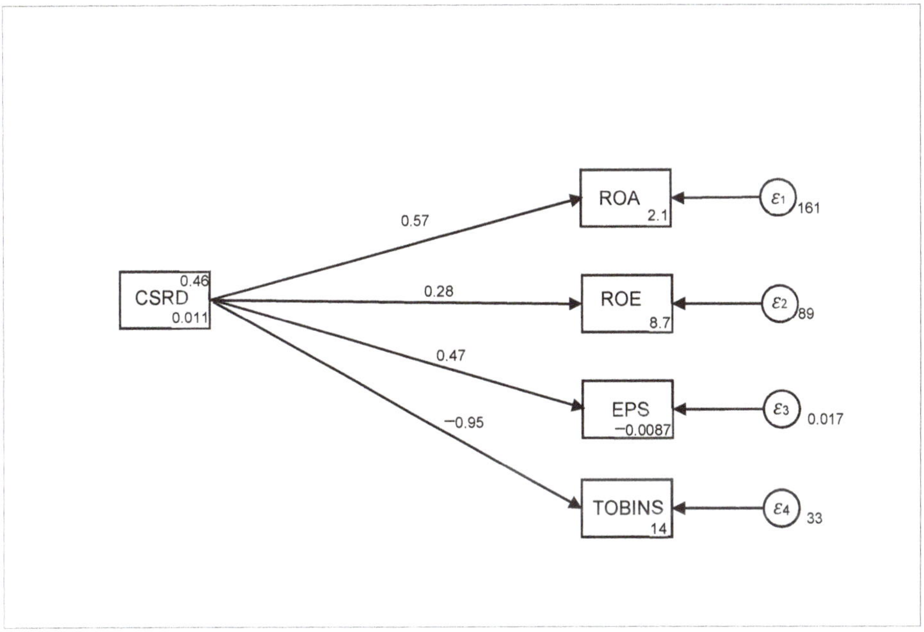

Figure 3. Result of corporate social responsibility disclosure and bank performance.

Table 3. Result of corporate social responsibility disclosure and bank performance.

Structural	Coefficient	P > Z
ROA < CSRD	0.5744962	0.068
ROE < CSRD	0.2805739	0.0000
EPS < CSRD	0.4730511	0.0000
TOBIN'S < CSRD	−0.9463632	0.0000

Figure 4 and Table 4 shows the result of the last condition of the Causal Steps Method, which confirmed that corporate social responsibility disclosure mediates the relationship between corporate governance and bank performance. In addition, the effect of corporate governance on bank performance increases through corporate social responsibility as a mediator. Thus, the results support the mediating hypothesis that corporate social responsibility disclosure mediates the relationship between corporate governance and bank performance.

Table 4. Result of mediating role of corporate social responsibility disclosure between corporate governance and bank performance.

CSRD	Coefficient	P > Z	Standard Err.
CG	1	0.0000	0.0095563
BANK PERFORMANCE	Coefficient	P > Z	Standard Err.
ROA	0.568037	0.961	11.72002
ROE	−2.196829	0.377	2.485405
EPS	0.4008096	0.001	0.1203536
TOBIN'S Q	−2.976764	0.581	5.391648
CG WITH ROA	0.1302887	0.988	58.77497
CG WITH ROE	0.4992941	0.047	0.2514511
CG WITH EPS	1.455955	0.103	0.8928138
CG WITH TOBIN'S Q	0.4092057	0.264	0.3660848

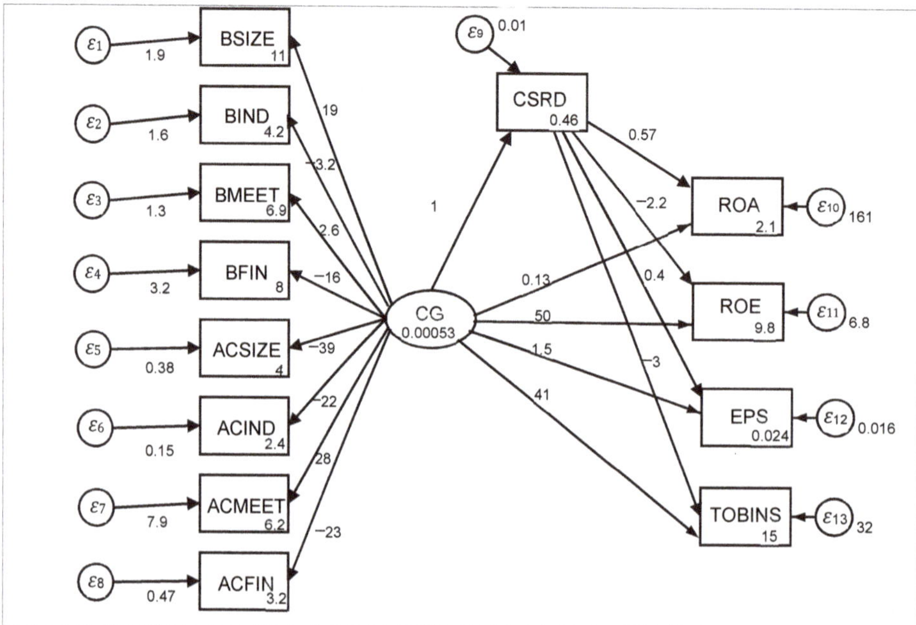

Figure 4. Result of mediating role of corporate social responsibility disclosure between corporate governance and bank performance.

According to Table 4, the mediating role of CSRD between corporate governance and bank performance since the p-value is significant.

Sensitivity Analysis (Sobel Test)

This section includes additional tests to see if the main finding of mediation models is consistent with other findings. In addition, this study uses the Sobel Test to confirm the result of the Casual Step Method. The following is the Sobel Test equation:

$$Z = \frac{ab}{\sqrt{(b^2 SE_a^2) + (a^2\, SE_b^2)}} \quad (7)$$

According to Table 5, the relationship between corporate governance and EPS is mediated by CSRD. p-values with one-tailed probability and two-tailed probability are significant but are insignificant for other variables.

Table 5. Result of mediating role of corporate social responsibility disclosure between corporate governance and bank performance by Sobel test.

Sobel Test	Coef.	Standard Error	One-Tailed Probability	Two-Tailed Probability
CG- CSRD-ROA	1 0.568037	0.0095563 11.72002	0.48067214	0.96134429
CG CSRD-ROE	1 −2.19682	0.0095563 2.485405	0.18838849	0.37677698
CG -CSRD-EPS	1 0.4008096	0.0095563 0.1203536	0.00043641	0.00087283
CG-CSRD–TOBIN'S Q	1 −2.976764	0.0095563 5.391648	0.29044410	0.58088821

5. Conclusions

More empirical research is required for improving the existing financial governance framework, before recommending rigor. Empirical evidence, for example, is required to evaluate the effect of each corporate governance variable on the financial performance of banks and their associated social responsibility disclosures. Furthermore, according to agency theory, corporate governance might be changed to make managers accountable to all stakeholders rather than just a small number of shareholders. The revisions have an impact on agency theory. The empirical findings of the study disclosed that corporate social responsibility disclosures act as mediators when bank performance is measured by earning per share, thus the hypothesis (H1) is affirmed. Moreover, several corporate governance measures have a significant impact on bank performance when corporate social responsibility disclosure is taken as a mediator variable. In the future, there are various potential prospects for more research and enhancements to be explored. The Jordanian Central Bank created the Bank Corporate Governance Code in order to strengthen the Jordanian banking system; this could serve as a point to begin investigation.

Author Contributions: Conceptualization, I.A.; methodology, M.B.M.S.; writing—original draft preparation, I.A., M.B.M.S. and S.A/L.R.; writing—review and editing, I.A. and M.B.M.S. All authors have read and agreed to the published version of the manuscript.

Funding: This research received no external funding.

Institutional Review Board Statement: Not applicable.

Informed Consent Statement: Not applicable.

Data Availability Statement: Not applicable.

Conflicts of Interest: The authors declare no conflict of interest.

References

1. Qaderi, S.A.; Alhmoud, T.R.; Abdulraheem Ghaleb, B.A. Audit committee features and CSR disclosure: Additional evidence from an emerging market. *Int. J. Financ. Res.* **2020**, *11*, 226–237. [CrossRef]
2. Daas, A.; Alaraj, R. The complementarity between corporate social responsibility disclosure and institutional investor in Jordan. *Int. J. Islamic Middle East. Financ. Manag.* **2019**, *12*, 191–215. [CrossRef]
3. Al-marayat, A.M.S. Corporate Social Responsibility Effective Determinants of the Banking Sector in Jordan. *RJFA* **2019**, *10*, 36–44. [CrossRef]
4. Reverte, C.; Gómez-Melero, E.; Cegarra-Navarro, J. The influence of corporate social responsibility practices on organizational performance: Evidence from Eco-Responsible Spanish firms. *J. Clean. Prod.* **2016**, *112*, 2870–2884. [CrossRef]
5. Malik, M.S.; Kanwal, L. Impact of corporate social responsibility disclosure on financial performance: Case study of listed pharmaceutical firms of Pakistan. *J. Bus. Ethics* **2018**, *150*, 69–78. [CrossRef]
6. Maqbool, Z. Corporate social responsibility and financial performance: An empirical analysis of Indian banks. *Future Bus. J.* **2018**, *4*, 84–93. [CrossRef]
7. Sial, M.S.; Chunmei, Z.; Khan, T.; Nguyen, V.K. Corporate social responsibility, firm performance and the moderating effect of earnings management in Chinese firms. *Asia-Pac. J. Bus. Adm.* **2018**, *10*, 184–199. [CrossRef]
8. Laskar, N.; Chakraborty, T.K.; Maji, S.G. Corporate sustainability performance and financial performance: Empirical evidence from Japan and India. *Manag. Labour Stud.* **2017**, *42*, 88–106. [CrossRef]
9. Allegrini, M.; Greco, G. Corporate boards, audit committees and voluntary disclosure: Evidence from Italian listed companies. *J. Manag. Gov.* **2013**, *17*, 187–216. [CrossRef]
10. Yusoff, W.F.W.; Adamu, M.S. The relationship between corporate social responsibility and financial performance: Evidence from Malaysia. *Int. Bus. Manag.* **2016**, *10*, 345–351.
11. Al-Samman, E.; Al-Nashmi, M.M. Effect of corporate social responsibility on nonfinancial organizational performance: Evidence from Yemeni for-profit public and private enterprises. *Soc. Responsib. J.* **2016**, *12*, 247–262. [CrossRef]
12. Weshah, S.R.; Dahiyat, A.A.; Abu Awwad, M.R.; Hajjat, E.S. The impact of adopting corporate social responsibility on corporate financial performance: Evidence from Jordanian banks. *Interdiscip. J. Contemp. Res. Bus.* **2012**, *4*, 34–44.
13. Khan, B.; Tariq, R. Corporate social responsibility impact on financial performance of Islamic and conventional banks: Evidence from Asian countries. *Res. J. Financ. Account.* **2017**, *8*, 20–28.
14. Babalola, Y.A. The impact of corporate social responsibility on firms' profitability in Nigeria. *Eur. J. Econ. Financ. Adm. Sci.* **2012**, *45*, 39–50.

15. Mangantar, M. The influence of corporate social responsibility and corporate governance on banking financial performance. *Eur. Res. Stud. J.* **2019**, *22*, 95–105. [CrossRef]
16. Creswell, J.W.; Tashakkori, A. *Differing Perspectives on Mixed Methods Research*; Sage Publications Sage CA: Los Angeles, CA, USA, 2007.
17. Tashakkori, A.; Teddlie, C. Issues and dilemmas in teaching research methods courses in social and behavioural sciences: US perspective. *Int. J. Soc. Res. Methodol.* **2003**, *6*, 61–77. [CrossRef]
18. Al Fadli, A.; Sands, J.; Jones, G.; Beattie, C.; Pensiero, D. Board Gender Diversity and CSR Reporting: Evidence from Jordan. *Australas. Account. Bus. Financ. J.* **2019**, *13*, 29–52. [CrossRef]
19. Baron, R.M.; Kenny, D.A. The moderator–mediator variable distinction in social psychological research: Conceptual, strategic, and statistical considerations. *J. Personal. Soc. Psychol.* **1986**, *51*, 1173. [CrossRef]
20. Sobel, M.E. Asymptotic confidence intervals for indirect effects in structural equation models. *Sociol. Methodol.* **1982**, *13*, 290–312. [CrossRef]
21. Başar, B.D.; Bouteska, A.; Büyükoğlu, B.; Ekşi, I.H. The effect of corporate governance on bank performance: Evidence from Turkish and some MENA countries banks. *J. Asset Manag.* **2021**, *22*, 153–162. [CrossRef]
22. Nour, A.I.; Ahmad Sharabati, A.A.; Mahmoud Hammad, K. Corporate governance and corporate social responsibility disclosure. *Int. J. Sustain. Entrep. Corp. Soc. Responsib.* **2020**, *5*, 20–41. [CrossRef]
23. Rehman, Z.U.; Zahid, M.; Rahman, H.U.; Asif, M.; Alharthi, M.; Irfan, M.; Glowacz, A. Do corporate social responsibility disclosures improve financial performance? A perspective of the Islamic banking industry in Pakistan. *Sustain. Digit. Transform. Fintech* **2020**, *12*, 3302. [CrossRef]

proceedings

Proceeding Paper

Industrial Revolution (IR) 4.0: Opportunities and Challenges in Online Business †

Norhidayah Ali [1,*], Zuraidah Mohamed Isa [1], Suhaida Abu Bakar [2], Fathiyah Ahmad @ Ahmad Jali [3] and Sarah Shaharruddin [4]

1 Faculty of Business and Management, Universiti Teknologi MARA Kedah Branch, Merbok 08400, Kedah, Malaysia
2 Academy of Contemporary Islamic Studies, Universiti Teknologi MARA Kedah Branch, Merbok 08400, Kedah, Malaysia
3 Academy of Language Studies, Universiti Teknologi MARA Kedah Branch, Merbok 08400, Kedah, Malaysia
4 School of Business Management, Universiti Utara Malaysia, Sintok 06010, Kedah, Malaysia
* Correspondence: norhidayah@uitm.edu.my
† Presented at the International Academic Symposium of Social Science 2022, Kota Bharu, Malaysia, 3 July 2022.

Abstract: This article aims to foster an understanding of Industrial Revolution (IR) 4.0 and the opportunities and challenges it creates for online business. This study employed a qualitative approach using library research. Google Scholar and ResearchGate were used to collect and search past studies. The article selection process focused on relevant recent research and used keywords such as Online Business, Industrial Revolution (IR) 4.0, and Opportunities and Challenges in Online Business. Approximately 80 articles were found in the Emerald Insight database. The ProQuest database contained 20 articles, 100 articles were found in the Google Scholar database, and 10 articles were found in the ResearchGate database. Following the screening, 24 publications were chosen for further discussion. It was discovered that few studies had been conducted on this subject. The results revealed that internet-based companies will be assisted in developing their smartness and efficiency through IR 4.0, also known as the new industrial revolution. Organizations will be able to use data in real time within supply chains and economies that also operate in real time. They will become more sustainable and their operating conditions will improve, earning them the contemporary consumers' trust and loyalty as they offer opportunities for personalization. Online companies were anticipated to grow tremendously because not only were easier and more convenient transactions permitted, but also significantly improved product offerings were made feasible by the IR 4.0 system. Online businesses should seize these opportunities and overcome the challenges to maximize their business performance. When examining the potential advancements, researchers have foreseen that IR 4.0 features both opportunities and challenges, although the opportunities to be gained far outweigh the challenges. Thus, further research should be conducted to gauge the impacts of the opportunities and challenges associated with IR 4.0, and a proper course of action should be suggested.

Keywords: online business; Industrial Revolution (IR) 4.0; opportunities and challenges in online business

Citation: Ali, N.; Mohamed Isa, Z.; Abu Bakar, S.; Ahmad @ Ahmad Jali, F.; Shaharruddin, S. Industrial Revolution (IR) 4.0: Opportunities and Challenges in Online Business. *Proceedings* **2022**, *82*, 85. https://doi.org/10.3390/proceedings2022082085

Academic Editor: Mohamad Rahimi Mohamad Rosman

Published: 26 September 2022

1. Introduction

An unprecedented technological revolution in human history is the fourth industrial revolution; dubbed IR 4.0, it will quickly advance into combinations of digitalization technologies in fields such as engineering, physics, and biology, thereby creating completely new possibilities. Through IR 4.0, a context will be created whereby the production chain's worldwide processes occur both online and physically, blending together into a flexible and seamless system. Not only does IR 4.0 refer to processes and devices that are smart and connected, it also involves a far more expansive mentality and affects profoundly the worldwide structures of politics, society, and economics. Major advantages and a considerable effect on global economics are likely to result from IR 4.0, while it is simultaneously influencing the realization of a host of innovations in various fields, including quantum computing, gene sequencing, renewable energies, and nanotechnology [1]. After the conceptualization of IR 4.0 was made public in 2011, the concept grew and developed to the extent that some of the theories have become actual applications. The wide-ranging ways that this new industrial revolution has been applied in so many areas has affected approximately everyone, although differently. National and international developments are taking place as people become acclimatized to the realization of IR 4.0. Growing evidence indicates that while innovations have been introduced, there are clear and robust connections between the advancement of technology, government policy, and the market [2].

Academics, business owners, governmental agencies, and public groups continually discuss the issues related to the now worldwide Industry 4.0. Economists, for example, are paying increasing attention to the impacts that the concept of Industry 4.0 is having on international and state economic systems, specific industrial fields, work, and capital markets. Recent years have witnessed a significant change in the worldwide industrial context due to the developments in technology and innovative production processes. Despite the emergence of Industry 4.0 as a concept, this subject has been underexplored by academics. Thus, it has not yet been comprehensively defined, nor has consensus on the topic been achieved [3].

Rapid changes are occurring within the landscapes of many businesses and these extend across the whole value chain. This encompasses research and development (R&D), production, logistics, and customer services, among others, and causes the costs of transactions and deliveries to reduce considerably. Manufacturing is likely to be affected substantially by IR 4.0, which will alter considerably the processes of production, as well as cause real-world and digitized features to converge, thus creating the Internet of Things (IoT), or a universal internet [4]. This development will be controlled in domestic environments despite encompassing every factory-based activity via the internet, because changes to the manufacturing methods can be achieved with the new forms of technology connecting the physical and online environments. All aspects of industry—from the infrastructure of manufacturing to healthcare—will be impacted through this IoT-led process of control. For example, Industry 4.0 can reduce considerably the costs of transactions and deliveries within the commercial field. Secondly, the characteristics of IR 4.0 mean that in terms of investment, technology should be a highly rewarding sector, with substantial future investments potentially enhancing the digitalization and internet segments in particular. Furthermore, an organization can gain a major advantage through IR 4.0, since this concept features the capacity to analyze data in real time. Organizations can also become increasingly visible, monitor activities autonomously, as well as increase their production levels and competitiveness. Organizations, sectors, and countries all rely on innovation. Numerous aspects will potentially be greatly enhanced through IR 4.0, while workplaces will be transformed if it is implemented [5].

Although innovations in technology regularly enhance productivity and make people more prosperous, these can alter at a pace that creates substantial pressures as the labor resources adapt. Considerable injustices could arise from IR 4.0, while it risks dismantling the labor market. If the whole economy features automated replacements for human staff, widespread redundancies will occur. As IR 4.0 progresses, factory staff will adopt modified work roles with new demands, while they will operate in much-changed working environments or organizations compared to those of today. The necessity of finding innovative and successful methods of production, as well as dealing with issues such as uncertain security, aging populations, and climate change, will lead to the robust growth of IR 4.0. Internet-based companies and their proprietors are likely to face both difficulties and opportunities through this revolution. Thus, the objective of this paper is to examine the literature pertaining to IR 4.0 within online business, as well as the business-related opportunities and challenges associated with IR 4.0.

2. Research Methodology

This study used a qualitative method based on library research. Researchers conduct several tasks at this stage of the methodology, including collecting, identifying, analyzing, formulating, and critically evaluating relevant previous studies. Emerald Insight, ProQuest, Google Scholar, and ResearchGate were used to collect and search past studies. The article selection procedure focused on relevant previous studies and utilized keywords such as Online Business, Industrial Revolution (IR) 4.0, and Opportunities and Challenges in Online Business. Approximately 80 articles were found in the Emerald Insight database. The ProQuest database contained 20 articles, 100 articles were found in the Google Scholar database, and 10 articles were found in the ResearchGate database. Following the screening, 24 publications were chosen for further discussion. It was discovered that few studies had focused on this topic. Information from other databases, such as websites and periodicals, was also retrieved. Data was extracted at two levels, namely dimensions and research findings. Through this methodology, it is hoped that this study will assist other researchers to gain a deeper understanding of the topic and draw a comprehensive conclusion.

3. Online Business

Nowadays, most human activities, including those related to business are being supported through technologies. Most businesses employ technologies to connect with almost all their business partners, including their customers. Since the end of 2019, online business has expanded due to the COVID-19 outbreak, during which the world has undergone quarantines, lockdowns, and standard operating procedures (SOP). The ability of organizations and individuals to engage across the globe is perhaps the most important development in an increasingly digitalized landscape [6]. Furthermore, it has been noted that businesses in almost all industries are becoming increasingly aware of the benefits of expanding their services beyond their four walls, practicing contactless transactions, and conducting operations with the use of recent technology. In addition, online businesses have exploited and should exploit technology via the Fourth Industrial Revolution (4IR or IR 4.0) to connect their business systems directly to customers, employees, and suppliers. Machine-to-machine communications (M2Ms) can occur on a large scale with the use of IR 4.0, including features such as fabrication and manufacturing using newly developed forms of smart technology. The Internet of Things (IoT) and industrial Internet of Things (IIoT) have experienced integration, enabling communications to improve, automation to increase, and production to gain far more flexibility, making it possible for products to be highly customized [7]. Online business would exponentially improve due to not only effortless and convenient transactions but also the considerably enhanced product offerings enabled by IR 4.0-related manufacturing and production systems.

4. IR 4.0 and Online Business

One feature of the age of cyber-physical systems is IR 4.0, within which smart devices, systems of storage, and production facilities can become networked. Thus, information can be exchanged autonomously, activities can be initiated, and each part of the network controlled by the others [8]. IR 4.0 has been described as the most recent trend in manufacturing technology automation and data interchange. Furthermore, three key aspects of IR 4.0 have been identified, as follows: vertical and horizontal value chains have been digitalized and increasingly integrated; products and services have been digitalized; and new business models have included innovation and digitalization [9].

Four key implications of IR 4.0 for businesses have been introduced: the expectations of customers, enhancing products, collaborating on innovations, and types of organizations [1]. The demands of consumers are undergoing dramatic shifts, which companies must accommodate if their transparency is to increase and new customer behavioral patterns are to be recognized. The connections between a business organization and its partners will be revolutionized by IR 4.0, so greater customization will come to be expected. Businesses can now provide clients with digital previews of products so that customized orders can be placed. As a result, Malaysian businesses are being forced to rethink and change the strategies they use to deliver products and services [10].

In recent years, the number of entrepreneurs has grown. The competition platform has also shifted from traditional to digital, allowing companies and businesses to acquire a larger market share by utilizing new technology capabilities. Online business is an example of a digital means of doing business. It also acts as a viable entry point into the realm of cross-border business. The relevance of digitization and the Internet of Things (IoT) to businesses has been underlined by the fourth industrial revolution (IR 4.0). In this context, IR 4.0 allows entrepreneurs to venture into more online business opportunities. Online entrepreneurs can explore other potential jobs and businesses, such as online retail consignor, content creator, and social media influencer [8].

Although it is fairly challenging to transition to IR 4.0, this must be done properly. Online business is one of the opportunities that can be pursued as it allows someone to start an online operation that links them with individuals worldwide. Entrepreneurs can now start their own firms through virtual organizations and no longer need to rely on jobs supplied by the government or private institutions. Through virtual organizations, online entrepreneurs may also be able to offer job opportunities to others.

5. IR 4.0 and Opportunities in Online Business

In the context of business operation, embracing IR 4.0 enables business operations to improve by creating, adopting, and integrating technological solutions. Some businesses may be concerned about the cost and complexity of using IR 4.0 in their online business. However, the adoption of IR 4.0 has created greater opportunities for businesses in terms of increased productivity, improved product quality, lower operating costs, and achieving a competitive advantage. This revolution will increase organizational efficiency by ensuring agility, adaptation, and alignment with other organizations, which benefits businesses by enabling them to acquire a competitive edge [6]. Integrating IR 4.0 technology into any business may require time and effort, but this may be highly rewarding for businesses as IR 4.0 can be used to manage their daily operations and business decisions. Furthermore, IR 4.0 allows businesses to market and compete both locally and globally [11]. It has also been remarked that IR 4.0 can fulfill the customer's needs and meet the demand for company sustainability.

Customers' needs can be fulfilled because of the substantial time reductions, i.e., from when a customer orders a product until it reaches them. The old ways of making phone calls, sending emails, and completing forms are being eliminated from the customer buying process. By simply using technology, the completion time becomes faster and easier, which is convenient for businesses and their customers. In addition, information is easily restored and retrieved, especially for any subsequent orders. Thus, by providing better and more convenient services using IR 4.0 technology, businesses can be more efficient and competitive, resulting in more sustainable and improved organizations. For instance, through the use of artificial intelligence (AI), customer choice predictions and higher profits can be achieved through the consumer insights that can be optimized by company marketers [12].

Moreover, business performance can be improved. Businesses will perform better during IR 4.0. Another benefit they will encounter is the optimization of production [13]. Using nine major technological advances—such as autonomous robots and the Internet of Things (IoT)—IR 4.0 can increase automation and reduce the reliance on people, who are susceptible to making errors. Businesses may utilize big data and analytics to manage large amounts of data, develop insights, and formulate business plans [14]. It has also been noted that machine learning assists businesses to gain an understanding of consumer behavior and be more proactive in adapting their marketing tactics. Using IR 4.0 allows businesses to focus on their core operations. It can also enhance the capacity to boost efficiency and manufacture a wide range of higher-quality items more quickly. Another advantage of IR 4.0 is that it reduces the operating costs due to its higher level of automation, for which fewer personnel are needed, resulting in less waste and greater efficiency. This aligns with a study which found that IR 4.0 may help to reduce business operational costs [15].

Although all modern-day companies and organizations are unique, each confronts one major difficulty: all must be able to connect to and access up-to-the-minute details involving procedures, partnerships, the items produced, and people. As such, Industry 4.0 plays a far greater role than merely investing in new technologies and tools to boost online business efficiency—it is about transforming the entire way that online businesses run and grow. However, the effects of the fourth industrial revolution are difficult to anticipate, but expectations should be posited in relation to new, unknown inventions that might alter the perceptions of future online businesses. In conjunction with this, the digital factory and intelligent supply chain are possible outcomes.

From the perspective of the supply chain, digitalized factories and smart supply chains form systems of great flexibility, which create automated performance improvements within larger networks, regulate themselves, and learn about changes in actual or near-real time [16]. Additionally, various applications will potentially be possible for any internet-based business. For instance, real and digitalized environments could be integrated. They could also optimize by reducing their use of materials, power, and a human workforce; become more transparent by viewing demand–supply process information in real time; and allow activities to be monitored and controlled to enable their use in internet-based businesses.

Moreover, IR 4.0 will greatly facilitate all internet-based business operations. Digitalizing, automating, making generally autonomous, decentralizing, and personifying the ways to order, store, produce, and distribute products, as well as conduct consumer relations, will enable extensive horizontal and vertical process integration within the digitalized factories and smart supply chains. Table 1 lists the overall advantages of these innovation-based measures [17].

Table 1. Benefits of online business in the IR 4.0 environment.

IR 4.0 and Benefits for Online Business
- Businesses better organized and managed. - Production procedures better planned and monitored. - Savings made on raw materials, other items, energy use, and human laborers. - Shutdown time and manufacturing bottlenecks eliminated. - Manufacturing procedures continuously transparent. - Irrespective of location, every level and unit of the company has ready access to information in real time. - Transport costs saved; logistical improvements made. - Manufacturing procedures can be controlled from anywhere across the world. - Total production costs are minimized. - The ability to manufacture intelligent products. - R&D and innovation procedures accelerated by using worldwide digital platforms and open innovation methods to create new forms of ecosystems. - Consumer preference can dictate the manufacture of items that feature personalization. - Less likely that products will be "missed". - Alterations in the demand from the market can be responded to flexibly. - Consumer interaction occurs during the full cycle of product development - Operations become more productive and efficient. - Companies become more competitive. - Operations have a less negative environmental effect. - The tasks and actions undertaken become more attractive. - Human needs can be better met by enhancing the educational and training levels. - Society enjoys increased well-being. - More free time.

6. IR 4.0 and Challenges in Online Business

The normal view of any revolution is that challenges will definitely be encountered. From the first to the third industrial revolutions, the world has faced numerous challenges in ensuring that these revolutions keep pace with the current developments and societal will. Industrial Revolution 4.0 is not without its obstacles, which must be handled by all the parties involved if this new industrial technological transition is to be executed optimally at all levels.

Online businesses are no exception as they also face multiple challenges in the IR 4.0 era [18]. Among these are a lack of technical skills, interoperability, and handling data growth [19]. Furthermore, it has been noted that workforce requirements are constantly changing. Business models can only successfully deploy new technology and maintain operations with the correct people and skills. Therefore, taking these developments into account is challenging. A further issue relates to interoperability, whereby protocol, component, product, and system cannot be separated. Interoperability limits companies' ability to innovate.

Since vendors cannot be replaced by others, the components of a system cannot easily be upgraded, which is a drawback of interoperability. Thus, this is another challenging factor to consider. This article notes other difficulties [19], such as the capacity to handle the growth in data: depending increasingly on artificial intelligence means the faster generation of greater volumes of data and its presentation in various forms. Addressing this is likely to overwhelm many businesses. Systems of artificial intelligence need to feature greater simplicity so that such extensive amounts of data can be managed. Furthermore, any algorithm needs to be able to combine various forms of data covering different ranges of time. Therefore, a further issue needs to be taken into consideration. Although AI is expected to supersede human capacities in every walk of life between 2020 and 2060, this is also considered a warning sign and may create some obstacles for humanity in the future [20]. Moreover, it has also been stated that AI machines could potentially be used irresponsibly and harmfully because they might attract criminals [21]. Therefore, there is no guarantee that technologies can be fully trusted without humans playing a role.

The internet is an important medium in businesses, especially online operations. However, cyber security requires serious attention as it could create some potential drawbacks for businesses. For example, cyber risk is one of the challenges that should be addressed in any assessment of an organization's IR 4.0 readiness. Contingency plans must be established to prevent exposure to cyber attacks such as from hackers, virus transmission, data breaches, and cyber extortion because cyber risk is caused by online networks, online traffic, and personal information being kept on the internet [6]. The safety and security of systems is also a concern. Unauthorized access causes technological difficulties during production since confidential data could be collected and disseminated without authorization, while the ability to adjust the production process content may be lost due to security breaches. This will endanger production, risk the loss of the consumer's trust, cause significant financial losses, and compromise the organization's reputation. To limit the dangers of misuse and illegal access, companies must maintain contingencies or a recovery backup plan [22].

Another factor that contributes to the challenges of Industry 4.0 is the lack of knowledge of how to adapt to IR 4.0 [23]. This knowledge gap is also a significant reason for the current employees' capacity to perform certain tasks. Employees possess a wide range of competencies and skill sets, but it has been revealed that they do not usually possess the amount of knowledge required for the adoption of Industry 4.0. It has been found that they need to know more about IR 4.0 and gain a better knowledge of what it implies. Understanding and knowing about the concept of IR 4.0, as well as planning for it, are critical for success. However, this requires new skills and continuous learning. It is recommended that training is conducted to teach people how to perform their job tasks smoothly and facilitate the change process [24]. Table 2 summarizes the challenges of IR 4.0 in the context of online business.

Table 2. Challenges to online business in the IR 4.0 environment.

IR 4.0 and Challenges for Online Business
- Lack of technical skills.
- Lack of interoperability.
- Lack of skills in handling data growth.
- Workforce requirements are constantly changing.
- Protocol, component, product, and system not separated.
- Interoperability limits companies' ability to innovate.
- Interoperability limits the possibilities of upgrading system components.
- Lack of contingency plans for cyber risks.
- Unauthorized access causes technological difficulties during production.
- Confidential data is collected and disseminated without authorization.
- The ability to adjust the production process content is lost.
- The lack of knowledge about adaptation to IR 4.0.

7. Conclusions

In summary, this preliminary study obtained various insights on IR 4.0, its opportunities, and its challenges in relation to online business. The study included an introduction and covered online business, IR 4.0, and the associated opportunities and challenges. The findings from the existing literature show that IR 4.0 has created opportunities and challenges. Internet-based companies will be assisted in developing their smartness and efficiency through IR 4.0, also known as the new industrial revolution. Organizations will be able to use data in real time within supply chains and economies that also operate in real time. They will become more sustainable, and their operating conditions will improve, earning them contemporary consumers' trust and loyalty as the companies offer opportunities for personalization. Nevertheless, it is crucial to remember that the advancements involved in this industrial revolution will impact the reality of the future. Moreover, the developments will impact trends of demography, environmental conditions, and the geopolitical sphere, creating new patterns and behaviors within societies and cultures. Therefore, the ability to be prepared in advance for these scenarios should also be considered, as the reduction of the incoming dangers will be critical for future and current generations. Despite the challenges that must be faced by proprietors, these are outweighed by the opportunities to sustain their business and improve performance. To achieve both of these, online businesses should take advantage of the opportunities and overcome the challenges. Further research could be conducted to foresee IR 4.0 impacts on online businesses to enable owners to further improve their business operations while adapting to this industrial revolution. Meanwhile, future studies could investigate the perceptions of internet-based business among the older generations, as well as ways it might be beneficial for them. IR 4.0 has been linked with causing various contemporary problems, including rising job losses; the stratification of society; security threats associated with cyber-attacks; infractions of the rules of privacy, ethics, and society; the dangers of new wars based on invention; as well as a widening rich–poor division in national terms. The truth of these contentions could be investigated in future research.

Author Contributions: Conceptualization, N.A., Z.M.I., S.A.B., F.A.A.J. and S.S.; methodology, N.A. and Z.M.I.; investigation, N.A., Z.M.I., S.A.B., F.A.A.J. and S.S.; writing—original draft preparation, N.A., Z.M.I., S.A.B., F.A.A.J. and S.S.; writing—review and editing, N.A., Z.M.I., S.A.B., F.A.A.J. and S.S.; visualization, N.A., Z.M.I., S.A.B., F.A.A.J. and S.S.; project administration, N.A., Z.M.I., S.A.B., F.A.A.J. and S.S.; funding acquisition, N.A., Z.M.I., S.A.B., F.A.A.J. and S.S. All authors have read and agreed to the published version of the manuscript.

Funding: This research was funded by UiTM Kedah Branch under *Geran Dalaman Negeri* (GDN), grant number 600-UiTMKDH (PJI.5/4/1) (June 2020).

Institutional Review Board Statement: Not applicable.

Informed Consent Statement: Not applicable.

Data Availability Statement: Not applicable.

Acknowledgments: The authors are grateful to UiTM Kedah Branch for awarding the *Geran Dalaman Negeri (GDN) Tahun 2020* for the research entitled *"Kajian Perniagaan Online Patuh Syariah dalam kalangan siswapreneur di Institut Pengajian Tinggi Awam Malaysia"* with the project identification number 600-UiTMKDH (PJI.5/4/1) (June 2020).

Conflicts of Interest: The authors declare no conflict of interest.

References

1. Schwab, K. The Fourth Industrial Revolution: What It Means, How to Respond, 2016. World Economic Forum. Available online: https://www.weforum.org/about/the-fourth-industrial-revolution-by-klaus-schwab (accessed on 24 October 2021).
2. Yang, F.; Gu, S. Industry 4.0, a revolution that requires technology and national strategies. *Complex Intell. Syst.* **2021**, *7*, 1311–1325. [CrossRef]
3. Pereira, A.; Romero, F. A review of the meanings and the implications of the industry 4.0 concept. *Procedia Manuf.* **2017**, *13*, 1206–1214. [CrossRef]
4. Rayhan, D.S.A.; Ahmad, F. Possible technological innovations for Industry 4.0 practice in furniture industry of Bangladesh. *Rev. Ind. Eng. Lett.* **2021**, *7*, 1–18. [CrossRef]
5. Abdelmajied, F.Y. Industry 4.0 and Its Implications: Concept, Opportunities, and Future Directions. In *Supply Chain—Recent Advances and New Perspectives in the Industry 4.0 Era*; Bányai, T., Bányai, A., Kaczmar, I., Eds.; Intechopen: London, UK, 2022. [CrossRef]
6. Stouthuysen, K.A. Perspective on "The building of online trust in e-business relationships". *Electron. Commer. Res. Appl.* **2020**, *40*, 100929. [CrossRef]
7. Moore, M. What Is Industry 4.0? *Everything You Need to Know*. 2019. TechRadar. Available online: https://www.techradar.com/sg/news/what-is-industry-40-everything-you-need-to-know (accessed on 5 August 2021).
8. Nayan, S.; Shafie, L.A.; Chulan, M.; Zakaria, F.; Nayan, S. Graduate employability and digital entrepreneurship in the Era of IR 4.0. *J. Intelek* **2021**, *16*, 158–165. [CrossRef]
9. Petrillo, A.; De Felice, F.; Cioffi, R.; Federico Zomparelli, F. Fourth industrial revolution: Current practices, challenges, and opportunities. In *Digital Transformation in Smart Manufacturing*; Petrillo, A., Cioffi, R., De Felice, F., Eds.; IntechOpen: London, UK, 2018; pp. 1–20. [CrossRef]
10. Balakrishnan, B.; Othman, Z.; Ahmad Zaidi, M.F. Review of IR4.0 readiness and adoption in Malaysian manufacturing sector. *Int. J. Bus. Econ.* **2021**, *3*, 24–35.
11. Hamidi, S.R.; Aziz, A.A.; Shuhidan, S.M.; Abdul Aziz, A.; Mokhsin, M. SMEs Maturity Model Assessment of IR4.0 Digital Transformation. In Proceedings of the 7th International Conference on Kansei Engineering and Emotion Research 2018, Kuching, Malaysia, 19–22 March 2018; Lokman, A., Yamanaka, T., Lévy, P., Chen, K., Koyama, S., Eds.; Springer: Singapore, 2018; pp. 721–734.
12. Eabrasu, M. What if? Fine-tuning the expectations of business simulation technology through the lens of philosophical counterfactual analysis. *Organization*, 2021; *Online First*. [CrossRef]
13. Maskuriy, R.; Selamat, A.; Maresova, P.; Krejcar, O.; David, O.O. Industry 4.0 for the construction industry: Review of management perspective. *Economies* **2019**, *7*, 68. [CrossRef]
14. Shan, S.T.; Kee, D.M.H. The readiness of small and medium enterprises for the Industrial Revolution 4.0. *Glob. J. Bus. Soc. Sci. Rev.* **2019**, *7*, 217–223.
15. Oláh, J.; Aburumman, N.; Popp, J.; Khan, M.A.; Haddad, H.; Kitukutha, N. Impact of industry 4.0 on environmental sustainability. *Sustainability* **2020**, *12*, 4674. [CrossRef]
16. Burke, R.; Mussomeli, A.; Laaper, S. *The Smart Factory*; Deloitte University Press: Westlake, TX, USA, 2017.
17. Gregor, M.; Medvecky, S.; Matuszek, J.; Stefanik, A. Digital factory. *J. Autom. Mob. Robot. Intell. Syst.* **2009**, *3*, 123–132.
18. Mohamad, S. *Revolusi Industri 4.0: Cabaran dan Peluang. Perspektif*; Terengganu Strategic & Integrity Institute TSIS: Kuala Terengganu, Malaysia, 2020; Available online: https://www.tsis.my/wpcontent/uploads/2020/03/PERSPEKTIF-Revolusi-Industri-4.0-Cabaran-danPeluang.pdf (accessed on 2 April 2020).
19. Stefanini Group. The Fourth Industrial Revolution: Industry 4.0 Challenges and Opportunities for Your Business. 2021. Available online: https://stefanini.com/en/trends/news/the-fourth-industrial-revolution-industry-4-0-challenges-and-opp (accessed on 28 October 2021).
20. Helbing, D.; Frey, B.S.; Gigerenzer, G.; HAfen, E.; Hagner, M.; Hofstetter, Y.; van den Hoven, J.; Zicari, R.V. Will democracy survive big data and artificial intelligence? In *Towards Digital Enlightenment*; Helbing, D., Ed.; Springer: Cham, Switzerland, 2019; pp. 73–98.
21. Winfield, A.F.; Jirotka, M. Ethical governance is essential to building trust in robotics and artificial intelligence systems. *Philos. Trans. R. Soc. A Math. Phys. Eng. Sci.* **2018**, *376*, 20180085. [CrossRef]
22. Balasingham, K. Industry 4.0: Securing the Future for German Manufacturing Companies. Master's Thesis, University of Twente, Enschede, The Netherlands, 2016. Available online: https://essay.utwente.nl/70665/1/Balasingham_BA_MA.pdf (accessed on 20 October 2021).

23. Shahroom, A.A.; Hussin, N. Industrial Revolution 4.0 and education. *Int. J. Acad. Res. Bus. Soc. Sci.* **2018**, *8*, 314–319. [CrossRef]
24. Kamruzzaman, M.; Hamid, M. The mediating role of knowledge transfer between the relationship of training and accepting new technology in the era of industry revolution 4.0. *IJPSAT Int. J. Progress. Sci. Technol.* **2021**, *24*, 489–496.

proceedings

MDPI

Proceeding Paper

The Comparison of Evaluation on User Experience and Usability of Mobile Banking Applications Using User Experience Questionnaire and System Usability Scale †

Nina Setiyawati [1] and Dwi Hosanna Bangkalang [2,*]

[1] Department of Informatics Engineering, Universitas Kristen Satya Wacana, Salatiga 50715, Central Java, Indonesia
[2] Department of Information System, Universitas Kristen Satya Wacana, Salatiga 50715, Central Java, Indonesia
* Correspondence: dwihosanna.bangkalang@uksw.edu
† Presented at the International Academic Symposium of Social Science 2022, Kota Bharu, Malaysia, 3 July 2022.

Abstract: The digitalization in the banking sector is improving rapidly as shown in the 46.72% increase in the number of digital banking system users in Indonesia, with significant increase in the number of mobile banking users. This enthusiasm should be followed by proper development of user experience (UX) and usability of the applications. User experience is a factor that affects the continuity intention to use an application which later also affects customer satisfaction. Usability also positively affects customer satisfaction. In this study, the UX and usability of four mobile banking applications in Indonesia were measured using the User Experience Questionnaire (UEQ) and System Usability Scale (SUS). The outcomes of the measurement of each application were then presented and compared to analyse the UX and usability of the application. Of four applications, the minimum rating was NEUTRAL, and the usability was regarded as GOOD. These results indicate that users have accepted the usability of the mobile banking applications that they used.

Keywords: user experience; usability; mobile banking; user experience questionnaire; system usability scale

check for updates

Citation: Setiyawati, N.; Bangkalang, D.H. The Comparison of Evaluation on User Experience and Usability of Mobile Banking Applications Using User Experience Questionnaire and System Usability Scale. *Proceedings* **2022**, *82*, 87. https://doi.org/10.3390/proceedings2022082087

Academic Editor: Mohamad Rahimi Mohamad Rosman

Published: 28 September 2022

1. Introduction

The development of digital banking systems in Indonesia is rapidly increasing as both the COVID-19 pandemic and Regulation No.12/POJK.03/2018 issued by the Financial Services Authority on the Implementation of Digital Banking Services by Commercial Banks accelerate it. The regulation encourages banking service efficiency using advancements of IT, allowing banks to provide faster and easier quality service that fits customers' needs [1]. The number of digital bank service users in Indonesia has increased by 46.72% in 2021, with the highest increase found in mobile banking service [2–4]. The increase in the number of users is then grouped into the increases in transaction volume using mobile banking and total downloads of mobile banking applications (acquisition of new users).

This positive trend should be followed by good user experience and usability of banking applications. User experience (UX) deals with user-related factors (users' tendencies, expectations, needs, motivations, and moods), system-design-related factors (complexity, purpose, usability, and functionality) and the context or environment where the interaction takes place [5]. UX covers the overall acceptance of a product or service that includes various aspects, including the pragmatic and hedonic of a product [6] that deals with the ease of use of an application that fits users' feelings and emotions when using the application [7]. Usability is a quality attribute that assesses the ease of the system interface of an application [8]. In a broader perspective, usability is the extent to which a product or system can be used by certain users to achieve certain goals in effective and efficient ways that it makes users satisfied in certain contexts of use [8,9]. Several studies have been carried out to examine how user experience and the usability of a product affect customer satisfaction and/or continuous intention. Lei and Lee (2020) verified that user experience is an important factor influencing users' intention to use mobile game applications continuously [10]. Similarly, Mantala and Firdaus (2016) also found that better user experience was positively associated with higher customer satisfaction [11]. Febrian et al. (2021) also emphasized that customer experience had a positive and significant effect on customer satisfaction [12]. Those studies illustrated the importance of user experience in relation to customer satisfaction and/or continuous intention. The relationship between user experience, customer satisfaction, continuous intention, and customer loyalty is shown in Figure 1.

Figure 1. The relationship among user experience, customer loyalty, and continuous intention.

Usability relates to the ease of use, learning, and user satisfaction which is one of the crucial attributes in designing a product. Alshira'H (2020) explains that usability has a positive role in user satisfaction of e-government websites in Jordan [13]. Besides having a positive effect on user satisfaction, usability also affects user experience in accessing a system [14]. Aspects of usability such as simplicity and ease of use make the human–computer interaction more efficient for the gap between the human and the computer, and the software becomes closer with good interface, making the use of an application more effective.

In this study, we evaluated the UX and usability of several mobile banking applications in Indonesia based on UEQ and SUS metrics. Concerning the massive growth of mobile banking usage and increasing bank transactions, it is necessary to identify users' perceptions of usability and UX flow in mobile banking applications. Users' perceptions can be used as a basis for the development and maintenance of mobile banking applications. In addition, it is important to obtain data on UX and usability can support the projection and direction of customer satisfaction and/or continuous intention.

2. Method

The flow of this study is presented in Figure 2. UEQ and SUS questionnaires were distributed to mobile banking users. There were four mobile banking applications set as the objects of this study. At least 80 users of each banking application participated in this study. Respondents' demographic data include: (1) respondents are users of one of the four mobile banking services; (2) male and female; (3) aged between 20–40 years. The aspects measured in UEQ and SUS questionnaires are shown in Figures 3 and 4.

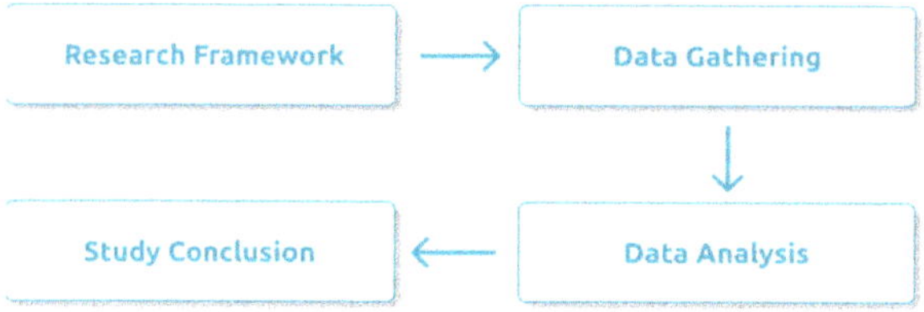

Figure 2. Method of the study.

The User Experience Questionnaire (UEQ) consisted of 26 items measuring user experience, including [15,16]:

1. Attractiveness (users' general perception);
2. Perspicuity which relates to how easy an application is to be used;
3. Efficiency (the amount of time required in completing a task);
4. Dependability (whether users have control over the application);
5. Stimulation (the extent to which users are motivated to use the product;
6. Novelty (innovation in a system, service and product).

Of 6 aspects of UEQ, attractiveness is a pure valence dimension. Meanwhile, Perspicuity, Efficiency and Dependability are pragmatic quality (goal-directed) aspects, and Stimulation and Novelty are aspects of hedonic quality (not goal-directed) [17].

The System Usability Scale (SUS) was employed to measure the usability attributes in mobile banking applications, including effectiveness, efficiency, satisfaction, easiness to learn, easiness to remember and few errors. The SUS questionnaire consisted of 10 items [18,19] containing positive and negative tone items.

In the data-gathering process, the questionnaires were distributed through Google forms, which results were then analysed based on UEQ and SUS metrics. The score of the answers for each UEQ question ranged from 1 to 7 according to the indicators of each aspect. Each SUS question is expressed in a 5-point Likert scale.

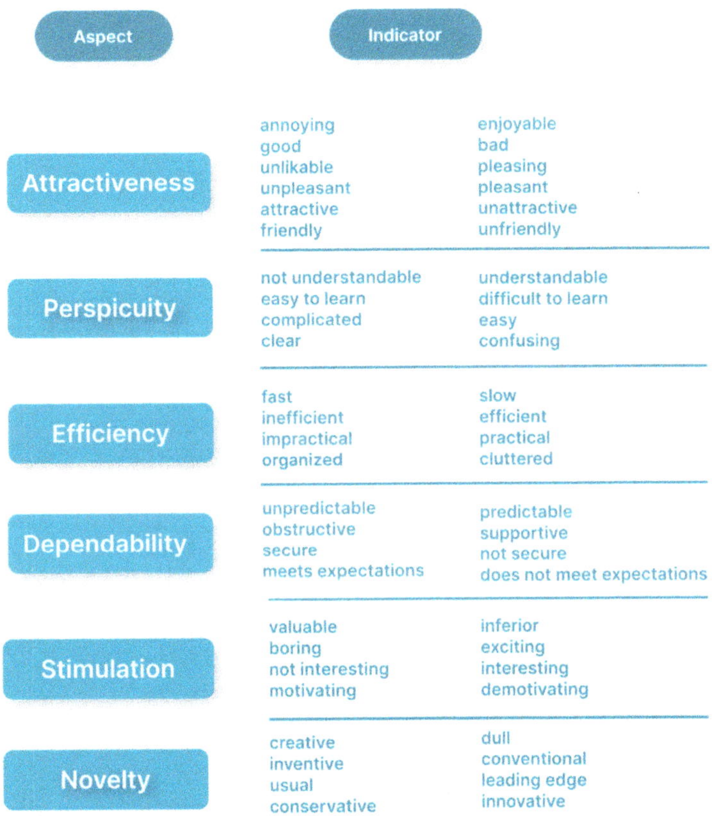

Figure 3. The indicators of user experience questionnaire.

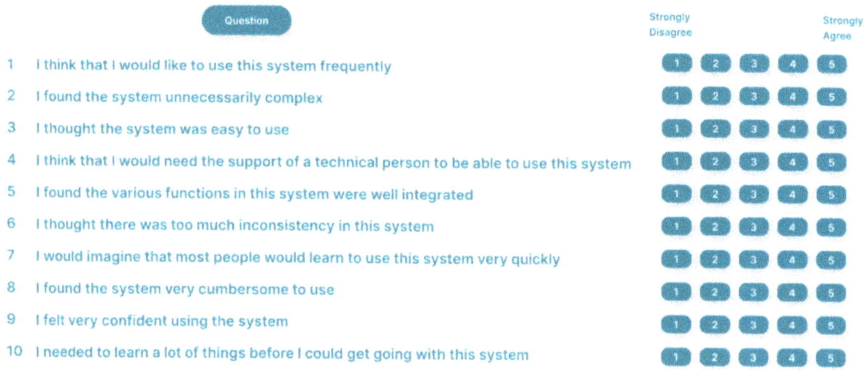

Figure 4. Items of the system usability scale. Adapted from [20].

3. Results and Discussions

Questionnaires were filled out by respondents, and results were processed using UEQ and SUS metrics. Table 1 presents the number of respondents of each mobile banking application.

Table 1. The number of respondents of each mobile banking application.

Mobile Banking Application	Number
BCA Mobile	88
Octo Mobile	80
BNI Mobile	94
Livin	100

Figure 5 shows that all mobile banking applications have slightly different values in functionality and interaction concepts. *t*-Test calculation at Alpha level 0.05 indicates "Significant Differences" in the Perspicuity and Efficiency aspects of BNI Mobile, Livin and Octo Mobile. BNI Mobile's Perspicuity compared with the one of BCA Mobile also shows a "Significant Difference", whereas the results of other comparisons show "No Significant Difference". Table 2 presents the interpretation of the UEQ scale for each mobile banking application, with an average value of −0.8 to 0.8 (neutral), value > 0.8 (positive), and an average value < −0.8 (negative) [17,21,22].

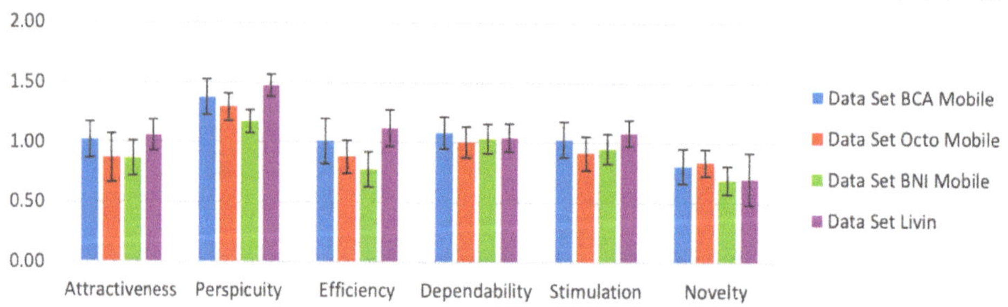

Figure 5. UEQ comparison on the mobile banking application.

Table 2. The interpretation of the mobile banking application UX.

Scale	BCA Mobile	Octo Mobile	BNI Mobile	Livin
Attractiveness	Positive	Positive	Positive	Positive
Perspicuity	Positive	Positive	Positive	Positive
Efficiency	Positive	Positive	Neutral	Positive
Dependability	Positive	Positive	Positive	Positive
Stimulation	Positive	Positive	Positive	Positive
Novelty	Positive	Positive	Neutral	Neutral

Figure 6 presents the usability of mobile banking applications measured using SUS. Raw SUS scores are converted into percentiles which indicate the extent to which mobile banking applications included in this study compare with other sets of data in the database. The average score (at the 50th percentile) is 68, implying that scores greater than 68 are above average and scores lower than 68 are below the average. SUS scores can be interpreted based on several measures, as shown in Figure 7 [23]. Adjective ratings describe the usability of a system using a phrase to make the usability value more meaningful. Acceptability ranges are another indicator as to whether a system is accepted. Conversely, the grade scale ranges from grade A for superior performance to F for failing performance.

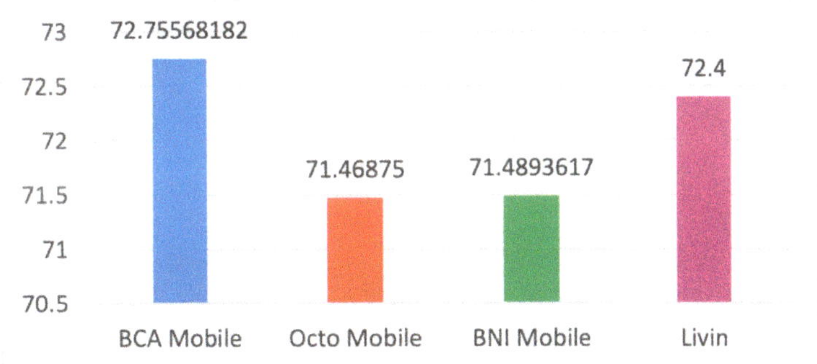

Figure 6. Results of usability measurement using SUS.

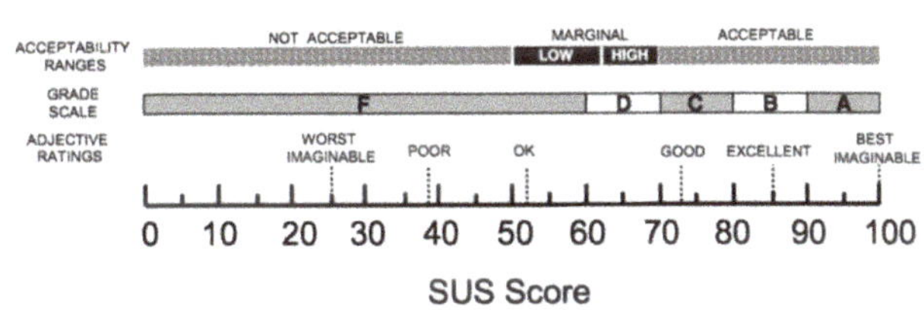

Figure 7. SUS score interpretation.

In the interpretation using the adjective ratings as shown in Figure 7 shows that: (a) the adjective ratings of all mobile applications are GOOD; (b) the grade of the mobile banking applications is C+, except BCA Mobile which obtained B- grade; (*c) all applications are acceptable.

Based on the UX and usability evaluation of mobile banking applications, all applications have a neutral and positive UX design, as well as good usability. Users also accept the usability of the four mobile banking applications.

4. Conclusions

This study measured the UX and usability of four mobile banking applications in Indonesia using UEQ and SUS metrics. The minimum category for every UX aspect of the four mobile banking applications is NEUTRAL with GOOD usability. These results show that the general impression of users on the mobile banking application is positive. In addition, users have "accepted" the usability of the mobile banking applications that they used. BNI Mobile and Livin were rated NEUTRAL in terms of Novelty. Therefore, the Novelty aspect of BNI Mobile and Livin needs to be reviewed and improved for more positive evaluation. The T-Test showed that the Perspicuity and Efficiency aspects of BNI Mobile with Livin and Octo Mobile with Livin have a "Significant Difference". Likewise, BNI Mobile's Perspicuity when compared with that of BCA Mobile is significantly different which can be further improved.

Author Contributions: Conceptualization, N.S. and D.H.B.; methodology, N.S. and D.H.B.; software, N.S.; validation, N.S. and D.H.B.; formal analysis, N.S.; investigation, N.S.; resources, N.S.; data curation, N.S. and D.H.B.; writing—original draft preparation, N.S.; writing—review and editing, N.S. and D.H.B.; visualization, N.S. and D.H.B.; supervision, N.S.; project administration, N.S.; funding acquisition, N.S. All authors have read and agreed to the published version of the manuscript.

Funding: This research received no external funding.

Institutional Review Board Statement: Not applicable.

Informed Consent Statement: Not applicable.

Data Availability Statement: Not applicable.

Acknowledgments: Paper presented during a panel at the 2022 International Academic Symposium of Social Science (IASSC 2022).

Conflicts of Interest: The authors declare no conflict of interest.

References

1. OJK. Digital Finance Innovation Road Map And Action Plan 2020–2024. Available online: http://www.digifin.org/digital-finance-innovation/ (accessed on 24 April 2022).
2. Walfajri, M.; Handojo, B.I. Transaksi Digital Banking Meningkat 46,72% yoy Sampai September 2021. *NewsSetup*. Available online: https://newssetup.kontan.co.id/news/bi-transaksi-digital-banking-meningkat-4672-yoy-sampai-september-2021 (accessed on 24 April 2022).
3. Ariesta, A. Transaksi Mobile dan Internet Banking Melonjak 300 Persen di 2021. *IDX Channel*. Available online: https://www.idxchannel.com/banking/transaksi-mobile-dan-internet-banking-melonjak-300-persen-di-2021 (accessed on 24 April 2022).
4. Bayu, D.J.; Ridhoi, M.A. Layanan Perbankan Digital Makin Sering Digunakan Saat Pandemi. Available online: https://databoks.katadata.co.id/datapublish/2020/11/18/layanan-perbankan-digital-makin-sering-digunakan-saat-pandemi (accessed on 24 April 2022).
5. Berni, A.; Borgianni, Y. From the Definition of User Experience to a Framework to Classify Its Applications. In Proceedings of the Design Society, Gothenburg, Sweden, 16–20 August 2021; Cambridge University Press: Cambridge, UK; Volume 1, pp. 1627–1636. [CrossRef]
6. Yazid, M.A.; Jantan, A.H. User experience design (UXD) of mobile application: An implementation of a case study. *J. Telecommun. Electron. Comput. Eng.* **2017**, *9*, 197–200.
7. Putra, D.O.; Setiawan, A. The Importance of User Experience Analysis in the Design of an Education Information System Application. In Proceedings of the 1st Borobudur International Symposium on Humanities, Economics and Social Sciences (BIS-HESS 2019), Magelang, Indonesia, 30 October 2019; Atlantis Press: Dordrecht, The Netherlands; 2020; Volume 436, pp. 1208–1211. [CrossRef]
8. Weichbroth, P. Usability of Mobile Applications: A Systematic Literature Study. *IEEE Access* **2020**, *8*, 55563–55577. [CrossRef]
9. Ergonomics of Human-System Interaction—Part 11: Usability: Definitions and Concepts. Available online: https://www.iso.org/obp/ui/#iso:std:iso:9241:-11:ed-2:v1:en (accessed on 22 April 2022).
10. Lei, B.; Lee, J. Analysis on continuous usage intention of chinese mobile games from the perspective of experiential marketing and network externality. *J. Inf. Technol. Appl. Manag.* **2020**, *27*, 197–224. [CrossRef]
11. Mantala, R.; Firdaus, M.R. Pengaruh customer experience terhadap customer satisfaction pada pengguna smartphone android (studi pada mahasiswa politeknik negeri banjarmasin). *J. Wawasan Manaj.* **2016**, *4*, 153–164.
12. Febrian, D.; Simanjuntak, M.; Hasanah, N. The Effect of Benefits Offered and Customer Experience on Re-use Intention of Mobile Banking through Customer Satisfaction and Trust. *J. Keuang. Perbank.* **2021**, *25*, 551–569. [CrossRef]
13. Badran, O.; Al-Haddad, S. The impact of software user experience on customer satisfaction. *J. Manag. Inf. Decis. Sci.* **2018**, *21*, 1–20.

14. Mohamad Salleh, M.A.; Salman, A.; Ali, M.N.S.; Hashim, H. The Importance of Usability Features in Enhancing Online Communication Satisfaction. *Malays. J. Commun.* **2016**, *32*, 1–15. [CrossRef]
15. Hartzani, A.G. Evaluasi User Experience Pada Dompet Digital Ovo Menggunakan User Experience Questionnaire (UEQ). Universitas Islam Negeri Syarif Hidayatullah. Available online: https://repository.uinjkt.ac.id/dspace/bitstream/123456789/57729/1/AL%20GHIFARI%20HARTZANI-FST.pdf (accessed on 11 April 2022).
16. Izabal, S.V.; Aknuranda, I.; Az-zahra, H.M. Evaluasi dan Perbaikan User Experience Menggunakan User Experience Questionnaire (UEQ) dan Focus Group Discussion (FGD) pada Situs Web FILKOM Apps Mahasiswa Fakultas Ilmu Komputer Universitas Brawijaya. *J. Pengemb. Teknol. Inf. Ilmu Komput. Univ. Brawijaya* **2018**, *2*, 3224–3232.
17. Schrepp, M.; Hinderks, A.; Thomaschewski, J. Applying the user experience questionnaire (UEQ) in different evaluation scenarios. In Proceedings of the International Conference of Design User Experience, and Usability, Crete, Greece, 22–27 June 2014; Springer: Berlin/Heidelberg, Germany, 2014; pp. 383–392. [CrossRef]
18. digital.gov. System Usability Scale (SUS): Improving Products Since 1986. Available online: https://digital.gov/2014/08/29/system-usability-scale-improving-products-since-1986/ (accessed on 16 May 2022).
19. Sidik, A. Penggunaan system usability scale (SUS) sebagai evaluasi website berita mobile. *J. Ilm. Technol.* **2018**, *9*, 83–88. [CrossRef]
20. Lewis, J.R.; Sauro, J. Item benchmarks for the system usability scale. *J. User Exp.* **2018**, *13*, 158–167.
21. Rauschenberger, M.; Schrepp, M.; Perez-Cota, M.; Olschner, S.; Thomaschewski, J. Efficient Measurement of the User Experience of Interactive Products. How to use the User Experience Questionnaire (UEQ). Example: Spanish Language Version. *Int. J. Interact. Multimed. Artif. Intell.* **2013**, *2*, 39–45. [CrossRef]
22. Hinderks, A.; Winter, D.; Schrepp, M.; Thomaschewski, J. Applicability of user experience and usability questionnaires. *J. Univers. Comput. Sci.* **2019**, *25*, 1717–1735.
23. Bangor, A.; Kortum, P.; Miller, J. Determining what individual SUS scores mean: Adding an adjective rating scale. *J. User Exp.* **2009**, *4*, 114–123.

proceedings

MDPI

Proceeding Paper

Could Lean Practices and the Theory of Inventive Problem Solving (TRIZ) Improve the Entrepreneurial Ecosystem of Small- and Medium-Sized Enterprises? †

Azyyati Anuar [1,*], Mohd Zulfabli Hasan [2], Mohamad Fazali Ghazali [3], Law Kuan Kheng [1], Daing Maruak Sadek [4] and Nurul Fadly Habidin [5]

1 Faculty of Business and Management, UiTM Cawangan Kedah, Merbok 08400, Kedah, Malaysia
2 Politeknik Tunku Syed Sirajuddin Pauh Putra, Arau 02600, Perlis, Malaysia
3 Kulliyyah Muamalat & Sains Pengurusan, Universiti Islam Antarabangsa Sultan Abdul Halim Mu'adzam Shah, Kuala Ketil 09300, Kedah, Malaysia
4 Academy of Contemporary Islamic Studies, UiTM Cawangan Kedah, Merbok 08400, Kedah, Malaysia
5 Fakulti Pengurusan dan Ekonomi, Universiti Pendidikan Sultan Idris, Tanjung Malim 36900, Perak, Malaysia
* Correspondence: azyyati@uitm.edu.my

† Presented at the International Academic Symposium of Social Science 2022, Kota Bharu, Malaysia, 3 July 2022.

Abstract: The concept of a Lean Business Model (LBM) is derived from the Business Model Canvas (BMC), which is primarily focused on well-established businesses. However, in a competitive environment typified by Schumpeterian waves of creative destruction, new businesses in the entrepreneurial ecosystem face the strategic challenge of constantly adapting and evolving alongside well-established companies. This proposes that the threat of newcomer disruption may affect all types of organisations, including small- and medium-sized enterprises (SMEs), particularly small and young businesses. The presence of SMEs is one of the most crucial factors in the growth of Malaysia's economy. Although SMEs contribute significantly to the economy, do they provide a better entrepreneurial ecosystem? This study aims to conceptualise how the integration of lean business and the Theory of Inventive Problem Solving (TRIZ) might improve the entrepreneurial ecosystem in Malaysia. The development of this model will be conducted using a triangulation of qualitative data sources comprising document analysis, interviews, and archival records. The validity of the findings will be examined based on the dimensions of trustworthiness to ensure that data obtained are accurate and reliable. The anticipated results of this study may show the impact of a new hybrid between lean business and the TRIZ model on the Malaysian SMEs entrepreneurial ecosystem. This new model will provide fresh insights into ways to transform the entrepreneurial ecosystem in line with the National Entrepreneurial Policy (NEP) 2030. Furthermore, the SMEs can be strengthened for the purpose of expanding their proportion of Gross Domestic Product (GDP) and exports, not just locally but also globally.

Keywords: entrepreneurial; ecosystem; lean practices; small–medium enterprises; TRIZ

check for updates

Citation: Anuar, A.; Zulfabli Hasan, M.; Fazali Ghazali, M.; Kuan Kheng, L.; Maruak Sadek, D.; Fadly Habidin, N. Could Lean Practices and the Theory of Inventive Problem Solving (TRIZ) Improve the Entrepreneurial Ecosystem of Small- and Medium-Sized Enterprises? *Proceedings* **2022**, *82*, 89. https://doi.org/10.3390/proceedings2022082089

Academic Editor: Mohamad Rahimi Mohamad Rosman

Published: 30 September 2022

1. Introduction

Governments in some nations have focused on creating an entrepreneurial ecosystem to create high-potential entrepreneurs globally. Entrepreneurial ecosystems are distinct networks of interdependent individuals and relationships that, either directly or indirectly, assist in the emergence and expansion of new businesses [1]. The ecosystem's strength lies in its ability to allow many entities to share their knowledge, resources, and learning opportunities [2]. In fact, the introduction of the National Entrepreneurship Policies will intensify the need for necessary skills by enabling entrepreneurs to adapt to changing market conditions and disruptive technologies [3]. Research into how entrepreneurial firms network and what entrepreneurial ecosystems can offer is necessary to comprehend how ecosystems function and reap their potential benefits for businesses [4]. The major challenge for governments around the world, however, is the shift from "traditional economy" to "new economy." Entrepreneurs must, therefore, have "21st-century skills", embrace business digitisation, and collaborate with other entrepreneurs in order to survive in this dynamic environment, particularly in small-to-medium-sized businesses.

Small- and medium-sized enterprises (SMEs) have begun implementing lean practices after realising they are one of the main drivers of global economic growth [5]. Lean management has been adopted by the majority of businesses across many industries to boost operational efficiency. However, a lot of them struggle with implementation due to various obstacles or difficulties, which have resulted in failures that can impede the lean implementation process [6,7]. Some of the obstacles to why the goal of lean practices cannot be reached are due to reasons such as cultural, human, and geographic factors [7]. In particular, decay and a return to the original way of doing business occur in 70% of lean implementations [8].

Although lean implementation efforts help businesses save time and money, working under these lean principles has a negative impact on work quality, since it prevents employees from expressing themselves [9]. A study conducted by [7] found 24 barriers (which are not specific to SMEs) in lean implementation. The main obstacles to the implementation of lean in Indian SMEs, according to [10], include lack of management commitment, leadership, and resources. However, identifying all the barriers to lean management has given a negative impact on employees' job satisfaction. Thus, lean business models (LBMs) are a collection of practical strategies that were developed to address this need. LBMs support business owners in the process of validating their assumptions through market testing and early customer feedback [1]. It was proven in this study that when LBM was applied in digital entrepreneurship, the results demonstrated how a business model canvas (BMC) can benefit from lean principles in a digital context, and how these concepts can be integrated into an original lean framework to experiment with, validate, and then improve a business model.

Nevertheless, to ensure LBM has a strong capability to change the entrepreneurial ecosystem in Malaysian SMEs, the introduction of TRIZ comes into place, which can provide a systematic approach to decision-makers in solving such SMEs problems. In current years, TRIZ has appeared as a valuable tool for inventing and solving technological challenges, including a wide range of complexities [11]. As stated by [12], both lean and TRIZ are aiming to maximise the utilisation of available resources. The purpose of lean is to remove waste because waste indicates inefficiencies and counterproductive behaviours in the system. Indeed, the basic concepts of lean, namely waste reduction, increased value, and improved customer satisfaction, are relevant to SMEs, not only to large enterprises [5]. In TRIZ, the problem-solution frequently makes use of a resource that was previously considered a nuisance or a waste. The idea of TRIZ is based on a substantial study that examined hundreds of thousands of patents in a variety of sectors to uncover broad trends in inventive solutions as well as the distinctive qualities of the issues that these inventions have solved [13].

Similarly, the concept of lean in business, using TRIZ, can indeed help to improve the operational performance in the business setting [12]. However, how far can lean and TRIZ contribute to enhancing Malaysian SMEs entrepreneurial ecosystem? Will it contribute to a more dynamic and effective SME landscape, particularly in light of the recent unprecedented pandemic crisis? In light of the shortcomings of past studies, it appears this study is necessary. Therefore, this research aims to conceptualise how the integration of LBM and TRIZ might improve the entrepreneurial ecosystem in Malaysia. With this hybrid model, it contributes to the novelty of this research, whereby there is a dearth of research investigating LBM, TRIZ, and the entrepreneurial ecosystem simultaneously. This paper begins with descriptions of the entrepreneurial ecosystem, lean business, TRIZ, and using TRIZ in lean, and then continues with methods. Finally, conclusions and an outline of a potential future research agenda for the area are presented.

2. Literature Review

2.1. Entrepreneurial Ecosystem

A stable environment and working conditions are necessary for the growth of an entrepreneurial ecosystem [14]. This calls for a high level of engagement and collaboration readiness among all members of an entrepreneurial ecosystem. Supported by [15], entrepreneurial activities and entrepreneurs develop within a highly integrated and complex system with a variety of on-screen characters to move the entrepreneurial ecosystem. The networks by which small business owners and entrepreneurs interact with other fictional characters have thus been precisely described and explained by the term entrepreneurial ecosystem.

The entrepreneurial ecosystem has long been a topic of interest for researchers from financial geology, finance, business, and other disciplines, many of whom have attempted to justify the reasons and mechanisms by which some areas support more significant startup development than others. Moreover, the phrase, "entrepreneurial ecosystem" has been around for more than 20 years [16], but it started gaining widespread recognition with articles such as "How to Start an Entrepreneurial Revolution" by [17] and "Startup Communities" by [18,19]. Since then, a wide range of organisations and researchers have added to our understanding of entrepreneurial ecosystems and the components that make them up. An entrepreneurial ecosystem, according to [20], is made up of three elements: opportunities, talented people, and assets. [17] has broaden this definition by including governance, culture, capital markets, and clients with an open mind.

In his subsequent work, he has further discovered that the entrepreneurial ecosystem is "a set of interconnected components that cultivate entrepreneurial development such as leadership, culture, capital, markets, human aptitudes, and bolster" [21]. Utilising the five Cs: capital (financial asset), capability (business visionary and proprietor skillset), connection (resource and relationship network), culture (local communities' recognition and support of business enterprise), and climate (administrative, financial improvement, and approach environment), the Center for Country Business enterprise completes an environment [22]. The proximity of large businesses, colleges, and advantageous suppliers is another factor that is thought to enhance an entrepreneurial ecosystem [23]. The presence or absence of these elements, as well as how well they work together, distinguishes different environments from one another and could have an impact on how robust an entrepreneurial ecosystem can be [24].

2.2. Lean Manufacturing (LM)

Krafcik, a researcher with the MIT International Motor Vehicle Program, coined the term "lean manufacturing", after studying various international automotive practices [25]. In his point of interest paper, Krafcik has presented the term "lean" to depict a generation framework that employs fewer assets of the whole thing compared to mass generation [26]. In any case, numerous researchers characterised lean in unexpected ways. Subsequently, [27] examined lean perspectives and summarised the terminology that had been used to describe lean. Based on the survey findings, lean has been described as a method, a plan, a collection of standards, a group of instruments and strategies, a method, a conception, a way of thinking, a practice, a framework, a plan, a fabricating worldview, and a demonstration [26].

To produce leading conceivable value and diminish non-value, organisations need to include exercises and producers connected to distinctive lean manufacturing (LM) standards, tools, and methods. In any case, numerous organisations discover it is challenging and troublesome to do that effectively. Within the setting of SMEs, LM activities posit advanced challenges. The concept of LM has been broadly connected to benefit the industry as a whole for a long time. The central thought of LM is to construct a shared belief and special way of working that exceedingly guarantee responsiveness to customers' requests while continually cutting fetches and disposing of squanders all through the organisation [27]. Furthermore, ref. [28] have acknowledged that SMEs are still hesitant to use LM despite the potential benefits and tangible results. Many companies are concerned that implementing LM will waste time and money.

Some researchers claim that large companies are more likely than SMEs to implement complete LM programmes [29–31]. In addition, ref. [28] also have recognised the fact that SMEs are still unsure of the value of LM execution and the tangibleness of the outcomes and benefits they might realise. The majority of these businesses worry that adopting LM will cost them time and money. According to [32], the successful selection of LM ventures in SMEs necessitates prior associational stability and affirmation of economics, human resources, courses, special activities, crucial arranging, and compensation mechanisms. [33] has asserted that SMEs should begin using less expensive and fundamental LM tools such as 5S, Kaizen, and visual control. Once that is accomplished, they can move on to more unconventional tools such as Kanban and small part sizes. He made it clear that the workforce practices and human mentality are what influence LM practices the most. In this way, the SMEs must start by increasing their staff members' awareness of the needs and nature of LM.

Several earlier exploration studies have shown how lean practices affect business performance. According to [34–36], lean emphasises throughput development, waste lessening, and resource proficiency, which improves business performance through cost reduction. Waste reduction leads to increased productivity and lowers costs for the business, which raises returns on assets [37]. A company's quality reputation can also increase market acceptance and turnover [38]. To support a business' cost-effectiveness and sustainable growth, total quality management (TQM), Kaizen, just-in-time (JIT), and other lean practices are implemented [34,37]. One TQM practice that has an indirect relationship with financial performance in SMEs is lean implementation [39]. In a previous study, ref. [40] looked at the effects of lean practices on the financial, market, and operational performance of manufacturing SMEs. The results supported the notion that combined lean practices improve the performance of SMEs as a whole.

2.3. *Lean Business*

Similar to LM, the lean business concept develops business procedures that produce more value for clients while speeding up performance, reducing waste, and balancing material, informational, and monetary flows. Lean business, in its most basic definition, is a company that maximises value while reducing waste. Although it takes a lot of effort and time to complete, the outcome is a certain success. Moreover, lean startup techniques are receiving more and more attention in the entrepreneurial communities [41,42]. Lean startup techniques have also been incorporated into some entrepreneurial programmes [43]. Conceptual essays have explained how lean startup fits into recent and past academic debates [44].

Additionally, it is evident that effective lean implementation is complex and multifaceted, incorporating a variety of elements that have an impact on success, including the implementation strategies employed and the tools and techniques utilised, as well as the organisational support and managerial skills observed [45]. Lean is a very broad concept that can be applied everywhere. However, it still rests on five guiding principles: value, value stream, flow, pull, and perfection [46]. It is crucial to note that this study did not focus solely on the effects of lean in organisations. However, its goal is to imagine how LBM and TRIZ incorporated might enhance Malaysia's entrepreneurial ecosystem.

2.4. *TRIZ Approach*

One of the most effective invention methodologies is TRIZ (Theory for Inventive Problem Solving), a method with a scientific foundation and empirical underpinnings that was initially developed for the analysis of the global patent database [47]. TRIZ (pronounced "trees"), is a collection of problem-solving methods. TRIZ is based on several universally applicable innovation principles and techniques. The TRIZ method has been used for decades by large multinationals such as Hewlett-Packard, Boeing, and Samsung to develop new products, optimise processes, and better understand market developments and trends [48]. TRIZ has developed into a term that encompasses a wide range of creative ideas, methods, and tools that are frequently applied to the resolution of challenging issues. According to [49], the development of TRIZ was prompted by the failure of alternative approaches that were already in use to provide suitable answers for designing and creating effective technological functionalities of both small- and large-scale magnitudes.

In the 1950s, Genrich Altshuller, a Soviet naval patent clerk, created the foundation for TRIZ. Altshuller realised as he organised the patents that there were only so many ways that problems could be solved, and problems in various industries had been solved in similar ways. After sifting through countless patents, he discovered that each of the inventions he saw in these patents could be explained by one (or more) of 40 principles. The contradiction that had existed in the technology's development was also resolved by each of these principles, he observed (i.e., an item has to be strong but light). A total of 38 potential attributes that might result in paradoxes were revealed by Altshuller. He created a method for solving problems based on these realisations [12].

2.5. *Using TRIZ in Lean*

Lean can benefit from the addition of TRIZ. In the context of lean, TRIZ can be used to identify solutions to problems that might not otherwise be discovered [12]. Finding solutions that make use of resources that are already on hand but might otherwise be considered waste (or "muda" in lean) is a strong suit of TRIZ. Furthermore, by examining a specific process and its function within a larger system and figuring out how it can be used ideally for both to be balanced, TRIZ's ideal final result could aid in the development of future state maps. Finally, TRIZ's problem-defining techniques, such as removing jargons, could be very helpful in developing the current state map, because there are many aspects of any problem situation that have not been sufficiently explored in terms of context and definition. The "9-Window" exercise from TRIZ, for instance, could help put a company's position in the market into context and show others how crucial it is to collaborate with the "super system" of distributors and the "subsystem" of suppliers. Additionally, this exercise aids in outlining the company's past and future. This background information would be very helpful in creating a current state map because it could provide a quick overview of the company to those who are unfamiliar with it [12]. To strengthen the entrepreneurial ecosystem in Malaysian SMEs, this research tries to incorporate the lean business model and TRIZ approach.

3. Methodology

The present research aims to develop a lean business model using the TRIZ approach with the purpose to improve the entrepreneurial ecosystem in Malaysian SMEs. The study will use a qualitative approach (subjectivism) that can address the research questions to meet its goal. It was decided that using an interpretive description of a phenomenon and subjective nature of reality will be the best approach. This study will be conducted using the case study method by [50], to better understand the nature of a qualitative approach. The research design for this study is anticipated to be conducted in a number of ways, including: (1) case study protocol; (2) data collection method; (3) determining the unit of analysis and sampling method; (4) qualitative data analysis; and (5) validity and reliability.

This research will use purposive sampling to obtain all information for the entrepreneurial ecosystem from a group of experts who are working in various SMEs, specifically those who are practicing lean and TRIZ in the organisations. Indeed, they are most readily available and are in the best position to provide the information required. This sampling technique is most appropriate when the topic entails the use of small samples and the subjects are best selected according to the judgment of the researchers [51]. Data will be collected from six companies located in the northern region of Peninsular Malaysia, which includes Kedah and Penang. There are two companies from Kedah state, namely the Al Haddad Mfg. Sdn Bhd and Ideal Healthcare Sdn. Bhd. Another four companies are located in Penang, namely the CG Global

Profastex Manufacturing Sdn Bhd, Herbagus Sdn Bhd, Dee Noon Corporation Sdn Bhd, and Puncak Bumi Utama. Thus, the researcher will gather information from well-known people who have extensive business experience in SMEs. A completely different circumstance pertains to the case study's data collection. The researcher needs to accommodate the interviewees' schedule and availability when interviewing key informants. The nature of the interview is much more semi-structured, and the subject may not always cooperate by answering the interviewer's questions.

The participants involved during the interview session will consist of SMEs entrepreneurs and members of TRIZ organisations and lean units. The interview will be conducted face-to-face, where the interviewer will seek out views, ideas, and opinions on the principles of and understanding of the notion of this principle from the business owners of SMEs. The face-to-face interview is notionally effective to accumulate the information and also to upsurge the collaboration between respondents and interviewers [52]. Furthermore, it has long been asserted that face-to-face interviews offer a stronger foundation for developing a strong rapport between the researcher and the participant.

4. Conclusions and Policy Implications

This study aimed to conceptualise how the integration of LBM and TRIZ might improve the entrepreneurial ecosystem in Malaysia. It is anticipated that a company that adopts both lean and TRIZ will reap the greatest benefits, increase the size of the market, and strengthen its overall competitiveness and sustainability for entrepreneurial ecosystems in the future. Indeed, policymakers, government, and corporate sectors need to play their role to ensure that these are in line with global needs, technology, and developments. One of the essential elements of the entrepreneurial ecosystem is policy, which directs the establishment and implementation of research institutions, regulatory mechanisms, and laws to set the direction and vision of entrepreneurship in the nation. In this regard, effective mechanisms should be pioneered to form cooperation between ministries, agencies, and enforcement bodies to make room for the trial of new business models to be in line with the development of today's Internet and communication technologies as well as the evolution of new commerce globally. Future research will look empirically at how the integration of LBM and TRIZ can improve the entrepreneurial ecosystem in various sectors such as education, food, construction, and others. Moreover, it is suggested to conduct this research quantitatively by focusing on the respondents from multi-national companies in Malaysia, or it can be other countries as well.

Author Contributions: Conceptualisation, A.A.; methodology, L.K.K.; writing—original draft preparation, A.A., M.Z.H., M.F.G., L.K.K., and D.M.S.; writing—review and editing, A.A. and N.F.H.; funding acquisition, A.A. All authors have read and agreed to the published version of the manuscript.

Funding: This research was funded by UITM KEDAH BRANCH, grant number 600-UITMKDH (PJI.5/4/1(2/2021)).

Institutional Review Board Statement: Not Applicable.

Informed Consent Statement: Not applicable.

Data Availability Statement: Not applicable.

Acknowledgments: The researchers would like to acknowledge the UiTM Kedah branch for the financial funding of this research through Geran Dana Penyelidikan Khas (DPK).

Conflicts of Interest: The authors declare no conflict of interest.

References

1. Balocco, R.; Cavallo, A.; Ghezzi, A.; Berbegal-Mirabent, J. Lean business models change process in digital entrepreneurship. *Bus. Process Manag. J.* **2019**, *25*, 1520–1542. [CrossRef]
2. Muldoon, J.; Bauman, A.; Lucy, C. Entrepreneurial ecosystem: Do you trust or distrust? *J. Enterprising Communities People Places Glob. Econ.* **2018**, *12*, 158–177. [CrossRef]

3. National Entrepreneurship Policy-Prime Minister's Office of Malaysia [Internet]. Gov.my. Available online: https://www.pmo.gov.my/2019/08/national-entrepreneurship-policy/ (accessed on 30 June 2022).
4. Gueguen, G.; Delanoë-Gueguen, S.; Lechner, C. Start-ups in entrepreneurial ecosystems: The role of relational capacity. *Manag. Decis.* **2021**, *59*, 115–135. [CrossRef]
5. Yadav, V.; Jain, R.; Mittal, M.L.; Panwar, A.; Lyons, A. The impact of lean practices on the operational performance of SMEs in India. *Ind. Manag. Data Syst.* **2019**, *119*, 317–330. [CrossRef]
6. Chaple, A.P.; Narkhede, B.E.; Akarte, M.M.; Raut, R. Modeling the lean barriers for successful lean implementation: TISM approach. *Int. J. Lean Six Sigma* **2018**, *12*, 98–119. [CrossRef]
7. Jadhav, J.R.; Mantha, S.S.; Rane, S.B. Exploring barriers in lean implementation. *Int. J. Lean Six Sigma* **2014**, *5*, 122–148. [CrossRef]
8. Schipper, T.; Swets, M. *Innovative Lean Development: How To Create, Implement And Maintain A Learning Culture Using Fast*; CRC Press: Boca Raton, FL, USA, 2010.
9. Minh, K.S.; Zailani, S.; Iranmanesh, M.; Heidari, S. Do lean manufacturing practices have negative impact on job satisfaction? *Int. J. Lean Six Sigma* **2018**, *5*, 122–148. [CrossRef]
10. Yadav, V.; Jain, R.; Mittal, M.L.; Panwar, A.; Sharma, M.K. An appraisal on barriers to implement lean in SMEs. *J. Manuf. Technol. Manag.* **2019**, *30*, 195–212. [CrossRef]
11. Jafari, M.; Akhavan, P.; Zarghami, H.R.; Asgari, N. Exploring the effectiveness of inventive principles of TRIZ on developing researchers' innovative capabilities: A case study in an innovative research center. *J. Manuf. Technol. Manag.* **2013**, *24*, 747–767. [CrossRef]
12. Bligh, A. *The Overlap Between TRIZ And Lean*; University of Rhode Island: South Kingstown, RI, USA, 2006.
13. Altshuller, G.; Al'tov, G.; Altov, H. *And Suddenly The Inventor Appeared: TRIZ, The Theory of Inventive Problem Solving*; Technical Innovation Center, Inc.: Worcester, MA, USA, 1996.
14. Allaoui, A.; Benmoussa, R. Employees' attitudes toward change with Lean Higher Education in Moroccan public universities. *J. Organ Chang. Manag.* **2020**, *33*, 253–288. [CrossRef]
15. Cowell, M.; Lyon-Hill, S.; Tate, S. It takes all kinds: Understanding diverse entrepreneurial ecosystems. *J. Enterprising Communities: People Places Glob. Econ.* **2018**, *12*, 198. [CrossRef]
16. Dubin, P. The influence of motivations and environment on business start-ups: Some hints for public policies. *J. Bus. Ventur.* **1989**, *4*, 11–26. [CrossRef]
17. Isenberg, D.J. How to start an entrepreneurial revolution. *Harv. Bus. Rev.* **2010**, *88*, 40–50.
18. Feld, B. *Startup Communities. Building an Entrepreneurial Ecosystem in Your City*; John Wiley & Sons: Hoboken, NJ, USA, 2012; Volume 50, pp. 1285–1307.
19. Stam, F.C.; Spigel, B. *Entrepreneurial Ecosystems*; USE Discussion Paper Series; USE: Utrecht, The Netherlands, 2016; p. 16.
20. Ahmad, N.; Hoffmann, A. *A Framework for Addressing and Measuring Entrepreneurship*; Entrepreneurship Indicators Steering Group: Paris, France, 2007.
21. Atiase, V.Y.; Mahmood, S.; Wang, Y.; Botchie, D. Developing entrepreneurship in Africa: Investigating critical resource challenges. *J. Small Bus. Enterp. Dev.* **2017**, *25*, 644–666. [CrossRef]
22. Macke, D.W.; Markley, D.M.; Fulwider, J.M. *Energizing Entrepreneurial Communities: A Pathway to Prosperity*; Center for Rural Entrepreneurship: Kearney, NE, USA, 2014.
23. Gertler, M.S. Rules of the game: The place of institutions in regional economic change. *Reg. Stud.* **2010**, *44*, 1–5. [CrossRef]
24. Colin, N. What Makes an Entrepreneurial Ecosystem. The Family Papers# 000 13, Out. 2015. Available online: https://salon.thefamily.co/what-makes-an-entrepreneurial-ecosystem-815f4e049804 (accessed on 30 June 2022).
25. Krafcik, J.F. Triumph of the lean production system. *Sloan Manag. Rev.* **1988**, *30*, 41–52.
26. Hasan, M.Z.; Saad, M.N.; Iteng, R. Ethical Climates as a moderator on the Relationship between Lean Manufacturing Practices and Manufacturing Performance using PLS-SEM. *Int. Res. J. Eng. Technol.* **2018**, *6*, 6740–6750.
27. Bhamu, J.; Sangwan, K.S. Lean manufacturing: Literature review and research issues. *Int. J. Oper. Prod. Manag.* **2014**, *4*, 876–940. [CrossRef]
28. Achanga, P.; Shehab, E.; Roy, R.; Nelder, G. Critical success factors for lean implementation within SMEs. *J. Manuf. Technol. Manag.* **2006**, *17*, 460–471. [CrossRef]
29. Bakås, O.; Govaert, T.; Van Landeghem, H. Challenges and success factors for implementation of lean manufacturing in European SMES. In Proceedings of the 13th International conference on the Modern Information Technology in the Innovation Processes of the Industrial Enterprise (MITIP 2011), Trondheim, Norway, 22–24 June 2011; Volume 1.
30. Rose, A.M.; Md Deros, B.; Nordin, N. Lean manufacturing best practices in SMEs. In Proceedings of the 2011 International Conference on Industrial Engineering and Operations Management, Kuala Lumpur, Malaysia, 22–24 January 2011.
31. Shah, R.; Ward, P.T. Lean manufacturing: Context, practice bundles, and performance. *J. Oper. Manag.* **2003**, *21*, 129–149. [CrossRef]
32. Hayes, B.J. Assessing for Lean Six Sigma Implementation and Success. 2000. Available online: https://www.isixsigma.com/implementation/basics/assessing-lean-six-sigma-implementation-and-success/ (accessed on 23 June 2022).
33. Nguyen, D.M. A new application model of lean management in small and medium sized enterprises. *Int. J. Simul. Model* **2015**, *14*, 289–298. [CrossRef]

34. Martínez-Jurado, P.J.; Moyano-Fuentes, J. Lean management, supply chain management and sustainability: A literature review. *J. Clean. Prod.* **2014**, *85*, 134–150. [CrossRef]
35. Khanchanapong, T.; Prajogo, D.; Sohal, A.S.; Cooper, B.K.; Yeung, A.C.; Cheng, T.C. The unique and complementary effects of manufacturing technologies and lean practices on manufacturing operational performance. *Int. J. Prod. Econ.* **2014**, *153*, 191–203. [CrossRef]
36. Bortolotti, T.; Danese, P.; Flynn, B.B.; Romano, P. Leveraging fitness and lean bundles to build the cumulative performance sand cone model. *Int. J. Prod. Econ.* **2015**, *162*, 227–241. [CrossRef]
37. Yang, M.G.; Hong, P.; Modi, S.B. Impact of lean manufacturing and environmental management on business performance: An empirical study of manufacturing firms. *Int. J. Prod. Econ.* **2011**, *129*, 251–261. [CrossRef]
38. Kaynak, H. The relationship between total quality management practices and their effects on firm performance. *J. Oper. Manag.* **2003**, *21*, 405–435. [CrossRef]
39. Sajan, M.; Shalij, P.; Ramesh, A. Lean manufacturing practices in Indian manufacturing SMEs and their effect on sustainability performance. *J. Manuf. Technol. Manag.* **2017**, *28*, 772–793.
40. Valente, C.M.; Sousa, P.S.A.; Moreira, M.R.A. Assessment of the lean effect on business performance: The case of manufacturing SMEs. *J. Manuf. Technol. Manag.* **2019**, *31*, 501–523. [CrossRef]
41. Welter, C.; Scrimpshire, A.; Tolonen, D.; Obrimah, E. The road to entrepreneurial success: Business plans, lean startup, or both? *N. Engl. J. Entrep.* **2021**, *24*, 21–42. [CrossRef]
42. Ries, E. *The Lean Startup: How Today's Entrepreneurs Use Continuous Innovation to Create Radically Successful Businesses*; Crown Business: New York, NY, USA, 2011.
43. Bliemel, M.J. Getting entrepreneurship education out of the classroom and into students' heads. *Entrep. Res. J.* **2013**, *4*, 237–260. [CrossRef]
44. Levinthal, D.; Contigiani, A. Situating the construct of lean startup: Adjacent 'conversations' and possible future directions. *Ind. Corp. Chang.* **2019**, *28*, 551–564. [CrossRef]
45. Connor, D.O.; Cormican, K. Leading from the middle: How team leaders implement lean success factors. *Int. J. Lean Six Sigma* **2021**, *13*, 253–275. [CrossRef]
46. Womack, J.P.; Jones, D.T. Lean thinking-banish waste and create wealth in your corporation. *J. Oper. Res. Soc.* **1997**, *48*, 1148. [CrossRef]
47. Ikovenko, S.; Bradle, J. TRIZ as a lean thinking tool. In Proceedings of the 4th TRIZ Future Conference, Florence, Italy, 3–5 November 2004.
48. Mulder, P. TRIZ Method by Genrich Altshuller [Internet]. Toolshero. 2022. Available online: https://www.toolshero.com/problem-solving/triz-method/ (accessed on 23 June 2022).
49. Mazur, G. Theory of Inventive Problem-Solving (TRIZ). 1995. Available online: http://www.mazur.net/triz/ (accessed on 23 June 2022).
50. Yin, R.K. *Case Study Research and Applications: Design and Methods*; Sage: Los Angeles, CA, USA, 2018.
51. Saunders, M.; Lewis, P.H.; Thornhill, A.D. *Business Students*, 4th ed.; Pearson Education Limited: London, UK, 2007.
52. Sekaran, U.; Bougie, R. *A Skill-Building Approach*; John Wiley & Sons, Inc.: New York, NY, USA, 2013.

proceedings

MDPI

Proceeding Paper

The Organizational Justice and Organizational Reputation Attracting Digital Natives with High Self-Esteem †

Hasnun Anip Bustaman [1,*], Azni Zarina Binti Taha [2], Mohammad Nazri Bin Mohd Nor [2], Muhammad Zia Aslam [2] and Mohammed Mustafa Mohammed Yousif [3]

1 Faculty of Business Management, MARA University of Technology, Bukit Ilmu, Machang 18500, Malaysia
2 Faculty of Business and Accountancy, Department of Business Strategy and Policy, University of Malaya, Kuala Lumpur 50603, Malaysia
3 College of Business Studies, Sudan University of Science and Technology, Khartoum P.O. Box 407, Sudan
* Correspondence: hasnu380@uitm.edu.my
† Presented at the International Academic Symposium of Social Science 2022, Kota Bharu, Malaysia, 3 July 2022.

Abstract: The present study seeks an understanding of digital natives' preference to join an organization based on the factors of organizational justice, organizational reputation and self-esteem. The study conceptualizes organizational reputation as a mediator between organizational justice and digital native attraction, while self-esteem is a moderator. Three-hundred and twenty-seven digital natives undergoing internships at reputable organizations comprise the selected sample for the present study. The result shows that organizational reputation mediated the relationship between organizational justice and digital native attraction, and it is stronger for digital natives with high self-esteem. The findings imply that organizations have no other choices but to improve their organizational reputation through organizational justice, which will attract digital natives with high self-esteem, while the government and related parties cooperate with each other to develop digital native self-esteem to a high level, which in turn will attract digital natives to apply for the job.

Keywords: digital native attraction; third-party organizational justice; organizational reputation; self-esteem and recruitment

Citation: Bustaman, H.A.; Taha, A.Z.B.; Nor, M.N.B.M.; Aslam, M.Z.; Yousif, M.M.M. The Organizational Justice and Organizational Reputation Attracting Digital Natives with High Self-Esteem. *Proceedings* **2022**, *82*, 112. https://doi.org/10.3390/proceedings2022082112

Academic Editor: Mohamad Rahimi Mohamad Rosman

Published: 27 October 2022

1. Introduction

Digitalization is part of our socialization, religiosity, modes of exercise, and work. The avoidance of digitalization in our lives is not an option anymore [1]. Moreover, the digitalization wave is set to expand even faster than ever imagined, ultimately becoming a new global trend [2]. Elon Musk, CEO of Tesla (a globally leading company in digitalization), has warned that the world should be ready to face frightening outcomes from the next generation of artificial intelligence [3], and the drive to bring digitalization to the next level. In today's world, the development of the younger generation is greatly affected by digitalization. Young people are known as some of the heaviest users of technology. Technology has a great influence, and they are constantly surrounded by it. In other words, they are the generation who rely most heavily on technology for learning, communicating and entertainment, and so scholars have referred to them as digital natives [4].

Recent studies have shown that digital natives as organizational outsiders pay attention to organizational justice; for instance, the way hotel staff are treated by management has a significant effect on the guests' feedback to the organization. For examples, Ref. [5] found that the way employers manage their employees in a hotel significantly predicts customers' responses to the company. The presence of social media makes organizations more vulnerable to outsiders' responses [6]. One study discovered that the organizational justice gossip that spreads beyond the company's control, through social media, can reach customers and affect outsiders' responses [7]. In a nutshell, organizational justice, as a universal concept, is a significant organizational characteristic to outsiders; thus, the concept implies there will be influences on digital natives, who are surrounded by social media all the time, in choosing an organization.

As for Malaysian digital natives, most of them rate a company's reputation as the main criterion when making a decision [8]. Digital natives should be concerned about a company's reputation, because it signifies the "bonding signals" that transmit information about an organization's positive brand image in the absence complete information [9]. This means that digital natives with limited knowledge of an organization will make decisions to apply and accept job offers based on the reputation of a company, because it can spread a positive value to them. The positive value associated to a company will be shared through social media, emotionally attracting digital natives to apply for jobs with them. In another study, employer reputation was found to be more important than the digital native's employer knowledge in attracting them to apply for a job at an organization [10,11]. For example, an organization's good reputation will affect digital natives, because it makes them perceive the company as a strong potential employer. Ref. [12] mentioned that if peers rated an organization positively, then this perception will act to strongly attract an individual to an organization. Therefore, we assume that an organization's reputation does have a significant impact on digital native attraction.

Finally, the present study seeks to understand how self-esteem plays a role in the relationship between organizational reputation and digital natives' attraction. Long ago, [13] found that an organization with positive identity will attract digital natives who having low self-esteem. Ref. [12] acknowledged the finding, and translated it into a proposition. The proposition articulates the relationship between organizational reputation and digital native attraction, moderated by self-esteem. The scholars subdivided the organizational reputation characteristics into such categories as reward system, the size of the company, the management system (whether centralized or not), and the organization's location; thus, they did not study organizational reputation per se [14]. This relationship has never been tested explicitly, but manipulative factors have been used to generate propositions. Therefore, building from these arguments, the present study proposes to study how self-esteem intercepts with organizational reputation and digital native attraction using a moderated mediation model.

2. Literature Review

2.1. Digital Natives Attraction Related to Organizational Justice

A digital native's attraction refers to the digital native's overall evaluation of the organization's attractiveness [15]. In other words, it is a concept that helps us understand why digital natives are attracted to a company [16]. Typically, scholars use the definition of the perceptions of digital natives set out in [17]. Other scholars state that the digital native's attitude or general positive emotion towards an organization is the meaning of this attraction [18]. Later, researchers described digital natives' attraction as reflected in individuals' affective and attitudinal thoughts about companies as potential places for employment [19,20]. All in all, digital native attraction can be understood as digital natives' perception of a company as an attractive place to work, or as a reference to digital native's fascination in pursuing employment with an organization.

In studies on digital natives' attraction, researchers outlined the predictors as "job and organizational characteristics, recruiter characteristics, perceptions of the recruitment process, perceived fit, perceived alternatives and hiring expectancies" [15,21]. Over the past 50 years, digital natives' attraction has been studied extensively, and so in our study, reviewing 71 reputable papers is the first step in listing the predictors. The findings show that digital native attraction should be understood through the organizational working environment.

There are varieties of work environment that have been tested; thus, it is essential that a study be focused. This study focuses on the one of the more renowned aspects of a work environment, which is organizational justice, mainly because the subject has recently received more attention from scholars following the finding that it is related to third-party individuals' perceptions [22]. Previously, studies on organizational justice have been aimed at understanding those who are involved with the justice situation, but rather being exclusively focused on this, they have extended beyond it to understand how the concept affects third-party individuals' perception. A recent meta-analytical review released has shown the relevance of and scholarly interest in the concept [23]. Therefore, the organizational justice of the work environment has become an interest of the present study. This interest has been channeled to investigate relevant theory to better understand the concept and to help develop new guidelines.

A theory related to discussions of third-party perception that is lacking complete information is signaling theory. The theory suggests that the organizational work environment is the initial information that is given to describe the situation of working with the organization. For this reason, the organizational work environment is understood to signal information about the working conditions in that organization [24], as understood via signaling theory. For example, the organizational working environment, such as its romance policy [25], and its approach to flexitime, flexplace [26] and corporate social performance [24], provide information about the organization's working conditions. Likewise, organizational justice represents an organizational working environment by signaling information about the organizational working conditions. The organizational working environment signifies certain organizational moral values and norms [23]. As [27] noted, digital natives would prefer to join a company they perceive as having positive values, moral and norms. The present study considers that organizational justice signals certain positive values, moral and norms that will affect the digital native's attraction to apply for a job. Hence, informed by signaling theory, the present study predicts that organizational justice is an aspect of the organizational working environment that is significant to attracting digital natives to apply job at that company. Thus, the hypothesis has been developed:

Hypothesis 1 (H1). *Perceived organizational justice is related to digital native attraction.*

2.2. *The Role of Organizational Reputation in Attaracting Digital Native*

Over the last two decades, organizational reputation has been frequently associated with digital native attractiveness [27]. In facts, the concept is the strongest predictor of digital natives attraction [28]. Moreover, the organizational reputation, defined by scholars as a universal, has appeared in many studies across time and countries [11,29–31]. However, researchers' interest in the organizational reputation as associated with digital native attraction is far from being exhausted, and in some places has massively increased due to the digitalization organizations. In some studies (on, e.g., Corporate Social Responsibility, pro-environment, flexible working hours and others) related to organizational reputation, the concept has received a very positive response from researchers [32,33].

Previously, organizational reputation has been directly related to digital natives attraction [34]. As time goes by, some organizational characteristics have been associated with organizational reputation [33] in understanding the attractiveness of a job. In other words, organizational reputation has become a mediator between organizational characteristics and digital native attraction. For example, organizational characteristics such as organizational pro-environmentalism [35], corporate social performance [33] and media richness [36] have been associated with organizational reputation in understanding digital native attraction towards a job. According to [14], a study would contribute a better understanding by providing a mediator in the direct relationship.

The present study found that organizational justice relates to organizational reputation [23], while reputation has a strong influence on digital native attraction. In other words, organizational justice associates to organizational reputation, then the relationship anticipates an affect on digital native attraction. Thus, the present study anticipates that the relationship between organizational justice and digital native attraction is mediated by organizational reputation [37–39]. Scholars have drawn out the mediator effect of organizational reputation, but have not examined the particular mechanisms, e.g., [32,33,35,40]. Thus, the hypothesis is developed:

Hypothesis 2 (H2). *The relationship between organizational justice and digital native attraction is mediated by organizational reputation.*

2.3. Self-Esteem and Digital Native Attraction

Organizational reputation as related to digital natives underlies the individual self-concept. Particularly, social identity theory states that people generally endeavour for a positive self-concept, and that can be generated from organizational membership [41]. In other words, a person may have a huge desire to associate themselves with a positively perceived organization, because the association will help them to derive a positive self-concept, as underlying social identity theory.

Studies have discovered that positive self-concept is something that people strive for, e.g., [11,33,35,36]; however, some studies found an insignificant relationship. A study on determining the positive or negative effect of the reputation of an organization found that both are significant to the organization [42]. A good reputation is not always consistently associated with positive value; thus, people do not strive to assimilate with this positive self-concept. Ref. [43] demonstrated that the relationship between organizational reputation and perceived suitability is not significant. Ref. [44] supported Falkenreck's study, as they found an insignificant relationship between organizational reputation and digital native attraction. Ref. [45] suggested introducing a moderator variable in the unpredictably weak or inconsistent results between an antecedent and a dependent variable. Thus, the present study introduces self-esteem as a moderator between organizational reputation and digital native attraction.

Ref. [13] mentioned that individuals are different in defining their self-concept, since each individual's characteristics are different; thus, the inconsistent results of the recent study are not unexpected. Ref. [46] supported Turban and Keon's finding that individual differences are influenced by environmental cues associated with self-esteem. Self-esteem is defined as an individual's set of thoughts and feelings about his or her own worth and importance [47]. Self-esteem is the most commonly studied moderator of self-concept in digital native attraction, which explains the differences in how individuals associate themselves to organizations' characteristic [12,13,46].

The present study refers to the behavioral plasticity hypotheses in understanding how individuals differ in terms of self-esteem response to different environments [48]. The behavioral plasticity hypotheses propose individual's differences in terms of self-esteem respond to external factors based on low or high self-esteem. According to [49], because people with low self-esteem tend to be more sensitive to external factors, the scholar hypothesized that they are more behaviorally plastic (reactive) than those with high self-esteem. In other words, the level of susceptibility to environmental and social cues is higher in people with low self-esteem compared to those with high self-esteem [49,50]. A study has been executed on organizational attractiveness [13] to empirically investigate the hypotheses that digital natives with low self-esteem are more inclined to choose organizations with a positive image, such as large companies, than high-self-esteem digital natives. The reason is they (low-self-esteem individuals) would like to associate themselves with a reputable company [51,52], as the association will help them to derive a positive self-concept from the positive identity of the organization. We posit that digital natives with low (versus high) self-esteem have a stronger attraction to reputable organizations.

Hypothesis 3 (H3). *The relationship between organizational reputation and digital native attraction is moderated by self-esteem.*

Hypothesis 4 (H4). *The indirect effect of organizational justice on digital native attraction via organizational reputation is moderated by self-esteem.*

The present study illustrated the four (4) study hypotheses in the hypothesized theoretical model (see Figure 1)

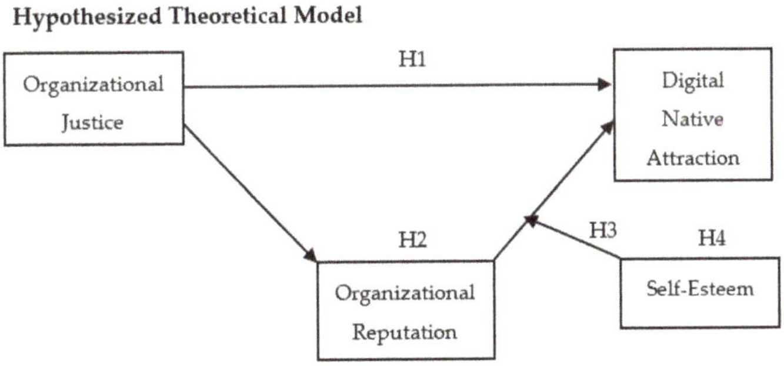

Figure 1. Model of Organizational Justice Inference and Digital Native Attraction (MOJA).

3. Methodology

The present study uses a quantitative method to investigate the role of organizational justice, organizational reputation and self-esteem in digital native attraction. The sample in the study consists of 327 accounting and financial services interns. In terms of the demographical composition, the sample were 22% male and 55% non-Malays, including Chinese, Indian and other. The present study purposely concentrated on two significant courses, which were accounting with 56% and financial services with 44%, according to the Critical Occupational List (COL), which lists difficult-to-fill vacancies. For the record, our participants are from the big four accounting firms in Malaysia, which consist of Ernst & Young, KPMG, Deloitte and Pricewaterhouse Coopers, with percentages of 32%, 23%, 22% and 23%, respectively.

3.1. Design and Procedure

The present study's design and the procedure for data collection strictly follow the purposive sampling method. This method requires a study be confined to the problem it faces and the purpose of the study in selecting participants. The problem that drives the present study is the difficulty experienced in organizations in hiring the right candidate, for the reason that digital natives exhibit a choosiness attitude. A study [53] discovered numerous business sectors facing difficulties in filling job vacancies in organizations. The sector listed these under the Critical Occupational List (COL). The accounting and financial sector has been categorized as the most critical sector by the Manpower Group in facing the problem. Thus, the present study confines the selection of participants to digital natives undertaking internships at the big four accounting firms in Malaysia, undoubtedly involving the accounting and financial services. The big four firms were chosen because the procedures for hiring interns at the firms are relatively strict; they only hire those who possess competency and are of a high academic quality. Consequently, the present study's findings are more fascinating, as they are related to competent digital natives.

The participants in the present study answered the questionnaire after three months spent in an organization as an intern. They expressed their answers by referring to the organization with which they are doing their internship program. The present study states that the period employed was sufficient to provide participants with the information to answer the questionnaire. Compared to information from the vignette version, or a policy capturing study, which involves using identical information pertaining to the organization, the participants' experience provides the best information with which to answer the present questionnaire. Next, the participants answer the question as to whether they are in an organization or in university. For those who answer the questionnaire while they are at the organization, they receive the questionnaire through their organization. Upon approval from the organization, the questionnaire is handed to the present study participant. Meanwhile, interns who are attending university courses, such as data analysis, problem statement development and others, will receive the present study questionnaire. This means that they answer the questionnaire while at university; however, they refer to their internship organization in providing the answers. A few questions related to their demographic are provided to get details on the participant involved.

3.2. Measures

Digital natives' attraction was measured using five items developed by [18]. Organizational justice was measured using twenty items developed by [47]. As regards organizational reputation, the participant expresses their perception based on the experiences they have had during the internship program. The information provides them with some understanding with which to rate the organization as being reputable or not. The expressions were derived after they reviewed five items with a Cronbach's alpha of 0.83, as developed by [54]. Finally, self-esteem was measured using eight items adapted from [47], with a Cronbach's alpha of 0.83. The present study adopts a similar scale for all variables, which is 1 = strongly disagree, and 7 = strongly agree.

4. Data Analysis and Result

The present study adopted a model called the moderated mediation model, wherein the mediator (organizational reputation) and the moderator (self-esteem) are tested simultaneously with an independent variable (organizational justice) and a dependent variable (digital native attraction). To perform the test, four procedures have to be followed. The first step is to assess the relationship of organizational justice with organizational reputation. The present study found no straddle (0) when referring to the CI bias correlation (LL = 0.911, UL = 1.016); thus, the result is significant [55,56]. The second step is the assessment of the relationship between organizational reputation and digital native attraction, and the results were also significant for no straddle (0) (LL = 0.684, UL = 1.045). The second last step is to assess the moderator of the present study, as the direct relationship of organizational justice was shown to be significant by referring to ($b_{2i} = -0.297$, $t = 6.731$) (LL = 0.211, UL = 0.385). Previous assessments have stated that the eligibility of the present study depends on the last step. The final step is to test the interaction of self-esteem and organizational reputation to determine the relationship between organizational justice and digital native interaction. The results demonstrate a significant relationship ($b_{3i} = -0.105$, $t = -2.196$) (LL = −0.200, UL = −0.011); however, the impact is negative ($p < 0.05$) (see Table 1).

Table 1. Moderated Mediation Regression Result of Organizational Justice.

Equation	T Statistics	Std Error	LL	UL	
ai	36.1031 **	0.0267	0.9105	1.0155	H2 Supported
bi	−2.8804 **	0.0830	−0.4024	−0.0758	
c′	9.4173 **	0.0918	0.6839	1.0451	H1 Supported
b2i	6.7312 **	0.0443	0.2108	0.3849	
b3i	−2.1962 *	0.0479	−0.1995	−0.0110	H3 Supported

Note: ** $p < 0.01$ and * $p < 0.05$.

In term of hypotheses, the present study developed its own based on the four procedures of the moderated mediation model. Thus, the hypotheses can be seen in Table 1, and Table 2 contains the moderated mediation regression result of digital native attraction. The first hypothesis of the present study in terms of the relation between organizational justice and digital native attraction is referred to via the c' value. The value of c' demonstrates the positive significant relationship with no straddle (0) (LL = 0.911, UL = 1.016) [55,56]; thus, Hypothesis 1 is supported.

Table 2. Moderated Mediation Indirect Effect of Organizational Justice.

	Self-Esteem	Effect	BootSE	BootLLCI	BootULCI	
Indirect Effect	Low	−0.1572	0.0901	−0.3289	−0.0246	
	Medium	−0.2302	0.0839	−0.3943	−0.0600	
	High	−0.3033	0.0863	−0.4718	−0.1324	
Mediated Moderation		−0.1013	0.0377	−0.1816	−0.0311	H4 not Supported

The next hypothesis assesses the relation of organizational justice to digital native attraction, mediated by organizational reputation. To assess the relationship, the present study strictly adheres to the suggestions of [45]. First, the independent variable is regressed to the outcome, followed by the mediator being regressed to the outcome variable, and the last step is the independent variable being regressed to the mediator. In case of all three steps are significant, the second hypothesis is supported. The present assessment of the first step (c' = 0.845, t = 9.417) (LL = 0.6839, UL = 1.0451) shows a significant relationship. The second step yields the same result (b_{1i} = −0.239, t = 9.417) (LL = 0.684, UL = 1.045). The final step also showed a significant relationship (a_i = 0.963, t = 36.103) (LL = 0.911, UL = 1.016), meaning Hypothesis 2 is supported.

The third hypothesis describes self-esteem as a moderator of organizational reputation regressed to digital native attraction. The present study refers to (b_{3i} = −0.105, t = −2.196) (LL = −0.200, UL = −0.011), as suggested by [45]. The results show that it does not straddle 0 in between the lower level and the upper level (LL = −0.200, UL = −0.011), and the t-value also supports this, which means self-esteem significantly impacts the relationship [55,56]; thus, Hypothesis 3 is supported.

The final step is crucial for the present study, following the assessment of the moderated mediation model, which is the main purpose of the study. The continuous variable is transformed to be mean-centered, following the recommendations of [57], and the present study assumes normal distribution. Table 2 shows a conditional indirect effect at three different levels, concluding with a moderated mediation model. The level describes the conditions in detail, and the moderated mediation assessment is used to show the significance of the organizational reputation and self-esteem in the present study. The assessment of the moderated mediation model found significance, with reference to IE = −0.101, se = 0.037, 95% CI (−0.182, −0.031), with no straddle (0) in between the lower lever and the upper level; however, the result is negative. Thus, Hypothesis 4 is not supported. The details of the condition's indirect effects need to be reviewed. The low IE = −0.157, se = 0.090, 95% CI (−0.329, −0.246), medium IE = −0.230, se = 0.084, 95% CI (−0.394, −0.060) and high IE = −0.303, se = 0.086, 95% CI (−0.472, −0.132) showed that all conditions' indirect effects are significant. An interesting conclusion is generated by these results, as detailed in the next paragraph.

Hypothesis 4 of the present study is not supported, despite the significant moderated mediation model achieved. This means that organizational justice, organizational reputation and self-esteem are still relevant to digital native attraction; however, the presence of low self-esteem does not strengthen the relationship, as hypothesized. Likewise, the presence of low self-esteem weakens the relationship, whereas the higher the reputation an organization achieves, the weaker the attraction for those who have low self-esteem. This finding necessitates the further investigation of the impact of self-esteem on the proposed model. The two levels of self-esteem are intensely scrutinized, and the discussion is given in the next paragraph.

The two levels, namely, low and high self-esteem, have been reviewed to develop a clearer picture. The results show both levels make a significant negative contribution to digital native attraction (see Table 2). Nevertheless, the straight-line position for people with high self-esteem is higher than that of people with low self-esteem (see Figure 2); similarly, people with high self-esteem show greater digital native attraction. In addition, Figure 2 has demonstrated that organizations go far in establishing a stronger reputation, while the changes made to improve attractiveness to the organization are diminished, and the magnitude of the change is even higher for people with low self-esteem compared to people with high self-esteem, as can be seen from the steepness of the slope. Therefore, the present study has revealed that digital natives with high self-esteem are more attracted to reputable organizations than those with low self-esteem, and this correlation is even stronger for organizations with better reputations.

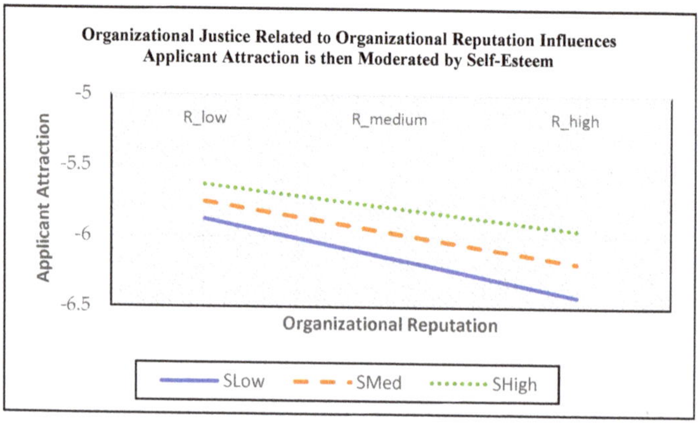

Figure 2. Moderated Mediation Interaction of Organizational Justice.

The intriguing part is the negative impact, despite the significant relationship, of the moderated mediation model. The present study finds that today, the Malaysian labor market is relatively challenging in terms of the competitiveness for jobs. The difficulty of getting a job explains the result of this study. Psychologically, there will be a negative impact when people feel it is impossible to get a job in the extremely competitive labor market, and this experience is even worse for people with low self-esteem. The impact is lesser for people with high self-esteem, as explained by the higher level of self-confidence they have in competing with other people. A high confidence level (high self-esteem) contributes to their likelihood to apply a job at a reputable company, which is stronger compared to those with low self-esteem, who are less self-confident. Therefore, the present study finds that the competitiveness of the labor market plays a substantial role in influencing individuals with different levels of self-esteem in responding to organizational attractiveness.

Author Contributions: Conceptualization, H.A.B., A.Z.B.T., M.N.B.M.N., M.Z.A. and M.M.M.Y.; methodology, A.Z.B.T. and M.M.M.Y.; validation, M.Z.A. and M.M.M.Y.; formal analysis, H.A.B., A.Z.B.T. and M.M.M.Y.; investigation, H.A.B., A.Z.B.T. and M.N.B.M.N.; resources, H.A.B., A.Z.B.T. and M.N.B.M.N.; data curation, M.Z.A. and M.M.M.Y.; writing—original draft preparation, H.A.B.; writing—review and editing, H.A.B.; supervision, A.Z.B.T. and M.N.B.M.N.; project administration, H.A.B., A.Z.B.T., M.N.B.M.N., M.Z.A. and M.M.M.Y. All authors have read and agreed to the published version of the manuscript.

Funding: This research received no external funding.

Institutional Review Board Statement: Not applicable.

Informed Consent Statement: Not applicable.

Data Availability Statement: Not applicable.

Acknowledgments: Paper presented during a panel organized by UiTM in the context of the Social Science conference, organized and hosted by the Faculty Information Technology, during the IASSc 2022.

Conflicts of Interest: The authors declare no conflict of interest.

References

1. Colbert, A.; Yee, N.; George, G. The digital workforce and the workplace of the future. *Acad. Manag. J.* **2016**, *59*, 731–739. [CrossRef]
2. Mesquita, A.; Oliveira, L.; Sequeira, A. The Future of the Digital Workforce: Current and Future Challenges for Executive and Administrative Assistants. In *World Conference on Information Systems and Technologies*; Springer: Cham, Switzerland, 2019.
3. Lipscomb, S.T.; Chandler, K.D.; Abshire, C.; Jaramillo, J.; Kothari, B. Early childhood teachers' self-efficacy and professional support predict work engagement. *Early Child. Educ. J.* **2022**, *50*, 675–685. [CrossRef] [PubMed]
4. Prensky, M. Digital Natives, Digital Immigrants Part 2: Do They Really Think Differently? *Horizon* **2001**, *9*, 1–6. [CrossRef]
5. Zoghbi-Manrique-de-Lara, P.; Aguiar-Quintana, T.; Suárez-Acosta, M.A. A justice framework for understanding how guests react to hotel employee (mis)treatment. *Tour. Manag.* **2013**, *36*, 143–152. [CrossRef]
6. Kulik, C.T.; Pepper, M.B.; Shapiro, D.L.; Cregan, C. The electronic water cooler: Insiders and outsiders talk about organizational justice on the internet. *Communic. Res.* **2012**, *39*, 565–591. [CrossRef]
7. Beersma, B.; Van Kleef, G.A. Why people gossip: An empirical analysis of social motives, antecedents, and consequences. *J. Appl. Soc. Psychol.* **2012**, *42*, 2640–2670. [CrossRef]
8. Mynewshub. *Factors Undergraduate Choose Employment*; Mynewshub: Kuala Lumpur, Malaysia, 2016.
9. Graham, M.E.; Cable, D.M. Consideration of the incomplete block design for policy-capturing research. *Organ. Res. Methods* **2001**, *4*, 26–45. [CrossRef]
10. Turban, D.B.; Cable, D.M. Firm reputation and applicant pool characteristics. *J. Organ. Behav.* **2003**, *24*, 733–751. [CrossRef]
11. Cable, D.M.; Turban, D.B. The value of organizational reputation in the recruitment context: A brand-equity perspective. *J. Appl. Soc. Psycol.* **2003**, *33*, 2244–2266. [CrossRef]
12. Cable, D.M.; Turban, D.B. Establishing the dimensions, sources and value of job seekers' employer knowledge during recruitment. In *Research in Personnel and Human Resources Management*; Emerald Group Publishing Limited: Bingley, UK, 2001; Volume 20, pp. 115–163.
13. Turban, D.B.; Keon, T.L. Organizational attractiveness—An interactionist perspective. *J. Appl. Psychol.* **1993**, *78*, 184–193. [CrossRef]
14. Lievens, F.; Slaughter, J.E. Employer image and employer branding: What we know and what we need to know. *Annu. Rev. Organ. Psychol. Organ. Behav.* **2016**, *3*, 407–440. [CrossRef]
15. Chapman, D.S.; Uggerslev, K.L.; Carroll, S.A.; Piasentin, K.A.; Jones, D.A. Applicant attraction to organizations and job choice: A meta-analytic review of the correlates of recruiting outcomes. *J. Appl. Psychol.* **2005**, *90*, 928–944. [CrossRef]
16. Ehrhart, K.H.; Ziegert, J.C. Why are individuals attracted to organizations? *J. Manag.* **2005**, *31*, 901–919. [CrossRef]
17. Rau, B.L.; Hyland, M.A.M. Role conflict and flexible work arrangements: The effects on appll6icant attraction. *Pers. Psychol.* **2002**, *55*, 111–136. [CrossRef]
18. Highhouse, S.; Lievens, F.; Sinar, E.F. Measuring attraction to organizations. *Educ. Psychol. Meas.* **2003**, *63*, 986–1001. [CrossRef]
19. Aiman-smith, L.; Bauer, T.N.; Cable, D.M. Are you attracted? Do you intend to pursue? A recruiting policy-capturing study. *J. Bus. Psychol.* **2001**, *16*, 219–237. [CrossRef]
20. Ong, L. Employer branding and its influence on potential job applicants. *Aust. J. Basic Appl. Sci.* **2011**, *5*, 1088–1092.
21. Uggerslev, K.L.; Fassina, N.E.; Kraichy, D. Recruiting through the stages: A meta-analytic test of predictors of applicant attraction at different stages of the recruiting process. *Pers. Psychol.* **2012**, *65*, 597–660. [CrossRef]
22. Crawshaw, J.R.; Cropanzano, R.; Bell, C.M.; Nadisic, T. Organizational justice: New insights from behavioural ethics. *Hum. Relat.* **2013**, *66*, 885–904. [CrossRef]
23. Skarlicki, D.P.; O'Reilly, J.; Kulik, C.T. Third-party reactions to employee (mis)treatment: A justice perspective. In *The Oxford Handbook of Justice in the Workplace*, 1st ed.; Cropanzano, R.S., Ambrose, M.L., Eds.; Elsevier Ltd.: London, UK, 2015; Volume 26, pp. 183–229.
24. Zhang, L.; Gowan, M.A. Corporate social responsibility, applicants' individual traits, and organizational attraction: A person-organization fit perspective. *J. Bus. Psychol.* **2012**, *27*, 345–362. [CrossRef]
25. Pierce, C.A.; Karl, K.A.; Brey, E.T. Role of workplace romance policies and procedures on job pursuit intentions. *J. Manag. Psychol.* **2012**, *27*, 237–263. [CrossRef]
26. Thompson, R.J.; Payne, S.C.; Taylor, A.B. Applicant attraction to flexible work arrangements: Separating the influence of flextime and flexplace. *J. Occup. Organ. Psychol.* **2015**, *88*, 726–749. [CrossRef]
27. Turban, D.B.; Greening, D.W. Corporate social performance and organizational attractiveness to prospective employees. *Acad. Manag. J.* **1997**, *40*, 658–672. [CrossRef]
28. Lis, B. The relevance of corporate social responsibility for a sustainable human resource management: An analysis of organizational attractiveness as a determinant in employees' selection of a (potential) employer. *Manag. Rev.* **2012**, *23*, 279–295. [CrossRef]
29. Anderson, M.H.; Haar, J.M.; Gibb, J.L. Personality trait inferences about organizations and organizational attraction: An organizational-level analysis based on a multi-cultural sample. *J. Manag. Organ.* **2010**, *16*, 140–150. [CrossRef]
30. Rindova, V.P.; Williamson, I.O.; Petkova, A.P.; Sever, J.M. Being good or beeing known: An empirical examination of the acendents, and consequences of organizational reputation. *Acad. Manag. J.* **2005**, *48*, 1033–1049. [CrossRef]

31. Williamson, I.O.; King, J.E.; Lepak, D.; Sarma, A. Firm reputation, recruitment web sites, and attracting applicants. *Hum. Resour. Manag.* **2010**, *49*, 669–687. [CrossRef]
32. Duarte, A.P.; Gomes, D.R.; das Neves, J.G. Tell me your socially responsible practices, I will tell you how attractive for recruitment you are the impact of perceived CSR on organizational attractiveness. *Tékhne* **2014**, *12*, 22–29. [CrossRef]
33. Jones, D.A.; Willness, C.R.; Madey, S. Why are job seekers attracted by corporate social performance? Experimental and field tests of three signal-based mechanisms. *Acad. Manag. J.* **2014**, *57*, 383–404. [CrossRef]
34. Turban, D.B.; Forret, M.L.; Hendrickson, C.L. Applicant attraction to firms: Influences of organization reputation, job and organizational attributes, and recruiter behaviors. *J. Vocat. Behav.* **1998**, *44*, 24–44. [CrossRef]
35. Behrend, T.S.; Baker, B.A.; Thompson, L.F. Effects of pro-environmental recruiting messages: The role of organizational reputation. *J. Bus. Psychol.* **2009**, *24*, 341–350. [CrossRef]
36. Baum, M.; Kabst, R. The effectiveness of recruitment advertisements and recruitment websites: Indirect and interactive effects on applicant attraction. *Hum. Resour. Manag.* **2014**, *53*, 353–378. [CrossRef]
37. Shrout, P.E.; Bolger, N. Mediation in experimental and nonexperimental studies: New procedures and recommendations. *Psychol. Methods* **2002**, *7*, 422–445. [CrossRef]
38. Mackinnon, D.P.; Krull, J.L.; Lockwood, C.M. Equivalence of the Mediation, Confounding and Suppression Effect. *Prev. Sci.* **2000**, *1*, 173–181. [CrossRef] [PubMed]
39. Edwards, J.R.; Lambert, L.S. Methods for integrating moderation and mediation: A general analytical framework using moderated path analysis. *Psychol. Methods* **2007**, *12*, 1–22. [CrossRef]
40. Banks, G.C.; Kepes, S.; Joshi, M.; Seers, A. Social identity and applicant attraction: Exploring the role of multiple levels of self. *J. Organ. Behav.* **2015**, *37*, 326–345. [CrossRef]
41. Glavas, A.; Godwin, L.N. Is the perception of 'goodness' good enough? Exploring the relationship between perceived corporate social responsibility and employee organizational identification. *J. Bus. Ethics* **2013**, *114*, 15–27. [CrossRef]
42. Brooks, M.E.; Highhouse, S.; Russell, S.S.; Mohr, D.C. Familiarity, ambivalence, and firm reputation: Is corporate fame a double-edged sword? *J. Appl. Psychol.* **2003**, *88*, 904–914. [CrossRef]
43. Falkenreck, C. *Reputation Transfer to Enter New B-to-B Markets: Measuring and Modelling Approaches*, 1st ed.; Springer Science & Business Media: Berlin/Heidelberg, Germany, 2009.
44. Waight, P.; Chow, J. Employer Branding in Australia: A content analysis of recruitment advertising in the mining and higher education industries. In Proceedings of the 23rd ANZAM Conference 2009: Sustainable Management and Marketing, Melbourne, Australia, 1–4 December 2009; Promaco Conventions Pty Ltd.: Bateman, WA, Australia, 2009.
45. Baron, R.M.; Kenny, D.A. The moderator-mediator variable distinction in social the moderator-mediator variable distinction in social psychological research: Conceptual, strategic, and statistical considerations. *J. Pers. Soc. Psychol.* **1986**, *51*, 1173–1182. [CrossRef] [PubMed]
46. Highhouse, S.; Thornbury, E.E.; Little, I.S. Social-identity functions of attraction to organizations. *Organ. Behav. Hum. Decis. Process* **2007**, *103*, 134–146. [CrossRef]
47. Rosenberg, M. *Rosenberg Self-Esteem Scale, Society and the Adoslescent Self-Image*; University Press: Princeton, NJ, USA; New York, NY, USA, 1965.
48. Pierce, J.L.; Gardner, D.G. Self-esteem within the work and organizational context: A review of the organization-based self-esteem literature. *J. Manag.* **2004**, *30*, 591–662. [CrossRef]
49. Brockner, J. *Self-Esteem at Work*; Lexington Books: Lexington, MA, USA, 1988.
50. Campbell, J.D. Self-esteem and clarity of the self-concept. *J. Pers. Soc. Psychol.* **1990**, *59*, 538–549. [CrossRef]
51. Ashforth, B.E.; Mael, F. Social identity theory and the organization. *J. Bus. Ethics* **1989**, *14*, 20–39. [CrossRef]
52. Dutton, J.E.; Dukerich, J.M.; Harquail, C.V. Organizational images and member identification. *Adm. Sci. Q.* **1994**, *39*, 239–263. [CrossRef]
53. ManpowerGroup. *Talent Shortage*; ManpowerGroup: Milwaukee, WI, USA, 2015.
54. Lievens, F.; Highhouse, S. The relation of instrumental and symbolic attributes to a company's attractiveness as an employer. *Pers. Psychol.* **2003**, *56*, 75–102. [CrossRef]
55. Preacher, K.J.; Hayes, A.F. SPSS and SAS procedures for estimating indirect effects in simple mediation models. *Behav. Res. Methods Instrum. Comput.* **2004**, *36*, 717–731. [CrossRef] [PubMed]
56. Preacher, K.J.; Hayes, A.F. Asymptotic and resampling strategies for assessing and comparing indirect effects in multiple mediator models. *Behav. Res. Methods* **2008**, *40*, 879–891. [CrossRef]
57. Aiken, L.S.; West, S.G. *Multiple Regression: Testing and Interpreting Interactions*; Sage: Newbury Park, CA, USA, 1996.

proceedings

MDPI

Proceeding Paper

Investment Education: Understanding Portfolio Optimization †

Ahmad Bukhari Mohd Yasin, Khairu Amin Ismail and Zulkifli Mohamed *

Faculty of Business and Administration, Universiti Teknologi MARA (UiTM), Shah Alam 40450, Malaysia
* Correspondence: zulkifli030@uitm.edu.my
† Presented at the International Academic Symposium of Social Science 2022, Kota Bharu, Malaysia, 3 July 2022.

Abstract: Investing in a stock market is very challenging. Investors need to understand the nature of stock-price movement and various factors that influence this fluctuation. Being unable to read the market's movements will expose investors to uncertainty in investment returns and can lead to capital loss. Investors need to manage investment risks well. When managing an investment, risk can be managed by investing in a portfolio of assets rather than in a single asset. However, how does a portfolio of assets work in reducing investment risk? This can be explained by a complex mathematical approach. Therefore, this study aimed to provide an exposition to investors and students about managing a portfolio's risk and the importance of asset allocation. A case study of two risky-asset portfolios has been used in the construction of efficient portfolios using a mean–variance model. As a result, students and investors can see the risk–return relationship in a portfolio's efficient management and the importance of risk-free assets in the generation of a portfolio's rate of return.

Keywords: investment; portfolio; diversification; optimization

Citation: Yasin, A.B.M.; Ismail, K.A.; Mohamed, Z. Investment Education: Understanding Portfolio Optimization. *Proceedings* **2022**, *82*, 113. https://doi.org/10.3390/proceedings2022082113

Academic Editor: Mohamad Rahimi Mohamad Rosman

Published: 28 October 2022

1. Introduction

We are currently living in a very challenging world. The Fourth Industrial Revolution and the COVID-19 pandemic have really affected human life during this century. Digital technologies and the Internet have effected tremendous changes in various areas of human life. It has changed the way of communication, working, doing business and transaction, get a medical treatment, teaching and learning in schools and universities. It has also affected the way investors analyze the stock market and make investment decisions. These situations have put investors in uncertain environments. Making investment decisions in a vague and uncertain environment is not an easy task. Investors need to take into account many considerations that influence assets' returns [1].

Uncertainty in a stock market is an issue that has long been raised by many scholars. In the current scenario characterized by the COVID-19 pandemic and the advent of disruptive technologies, the situation has been worsened. Business operations and social life have been influenced by digital technologies. Technology changes very quickly and frequently. It has changed the way people communicate, how businesses operate, marketing media, the banking process, the transaction system, etc. These changes have influenced the state of volatility in business sales and turnover since the environment has become more uncertain, and businesses are operating in a more complicated environment. Decisions must be made in an environment of ambiguity. Stock market movements have become more unpredictable, uncertain, and volatile. The environment can be simplified as one characterized by volatility, uncertainty, complexity, and ambiguity (VUCA).

In portfolio investment, stock market fluctuation is a normal situation. Investors need to manage the stock market's risks well. Stock market risks can be classified into systematic and unsystematic risks. A VUCA business environment can be classified as a systematic risk which cannot be eliminated but which can be reduced. Systematic risk can be minimized by using portfolio diversification. By using an appropriate approach of portfolio diversification, investors can minimize their portfolio risk, and at the same time, they can maximize their portfolios' rates of return. This can be proven using a complicated mathematical approach. Therefore, this study proposes a simplified method for developing understanding in investors and students for managing a portfolio's risk.

2. Literature Review

Investment plays very important roles in asset management. It cannot be managed by siloing, and it requires an expert to manage it for the purpose of obtaining diversification benefits. According to Stewart, Piros & Heisler, portfolio management is becoming increasingly more sophisticated due to the ongoing advancement of theory and the growing complexity of practice led by a number of trends, including:

i. Advances in modern portfolio theory,
ii. More complexity in instruments,
iii. Increased demands on performance,
iv. Increased client sophistication,
v. Rising retirement costs, and the growing trend toward individual responsibility for those costs, and
vi. Dramatic growth in assets under management.

For the past few decades, most investors have concentrated their money in single-quality stock investments [2]. Now, however, due to the above-mentioned changes, portfolio diversification is one of the strategies used to minimize investment risks. Investing in the stock market has become an important medium for retail and institutional investors, either to create additional wealth or to protect themselves from inflation. In today's fast-changing environment, it is difficult for investors to make investment decisions, especially when in conjunction with volatile supply and demand, an uncertain business environment, complex data, and ambiguous information. The utilization of technology has become a requirement for any investment decision. This situation is consistent with that which [2] mentioned—that today's uncertain and volatile market conditions require investors to utilize information technology (IT) to drive future investment decisions through correct analysis and judgment.

Through the application of the appropriate investment management and diversification strategies, investment risk can be minimized. Asset allocation between risky assets and risk-free assets can be utilized in order to manage a portfolio's risk level. According to [3], studies of stocks on the US stock market have found that the correlation structure of return is largely explained by the 'single index' model. That is, there is one dominant, causal agent that operates as a single factor or "market factor" in the correlation structure of a stock return. Other factors, such as the industry group, are present, but they have been found to be of relatively minor importance.

In modern portfolio theory, asset correlation is very important in determining a portfolio's risk level. Negative correlation assets are needed in order to minimize the portfolio's risk. Moreover, investors also can use risk-free assets in their portfolios in order to minimize the portfolio's risk. Inclusion of risk-free assets in a portfolio is able to reduce the amount of total risk since the asset variance is zero.

In this exploration, the study sample was taken from Bursa Malaysia. The data were obtained from the web-based DataStream, which is composed of a huge number of companies which Bursa Malaysia prepared for selection. The following steps can guide students in the process of determining efficient frontiers and capital-allocation lines.

3. Methodology

To simplify the discussion, this study only focuses on a two-company investment portfolio scenario. The study included 30 companies listed in the FTSE Bursa Malaysia Kuala Lumpur Composite Index as the sample. The companies selected for the study were Top Glove Corporation Berhad and Genting Malaysia Berhad. The two companies were chosen because their combination had the lowest correlation value (-0.12) of any of the sample companies' combinations. A low correlation value is the main selection criterion as it can theoretically provide the optimum diversification strategy. The study covers 13 years of monthly data, ranging from 24 May 2009 to 24 May 2022. The data-analysis technique is broken down into three stages: First, calculating the efficient frontier, then calculating the capital allocation line, and finally, determining the optimal portfolio construction by looking at the tangent between the efficient frontier line and the capital allocation line. The formulas below were taken from [4,5].

3.1. Efficient Frontier Line

The portfolio return and portfolio standard deviation must be computed before the efficient frontier can be constructed.

3.1.1. Portfolio Return

To calculate the portfolio return, one must start with the calculation of a single stock return:

$$R_i = \frac{P_1 - P_0 + D}{P_0} \tag{1}$$

$R_i = Single\ stock\ return$
$P_n = Share\ price\ at\ time\ n$
$D = Dividend$

The average return (mean) is then calculated:

$$\overline{R_i} = \frac{\sum R_i}{n} \tag{2}$$

The following is the final step in determining the portfolio return:

$$\overline{R}_p = \sum_{i=1}^{n} W_i R_i \tag{3}$$

$W_i = the\ percent\ of\ the\ portfolio\ in\ asset\ i$
$R_i = rate\ of\ return\ for\ asset$

3.1.2. Portfolio Standard Deviation

The process of calculating the portfolio standard deviation begins with calculating the single-asset standard deviation:

$$\sigma_i = \sqrt{\frac{\sum (R_i - \overline{R}_i)^2}{n-1}} \tag{4}$$

$\sigma_i = Standard\ deviation\ of\ asset\ i$

The next step is calculating the covariance between the two assets:

$$Cov_{i,j} = \frac{\sum (R_i - \overline{R}_i)(R_j - \overline{R}_j)}{n-1} \tag{5}$$

$Cov_{i,j} = Covariance\ between\ asset\ i\ and\ j$

The last step is calculating the portfolio standard deviation

$$\sigma_p = \sqrt{\sum_{i=1}^{n} w_i^2 \sigma_i^2 + \sum_{i=1}^{n} \sum_{i=1}^{n} w_i w_j Cov_{ij}} \tag{6}$$

$\sigma_p = standard\ deviation\ of\ the\ portfolio$
$w_i = proportion\ of\ value\ in\ the\ portfolio$
$\sigma_i^2 = Variance\ for\ asset\ i$
$Cov_{ij} = Covariance\ between\ rates\ of\ return\ for\ asset\ i\ and\ j$

Various combinations of percentage capital allocation (weightage) between Top Glove Corporation Berhad and Genting Malaysia Berhad have been considered in constructing the efficient frontier.

3.2. Capital Allocation Line (CAL)

The second process is calculating the capital allocation line. The capital allocation line can be calculated based on the following formula below:

$$R_c = R_f + \sigma_c \frac{R_p - R_f}{\sigma_p} \tag{7}$$

$R_c = Return\ on\ portfolio\ combined\ with\ risk\ free\ asset$
$R_f = Risk\ free\ rate$
$\sigma_c = Standard\ deviation\ on\ portfolio\ combined\ with\ risk\ free\ asset$
$R_p = Return\ on\ portfolio$
$\sigma_p = Standard\ deviation\ of\ portfolio$
$\frac{R_p - R_f}{\sigma_p}$ is also known as the Sharpe ratio

3.3. Combining the Efficient Frontier and CAL

The final step is to determine the optimum portfolio. The efficient frontier and CAL are plotted on the same graph with the portfolio standard deviation as the x-axis and the portfolio return as the y-axis. The model suggests that the optimum portfolio is achieved at a point where the CAL line intersects with the efficient frontier line, and at this point, the Sharpe ratio is at the maximum level.

Investors can follow the above-presented steps in order to develop an understanding of the efficient frontier and optimum portfolio, which will provide them with the lowest portfolio risk and the highest return at any point of risk and return.

4. Results and Discussion

As a result, the efficient frontier for the two-asset portfolio was derived as presented in Figure 1, which shows the efficient frontier line and CAL plotted on the same graph.

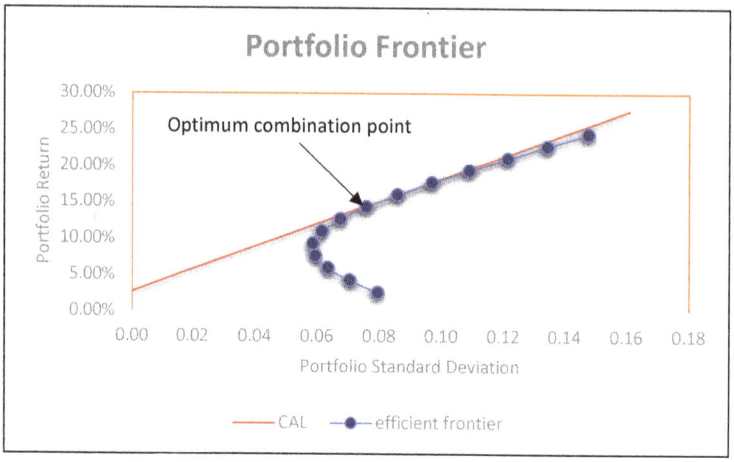

Figure 1. Portfolio frontier for a two-asset portfolio.

Figure 1 shows the optimum combination for an investment portfolio containing Top Glove Corporation Berhad and Genting Malaysia Berhad. The optimum combination is found at the tangent where the efficient frontier line and the CAL intersect. At this point, the Sharpe ratio is also at the maximum value. At the optimum combination point, the model proposes that 70% of the capital should be invested in Top Glove Corporation Berhad and 30% should be allocated to Genting Malaysia Berhad. At this level, the return of the proposed portfolio will 16%, while the risk faced by investors is at 8.6%.

Although this study proposed a 70% capital allocation to Top Glove Berhad and a 30% allocation to Genting Malaysia Berhad, it should be read with extra care. First, the study used historical data based on 13 years of market performance. However, since the emergence of the COVID-19 pandemic in December 2019 and the beginning of the Russian invasion of Ukraine on 24 February 2022, the economy and the market have been very volatile. Many profit-making industries have been faced with a gloomy future. Second, the model only looking at and is restricted to the Markowitz model, and the reader should also look into other investment models. Third, one of the objectives of this study is to show how students can use complex theory and turn it into something practical; therefore, the discussion is confined to the Markowitz model alone.

5. Conclusions

Portfolio management is very important, and it requires careful asset selection in order to minimize the portfolio's risk. The aim of the asset selection in portfolio management is to minimize the investment and maximize the portfolio return. Portfolio optimization is achieved when all of the portfolio lies on the efficient frontier curve. In the case of risk-free assets present in the portfolio, the capital allocation line was used. By using this approach, investors are guided in the construction of the most efficient portfolio for their investments.

Author Contributions: Conceptualization, Z.M.; methodology, A.B.M.Y.; software, K.A.I.; validation, Z.M., A.B.M.Y. and K.A.I.; formal analysis, A.B.M.Y.; investigation, K.A.I.; resources, Z.M.; data curation, A.B.M.Y.; writing—original draft preparation, Z.M.; writing—review and editing, K.A.I.; visualization, A.B.M.Y.; supervision, Z.M.; project administration, Z.M. All authors have read and agreed to the published version of the manuscript.

Funding: This research received partial funding from General Training Allocation, Universiti Teknologi MARA, Kelantan Branch.

Institutional Review Board Statement: Not applicable.

Informed Consent Statement: Not applicable.

Data Availability Statement: Not applicable.

Acknowledgments: Great appreciation goes to the Universiti Teknologi MARA Kelantan Branch Administration Team for the coorporation and support the research project. Thank you, and congratulations to the organizing committee of the International Annual Social Science Conference (IASSC) 2022.

Conflicts of Interest: The authors declare no conflict of interest.

References

1. Mohamed, Z.; Mohamad, D.; Samat, O. A fuzzy approach to portfolio selection. *Sains Malays.* **2009**, *38*, 895–899.
2. Zulkifli, M.; Basaruddin Shah, B.; Norzaidi, M.D.; Chong, S.C. Portfolio diversification: The role of information technology in future investment decision-making. *Int. J. Electron. Financ.* **2008**, 2, 451–468. [CrossRef]
3. Hunter, J.E.; Coggin, T.D. An analysis of the diversification benefit from international equity investment. *J. Portf. Manag.* **1990**, *17*, 33. [CrossRef]
4. Reilly, F.K.; Brown, K.C. *Investment Analysis and Portfolio Management*, 10th ed.; South-Western Cangage Learning: Mason, OH, USA, 2011.
5. Scott, D.S.; Christopher, D.P.; Jeffrey, C.H. *Portfolio Management: Theory and Practice*, 2nd ed.; John Wiley & Sons, Inc.: Hoboken, NJ, USA, 2011.

proceedings

MDPI

Proceeding Paper

Impact of Social Care on Elderly Well-Being in Malaysia: A Mediating Role of the *Waqf* Fund †

Siti Nurhanis Mohd Fadzil [1], Idris Osman [2,*], Shafinar Ismail [2], Maryam Jameelah Mohd Hashim [3] and Mohd Rahim Khamis [3]

1 Faculty of Business and Management, Universiti Teknologi MARA, Shah Alam Campus, Shah Alam 40450, Malaysia
2 Faculty of Business and Management, Universiti Teknologi MARA, Cawangan Melaka Kampus Bandaraya Melaka, Melaka 75350, Malaysia
3 Faculty of Business and Management, Universiti Teknologi MARA, Selangor Branch, Puncak Alam Campus, Puncak Alam 42300, Malaysia
* Correspondence: idris424@uitm.edu.my
† Presented at the International Academic Symposium of Social Science 2022, Kota Bharu, Malaysia, 3 July 2022.

Abstract: The aim of this study is to investigate the impact of social care on elderly well-being in Malaysia, and the *waqf* fund acts as a mediating role in this relationship. A total of 150 responses were analyzed, and PLS-SEM was used to validate the research model and hypotheses. The findings revealed that social care and the *waqf* fund were proven to have a considerable impact on the well-being of the elderly. This study provided major contributions by demonstrating that the *waqf* fund has an influence on the well-being of the elderly both directly and indirectly through social care. Social care assistance is an important factor in the elderly citizens' perceptions of their quality of life. The *waqf* fund, as a financial tool infused with charity, social contribution, and mutual collaboration, has considerable potential to be used as a method to safeguard the well-being of the elderly.

Keywords: well-being; elderly; social care; *waqf* fund; Malaysia

Citation: Fadzil, S.N.M.; Osman, I.; Ismail, S.; Hashim, M.J.M.; Khamis, M.R. Impact of Social Care on Elderly Well-Being in Malaysia: A Mediating Role of the *Waqf* Fund. *Proceedings* **2022**, *82*, 114. https://doi.org/10.3390/proceedings2022082114

Academic Editor: Mohamad Rahimi Mohamad Rosman

Published: 8 November 2022

1. Introduction

The ageing population happens when 7 percent of the population is 65 years old or older. Statistics have shown that the senior population across the globe is currently growing at a rapid rate, eventually leading to the ageing phenomena [1,2]. In Malaysia, the proportion of elderly people is growing, too, where there were 2.2 million senior individuals in 2019, and the figure is predicted to increase to 3.6 million by 2030. With respect to ageing population, chronic diseases, physical disabilities, mental illnesses, and other co-morbidities are becoming more widespread [3]. Thus, health needs and issues related to the elderly, including social issues and senior mistreatment, cannot be considered in isolation. In addition, psycho-emotional concerns (isolation, mental tension, difficulties staying busy) and financial restraints have a substantial impact on the elderly's quality of life [4,5].

Satisfaction with social care is a crucial determinant for elderly people's general well-being [3]. Older people's care is organised as part of social service delivery in Sweden and Finland, for example, with an increasing emphasis on delivering integrated social and healthcare services [6,7]. Even if they are unable to drive, public transit allows the elderly to access goods, services, occupations, and other activities, allowing them to maintain an active lifestyle [8]. Social needs are one of the most significant human needs for the elderly. When social requirements are not addressed, loneliness and social isolation may develop [9]. Loneliness and social isolation have been shown to be linked to sickness and death [10]. There is a large body of evidence that shows how satisfied social needs can safeguard physical and mental health [11]. It has been proven that meeting social needs protects against diseases and depression, while also improving self-esteem, life satisfaction and well-being [12].

According to [13], social care can help the elderly feel less depressed and lonely. Improved social care helps to connect the elderly with their environment and almost helps to sustain social care in creating well-being. Earlier studies revealed the factors that contribute to differences in social involvement and loneliness among the elderly, but developing a livable and healthy social care system for them remains a serious challenge. Despite various studies on social care and the elderly, the number of elderly individuals feeling loneliness and abused has been steadily increasing for several years. In addition, maintaining full necessities, amenities, and safety for the elderly requires large resources [14]. As a result, academics advocate for the above-mentioned uses of *waqf* funding. The *waqf* fund may be used to provide additional financial help to the elderly [15]. According to [16–18], the *waqf* fund can also be utilised to cover administrative, maintenance, and service costs. Based on these identified gaps, therefore, this study aims to investigate the impact of social care on elderly well-being and the mediating effect of the *waqf* fund on the relationship between social care and elderly well-being.

2. Literature Review

A sense of health, happiness, and riches, as well as strong mental health, a high degree of life satisfaction, a sense of meaning or purpose, and the ability to manage with stress, are all considered to be indicators of well-being [19,20]. Individual or community well-being has been described in a variety of ways, including standard of living and quality of life [21]. Effective ageing, according to activity theory, also known as normal ageing theory and public ageing theory, happens when older individuals stay active and retain social ties. As a result, maintaining a healthy well-being that encompasses health, social care, and daily surroundings, as well as the ability to express oneself, is vital [22].

2.1. The Relationship between Social Care and Elderly Well-Being

When defining senior individuals using accessibility terms, previous research has highlighted public transportation [23]. Methods for mapping and assessing elderly people's access to subway stations are proposed by [24]. As people get older, their driving habits change. Furthermore, the most important elements that positively influence the four dimensions of elderly well-being are care services (daily care services and cleaning services) and the environment (space, barrier-free design, fire safety, and support facilities). According to [25], increasing the availability of social services and health facilities in the community is one way of improving the well-being of the elderly. Many ageing issues are caused by unhealthy lifestyles, and various important programs might be investigated [26]. Ref. [27] also advocates for the development of dependable, low-cost, private, easy-to-use, and portable assisted information technology for the elderly. Therefore, it is hypothesized that:

H1. *Social care is positively related to elderly well-being.*

2.2. The Relationship between Social Care and the Waqf Fund

The potential of *waqf* to expand elderly care is undeniable if managed professionally and in accordance with Islamic work culture, and institutions must work together to promote the country's *waqf* fund [28]. *Waqf* of services and expertise is a sort of *waqf* of energy that is allowed to play a role and provide expertise for free under Islamic law. Doctors, for example, provide their time, knowledge, and services to care for the health and nutrition of the old; volunteers donate their time to care for the elderly [29]. Donors' monetary *waqf* can also be used to assist with the construction of geriatric homes, health clinics, and other services for the elderly [30]. As a majority of Malaysian employees have no savings other than the mandated contributions to the Employees Provident Fund (EPF), the ageing population is a worry [18]. Refs. [16,18] propose a *Waqf*-based pension model in which *waqf* money is used to build community infrastructure such as health care, education, housing, transportation, and religious institutions. Therefore, the following hypothesis is developed:

H2. *Social care and waqf fund are significantly related.*

2.3. The Relationship between the Waqf Fund and Elderly Well-being

The value of *waqf* as a financial tool packed with charity, social responsibility, and cooperation has a significant potential for use as a method to protect the elderly [31]. The Islamic economic system relies heavily on the *waqf* [32]. *Waqf*'s general objective is to provide spiritual reward and humanitarian assistance to those in need, as well as to protect the heirs and descendants of those who suffer poverty and misfortune [1]. This *waqf* has a huge impact on the neighbourhood and the country, since it promotes economic growth, education, and social development [33]. The most successful strategies of accomplishing university development in Malaysia. According to [26] a prior study, most researchers are found to have used *waqf* as an independent variable in their research. Nevertheless, despite *Waqf* being an effective financial instrument for protecting the elderly [18], few studies have used the *waqf* fund as a mediating variable [17,34]. Money granted by the Malaysian government, which is said to have contributed to the growth of *waqf* institutions and various initiatives for the benefit and welfare of the elderly, is evidence of this [28]. *Waqf* institutions provide superior wellness services in the areas of health care, education, social well-being, the environment, and other community-based activities [28,35]. Thus, we hypothesized that:

H3. *Waqf fund and elderly well-being are significantly related;*

H4. *Waqf fund has mediated in the relationship between social care and elderly well-being.*

Herewith, Figure 1 demonstrate the research framework that explicate the relationship between the exogenous, endogenous and mediator variables.

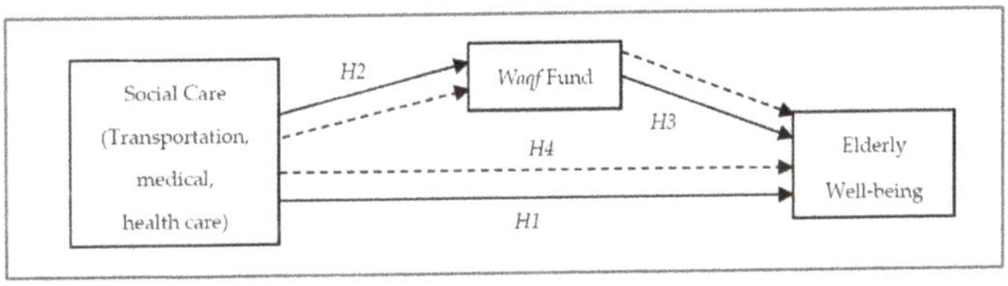

Figure 1. Research Framework.

3. Methodology

Data for the study were obtained from Malaysian seniors aged 60 and above to put the conceptual model to the test. This is the most efficient approach for gathering samples from a big population using the statistical analysis computations in G*Power version 3.1.9.2 [36]. In a senior citizen assistance facility, 200 elderly folks were approached privately. The Department of Statistics Malaysia (DOSM) website, version 2021, provided the list of aged care facilities. The survey was given to 200 senior citizens who had agreed to participate in the study. As a result, 150 questionnaires were completed and made available for use.

Quantitative measurements were adapted using a self-administered survey. The responses were scored on a five-point Likert scale, with 1 being the most strongly disagreed and 5 being the strongly agreed. Five social care items were taken from [28], three *waqf* fund items from [37], and five elderly well-being items from [28]. The hypotheses given in this paper were investigated using PLS-SEM in two stages: (1) measurement model evaluation and (2) structural model evaluation [38]. In addition, the model's route associations were built using a bootstrapping approach, which is a non-parametric PLS test. Ref. [17] defines a bootstrapped model as one in which the original sample is replaced with repeated random sampling in order to collect standard errors for hypothesis testing and produce a bootstrapped model.

4. Results

Prior to performing the quantitative analysis, the dataset was first inspected and vetted. A reliability test was performed on the total measurement scales, and descriptive statistics were reported using IBM SPSS version 26. To assess whether the data are regularly distributed, a normality test incorporating the metrics of kurtosis and skewness should be performed [38]. In this study, skewness values vary from −0.161 to 1.077, and kurtosis values range from −1.515 to 0.756. The variables were all within acceptable limits. With negative and positive skewness and kurtosis values, all variables were found to be regularly distributed.

4.1. Reflective Measurement Model

To test the hypotheses mentioned in this paper, PLS-SEM was utilised in two steps: (1) measurement model evaluation and (2) structural model evaluation [39]. Table 1 shows that the outer loading ranges from 0.558 to 0.958. Items with an outer loading value of less than 0.4 should normally be rejected due to the lack of explanation provided to the model, and values more than 0.70 should be preferred [40]. The items' internal consistency and accuracy were then assessed using composite reliability, which yielded a score of greater than 0.70 [28]. The composite dependability values ranged from 0.892 to 0.929, showing a high level of internal consistency. The Cronbach's alpha value was also greater than 0.50, which was deemed acceptable and sufficient [41]. The readings were in the range of 0.846 to 0.886, indicating that they were adequate. Figure 2 shows the measurement model of this study.

Table 1. Results of Reflective Measurement Model Assessment.

Latent Variables	Indicators	Outer Loadings	Composite Reliability	AVE	Cronbach Alpha
Social Care (SC)	SC1	0.867	0.915	0.686	0.883
	SC2	0.883			
	SC3	0.883			
	SC4	0.859			
	SC5	0.625			
	SC1	0.867			
Waqf Fund (WF)	WF1	0.958	0.929	0.815	0.886
	WF2	0.945			
	WF3	0.797			
Elderly Well-being (EWB)	EW1	0.890	0.892	0.633	0.846
	EW2	0.933			
	EW3	0.924			
	EW4	0.558			
	EW5	0.583			

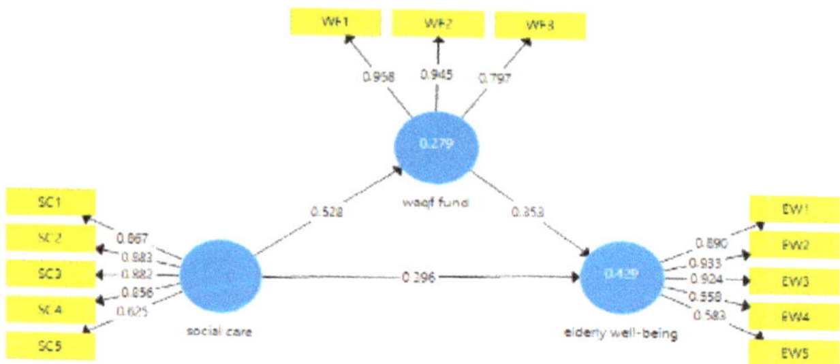

Figure 2. Measurement Model.

The average variance extracted (AVE) was utilised to determine the convergent validity of each construct. The AVE must be at least 0.5, indicating that there is a large amount of shared variance among multiple items measuring the same variable [39]. AVE values between 0.633 and 0.815 show good convergent validity. To ensure that the constructs were not too closely connected, they were also examined for discriminant validity [40]. Its purpose is to determine how distinct one construct's indicators are from those of another build. This was accomplished using Fornell-Larcker criteria, cross-loadings, and the HTMT criterion.

The HTMT criterion was employed to examine discriminant validity, as recommended by [42]. It calculates the average correlation of the indicator across all constructions. The HTMT values should be less than 0.85, with a more lenient 0.90 criterion [7,43]. As indicated in Table 2, the HTMT findings were all lower than the stricter threshold of 0.639, showing that the respondents understood that all five concepts are distinct.

Table 2. Heterotrait-Monotrait Ratio (HTMT).

	Elderly Well-Being	Social Care	*Waqf* Fund
Elderly Well-Being			
Social Care	0.625		
Waqf Fund	0.639	0.567	

4.2. Structural Model

The structural model aims to identify which latent constructs influence the values of the model's other latent constructs [38]. SmartPLS 3.3 was used to create a path diagram that depicted the model's different linkages in this investigation. The value of the variance inflation factor (VIF) was determined, and it should be less than 5. Collinearity concerns are indicated by a VIF value greater than 5. All of the results were less than 5, indicating that there was no strong evidence of multicollinearity.

Table 3 shows that the findings demonstrate that social care has a substantial positive link with elderly well-being (β = 0.396; t = 5.062; *p*-value = 0.000). The H2 indicates that social care and the *waqf* fund are linked (β = 0.528; t = 9.235; *p*-value = 0.000). The results of this study support the third hypothesis (H3) that claims a link exists between the *waqf* fund and the well-being of the elderly (β = 0.353; t = 4.033; *p*-value = 0.000). As a result, H3 has a statistical significance and is supported by the research findings.

Table 3. Path Coefficient (β), T-Value and Significance Level.

	Path Coefficient (β)	T Values	*p* Values	Results
SC → EW	0.396	5.062	0.000	Supported
SC → WF	0.528	9.235	0.000	Supported
WF → EW	0.353	4.033	0.000	Supported

Figure 3 shows the structural model of the study. When assessing mediation, the bootstrapping approach is used to determine the significance of indirect effects and their magnitude [38,44]. Instead of utilising *p*-values to test the significance of a hypothesis, Ref. [39] determined that a combination of criteria such as *p*-values, confidence intervals, and effect sizes should be employed. If the confidence interval does not straddle 0, it is assumed that there is a considerable mediation [45]. Before investigating the mediation, the direct effects of H1, H2, and H3 were examined.

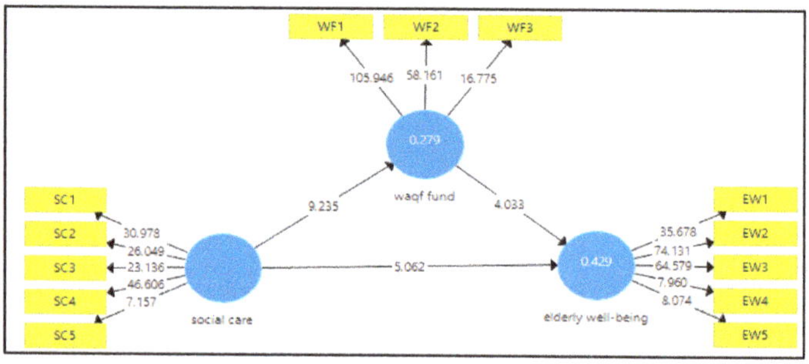

Figure 3. Structural Model.

The relationships between social care and *waqf* fund ($\beta = 0.396, p = 0.000$) and between *waqf* fund and social care ($\beta = 0.528$, $p = 0.000$) were both significant factors in elderly well-being, whereas the relationship between *waqf* fund and elderly ($\beta = 0.353, p = 0.000$) was also significant. As a result, all the hypotheses were supported as indicated in Table 4.

Table 4. Hypotheses Testing Direct Effect.

Relationships	Std. Beta	Std. Dev.	t-Values	*p*-Values	BCI LL	BCI UL	f2
SC→ EW	0.396	0.078	5.062	0.000	0.251	0.557	0.198
SC→ WF	0.528	0.057	9.235	0.000	0.417	0.643	0.338
WF→ EW	0.353	0.088	4.033	0.000	0.180	0.518	0.157

This study follows [46] in bootstrapping the indirect effect to evaluate the mediation hypotheses. Table 5 state that social care, the *waqf* fund, and elderly well-being have a substantial mediation connection ($\beta = 0.187, p = 0.000$). In the bias-corrected confidence intervals, no intervals crossed zero. As a result, H4 was supported. Social care, the *waqf* fund, and elderly well-being have a substantial mediation connection. In the bias-corrected confidence intervals, no intervals crossed a zero.

Table 5. Hypotheses Testing Indirect Effect.

Relationship	Std. Beta	Std. Dev.	t-Values	*p*-Values	BCI LL	BCI UL
SC → WF → EW	0.187	0.051	3.628	0.000	0.092	0.294

5. Conclusions

Overall, the current study's findings highlighted the relevance of social care and the *waqf* fund in enhancing the well-being of the elderly. Contentment with social care, according to previous studies, is crucial for older people's overall well-being [3]. According to [47], the *waqf* fund has significant benefits and impacts on the community and country, since it promotes economic growth, education, and social development. This research contributed significantly by establishing that *waqf* funds have a direct and indirect impact on the well-being of the elderly through social care. The impact of social care assistance on senior individuals' perceptions of their quality of life is important.

The variable degree of functioning determines the subjective assessment of functioning in different qualities of life dimensions. Adequate elder care, whether offered by children or by professionals, stops them from feeling lonely and depressed while also providing them with a sense of community and social life which nourishes and energises them. The elderly will benefit from increased service and care. As a result, social care is an important factor in predicting elderly satisfaction. Furthermore, the potential for developing a *waqf* fund as a financial tool infused with compassion, social commitment, and teamwork as a mechanism to safeguard the elderly is enormous.

Additionally, with the implementation of appropriate legislation and regulations, the well-being of Malaysia's senior citizens has the potential to be considerably improved and enhanced. This would subsequently lead towards long-term economic and social benefits, especially when it comes to societal challenges like abuse and neglect. As this research looked into the factors that influence senior well-being in order to contribute to the creation of social and economic values that benefit the aged, the efforts were hampered by a lack of professional help and service barriers, highlighting the reality that many services and care delivery models are still not designed with older people in mind. As a result of these findings, tailoring interventions and support services to the requirements of older persons is crucial.

There are certain flaws in this study that can be addressed in future research. The study's initial purpose was to investigate the well-being of the elderly, and it covered topics such as social care and the role of the *waqf* fund as a mediator. Other social elements, such as social relationships and social activities, can be investigated as future sources of happiness and well-being, particularly for elderly people who require social care and assistance. Second, only elderly Malaysians from the states of Johor, Selangor, Kuala Lumpur, and Perak were included in the current sample study. As a result, the findings are not applicable to other regions of Malaysia. Hence, researchers and industry experts are proposed to do research on elderly well-being from various locations (e.g., the Northern Region, the Central Region, the East Coast Region, and Sabah, Sarawak, and Labuan).

Author Contributions: Conceptualization, S.N.M.F. and I.O.; methodology, S.I., M.J.M.H. and M.R.K.; software, S.N.M.F. and M.J.M.H.; validation, S.N.M.F. and I.O.; formal analysis, S.N.M.F. and I.O.; resources, S.N.M.F. and I.O.; data curation, S.N.M.F. and I.O.; writing—original draft preparation, S.N.M.F. and I.O.; writing—review and editing, S.N.M.F. and I.O.; visualization, S.N.M.F. and I.O.; supervision, I.O.; project administration, I.O.; funding acquisition, I.O., S.I., M.J.M.H. and M.R.K. All authors have read and agreed to the published version of the manuscript.

Funding: This research was funded by the Fundamental Research Grant Scheme (FRGS) (600-IRMI/FRGS 5/3 (248/2019)).

Institutional Review Board Statement: Not applicable.

Informed Consent Statement: Not applicable.

Data Availability Statement: Not applicable.

Acknowledgments: Panel and reviewers of the International Academic Symposium of Social Science 2022, organized by Universiti Teknologi MARA, Kelantan branch.

Conflicts of Interest: The authors declare no conflict of interest.

References

1. Abdul Hamid, A.R.; Muhamad, N.; Zakaria, R.; Aminuddin, E.; AAlwee, A.A.; Tok, A. The challenges of the ageing population on the Malaysian construction industry. *J. Phys. Conf. Ser.* **2018**, *1049*, 1. [CrossRef]
2. Amarya, S.; Singh, K.; Sabharwal, M. Ageing process and physiological changes. In *Gerontology*; D'Onofrio, G., Greco, A., Sancarlo, D., Eds.; IntechOpen: London, UK, 2018. [CrossRef]
3. Abdi, S.; Spann, A.; Borilovic, J.; de Witte, L.; Hawley, M. Understanding the care and support needs of older people: A scoping review and categorisation using the WHO international classification of functioning, disability and health framework (ICF). *BMC Geriatr.* **2019**, *20*, 23. [CrossRef] [PubMed]
4. Manfredini, F.; Di Rosa, C. Measuring spatial accessibility for elderly: An application to subway stations in Milan. *TeMA J. Land Use, Mobil. Environ.* **2018**, *2018*, 85–94. [CrossRef]
5. Sapuan, M.M.; Mohd Zeni, N.A.; Md Hashim, S.L. Elevation of waqf industry for social wellbeing in Malaysia: Issues and challenges. *Int. J. Adv. Biotechnol. Res.* **2017**, *8*, 312–318.
6. Bäck, M.A.; Calltorp, J. The Norrtaelje model: A unique model for integrated health and social care in Sweden. *Int. J. Integ. Care* **2015**, *15*, e016. [CrossRef]
7. Keller, D.; McGee, M.F.; Goyal, S.; Nobel, T.; Ermlich, B.O.; Cheruvu, V.; Delaney, C. Construct validation and comparison of a novel postoperative quality-of-life metric and the short form-36 in colorectal surgery patients. *Surgery* **2013**, *154*, 690–696. [CrossRef]

8. Sabit, M.T. *An Ideal Mechanism for the Development of Waqf Properties*, 2nd ed.; Universiti Teknologi Malaysia Press: Johor, MA, USA, 2006.
9. Abdul Shukor, S.; Anwar, I.F.; Aziz, S.A.; Sabri, H. Muslim attitude towards participation in cash waqf: Antecedents and consequences. *Int. J. Bus. Soc.* **2017**, *18*, 193–204.
10. Rendón, R.C.; Angulo, Y.L.; Villalobos, M.V.P.; Mujica, A.D. Perceived social support and its effects on changes in the affective and eudaimonic well-being of Chilean University Students. *Front. Psychol.* **2020**, *11*, 590513. [CrossRef]
11. Othman, Y.H.; Abd Latib, M.F.; Ahmad, M.N.; Hasnan, R. The role of Islamic social finance in reviving the economy during COVID-19 pandemic crisis. *Int. J. Muamalat* **2020**, *4*, 147–152. [CrossRef]
12. Shrestha, P.; Millonig, A.; Hounsell, N.B.; McDonald, M. Review of public transport needs of older people in European context. *Popul. Ageing* **2017**, *10*, 343–361. [CrossRef]
13. Chen, L.; Alston, M.; Guo, W. The influence of social support on loneliness and depression among older elderly people in China: Coping styles as mediators. *J. Community Psychol.* **2019**, *47*, 1235–1245. [CrossRef] [PubMed]
14. Thielke, S.; Harniss, M.; Thompson, H.; Patel, S.; Demiris, G.; Johnson, K. Maslow's hierarchy of human needs and the adoption of health-related technologies for older adults. *Ageing Int.* **2017**, *37*, 470–488. [CrossRef]
15. Ab Rahman, F.; Abd Murad, F.A.H.; Abdullah Thaidi, H.A.; Ab Rahman, A. Role of waqf institution in the sustainable management of senior citizens in Malaysia. *Perdana Int. J. Acad. Res.* **2020**, *8*, 20–29.
16. Folk, Y.; Beh, L.S.; Baranovich, D.L. Financial education: Determinant of retirement planning in Malaysia. *J. Bus. Manag. Econ.* **2012**, *3*, 069–078.
17. Hair, J.F.; Ringle, C.M.; Sarstedt, M. PLS-SEM: Indeed, a Silver Bullet. *J. Mark. Theory Pract.* **2011**, *18*, 139–152. [CrossRef]
18. Steverink, N.; Lindenberg, S. Which social needs are important for subjective well-being? What happens to them with aging? Am. *Psychol. Assoc.* **2006**, *21*, 281–290. [CrossRef]
19. Davis, S. What Is Well-Being? Definition, Types, and Well-Being Skills. Available online: https://www.psychologytoday.com/us/blog/click-here-happiness/201901/what-is-well-beingdefinition-types-and-well-being-skills (accessed on 1 August 2019).
20. Taber, K. The use of cronbach's alpha when developing research instruments in science education. *Res. Sci. Educ.* **2017**, *48*, 1–24. [CrossRef]
21. Hubley, A.M. Discriminant validity. In *Encyclopedia of Quality of Life and Well-Being Research*; Michalos, A.C., Ed.; Springer: Dordrecht, The Netherlands, 2014.
22. Paivi, N.; Seppänen, M.; Begum, S. (Eds.) *Ageing, Wellbeing and Climate Change in the Arctic: An Interdisciplinary Analysis*; Routledge: England, UK, 2015; p. 274.
23. Minhat, H.S. Social Care for Older Persons in Malaysia. Universiti Putra Malaysia. 2018. Available online: https://www.agediversity.org/wp-content/uploads/2018/09/halima-minhat-upm.pdf (accessed on 15 December 2020).
24. Kline, R.B. *Principles and Practice of Structural Equation Modelling*; Guilford Press: New York, NY, USA, 2011.
25. McCombes, S. How to Create a Research Design. Available online: https://www.scribbr.com/research-process/research-design (accessed on 1 August 2019).
26. Eliasi, L.G.; Rasi, H.A.; Tavakoli, A. Factors affecting quality of life among elderly population in Iran. *Humanit. Soc. Sci.* **2017**, *5*, 26–30. [CrossRef]
27. Ng, S.T.; Tey, N.P.; Asadullah, M.N. What matters for life satisfaction among the oldestold? Evidence from China. *PLoS ONE* **2017**, *12*, e0171799. [CrossRef]
28. Preacher, K.J.; Hayes, A.F. SPSS and SAS procedures for estimating indirect effects in simple mediation models. *Behav. Res. Methods Instrum. Comput.* **2004**, *36*, 717–731. [CrossRef]
29. Ab Rahman, F.; Amanullah, M. Challenges and problems facing the application of temporary waqf in selected states of Malaysia: New issues and their guiding rules. *J. King Abdulaziz Univ. Islamic Econ.* **2017**, *30*, 291–314. [CrossRef]
30. Hassan, T.; Yusoff, A.; Ahmad Asnawi, A.W.; Adnan, I.S. Takaful, zakat and waqf instruments in ensuring holistic senior citizens program development in Malaysia. *J. Islamic Financ.* **2020**, *9*, 79–88.
31. Oladeji, D. Family care, social services, and living arrangements factors influencing psychosocial well-being of elderly from selected households in Ibadan, Nigeria. *Edu. Res. Int.* **2011**, *2011*, 421898. [CrossRef]
32. Budiman, M.A. The significance of waqf for economic development. *Equilibrium* **2014**, *2*, 19–34. [CrossRef]
33. Ubaidillah, M.A.; Ismail Mohd Noor, M.A. The waqf integrated financial instrument of pension model in Malaysian social security: A conceptual proposition. *JMIFR* **2020**, *17*, 33–55. [CrossRef]
34. Amiruddin, M.M.; Shamad, Y. Waqf conflict resolution through mediation (Islamic and Bugis norm perspective). *Li Falah-J. Studi Ekon. Dan Bisnis Islam* **2019**, *4*, 79–95. [CrossRef]
35. Shahriar, S.M.; Alam, M.M.; Said, J.; Monzur-E-Elahi, M. Waqf as a tool for rendering social welfare services in the social entrepreneurship context. *Glob. J. Al-Thaqafah* **2019**, 87–98. [CrossRef]
36. Kroger, T.; Leinonen, A. Transformation by stealth: The retargeting of home care services in Finland. *Health Soc. Care Community* **2012**, *20*, 319–327. [CrossRef]
37. Schönfelder, W.; Eggebø, H.; Munkejord, M.C. Social care for older people–a blind spot in the Norwegian care system. *Soc. Work Health Care* **2020**, *59*, 631–649. [CrossRef]
38. Hair, J.F.; Black, W.; Babin, B.J.; Anderson, R.E. *Multivariate Data Analysis: A Global Perspective*, 7th ed.; Pearson: London, UK, 2010.

39. Hahn, E.D.; Ang, S.H. From the editors: New directions in the reporting of statistical results in the Journal of World Business. *J. World Bus.* **2017**, *52*, 125–126. [CrossRef]
40. Hinton, P.R.; Brownlow, C.; McMurray, I.; Cozens, B. *SPSS Explained*; Routledge: London, UK, 2004. [CrossRef]
41. Shrivastava, S.R.B.L.; Shrivastava, P.S.; Ramasamy, J. Health-care of elderly: Determinants, needs and services. *Int. J. Prev. Med.* **2013**, *4*, 1224–1225. [PubMed]
42. Henseler, J.; Ringle, C.M.; Sinkovics, R.R. The use of partial least squares path modeling in international marketing. *Adv. Int. Mark.* **2009**, *20*, 277–320.
43. Gold, A.; Malhotra, A.; Segars, A. Knowledge management: An organizational capabilities perspective. *J. Manag. Inf. Syst.* **2001**, *18*, 185–214. [CrossRef]
44. Cepeda, G.; Nitzl, C.; Roldán, M.J.L. *Mediation Analyses in Partial Least Squares Structural Equation Modeling: Guidelines and Empirical Examples*; Springer: Berlin/Heidelberg, Germany, 2018.
45. Cohen, J. *Statistical Power Analysis for the Behavioral Sciences*, 2nd ed.; Lawrence Erlbaum Associates: Hillsdale, NJ, USA, 1988.
46. Poudel, A.; Gurung, B.; Khanal, G.P. Perceived social support and psychological wellbeing among Nepalese adolescents: The mediating role of self-esteem. *BMC Psychol.* **2020**, *8*, 43. [CrossRef]
47. Yakubu, A.; Abdul Aziz, A.H. The role of waqf in economic sustainability and poverty reduction. *Int. J. Acad. Res. Bus. Soc. Sci.* **2019**, *9*, 708–714.